FUNDAMENTALS
OF QUALITY CONTROL
AND IMPROVEMENT

FUNDAMENTALS OF QUALITY CONTROL AND IMPROVEMENT

Third Edition

AMITAVA MITRA
Auburn University
College of Business
Auburn, Alabama

WILEY

A JOHN WILEY & SONS, INC., PUBLICATION

Published by John Wiley & Sons, Inc., Hoboken, New Jersey.
Published simultaneously in Canada.

For general information on our other products and services or for technical support, please contact our Customer Care Department within the United States at (800) 762-2974, outside the United States at (317) 572-3993 or fax (317) 572-4002.

Wiley also publishes its books in a variety of electronic formats. Some content that appears in print may not be available in electronic formats. For more information about Wiley products, visit our web site at www.wiley.com.

Library of Congress Cataloging-in-Publication Data:

Mitra, Amitava.
   Fundamentals of quality control and improvement / Amitava Mitra. – 3rd ed.
       p. cm.
   Includes index.
   ISBN 978-0-470-22653-7 (cloth)
1.   Quality control – Statistical methods.   I. Title.
   TS156 . M54 2008
   658.4'0.13–dc22
                                                        2007036433

Printed in the United States of America

10 9 8 7 6 5 4 3

To the memory of my father, and my mother,
who instilled the importance of
an incessant inquiry for knowledge —
and whose inspiration transcends mortality

CONTENTS

Appendixes 665

Index 687

PREFACE

This book covers the foundations of modern methods of quality conrol and improvement that are used in the manufacturing and service industries. Quality is key to surviving tough competition. Consequently, business needs technically competent people who are well-versed in statistical quality control and improvement. This book should serve the needs of students in business and management and students in engineering, technology, and related disciplines. Professionals will find this book to be a valuable reference in the field.

An outgrowth of many years of teaching, research, and consulting in the field of quality assurance and statistical process control, the methods discussed in this book apply statistical foundations to real-world situations. Mathematical derivations and proofs are kept to a minimum to allow a better flow of material. Although an introductory course in statistics would be useful to a reader, the foundations of statistical tools and techniques discussed in Chapter 4 should enable students without a statistical background to understand the material.

Prominently featured are many real-world examples. For each major concept, at least one example demonstrates its application. Furthermore, case studies at the end of some chapters enhance student understanding of pertinent issues.

The book is divided into five parts. Part I, which deals with the philosophy and fundamentals of quality control, consists of three chapters. Chapter 1 is an introduction to quality control and the total quality system. In addition to introducing the reader to the nomenclature associated with quality control and improvement, it provides a framework for the systems approach to quality. Discussions of quality costs and their measurement, along with activity-based costing, are presented. In Chapter 2 we examine philosophies of such leading experts as Deming, Crosby, and Juran. Deming's 14 points for management are analyzed, and the three philosophies are compared. Features of quality in the service sector are introduced. Chapter 3 covers quality management practices, tools, and standards. Topics such as total quality management, balanced scorecard, quality function deployment, benchmarking, failure mode and effects criticality analysis, and tools for quality improvement are presented. We also discuss the criteria for the Malcolm Baldrige National Quality Award in the United States and the International Organization for Standardization (ISO) 9000 standards.

Part II deals with the statistical foundations of quality control and consists of two chapters. Chapter 4 offers a detailed coverage of statistical concepts and techniques in quality control and improvement. It present a thorough treatment of inferential statistics.

Depending on the student's background, only selected sections of this chapter will need to be covered.

Chapter 5 covers some graphical methods of empirical distributions. Identification of the population distribution using probability plotting along with the several transformations to achieve normality are presented. Analysis of count data, including contingency table analysis and measures of association, are discussed. Finally, some common sampling designs and determination of an appropriate sample size are features of this chapter.

The field of statistical quality control consists of two areas: statistical process control and acceptance sampling. Part III deals with statistical process control and consists of four chapters. Chapter 6 provides an overview of the principles and use of control charts. A variety of control charts for variables are discussed in detail in Chapter 7. In additon to charts for the mean and range, those for the mean and standard deviation, individual units, cumulative sum, moving average, geometric moving average, and trends are presented. Control charts for attributes are discussed in Chapter 8. Charts such as the p-chart, np-chart, c-chart, u-chart, and U-chart are presented. The topic of process capability analysis is discussed in Chapter 9. The ability of a process to meet customer specifications is examined in detail. Process capability analysis procedures and process capability indices are also treated in depth. The chapter covers proper approaches to setting tolerances on assemblies and components. Part III should form a core of material to be covered in most courses.

Part IV deals with acceptance sampling procedures and cosists of one chapter. Methods of acceptance of a product based on information from a sample are described. Chapter 10 presents acceptance sampling plans for attributes and variables. Lot-by-lot attribute and variable sampling plans are described. With the emphasis on process control and improvement, sampling plans do not occupy the forefront. Nevertheless, they are included to make the discussion complete.

Part V deals with product and process design and consists of two chapters. With the understanding that quality improvement efforts are generally being moved further upstream, these chapters constitute the backbone of current methodology. Chapter 11 deals with reliability and explores the effects of time on the proper functioning of a product. Chapter 12 provides the fundamentals of experimentals design and the Taguchi method. Different designs, such as the completely randomized design, randomized block design, and Latin square design are presented. Estimation of treatment effects using factorial experiments is included. This chapter also provides a treatment of the Taguchi method for design and quality improvement; the philosophy and fundamentals of this method are discussed. Various sections of Part V could also be included in the core material for a quality control course.

This book may serve as a text for an undergraduate or graduate course for students in business and management. It may also serve the needs of students in engineering, technology, and related disciplines. For a one-semester or one-quarter course, Part I, selected portions of Part II, selected portions of Part III, and selected portions of Part V could be covered. For a two-semester or two-quarter course, all of Parts II, III, and V, along with portions from Part IV, could be covered as well.

CHANGES IN THE THIRD EDITION

Some major changes have been made in the third edition. First, the theme of applications of quality control and improvement in the service sector has been integrated throughout the book. Second, the use of a computer software package, Minitab, is demonstrated. Third, new case studies are introduced in some of the chapters. Some of the initial chapters also include profiles of Malcolm Baldrige National Quality Award winners that demonstrate best-business practices and their performance achievements. Fourth, many of the chapter-end exercises have been revised. Fifth, some new material has been included in appropriate chapters as discussed below.

Chapter 1 has been revamped completely. Concepts and applications of activity-based costing are included. The discussion of quality costs and the impact of various factors, such as product development, have been expanded. The effects of first-pass yield and inspection on unit costs are elaborated. Chapter 2 now includes a discussion of the unique character-istics of a model for quality in the service sector. Ideas on the balanced scorecard and failure mode effects and criticality analysis have been added to Chapter 3. Multivariable charts, matrix plots, and surface plots are included as part of the tools for quality improvement. The gamma and lognormal distributions as well as the paired-difference t-test are now discussed in Chapter 4. Chapter 5 has been modified intensively to include validation of distributional assumptions, since many statistical tests and inferences are based on a population distri-butional assumptions. Special attention is devoted to transformations to achieve normality. Expanded coverage of sampling and sample size determination are now featured. Analysis of count data is an added section. Control charts for short production runs and added discussion of multivariate control charts, including the generalized variance chart, are part of Chapter 7.

A section on control charts for highly conforming processes is now included in Chapter 8. Some new process capability measures, C_{pmk} and C_{pq}, along with making inferences on capability indices, are included in Chapter 9. The discussion on gage repeatability and reproducibility has been expanded, with capability analysis for nonnormal distributions being featured. Setting tolerances on nonlinear combinations of random variables are part of Chapter 9. The use of Bayes' rule in updating prior probabilities and the use of sample information to make optimal decisions are illustrated in Chapter 10. A brief discussion of availability is now provided in Chapter 11. Finally, accumulation analysis using attribute data is discussed in Chapter 12.

ACKNOWLEDGMENTS

Many people have contributed to the development this book, and thanks are due to them. Modern trends in product/process quality through design and improvement, as well as discussions and questions from undergraduate and graduate classes over the years, have shaped this book. Applications encountered in a consulting environment have provided a scenario for examples and exercises. Input from faculty and professional colleagues, here and abroad, has facilitated composition of the material. Constructive comments from the reviewers have been quite helpful. Many of the changes in the third edition are based on input from those who have used the book as well as from reviewers.

The manuscript preparation center of the College of Business at Auburn University did a remarkable job for which Margie Maddox is to be congratulated. I would like to thank

Minitab, Inc. (Quality Plaza, 1829 Pine Hall Road, State College, PA 16801-3008) for its assistance in providing software support. My editor, Steve Quigley, is to be commended for his patience and understanding.

Learning is a never-ending process. It takes time and a lot of effort. So does writing and revising a book. That has been my reasoning to my wife, Sujata, and son, Arnab. I believe they understand this—my appreciation to them. Their continual support has provided an endless source of motivation.

PART I

PHILOSOPHY AND FUNDAMENTALS

1

INTRODUCTION TO QUALITY CONTROL AND THE TOTAL QUALITY SYSTEM

1-1 INTRODUCTION AND CHAPTER OBJECTIVES

Wa. . .ah, Wa. . .ah, Wa. . .ah. It was about 2 A.M. The resident physician had just delivered the newborn and the proud parents were ecstatic. Forgotten in an instant, as she held the infant close to her, was the pain and suffering of the mother. "Wouldn't it be nice if one could share this moment, right now, with family and loved ones who are far away?", thought the parents. As if through extrasensory perception, the physician understood this desire. In a moment, the digital camera that was already in place started streaming the video to the cell phones and Web addresses of those designated. Now, both parties could hear and see each other as they shared this unique and rare moment in a bond of togetherness. The developments of the twenty-first century and the advances in quality make this possible.

The objectives of this chapter are, first, to define quality as it relates to the manufacturing and service sector, to introduce the terminology related to quality, and to set up a framework

Fundamentals of Quality Control and Improvement, Third Edition, By Amitava Mitra
Copyright © 2008 John Wiley & Sons, Inc.

for the design and implementation of quality. Of importance will be the ability to identify the unique needs of the customer, which will assist in maintaining and growing market share. A study of activity-based product costing will be introduced along with the impact of quality improvement on various quality-related costs. The reader should be able to interpret the relationships among quality, productivity, long-term growth, and customer satisfaction.

1-2   EVOLUTION OF QUALITY CONTROL

The quality of goods produced and services rendered has been monitored, either directly or indirectly, since time immemorial. However, using a quantitative base involving statistical principles to control quality is a modern concept.

The ancient Egyptians demonstrated a commitment to quality in the construction of their pyramids. The Greeks set high standards in arts and crafts. The quality of Greek architecture of the fifth century B.C. was so envied that it profoundly affected the subsequent architectural constructions of Rome. Roman-built cities, churches, bridges, and roads inspire us even today.

During the Middle Ages and up to the nineteenth century, the production of goods and services was confined predominantly to a single person or a small group. The groups were often family-owned businesses, so the responsibility for controlling the quality of a product or service lay with that person or small group—Those also responsible for producing items conforming to those standards. This phase, comprising the time period up to 1900, has been labeled by Feigenbaum (1983) the *operator quality control period*. The entire product was manufactured by one person or by a very small group of persons. For this reason, the quality of the product could essentially be controlled by a person who was also the operator, and the volume of production was limited. The worker felt a sense of accomplishment, which lifted morale and motivated the worker to new heights of excellence. Controlling the quality of the product was thus embedded in the philosophy of the worker because pride in workmanship was widespread.

Starting in the early twentieth century and continuing to about 1920, a second phase evolved, called the *foreman quality control period* (Feigenbaum 1983). With the Industrial Revolution came the concept of mass production, which was based on the principle of specialization of labor. A person was responsible not for production of an entire product but rather, for only a portion of it. One drawback of this approach was the decrease in the workers' sense of accomplishment and pride in their work. However, most tasks were still not very complicated, and workers became skilled at the particular operations that they performed. People who performed similar operations were grouped together. A supervisor who directed that operation now had the task of ensuring that quality was achieved. Foremen or supervisors controlled the quality of the product, and they were also responsible for operations in their span of control.

The period from about 1920 to 1940 saw the next phase in the evolution of quality control. Feigenbaum (1983) calls this the *inspection quality control period*. Products and processes became more complicated, and production volume increased. As the number of workers reporting to a foreman grew in number, it became impossible for the foreman to keep close watch over individual operations. Inspectors were therefore designated to check the quality of a product after certain operations. Standards were set, and inspectors compared the quality of the item produced against those standards. In the event of discrepancies between a standard and a product, deficient items were set aside from those that met the standard. The nonconforming items were reworked if feasible, or were discarded.

During this period, the foundations of statistical aspects of quality control were being developed, although they did not gain wide usage in U.S. industry. In 1924, Walter A. Shewhart of Bell Telephone Laboratories proposed the use of statistical charts to control the variables of a product. These came to be known as *control charts* (sometimes referred to as *Shewhart control charts*). They play a fundamental role in statistical process control. In the late 1920s, H. F. Dodge and H. G. Romig, also from Bell Telephone Laboratories, pioneered work in the areas of acceptance sampling plans. These plans were to become substitutes for 100% inspection.

The 1930s saw the application of acceptance sampling plans in industry, both domestic and abroad. Walter Shewhart continued his efforts to promote to industry the fundamentals of statistical quality control. In 1929 he obtained the sponsorship of the American Society for Testing and Materials (ASTM), the American Society of Mechanical Engineers (ASME), the American Statistical Association (ASA), and the Institute of Mathematical Statistics (IMS) in creating the Joint Committee for the Development of Statistical Applications in Engineering and Manufacturing.

Interest in the field of quality control began to gain acceptance in England at this time. The British Standards Institution Standard 600 dealt with applications of statistical methods to industrial standardization and quality control. In the United States, J. Scanlon introduced the *Scanlon plan*, which dealt with improvement of the overall quality of worklife (Feigenbaum 1983). Furthermore, the U.S. Food, Drug, and Cosmetic Act of 1938 had jurisdiction over procedures and practices in the areas of processing, manufacturing, and packing.

The next phase in the evolution process, called the *statistical quality control phase* by Feigenbaum (1983), occurred between 1940 and 1960. Production requirements escalated during World War II. Since 100% inspection was often not feasible, the principles of sampling plans gained acceptance. The American Society for Quality Control (ASQC) was formed in 1946, subsequently renamed the American Society for Quality (ASQ). A set of sampling inspection plans for attributes called MIL-STD-105A was developed by the military in 1950. These plans underwent several subsequent modifications, becoming MIL-STD-105B, MIL-STD-105C, MIL-STD-105D, and MIL-STD-105E. Furthermore, in 1957, a set of sampling plans for variables called MIL-STD-414 was also developed by the military.

Although suffering widespread damage during World War II, Japan embraced the philosophy of statistical quality control wholeheartedly. When W. Edwards Deming visited Japan and lectured on these new ideas in 1950, Japanese engineers and top management became convinced of the importance of statistical quality control as a means of gaining a competitive edge in the world market. J. M. Juran, another pioneer in quality control, visited Japan in 1954 and further impressed on them the strategic role that management plays in the achievement of a quality program. The Japanese were quick to realize the profound effects that these principles would have on the future of business, and they made a strong commitment to a massive program of training and education.

Meanwhile, in the United States, developments in the area of sampling plans were taking place. In 1958, the Department of Defense (DOD) developed the Quality Control and Reliability Handbook H-107, which dealt with single-level continuous sampling procedures and tables for inspection by attributes. Revised in 1959, this book became the Quality Control and Reliability Handbook H-108, which also covered multilevel continuous sampling procedures as well as topics in life testing and reliability.

The next phase, *total quality control*, took place during the 1960s (Feigenbaum 1983). An important feature during this phase was the gradual involvement of several departments and management personnel in the quality control process. Previously, most of these activities were dealt with by people on the shop floor, by the production foreman, or by people from the inspection and quality control department. The commonly held attitude prior to this period was that quality control was the responsibility of the inspection department. The 1960s, however, saw some changes in this attitude. People began to realize that each department had an important role to play in the production of a quality item. The concept of *zero defects*, which centered around achieving productivity through worker involvement, emerged during this time. For critical products and assemblies – [e.g., missiles and rockets used in the space program by the National Aeronautics and Space Administration (NASA)] this concept proved to be very successful. Along similar lines, the use of **quality circles** was beginning to grow in Japan. This concept, which is based on the participative style of management, assumes that productivity will improve through an uplift of morale and motivation, achieved in turn, through consultation and discussion in informal subgroups.

The advent of the 1970s brought what Feigenbaum (1983) calls the *total quality control organizationwide phase*, which involved the participation of everyone in the company, from the operator to the first-line supervisor, manager, vice president, and even the chief executive officer. Quality was associated with every person. As this notion continued in the 1980s, it was termed by Feigenbaum (1983) the **total quality system**, which he defines as follows: "A quality system is the agreed on companywide and plantwide operating work structure, documented in effective, integrated technical and managerial procedures, for guiding the coordinated actions of the people, the machines, and the information of the company and plant in the best and most practical ways to assure customer quality satisfaction and economical costs of quality."

In Japan, the 1970s marked the expanded use of a graphical tool known as the **cause-and-effect diagram**. This tool was introduced in 1943 by K. Ishikawa and is sometimes called an **Ishikawa diagram**. It is also called a **fishbone diagram** because of its resemblance to a fish skeleton. This diagram helps identify possible reasons for a process to go out of control as well as possible effects on the process. It has become an important tool in the use of control charts because it aids in choosing the appropriate action to take in the event of a process being out of control. Also in this decade, G. Taguchi of Japan introduced the concept of quality improvement through statistically designed experiments. Expanded use of this technique has continued in the 1990s as companies have sought to improve the design phase.

In the 1980s, U.S. advertising campaigns placed quality control in the limelight. Consumers were bombarded with advertisements related to product quality, and frequent comparisons were made with those of competitors. These promotional efforts tried to point out certain product characteristics that were superior to those of similar products. Within the industry itself, an awareness of the importance of quality was beginning to evolve at all levels. Top management saw the critical need for the marriage of the quality philosophy to the production of goods and services in all phases, starting with the determination of customer needs and product design and continuing on to product assurance and customer service.

As computer use exploded during the 1980s, an abundance of quality control software programs came on the market. The notion of a total quality system increased the emphasis on vendor quality control, product design assurance, product and process quality audit, and related areas. Industrial giants such as the Ford Motor Company and General Motors Corporation adopted the quality philosophy and made strides in the implementation of statistical quality control methods. They, in turn, pressured other companies to use quality control techniques. For example, Ford demanded documentation of statistical process control from its vendors. Thus, smaller companies that had not used statistical quality control methods previously were forced to adopt these methods to maintain their contracts. The strategic importance of quality control and improvement was formally recognized in the United States through the **Malcolm Baldrige National Quality Award** in 1987.

The emphasis on customer satisfaction and continuous quality improvement globally created a need for a system of standards and guidelines that support the quality philosophy. The International Organization for Standardization (ISO) developed a set of standards, **ISO 9000–9004**, in the late 1980s. The American National Standards Institute (ANSI) and ASQC brought their standards in line with the ISO standards when they developed the ANSI/ASQC Q90–Q94 in 1987, which was subsequently revised in 1994 to ANSI/ASQC Q9000–Q9004, and further in 2000, to ANSI/ISO/ASQ Q9000–2000. The ISO 9000–9004 standards were also revised in 1994 and 2000.

Beginning with the last decade of the twentieth century and continuing on to the current century, the world has seen the evolution of an era of information technology. This is the major revolution since the Industrial Revolution of the late eighteenth century. The twenty-first century is undergoing its revolution in information technology digitally, using wireless technology. Such advances promote the maintenace and protection of information quality while delivering data in an effective manner. Further, advances in computational technology have made it feasible to solve, in a timely fashion, complex and/or large-scale problems to be used for decision making. Moreover, the Internet is part and parcel of our everyday lives. Among a multitude of uses, we make travel arrangements, purchase items, look up information on a variety of topics, and correspond. All of these activities are conducted on a real-time basis, thus raising expectations regarding what constitutes timely completion. On receiving an order through the Internet, service providers will be expected to conduct an error-free transaction, for example, either assemble the product or provide the service, receive payment, and provide an online tracking system for the customer to monitor. Thus, the current century will continue to experience a thrust in growth of quality assurance and improvement methods that can, using technology, assimilate data and analyze them in real time and with no tolerance for errors.

1-3 QUALITY

The notion of **quality** has been defined in different ways by various authors. Garvin (1984) divides the definition of quality into five categories: transcendent, product-based, user-based, manufacturing-based, and value-based. Furthermore, he identifies a framework of eight attributes that may be used to define quality: performance, features, reliability, conformance, durability, serviceability, aesthetics, and perceived quality. This frequently used definition is attributed to Crosby (1979): "Quality is conformance to requirements or specifications." A more general definition proposed by Juran (1974) is as follows: "Quality is fitness for use."

In this book we adopt, the latter definition and expand it to cover both the **manufacturing** and **service** sectors. The service sector accounts for a substantial segment of our present economy; it is a major constituent that is not to be neglected. Projections indicate that this proportion will expand even further in the future. Hence, quality may be defined as follows: The quality of a product or service is the fitness of that product or service for meeting or exceeding its intended use as required by the customer.

So, who is the driving force behind determining the level of quality that should be designed into a product or service? *The customer*! Therefore, as the needs of customers change, so should the level of quality. If, for example, customers prefer an automobile that gives adequate service for 15 years, then that is precisely what the notion of a quality product should be. Quality, in this sense, is not something that is held at a constant universal level. In this view, the term *quality* implies different levels of expectations for different groups of consumers. For instance, to some, a quality restaurant may be one that provides extraordinary cuisine served on the finest china with an ambience of soft music. However, to another group of consumers, the characteristics that comprise a quality restaurant may be quite different: excellent food served buffet style at moderate prices until the early morning hours.

Quality Characteristics

The preceding example demonstrates that one or more elements define the intended quality level of a product or service. These elements, known as **quality characteristics**, can be categorized in these groupings: **Structural characteristics** include such elements as the length of a part, the weight of a can, the strength of a beam, the viscosity of a fluid, and so on; **sensory characteristics** include the taste of good food, the smell of a sweet fragrance, and the beauty of a model, among others; **time-oriented characteristics** include such measures as a warranty, reliability, and maintainability; and **ethical characteristics** include honesty, courtesy, friendliness, and so on.

Variables and Attributes

Quality characteristics fall into two broad classes: variables and attributes. *Characteristics that are measurable and are expressed on a numerical scale* are called *variables*. The waiting time in a bank before being served, expressed in minutes, is a variable, as are the density of a liquid in grams per cubic centimeter and the resistance of a coil in ohms.

Prior to defining an attribute, we should defined a nonconformity and a nonconforming unit. A *nonconformity* is a *quality characteristic that does not meet its stipulated specifications*. Let's say that the specification on the fill volume of soft drink bottles is 750 ± 3 milliliters (mL). If we have a bottle containing 745 mL, its fill volume is a nonconformity. A *nonconforming unit* has *one or more nonconformities such that the unit is unable to meet the intended standards and is unable to function as required*. An example of a nonconforming unit is a cast iron pipe whose internal diameter and weight both fail to satisfy specifications, thereby making the unit dysfunctional.

A quality characteristic is said to be an **attribute** if it is classified as *either conforming or nonconforming to a stipulated specification*. A quality characteristic that cannot be measured on a numerical scale is expressed as an attribute. For example, the smell of a cologne is characterized as either acceptable or is not; the color of a fabric is either acceptable or is not. However, there are some variables that are treated as attributes because it is simpler to measure them this way or because it is difficult to obtain data on them. Examples in this category are numerous. For instance, the diameter of a bearing is, in theory, a variable. However, if we measure the diameter using a go/no-go gage and classify it as either conforming or nonconforming (with respect to some established specifications), the characteristic is expressed as an attribute. The reasons for using a go/no-go gage, as opposed to a micrometer, could be economic; that is, the time needed to obtain a measurement using a go/no-go gage may be much shorter and consequently less expensive. Alternatively, an inspector may not have enough time to obtain measurements on a numerical scale using a micrometer, so such a classification of variables would not be feasible.

Defects

A **defect** is associated with a quality characteristic that does not meet certain standards. Furthermore, the severity of one of more defects in a product or service may cause it to be unacceptable (or defective). The modern term for *defect* is *nonconformity*, and the term for *defective* is *nonconforming item*. The American National Standards Institute, the International Organization for Standardization, and the American Society for Quality provide a definition of a defect in ANSI/ISO/ASQ Standard A8402 (ASQ 1994).

Standard or Specification

Since the definition of quality involves meeting the requirements of the customer, these requirements need to be documented. A **standard**, or a **specification**, refers to *a precise statement that formalizes the requirements of the customer; it may relate to a product, a process, or a service*. For example, the specifications for an axle might be 2 ± 0.1 centimeters (cm) for the inside diameter, 4 ± 0.2 cm for the outside diameter, and 10 ± 0.5 cm for the length. This means that for an axle to be acceptable to the customer, each of these dimensions must be within the values specified. Definitions given by the National Bureau of Standards (NBS, 2005) are as follows:

- *Specification:* a set of conditions and requirements, of specific and limited application, that provide a detailed description of the procedure, process, material, product, or service for use primarily in procurement and manufacturing. Standards may be referenced or included in a specification.

- *Standard:* a prescribed set of conditions and requirements, of general or broad application, established by authority or agreement, to be satisfied by a material, product, process, procedure, convention, test method; and/or the physical, functional, performance, or conformance characteristic thereof. A physical embodiment of a unit of measurement (for example, an object such as the standard kilogram or an apparatus such as the cesium beam clock).

Acceptable bounds on individual quality characteristics (say, 2 ± 0.1 cm for the inside diameter) are usually known as **specification limits**, whereas the document that addresses the requirements of all the quality characteristics is labeled the *standard*.

Three aspects are usually associated with the definition of quality: quality of design, quality of conformance, and quality of performance.

Quality of Design

Quality of design deals with the stringent conditions that a product or service must minimally possess to satisfy the requirements of the customer. *It implies that the product or service must be designed to meet at least minimally the needs of the consumer.* Generally speaking, the design should be the simplest and least expensive while still meeting customer's expectations. Quality of design is influenced by such factors as the type of product, cost, profit policy of the firm, demand for product, availability of parts and materials, and product safety. For example, suppose that the quality level of the yield strength of steel cables desired by the customer is 100 kg/cm^2 (kilograms per square centimeter). When designing such a cable, the parameters that influence the yield strength would be selected so as to satisfy this

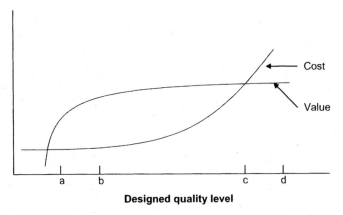

FIGURE 1-1   Cost and value as a function of designed quality.

requirement at least minimally. In practice, the product is typically overdesigned so that the desired conditions are exceeded. The choice of a safety factor (k) normally accomplishes this purpose. Thus, to design a product with a 25% stronger load characteristic over the specified weight, the value of k would equal 1.25, and the product will be designed for a yield strength of $100 \times 1.25 = 125 \, \text{kg/cm}^2$.

In most situations, the effect of an increase in the design quality level is to increase the cost at an exponential rate. The value of the product, however, increases at a decreasing rate, with the rate of increase approaching zero beyond a certain designed quality level. Figure 1-1 shows the impact of the design quality level on the cost and value of the product or service. Sometimes, it might be of interest to choose a design quality level b, which maximizes the differences between value and cost given that the minimal customer requirements a are met. This is done with the idea of maximizing the return on investment. It may be observed from Figure 1-1 that for a designed quality level c, the cost and value are equal. For any level above c (say, d) the cost exceeds the value. This information is important when a suitable design level is being chosen.

Quality of Conformance

Quality of conformance implies that *a manufactured product or a service rendered must meet the standards selected in the design phase.* With respect to the manufacturing sector, this phase is concerned with the degree to which quality is controlled from the procurement of raw material to the shipment of finished goods. It consists of the three broad areas of defect prevention, defect finding, and defect analysis and rectification. As the name suggests, defect prevention deals with the means to deter the occurrence of defects and is usually achieved using statistical process control techniques. Locating defects is conducted through inspection, testing, and statistical analysis of data from the process. Finally, the causes behind the presence of defects are investigated, and corrective actions are taken.

Figure 1-2 shows how quality of design, conformance, and performance influence the quality of a product or service. The quality of design has an impact on the quality of conformance. Obviously, one must be able to produce what was designed. Thus, if the design specification for the length of iron pins is 20 ± 0.2 mm (millimeters), the question that must be addressed is how to design the tools, equipment, and operations such that the manufactured product will meet the design specifications. If such a system of production can be

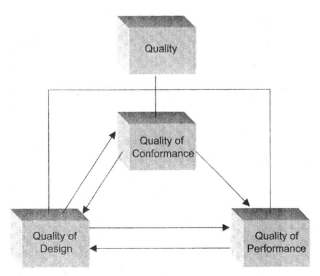

FIGURE 1-2   The three aspects of quality.

achieved, the conformance phase will be capable of meeting the stringent requirements of the design phase. On the other hand, if such a production system is not feasibly attained (e.g., if the process is only capable of producing pins with a specification of 20 ± 0.36 mm), the design phase is affected. This feedback suggests that the product be redesigned because the current design cannot be produced using the existing capability. Therefore, there should be a constant exchange of information between the design and manufacturing phases so that a feasible design can be achieved.

Quality of Performance

Quality of performance is concerned with *how well a product functions or service performs when put to use. It measures the degree to which the product or service satisfies the customer.* This is a function of both the quality of design and the quality of conformance. Remember that the final test of product or service acceptance always lies with the customers. Meeting or exceeding their expectations is the major goal. If a product does not function well enough to meet these expectations, or if a service does not live up to customer standards, adjustments need to be made in the design or conformance phase. This feedback from the performance to the design phase, as shown in Figure 1-2, may prompt a change in the design because the current design does not produce a product that performs adequately.

1-4   QUALITY CONTROL

Quality control may generally be defined as *a system that maintains a desired level of quality,* through feedback on product/service characteristics and implementation of remedial actions, in case of a deviation of such characteristics from a specified standard. This general area may be divided into three main subareas: off-line quality control, statistical process control, and acceptance sampling plans.

Off-Line Quality Control

Off-line quality control procedures deal with measures to select and choose controllable product and process parameters in such a way that the deviation between the product or process output and the standard will be minimized. Much of this task is accomplished through product and process design. The goal is to come up with a design within the constraints of resources and environmental parameters such that when production takes place, the output meets the standard. Thus, to the extent possible, the product and process parameters are set before production begins. Principles of experimental design and the Taguchi method, discussed in a later chapter, provide information on off-line process control procedures.

Statistical Process Control

Statistical process control involves comparing the output of a process or service with a standard and taking remedial actions in case of a discrepancy between the two. It also involves determining whether a process can produce a product that meets desired specifications or requirements.

For example, to control paperwork errors in an administrative department, information might be gathered daily on the number of errors. If the number observed exceeds a specified standard, then on identification of possible causes, action should be taken to reduce the number of errors. This may involve training the administrative staff, simplifying operations if the error is of an arithmetic nature, redesigning the form, or taking other appropriate measures.

Online statistical process control means that information is gathered about the product, process, or service while it is functional. When the output differs from a determined norm, corrective action is taken in that operational phase. It is preferable to take corrective action on a real-time basis for quality control problems. This approach attempts to bring the system to an acceptable state as soon as possible, thus minimizing either the number of unacceptable items produced or the time over which undesirable service is rendered. Chapters 6 to 9 cover the background and procedures of online statistical process control methods.

One question that may come to mind is: Shouldn't all procedures be controlled on an off-line basis? The answer is "yes," to the extent possible. The prevailing theme of quality control is that quality has to be designed into a product or service; it cannot be inspected into it. However, despite taking off-line quality control measures, there may be a need for online quality control, because variation in the manufacturing stage of a product or the delivery stage of a service is inevitable. Therefore, some rectifying measures are needed in this phase. Ideally, a combination of off-line and online quality control measures will lead to a desirable level of operation.

Acceptance Sampling Plans

Acceptance sampling plans involve inspection of a product or service. When 100% inspection of all items is not feasible, a decision has to be made as to how many items should be sampled or whether the batch should be sampled at all. The information obtained from the sample is used to decide whether to accept or reject the entire batch or lot. In the case of attributes, one parameter is the acceptable number of nonconforming items in the sample.

If the number of nonconforming items observed is less than or equal to this number, the batch is accepted. This is known as the *acceptance number*. In the case of variables, one parameter may be the proportion of items in the sample that are outside the specifications. This proportion would have to be less than or equal to a standard for the lot to be accepted. *A plan that determines the number of items to sample and the acceptance criteria of the lot, based on meeting certain stipulated conditions (such as the risk of rejecting a good lot or accepting a bad lot), is known as an acceptance sampling plan.*

Let's consider a case of attribute inspection where an item is classified as conforming or not conforming to a specified thickness of 12 ± 0.4 mm. Suppose that the items come in batches of 500 units. If an acceptance sampling plan with a sample size of 50 and an acceptance number of 3 is specified, the interpretation of the plan is as follows. Fifty items will be randomly selected by the inspector from the batch of 500 items. Each of the 50 items will then be inspected (say, with a go/no-go gage) and classified as conforming or not conforming. If the number of nonconforming items in the sample is 3 or less, the entire batch of 500 items is accepted. However, if the number of nonconforming items is greater than 3, the batch is rejected. Alternatively, the rejected batch may be screened; that is, each item is inspected and nonconforming ones are removed. Acceptance sampling plans for attributes and variables are discussed in Chapter 10.

1-5 QUALITY ASSURANCE

Quality is not just the responsibility of one person in the organization—this is the message. Everyone involved directly or indirectly in the production of an item or the performance of a service is responsible. Unfortunately, something that is viewed as everyone's responsibility can fall apart in the implementation phase and become no one's responsibility. This behavior can create an ineffective system where the quality assurances exist only on paper. Thus, what is needed is *a system that ensures that all procedures that have been designed and planned are followed.* This is precisely the role and purpose of the quality assurance function.

The objective of the quality assurance function is to have in place a formal system that continually surveys the effectiveness of the quality philosophy of the company. The quality assurance team thus audits the various departments and assists them in meeting their responsibilities for producing a quality product.

Quality assurance may be conducted, for example, at the product design level by surveying the procedures used in design. An audit may be carried out to determine the type of information that should be generated in the marketing department for use in designing the product. Is this information representative of the customer's requirements? If one of the customer's key needs in a food wrap is that it withstand a certain amount of force, is that information incorporated in the design? Do the data collected represent that information? How frequently are the data updated? Are the forms and procedures used to calculate the withstanding force adequate and proper? Are the measuring instruments calibrated and accurate? Does the design provide a safety margin? The answers to all of these questions and more will be sought by the quality assurance team. If discrepancies are found, the quality assurance team will advise the relevant department of the changes that should be adopted. This function acts as a watchdog over the entire system.

1-6 QUALITY CIRCLES AND QUALITY IMPROVEMENT TEAMS

A **quality circle** is typically *an informal group of people that consists of operators, supervisors, managers, and so on, who get together to improve ways to make a product or deliver a service.* The concept behind quality circles is that in most cases, the persons who are closest to an operation are in a better position to contribute ideas that will lead to an improvement in it. Thus, improvement-seeking ideas do not come only from managers but also from all other personnel who are involved in the particular activity. A quality circle tries to overcome barriers that may exist within the prevailing organizational structure so as to foster an open exchange of ideas.

A quality circle can be an effective productivity improvement tool because it generates new ideas and implements them. Key to its success is its participative style of management. The group members are actively involved in the decision-making process and therefore develop a positive attitude toward creating a better product or service. They identify with the idea of improvement and no longer feel that they are outsiders or that only management may dictate how things are done. Of course, whatever suggestions that a quality circle comes up with will be examined by management for feasibility. Thus, members of the management team must understand clearly the workings and advantages of the action proposed. Only then can they evaluate its feasibility objectively.

A **quality improvement team** is another means of identifying feasible solutions to quality control problems. Such teams are typically cross-functional in nature and involve people from various disciplines. It is not uncommon to have a quality improvement team with personnel from design and development, engineering, manufacturing, marketing, and servicing. A key advantage of such a team is that it promotes cross-disciplinary flow of information in real time as it solves the problem. When design changes are made, the feasibility of equipment and tools in meeting the new requirements must be analyzed. It is thus essential for information to flow between design, engineering, and manufacturing. Furthermore, the product must be analyzed from the perspective of meeting customer needs. Do the new design changes satisfy the unmet needs of customers? What are typical customer complaints regarding the product? Including personnel from marketing and servicing on these teams assists in answering these questions.

The formation and implementation of quality improvement teams is influenced by several factors. The first deals with selection of team members and its leader. Their knowledge and experience must be relevant to the problem being addressed. People from outside the operational and technical areas can also make meaningful contributions; the objective is to cover a broad base of areas that have an impact. Since the team leader has the primary responsibility for team facilitation and maintenance, he or she should be trained in accomplishing task concerns as well as people concerns, which deal with the needs and motivation of team members.

Team objectives should be clearly defined at the beginning of any quality improvement team project. These enable members to focus on the right problem. The team leader should prepare and distribute an agenda prior to each meeting. Assignments to individual members or subgroups must be clearly identified. Early in the process, the team leader should outline the approach, methods, and techniques to be used in addressing the problem. Team dynamics deals with interactions among members that promote creative thinking and is vital to the success of the project. The team leader plays an important role in creating this climate for creativity. He or she must remove barriers to idea generation and must encourage differing points of view and ideas. All team members should be encouraged to contribute their ideas or to build on others.

Regular feedback on the results and actions taken at meetings is important. It keeps the team on track, helps eliminate the personal bias of members, if any, and promotes group effort. Such reviews should ensure that all members have been assigned specific tasks; this should be documented in the minutes. Progress should be reviewed systematically, the objective being to come up with a set of action plans. This review is based on data collected from the process, which is analyzed through basic quality improvement tools (some of which are discussed in Chapters 3 and 5). Based on the results of the analysis, action plans can be proposed. In this way, team recommendations will not be based on intuition but on careful analysis.

1-7 CUSTOMER NEEDS AND MARKET SHARE

For the manufacturing or service sector, satisfying the customers—both internal and external—is fundamental to growth and improving market share. An important aspect of the *quality of design* phase deals with identification of customer needs and wants. These customer needs may be grouped into the three broad categories of *critical to quality, critical to delivery,* and *critical to cost.* Not all needs are of equal importance to the customer. Moreover, some are expressed while others are taken for granted.

Kano Model

Noriaki Kano, a Japanese consultant, developed a model relating design characteristics to customer satisfaction (Cohen 1995). Customer needs or expectations can be divided into three *prioritized* categories: *basic needs* (dissatisfiers); *performance needs* (satisfiers); and *excitement needs* (delighters). Basic needs are those that are taken for granted by the customer. Meeting these needs may not steeply increase customer satisfaction; but not meeting them will definitely cause dissatisfaction. For example, in a city public library, it is taken for granted that current editions of popular magazines will be available. Not having them will lead to dissatisfied consumers.

Performance needs are those that the consumer expects. Thus, the better these are met, the more satisfied the customer. Typically, customer satisfaction increases as a linear function of the degree to which such needs are met. Ease of checking out a book or video at a city library could be one such need. Excitement needs, also known as delighters, are those that surprise the customer unexpectedly. The consumer does not necessarily expect these and hence may not express them. So, when they are met, it increases customer satisfaction in an exponential manner. For example, if the city library offered free consultation on tax-form preparation, customers might be delighted beyond bounds.

Figure 1-3 shows the Kano model, relating the degree of meeting customer needs and customer satisfaction. Note the three curves associated with basic, performance, and excitement needs and their relative impact on increasing customer satisfaction. Basic and excitement needs are usually not identifiable from customer surveys. Satisfying basic needs may prevent customer loss but not necessarily promote growth. Survey data are typically used to address performance needs and the degree to which improvement in these needs is necessary in order to grow market share linearly, to a certain extent. Excitement needs, not generally expressed by consumers in surveys, demand a major source of attention for organizations seeking market share growth. These needs, if incorporated in the design phase, will distinguish the company from its competitors.

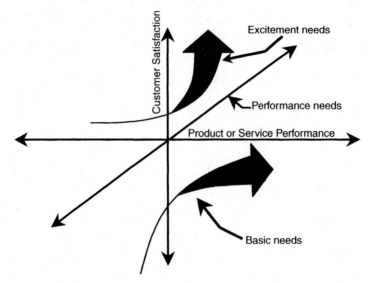

FIGURE 1-3 Kano model.

1-8 BENEFITS OF QUALITY CONTROL AND THE TOTAL QUALITY SYSTEM

The goal of most companies is to conduct business in such a manner that an acceptable rate of return is obtained by the shareholders. What must be considered in this setting is the short-term goal versus the long-term goal. If the goal is to show a certain rate of return this coming year, this may not be an appropriate strategy because the benefits of quality control may not be realized immediately. However, from a long-term perspective, a quality control system may lead to a rate of return that is not only better but is also sustainable.

One of the drawbacks of the manner in which many U.S. companies operate is that the output of managers is measured in short time frames. It is difficult for a manager to show a 5% increase in the rate of return, say, in the quarter after implementing a quality system. Top management may then doubt the benefits of quality control.

The advantages of a quality control system, however, become obvious in the long run. First and foremost is the improvement in the quality of products and services. Production improves because a well-defined structure for achieving production goals is present. Second, the system is continually evaluated and modified to meet the changing needs of the customer. Therefore, a mechanism exists for rapid modification of product or process design, manufacture, and service to meet customer requirements so that the company remains competitive. Third, a quality control system improves productivity, which is a goal of every organization. It reduces the production of scrap and rework, thereby increasing the number of usable products. Fourth, such a system reduces costs in the long run. The notion that improved productivity and cost reduction do not go hand in hand is a myth. On the contrary, this is precisely what a quality control system does achieve. With the production of few nonconforming items, total costs decrease, which may lead to a reduced selling price and thus increased competitiveness. Fifth, with improved productivity, the lead time for producing parts and subassemblies is reduced, which results in improved delivery dates. One again, quality control keeps customers satisfied. Meeting or exceeding their needs on a timely basis helps sustain a good relationship. Last, but not least, a quality control system maintains an "improvement" environment where everyone strives for improved quality and productivity. There is no end to this process—there is always room for improvement. A company that adopts this philosophy and uses a quality control system to help meet this objective is one that will stay competitive.

Total Quality System

Quality is everyone's responsibility. This means that comprehensive plans should be developed to show the precise responsibilities of the various units, procedures should be defined to check their conformance to the plans, and remedial measures should be suggested in the event of discrepancies between performance and standard. The quality assurance function, as defined earlier, monitors the system.

The **systems approach** to quality integrates the various functions and responsibilities of the various units and provides a mechanism to ensure that organizational goals are being met through the coordination of the goals of the individual units. The ISO, in conjunction with ANSI and ASQ, has developed standards ANSI/ISO/ASQ 9000–9004 (ASQ 2004) that describe quality systems.

In this book we focus on the analytical tools and techniques within the context of a total quality system. An overview of the chapter contents in the book follows. A foundation, along with appropriate terminology, is presented in this chapter. In Chapter 2 we introduce some quality philosophies developed by pioneers in the field. Further, similarities and differences between quality in the manufacturing and service sectors are delineated in this chapter. Quality management practices and their associated standards, developed by various organizations (ISO/ANSI/ASQ), which define acceptable norms, are the focus of Chapter 3, where the six sigma metric and methodology are discussed. Chapter 4 covers the fundamentals of statistical concepts and techniques used in quality control. In Chapter 5 we present some statistical techniques for quality analysis and improvement. The idea of process control through control charts, which is one of the primary quality control tools, is covered in Chapters 6 to 9. The fundamental principles of control charts are introduced in Chapter 6. Chapter 7 focuses on control charts for variables, while those for attributes are covered in Chapter 8. Statistical methods for determining whether a process is capable of producing items that conform to a standard are described in Chapter 9. These methods involve process capability analysis. The topics of acceptance sampling plans for attributes and variables are given in Chapter 10. Statistical methods dealing with life testing and reliability are covered in Chapter 11; these techniques concern the performance of a product over a period of time. Designing experiments for use in systematically analyzing and guiding process parameter settings is covered in Chapter 12. Some fundamental concepts of the Taguchi method of off-line quality control are also presented in Chapter 12. Since more than 70% of the gross national product comes from the service sector, techniques for monitoring quality that have been used primarily in the manufacturing sector are also warranted here. Such service-sector applications are integrated in the various chapters. Finally, computers play a fundamental role in quality control, and their use will expand even more in the years to come. The use of computer software (Minitab) for a variety of statistical techniques in quality control and improvement is integrated throughout.

1-9 QUALITY AND RELIABILITY

Reliability refers to the ability of a product to function effectively over a certain period of time. Reliability is related to the concept of quality of performance. Since the consumer has the ultimate say on the acceptability of a product or service, the better the performance over a given time frame, the higher the reliability and the greater the degree of customer satisfaction. Achieving desirable standards of reliability requires careful analysis in the product design phase. Analysis of data obtained on a timely basis during product performance keeps the design and production parameters updated so that the product may continue to perform in an acceptable manner. Reliability is built in through quality of design.

The product is often overdesigned so that it more than meets the performance requirements over a specified time frame. For example, consider the quality of a highway system where roads are expected to last a minimum time period under certain conditions of use. Conditions of use may include the rate at which the road system is used, the weight of vehicles, and such atmospheric conditions as the proportion of days that the temperature exceeds a certain value. Suppose that the performance specifications require the road system to last at least 20 years. In the design phase, to account for the variation in the uncontrollable parameters, the roads might be designed to last 25 years. This performance level may be achieved through properly selected materials and the thickness of the concrete and tar layers.

1-10   QUALITY IMPROVEMENT

Efforts to reduce both the variability of a process and the production of nonconforming items should be ongoing because quality improvement is a *never-ending* process. Whereas process control deals with identification and elimination of special causes (those for which an identifiable reason can be determined) that force a system to go out of control (e.g., tool wear, operator fatigue, poor raw materials), **quality improvement** relates to the detection and elimination of common causes. **Common causes** are inherent to the system and are always present. Their impact on the output may be uniform relative to that of special causes. An example of a common cause is the variability in a characteristic (say, a diameter) caused by the inherent capability of the particular equipment used (say, a milling machine). This means that all other factors held constant, the milling machine is unable to produce parts with exactly the same diameter. To reduce the inherent variability of that machine, an alternative might be to install a better or more sophisticated machine. **Special causes** are controllable mainly by the operator, but common causes need the attention of management. Therefore, quality improvement can take place only through the joint effort of the operator and management, with the emphasis primarily on the latter. For instance, a decision to replace the milling machine must be made by management. Another example could be the inherent variation in the time to process purchase orders. Once special causes have been eliminated, ways in which the average time or variability could be reduced could be through changes in the procedure/process, which requires management support. Eliminating or reducing the impact of some of the common causes results in improved process capability, as measured by less variation of the output.

Most quality control experts agree that common causes account for at least 90% of the quality problems in an organization. The late W. Edwards Deming, the noted authority on quality, strongly advocated this belief. He concluded that management alone is responsible for common-cause problems and, hence, only management can define and implement remedial actions for these problems. The operator has no control on nonconforming product or service in a majority of the instances. Therefore, if a company is interested in eliminating the root causes of such problems, management must initiate the problem-solving actions.

Quality improvement should be the objective of all companies and individuals. It improves the rate of return or profitability by increased productivity and by cost reduction.

It is consistent with the philosophy that a company should continually seek to expand its competitive edge. It supports the principle that no deviation from a standard is acceptable, which is akin to the principle of the loss function developed in the Taguchi methods (Taguchi 1986; Taguchi and Wu 1979). So even if the product is within the specification limits, an ongoing effort should be made to reduce its variability around the target value.

Let's say that the specifications for the weight of a package of sugar are 2.00 ± 0.02 kg. If the output from the process reveals that all packages weigh between 1.98 and 2.02 kg, the process is capable and all items will be acceptable. However, not all of the packages weigh exactly 2.00 kg, the target value; that is, there is some variability in the weights of the packages. The Taguchi philosophy states that any deviation from the target value of 2.00 kg is unacceptable with the loss being proportional to the deviation. Quality improvement is a logical result of this philosophy.

Quality function deployment techniques, which incorporate the needs and priorities of a customer in designing a product or service, are demonstrated in Chapter 3. Some methods for quality improvement are discussed in Chapter 5. These include such graphical techniques as Pareto analysis, histograms, and cause-and-effect or fishbone diagrams. Additional techniques discussed in Chapter 9 deal with process capability analysis. Quality improvement through design may also be achieved through experimental design techniques and the Taguchi method; these are discussed in Chapter 12.

1-11 PRODUCT AND SERVICE COSTING

In costing a product or service, the broad categories of direct and indirect costs come into play. **Direct costs**, such as direct labor and materials, are a function of the number of units of the manufactured product or the number of customers serviced. On the contrary, **indirect costs** do not change with each unit produced or each customer served, such as machine setup for the same product, depreciation of building, property taxes, and so on. Accounting methods that use a system that allocates indirect costs adequately to the particular product or service are highly desirable. This is true especially when multiple products are produced or types of services are performed. Indirect costs should be distributed to products or services based on cause-and-effect relations or actual use.

Traditional accounting methods can lead to misleading product/service costs where indirect costs are allocated based on direct labor or direct material. However, the actual use of the resource is not necessarily a function of the direct labor or direct material cost. In such cases, a better estimate of product costing is arrived at by using activities that measure the degree of use of the particular resource. This is known as *activity-based costing*. The implicit assumption in traditional financial/cost accounting methods is that indirect costs are a relatively small proportion of the unit cost. In the new century, as product/service options, product complexity, and volume continue to grow, the method of allocation of indirect costs becomes important, since use of indirect resources is not necessarily similar for all types of product/service.

Activity-Based Costing

Activities are tasks performed by a specialized group or department, say the purchasing unit in an organization, also known as an *activity* or *cost center*. The types of transactions that generate costs are identified as *cost-drivers*. For instance, the number of purchase orders processed is a cost-driver. Whether the purchase order is for one product item or 50, it uses the same amount of the purchasing department resource. Thus, allocation of the indirect costs associated with use of the purchasing department should incorporate the number of purchase orders, not the number of direct labor hours or direct material costs in making the product, as this represents a better picture of the resource use by the product/ service.

Cost-drivers are typically categorized into four groups. **Unit-level costs** comprise those activities that are associated with each product/service unit. Direct labor and material, machining, and assembly costs are examples. **Batch-level costs** are based on activities that are performed once for each lot or batch of products and are not influenced by the number of units in the batch. A machine setup for an operation for a batch, or invoicing for a batch, are examples. Hence, the number of batches will influence the allocation of costs. The next level of cost driver is the **product/service-level cost** which is based on the type of product/service. In engineering design, if two products are being made, the resource spent on each product design will be an example of such a cost. Finally, there is the **production/service-sustaining cost level** which incorporates activities that use all other resources necessary to maintain operations. Building depreciation, insurance, and property taxes are examples in this category. These do not depend on the number of product/service units, batches, or product/service lines.

To determine the product/service unit cost, the following guidelines are used in activity-based costing. The direct costs are the unit-level costs that can be assigned to a product/ service unit. Once the batch-level costs are identified, based on the number of batches used, the allocation to a product/service unit is computed by dividing by the number of units in a particular batch. Similarly, product/service level costs, once identified, based on the number of types of products/services, can be spread over all product/service units of a certain type to include toward computation of the unit cost. Finally, the production/service-sustaining costs, which cannot be linked directly to units, batches, or product/service lines, can be allocated in a sequential manner. First, they are assigned to product/service lines, then to batches, and eventually to product/service units. Another approach for this cost category is to allocate directly to units based on direct labor hours.

Thus, the essence of activity-based costing is based on the proper identification of cost-drivers, nonfinancial measures such as the number of purchase orders processed or the number of types of product made. Although the number of cost-drivers used might better identify product/service unit costs, the drawback lies in the cost of obtaining information. Further, some assumptions include that the unit batch cost is not dependent on the batch size or the type of product/service. In actuality, it could be that purchase order processing times, and thereby costs, vary based on the nature of the product/service. The benefits of activity-based costing are in decision making rather than decision control. It leads to better pricing and product/service mix decisions, especially in situations in multiproduct/service organizations. With product/service volumes that vary greatly between product/service types, it provides a better representation of the use of common resources in the organization.

Example 1-1   Two types of microchips (A and B) are being manufactured, with microchip B being slightly more complex. There are two sizes of microchip A, A1 and A2. For each type and each size, microchips are manufactured in batch sizes of 100,000. Table 1-1 shows the production volume and direct costs/batch, while Table 1-2 displays the overhead costs for the past year. Using the traditional costing method (unit-based costing) of allocating overhead rate based on direct labor costs, calculate the cost per batch of each type of microchip.

TABLE 1-1   Production Volume and Direct Costs per Batch of Microchips

	Microchip A		Microchip B
	A1	A2	
Number of batches produced	500	800	1200
Cost/batch			
Direct labor ($)	500	600	800
Direct material ($)	2500	3200	3800
Processing ($)	1500	2000	3000

TABLE 1-2   Overhead Cost of Plant

Category	Cost ($ millions)
Setup and testing	$2.20
Product-line costs	
Microchip A	5.50
Microchip B	9.30
Other plant costs	4.50
Total overhead costs	$21.50

TABLE 1-3   Overhead Rate Using Unit-Based Allocation

	Microchip			Total
	A1	A2	B	
Number of batches produced	500	800	1200	
Direct labor cost per batch ($)	500	600	800	
Total direct labor cost ($ millions)	0.25	0.48	0.96	1.69
Total overhead ($ millions)				21.50
Overhead rate (%)				1272.19

Calculate the cost per batch using the activity-based costing method and compare with the figures calculated previously.

Solution   The unit-based allocation scheme is demonstrated first. The total direct labor cost is used, based on which the overhead rate is determined by using the ratio of the total overhead cost to the total direct labor cost. This common overhead rate is then applied to each microchip type and size to determine the overhead cost assignment. Subsequently, a unit cost is calculated. Table 1-3 shows the computation of the common overhead rate. Costs per batch of each microchip type and size are shown in Table 1-4 using the computed overhead rate in Table 1-3.

Next, the activity-based cost allocation scheme is used. The unit cost for each of the three activity-based cost drivers is calculated as follows:

- *Batch-related costs*:   Setup and testing: $2.20 million ÷ 2500 batches = $880 per batch
- *Product line–related costs*:   Microchip A: $5.50 million ÷ 1300 = $4231 per batch
  Microchip B: $9.30 million ÷ 1200 = $7750 per batch

TABLE 1-4   Cost per Batch Using Unit-Based Allocation

Cost Component	Microchip		
	A1	A2	B
Direct labor	$500	$600	$800
Direct material	2,500	3,200	3,800
Processing	1,500	2,000	3,000
Overhead (1272.19% of direct labor)	6,361	7,633	10,178
Total cost per batch	$10,861	$13,433	$17,778
Cost per microchip	$0.1086	$0.1343	$0.1778

TABLE 1-5   Cost per Batch Using Activity-Based Allocation

Cost Component	Microchip		
	A1	A2	B
Direct labor	$500	$600	$800
Direct material	2,500	3,200	3,800
Processing	1,500	2,000	3,000
Overhead			
Batch-related	880	880	880
Product-line related	4,231	4,231	7,750
Production-sustaining (266.27% of direct labor)	1,331	1,598	2,130
Total cost per batch	$10,942	$12,509	$18,360
Cost per microchip	$0.1094	$0.1251	$0.1836

- *Production-sustaining costs*:   Overhead rate per direct labor dollar $= \$4.50$ million $\div \$1.69$ million $= 266.27\%$

An assumption made in batch-related costs, in this example, is that setup and testing costs are quite similar for both types of microchips. Hence, the total setup and testing cost is averaged out over the total batches of both types of microchips. Using these computed values, Table 1-5 shows the costs per batch for each type and size of microchip using the activity-based allocation method.

From Tables 1-4 and 1-5, differences are observed between the two methods of costing in the cost per batch, and the corresponding cost per microchip, for each type and size of microchip. Some conclusions that can be drawn are as follows. The unit-based (traditional) costing method tends to over-cost the high-volume items within a product type. Also, since more complex products require more product-line costs, the activity-based costing method will make proportional allocations. However, the unit-based costing method, using direct labor as a measure of allocating overhead costs, will under-cost more complex products.

1-12   QUALITY COSTS

The value of a quality system is reflected in its ability to satisfy the customer. In this context, quality costs reflect the achievement or nonachievement of meeting product or service requirements, as determined from the perspective of the customer. These requirements may include design specifications of a product, operating instructions, government regulations, timely delivery, marketing procedures, and servicing commitments, among others.

The various components of quality costs are designated based on product/service conformance or nonconformance. The achievement of requirements, identified by product or service conformance, consists of a cost component, identified as prevention costs, while nonconformance consists of the cost components of appraisal and failure costs (Campanella 1999). To summarize, quality costs may be interpreted as the difference between the actual cost and the reduced cost if products and services were all conforming. The four major categories of quality costs are discussed here.

Prevention Costs

Prevention costs are incurred in planning, implementing, and maintaining a quality system to prevent poor quality in products and services. They include salaries and developmental costs for product design, process and equipment design, process control techniques (through such means as control charts), information systems design, and all other costs associated with making the product right the first time. Also, costs associated with education and training are included in this category. Other such costs include those associated with defect cause and removal, process changes, and the cost of a quality audit. Prevention costs increase with the introduction of a quality system and, initially, may be a significant proportion of the total quality costs. However, the rate of increase slows with time. Even though prevention costs increase, they are more than justified by reductions in total quality costs due to reductions in internal and external failure costs.

Appraisal Costs

Appraisal costs are those associated with measuring, evaluating, or auditing products, components, purchased materials, or services to determine their degree of conformance to the specified standards. Such costs include dealing with the inspection and testing of incoming materials as well as product inspection and testing at various phases of manufacturing and at final acceptance. Other costs in this category include the cost of calibrating and maintaining measuring instruments and equipment and the cost of materials and products consumed in a destructive test or devalued by reliability tests. Appraisal costs typically occur during or after production but before the product is released to the customer. Hence, they are associated with managing the outcome, whereas prevention costs are associated with managing the intent or goal. Appraisal costs normally decline with time as more non-conformities are prevented from occurring.

Internal Failure Costs

Internal failure costs are incurred when products, components, materials, and services fail to meet quality requirements prior to the transfer of ownership to the customer. These costs would disappear if there were no nonconformities in the product or service. Internal failure costs include scrap and rework costs for the materials, labor, and overhead associated with production. The cost of correcting nonconforming units, as in rework, can include such additional manufacturing operations as regrinding the outside diameter of an oversized part. If the outside diameter were undersized, it would not be feasible to use it in the finished product, and the part would become scrap. The costs involved in determining the cause of failure or in reinspecting or retesting reworked products are other examples from this category. The cost of lost production time due to nonconformities must also be considered (e.g., if poor quality of raw materials requires retooling of equipment). Furthermore, *downgrading costs,* the revenue lost because a flawed product has to be sold at a lower price, constitutes another component. As a total quality system is implemented and becomes effective with time, internal failure costs will decline. Less scrap and rework will result as problems are prevented.

External Failure Costs

External failure costs are incurred when a product does not perform satisfactorily after ownership is transferred to the customer or services offered are nonconforming. If no non-conforming units were produced, this cost would vanish. Such costs include those due to customer complaints, which include the costs of investigation and adjustments, and those associated with receipt, handling, repair, and replacement of nonconforming products. Warranty charges (failure of a product within the warranty time) and product liability costs (costs or awards as an outcome of product liability litigation) also fall under this category. A reduction in external failure costs occurs when a quality control system is implemented successfully.

Hidden Failure Costs

The measurable components of failure costs include those associated with scrap, rework, or warranty, which are easily tracked by accounting systems. A significant segment of the failure costs are "hidden." These include management and engineering time associated with cause identification and determination of remedial actions associated with failures. Line downtime, the necessity to carry increased inventory, the decrease in available capacity, and orders lost due to poor quality are examples of costs not easily tracked by accounting systems. Hence, what is typically reported as failure costs is but a minute portion of the true failure costs.

Quality Costs Data Requirements

Quality costs should be monitored carefully. Because indirect costs are as important as such direct costs as raw material and labor, well-defined accounting procedures should be set up to determine realistic quality cost estimates. Consider the case where quality cost data cross departmental lines. This occurs, for example, when a quality control supervisor in a staff position identifies the reason for scrap or rework, and a machine operator conducts an extra operation to rework those items. Similarly, should rework or scrap inspire a change in the product design, the redesign time is assigned to quality costs.

Figure 1-4 shows the data requirements at various management levels. Data are collected for each product line or project and distributed to each level of management. The needs are somewhat different at each level. Top management may prefer a summary of the total quality costs, broken down into each of the four categories, at the division or plant level. On the other hand, line management or supervisors may want a summary of the direct costs, which include labor and material costs, as it relates to their area.

This means that if a change is made in product or process design, it is possible for one or more quality cost categories to be affected. The time spent by the design engineer would be allocated, costwise, to prevention cost. On the other hand, if the design calls for new

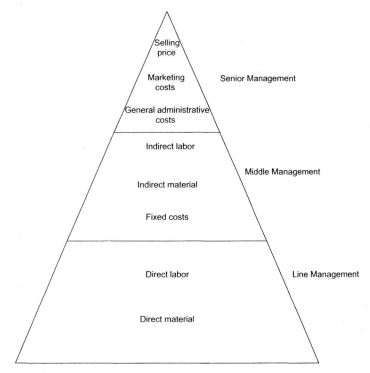

FIGURE 1-4   Quality costs data requirements at different management levels.

inspection equipment, that would be allocated to appraisal cost. Thus, a costwise breakdown into the four categories of prevention, appraisal, internal failure, and external failure is incorporated into the accounting system for the variety of functions performed by management and operators. Information from cost systems must be used to identify *root causes* associated with failures. Only when *remedial actions* are taken to prevent these from occurring will the true benefits of a cost system be reaped.

Process Cost Approach

In a conventional approach, the normal production costs associated with running a process (e. g., direct material, labor, and overhead costs) or providing a service (salaries and wages of personnel, shipment or delivery costs) may not be included in the quality costs of conformance (i.e., prevention). Costs of nonconformance typically include internal and external failure costs. It is possible that greater cost-saving opportunities might lie in reducing the cost of conformance. In this regard, the process-cost approach could be helpful by eliminating non-value-added activities or combining process steps, to reduce the cost of conformance. Such an approach may be effective where quality improvement efforts have reached maturation. It has the advantage of tackling costs associated with efficiency as well as quality and may reduce the "hidden" quality costs discussed previously.

Example 1-2   A company manufacturing a chemical compound has identified the following cost components (per kilogram): direct labor, $2.00; material, $5.00; energy, $0.50; overhead, 30% of direct labor and material. The process is operating at a yield rate of 96%. A quality improvement team, through analysis, has been able to increase yield to 97%. Find the cost per kilogram of product before and after this improvement. On further study, the team analyzed the process and was able to eliminate several non-value-added steps. Direct labor and material costs were reduced by 25%. Calculate the cost per kilogram of product after these process changes are made. By what percentage have total costs been reduced? What is the percentage increase in capacity?

Solution   The conformance costs include the process costs, comprising direct labor, material, energy, and overhead costs. Overhead costs are $2.10/kg, leading to total conformance costs of $9.60/kg. The total cost (conformance and nonconformance) is $10/kg (9.60/ 0.96), implying a nonconformance cost of $0.40/kg. By improving yield to 97%, the total cost/ kg is $9.90. Thus, the nonconformance cost has been reduced to $0.30/kg. On elimination of non-value-added steps in the process, direct labor and material costs are $5.25/kg. With overhead costs of $1.58 (5.25 × 0.3), the process costs are $7.33/kg. Since the yield rate is now 97%, the total cost of the product is $7.56/kg, implying a nonconformance cost of $0.23/kg. Total costs have been reduced by 24.4% [(10.00 − 7.56)/10.00]. Relative level in capacity 0.97/0.96 = 1.010, indicating a 1% increase in capacity. The reduction in total costs, by analyzing the process, has made a major impact.

1-13   MEASURING QUALITY COSTS

The magnitude of quality costs is important to management because such indices as return on investment are calculated from it. However, for comparing quality costs over time, magnitude may not be the measure to use because conditions often change from one quarter to the next. The number of units produced may change, which affects the direct costs of labor and materials, so the total cost in dollars may not be comparable. To alleviate this situation, a measurement base that accounts for labor hours, manufacturing costs, sales dollars, or units produced could be used to produce an index. These ideas are discussed here.

1. *Labor-based index.* One commonly used index is the quality costs per direct-labor hour. The information required to compute this index is readily available, since the accounting department collects direct-labor data. This index should be used for short periods because over extended periods, the impact of automation on direct-labor hours may be significant. Another index lists quality costs per direct-labor dollar, thus eliminating the effect of inflation. This index may be most useful for line and middle management.

2. *Cost-based index.* This index is based on calculating the quality costs per dollar of manufacturing costs. Direct-labor, material, and overhead costs make up manufacturing costs, and the relevant information is readily available from accounting. This index is more stable than the labor base index because it is not significantly affected by price fluctuations or changes in the level of automation. For middle management, this might be an index of importance.

3. *Sales-based index.* For top management, quality costs per sales dollar may be an attractive index. It is not a good measure for short-term analysis, but for strategic decisions, top management focuses on long-term outlook. Sales lag behind production and are subject to seasonal variations (e.g., increased sales of toys during Christmas). These variations have an impact in the short run. However, they smooth out over longer periods of time. Furthermore, changes in selling price also affect this index.

4. *Unit-based index.* This index calculates the quality costs per unit of production. If the output of different production lines is similar, this index is valid. Otherwise, if a company produces a variety of products, the product lines would have to be weighted and a standardized product measure computed. For an organization producing refrigerators, washers, dryers, and electric ranges, for example, it may be difficult to calculate the weights based on a standard product. For example, if 1 electric range is the standard unit, is a refrigerator 1.5 standard units of a product and a washer 0.9 standard unit? The other indexes should be used in such cases.

For all of these indexes, a change in the denominator causes the value of the index to change, even if the quality costs do not change. If the cost of direct labor decreases, which may happen because of improvement in productivity, the labor-based index increases. Such increases should be interpreted cautiously because they can be misconstrued as increased quality costs.

A sample monthly quality cost report is shown in Table 1-6. Monthly costs are depicted for individual elements in each of the major quality cost categories of prevention, appraisal,

TABLE 1-6   Sample Monthly Quality Cost Report

Cost Categories	Amount	Percentage of Total
Prevention costs		
Quality planning	15,000	
Quality control engineering	30,000	
Employee training	10,000	
Total prevention costs	55,000	26.31
Appraisal costs		
Inspection	6,000	
Calibration and maintenance of test equipment	3,000	
Test	2,000	
Vendor control	4,000	
Product audit	5,000	
Total appraisal costs	20,000	9.57
Internal failure costs		
Retest and troubleshooting	10,000	
Rework	30,000	
Downgrading expense	3,000	
Scrap	16,000	
Total of internal failure costs	59,000	28.23
External failure costs		
Manufacturing failures	15,000	
Engineering failures	20,000	
Warranty charges	40,000	
Total external failure costs	75,000	35.89
Total quality cost	209,000	

Bases		Ratios	
Direct labor	800,000	Internal failure to labor	7.375%
Manufacturing cost	2,000,000	Internal failure to manufacturing	2.950%
Sales	3,800,000	Total quality costs to sales	5.500%

internal failure, and external failure. Observe that external failure and internal failure are the top two cost categories, comprising 35.89 and 28.23%, respectively, of the total cost. Comparisons of costs with different bases are also found in Table 1-6. Data on direct-labor cost, manufacturing cost that includes direct-labor, material, and overhead costs, and sales are shown in the table. Cost of goods sold includes manufacturing costs plus selling costs. Internal failure costs, for example, are 7.375% of direct-labor costs and 2.950% of manufacturing costs. Total quality costs are 5.5% of sales. From the information in the table, it seems that management needs to look into measures that will reduce internal and external failure costs, which dominate total quality costs. Perhaps better planning and design, and coordination with manufacturing, will reduce the external failure costs.

Impact of Quality Improvement on Quality Costs

The traditional notion of determining the optimal level of quality based on minimizing total quality costs is based on a static concept. It also does not incorporate the impact of quality improvement, an ideology that is an integral fabric of organizations, on the various quality cost functions. In fact, improving the level of quality of conformance and decreasing quality costs are not conflicting objectives. Rather, striving for improvements in both is complementary in nature and hence, can be achieved.

Minimization of total quality costs to determine the optimal operational level of quality using a traditional static concept is shown in Figure 1-5. In this case, prevention costs increase at an exponential rate with an improvement in the level of quality. Appraisal costs, however, may not increase rapidly with the level of quality. The combined prevention and appraisal cost function is dominated by the prevention costs, leading to the shape of the function in Figure 1-5. On the contrary, as the level of quality improves, a decline in the internal and external failure costs takes place, demonstrated by the nonlinear decay function. The total quality cost function, the sum of the prevention and appraisal costs and the internal failure and external failure costs, is also shown in Figure 1-5. The minimization of the total quality cost function leads to an optimal quality level (q_0).

We now discuss the more appropriate dynamic concept of analyzing quality costs and the manner in which the analysis is affected by the continuous quality improvement philosophy.

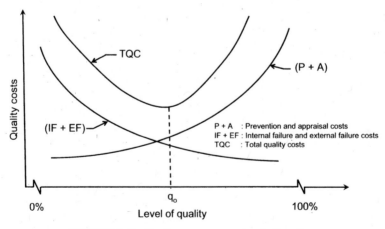

FIGURE 1-5   Quality costs versus level of quality.

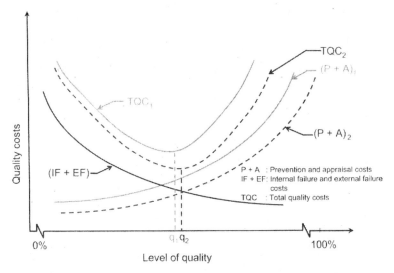

FIGURE 1-6   Dynamic concept of the impact of quality improvement.

First, with continuous improvement, not only is there a reduction in the unit cost of the product or service, but also a change in the shape of the prevention and appraisal cost function. Usually, the rate of increase of this function with the level of quality will be smaller than in the original situation. Ignoring, for the present, the other impacts of quality improvement, Figure 1-6 shows the shifted prevention and appraisal cost function. Note that the optimum level of quality desired improves (from q_1 to q_2). Rationalizing along these lines, the target level of quality to strive for, in the long run, should be total conformance.

Improvements in technology and advances in knowledge will initially affect the prevention and appraisal cost function, shifting it to the right with a reduction in slope. Although such advancements start out in *incremental steps* (i.e., the *Kaizen* concept of continuous improvement), after the achievement of a certain quality level, management must focus on *technological breakthroughs* to further improve quality. Such major innovations may lead to a reduction in the rate of change in the level of the prevention and appraisal cost function and consequently, a *change in the slope* of the prevention and appraisal cost function. The shape of the prevention and appraisal cost function changes from concave to convex after a certain level of quality (inflection point). Due to such a change, the shape of the total quality cost function will also change and will show a decreasing trend with the level of quality. Figure 1-7 demonstrates this impact. The target level of quality is 100% conformance.

Let us now discuss some of the other effects of continuous quality improvement that occur in a dynamic situation. Improvement in the performance of a product or service leads to improved customer satisfaction. As a result, market share improves. In a static situation, this increase in market share is not incorporated. Increased market share results in better overall company performance and return to stakeholders, necessitating a change to improved quality levels. An indirect impact of improved customer satisfaction is the ability of the manufacturer or service provider to charge a higher unit price, which leads to further improvement in profitability. In a static situation, the provider does not have the ability to increase unit price through quality improvement. To summarize, if we denote profitability as being proportional to the product of the market share and the difference between unit price and unit cost, quality

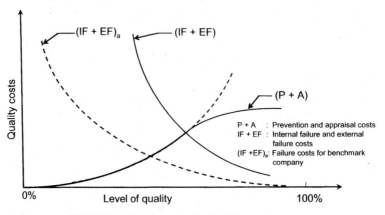

FIGURE 1-7   Target level of 100% conformance.

improvement affects all three components on a dynamic basis. It increases market share, decreases unit cost, and may also lead to increases in unit prices, thereby affecting profitability from three angles.

The focus of organizations should be on reduction of failure costs, internal and external. External failure is damaging. It leads to customer dissatisfaction, brand switching, and market share reduction. Although such opportunity costs are difficult to measure, they are significant. With an improvement in the level of quality, all other factors being constant, total failure costs should decrease. However, external failure costs are also influenced by the relative position of the organization relative to its competitors' offering. Thus, if a company's product lags significantly behind that of the benchmark or leader, the customer is likely not to be satisfied with the product, leading to increased external failure costs. Assuming that external failure costs are the more dominant of the two failure costs, Figure 1-7 also shows the total failure cost curve for the benchmark company in the industry. Whereas the prevention and appraisal cost function is influenced by actions and policies adopted by the company, the failure cost function is affected as well by the company and its competitors and by customer preferences. This demonstrates that managing external failure costs is much more encompassing. Even if the company maintains, internally, the same level of quality, external failure costs may go up, given the competitive environment of business. Further, customer preferences are dynamic in nature. It is not sufficient to improve only the manufacture of a chosen product. Keeping up with the needs and expectations of the customer is imperative.

1-14 MANAGEMENT OF QUALITY

Depending on the nature of the business (i.e., manufacturing, assembly, or service; range of product or service offerings; and degree of outsourcing), the management function of quality may employ appropriate models. Regardless, meeting and exceeding customer needs must be the central theme in all these models.

The first model describes the concept of *supply chain management*. Here, companies link to form partnerships with external organizations in order to leverage their strategic positioning as well as to improve operational efficiency. Consider Figure 1-8, which demonstrates a manufacturing or assembly situation conducted by the original equipment

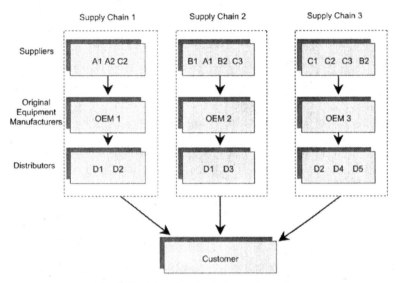

FIGURE 1-8   Supply chain configuration.

manufacturer (OEM). Based on the *core competencies* of the OEM, the OEM selects suppliers that can address their non-core competencies. Components or subassemblies could be obtained from suppliers which are not necessarily unique to a given OEM. The same supplier may serve multiple OEMs. Similarly, an OEM may market its products through one or more distributors, who also may not necessarily be unique to an OEM. The customer buys the products from the distributors. Thus, we have a situation in which supply chains rather than OEMs compete with each other. Quality and cost associated with the product are influenced by that of the suppliers, OEMs, and distributors considered collectively.

A second model describes the situation where the same suppliers, labeled in a tiered fashion that follows the process of assembly, feed all the OEMs. In this case, the OEMs are usually limited in number. Consider the automobile industry, where there are only a handful of OEMs: for example, Ford, General Motors, DaimlerChrysler, Toyota, Honda, and Hyundai. Since the same types of components are needed for each OEM, a tier 1 supplier producing a thermostatic control system to regulate engine temperature could conceivably supply all the OEMs. Similarly, at the tier 2 level, the components to produce the thermostatic control system could be manufactured by dedicated suppliers that focus on making only certain parts. Hence, parts produced by suppliers A and B in tier 2 are used to make a component or subassembly in tier 1. Information on individual parts and/or subassemblies could be monitored through a central "infomediary". The OEMs would draw parts and components using such information. As before, the customer buys from the distributors. Figure 1-9 shows such a tiered supply chain structure.

In considering the OEM as the organizational unit that attempts to maximize the performance of the associated supply chain, which involves its suppliers, distributors, customer representatives, and employees, an enterprise-wide concept could be incorporated. Data would be collected from the various facets of the organization in order to develop and monitor measures of quality and costs. Figure 1-10 shows the information needs for an enterprise-wide system. The quality of the product and/or service will be influenced by all of the contributing units, making management's task quite encompassing.

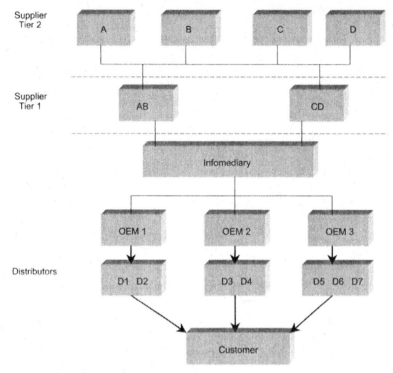

FIGURE 1-9   Tiered supply chain.

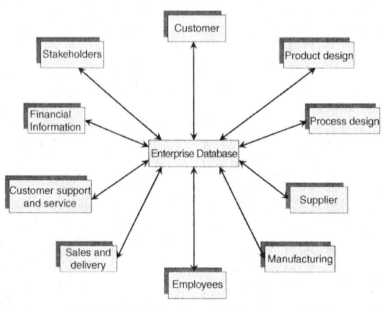

FIGURE 1-10   Enterprise-wide information needs.

1-15 QUALITY AND PRODUCTIVITY

A misconception that has existed among businesses (and is hopefully in the process of being debunked) is the notion that quality decreases productivity. On the contrary, the relationship between the two is positive: Quality *improves* productivity. Making a product right the first time lowers total costs and improves productivity. More time is available to produce defect-free output because items do not have to be reworked and extra items to replace scrap do not have to be produced. In fact, doing it right the first time increases the available capacity of the entire production line. As waste is reduced, valuable resources—people, equipment, material, time, and effort—can be utilized for added production of defect-free goods or services. The competitive position of the company is enhanced in the long run, with a concomitant improvement in profits.

Effect on Cost

As discussed previously, quality costs can be grouped into the categories of prevention, appraisal, internal failure, and external failure. Improved productivity may affect each of these costs differently.

1. *Prevention and appraisal costs.* With initial improvements in productivity, it is possible that prevention and appraisal costs will increase. As adequate process control procedures are installed, they contribute to prevention and appraisal costs. Further-more, process improvement procedures may also increase costs in these two categories. These are thus called the *costs of conformance* to quality requirements. With time, a reduction in appraisal costs is usually observed. As process quality improves, it leads to efficient and simplified operations. This may yield further improvements in productivity.

2. *Internal and external failure costs.* A major impact of improved quality is a reduction in internal and external failure costs. In the long run, decreasing costs in these two categories usually offset the increase in prevention and appraisal costs. The total cost of quality thus decreases. Moreover, as less scrap and rework is produced, more time is available for productive output. The company's profitability increases. As external failures are reduced, customer satisfaction improves. Not only does this emphasis on quality reduce the tangible costs in this category (such as product warranty costs and liability suits), it also significantly affects intangible costs of customer dissatisfaction. Figure 1-11 shows how improved quality leads to reduced costs, improved productivity, increased customer satisfaction; and eventually, increased profits through improved competitive positioning.

As noted previously, management must focus on long-term profits rather than short-term gain. A reason cited frequently for not adopting a total quality system is management's emphasis on short-term profits. As is well known, short-term profits can be enhanced by postponing much-needed investment in process improvement equipment and methods, by reducing research and development, and/or by delaying preventive maintenance. These actions eventually hurt competitiveness and profitability.

Effect on Market

An improvement in quality can lead to increased market shares, improved competitive position, and increased profitability.

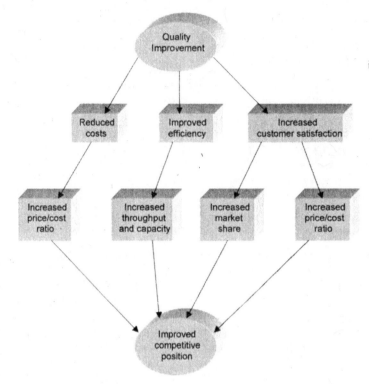

FIGURE 1-11   Impact of quality on competitive position.

1. *Market share.* With a reduction in external failure costs and improved performance of a product in its functional phase, a company is in a position to raise the satisfaction level of its customers, many of whom return to buy the product again. Satisfied customers spread the word about good quality, which leads to additional customers. Market share goes up as the quality level goes up.

2. *Competitive position.* All organizations want to stay competitive and to improve their market position, but simply improving quality or productivity may not be sufficient since competitors are doing the same. Organizations must monitor their relative position within the industry as well as the perception of customers. Efforts to improve quality are crucial in attaining these goals. Through process control and improvement and efficient resource utilization (reduced production of scrap and network), a firm can minimize its costs. So even if the selling price remains fixed, an improved price/ cost ratio is achieved. Alternatively, as quality improves, the firm may be able to charge a higher price for its product, although customer satisfaction and expectations ultimately determine price. In any event, an improved competitive position paves the way for increased profitability.

Example 1-3   Three independent operations are performed sequentially in the manufacture of a product. The first-pass yields (proportion conforming) for each operation are given by $p_1 = 0.90$, $p_2 = 0.95$, and $p_3 = 0.80$, respectively, as shown in Figure 1-12. The unit production costs for each operation are $u_1 = \$5$, $u_2 = \$10$, and $u_3 = \$15$, respectively.

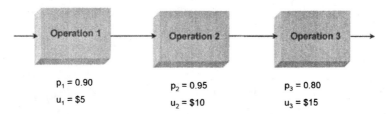

$p_1 = 0.90$ $p_2 = 0.95$ $p_3 = 0.80$
$u_1 = \$5$ $u_2 = \$10$ $u_3 = \$15$

FIGURE 1-12   Operations sequence in manufacturing.

(a) What is the unit cost *per conforming product?*

Solution   The first-pass yield (proportion conforming) at the completion of all three operations $= (0.90)(0.95)(0.80) = 0.684$. The total production cost of all three operations $30. Thus, if 1000 parts are manufactured at a cost of $30 each, only 684 of them are conforming. Thus, the unit cost per conforming product $= \$30/0.684 = \$43.86$. This is about a 46% increase over the unit production costs. As the level of quality is improved in the operations, the unit cost per conforming product may be reduced, its limiting value being $30, at which point the conformance rate for each operation is at 100%.

(b) Suppose, through quality improvement effort, the first-pass yield for each operation is improved to the following levels: $p_1 = 0.94$, $p_2 = 0.96$, $p_3 = 0.88$. Relative to part (a), determine how much improvement in *capacity* has taken place.

Solution   The first-pass yield at completion of all three operations now $= (0.94)(0.96)(0.88) = 0.794$. Relative level in capacity $= 0.794/0.684 = 1.161$, compared to the previous operation level, indicating an increase in available production capacity of 16.1%.

(c) Management is contemplating a 100% inspection process after either operation 1 or 2. Assume that the inspection process is completely reliable [i.e., all units are identified correctly (conforming or not)]. Unit inspection costs after operations 1 and 2 are $0.10 and $0.20, respectively. Nonconforming parts are not forwarded to subsequent operations. Find the unit cost per conforming product for each plan for the improved process.

Solution
Plan 1: Inspection only following operation 1   In this plan, all units following operation 1 will undergo inspection. However, only the conforming parts will be forwarded to operation 2. Suppose that 1000 parts are input to operation 1. The total production and inspection costs for all the operations are as follows: $(5 + 0.10)$ $(1000) + (0.94)(1000)(10 + 15) = \$28,600$. The number of conforming product units $= (1000)(0.94)(0.96)(0.88) = 794.112$. Hence, the unit cost per conforming product $= 28,600/794.112 = \$36.02$.

Plan 2: Inspection only following operation 2   Here, all units following operation will be forwarded to operation 2. After inspection on completion of operation 2, only the conforming units will be sent to operation 3. So the total production and inspection costs for all operations will be $(1000)(5) + (1000)(10 + 0.20) + (0.94)(0.96)(1000)(15) = \$28,736$. This leads to a unit cost per conforming product $= \$28,736/794.112 = \$36.19$. Management should recommend plan 1 if unit cost per conforming product is used as the selection criterion.

1-16   TOTAL QUALITY ENVIRONMENTAL MANAGEMENT

In recent years we have witnessed the emergence of many national and regional standards in the environmental management field. Some companies, of course, have long felt a social responsibility to operate and maintain safe and adequate environmental conditions, regardless of whether outside standards required it. Xerox Corporation is an example of a large corporation that takes its social obligations toward the environment seriously. The company has undertaken a major effort to reduce pollution, waste, and energy consumption. The quality culture is reflected in Xerox's protection of the environment; their motto is to reuse, remanufacture, and recycle. Company goals are aimed at creating waste-free products in waste-free factories using a "Design for the Environment" program. To support their environmental management program, Xerox uses only recyclable and recycled thermoplastics and metals.

With the concern for protection of the environment that is mandated in regional and national standards, standards need to be developed in environmental management tools and systems. British Standards Institute's BSI 7750 standards on environmental management is one such example; the European Union's (EU) eco-label and Eco-Management and Auditing Scheme (EMAS) are other examples. Both of these rely on consensus standards for their operational effectiveness. Similarly, in the United States, technical environmental standards under the sponsorship of the American Society for Testing and Materials (now ASTM international) have been published that address the testing and monitoring associated with emission and effluent pollution controls.

The International Organization for Standardization (ISO) based in Geneva, Switzerland, has long taken the lead in providing quality management standards. Its ISO 9000 standards have become a benchmark for quality management practices. U.S. companies have adopted ISO 9000 standards and have found this particularly beneficial in doing business with or trading in the European Union. Technical Committee (TC) 207 of the ISO has developed standards in the field of environmental management tools and systems; their document, ISO 14000: *An International Environmental Management Standard*, deals with developing management systems for day-to-day operations that have an impact on the environment (Marcus and Willig 1997; Parry 2000; Welch 1998). ISO 14000 consists of six standards. Three of these are in the category of organizational evaluation, focusing on environmental and business management systems: the Environmental Management System (ISO 14001) Environmental Performance Evaluation (ISO 14031), and Environmental Auditing (ISO 14010) standards. The other three explore the product development process and consist of Life Cycle Assessment (ISO 14040), Environmental Labeling (ISO 14020), and Product Standards (ISO 14060).

Environmental management began as a regulation-based and compliance-driven system. It has subsequently evolved into a voluntary environmental stewardship process whereby companies have undertaken a continuous improvement philosophy to set goals that go beyond the protection levels required by regulations. The ISO 14000 standards promote this philosophy with the objective of developing uniform environmental management standards that do not create unnecessary trade barriers. These standards are therefore not product standards. Also, they do not specify performance or pollutant/effluent levels. Specifically excluded from the standards are test methods for pollutants and setting limit values for pollutants or effluents.

Environmental management systems and environmental auditing span a variety of issues, including top management commitment to continuous improvement, compliance and pollution prevention, creating and implementing environmental policies, setting appropriate targets and achieving them, integrating environmental considerations in operating procedures, training employees on their environmental obligations, and conducting audits of the environmental management system.

One of the major corporations taking a lead in adoption of environmental management systems is the Ford Motor Company. Not only are all its plants around the world certified in ISO 14001, they have achieved significant improvements in air pollution, utilization of energy resources, recycling, waste disposal, and water treatment. Major recycling efforts have kept solid waste out of landfills. Ford has promoted expansion of environmental consciousness by providing monetary incentives to suppliers who assist in their waste reduction effort.

Several benefits will accrue from the adoption of environmental management system standards. First and foremost is the worldwide focus on environmental management that the standards will help to achieve. This promotes a change in the corporate culture. At the commercial level, ISO 14000 will have an impact on creating uniformity in national rules and regulations, labels, and methods. They will minimize trade barriers and promote a policy that is consistent. Not only will the standard help in maintaining regulatory compliance, it will help create a structure for moving beyond compliance. Management commitment and the creation of a system that reflects the goal to maintain a self-imposed higher standard will pave the way for continuous improvement in environmental management.

1-17 PROFILE OF A COMPANY: THE BAMA COMPANIES, INC.[*]

The Bama Companies, Inc. is a privately held corporation that began in 1927 and has grown into a leading developer and manufacturer of frozen ready-to-use food products served worldwide by quick service and casual dining restaurant chains such as McDonald's and Pizza Hut. From four production facilities in Tulsa, Oklahoma and two in Beijing, China, Bama's 1100 employees generate over $200 million a year in revenues. The company's three main product categories—handheld pies, biscuits, and pizza crust—account for 92% of revenues. The company is a 2004 winner of the Malcolm Baldrige National Quality Award in the manufacturing category (NIST 2005).

Company History and Vision

Whereas overall sales in the frozen baked goods industry have remained flat since 1999, Bama's sales have grown 72%. At the Bejing plant, production leaped from 600,000 pies in 1993 to 90 million in 2004. In an industry dominated by companies several times its own size, Bama's agility, its unique approach to product innovation, and its System View pricing strategy (it has not raised prices for its handheld pies and biscuits since 1996) give it tremendous leverage in the marketplace.

The company is rooted in its original guiding principles; keep your eyes on quality and remember that people make a company. Yet the way that Bama applies these principles in today's competitive business environment is anything but traditional. The company's stated vision is to "Create and Deliver Loyalty, Prosperity and Fun for All, While Becoming a

* Adapted from the Baldrige National Quality Program, National Institute of Standards and Technology, U.S. Department of Commerce, Gaithersburg, MD, 2005, www.nist.gov.

Billion Dollar Company." Bama sees itself and its mission as "People Helping People Be Successful."

Eyes on Quality

In its endless quest for improvement, Bama uses a battery of advanced strategies and tools, including the Bama Quality Management System, based on the quality improvement philosophies of W. Edwards Deming and the company's own performance excellence model. The Bama Excellence System provides a framework for all decision making. A Principle Centered Bama Culture provides a context for creating and measuring excellence. Using six sigma methodologies since 2000, Bama has dramatically improved processes throughout the company. Total savings from six sigma improvements is over $17 million since 2001.

The Future Looks Bright

In 1999, Bama utilized a strategic planning process called Prometheus to develop a company Future Picture, a high-level view of the company as it wants to be in 2010. The future that Bama envisions includes billion-dollar sales, recognition of the company's world class quality, being first-choice supplier in all its target markets, and providing employees and other stakeholders with unparalleled personal and financial opportunities. To help achieve these goals but maintain its small company culture, the company focuses on five strategic outcomes: (1). people—create and deliver loyalty, prosperity, and fun; (2). learning and innovation; (3). continuous improvement; (4). being the customer's first choice; and (5). value-added growth.

Bama uses it centers of gravity (short-term action plans) and a balanced scorecard to assess progress toward meeting these outcomes. The plans and scorecard support the company's decision-making process at all levels and are posted throughout its facilities, allowing all employees to see at a glance how their unit is performing against goals. The senior management team reviews the information at weekly and monthly meetings.

Building long-term relationships with suppliers and customers also helps Bama stay at the top of its game. Most of its key suppliers have been partners for 10 years or more, two have worked with the Bama Companies for three decades. Relationships with customers are just as enduring. The McDonald's system has been a Bama customer for 37 years and Pizza Hut for 11 years. Through these long-term relationships, Bama understands its customers, their customers' markets, and what their customers need to succeed. The company has tailored its services to meet customer requirements in critical areas such as assured supply, precision manufacturing, and value pricing. Since 2001, Bama has achieved 98% on-time delivery of products to customers, with 99% of orders filled completely on the initial shipment. Customer satisfaction for the company's major national accounts has increased from 75% in 2001 to 100% in 2004, considerably higher than the food manufacturing benchmark of 85%.

Innovation and Quality Improvement

Bama's manufacturing capabilities and customer knowledge have positioned the company not just as a manufacturer and supplier, but also as a designer of innovative food products. Developing new and innovative products is a major reason for Bama's growing market share and sales growth rate and is also helping its customers increase their own market share. In recognition of the innovation Bama brings, Bama has received Pizza Hut's Innovator of the Year award in three consecutive years: 2001, 2002, and 2003.

To manage its innovation initiatives, the company developed a business opportunity management process to coordinate the activities required to get a product from the idea stage to market. As a result, from 2000 to 2004, sales from new and innovative products have grown from less than 0.5% of total sales to almost 25%; and sales per employee grew from $175,000 to $205,000. This exceeds the 2003 *Industry Week* benchmark by $40,000.

People Make the Company

Bama is committed to the success of everyone associated with its business: customers, employees, and the community. The company's *people assurance system* ensures that each employee is well trained, fully informed, and empowered, and centers on helping all employees develop their potential and achieve personal success. Bama encourages employees to seek a college education by providing tuition reimbursement.

Satisfaction and loyalty of their people is key, and the company shares its success with employees when certain financial measures are met. Since 2001, profit-sharing payments have averaged around $3000 per year for each employee. A promote-from-within philosophy offers every qualified employee opportunities for advancement and the opportunity to be considered for job openings. This commitment to employees is paying off. Bama's 14% employee turnover rate is well below the average rate of 20% in the Tulsa area.

Bama also is committed to its community. A full-time leader of community development directs all of Bama's charitable and volunteer efforts, and a volunteer coordinator matches community needs with corporate resources. Employees are given paid time off to volunteer to work on corporate-sponsored projects. The number of hours that Bama employees donated to organizations such as Meals on Wheels, Habitat for Humanity, domestic violence intervention services, emergency infant services, and others increased from 500 in 2000 to nearly 7000 in 2004. Bama is the third largest contributor to the Tulsa Area United Way (Manufacturing Division), contributing $150,000 in 2004 alone. In addition, the company contributes annually an average of 6% of its pretax income (over $2.6 million since 2000) to local organizations that provide essential social, educational, cultural, and health services.

SUMMARY

In this chapter we examine the detailed framework on which the concept of the total quality system is based. We introduce, some of the basic terminology and provide an overview of the design, conformance, and implementation phase of the quality concept. We trace the evolution of quality control and present specifics on the benefits of quality control, who is responsible for it, and how it is to be adopted. The importance of the various types of needs of the customer is discussed with emphasis on those that focus on increasing market share. Such concepts apply to the manufacturing and service sectors. In costing products and services, the concept of activity-based costing is demonstrated. The subject of quality costs is explored thoroughly and the trade-offs that take place among the cost categories upon successful implementation of a total quality system are presented. A critical consideration in the entire scheme is management of the quality function. With the quality of the end product or service being influenced by entities such as outside suppliers, neither monitoring of the quality function nor of informational needs is restricted to within the organization. An important outcome of improvement in quality is an increase in productivity, capacity, and market share, along with a decrease in costs. All of this leads to improved profitability.

KEY TERMS

acceptance sampling plans
attributes
cause-and-effect diagram
causes
  common or chance
  special or assignable
costing
  activity-based
costs
  batch-level
  product/service-level
  production/service-sustaining
  process
customer
  needs
  satisfaction
defect
fishbone diagram
inspection
Ishikawa diagram
Kano model
management of quality
market share
nonconforming unit
nonconformity
off-line quality control
product design
productivity
quality
  quality of conformance
  quality of design

quality of performance
responsibility for quality
quality assurance
quality characteristic
quality circles
quality control benefits
quality cost measurements
  cost-based index
  labor-based index
  sales-based index
  unit-based index
quality costs
  appraisal costs
  external failure costs
  hidden failure costs
  prevention costs
  internal failure costs
quality improvement
reliability
specification
  specification limits
standard
statistical process control
  online statistical process control
supply chain management
systems approach
total quality environmental management
total quality system
variables
zero-defects program

EXERCISES

Discussion Questions

1-1 Consider the following organizations. How would you define quality in each context? Specify attributes/variables that may measure quality. How do you integrate these measures? Discuss the ease or difficulties associated with obtaining values for the various measures.

(a) Call center for a company that sells computers

(b) Emergency services (i.e., ambulances) for a city or municipality

(c) Company making semiconductor chips

(d) Hospital

(e) Company that delivers mail/packages on a rapid basis

(f) Department store

(g) Bank

(h) Hydroelectric power plant

1-2 The senior management in an urban bank is committed to improving its services. Discuss the specifics of quality of design, conformance, and performance in this context. Elaborate on possible basic needs, performance needs, and excitement needs of the customer.

1-3 A travel agency is attempting to enter a market where several competitors currently exist. What are the various customer needs that they should address? How will quality be measured? As the company strives to improve its market share, discuss the impact on the various categories of quality costs.

1-4 Consider the hospitality industry. Describe special causes and common causes in this setting and discuss the role of quality control and quality improvement.

1-5 An OEM in the automobile industry is considering an improvement in its order-processing system with its tier 1 suppliers. Discuss appropriate measures of quality. What are some special and some common causes in this environment?

1-6 An intermodal logistics company uses trucks, trains, and ships to distribute goods to various locations. What might be the various quality costs in each of the categories of prevention, appraisal, internal failure, and external failure?

1-7 A quality improvement program has been instituted in an organization to reduce total quality costs. Discuss the impact of such a program on prevention, appraisal, and failure costs.

1-8 Classify each of the following into the cost categories of prevention, appraisal, internal failure, and external failure:
(a) Vendor selection
(b) Administrative salaries
(c) Downgraded product
(d) Setup for inspection
(e) Supplier control
(f) External certification
(g) Gage calibration
(h) Process audit

1-9 Discuss the indices for measuring quality costs. Give examples where each might be used.

1-10 Explain how it is feasible to increase productivity, reduce costs, and improve market share at the same time.

1-11 Explain why it is possible for external failure costs to go up even if the first-pass quality level of a product made by a company remains the same.

1-12   Discuss the impact of technological breakthrough on the prevention and appraisal cost and failure cost functions.

1-13   As natural resources become scarce, discuss the role of ISO 14000 in promoting good environmental management practices.

1-14   Discuss the processes through which supply chain quality may be monitored.

Problems

1-15   An assemble-to-order hardware company has two types of central processing units (CPUs), C1 and C2, and two types of display monitors, M1 and M2. Unit C2 is slightly more complex then C1, as is M2 compared to M1. The annual production volume and direct costs are shown in Table 1-7. Overhead costs for this past year are shown in Table 1-8. Assume that setup and testing costs are similar for both types of CPUs and monitors.

TABLE 1-7

	CPU		Monitor	
	C1	C2	M1	M2
Annual volume	10,000	15,000	18,000	20,000
Unit costs				
Direct labor ($)	80	140	120	200
Direct material ($)	60	100	80	120
Assembly ($)	40	60	60	100

TABLE 1-8

Category	Cost ($ millions)
Setup and testing	1.1
Product-line cost	
CPU C1	0.5
CPU C2	1.5
Monitor M1	0.8
Monitor M2	2.5
Other company costs	0.6

(a) Calculate the cost per unit of each product using the unit-based costing method by allocating overhead based on direct labor costs.
(b) Calculate the cost per unit of each product using the activity-based costing method.
(c) Discuss the unit costs calculated using these two methods.

1-16   For the hardware company in Exercise 1-15, suppose that setup and testing costs are different for CPUs and monitors. Annual costs for setup and testing are $0.4 million and $0.7 million, for CPUs and monitors, respectively. However, between two types of CPUs, these costs are basically similar, as is also the case for monitors.
(a) Calculate the cost per unit of each product using the activity-based costing method.
(b) Discuss these unit costs in relation to those in Exercise 1-15.

1-17   Consider the hardware company in Exercise 1-15. The company is contemplating outsourcing of its complex monitor M2. Assume that all other information remains as given in Exercise 1-15.
(a) What is the cost per unit of each product using activity-based costing assuming that M2 is not produced?
(b) A prospective supplier of monitor M2 has offered a unit price of $480. Should the company outsource the monitor M2 to this supplier? Discuss.

1-18   A pharmaceutical company has obtained the following cost information (per 1000 tablets) based on the production of a drug in the past year: material, $150; direct labor, $100; energy, $50; overhead, 40% of direct labor and material. Presently, the process yield rate is 94%.
(a) Find the cost per tablet of acceptable product.
(b) A team evaluating the entire process has suggested improvements that led to increasing the yield rate to 96%. What is the cost per tablet of conforming product and the percent increase in capacity?
(c) Process engineers have come up with an improved sequence of operations. Labor costs were reduced by 15% and energy costs by 20% from the original values. Find the cost per tablet of conforming product now, and calculate the percentage reduction in cost.

1-19   The following data (in $/m^3) was obtained from a company that makes insulation for commercial buildings: direct labor, 20; direct materials, 30; indirect labor and materials, 30% of direct labor; fixed expenses, 25; administrative costs, 25; selling costs, 10.
(a) Assuming a 100% first-pass yield, what should the selling price (per m^3) be such that a 10% profit margin, over the cost of goods sold, will be obtained?
(b) Suppose that the first-pass yield is 94%. If the selling price is kept the same as calculated in part (a), what is the profit margin?
(c) Through process improvements, first-pass yield has been improved to 98%. However, the capital expenditures necessary for such improvements is $150,000. If the selling price is kept the same as in part (a), what is the profit margin, ignoring additional capital expenditures?
(d) For the improved process in part (c), assuming that monthly demand is 5000 m^3, how long would it take for the company to break even on its added capital expenditures?
(e) Suppose that the company is able to sell product that does not meet first-pass quality criteria at a reduced price of $120/m^3. For the improved process in part (d), what is the break-even time now to recover added capital expenditures?

1-20 In the production of a part for a printer, four sequential operations are involved. Unit processing costs for the operations are $10, $6, $15, and $20, respectively. The first-pass yield for each operation is 0.95, 0.90, 0.95, and 0.85, respectively. Unit inspection costs after each operation are $0.50, $2.00, $3.00, and $5.00, respectively.
   (a) If no inspection is performed, what is the unit cost for an acceptable part?
   (b) Assume that an inspection process is able to identify all parts correctly. Suppose that inspection is conducted only after the first and second operations. Nonconforming parts are not forwarded to the next operation. What is the unit cost per acceptable part?
   (c) Suppose that inspection is conducted only after the third operation. Nonconforming parts are not forwarded to the next operation. What is the unit cost per acceptable part?
   (d) Based on the unit costs computed in parts (b) and (c), discuss where, in general, inspections should be conducted.

1-21 Suppose that the prevention and appraisal cost functions are given by $C_p = 50q^2$ and $C_a = 10q$, respectively, where q represents the degree of quality level ($0 < q < 1$). The cost of reworking a unit is $5, and the cost of a customer obtaining a nonconforming product is $85. Assume that these cost functions are linear in $(1 - q)$. What is the desirable operational level of quality for this static situation? Discuss the appropriateness of the cost functions.

1-22 Suppose that the prevention cost function is as given in Exercise 1-21. However, the unit appraisal cost is $2, with the cost function being linear in $1 - q$, implying a decrease in appraisal costs as quality improves. Further, the rework and external failure cost functions are given by $C_r = 5(1 - q)/q$ and $C_e = 85(1 - q)/q$, respectively. Construct the total cost function as a function of q and graph it for levels of q in the range 0.80 to 0.98. What is the desirable operational level of quality?

1-23 Consider Exercise 1-22. Suppose that market share is strongly influenced by the level of quality, with the revenue function given by $90q^2$. What is the net profit function? What is the minimum desirable quality level to break even?

REFERENCES

ASQ. (1994). *Quality Management and Quality Assurance Vocabulary*, ANSI/ISO/ASQ Standard A8402 Milwaukee, WI: American Society for Quality.

———(2000). *Quality Management Systems: Requirements*, ANSI/ISO/ASQ 9000 Milwaukee, WI: American Society for Quality.

———(2004). *Glossary and Tables for Statistical Quality Control*, 4th ed., Milwaukee, WI: American Society for Quality, Statistics Division.

Campanella, J., Ed. (1999). *Principles of Quality Costs: Principles, Implementation, and Use*, 3rd ed. Milwaukee, WI: American Society for Quality.

Cohen, L. (1995). *Quality Function Deployment: How to Make QFD Work for You*. Reading, MA: Addison-Wesley.

Crosby, P. B. (1979). *Quality Is Free*. New York: McGraw-Hill.

Feigenbaum, A. V. (1983). *Total Quality Control*. New York: McGraw-Hill.

Garvin, D. A. (1984). "What Does Product Quality Really Mean?" *Sloan Management Review*, 26 (1): 25–43.

Juran, J. M., Ed. (1974). *Quality Control Handbook*, 3rd ed. New York: McGraw-Hill.

Marcus, P. A., and J. T. Willig, Eds. (1997). *Moving Ahead with ISO 14000*. New York: Wiley.

NBS (2005). *Model State Laws and Regulations*, NBS Handbook 130. Gaithersburg, MD: U.S. Department of Commerce, National Bureau of Standards.

NIST (2005). *Malcolm Baldrige National Quality Award: Profiles of Winners*. Gaithersburg, MD: U.S. Department of Commerce, National Institute of Standards and Technology.

Parry, P. (2000). *The Bottom Line: How to Build a Business Case for ISO 14000*. Boca Raton, FL: St. Lucie Press.

Taguchi, G. (1986). *Introduction to Quality Engineering: Designing Quality into Products and Processes*. Tokyo, Japan: Asian Productivity Organization. Available in North America, U.K., and Western Europe from the American Supplier Institute, Inc., and UNIPUB/Kraus International Publications, White Plains, NY.

Taguchi, G. and Y. Wu (1979). *Introduction to Off-Line Quality Control*. Nagoya, Japan: Central Japan Quality Control Association.

Welch, T. E. (1998). *Moving Beyond Environmental Compliance*. Boca Raton FL: Lewis Publishers.

2

SOME PHILOSOPHIES AND THEIR IMPACT ON QUALITY

2-1 INTRODUCTION AND CHAPTER OBJECTIVES

Several people have made significant contributions in the field of quality control. In this chapter we look at the philosophies of three people: W. Edwards Deming, Philip B. Crosby, and Joseph M. Juran. Pioneers in the field of quality control, they are largely responsible for the global adoption and integration of quality assurance and control in industry.

Management commitment is key to a successful program in quality. This often requires a change in corporate culture. The idea of an ongoing quality control and improvement is now widely accepted. One of our objectives is to discuss some unique quality characteristics in the service industry. A second objective is to study the various philosophies on quality. In this chapter we examine Deming's philosophy in depth. Deming's **14 points for management** are fundamental to the implementation of any quality program. These points, which constitute a "road map," should be understood thoroughly by all who undertake the implementation of such programs. We also discuss Crosby's and Juran's philosophies and compare the three. The goal is the same in all three philosophies: creating and adopting a world-class quality business culture. Although the paths they describe are slightly different, companies should look closely at each approach before embarking on a quality program.

Fundamentals of Quality Control and Improvement, Third Edition, By Amitava Mitra
Copyright © 2008 John Wiley & Sons, Inc.

2-2 SERVICE INDUSTRIES AND THEIR CHARACTERISTICS

Today, service industries dominate our economy. The service sector accounts for more than 80% of jobs, and the number continues to grow. Quality improvement looms large in the ongoing success of this sector of the economy. However, major differences exist in the quality characteristics of manufacturing and service (Fitzsimmons and Fitzsimmons 1994; Patton 2005; Zeithaml 2000). Accordingly, both the measurement process and management's focus differ. In service industries, not only must the product meet the functional requirements of the customer, but employee behavior must also meet high standards. The total service concept is a combination of technical and human behavioral aspects, and the latter are much more difficult to quantify, measure, and control.

Let's consider the airline industry. A quantifiable goal is to transport people between two cities in a desirable time. Achieving this is dependent on aircraft design that enables certain speeds to be attained to cover the distance within the required time, and on proper scheduling of flights. Data on these factors are clearly quantifiable. However, customer satisfaction is often influenced by factors that are not so easy to quantify. For instance, the manner in which stewardesses and ticket agents treat customers is very important. Courteous and friendly, and warm and caring are not so obviously quantified. Thus, the manner in which service is performed is an important concern that might not be considered in manufacturing industries. Of course, we should realize that even manufacturing industries have to deal with service functions (e.g., payroll and accounting, customer relations, product service, personnel, purchasing, marketing). The importance of the service industry should not, therefore, be underestimated.

In this section we discuss quality characteristics unique to service industries. Fundamental differences between manufacturing and service are noted. The customer is, of course, the focal point of quality control and improvement, and customer feedback is essential. We find service industries in all facets of our society. Functions performed by service industries include education, banking, governmental services (such as defense, municipal services, and welfare), health care, insurance, marketing, personal services (such as hotels and motels), restaurants, traveling and tours, public utilities (including electricity, gas, and telephone service), and transportation (airlines, railroads, and buses). As shown in the preceding example from the airline industry, service industries provide both a tangible product *and* an intangible component that affects customer satisfaction.

Two parties are involved in providing a service. The one that assists or provides the service is the vendor, or company, and the party receiving the service is the vendee, or customer. Certain **service functions** are found in both the manufacturing and service sectors. In the manufacturing sector, these are staff functions and are preformed by staff personnel, rather than by line personnel. Staff personnel provide expertise to the operating departments and to customers to enable them to get the most value out of a product. Customer services and warranties are examples of this. In addition, clerical and administrative operations such as accounting, purchasing, payroll, and personnel are service functions that play a supportive role in a manufacturing organization. Research and development activities are also viewed as service functions, because their goal is to devise better product or process designs that will facilitate line operations.

Differences in the Manufacturing and Service Sectors

Basic differences in the manufacturing and service sectors are noted in Table 2-1. The manufacturing sector makes products that are tangible, whereas services have an associated

TABLE 2-1   Differences in the Manufacturing and Service Sectors

Manufacturing Sector	Service Sector
Product is tangible.	Service consists of tangible and intangible components.
Back orders are possible.	Services cannot be stored; if not used, they are lost.
Producer or company is the only party involved in the making of the product.	Producer and consumer are both involved in delivery of a service.
Product can be resold.	Services cannot be resold.
Customer usually provides formal specifications for the product.	Formal specifications need not be provided by the consumer. In fact, in monopolies involving public utilities (e.g., electricity, gas, telephone), federal and state laws dictate the requirements.
Customer acceptance of the product is easily quantifiable.	Customer satisfaction is difficult to quantify because a behavioral component is involved associated with the delivery of the service.
Ownership of a product changes hands at a specific point in time.	Rendering a service takes place over an interval of time.

intangible component: A caring attitude with a smile from servers leads to customer satisfaction. In manufacturing, in periods when product demand exceeds supply, back orders can eventually right the imbalance. However, services cannot usually be back-ordered; there is an associated time constraint. If a service is not provided within the necessary time frame, it cannot be used at a later time. An example of this is the empty seats in a mass transportation system from 10:00 to 11:00 A.M. These empty seats cannot be saved for use during the 5:00 to 6:00 P.M. rush hour.

Another distinguishing feature concerns the relationship of the provider and customer. In the health care industry, the doctor or nurse, the provider, interacts with the patient, the customer, to provide service. Responses from the patient influence how a service is delivered. In the manufacturing industries, the producer or manufacturing company alone influences the process through which a product is made. The customer affects the product in the sense that the product is designed to meet customer requirements, but once a satisfactory product has been achieved, the customer does not influence the product quality during production. Manufactured products can be resold; the same is not true of services.

Customers usually have a direct impact on creating formal product specifications in a manufacturing environment. Quality characteristics that influence customer satisfaction are identified and are incorporated into the product at the design phase. In some service industries, however, the customer does not provide *direct* input on the quality characteristics for services. Public utilities such as electricity, gas, and telephone are regulated by federal and state laws; thus, the services they provide and the prices they charge are mandated by governing bodies such as public service commissions. The customer's involvement is *indirect*: that is, they elect the public officials who make the regulations and they go before governmental committees to make their desires known, which may or may not influence the regulations.

The behavioral aspect associated with the delivery of services also differs for the manufacturing and service sectors. In manufacturing companies, the degree to which a product is accepted can be quantified: say, in terms of the proportion of unacceptable product. In service industries, the degree of customer satisfaction is not as easily quantified because of the human factors involved with delivery of a service. The behavioral traits of both the provider and the customer influence service delivery. Customer dissatisfaction can be the result of many intangible factors. On the other hand, in a manufacturing company, if a product is not accepted because a characteristic falls outside certain specifications, the reason for customer dissatisfaction can readily be found and remedial measures taken.

Service Quality Characteristics

In this subsection we consider features of quality characteristics in the service sector. The quality characteristics are grouped into four categories (see Table 2-2). Although exceptions to these groups exist, the categories generally summarize factors common to service functions and industries.

The quality of a service can be broken down into two categories: effectiveness and efficiency. **Effectiveness** deals with meeting the desirable service attributes that are expected by the customer. For example, the décor and available facilities in a hospital room, the quality and quantity of food served in a restaurant, and the types of checking and savings accounts available in a bank are related to service effectiveness. **Efficiency**, on the other hand, concerns the time required for the service to be rendered.

Human Factors and Behavioral Characteristics   Service quality is influenced by the attitude and behavior of the provider (Lefevre 1989; Normann 1991). Since providers and customers are *part* of the product (the service), their behavior affects the quality of the service.

TABLE 2-2   Service Quality Characteristics and Their Measures

Service Quality Characteristic	Measures of Service Quality
Human factors and behavioral characteristics	Number of customer complaints based on behavioral factors (or lack thereof) of persons involved in the service process
	Number of complimentary responses based on human traits in delivery of service
Timeliness characteristics	Waiting time in a bank prior to transaction
	Time to process a transaction
	Time to check in at an airport
	Waiting time before receiving baggage at an airport
	Time to hear from an insurance company regarding a payment
Service nonconformity characteristics	Number of errors per 1000 transactions in banks, insurance companies, and payroll departments
	Number of billing errors per 1000 accounts by utility companies
	Proportion of income tax returns prepared by an agency that have errors
Facility-related characteristics	Number of complaints due to:
	An uncomfortable bed in a hotel room
	Unavailability of a swimming pool in a hotel
	Insufficient legroom in an aircraft
	Inadequate temperature control in a convention meeting room
	Shabby appearance of a receptionist in a hotel or bank
	Lack of certain indoor activities (such as table tennis) in a recreation facility

Human factors thus include intensity, eagerness to help, thoughtfulness, complacency, courtesy, and so on. Some of these traits can be developed through adequate training; some are inherent in the person. Proper screening of employees and appropriate job assignment are ways to achieve desirable quality characteristics. A primary source of customer complaints is discourteous behavior.

On the other hand, the attitude of the customer is largely beyond the control of the service company. For instance, a customer's mood when purchasing a service can influence perceived quality; the quality may be good but it is not perceived as such because the customer is angry about something totally unrelated to the service or the provider. However, companies can influence customers' expectations through advertisement and reputation. Customers' mind-sets are often a function of what they expect to receive. Thus, the company affects the behavioral patterns of their customers by molding their expectations. If a bank advertises that in addition to providing its regular services, it now provides financial management services, customer expectations are raised. Customers will no longer be satisfied if questions related to financial management are not answered adequately. On the other hand, if the bank does not claim to provide financial management services, the customer will not expect such questions to be answered and will thus not be disappointed. Unfortunately, measurement of attitudes and behavioral characteristics is not as simple and well defined as for tangible criteria.

Timeliness Characteristics   A service that is not used in a given span of time cannot be stored for later use. A hospital with empty beds during certain days of a month cannot save them for use in the following month. Thus, the timeliness with which a service is performed is critical to customer satisfaction. How long did the customer have to wait before being served in a restaurant? How long did the customer have to wait in line to cash a check? Characteristics related to timeliness are categorized by the service phase with which they are associated. Categories might include the time to order the service, the waiting time before the service is performed, the time to serve, and the post-service time. These characteristics are much more amenable to measurement than are behavioral characteristics.

Service Nonconformity Characteristics   Nonconformity characteristics deal with deviation from target performance levels; a nonconformity is a deviation from the ideal level. Examples of such characteristics include the number of errors by bank employees in processing 100 vouchers, the number of errors by a data-entry operator per 1000 keystrokes, the number of billing errors per 100 accounts by a utility company, the number of complaints per 100 guests in a hotel, and so on. The target performance level for these examples is zero nonconformities. The goal of the service organization is to achieve the target level, thus meeting customer expectations, and then to exceed it through quality improvement measures. Quality characteristics in this category are well defined and are more readily measured than behavioral characteristics.

Facility-Related Characteristics   The physical characteristics of the facilities associated with a service and its delivery can affect customer satisfaction. The décor of a restaurant, the waiting area in a physician's office, and the availability of such amenities as a swimming pool or spa in a hotel are examples of quality characteristics for physical facilities that are involved in providing a service. The appearances of a waiter or waitress, a bank teller, or an insurance agent are attributes of employees performing a service. These characteristics are not as clearly defined and measurable as service nonconformity characteristics. They are, however, more quantifiable than behavioral characteristics.

Measuring Service Quality

In terms of ease of quantification and measurement, the categories identified in Table 2-2 can be ranked in the following order: service nonconformity, timeliness, facility-related, and human behavioral factors. Since the success of many service functions is determined predominantly by the interaction between the provider and the customer, measurement and evaluation of service quality is difficult because it is subjective. Defining the measurement unit itself is problematic.

People are not as predictable as equipment and facilities. The water temperature in a swimming pool can be stabilized by an adequate heating and recycling system, but the behavior of a check-in attendant is not always under the control of the company. Many factors influence employee behavior: family life, unforeseen personal events, and mental outlook, to name a few. Not only can these cause large performance variations but they are largely outside the influence of the company, and they cannot be predicted.

To counteract these performance variations in human behavior, procedures that generate representative statistics of performance can be devised. Randomly choosing samples of performance from the time interval under consideration is one way to eliminate bias. In situations where we know that behavioral patterns vary greatly based on the time period (e.g., if error rates are high in the first and eighth hours of an 8-hour workday), we can select a sampling plan that adequately reflects this. In this example, a stratified sampling plan for two strata is designed, one for the early morning and late afternoon periods, and one for the remainder of the day. If 20 samples are selected daily and a proportional sampling scheme is used, 5 samples (which comprise 25% of the total daily samples) would be randomly selected from the first stratum: that is, early morning (8 to 9 A.M.) and late afternoon (4 to 5 P.M.). The assumption is that based on an 8-hour day, the time interval covered by this stratum is 2 hours, or 25% of the total daily hours worked. The remaining 15 samples will be randomly selected from the second stratum, which represents the remaining time period.

Another difficulty is that significant differences exist between individuals. Thus, even though the scheme of stratified sampling is used to select appropriate samples that reflect a person's performance, it is not obvious whether this same scheme can be applied collectively to a group of persons. People vary in their peak performance periods: Some work best in the early morning, and others work best at night. If such differences can be identified, the sampling plan can be designed to reflect them.

Techniques for Evaluating Service Quality

As with manufacturing, achieving an appropriate design of a service system precedes any activities on control and improvement. However, in the service sector, ergonomic, anthropometric, and behavioral characteristics are important, as are the physical characteristics of the service systems and timeliness with which the service is provided. Understanding variability in service quality characteristics is important to the control and improvement of service quality. Certain sources of variability are similar to those encountered in the manufacturing sector, such as variation due to equipment, process, and environmental factors. Additionally, because of the extensive involvement of people, person-to-person variation or project-to-project variation can be significant in service functions. The motivation level of the individual affects quality of service. Thus, the quality of work performed by auditors may vary from one person to another. As in manufacturing, changing tasks may lead to performance variability in service functions because some people are more adept at doing certain things. A data-entry operator for a utility company may make fewer errors when reading a well-designed short form that shows power consumption. The same operator when entering personnel data from a more complex form is liable to make more errors.

Sampling techniques in service operations include the following:

1. *100% sampling*. When the cost of external errors or nonconformities is high, this sampling scheme is useful. The cost of sampling and inspection is high, but it is still cost-effective compared to having a nonconformity found by the customer. For example, errors in transactions that are worth more than $100,000 can cause customer dissatisfaction to a degree that seriously affects profitability.

2. *Convenience sampling*. Here samples are chosen by the ease of drawing them and are influenced by the subjectivity of the person performing the sampling. For example, one person will choose the thinnest files in inspecting insurance claims, and another will choose the ones on the top of the pile. In another example, if questionnaires are mailed to a group of people known to be satisfied and responsive, the inferences drawn from the study will certainly be distorted.

3. *Judgment sampling*. These samples are chosen based on expert opinion. This can also create a bias. Caution should be exercised in drawing statistical inferences from these samples even though summary measures can be computed. Let's say that to review the purchase order error rate, an expert recommends selecting vendors who have dealt with the company for over two years. The expert feels that because the vendors are familiar with the procedures involved in preparing purchase orders, the error rate may be low. Thus, the recommendation of the expert may be to draw a sample once a week from these categorized vendors. The bias may stem from not sampling from all vendors.

4. *Probability sampling*. This technique has a statistical basis and is preferable for most situations. In random sampling, each item has an equal chance of being selected. This can be accomplished using random number tables. Sampling of vouchers for auditing, where 20 vouchers are to be selected from a population of 500, is one example. An example for stratified random sampling, where large differences are expected between groups in the population, could be tax-return audits of the Internal Revenue Service. If the IRS first stratifies the returns based on the estimated income and then chooses tax returns as simple random samples from each group, all income categories will be represented in the audit.

2-3 MODEL FOR SERVICE QUALITY

The special features of service functions and service industries help to define the role for management. To see how this works, we use the service quality model shown in Figure 2-1. The key concept is this: Customer satisfaction is a function of the perceived quality of service, which is a measure of how actual quality compares to expected quality (Normann 1991). Internal and external factors affect customer perceptions of quality. External factors, which are not directly under the control of a service organization (shown by the dashed box in

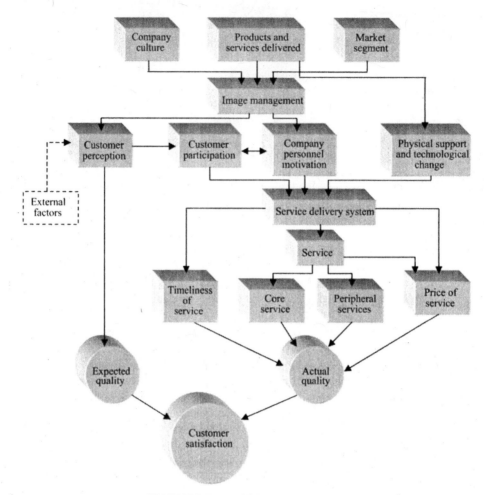

FIGURE 2-1   Model for service quality.

Figure 2-1, include the social values and lifestyles of customers, and knowledge of the service and of the services offered by the competitors. The image presented by the company also influences customer perceptions. Companies use various techniques for monitoring customer perceptions; this falls under the category of image management. Typical methods are annual company reports and quarterly marketing sales reports. A customer who reads about a high sales volume will perceive the company to be competent; that is, the company must be doing something right to have such a high sales volume.

Figure 2-2 shows external and internal factors that influence customer perceptions. Client management is important for improving the service and for meeting the changing needs of customers. Keeping abreast of customer needs through interviews, polls, and surveys and making changes in the facilities and services provided to meet these needs will facilitate retaining and expanding the customer base. Another important concept is creating a corporate culture that motivates employees. Motivated employees create favorable customer attitudes. This is a synergistic cycle: Satisfied customers cause employees to be further motivated.

In service industries, image management is central to retaining and attracting new customers. The corporate culture, the products and services delivered, and the market share

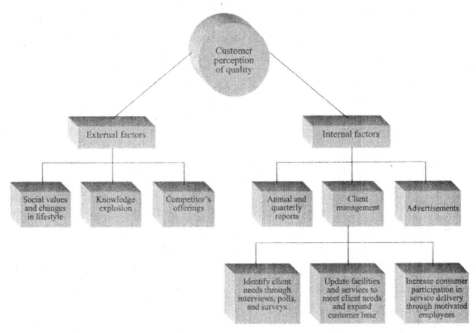

FIGURE 2-2   Factors that influence customer perception.

affect a company's reputation significantly. The company must advertise only what it can deliver. Building up expectations and failing to meet them creates dissatisfied customers and results in a decreased market share.

Creating a positive environment within the company often motivates employees and influences service delivery. In keeping with the philosophies of quality assurance and management, channels of open communication in a friendly atmosphere promote desirable employee behavior. It is difficult for employees to maintain a degree of warmth with the customer in service delivery if the same degree of congeniality does not exist between employees and management.

Service delivery is also influenced by customer participation, and customer participation is affected by customer expectations. A customer who does not expect to receive a free gift when buying a certain amount of merchandise will behave differently than one who does. Moreover, an employee's attitude may influence a customer's attitude, which in turn may have an effect on the employee's motivation, thereby closing the loop and continuing the cycle. Physical facilities and the manner in which they keep pace with the needs of the customer affect service functions and the efficiency and effectiveness of service delivery. A hospital with the latest technology in cancer diagnosis and treatment will attract patients who require such a service.

The service provided has two components: the core service and the peripheral services. The benefits received are both tangible and intangible. For a transportation company providing bus services between cities, the core service is the transport of customers between cities. The peripheral services are the comfort and safety of the bus stations, rest room and meal services, and transfer to and from main junctions in the city. Sometimes, peripheral services influence the customer to select one company over its competitors. A careful analysis of the entire range of services provided by the company is essential to maintaining and increasing market share.

Service quality is a function of several factors: timeliness of the service, the manner in which it is performed, the adequacy of the service (both core and peripheral services), and the price of service. Some service companies use a broad-based systems concept and provide several ancillary services. For example, an automobile dealer may also provide financing, insurance, maintenance contracts, and gasoline. A commitment to the customer in the post-sale period is helpful in retaining customers. Evidence of good service spreads through word of mouth. Customers are often a company's most important advertisers. As customer expectations change, so must the delivery of service.

2-4 W. EDWARDS DEMING'S PHILOSOPHY

Deming's philosophy emphasizes the role of management. Of the problems that industry faces, Deming said that over 85% can be solved only by management. These involve changing the system of operation and are not influenced by the workers. In Deming's world, workers' responsibility lies in communicating to management the information they possess regarding the system, for both must work in harmony. Deming's ideal management style is holistic: The organization is viewed as an integrated entity. The idea is to plan for the long run and provide a course of action for the short run. Too many U.S. companies in the past (and in the present) have focused on short-term gains.

Deming believed in the adoption of a total quality program and emphasized the never-ending nature of quality control in the quality improvement process. Such a program achieves the desired goals of improved quality, customer satisfaction, higher productivity, and lower total costs in the long run. He demonstrated that an improvement in quality inevitably leads to increased capacity and greater productivity. As experience demonstrates, these desirable changes occur over an extended period of time. Thus, Deming's approach is not a "quick fix" but rather, a plan of action to achieve long-term goals. He stressed the need for firms to develop a corporate culture where short-term goals such as quarterly profits are abandoned.

Deming's approach demands nothing less than a cultural transformation—it must become a way of life. The principles may be adapted and refined based on the experience of a particular organization, but they still call for total commitment. At the heart of this philosophy is the need for management and workers to speak a common language. This is the language of statistics—statistical process control. The real benefits of quality programs will accrue only when everyone involved understands their statistical underpinning. Therefore, Deming's fundamental ideas require an understanding and use of statistical tools and a change in management attitude. His 14 points, which we will look at shortly, identify a framework for action. This framework must be installed for the quality program to be successful. Management must commit to these points in thought, word, and deed if the program is to work.

Deming advocated certain key components that are essential for the journey toward continuous improvement. The following four components comprise the basis for what Deming called the **system of profound knowledge**:

1. *Knowledge of the system and the theory of optimization.* Management needs to understand that optimization of the *total system* is the objective, not necessarily the optimization of individual subsystems. In fact, optimizing subsystems can lead to a suboptimal total system. The total system consists of all constituents: customers,

employees, suppliers, shareholders, the community, and the environment. A company's long-term objective is to create a win–win situation for all its constituents.

2. *Knowledge of the theory of variation.* All processes exhibit variability, the causes of which are of two types: special causes and common causes. **Special causes** of variation are external to the system. It is the responsibility of operating personnel and engineering to eliminate such causes. **Common causes**, on the other hand, are due to the inherent design and structure of the system. They define the system. It is the responsibility of management to reduce common causes. A system that exists in an environment of common causes only is said to be *stable* and *in control*. Once a system is considered to be in control, its capability can be assessed and predictions on its output made.

3. *Exposure to the theory of knowledge.* Information, by itself, is not knowledge. Knowledge is evidenced by the ability to make predictions. Such predictions are based on an underlying theory. The underlying theory is supported or invalidated when the outcome observed is compared to the value predicted. Thus, experience and intuition are not of value to management unless they can be interpreted and explained in the context of a theory. This is one reason why Deming stressed a data analysis–oriented approach to problem solving where data are collected to ascertain results. The results then suggest what remedial actions should be taken.

4. *Knowledge of psychology.* Managing people well requires a knowledge of psychology because it helps us understand the behavior and interactions of people and the interactions of people with their work environment. Also required is a knowledge of what motivates people. People are motivated by a combination of intrinsic and extrinsic factors. Job satisfaction and the motivation to excel are intrinsic. Reward and recognition are extrinsic. Management needs to create the right mix of these factors to motivate employees.

Extended Process

The **extended process** envisioned by Deming expands the traditional organizational boundaries to include suppliers, customers, investors, employees, the community, and the environment. Figure 2-3 shows an extended process. An organization consists of people, machines, materials, methods, and money. The extended process adds a key entity—the customer. An organization is in business to satisfy the consumer. This should be its primary goal. Goals such as providing the investors with an acceptable rate of return are secondary. Achieving the primary goal—customer satisfaction—automatically causes secondary goals to be realized. This primary goal is especially relevant to a service organization; here the customer is more obviously central to the success of the organization. A prime example is the health care industry, where, in addition to the primary customers, the patients, are physicians, nurses, employees, the federal government, health-maintenance organizations, and insurance companies.

The community in which an organization operates is also part of this extended process. This community includes consumers, employees, and anyone else who is influenced by the operations of the company, directly or indirectly. An accepting and supportive community makes it easier for the company to achieve a total quality program. Community support ensures that there is one less obstacle in the resistance to the changes proposed by Deming.

Vendors are another component of the extended process. Because the quality of raw materials, parts, and components influences the quality of the product, efforts must be

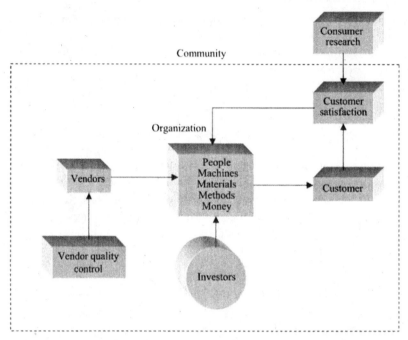

FIGURE 2-3   Extended process in Deming's philosophy.

made to ensure that vendors supply quality products. In Deming's approach, a long-term relationship between the vendor and the organization is encouraged, to their mutual benefit.

Deming's 14 Points for Management

The focus of Deming's philosophy (Deming, 1982,1986) is management. Since a major proportion of problems can be solved by management, Deming noted that management cannot "pass the buck." Only a minority of problems can be attributed to suppliers or workers, so in Deming's view, what must change is the fundamental style of management and the corporate culture (Fellers 1992; Gitlow and Gitlow 1987).

In Deming's ideal organization, workers, management, vendors, and investors are a team. However, without management commitment, the adoption and implementation of a total quality system will not succeed. It is management that creates the culture of workers' "ownership" and their investment in the improvement process. Managers create the **corporate culture** that enables workers to feel comfortable enough to recommend changes. Management develops long-term relationships with vendors. And finally, it is managers who convince investors of the long-term benefits of a quality improvement program. A corporate culture of trust can only be accomplished with the blessings of management.

Deming's Point 1   *Create and publish to all employees a statement of the aims and purposes of the company or other organization. The management must demonstrate constantly their commitment to this statement.*[*]

---

*Deming's 14 points (January 1990 revision) are reprinted from *Out of the Crisis,* by W. Edwards Deming, published by MIT Center for Advanced Engineering Study, Cambridge, MA. Copyright © 1986 by The W. Edwards Deming Institute.

This principle stresses the need to create the long-term strategic plans that will steer a company in the right direction. Mission statements should be developed and expressed clearly so that not only everyone in the organization understands them but also vendors, investors, and the community at large. Mission statements address such issues as the continued improvement of quality and productivity, competitive position, stable employment, and reasonable return for the investors. It is not the intent of such statements to spell out the finances required; however, the means to achieve the goals should exist. To develop strategic plans, management must encourage input from all levels. If the process through which the plan is developed is set up properly, companywide agreement with, and commitment to, the strategic plan will be a natural outcome.

Product Improvement Cycle   Committing to a specific rate of return on investment should not be a *strategic* goal. In this philosophy, the customers—not the investors—are the driving force in the creation of strategic goals. The old approach of designing, producing, and marketing a product to customers without determining their needs is no longer valid. Instead, the new approach is a four-step cycle that is customer-driven. Figure 2-4 shows this cycle, which includes designing a customer needs-based product, making and testing it, selling it, determining its in-service performance with market research, and using this information to start the cycle again. This approach integrates the phases of quality of design, conformance, and performance (discussed in Chapter 1).

Determining customer needs in clearly understood terms is central to improving product quality. For instance, noting that a customer prefers a better refrigerator that is less noisy and can produce ice quickly is not sufficient. What noise level (stated in decibels) is acceptable? How fast (in minutes) does the customer expect the ice to be produced? Only when specific attributes are quantified can the product be made better.

Constancy and Consistency of Purpose   Foresight is critical. Management must maintain a **constancy of purpose**. This implies setting a course (e.g., all departments within the

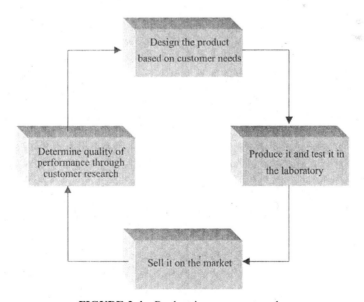

FIGURE 2-4   Product improvement cycle.

organization will pursue a common objective of quality improvement) and keeping to it. Too often, a focus on short-term results such as weekly production reports or quarterly profits deters management from concentrating on the overall direction of the company. Actions that create a profit now have a negative impact on profits 10 years from now.

Management must be innovative. They must allocate resources for long-term planning, including consumer research and employee training and education that addresses not only work performance but also the new philosophies. They must ensure that there are resources available to cover the new costs of determining quality of performance and for new methods of production or changes in equipment. The priorities are research and education and a constant improvement in the product or service.

In addition to constancy of purpose, there should be a **consistency of purpose**. This means that the company should not digress from long-term objectives. For example, are all units of the company working toward the company goal to improve quality synergistically? Or are they working for a departmental subgoal, which may be to increase production? Management should also accept the fact that variability exists and will continue to exist in any operation. What they must try to do is determine ways in which this variation can be reduced. And this is a never-ending process. As Deming put it, "Doing your best is not good enough. You have to know what to do. Then do your best."

Deming's Point 2   *Learn the new philosophy, top management and everybody.*

The new attitude must be adopted by everyone. Quality consciousness must be everything to everyone. Previously acceptable levels of defects should be abandoned; the idea that improvement is a never-ending process must be embraced wholeheartedly.

Human beings are resistant to change. Managers who have been successful under the old system where certain levels of defects were acceptable may find it difficult to accept the new philosophy. Overcoming this resistance is a formidable task, and it is one that only management can accomplish. The idea is not only continually to reduce defects but also to address the needs of the customer.

Declaring any level of defect to be acceptable promotes the belief that defects *are* acceptable. Say a contract specifies that a defective rate of 4 units in 1000 is acceptable; this *ensures* that 0.4% will be defective. This philosophy must be abandoned and a no-defects philosophy adopted in its place.

Deming's Point 3   *Understand the purpose of inspection, for improvement of processes and reduction of cost.*

Quality has to be *designed* into the product; it cannot be *inspected* into it. Creating a "design for manufacturability" is imperative because producing the desired level of quality must be feasible. Inspection merely separates the acceptable from the unacceptable. It does not address the root cause of the problem: that is, what is causing the production of nonconformities and how to eliminate them. The emphasis is on defect prevention, not on defect detection.

The production of unacceptable items does not come without cost. Certain items may be reworked, but others will be scrapped. Both are expensive. The product's unit price increases, and the organization's competitiveness decreases. Market share and available capacity are inevitably affected.

Drawbacks of Mass Inspection   Mass inspection does not prevent defects. In fact, depending on mass inspection to ensure quality *guarantees* that defects will continue. Even 100% inspection will not eliminate all the defectives if more than one person is responsible for inspection. When several points of inspection are involved, it is only human to assume that others will find what you have missed. This is *inherent* in the mass inspection system. Inspector fatigue is another factor in 100% inspection. Thus, defects can only be prevented by changing the *process*.

Deming's Recommendation   If inspection must be performed, Deming advocated a plan that minimizes the total cost of incoming materials and thus the final product. His plan is an inspect-all-or-none rule. Its basis is statistical evidence of quality, and it is applied to a stable process. This is the *kp rule* (see Chapter 10). In situations where inspection is automated and does not require human intervention, 100% inspection may be used to sort nonconforming items, for rework or disposition. Analysis of defects may indicate the root cause in the process—which leads to identification of remedial actions.

Deming's Point 4   *End the practice of awarding business on the basis of price tag alone.*
    Many companies, as well as state and federal governments, award contracts to the lowest bidder as long as they satisfy certain specifications. This practice should cease. Companies should also review the bidders' approaches to quality control. What quality assurance procedures do the bidders use? What methods do they use to improve quality? What is the attitude of management toward quality? Answers to these questions should be used, along with price, to select a vendor, because low bids do not always guarantee quality.
    Unless the quality aspect is considered, the effective price per unit that a company pays its vendors may be understated and, in some cases, unknown. Knowing the fraction of nonconforming products and the stability of the process provides better estimates of the price per unit.
    Suppose that three vendors, A, B, and C, submit bids of $15, $16, and $17 per unit, respectively. If we award the contract on the basis of price alone, vendor A will get the job. Now let's consider the existing quality levels of the vendors.
    Vendor B has just started using statistical process control techniques and has a rather stable defect rate of 8%. Vendor B is constantly working on methods of process improvement. The effective price we would pay vendor B is $16/(1-0.08) = $17.39 per unit (assuming that defectives cannot be returned for credit).
    Vendor C has been using total quality management for some time and has a defect rate of 2%. Therefore, the effective price we would pay vendor C is $17/(1-0.02) = $17.35 per unit.
    Vendor A has no formal documentation on the stability of its process. It does not use statistical procedures to determine when a process is out of control. The outgoing product goes through sampling inspection to determine which ones should be shipped to the company. In this case, vendor A has no information on whether the process is stable, what its capability is, or how to improve the process. The effective price we would pay vendor A for the acceptable items is unknown. Thus, using price as the only basis for selection controls neither quality nor cost.
    A flagrant example where the lowest-bidder approach works to the detriment of quality involves urban municipal transit authorities. These agencies are forced to select the lowest bidder to comply with the policy set by the Urban Transit Authority of the United States. The poor state of affairs caused by complying with this policy is visible in many areas around the country.

Principles of Vendor Selection   Management must change the process through which **vendor selection** is conducted. The gap between vendor and buyer must be closed; they must work as a team to choose methods and materials that improve customer satisfaction.

In selecting a vendor, the total cost (which includes the purchase cost plus the cost to put the material into production) should be taken into account. A company adhering to Deming's philosophy buys not only a vendor's products but also its process. Purchasing agents play an important role in the extended process. Knowing whether a product satisfies certain specifications is not enough. They must understand the precise problems encountered with the purchased material as it moves through the extended process of manufacture, assembly, and eventual shipment to the consumer. The buyers must be familiar with statistical methods so as to assess the quality of the vendor's plant. They must be able to determine the degree of customer satisfaction with the products, tell what features are not liked, and convey all related information to the vendor. Such information enables the vendor to improve its product.

Another important principle involves reducing the number of suppliers. The goal is to move toward single-supplier items. Companies in the United States have had multiple vendors for several reasons, including a fear of price increases, a vendor's inability to meet projected increases in demand, and a vendor's lack of timeliness in meeting delivery schedules.

There are several disadvantages to this policy. First, it promotes a feeling of mistrust between buyer and vendor and thereby creates a short-term, price-dependent relationship between them. Furthermore, vendors have no motivation to change their process to meet the buyer's specifications. Price, not quality, is the driving factor because another vendor may be selected if their price is lower. A long-term commitment cannot exist in such a situation.

Other disadvantages involve cost. Increased paperwork leads to increased order preparation costs. Travel costs of the vendor to purchaser sites increase. Volume discounts do not kick in because order sizes are smaller when there is more than one vendor. Setup costs go up because the buyer's process changes when the incoming supplier changes. Machine settings may have to be adjusted along with tooling. In continuous process industries, such as chemical companies producing sulfuric acid, differences in raw materials may require changes in the composition mix. Multiple setup periods mean idle production and therefore reduced capacity. Also, training the people who work with vendors costs more with multiple vendors.

One major disadvantage is the increased variability in incoming quality, even if individual vendors' processes are stable. Figure 2-5 explains this concept. Suppose that we purchase from three vendors, A, B, and C, each quite stable and having a small dispersion as far as the quality characteristic of interest is concerned (say, density of a red color pigment in producing a dye). However, the combined incoming product of three good suppliers may turn out to be mediocre. This happens because of the intervendor variability.

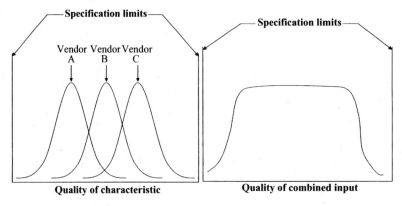

FIGURE 2-5   Mediocre incoming quality due to multiple vendors.

Many benefits are gained by moving to single-vendor items. The disadvantages mentioned in the preceding paragraphs can be eliminated when a long-term relationship with a quality-conscious vendor is developed. A long-term vendor can afford to change its process to meet the needs of the buyer because the vendor does not fear losing the contract. Such a relationship also permits open contract negotiations, which can also reduce costs.

Deming's Point 5   *Improve constantly and forever the system of production and service.*

In Deming's philosophy, companies move from defect detection to defect prevention and continue with process improvement to meet and exceed customer requirements on a never-ending basis. Defect prevention and process improvement are carried out by the use of statistical methods. Statistical training is therefore a necessity for everyone and should be implemented on a gradual basis.

Deming Cycle   The continuous cycle of process improvement is based on a scientific method originally called the *Shewhart cycle* after its originator, Walter A. Shewhart. He also developed control charts. In the 1950s, the Japanese renamed it the **Deming cycle**. It consists of four basic stages: *plan*, *do*, *check*, and *act* (the PDCA cycle). The Deming cycle is shown in Figure 2-6.

Plan Stage   In this stage (depicted in Figure 2-7), opportunities for improvement are recognized and defined operationally. A framework is developed that addresses the effect of controllable process variables on process performance. Since customer satisfaction is the focal point, the degree of difference between customer needs' satisfaction (as obtained through market survey and consumer research) and process performance (obtained as feedback information) is analyzed. The goal is to reduce this difference. Possible relationships between the variables in the process and their effect on outcome are hypothesized.

Suppose a company that makes paint finds that one major concern for customers is drying time; the preferred time is 1 minute. Feedback from the process says that the actual drying

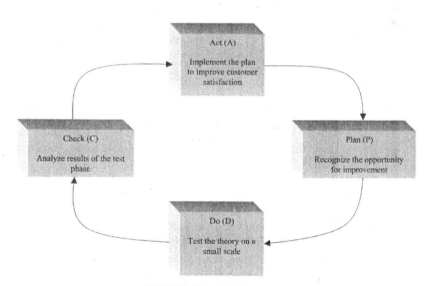

FIGURE 2-6   Deming cycle.

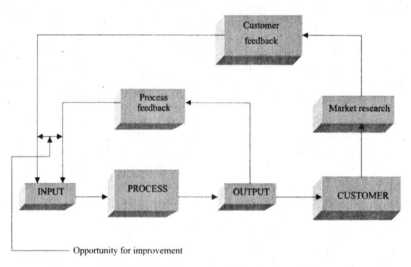

FIGURE 2-7   Plan stage of Deming's cycle.

time is 1.5 minutes. Hence, the opportunity for improvement, in operational terms, is to reduce the drying time by 0.5 minute.

The next task is to determine how to reduce drying time by 0.5 minute. The paint components, process parameter settings, and interaction between them are examined to determine their precise effect on drying time. Quality of design and quality of conformance studies are undertaken. Company chemists hypothesize that reducing a certain ingredient by 5% in the initial mixing process will reduce the drying time by 0.5 minute. This hypothesis is then investigated in the following stages.

Do Stage   The theory and course of action developed in the plan stage is put into action in the do stage. Trial runs are conducted in a laboratory or prototype setting. Feedback is obtained from the customer and from the process. At this stage, our paint company will test the proposed plan on a small scale. They reduce the ingredient by 5% and obtain the product.

Check Stage   Now the results are analyzed. Is the difference between customer needs and process performance reduced by the proposed action? Are there potential drawbacks relating to other quality characteristics that are important to the customer? Statistical methods will be used to find these answers. As our paint company attempts to reduce drying time by 0.5 minute, samples are taken from the modified output. The mean drying time and the variability associated with it are determined; the analysis yields a mean drying time of 1.3 minutes with a standard deviation of 0.2 minute. Prior to the modification, the mean was 1.5 minutes with a standard deviation of 0.3 minute. The results thus show a positive improvement in the product.

Act Stage   In the act stage, a decision is made regarding implementation. If the results of the analysis conducted in the check stage are positive, the plan proposed is adopted. Customer and process feedback will again be obtained after full-scale implementation. Such information will provide a true measure of the plan's success. If the results of the check stage show no significant improvement, alternative plans made be developed, and the cycle continues.

In our paint example, the proposed change in paint mix reduced the drying time, and the decision to change the mix is then implemented on a full scale. Samples produced by this new process are now taken. Is the mean drying time still 1.3 minutes with a standard deviation of 0.2 minute, as in the check stage? Or has full-scale implementation caused these statistics to change? If so, what are they? What can be done to further reduce the mean drying time and the standard deviation of the drying time?

Customers' needs are not constant. They change with time, competition, societal outlook, and other factors. In the paint example, do customers still desire a drying time of 1 minute? Do they have other needs higher in priority than the drying time? Therefore, the proposed plan is checked continuously to see whether it is keeping abreast of these needs. Doing so may require further changes; that is, the cycle continues, beginning once again with the plan stage.

Variability Reduction and Loss Function   Deming's philosophy calls for abandoning the idea that everything is fine if specifications are met. The idea behind this outmoded attitude is that there is no loss associated with producing items that are off-target but within specifications. Of course, just the opposite is true, which is the reason for striving for continual process improvement. Reducing process variability is an ongoing objective that minimizes loss.

Following the course set by Deming, in 1960, Genichi Taguchi of Japan formalized certain **loss functions**. He based his approach on the belief that economic loss accrues with any deviation from the target value. Achieving the target value wins high praise from the customer and yields no loss. Small deviations yield small losses. However, with larger deviations from the target, the losses increase in a nonlinear (say, a quadratic) relationship.

Figure 2-8 demonstrates the new loss function along with the old viewpoint. The new loss function ties into the idea that companies must strive for continual variability reduction; only then will losses be reduced. Any deviation from the customer target will not yield the fullest possible customer satisfaction. Losses may arise because of such problems as lost opportunities, warranty costs, customer complaint costs, and other tangible and intangible costs. There is even a loss associated with customers not praising the product even though they are not unhappy with it. It is important to get people to praise the product or service because this affects public perception of the product and hence of the company.

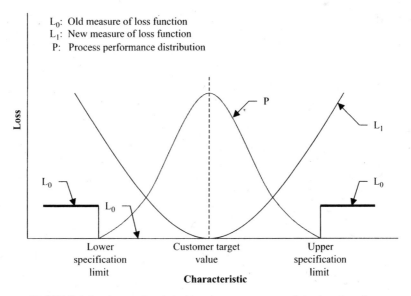

L_0: Old measure of loss function
L_1: New measure of loss function
P: Process performance distribution

FIGURE 2-8   Comparison of old and new measures of the loss function.

Deming's Point 6   *Institute training.*

Employee **training** is integral to proper individual performance in the extended process setting. If employees function in accordance with the goals of the company, an improvement in quality and productivity results. This in turn reduces costs and increases profits.

Employees are the fundamental asset of every company. When employees are hired, they should be carefully instructed in the company's goals in clear-cut operational terms. Merely stating that the company supports a total quality program is not sufficient. Instead, employees must know and buy into the company's long-term goals. Understanding these goals is essential to performing adequately. Employees' individual goals may not always be compatible with those of the company. For example, an employee's desire to produce 50 items per day may not be consistent with the company's goal of defect-free production. Instruction enables the employee to understand what his or her responsibilities are for meeting customers' needs.

Training must be presented in unambiguous operational terms. Employees must know exactly what is to be done and its importance in the entire process. Even the employee who performs only one operation of the many that a product goes through must understand the needs of the customer and the role of the supplier in the extended process. Statistical concepts and techniques play a central role in the Deming program. Consequently, employees must be trained in several statistical tools; these include flow diagrams, histograms, control charts, cause-and-effect diagrams, Pareto diagrams, scatter diagrams, and design of experiments. We examine these tools in detail later.

Deming's Point 7   *Teach and institute leadership.*

Supervisors serve as vital links between management and workers and have the difficult job of maintaining communication channels. Thus, they must understand both the problems of workers and top management's goals. Communicating management's commitment to quality improvement to the workers is a key function of supervisors. To be effective leaders, the supervisors must not think punitively but must think in terms of helping workers do a better job. Shifting to this positive attitude creates an atmosphere of self-respect and pride for all concerned.

Supervisors need to be trained in statistical methods; they are positioned to provide crucial leadership and instruction in these areas. By creating a supportive atmosphere, employee morale is improved and the achievement of the overall goal of quality improvement is facilitated. Supervisors are in the best position to identify common causes inherent in the system, causes for which the workers should not be blamed. It is management's responsibility to minimize the effects of common causes. Special causes, such as poor quality of an incoming raw material, improper tooling, and poor operational definitions should be eliminated first. Identification of these special causes can be accomplished through the use of control charts, which are discussed in later chapters. Supervisors often end up managing things (e.g., equipment) and not people. Such an approach overlooks the fundamental asset of an organization—people.

Deming's Point 8   *Drive out fear. Create trust. Create a climate for innovation.*

Functioning in an environment of fear is counterproductive, because employee actions are dictated by behavior patterns that will please supervisors rather than meet the long-term goals of the organization. The economic loss associated with fear in organizations is immense. Employees are hesitant to ask questions about their job, the methods involved in production, the process conditions and influence of process parameters, the operational definition of what is acceptable, and other such important issues. The wrong signal is given when a supervisor or manager gives the impression that asking these questions is a waste of time.

A fear-filled organization is wasteful. Consider an employee who produces a quota of 50 parts per day—without regard to whether they are all acceptable—just to satisfy the immediate supervisor. Many of these parts will have to be scrapped, leading to wasted resources and a less than optimal use of capacity. Fear can cause physical or physiological disorders, and poor morale and productivity can only follow. A lack of job security is one of the main causes of fear.

Creating an environment of trust is a key task of management. Only when this trust embraces the entire extended process—when workers, vendors, investors, and the community are included—can an organization strive for true innovation. As management starts to implement the 14 points, removing or reducing fear is one of the first tasks to tackle, because an environment of fear starts at the top. The philosophy of managing by fear is totally unacceptable; it destroys trust, and it fails to remove barriers that exist between different levels of the organization.

Deming's Point 9   *Optimize toward the aims and purposes of the company the efforts of teams, groups, staff areas.*

Organizational barriers (Figure 2-9) impede the flow of information. Internal barriers within organizations include barriers between organizational levels (e.g., between the supervisor and workers) and between departments (perhaps between engineering and production, or between product design and marketing). The presence of such barriers impedes the flow of information, prevents each entity in the extended process from perceiving organizational goals, and fosters the pursuit of individual or departmental goals that are not necessarily consistent with the organizational goals.

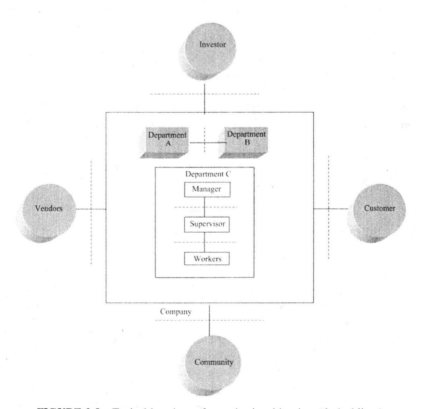

FIGURE 2-9 Typical locations of organizational barriers (dashed lines).

External barriers include those between the vendors and the company, the company and the customer, the company and the community, and the company and its investors. The very survival of the company may be in question if it does not incorporate the sentiments of the community in which it exists. The relationships of the company to its customers, the vendors, and community must be harmonious.

Poor communication is often a culprit in barrier creation. Perhaps top management fails to model open and effective communication. A fear-ridden atmosphere builds barriers. Another reason that barriers exist is the lack of cross-functional teams. Interdisciplinary teams promote communication between departments and functional areas. They also result in innovative solutions to problems. Employees must feel that they are part of the same team trying to achieve the overall mission of the company.

Breaking down barriers takes time; it requires changing attitudes. However, they can and do change when everyone involved is convinced of the advantages of doing so and of the importance of a team effort in achieving change. At the Ford Motor Company, for instance, this concept is found at every level of their design approval process. This process incorporates input from all related units, such as design, sales and marketing, advance product planning, vehicle engineering, body and assembly purchasing, body and assembly manufacturing, product planning, and others. Management emphasizes open lines of communication at all levels and among different departments. The reward system used to facilitate this process is based on teamwork rather than on an individual person's production.

Deming's Point 10   *Eliminate exhortations for the workforce.*

Numerical goals such as a 10% improvement in productivity set arbitrarily by management have a demoralizing effect. Rather than serving to motivate, such standards have the opposite effect on morale and productivity.

Consider, for example, an insurance company processing claims. The company tracks the average time to process a claim (in days) on a weekly basis, a plot of which is shown in Figure 2-10. The figure shows that this average hovers around 6 days. Suppose that management now sets a goal of 4 days for claims processing time. What are the implications of this? First, what is the rationale behind the goal of 4 days for claims processing time. Second, is management specifying ways to achieve the goal? If the answers to these questions are unsatisfactory, employees can only experience frustration when this goal is presented.

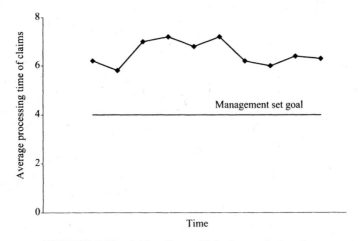

FIGURE 2-10   Arbitrarily established numerical goals.

If the process is stable, the employees have no means of achieving the goal unless management changes the process or the product. Management has to come up with a feasible course of action so that the desired goal can be achieved. Failing to do so will lower morale and productivity. Hence, goals should be set by management in a participative style, and procedures for accomplishment should be given.

Deming's Point 11   (a) *Eliminate numerical quotas for production. Instead, learn and institute methods for improvement.* (b) *Eliminate M.B.O. [management by objectives]. Instead, learn the capabilities of processes, and how to improve them.*

Work standards are typically established by someone other than those who perform the particular job in question. They are based on quantity without regard to quality. According to Deming, setting such work standards guarantees inefficiency and increases costs. The numerical quota defined by a work standard may derail implementing the Deming improvement cycle because people naturally strive to meet the quota rather than to produce acceptable goods. As such, numerical quotas actually promote the production of nonconforming items.

Another drawback of work standards is that they give no information about the procedure that might be used to meet the quota. Is the numerical value a feasible one? Usually, when determining work standards, an allowance is made for the production of nonconforming items, but this often simply ensures that a certain proportion of defectives will be produced. The company thus moves farther from the desirable goal of continuous improvement. Quotas provide no game plan for implementing a quality system.

A third drawback of the quota system is that it fails to distinguish between special causes and common causes when improvements in the process are sought. Consequently, workers may be penalized for not meeting the quota when it is really not their fault. As discussed previously, common causes can be eliminated only by management. Thus, improvements in process output cannot occur unless a conscientious effort is made by management. If the quota is set too high, very few workers will meet the objectives. This will lead to the production of more defective units by workers, because they will try to meet the numerical goal without regard for quality. Furthermore, workers will experience a loss of pride in their work, and worker morale and motivation will drop significantly.

Work standards are typically established through union negotiation and have nothing to do with the capability of the process. Changes in process capability are not pursued, so the standards do not reflect the potential of the current system. Workers who surpass a standard that has been set too high may be producing several defectives, and they may know it. They realize that they are being rewarded for producing nonconforming items—which is totally against Deming's philosophy. On the other hand, if quotas are set too low, productivity will be reduced. A worker may meet the quota with ease, but once the quota is achieved, he or she may have no motivation to exceed it; if management finds out that several people's output meets or exceeds the quota, the quota will probably be increased. This imposes an additional burden on the employee to meet the new quota, without the aid of improved methods or procedures.

The work standard system is never more than a short-term solution, if it is that. On the other hand, using control charts to analyze and monitor processes over time offers proper focus on long-term goals. Statistical methods are preferable over arbitrary work standards, because they help an organization stay competitive.

Deming's Point 12   *Remove barriers that rob people of pride of workmanship.*

A **total quality system** can exist only when all employees synergistically produce output that conforms to the goals of the company. Quality is achieved in all components of the extended process when the employees are satisfied and motivated, when they understand their role in the context of the organization's goals, and when they take pride in their work. It is management's duty to eliminate barriers that prevent these conditions from occurring. A direct effect of pride in workmanship is increased motivation and a greater ability for employees to see themselves as part of the same team—a team that makes good things happen.

Factors That Cause a Loss of Pride   Several factors diminish worker pride. First, management may not treat employees with dignity. Perhaps they are insensitive to workers' problems (personal, work, or community). Happy employees are productive, and vice versa. Happy employees don't need continuous monitoring to determine whether their output is acceptable.

Second, management may not be communicating the company's mission to all levels. How can employees help achieve the company's mission if they do not understand what the mission is?

Third, management may assign blame to employees for failing to meet company goals when the real fault lies with management. If problems in product output are caused by the system (such as poor-quality raw materials, inadequate methods, or inappropriate equipment), employees are not at fault and should not be penalized (even though the best employee might be able to produce a quality product under these circumstances). Assigning blame demoralizes employees and affects quality. As Deming noted, the problem is usually the system, not the people.

Focusing on short-term goals compounds these problems. Consider daily production reports. Different departments dutifully generate pertinent data, but the focus is wrong and they know it: Top management is "micromanaging" and not attending to long-term goals. Constant pressure to increase quantity on a daily basis does not promote the notion of quality. How many times have we heard of a department manager having to explain why production dropped today by, say, 50 units, compared to yesterday? Such a drop may not even be statistically significant. Inferences should be based on sound statistical principles.

Performance Classification Systems   Inadequate performance evaluation systems rob employees of their pride in workmanship. Many industries categorize their employees as excellent, good, average, fair, or unacceptable, and they base pay raises on these categorizations. These systems fail because there are often no clear-cut differences between categories, which leads inevitably to inconsistencies in performance evaluation. A person may be identified as "good" by one manager and "average" by another. That is not acceptable.

A major drawback is that management does not have a statistical basis for saying that there are significant differences between the output of someone in the "good" category and someone in the "average" category. For instance, if a difference in output between two workers were statistically insignificant (i.e., due to chance), it would be unfair to place the two workers in different categories. Figure 2-11 shows such a classification system composed of five categories. The categories numbered 1 through 5 (unacceptable through excellent) have variabilities that are not due to a fundamental difference in the output of the individuals. In fact, the employees may not even have a chance to improve their output because of system deficiencies. Thus, two employees may be rated by their supervisors as 3 (average) and 4 (good), respectively, with the employee rated as 4 considered superior in performance to the other. However, both of these employees may be part of the same distribution, implying that there is no statistically significant difference between them. With this particular system, whatever aggregate measure of evaluation is being used to lump employees into categories, there are no statistically significant differences between the

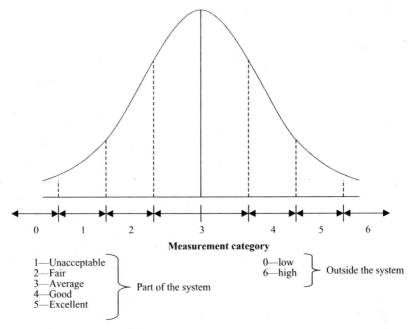

FIGURE 2-11 Improper classification system where categories are part of the same system.

values for categories 1, 2, 3, 4, or 5. The only two groups that may be considered part of another system, whose output therefore varies significantly from those in categories 1 to 5, are categories 0 and 6. Those placed in group 0 may be classified as performing poorly (so low that they are considered part of a different system). Similarly, those placed in group 6 may be classified as performing exceptionally well.

Note that the discussion above is based on the assumption that the numerical measure of performance being used to categorize individuals into groups is normally distributed as in Figure 2-11. Only those significant deviations from the mean (say, three or more standard deviations on either side) may be considered extraordinary, that is, not belonging to the same system that produces ratings of 1 to 5. Under these circumstances, category 0 (whose cutoff point is three standard deviations below the mean) and category 6 (whose cutoff point is three standard deviations above the mean) would be considered different from categories 1 to 5 because they are probably from a different distribution.

Thus, statistically speaking, the only categories should be the three groups corresponding to the original categories 0, 1 to 5, and 6. In Deming's approach, teamwork is considered extremely important. Consideration should be given to this point when conducting performance appraisal. The procedure should be designed so that promoting teamwork is a necessary criterion for placing someone in the highest category. Such team members should be rewarded through merit raises. For those placed in categories 1 to 5, monies set aside for investment in the system should be distributed evenly.

Deming's Point 13   *Encourage education and self-improvement for everyone.*

Deming's philosophy is based on continuous, long-term process improvement. To meet this goal, the organization's most important resource, its people, have to be motivated and adequately trained. This is the only way that the company can survive in today's highly competitive global business environment. Management must commit resources for education and retaining. This represents a sizable investment in both time and money. However, an educated workforce, with a focus on future needs, can safeguard improvements in quality and **productivity** and help a company maintain a competitive position.

Point 6, discussed previously, deals with training that enables new employees to do well. Point 13, on the other hand, addresses the need for ongoing and continual education and self-improvement for the entire organization. A workforce that is continually undergoing education and retraining will be able to keep up with cutting-edge technology. Today's workplace is changing constantly, and knowing that they are adapting to it gives employees a sense of security. A company that invests in its employees' growth and well-being will have a highly motivated workforce—a win–win situation for everyone. Education is an investment in the future.

Advantages of Education and Retraining   Investments in the workforce, such as educating and retraining employees at all levels, serve a multifold purpose. First, employees (who believe that if the company is willing to incur such expenditures for their benefit must be interested in their well-being) are likely to be highly motivated employees. Such a belief fosters the desire to excel at work.

Second, as employees grow with the company and their job responsibilities change, retraining provides a mechanism to ensure adequate performance in their new jobs. Such progressive action by the company makes it easier to adapt to the changing needs of the consumer.

Education should always include instruction on Deming's 14 points. Also, education and retraining applies to people in nonmanufacturing or service areas as well. Thus, employees in accounting, purchasing, marketing, sales, or maintenance, as well as those in banks, retail stores, federal and state government offices, health care, hospitality, information technology, or the transportation sector need this instruction to participate in the never-ending process of improvement. Industry should work with academia to aid in this process of education and retraining.

Deming's Point 14   *Take action to accomplish the transformation.*

Point 14 involves accepting the Deming philosophy and committing to seeing its implementation in the extended process. A structure must be created and maintained for the dissemination of the concepts associated with the first 13 points. Responsibility for creating this structure lies with top management. Besides being totally committed to the idea of quality, management must be visionary and knowledgeable as to its potential impacts, and they must be in it for the long run. For the implementation to succeed, people must be trained in statistics at all levels of the organization. A quality company requires statistical assistance at every level.

Deming's Deadly Diseases

Deming's 14 points for management provide a road map for continuous quality improvement. In implementing these points, certain practices of management, which Deming labels as deadly diseases or sins, must be eliminated. Most of **Deming's deadly diseases** involve a lack of understanding of variation. Others address management's failure to understand who (they and only they) can correct common causes in the system. The five deadly diseases, which Deming's 14 points seek to stamp out, are as follows:

Management by Visible Figures Only   This deadly disease is also known as "management by the numbers." Visible figures are such items as monthly production, amount shipped, inventory on hand, and quarterly profit. Emphasizing short-term visible figures doesn't always present an accurate picture of the financial state of an organization. Visible figures can easily be manipulated to show attractive quarterly profits when, in fact, the company's competitive position is in jeopardy. How many managers would deny that it is easy to show a large amount of production shipped at the end of the month and to hide the numbers of defectives returned? Cutting back research, training, and education budgets is amazingly tempting when quarterly profits are disappointing. Although these actions will show desirable short-term profits, they can seriously sabotage a company's long-term goals of remaining competitive and of improving constantly and forever. As Deming often stated, "he who runs his company on visible figures alone will in time have neither company nor figures."

Properly selected statistical data, unbiased and as objective as possible, is *the* quality assessment tool of the quality company. Some examples include data on customer satisfaction; employee satisfaction with, and perception of, the company; and employee morale and motivation. Often, however, the most pertinent data are unknown or difficult to measure. How does a company measure loss of market share due to customer dissatisfaction? How does a company measure loss of goodwill because of the unmet needs of its customers who decide to switch? How does it measure the damage caused by an unmotivated workforce? How does it measure the losses that accrue when management fails to create an atmosphere of innovation? These losses are difficult, if not impossible, to measure; we could call them the *invisible figures*. They are invisible and real, and they must be addressed in the transformation process.

Some of the Deming's 14 points for management seek to eliminate the deadly "visible figures only" disease. Consider point 5, which deals with constant, never-ending cycles of improvement. This suggests that a system should first be brought to a state of statistical control by eliminating the special or assignable causes. Subsequently, cross-disciplinary teams could be used to solve the common causes. Although variability will always be present in any system, it is important that common causes not be treated as special causes. Point 8 stresses eliminating fear and creating a climate of trust and innovation. Implementing this point promotes teamwork because teams obviously cannot function in fear-driven organizations. Then and only then can workers drop the "every person for himself or herself" attitude and work for the common good of the company. Once unrealistic numerical goals, which cannot be accomplished under the existing system, are eliminated, employee fear dissolves, and effective quality improvement teams can get to work.

Lack of Constancy of Purpose   This disease prevents organizations from making a transformation to the long-term approach. Top management must rid itself of myopia. Management cannot mouth a long-term vision but enforce a short-term approach to operational decisions. The slogan of the legendary founder of Wal-Mart, Sam Walton, applies here: "Management must walk their talk." Only then will employees be persuaded to pursue the quality work ethic.

Constancy of purpose ensures survival. A healthy company, doing everything it can to stay in for the long haul, provides a sense of security to its employees. No company can guarantee lifetime employment; however, when employees see that the company has a long-term vision that it lives by, and they see daily demonstrations of this vision, they also see stability.

Mission statements play an important role here. Unless the mission statement is propagated through all levels of the company and understood and adapted by all, it is a useless document. But when employees understand a company's mission statement and know that management and other employees believe in it *and* use it, it becomes the embodiment of the corporate **quality culture**. The wording key: The emphasis must be on the *process* and not on the end result. For example, a mission statement that talks about maximizing profits and return on investment does little to guide employees. Profits and such are an end result, not the means. The mission statement is the means, the vision, the road map writ large. Managing the process will automatically take care of the end result. Mission statements must be simple and provide vision and direction to *everyone*.

Effective mission statements are formulated only after considerable input from all constituencies, but top management has the ultimate say. Mission statements are usually nonquantitative. Here is an excerpt from General Electric Company's mission: *"Progress is our most important product."*

The mission is implemented bottom-up. New recruits should be hired to further the mission, and current employees must demonstrate their support for the mission. The wise company also makes its customers aware of its mission.

Performance Appraisal by the Numbers   In such systems, the focus is on the short term and on the outcome rather than on the process. One tremendous disadvantage of these systems is that they promote rivalry and internal competition. Employees are forced to go up against each other, and about 50% must be ranked as below average, whether or not their work is satisfactory. In our society, being ranked below average is a strong message, strong enough to cause depression and bitterness. It demoralizes and devastates. Whatever spirit of cooperation that may have existed among employees previously begins to erode.

The company must endeavor to stay competitive, even as it completes the transformation process. Such is the message of points 1 and 5. Numerical performance appraisals make this difficult because they encourage counterproductive behavior among employees. Rivalry within weakens the battle on rivalry without: that is, rivalry in the marketplace. For the company to stay in business, prosper, and create jobs, a concerted effort must be made by everyone to work as a team. Individual-centered performance, whose sole objective is to demonstrate one employee's superiority over another, sabotages this effort. Numerical performance appraisal thus has a detrimental effect on teamwork.

Synergism—effective teamwork—is attained only when employees share the common goal of improving companywide performance. It is possible for individual employees to dispatch their duties "by the book," yet fail to improve the competitive position of the company. Point 9 focuses on breaking down organizational barriers and promoting teamwork. Cross-functional teams, which focus on the process, are a means of continuous improvement. Obviously, successfully implementing such teams is difficult when employees are evaluated on individual performance.

Short-Term Orientation   Many companies choose less-than-optimal solutions because they focus on near-term goals and objectives. Decisions dictated by quarterly dividends or short-term profits can diminish a company's chances of long-term survival. This is in direct contradiction to points 1 and 2.

A reason often given for adopting a short-term approach is the need to satisfy shareholders. Companies are fearful of hostile takeovers and believe that demonstrating short-term profits can discourage this activity. Top-down pressure to create short-term profits causes many undesirable actions: Monthly or quarterly production may be manipulated, training programs may be cut, and other investments on employees may be curtailed.

Although such actions may defer costs and show increased short-term profits, they can have a devastating effect on a company's long-term survival. Moreover, the real impact of such schemes is not always tangible. The connection between a disheartened workforce and cutting back on quality control doesn't always manifest itself in the visible numbers.

Long-term partnerships can have a sizable impact on cost savings. IBM used to contract with about 200 carriers to move parts to different locations around the country. Thousands of invoices had to be processed annually, and stability among carriers was nonexistent as they struggled to stay in business despite fluctuating orders. By reducing the number of carriers over tenfold, IBM achieved significant savings in order-processing costs. Although the initial goal may have been to reduce invoicing headaches, in reality, relationships with a few selected vendors have positively affected IBM's competitive position. The concept of the extended process becomes a reality, as the vendor's technical problems are also that of the company's, and vice versa.

Linking executive salaries with annual bottom-line profit figures also promotes a short-term focus. In many companies, shareholders approve this policy, and when this prevails, senior managers sometimes postpone much-needed expenditures to demonstrate short-term gains. Solutions to this problem include basing compensation and bonus packages on five-year moving averages of corporate earnings.

Mobility of Management   A major cause of organizational instability is the short tenure of management. When the average tenure of a midlevel manager is four to five years, continuity is difficult, if not impossible. Getting acquainted with, and accustomed to, a quality culture requires major paradigm shifts, and this takes time. Frequent changes in management sabotage constancy of purpose. An atmosphere of trust is difficult to maintain when the "ground shifts" with each new change in management. To reduce mobility, top management can promote from within, institute job enrichment programs, and practice job rotation.

Companies must demonstrate to managers concern for their advancement. Recall that point 13 deals with providing education and self-improvement for everyone. By investing in management training and seminars that develop skills for expanded responsibilities, the company also gains loyalty from its managers. Attention to salaries relative to industry norms is important. When managers sense that the company is doing all it can to maintain salaries at competitive levels, they are often motivated to stay. Managers know that there are priorities, and they will stay if they are convinced that commitment to employees is the top one.

2-5 PHILIP B. CROSBY'S PHILOSOPHY

Philip B. Crosby founded Philip Crosby Associates in 1979. Prior to that, he was a corporate vice president for ITT, where he was responsible for worldwide quality operations. Crosby has a particularly wide-ranging understanding of the various operations in industry because he started as a line supervisor and worked his way up. Such firsthand experience has provided him with a keen awareness of what quality is, what the obstacles to quality are, and what can be done to overcome them. Crosby has trained and consulted with many people in manufacturing and service industries, and he has written many books (Crosby 1979, 1984, 1989) on quality management.

The Crosby approach begins with an evaluation of the existing quality system. His quality management grid (Crosby 1979) identifies and pinpoints operations that have potential for improvement. Table 2-3 is an example of a **quality management maturity grid**. The grid is divided into five stages of maturity, and six measurement categories aid in the evaluation process.

Four Absolutes of Quality Management

To demonstrate the meaning of quality, Crosby (1979) has identified four absolutes of quality management.

1. *Definition of quality.* Quality means conformance to requirements.
2. *System for achievement of quality.* The rational approach is prevention of defects.
3. *Performance standard.* The only performance standard is zero defects.
4. *Measurement.* The performance measurement is the cost of quality. In fact, Crosby emphasizes the costs of "unquality", such as scrap, rework, service, inventory, inspection, and tests.

14-Step Plan for Quality Improvement

Crosby's 14-step plan is discussed briefly here to help businesses implement a **quality improvement program**.[*] The reader who is interested in a detailed examination of these steps should consult the suggested references.

1. *Management commitment.* For quality improvement to take place, commitment must start at the top. The emphasis on defect prevention has to be communicated, and a quality policy that states the individual performance requirements needed to match customer requirements must be developed.
2. *Quality improvement team.* Representatives from each department or division form the quality improvement team. These individuals serve as spokespersons for each group they represent. They are responsible for ensuring that suggested operations are brought to action. This team brings all the necessary tools together.
3. *Quality measurement.* Measurement is necessary to determine the status of quality for each activity. It identifies the areas where corrective action is needed and where quality improvement efforts should be directed. The results of measurement, which are placed in highly visible charts, establish the foundation for the quality improvement program. These principles apply to service operations as well, such as counting the number of billing or payroll errors in the finance department, the number of drafting errors in engineering, the number of contract or order description errors in marketing, and the number of orders shipped late.
4. *Cost of quality evaluation.* The cost of quality (or rather unquality) indicates where corrective action and quality improvement will result in savings for the company. A study to determine these costs should be conducted through the comptroller's office, with the categories that comprise quality costs defined precisely. This study establishes a measure of management's performance.
5. *Quality awareness.* The results of the cost of nonquality should be shared with all employees, including service and administrative people. Getting everybody involved with quality facilitates a quality attitude.

*Adapted from P. B. Crosby, *Quality Is Free*, McGraw-Hill, New York, 1979.

TABLE 2-3 Crosby's Quality Management Grid

Stages of Maturity

Measurement Categories	Stage I: Uncertainty	Stage II: Awakening	Stage III: Enlightenment	Stage IV: Wisdom	Stage V: Certainty
Management understanding and attitude.	No comprehension of quality as a management tool. Tend to blame quality department.	Recognizing that quality management may be of value but not willing to provide money or time.	While going through an improvement program, learn about quality management.	Participating. Understand absolutes of quality management.	Consider quality management as essential part of company system.
Quality organization status.	Quality is hidden in manufacturing or engineering departments. Emphasis on appraisal and sorting.	A stronger quality leader is appointed, but the main emphasis is still on appraisal and moving the product.	Quality department reports to top management, all appraisal is incorporated.	Quality manager conducts as an officer of company, effective status reporting and preventive action. Involved with consumer affairs.	Quality manager on board of directors. Prevention is main concern.
Problem handling.	Problems are fought as they occur; no resolution; inadequate definition.	Teams are set up to attack major problems. Long-range solutions are not solicited.	Corrective action established. Problems are faced openly.	Problems are identified early. All functions are open to suggestions.	Except in the most unusual cases, problems are prevented.
Cost of quality as a percentage of sales.	Reported: unknown Actual: 20%	Reported: 3% Actual: 18%	Reported: 8% Actual: 12%	Reported: 6.5% Actual: 8%	Reported: 2.5% Actual: 2.5%
Quality improvement actions.	No organized activities. No understanding of such activities.	Trying obvious "motivational" short-range efforts.	Implementation of the 14-step program with thorough understanding.	Continuing the 14-step program.	Quality improvement is a continued activity.
Summary of company quality posture.	"We don't know why we have problems with quality."	"Is it always absolutely necessary to have problems with quality?"	"Through management commitment and quality improvement we are identifying and resolving our problems."	"Defect prevention is a routine part of our operation."	"We know why we do not have problems with quality."

Source: Adapted from Philip B. Crosby, *Quality Is Free*, McGraw-Hill, New York, 1979. Reproduced with permission of The McGraw-Hill Companies.

6. *Corrective action.* Open communication and active discussion of problems creates feasible solutions. Furthermore, such discussion also exposes other problems not identified previously and thus determines procedures to eliminate them. Attempts to resolve problems should be made as they arise. For those problems without immediately identifiable remedies, discussion is postponed to subsequent meetings. The entire process creates a stimulating environment of problem identification and correction.

7. *Ad hoc committee for the zero-defects program.* The concept of **zero defects** must be communicated clearly to all employees; everyone must understand that the achievement of such a goal is the company's objective. This committee gives credibility to the quality program and demonstrates the commitment of top management.

8. *Supervisor training.* All levels of management must be made aware of the steps of the quality improvement program. Also, they must be trained so they can explain the program to employees. This ensures propagation of the quality concepts from the chief executive officers to the hourly worker.

9. *Zero-defects day.* The philosophy of zero defects should be established companywide and should originate on one day. This ensures a uniform understanding of the concept for everyone. Management has the responsibility of explaining the program to the employees, and they should describe the day as signifying a "new attitude." Management must foster this type of quality culture in the organization.

10. *Goal setting.* Employees, in conjunction with their supervisors, should set specific measurable goals. These could be 30-, 60-, or 90-day goals. This process creates a favorable attitude for people ultimately to achieve their own goals.

11. *Error-cause removal.* The employees are asked to identify reasons that prevent them from meeting the zero-defects goal—not to make suggestions but to list the problems. It is the task of the appropriate functional group to come up with procedures for removing these problems. Reporting problems should be done quickly. An environment of mutual trust is necessary so that both groups work together to eliminate the problems.

12. *Recognition.* Award programs should be based on recognition rather than money and should identify those employees who have either met or exceeded their goals or have excelled in other ways. Such programs will encourage the participation of everyone in the quality program.

13. *Quality councils.* Chairpersons, team leaders, and professionals associated with the quality program should meet on a regular basis to keep everyone up to date on progress. These meetings create new ideas for further improvement of quality.

14. *Do it over again.* The entire process of quality improvement is continuous. It repeats again and again as the quality philosophy becomes ingrained.

2-6   JOSEPH M. JURAN'S PHILOSOPHY

Joseph M. Juran founded the Juran Institute, which offers consulting and management training in quality. Juran has worked as an engineer, labor arbitrator, and corporate director in the private sector and as a government administrator and a university professor in the public sector. He has authored many books on the subjects of quality planning, control, management, and improvement (Juran 1986, 1988a,b, 1989; Juran and Gryna 1993).

Like Deming, Juran visited Japan in the early 1950s to conduct training courses in quality management. He eventually repeated these seminars in over 40 countries on all continents. In the 1980s, Juran met the explosive demand for his services with offerings through the Juran

Institute. His books and videotapes have been translated into many languages, and he has trained thousands of managers and specialists. Juran believes that management has to adopt a unified approach to quality. Quality is defined as "fitness for use." The focus here is on the needs of the customer.

Certain nonuniformities deter the development of a unified process. One is the existence of multiple functions—such as marketing, product design and development, manufacture, and procurement—where each function believes itself to be unique and special. Second, the presence of hierarchical levels in the organizational structure creates groups of people who have different responsibilities. These groups vary in their background and may have different levels of exposure to the concepts of quality management. Third, a variety of product lines that differ in their markets and production processes can cause a lack of unity.

Juran proposes a universal way of thinking about quality, which he calls the **quality trilogy**: quality planning, quality control, and quality improvement. This concept fits all functions, levels of management, and product lines.

Quality Trilogy Process

The quality trilogy process starts with quality planning at various levels of the organization, each of which has a distinct goal. At the upper management level, planning is termed *strategic quality management*. Broad quality goals are established. A structured approach is selected in which management chooses a plan of action and allocates resources to achieve the goals. Planning at the middle management level is termed *operational quality management*. Departmental goals consistent with the strategic goals are established. At the workforce level, planning involves a clear assignment to each worker. Each worker is made aware of how his or her individual goal contributes to departmental goals.

After the planning phase, quality control takes over. Here, the goal is to run the process effectively such that the plans are enacted. If there are deficiencies in the planning process, the process may operate at a high level of chronic waste. Quality control will try to prevent the waste from getting worse. If unusual symptoms are detected sporadically, quality control will attempt to identify the cause behind this abnormal variation. Upon identifying the cause, remedial action will be taken to bring the process back to control.

The next phase of the trilogy process is **quality improvement**, which deals with the continuous improvement of the product and the process. This phase is also called the **quality breakthrough sequence**. Such improvements usually require action on the part of upper and middle management, who deal with such actions as creating a new design, changing methods or procedures of manufacturing, and investing in new equipment.

Table 2-4 shows an outline of the various steps involved in the quality planning, quality control, and quality improvement phases. Readers should consult the listed references for an elaborate treatment of the details of each phase.

Quality Planning

1. *Establish quality goals.* Goals, as established by the organization, are desired outcomes to be accomplished in a specified time period. The time period may be short-term or long-term.

2. *Identify customers.* Juran has a concept similar to Deming's extended process. Juran's includes vendors and customers. He stresses the importance of identifying the customer, that could be internal or external. In cases where the output form one department flows to another, the customer is considered internal.

TABLE 2-4 Universal Process for Managing Quality

Quality Planning	Quality Control	Quality Improvement
Establish quality goals	Choose control subjects	Prove the need
Identify customers	Choose units of measure	Identify projects
Discover customer needs	Set goals	Organize project teams
Develop product features	Create a sensor	Diagnose the causes
Develop process features	Measure actual performance	Provide remedies, prove that the remedies are effective
Establish process controls, transfer to operations	Interpret the difference	Deal with resistance to change
	Take action on the difference	Control to hold the gains

Source: Adapted from J.M. Juran and F.M. Gryna, *Quality Planning and Analysis*, 3rd ed., McGraw-Hill, New York, 1993. Reproduced with permission of The McGraw-Hill Companies.

3. *Discover customer needs.* Long-term survival of the company is contingent upon meeting or exceeding the needs of the customer. Conducting analysis and research, surveying clients and nonclients, and keeping abreast of the dynamic customer needs are a few examples of activities in this category.

4. *Develop product features.* With customer satisfaction as the utmost objective, the product or service should be designed to meet the customer requirements. As customer needs change, the product should be redesigned to conform to these changes.

5. *Develop process features.* While a product is designed based on a knowledge of customer needs, this step deals with the manufacturing process of that product. Methods must be developed, and adequate equipment must be available to make the product match its design specifications. For service organizations, effective and efficient processes that meet or exceed customer requirements are critical.

6. *Establish process controls, transfer to operations.* For manufacturing operations, bounds should be established on process variables for individual operations, that assist in making an acceptable product. Similarly, in the service setting, norms on operations such as time to complete a transaction must be adopted.

Quality Control

1. *Choose control subjects.* Product characteristics that are to be controlled in order to make the product conform to the design requirements should be chosen. For instance, a wheel's control characteristics may be the hub diameter and the outside diameter. Selection is done by prioritizing the important characteristics that influence the operation or appearance of the product and hence impact the customer.

2. *Choose units of measure.* Based on the quality characteristics that have been selected for control, appropriate units of measure should be chosen. For example, if the hub diameter is being controlled, the unit of measurement might be millimeters.

3. *Set goals.* Operational goals are created such that the product or service meets or exceeds customer requirements. For instance, a standard of performance for the hub diameter could be 20 ± 0.2 mm. A hub with a diameter in this range would be compatible in final assembly and would also contribute to making a product that will satisfy the customer.

4. *Create a sensor.* To collect information on the identified quality characteristics, automated equipment or individuals, who serve as auditors or inspectors, are integrated into the system. Databases that automatically track measurements on the quality characterstic (diameter of hub or processing time of purchase order) could also serve as sensors.

5. *Measure actual performance.* This phase of quality control is concerned with the measurement of the actual process output. Measurements are taken on the previously selected control subjects (or quality characteristics). Such measurements will provide information on the operational level of the process.

6. *Interpret the difference.* This involves comparing the performance of the process with the established goals. If the process is stable and capable, then any differences between the actual and the standard may not be significant.

7. *Take action on the difference.* In the event that a discrepancy is found between the actual output of the process and the established goal, remedial action needs to be taken. It is usually management's responsibility to suggest a remedial course of action.

Quality Improvement

1. *Prove the need.* Juran's breakthrough sequence tackles the chronic problems that exist because of a change in the current process; this task requires management involvement. First, however, management has to be convinced of the need for this improvement. Problems such as rework and scrap could be converted to dollar figures to draw management's attention. It would also help to look at problems as cost savings opportunities.

2. *Identify projects.* Because of the limited availability of resources, not all problems can be addressed simultaneously. Therefore, problems should be prioritized. A Pareto analysis is often used to identify vital problems. Juran's quality improvement process works on a project-by-project basis. A problem area is identified as a project, and a concerted effort is made to eliminate the problem.

3. *Organize project teams.* The organizational structure must be clearly established so projects can be run smoothly. Authority and responsibility are assigned at all levels of management to facilitate this. Top management deals with strategic responsibilities, and lower management deals with the operational aspects of the actions. Furthermore, the structure should establish precise responsibilities for the following levels: guidance of overall improvement program, guidance for each individual project, and diagnosis and analysis for each project.

4. *Diagnose the causes.* This is often the most difficult step in the whole process. It involves data gathering and analysis to determine the cause of a problem. The symptoms surrounding the defects are studied, and the investigator then hypothesizes causes for the symptoms. Finally, an analysis is conducted to establish the validity of the hypotheses. Juran defines a **diagnostic arm** as a person or group of persons brought together to determine the causes of the problem. The organization needs to enlist the right people and to ensure that the required tools and resources are available. This is accomplished through a **steering arm**.

5. *Provide remedies, prove that the remedies are effective.* Here, remedial actions are developed to alleviate the chronic problems. Remedies may deal with problems that are

controllable by management or those that are controllable by operations. Changes in methods or equipment should be considered by management and may require substantial financial investment. Frequently, the return on investment is analyzed. This is also the real test of the effectiveness of the remedies proposed. Can the suggested actions be implemented, and do they have the beneficial effect that has been hypothesized?

6. *Deal with resistance to change.* The breakthrough process requires overcoming resistance to change. Changes may be technological or social in nature. The proposed procedure may require new equipment, and operators may have to be trained. Management commitment is vital to the effective implementation of the changes. By the same token, social changes, which deal with human habits, beliefs, and traditions, require patience, understanding, and the participation of everyone involved.

7. *Control to hold the gains.* Once the remedial actions have been implemented and gains have been realized, there must be a control system to sustain this new level of achievement. In other words, if the proportion of nonconforming items has been reduced to 2%, we must make sure that the process does not revert to the former nonconformance rate. A control mechanism is necessary, for example, audits may be performed in certain departments. Such control provides a basis for further process improvement as the whole cycle is repeated.

2-7 THE THREE PHILOSOPHIES COMPARED

We have now briefly examined the quality philosophies of three experts: Deming, Crosby, and Juran. All three philosophies have the goal of developing an integrated total quality system with a continual drive for improvement. Although there are many similarities in these approaches, some differences do exist. A good discussion of these three philosophies may be found in an article by Lowe and Mazzeo (1986).

Definition of Quality

Let's consider how each expert defines **quality**. Deming's definition deals with a predictable uniformity of the product. His emphasis on the use of statistical process control charts is reflected in this definition. Deming's concern about the quality of the product is reflected in the quality of the process, which is the focal point of his philosophy. Thus, his definition of quality does not emphasize the customer as much as do Crosby's and Juran's. Crosby defines quality as conformance to requirements. Here, requirements are based on customer needs. Crosby's performance standard of zero defects implies that the set requirements should be met every time. Juran's definition of quality—fitness of a product for its intended use— seems to incorporate the customer the most. His definition explicitly relates to meeting the needs of the customer.

Management Commitment

All three philosophies stress the importance of top management commitment. Deming's first and second points (creating a constancy of purpose toward improvement and adopting the new philosophy) define the tasks of management. In fact, his 14 points are all aimed at management, implying that management's undivided attention is necessary to create a total quality system. Point 1 in Crosby's 14-step process deals with management commitment. He stresses the importance of management communicating its understanding and commitment. Crosby's philosophy is focused on the creation of a "quality culture," which can be attained through management commitment. Juran's quality planning, control, and improvement process seeks management support at all levels. He believes in quality improvement on a project basis. The project approach gets managers involved and assigns responsibilities to each. Thus, in all three philosophies, the support of top management is crucial.

Strategic Approach to a Quality System

Deming's strategy for top management involves their pursuing the first 13 points and creating a structure to promote the 13 points in a never-ending cycle of improvement (i.e., point 14). Crosby's approach to quality improvement is sequenced. His second step calls for the creation of quality improvement teams. Under Juran's philosophy, a quality council guides the quality improvement process. Furthermore, his quality breakthrough sequence involves the creation of problem-solving steering arms and diagnostic arms. The steering arm establishes the direction of the problem-solving effort and organizes priorities and resources. The diagnostic arm analyzes problems and tracks down their causes.

Measurement of Quality

All three philosophies view quality as a measurable entity, although in varying degrees. Often, top management has to be convinced of the effects of good quality in dollars and cents. Once they see it as a cost-reducing measure, offering the potential for a profit increase, it becomes easier to obtain their support. A fundamental aim of the quality strategy is to reduce and eliminate scrap and rework, which will reduce the cost of quality. A measurable framework for doing so is necessary. The total cost of quality may be divided into subcategories of prevention, appraisal, internal failure, and external failure. One of the difficulties faced in this setting is the determination of the cost of nonquality, such as customer nonsatisfaction. Notice that it is difficult to come up with dollar values for such concerns as customer dissatisfaction, which is one of Deming's concerns in deriving a dollar value for the total cost of quality. Crosby believes that quality is free; it is "unquality" that costs.

Never-Ending Process of Improvement

These philosophies share a belief in the never-ending process of improvement. Deming's 14 steps repeat over and over again to improve quality continuously. Deming's PDCA cycle (plan–do–check–act) sustains this never-ending process, as does Juran's breakthrough sequence. Crosby also recommends continuing the cycle of quality planning, control, and improvement.

Education and Training

Fundamental to quality improvement is the availability of an adequate supply of people who are educated in the philosophy and technical aspects of quality. Deming specifically referred to this in his sixth point, which talks about training all employees, and in his thirteenth point, which describes the need for retraining to keep pace with the changing needs of the customer. Deming's focus is on education in statistical techniques. Education is certainly one of Crosby's concerns as well; his eighth step deals with quality education. However, he emphasizes developing a quality culture within the organization so that the right climate exists. Juran's steps do not explicitly call for education and training. However, they may be implicit, because people must be knowledgeable to diagnose defects and determine remedies.

Eliminating the Causes of Problems

In Deming's approach, **special causes** refer to problems that arise because something unusual has occurred, and **common causes** refer to problems that are inherent to the system. Examples of special causes are problems due to poor quality from an unqualified vendor or use of an improper tool. With common causes, the system itself is the problem. Examples of common causes are inherent machine variability or worker capability. These problems are controllable only by management. Both Deming and Juran have claimed that about 85% of problems have common causes. Hence, only action on the part of management can eliminate them; that is, it is up to management to provide the necessary authority and tools to the workers so that the common causes can be removed.

At the heart of Deming's philosophy are the statistical techniques that identify special causes and common causes—especially statistical process control and control charts. Variations *outside* the control limits are attributed to special causes. These variations are worker-controllable, and the workers are responsible for eliminating these causes. On the other hand, variations *within* the control limits are viewed as the result of common causes. These variations require management action.

Juran's approach is similar to Deming's. In his view, special causes create **sporadic problems**, and common causes create **chronic problems**. Juran provides detailed guidelines for identifying sporadic problems. For example, he categorizes operator error as being inadvertent, willful, or due to inadequate training or improper technique. He also specifies how the performance standard of zero defects that Crosby promotes can be achieved. Crosby, of course, suggests a course of action for error cause removal in his eleventh step, whereby employees identify reasons for nonconformance.

Goal Setting

Deming was careful to point out that arbitrarily established numerical goals should be avoided. He asserted that such goals impede, rather than hasten, the implementation of a total quality system. Short-term goals based mainly on productivity levels without regard to quality are unacceptable. By emphasizing the never-ending quality improvement process, Deming saw no need for short-term goals. On the other hand, both Crosby and Juran call for setting goals. Crosby's tenth point deals with goal setting; employees (with guidance from their supervisors) are asked to set measurable goals for even short-term periods such as 30, 60, or 90 days. Juran recommends an annual quality improvement program with specified goals. He believes that such goals help measure the success of the quality projects undertaken in a given year. The goals should be set according to the requirements of the customer. Juran's approach resembles the framework of management by objectives, where performance is measured by achievement of stipulated numerical goals.

Structural Plan

Deming's 14-point plan emphasizes using statistical tools at all levels. Essentially a bottom-up approach, the process is first brought into a state of statistical control (using control charts) and then improved. Eliminating special causes to bring a process under control takes place at the lower levels of the organizational structure. As these causes are removed and the process assumes a state of statistical control, further improvements require the attention of upper-level management.

Crosby, on the other hand, takes a top-down approach. He suggests changing the management culture as one of the first steps in his plan. Once the new culture is ingrained, a plan for managing the transition is created.

Finally, Juran emphasizes quality improvement through a project-by-project approach. His concept is most applicable to middle management.

Because each company has its own culture, companies should look at all three approaches and select the one (or combination) that is most suited to its own setting.

SUMMARY

In this chapter we discuss the quality philosophies of Deming, Crosby, and Juran, with emphasis on Deming's 14 points for management. Many companies are adopting the Deming approach to quality and productivity. The quality philosophies of Crosby and Juran provide the reader with a broad framework of the various approaches that exist for management. All three approaches have the same goal, with slightly different paths.

Whereas Deming's approach emphasizes the importance of using statistical techniques as a basis for quality control and improvement, Crosby's focuses on creating a new corporate culture that deals with the attitude of all employees toward quality. Juran advocates quality improvement through problem-solving techniques on a project-by-project basis. He emphasizes the need to diagnose correctly the causes of a problem based on the symptoms observed. Once these causes have been identified, Juran focuses on finding remedies. Upon understanding all three philosophies, management should select one or a combination to best fit their own environment. These philosophies of quality have had a global impact.

The chapter introduces quality characteristics associated with the service industry, a dominant segment of the economy. While distinguishing the differences between the manufacturing and service sectors, the chapter presents key traits that are important to customer satisfaction in service organizations. A model for service quality is discussed. Customer satisfaction in a service environment is influenced by the levels of expected and actual quality, as perceived by the customer. Hence, factors that influence such perceptions are presented.

CASE STUDY: CLARKE AMERICAN CHECKS, INC.*

Headquartered in San Antonio, Texas, Clarke American supplies personalized checks, checking-account and bill-paying accessories, financial forms, and a growing portfolio of services to more than 4000 financial institutions in the United States. Founded in 1874, the company employs about 3330 people at 25 sites in 15 states. It received the Malcolm Baldrige National Quality Award in 2001 in the manufacturing category.

In addition to filling more than 50 million personalized check and deposit orders every year, Clarke American provides 24-hour service and handles more than 11 million calls

---

*Adapted from "Malcolm Baldrige National Quality Award, 2001 Award Recipient Profile," U.S. Department of Commerce, National Institute of Standards and Technology, Gaithersburg, MD, www.nist.gov.

annually. Clarke American competes in an industry that has undergone massive consolidation. Three major competitors have a 95% market share and vie for the $1.8 billion U.S. market for check-printing services supplied to financial institutions. Since 1996, Clarke American's market share has increased by 50%. Revenues were over $460 million in 2001.

The company is organized into a customer-focused matrix of three divisions and 11 processes. It is in a nearly continual state of organizational redesign, reflecting ever more refined segmentation of its partners and efforts to better align with these customers' requirements and future needs.

"First in Service"

In the early 1990s, when an excess manufacturing capacity in check printing triggered aggressive price competition, Clarke American elected to distinguish itself through service. Company leaders made an all-out commitment to ramp up the firm's "first in service" (FIS) approach to business excellence. Comprehensive in scope, systematic in execution, the FIS approach defines how Clarke American conducts business and how all company associates are expected to act to fulfill the company's commitment to superior service and quality performance.

FIS is the foundation and driving force behind the company's continuous improvement initiatives. It aligns Clarke American's goals and actions with the goals of its partners and the customers of these financial institutions. The company uses this single-minded organizational focus to accomplish strategic goals and objectives.

The company's key leadership team (KLT), consisting of top executives, general managers of business divisions, and vice presidents of processes, establishes, communicates, and deploys values, direction, and performance expectations. Leadership responsibilities go beyond task performance. KLT members are expected to be role models who demonstrate commitment and passion for performance excellence. Each year, associates evaluate the competencies of executives and general managers in 10 important leadership areas. Senior leaders also keep a scorecard to track their progress in implementing company strategies and in demonstrating key behaviors, a tool adopted from a previous Baldrige Award winner.

Goals, plans, processes, measures, and other vital elements of performance improvement are clearly documented and accessible to all. However, the company also places a premium on two-way, face-to-face communication. For example, KLT members and other senior local management lead monthly FIS meetings in their respective division or process. These two- to three-hour meetings are held at all Clarke American facilities, and all associates are expected to attend. Agenda items include competitive updates, reviews of company goals and direction, key-project progress reports, associate and team recognition, and question-and-answer periods.

Running and Changing the Business

Strategic and annual planning are tightly integrated through the company's goal deployment process. Ambitious long-term (three to five years) business goals help to guide the development of short-term (one to three years) objectives and the selection of priority projects necessary to accomplish them. This dual planning approach benefits from detailed analyses of information. Inputs include periodic studies of the industry, market and competitor analyses, benchmarking results, voice of the customer data, the annual associate opinion survey, partner and supplier scorecards, an independent study of customer satisfaction, and assessments of business performance and core competencies.

The goal deployment process has two main objectives: to identify "change the business" objectives and to ensure alignment of the entire company. Long-term *change the business* objectives aim for breakthrough business and performance improvements, on the order of 20% or better. To ensure alignment, measures are used to gage progress toward these goals and are tracked in the company's balanced scorecard.

Annual "run the business" objectives are also generated. These improvements in daily operations are deemed necessary to satisfy evolving customer requirements and to reach the company's "change the business" goals. A set of key performance indicators is developed to monitor efforts to accomplish short-term operational objectives. Goals are distilled into concrete targets for every process and at every level of the company. Progress reviews are also held at every level.

Accomplishment-Oriented Associates

Clarke American believes that empowered and accomplishment-oriented associates are its greatest competitive advantage, and it fully recognizes the correlation between high employee satisfaction and superior performance. From orientation and onward, associates are steeped in the company's culture and values: customer first, integrity and mutual respect, knowledge sharing, measurement, quality workplace, recognition, responsiveness, and teamwork. They are schooled regularly in the application of standardized quality tools, performance measurement, use of new technology, team disciplines, and specialized skills. In 2000, associates averaged 76 hours of training, more than the "best in class" companies tracked by the American Society for Training and Development.

Work teams and improvement teams carry out efforts to attain operational improvements spelled out in "run the business" goals. Cross-functional project teams attend to "change the business" initiatives. Sharing of knowledge across teams, a company value, is facilitated through systematic approaches to communication and a highly competitive team excellence award process.

Individual initiative and innovation are expected. Associates are encouraged to contribute improvement ideas under Clarke American's S.T.A.R. (suggestions, teams, actions, results) program. In 2001, more than 20,000 process improvement ideas saved the company an estimated $10 million. Since the program started in 1995, implementation rates for S.T.A.R. ideas have increased from below 20% to 70% in 2001. At the same time, financial rewards flowed back to associates, who averaged nearly $5000 in bonus and profit-sharing payouts.

Leveraging Technology

Since 1995, Clarke American has invested substantially in new technology, using it to improve performance and to deepen relationships with partners through its offerings of customer management solutions. For example, with a major information technology supplier, the company developed digital printing capability that enables it to provide faster and more customized products and services to financial institutions and their customers. New technology has led to major reductions in cycle time, errors, nearly complete elimination of hazardous materials, less waste, and dramatically improved quality.

Clarke American, its partners, and its associates are reaping the benefits of these and other improvements. Since 1997, company-conducted telephone surveys of partner organizations consistently show a 96% satisfaction rate. Partner loyalty ratings have increased from 41% in first quarter 2000 to 54% in third quarter 2001. In independent surveys, commissioned by the company every 18 months, Clarke American's customer-satisfaction scores for all three partner segments are trending upward and top those of its major competitors.

Among Clarke American associates, overall satisfaction has improved from 72% in 1996 to 84% in 2000, when survey participation reached 96%, comparable to the world-class benchmark. Rising associate satisfaction correlates with the 84% increase in revenue earned per associate since 1995. Annual growth in company revenues has increased from a rate of 4.2% in 1996 to 16% in 2000, compared to the industry's average annual growth rate of less than 1% over the five-year period.

Highlights

- Since 1996, Clarke American's market share has increased by 50% to its current 26%.
- In 2001, more than 20,000 ideas from Clarke American associates were implemented for a cost savings of an estimated $10 million.
- Since 1997, surveys of partner organizations consistently have shown a 96% satisfaction rate.
- In 2000, Clarke American associates averaged 76 hours of training, more than the "best in class" companies tracked by the American Society for Training and Development.

Questions for Discussion

1. Who is the customer? How does Clarke American ensure customer satisfaction?
2. Explain how the organization links long-term objectives, short-term objectives, and customer satisfaction.
3. Should top management set strategic plans by themselves? From which constituencies should they receive input?
4. How has the organization utilized the latest developments in technology?
5. Explain the significance of utilizing cross-functional teams in this context.
6. How does Clarke American incorporate the dynamic needs of its customers?
7. Discuss how the organization is committed to image management and customer perception.

KEY TERMS

chronic problems	Deming's deadly diseases
common causes	Deming's 14 points for management
consistency of purpose	diagnostic arm
constancy of purpose	effectiveness
corporate culture	efficiency
Crosby's 14-step plan for quality improvement	extended process
	Juran's philosophy
Crosby's philosophy	leadership
customer satisfaction	loss function
Deming cycle	organizational barriers

performance classification
process capability
product improvement cycle
productivity
quality
quality breakthrough sequence
quality culture
quality improvement
quality management maturity grid
quality trilogy
sampling
  convenience sampling
  judgment sampling
  100% sampling

probability sampling
  random sampling
service functions
service industries
service nonconformity
special causes
sporadic problems
steering arm
system of profound knowledge
total quality system
training
vendor selection
work standards
zero defects

EXERCISES

2-1   Who is the customer in health care? Describe some of the customer's needs.

2-2   Discuss some service nonconformity and behavioral characteristics in the following areas:
(a) Health care
(b) Call center
(c) Internal Revenue Service
(d) Airline industry

2-3   Refer to Exercise 2-2. For each situation, discuss the ease or difficulty of measuring service quality. What are some remedial measures?

2-4   Refer to Exercise 2-2. For each situation, explain what factors influence customer perception of quality and how they are to be managed.

2-5   The following companies are interested in conducting a market survey of their products/services. Explain the possible sampling techniques that they might choose.
(a) High-end automobiles
(b) Cooking range
(c) Cell phones
(d) Boutique clothes
(e) City municipal services
(f) State revenue department
(g) Home insurance company

2-6   Explain the notion of the extended process and its significance. Discuss this in the context of the following organizations:
(a) Hardware vendor (such as for computers and printers)
(b) Hospital
(c) Software company
(d) Entertainment industry

2-7   What are the reasons for mass inspection not being a feasible alternative for quality improvement?

2-8   Describe some characteristics for selecting vendors in the following organizations and the selection process to be followed.

(a) Supermarket

(b) Physician's office

(c) Fast-food restaurant

2-9   Explain the organizational barriers that prevent a company from adopting the quality philosophy. Describe some specific action plans to remove such barriers.

2-10   What are the drawbacks of some traditional performance appraisal systems, and how may they be modified?

2-11   What is the difference between quality control and quality improvement? Discuss the role of management in each of these settings.

2-12   Discuss the five deadly diseases in the context of Deming's philosophy of management. What remedial actions would you take?

2-13   Explain the drawbacks of a bottom-line management approach.

2-14   Discuss the dilemma management faces when they sacrifice short-term profits for long- run stability. What approach is recommended?

2-15   Explain Deming's system of profound knowledge with specifics in the following industries:

(a) Hospital

(b) Software company

2-16   What are some organizational culture issues that management must address as they strive for long-run stability and growth?

2-17   American Airlines, through the use of its SABRE reservation system, realized the potential of yield management. Through such a system, it monitored the status of its upcoming flights and competitors' flights continuously to make pricing and allocation decisions on unsold seats.

(a) Discuss the impact of this on customer satisfaction.

(b) Could it lead to customer dissatisfaction?

(c) What is a possible objective function to be optimized?

(d) Could such yield management practices be used in other industries? Explain through specifics.

2-18   American Express has access to the spending habits of its cardholders. How may it use this information to improve customer satisfaction? How may its retail customers use this information?

REFERENCES

Crosby, P. B. (1979). *Quality Is Free*. New York: McGraw-Hill.

———(1984). *Quality Without Tears: The Art of Hassle-Free Management*. New York: McGraw-Hill.

———(1989). *Let's Talk Quality*. New York: McGraw-Hill.

Deming, W. E. (1982). *Quality, Productivity, and Competitive Position*. Cambridge, MA: Center for Advanced Engineering Study, Massachusetts Institute of Technology.

———(1986). *Out of the Crisis*. Cambridge, MA: Center for Advanced Engineering Study, Massachusetts Institute of Technology.

Fellers, G. (1992). *The Deming Vision: SPC/TQM for Administrators*. Milwaukee, WI: ASQC Press.

Fitzsimmons, J. A., and M. J. Fitzsimmons (1994). *Service Management for Competitive Advantage*. New York: McGraw-Hill.

Gitlow, H. S., and S. J. Gitlow (1987). *The Deming Guide to Quality and Competitive Position*. Englewood Cliffs, NJ: Prentice Hall.

Juran, J. M. (1986). "The Quality Trilogy", *Quality Progress*, Aug., pp. 19–24.

———(1988a). *Juran on Planning for Quality*. New York: Free Press.

———(1988b). *Juran's Quality Control Handbook*. New York: McGraw-Hill.

———(1989). *Juran on Leadership for Quality: An Executive Handbook*. New York: Free Press.

Juran, J. M., and F. M. Gryna, Jr. (1993). *Quality Planning and Analysis: From Product Development Through Use*. 3rd ed., New York: McGraw Hill.

Lefevre, H. L. (1989). *Quality Service Pays: Six Keys to Success*. Milwaukee, WI: American Society for Quality Control.

Lowe, T. A., and J. M. Mazzeo (1986). "Three Preachers, One Religion", *Quality*, Sept. pp. 22–25.

NIST, (2001). "Malcolm Baldrige National Quality Award, 2001 Award Recipient Profile", U.S. Department of Commerce, National Institute of Standards and Technology, Gaithersburg, MD, www.nist.gov.

Normann, R. (1991). *Service Management: Strategy and Leadership in Service Business,* 2nd ed. New York: Wiley.

Patton, F. (2005). "Does Six Sigma Work in Service Industries?" *Quality Progress*, 38 (9): 55–60.

Zeithaml, V. A. (2000) "Service Quality, Profitability, and the Economic Worth of Customers: What We Know and What We Need to Learn", *Journal of the Academy of Marketing Science*, 28(1): 67–85.

3

QUALITY MANAGEMENT: PRACTICES, TOOLS, AND STANDARDS

3-1 INTRODUCTION AND CHAPTER OBJECTIVES

The road to a quality organization is paved with the commitment of management. If management is not totally behind this effort, the road will be filled with potholes, and the effort will drag to a halt. A keen sense of involvement is a prerequisite for this journey, because like any journey of import, the company will sometimes find itself in uncharted territory. Company policies must be carefully formulated according to principles of a quality program. Major shifts in paradigms may occur. Resources must, of course, be allocated to accomplish the objectives, but this by itself is not sufficient. Personal support and motivation are the key ingredients to reaching the final destination.

In this chapter we look at some of the quality management practices that enable a company to achieve its goals. These practices start at the top, where top management creates the road map, and continue with middle and line management, who help employees follow the map. With an ever-watchful eye on the satisfaction of the customer, the entire workforce embarks on an intensive study of product design and process design. Company policies on vendor selection are discussed. *Everything* is examined through the lens of quality improvement.

The prime objectives of this chapter are to provide a framework through which management accomplishes its task of quality assurance. Principles of total quality management are presented. Additional tools such as quality function deployment, which plays a major role in incorporating customer needs in products and processes, are discussed. Problems to address are investigated through Pareto charts and failure mode and effects

Fundamentals of Quality Control and Improvement, Third Edition, By Amitava Mitra
Copyright © 2008 John Wiley & Sons, Inc.

criticality analysis. Following this, root cause identification is explored through cause-and-effect diagrams. The study of all processes, be they related to manufacturing or service, typically starts with a process map that identifies all operations, their precedence relationships, the inputs and outputs for each operation along with the controllable and uncontrollable factors, and the designated ownership of each. A simpler version of the process map is a flowchart that shows the sequence of operations and decision points and assists in identifying value-added and non-value-added activities.

Finally, we consider the standards set out by the International Organization for Standardization (ISO): in particular, ISO 9000 standards. Organizations seek to be certified by these standards to demonstrate the existence of a quality management process in their company. We look at some prime examples of companies that foster quality and the award that gives them recognition for their efforts: the highly prestigious Malcolm Baldrige National Quality Award. Companies that win this award become the benchmarks in their industries.

3-2   MANAGEMENT PRACTICES

A company's strategic plan is usually developed by top management; they are, after all, responsible for the long-range direction of the company. A good strategic plan addresses the needs of the company's constituencies. First and foremost, of course, is the customer, who can be internal and/or external. The customer wants a quality product or service at the lowest possible cost. Meeting the needs of the shareholders is another objective. Shareholders want to maximize their return on investment. Top management has the difficult task of balancing these needs and creating a long-term plan that will accomplish them.

What management needs are specific practices that enable them to install a quality program. That is what this chapter is about, but first we need some terminology. In this context, the term total quality management (TQM) refers to a comprehensive approach to improving quality. According to the U.S. Department of Defense, TQM is both a philosophy and a set of guiding principles that comprise the foundation of a continuously improving organization. Other frequently used terms are synonymous to TQM; among them are *continuous quality improvement, quality management, total quality control*, and *company-wide quality assurance*.

Total Quality Management

Total quality management revolves around three main themes: the customer, the process, and the people. Figure 3-1 shows some basic features of a TQM model. At its core are the company vision and mission and management commitment. They bind the customer, the process, and the people into an integrated whole. A company's vision is quite simply what the company wants to be. The mission lays out the company's strategic focus. Every employee should understand the company's vision and mission so that individual efforts will contribute to the organizational mission. When employees do not understand the strategic focus, individuals and even departments pursue their own goals rather than those of the company, and the company's goals are inadvertently sabotaged. The classic example is maximizing production with no regard to quality or cost.

Management commitment is another core value in the TQM model. It must exist at all levels for the company to succeed in implementing TQM. Top management envisions the

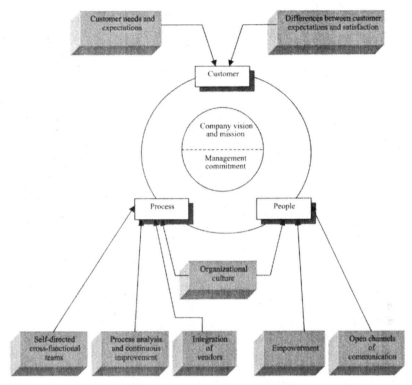

FIGURE 3-1   Features of a TQM model.

strategy and creates policy. Middle management works on implementation. At the operational level, appropriate quality management tools and techniques are used.

Satisfying customer needs and expectations is a major theme in TQM—in fact, it is the driving force. Without satisfied customers, market share will not grow and revenue will not increase. Management should not second-guess the customer. For example, commercial builders should construct general merchandise stores only after they have determined that there is enough customer interest to support them. If consumers prefer specialty stores, specialty stores should be constructed. Direct feedback using a data-driven approach is the best way to identify customer expectations and needs. A company's strategic plan must conform to these needs.

A key principle in quality programs is that customers are both internal and external. The receiving department of a processed component is a customer of that processing unit. Feedback from such internal customers identifies problem areas before the product reaches its finished stage, thus reducing the cost of scrap and rework.

Customer expectations can, to some extent, be managed by the organization. Factors such as the quality of products and services and warranty policies offered by the competitor influence customer expectations directly. The company can, through truthful advertising, shape the public's expectations. For example, if the average life of a lawn mower under specified operating conditions is 15 years, there is no reason to exaggerate it. In service operations, customers know which companies are responsive and friendly. This doesn't need advertising. Customer surveys can help management determine discrepancies between expectations and satisfaction. Taking measures to eliminate discrepancies is known as **gap analysis**.

The second theme in TQM is the process. Management is responsible for analyzing the process to improve it continuously. In this framework, vendors are part of the extended process, as advocated by Deming. As discussed earlier, integrating vendors into the process improves the vendors' products, which leads to better final products. Because problems can and do span functional areas, self-directed cross-functional teams are important for generating alternative feasible solutions—the process improves again. Technical tools and techniques along with management tools come in handy in the quest for quality improvement. Self-directed teams are given the authority to make decisions and to make appropriate changes in the process.

The third theme deals with people. Human "capital" is an organization's most important asset. **Empowerment**—involving employees in the decision-making process so that they take ownership of their work and the process—is a key factor in TQM. It is people who find better ways to do a job, and this is no small source of pride. With pride comes motivation. There is a sense of pride in making things better through the elimination of redundant or non-value-added tasks or combining operations. In TQM, managing is empowering.

Barriers restrict the flow of information. Thus, open channels of communication are imperative, and management had better maintain these. For example, if marketing fails to talk to product design, a key input on customer needs will not be incorporated into the product. Management must work with its human resources staff to empower people to break down interdepartmental barriers. From the traditional role of management of coordinating and controlling has developed the paradigm of coaching and caring. Once people understand that they, and only they, can improve the state of affairs, and once they are given the authority to make appropriate changes, they will do the job that needs to be done. There originates an intrinsic urge from within to do things better. Such an urge has a force that supersedes external forms of motivation.

Linking the human element and the company's vision is the fabric we call **organizational culture**. Culture comprises the beliefs, values, norms, and rules that prevail within an organization. How is business conducted? How does management behave? How are employees treated? What gets rewarded? How does the reward system work? How is input sought? How important are ethics? What is the social responsibility of the company? The answers to these and many other questions define an organization's culture. One culture may embrace a participative style of management that empowers its employees and delights its customers with innovative and timely products. Another culture may choose short-term profit over responsibility to the community at large. Consider, for example, the social responsibility adopted by the General Electric Company. The company and its employees made enormous contributions to support education, the arts, the environment, and human services organizations worldwide.

Vision and Quality Policy

A company's **vision** comprises its values and beliefs. Their vision is what they want it to be, and it is a message that every employee should not only hear, but should also believe in. Visions, carefully articulated, give a coherent sense of purpose. Visions are about the future, and effective visions are simple and inspirational. Finally, it must be motivational so as to evoke a bond that creates unison in the efforts of persons working toward a common organizational goal. From the vision emanates a **mission statement** for the organization that is more specific and goal oriented.

A service organization, IBM Direct, is dedicated to serving U.S. customers who order such IBM products as ES/9000 mainframes, RS/6000 and AS/400 systems, connectivity networks, and desktop software. Their vision for customer service is "to create an environment for customers where conducting business with IBM Direct is considered an enjoyable, pleasurable and satisfying experience." This is *what* IBM Direct wants to be. Their mission is "to act as the focal point for post-sale customers issues for IBM Direct customers. We must address customer complaints to obtain timely and complete resolutions. And, through root cause analysis, we must ensure that our processes are optimized to improve our customer satisfaction." Here again, the mission statement gets specific. This is *how* they will get to their vision. Note that no mention is made of a time frame. This issue is usually dealt with in goals and objectives.

Framed by senior management, a **quality policy** is the company's road map. It indicates what is to be done, and it differs from procedures and instructions, which address how it is to be done, where and when it is to be done, and who is to do it. A beacon in TQM leadership, Xerox Corporation is the first major U.S. corporation to regain market share after losing it to Japanese competitors. Xerox attributes its remarkable turnaround to its conversion to TQM philosophy. The company's decision to rededicate itself to quality through a strategy called *Leadership Through Quality* has paid off. Through this process, Xerox created a participatory style of management that focuses on quality improvement while reducing costs. It encouraged teamwork, sought more customer feedback, focused on product development to target key markets, encouraged greater employee involvement, and began competitive benchmarking. Greater customer satisfaction and enhanced business performance are the driving forces in their quality program, the commitment to which is set out in the Xerox quality policy: "Quality is the basic business principle at Xerox."

Another practitioner of TQM, the Eastman Chemical Company, manufactures and markets over 400 chemicals, fibers, and plastics for over 7000 customers around the world. A strong focus on customers is reflected in its vision: "to be the world's preferred chemical company." A similar message is conveyed in its quality goal: "to be the leader in quality and value of products and services." Its vision, values, and goals define Eastman's quality culture. The company's quality management process is set out in four directives: "focus on customers; establish vision, mission, and indicators of performance; understand, stabilize, and maintain processes; and plan, do, check, act for continual improvement and innovation."

Eastman Chemical encourages innovation and provides a structured approach to generating new ideas for products. Cross-functional teams help the company understand the needs of both its internal and external customers. The teams define and improve processes, and they help build long-term relationships with vendors and customers. Through the Eastman Innovative Process, a team of employees from various areas—design, sales, research, engineering, and manufacturing—guides an idea from inception to market. People have ownership of the product and of the process. Customer needs and expectations are addressed though the process and are carefully validated. One outcome of the TQM program has been the drastic reduction (almost 50%) of the time required to launch a new product. Through a program called Quality First, employees team with key vendors to improve the quality and value of purchased materials, equipment, and services. Over 70% of Eastman's worldwide customers have ranked the company as their best supplier. Additionally, Eastman has received an outstanding rating on five factors that customers view as most important: product quality, product uniformity, supplier integrity, correct delivery, and reliability. Extensive customer surveys led the company to institute a no-fault return policy on its plastic products. This policy, believed to be the only one of its kind in the chemical industry, allows customers to return any product for any reason for a full refund.

Balanced Scorecard

The Balanced Scorecard (BSC) is a management system that integrates measures derived from the organization's strategy. It integrates measures related to tangible as well as intangible assets. The focus of BSC is on accomplishing of the company's mission through the development of a communication and learning system. It translates the mission and strategy to objectives and measures that span four dimensions: learning and growth, internal processes, customers, and financial (Kaplan and Norton 1996). Whereas traditional systems have focused only on financial measures (such as return on investment), which is a short-term measure, BSC considers all four perspectives from a long-term point of view. So, for example, even for the financial perspective, it considers measures derived from the business strategy, such as sales growth rate or market share in targeted regions or customers. Figure 3-2 shows the concept behind the development of a balanced scorecard.

Measures in the learning and growth perspective that serve as drivers for the other three perspectives are based on three themes. First, employee capabilities, which include employee satisfaction, retention, and productivity, are developed. Improving satisfaction typically improves retention and productivity. Second, development of information systems capabilities is as important as the system for procuring raw material, parts, or components. Third, creation of a climate for growth through motivation and empowerment is an intangible asset that merits consideration.

For each of the four perspectives, diagnostic and strategic measures could be identified. Diagnostic measures relate to keeping a business in control or in operation (similar to the concept of quality control). On the contrary, strategic measures are based on achieving competitive excellence based on the business strategy. They relate to the position of the company relative to its competitors and information on its customers, markets, and suppliers.

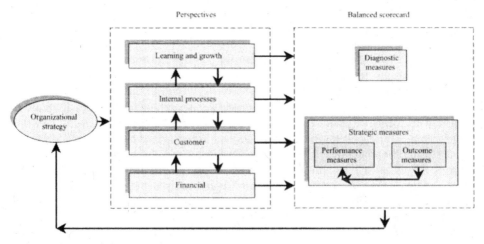

FIGURE 3-2   Balanced scorecard.

Strategic measures could be of two types: outcome measures and performance measures. *Outcome measures* are based on results from past efforts and are lagging indicators. Examples are return on equity or employee productivity. *Performance measures* reflect the uniqueness of the business strategy and are lead indicators, examples of which are sales growth rate by segment or percentage revenue from new products. Each performance measure has to be related to an outcome measure through a cause-and-effect type of analysis, which will therefore reflect the financial drivers of profitability. It could also identify the specific internal processes that will deliver value to targeted customers if the company strategy is to expand its market share for a particular category of customers.

In the learning and growth perspective, employee satisfaction, a strategic lag indicator, could be measured on an ordinal scale of 1 to 5. Another lag indicator could be the revenue per employee, a measure of employee productivity. A performance measure, a lead indicator, could be the strategic job coverage ratio, which is the ratio of the number of employees qualified for strategic jobs to the organizational needs that are anticipated. This is a measure of the degree to which the company has re-skilled its employees. Under motivation, an outcome measure could be the number of suggestions per employee or the number of suggestions implemented.

When considering the internal processes perspective, one is required to identify the critical processes that will enable the meeting of customer or shareholder objectives. Based on the expectations of specific external constituencies, they may impose demands on internal processes. Cycle time, throughput, and costs associated with existing processes are examples of diagnostic measures. In the strategic context, a business process for creating value could include innovation, operations, and post-sale service. Innovation may include basic research to develop new products and services or applied research to exploit existing technology. Time to develop new products is an example of a strategic outcome measure. Under post-sale service, measures such as responsiveness (measured by time to respond), friendliness, and reliability are applicable.

The strategic aspect of the customer perspective deals with identifying customers to target and the corresponding market segments. For most businesses, core outcome measures are market share, degree of customer retention, customer acquisition, customer satisfaction, and customer profitability from targeted segments. All of these are lagging measures and do not indicate what employees should be doing to achieve desired outcomes. Thus, under performance drivers (leading indicators), measures that relate to creating value for the customer are identified. These may fall in three broad areas: product/service attributes, customer relationship, and image and reputation. In association with product/service attributes, whereas lead time for existing products may be a diagnostic measure, time to serve targeted customers (e.g., quick check-in for business travelers in a hotel) is a strategic performance measure. Similarly, while quality of product/services (as measured by, say, defect rates) is considered as a "must," some unique measures such as service guarantees (and the cost of such), which offer not only a full refund but a premium above the purchase price, could be a performance measure.

Under the financial perspective, the strategic focus is dependent on the stage in which the organization currently resides (i.e., infancy, dominancy, or maturity). In the infancy stage, companies capitalize on significant potential for growth. Thus, large investments are made in equipment and infrastructure which may result in negative cash flow. Sales growth rate by product or market segment could be a strategic outcome measure. In the dominancy phase, where the business dwells mainly on the existing market, traditional measures such as gross margin or operating income are valid. Finally, for those in the mature stage, a company may not invest in new capabilities with a goal of maximizing cash flow. Unit costs could be a measure. For all three phases, some common themes are revenue growth, cost reduction, and asset utilization (Kaplan and Norton 1996).

Several features are to be noted about the balanced scorecard. First, under strategic measures, the link between performance measures and outcome measures represent a cause-and-effect relationship. However, based on the outcome measures observed and a comparison with the strategic performance expected, this feedback may indicate a choice of different performance measures. Second, all of the measures in the entire balanced scorecard represent a reflection of a business's performance. If such a performance does not match the performance expected based on the company strategy, a feedback loop exists to modify the strategy. Thus, the balanced scorecard serves an important purpose in linking the selection and implementation of an organization's strategy.

Performance Standards

One intended outcome of a quality policy is a desirable level of performance: that is, a defect-free product that meets or exceeds customer needs. Even though current performance may satisfy customers, organizations cannot afford to be complacent. Continuous improvement is the only way to stay abreast of the changing needs of the customer. The tourism industry, for instance, has seen dramatic changes in recent years; options have increased and customer expectations have risen. Top-notch facilities and a room filled with amenities are now the norm and don't necessarily impress the customer. Meeting and exceeding consumer expectations is no small challenge. Hyatt Hotels Corporation has met this challenge head-on. Their "In Touch 100" quality assurance initiative provides a framework for their quality philosophy and culture. Quality at Hyatt means consistently delivering products and services 100% of the time. The In Touch 100 program sets high standards—standards derived from guest and employee feedback—and specifies the pace that will achieve these standards every day. The core components of their quality assurance initiative are standards, technology, training, measurements, recognition, communication, and continuous improvement (Buzanis 1993).

Six Sigma Quality   Although a company may be striving toward an ultimate goal of zero defects, numerical standards for performance measurement should be avoided. Setting numerical values that may or may not be achievable can have an unintended negative emotional impact. Not meeting the standard, even though the company is making significant progress, can be demoralizing for everyone. Numerical goals also shift the emphasis to the short term, as long-term benefits are sacrificed for short-term gains.

So, the question is: How *do* we measure performance? The answer is: by making continuous improvement the goal, and then measuring the trend (not the numbers) in improvement. This is also motivational. Another effective method is benchmarking; this involves identifying high-performance companies or intra-company departments and using their performance as the improvement goal. The idea is that although the goals may be difficult to achieve, others have shown that it can be done.

Quantitative goals do have their place, however, as Motorola, Inc. has shown with its concept of **six sigma quality**. Sigma (σ) stands for the *standard deviation*, which is a measure of variation in the process. Assuming that the process output is represented by a normal distribution, about 99.73% of the output is contained within bounds that are three standard deviations (3σ) from the mean. As shown in Figure 3-3, these are represented as the lower and

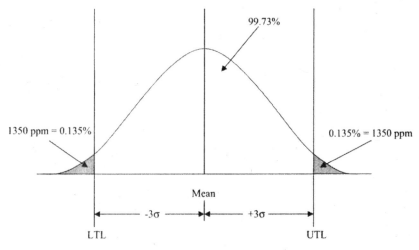

Legend: LTL: lower tolerance limit
UTL: upper tolerance limit
ppm: parts per million

FIGURE 3-3   Process output represented by a normal distribution.

upper tolerance limits (LTL and UTL). The normal distribution is characterized by two parameters: the mean and the standard deviation. The *mean* is a measure of the location of the process. Now, if the product specification limits are three standard deviations from the mean, the proportion of nonconforming product is about 0.27%, which is approximately 2700 parts per million (ppm); that is, the two tails, each 1350 ppm, add to 2700 ppm. On the surface, this appears to be a good process, but appearances can be deceiving. When we realize that most products and services consist of numerous processes or operations, reality begins to dawn. Even though a single operation may yield 97.73% good parts, the compounding effect of out-of-tolerance parts will have a marked influence on the quality level of the finished product. For instance, for a product that contains 1000 parts or has 1000 operations, an average of 2.7 defects per product unit is expected. The probability that a product contains no defective parts is only 6.72% ($e^{-2.7}$, using the Poisson distribution discussed in a later chapter)! This means that only about 7 units in 100 will go through the entire manufacturing process without a defect (rolled throughput yield)—not a desirable situation.

For a product to be built virtually defect-free, it must be designed to tolerance limits that are significantly *more* than $\pm 3\sigma$ from the mean. In other words, the process spread as measured by $\pm 3\sigma$ has to be significantly less than the spread between the upper and lower specification limits (USL and LSL). Motorola's answer to this problem is six sigma quality; that is, process variability must be so small that the specification limits are six standard deviations from the mean. Figure 3-4 demonstrates this concept. If the process distribution is stable (i.e., it remains centered between the specification limits), the proportion of non-conforming product should be only about 0.001 ppm on each tail.

In real-world situations, the process distribution will not always be centered between the specification limits; process shifts to the right or left are not uncommon. It can be shown that even if the process mean shifts by as much as 1.5 standard deviations from the center, the proportion nonconforming will be about 3.4 ppm. Comparing this to a 3σ capability of 2700 ppm demonstrates the improvement in the expected level of quality from the process. If we consider the previous example for a product containing 1000 parts and we design it for

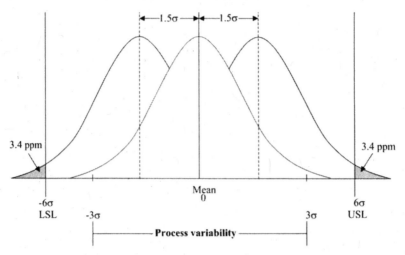

FIGURE 3-4 Six-sigma capability.

6σ capability, an average of 0.0034 defect per product unit (3.4 ppm) is expected instead of the 2.7 defects expected with 3σ capability. The cumulative yield (rolled throughput yield) from the process will thus be about 99.66%, a vast improvement over the 6.72% yield in the 3σ case.

Establishing a goal of 3σ capability is acceptable as a starting point, however, because it allows an organization to set a baseline for improvement. As management becomes more process oriented, higher goals such as 6σ capability become possible. Such goals may require fundamental changes in management philosophy and the organizational culture.

Although the previous description of six sigma has defined it as a metric, on a broader perspective six sigma may also be viewed as a philosophy or a methodology for continuous improvement. When six sigma is considered as a philosophy, it is considered as a strategic business initiative. In this context, the theme of identification of customer needs and ways to satisfy the customer is central. Along the lines of other philosophies, continuous improvement is integral to six sigma as well.

As a methodology, six sigma may be viewed as the collection of the following steps or phases: *define, measure, analyze, improve,* and *control.* Within each phase there are certain tools that could be utilized. Some of the tools are discussed in this chapter. In the *define phase*, customer needs are translated to specific attributes that are critical to meeting such needs. Typically, these are categorized in terms of critical to quality, delivery, or cost. Identification of these attributes will create a framework for study. For example, suppose that the waiting time of customers in a bank is the problem to be tackled. The number of tellers on duty during specific periods in the day is an attribute that is critical to reducing the waiting time, which might be a factor of investigation.

The *measure phase* consists of identifying metrics for process performance. This includes the establishment of baseline levels as well. In our example, a chosen metric could be the average waiting time prior to service, in minutes, while the baseline level could be the current value of this metric, say 5 minutes. Key process input and output variables are also identified. In the define phase, some tools could be the process flowcharts or its detailed version, the process map. To identify the vital few from the trivial many, a Pareto chart could be appropriate. Further, to study the various factors that may affect an outcome, a cause-and-effect diagram may be used. In the measure phase, one first has to ensure that the measurement system itself is stable. The technical name for such a study is *gage repeatability and reproducibility*. Thereafter, benchmark measures of process capability could be utilized, some of which are defects per unit of product, parts per million nonconforming, and rolled throughout yield, representing the proportion of the final product that has no defects. Other measures of process capability are discussed later in a separate chapter.

In the *analyze phase*, the objective is to determine which of a multitude of factors affects the output variable (s) significantly, through analysis of collected data. Tools may be simple graphical tools such as scatterplots and multivari charts. Alternatively, analytical models may be built linking the output or response variable to one of more independent variables through regression analysis. Hypothesis testing on selected parameters (i.e., average waiting time before and after process improvement) could be pursued. Analysis-of-variance techniques may be used to investigate the statistical significance of one or more factors on the response variable.

The *improve phase* consists of identifying the factor levels of significant factors to optimize the performance measure chosen, which could be minimize, maximize, or achieve a goal value. In our example, the goal could be to minimize the average waiting time of customers in the bank subject to certain resource or other process constraints. Here, concepts in design of experiments are handy tools.

Finally, the *control phase* deals with methods to sustain the gains identified in the preceding phase. Methods of statistical process control using control charts, discussed extensively in later chapters, are common tools. Process capability measures are also meaningful in this phase. They may provide a relative index of the degree to which the improved product, process, or service meets established norms based on customer requirements.

3-3 QUALITY FUNCTION DEPLOYMENT

Quality function deployment (QFD) is a planning tool that focuses on designing quality into a product or service by incorporating customer needs. It is a systems approach involving cross-functional teams (whose members are not necessarily from product design) that looks at the complete cycle of product development. This quality cycle starts with creating a design that meets customer needs and continues on through conducting detailed product analyses of parts and components to achieve the desired product, identifying the processes necessary to make the product, developing product requirements, prototype testing, final product or service testing, and finishing with after-sales troubleshooting.

QFD is customer driven and translates customers' needs into appropriate technical requirements in products and services. It is proactive in nature. Also identified by other names—*house of quality, matrix product planning, customer-driven engineering*, and *decision matrix*—it has several advantages. It evaluates competitors from two perspectives, the customer's perspective and a technical perspective. The customer's view of competitors provides the company with valuable information on the market potential of its products. The technical perspective, which is a form of benchmarking, provides information on the relative performance of the company with respect to industry leaders. This analysis identifies the degree of improvements needed in products and processes and serves as a guide for resource allocation.

QFD reduces the product development cycle time in each functional area, from product inception and definition to production and sales. By considering product and design along with manufacturing feasibility and resource restrictions, QFD cuts down on time that would otherwise be spent on product redesign. Midstream design changes are minimized, along with concerns on process capability and post-introduction problems of the product. This results in significant benefits for products with long lead times, such as automobiles. Thus, QFD has been vital for the Ford Motor Company and General Motors in their implementation of total quality management.

Companies use QFD to create training programs, select new employees, establish supplier development criteria, and improve service. Cross-functional teams have also used QFD to show the linkages between departments and thereby have broken down existing barriers of communication. Although the advantages of QFD are obvious, its success requires a significant commitment of time and human resources because a large amount of information is necessary for its startup.

QFD Process

Figure 3-5 shows a QFD matrix, also referred to as the **house of quality**. The objective statement delineates the scope of the QFD project, thereby focusing the team effort. For a space shuttle project, for example, the objective could be to identify critical safety features. Only one task is specified in the objective. Multiple objectives are split into separate QFDs in order to keep a well-defined focus.

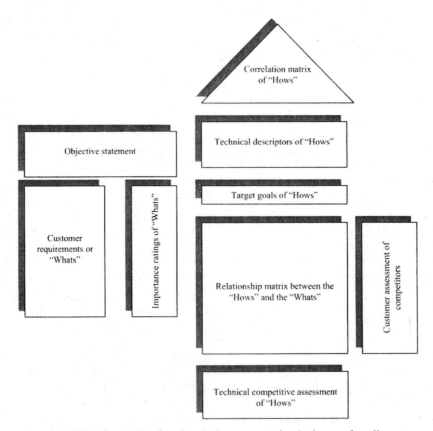

FIGURE 3-5   Quality function deployment matrix: the house of quality.

TABLE 3-1   Importance Rating of Credit-Card Customer Requirements

Customer Requirement ("Whats")	Importance Rating
Low interest rate	2
Error-free transactions	5
No annual fee	1
Extended warranty at no additional cost	3
Customer service 24 hours a day	4
Customers' advocate in billing disputes	4

The next step is to determine customer needs and wants. These are listed as the "whats" and represent the individual characteristics of the product or service. For example, in credit-card services, the "whats" could be attributes such as a low interest rate, error-free transactions, no annual fee, extended warranty at no additional cost, customer service 24 hours a day, and a customers' advocate in billing disputes. The list of "whats" is kept manageable by grouping similar items. On determination of the "whats" list, a customer importance rating that prioritizes the "whats" is assigned to each item. Typically, a scale of 1 to 5 is used, with 1 being the least important. Multiple passes through the list may be necessary to arrive at ratings that are acceptable to the team. The ratings serve as weighting factors and are used as multipliers for determining the technical assessment of the "hows." The focus is on attributes with high ratings because they maximize customer satisfaction. Let's suppose that we have rated attributes for credit-card services as shown in Table 3-1. Our ratings thus imply that our customers consider error-free transactions to be the most important attribute, and the least important to be charging no annual fee.

The customer plays an important role in determining the relative position of an organization with respect to that of its competitors for each requirement or "what." Such a comparison is entered in the section on "customer assessment of competitors." Thus, customer perception of the product or service is verified, which will help identify strengths and weaknesses of the company. Different focus groups or surveys should be used to attain statistical objectivity. One outcome of the analysis might be new customer requirements, which would then be added to the list of "whats," or the importance ratings might change. Results from this analysis will indicate what dimensions of the product or service the company should focus on. The same rating scale that is used to denote the importance ratings of the customer requirements is used in this analysis.

Consider, for example, the customer assessment of competitors shown in Table 3-2, where A represents our organization. The ratings are average scores obtained from various samples of consumers. The three competitors (companies B, C, and D) are our company's competition, so the maximum rating scores in each "what" will serve as benchmarks and thus the acceptable standard towards which we will strive. For instance, company C has a rating of 4 in the category "customer service 24 hours a day" compared to our 2 rating; we are not doing as well in this "what." We have identified a gap in a customer requirement that we consider important. To close this gap we could study company C's practices and determine whether we can adopt some of them. We conduct similar analyses with the other "whats," gradually implementing improved services. Our goal is to meet or beat the circled values in Table 3-2, which represent best performances in each customer requirement. That is, our goal is to become the benchmark.

Coming up with a list of technical descriptors—the "hows"—that will enable our company to accomplish the customer requirements is the next step in the QFD process.

TABLE 3-2   Customer Assessment of Competitors

Customer Requirements ("Whats")	Competitive Assessment of Companies			
	A	B	C	D
Low interest rate	3	2	④	2
Error-free transactions	4	⑤	3	3
No annual fee	⑤	⑤	2	3
Extended warranty at no additional cost	2	2	1	④
Customer service 24 hours a day	2	2	④	3
Customers' advocate in billing disputes	④	2	3	3

Multidisciplinary teams whose members originate in various departments will brainstorm to arrive at this list. Departments such as product design and development, marketing, sales, accounting, finance, process design, manufacturing, purchasing, and customer service are likely to be represented on the team. The key is to have a breadth of disciplines in order to "capture" all feasible "hows." To improve our company's ratings in the credit-card services example, the team might come up with these "hows": software to detect errors in billing, employee training on data input and customer services, negotiations and agreements with major manufacturers and merchandise retailers to provide extended warranty, expanded scheduling (including flextime) of employee operational hours, effective recruiting, training in legal matters to assist customers in billing disputes, and obtaining financial management services.

Target goals are next set for selected technical descriptors or "hows." Three symbols are used to indicate target goals: ↑ (maximize or increase the attained value), ↓ (minimize or decrease the attained value), and ⊙ (achieve a desired target value). Table 3-3 shows how our team might define target goals for the credit-card services example. Seven "hows" are listed along with their target goals. As an example, for how 2, creating a software to detect billing errors, the desired target value is zero: that is, no billing errors. For how 1, it is desirable to maximize or increase the effect of employee training to reduce input errors and

TABLE 3-3   Target Goals of Technical Descriptors

"Hows"	1	2	3	4	5	6	7
Target goals	⊙	↑	↑	⊙	↑	↑	↑

Legend	
Number	Technical descriptors or "hows"
1	Software to detect billing errors
2	Employee training on data input and customer services
3	Negotiations with manufacturers and retailers (vendors)
4	Expanded scheduling (including flextime) of employees
5	Effective recruiting
6	Legal training
7	Financial management services
Symbol	Target goal
↑	Maximize or increase attained value
↓	Minimize or decrease attained value
⊙	Achieve a target value

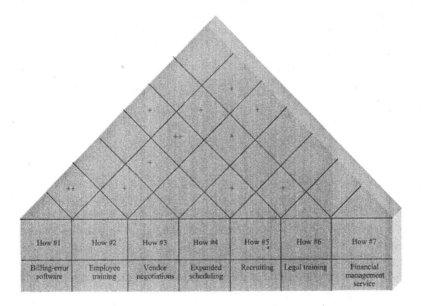

Legend ++   Strong positive relationship
      +   Positive relationship
      -       Negative relationship
     - -   Strong negative relationship

FIGURE 3-6   Correlation matrix of "Hows."

interact effectively with customers. Also, for how 4, the target value is to achieve customer service 24 hours a day. If measurable goals cannot be established for a technical descriptor, it should be eliminated from the list and the inclusion of other "hows" considered.

The correlation matrix of the relationship between the technical descriptors is the "roof" of the house of quality. In the correlation matrix shown in Figure 3-6, four levels of relationship are depicted: strong positive, positive, negative, and strong negative. These indicate the degree to which the "hows" support or complement each other or are in conflict. Negative relationships may require a trade-off in the objective values of the "hows" when a technical competitive assessment is conducted. In Figure 3-6, which correlates the "hows" for our credit-card services example, how 1, creating a software to detect billing errors, has a strong positive relationship (++) with how 2, employee training on data input and customer services. The user friendliness of the software will have an impact on the type and amount of training needed. A strong positive relationship indicates the possibility of synergistic effects. Note that how 2 also has a strong positive relationship with how 5; this indicates that a good recruiting program in which desirable skills are incorporated into the selection procedure will form the backbone of a successful and effective training program.

Following this, a technical competitive assessment of the "hows" is conducted along the same lines as the customer assessment of competitors we discussed previously. The difference is that instead of using customers to obtain data on the relative position of the company's "whats" with respect to those of the competitors, the technical staff of the company provides the input on the "hows." A rating scale of 1 to 5, as used in Table 3-2, may be used. Table 3-4 shows how our company's technical staff has assessed technical competitiveness for the "hows" in the credit-card services example. Our three competitors, companies B, C, and D, are reconsidered. For how 1 (creating a software to detect billing

TABLE 3-4   Technical Competitive Assessment of "Hows"

Company	Technical Descriptors ("Hows")						
	1	2	3	4	5	6	7
A	4	3	2	3	4	④	⑤
B	⑤	3	1	④	1	2	3
C	3	⑤	2	2	⑤	3	2
D	2	2	④	1	3	3	4

errors), our company is doing relatively well, with a rating of 4, but company B, with its rating of 5, is doing better; company B is therefore the benchmark against which we will measure our performance. Similarly, company C is the benchmark for how 2; we will look to improve the quality and effectiveness of our training program. The other assessments reveal that we have room to improve in hows 3, 4, and 5, but in hows 6 and 7 we are the benchmarks. The circled values in Table 3-4 represent the benchmarks for each "how."

The analysis shown in Table 3-4 can also assist in setting objective values, denoted by the "how muches," for the seven technical descriptors. The achievements of the highest-scoring companies are set as the "how muches," which represent the minimum acceptable achievement level for each "how." For example, for how 4, since company B has the highest rating, its achievement level will be the level that our company (company A) will strive to match or exceed. Thus, if company B provides customer service 16 hours a day, this becomes our objective value. If we cannot achieve these levels of "how muches," we should not consider entering this market because our product or service will not be as good as the competition's.

In conducting the technical competitive assessment of the "hows," the probability of achieving the objective value (the "how muches") is incorporated in the analysis. Using a rating scale of 1 to 5, 5 representing a high probability of success, the absolute scores are multiplied by the probability scores to obtain weighted scores. These weighted scores now represent the relative position within the industry and the company's chances of becoming the leader in that category.

The final step of the QFD process involves the relationship matrix located in the center of the house of quality (see Figure 3-5). It provides a mechanism for analyzing how each technical descriptor will help in achieving each "what." The relationship between a "how" and a "what" is represented by the following scale: 0 ≡ no relationship; 1 ≡ low relationship; 3 ≡ medium relationship; 5 ≡ high relationship. Table 3-5 shows the relationship matrix for the credit-card services example. Consider, for instance, how 2 (employee training on data input and customer services). Our technical staff believes that this "how" is related strongly to providing error-free transactions, so a score of 5 is assigned. Furthermore, this "how" has a moderate relationship with providing customer service 24 hours a day and serving as customers' advocate in billing disputes, so a score of 3 is assigned for these relationships. Similar interpretations are drawn from the other entries in the table. "Hows" that have a large number of zeros do not support meeting the customer requirements and should be dropped from the list.

The cell values, shown in parentheses in Table 3-5, are obtained by multiplying the rated score by the importance rating of the corresponding customer requirement. The absolute score for each "How" is calculated by adding the values in parentheses. The relative score is merely a ranking of the absolute scores, with 1 representing the most

TABLE 3-5   Relationship Matrix of Absolute and Relative Scores

Customer Requirements ("Whats")	Importance Ratings	Technical Descriptors ("Hows")						
		1	2	3	4	5	6	7
Low interest rate	2	0 (0)	0 (0)	5 (10)	0 (0)	0 (0)	0 (0)	5 (10)
Error-free transactions	5	5 (25)	5 (25)	0 (0)	3 (15)	5 (25)	0 (0)	0 (0)
No annual fee	1	0 (0)	0 (0)	3 (3)	0 (0)	0 (0)	0 (0)	5 (5)
Extended warranty	3	0 (0)	1 (3)	5 (15)	0 (0)	0 (0)	3 (9)	3 (9)
Customer service 24 hours a day	4	1 (4)	3 (12)	0 (0)	5 (20)	5 (20)	3 (12)	0 (0)
Customers' advocate in billing disputes	4	1 (4)	3 (12)	5 (20)	0 (0)	3 (12)	5 (20)	1 (4)
Absolute score		33	52	48	35	57	41	28
Relative score		6	2	3	5	1	4	7
Technical competitive assessment		5	5	4	4	5	4	5
Weighted absolute score		165	260	192	140	285	164	140
Final relative score		4	2	3	6.5	1	5	6.5

important. It is observed that how 5 (effective recruiting) is most important because its absolute score of 57 is highest.

The analysis can be extended by considering the technical competitive assessment of the "hows." Using the rating scores of the benchmark companies for each technical descriptor—that is, the objective values (the "how muches") from the circled values in Table 3-4—our team can determine the importance of the "hows." The weighted absolute scores in Table 3-5 are found by multiplying the corresponding absolute scores by the technical competitive assessment rating. The final scores demonstrate that the relative ratings of the top three "hows" are the same as before. However, the rankings of the remaining technical descriptors have changed. Hows 4 and 7 are tied for last place, each with an absolute score of 140 and a relative score of 6.5 each. Management may consider the ease or difficulty of implementing these "hows" in order to break the tie.

Our example QFD exercise illustrates the importance of teamwork in this process. An enormous amount of information must be gathered, all of which promotes cross-functional understanding of the product or service design system. Target values of the technical descriptors or "hows" are then used to generate the next level of house of quality diagram, where they will become the "whats." The QFD process proceeds by determining the technical descriptors for these new "whats." We can therefore consider implementation of the QFD process in different phases. As Figure 3-7 depicts, QFD facilitates the translation of customer requirements into a product whose features meet these requirements. Once such a product design is conceived, QFD may be used at the next level to identify specific

FIGURE 3-7   Phases of use of QFD.

characteristics of critical parts that will help in achieving the product designed. The next level may address the design of a process in order to make parts with the characteristics identified. Finally, the QFD process identifies production requirements for operating the process under specified conditions. Use of quality function deployment in such a multi-phased environment requires a significant commitment of time and resources. However, the advantages—the spirit of teamwork, cross-functional understanding, and an enhanced product design—offset this commitment.

3-4 BENCHMARKING AND PERFORMANCE EVALUATION

The goal of continuous improvement forces an organization to look for ways to improve operations. Be it a manufacturing or service organization, the company must be aware of the best practices in its industry and its relative position in the industry. Such information will set the priorities for areas that need improvement.

Organizations benefit from innovation. Innovative approaches cut costs, reduce lead time, improve productivity, save capital and human resources, and ultimately, lead to increased revenue. They constitute the breakthroughs that push product or process to new levels of excellence. However, breakthroughs do not happen very often. Visionary ideas are few and far between. Still, when improvements come, they are dramatic and memorable. The development of the computer chip is a prime example. Its ability to store enormous amounts of information in a fraction of the space that was previously required has revolutionized our lives. Figure 3-8 shows the impact of innovation on a chosen quality measure over time. At times a and b innovations occur as a result of which steep increases in quality from x to y and y to z are observed.

Continuous improvement, on the other hand, leads to a slow but steady increase in the quality measure. Figure 3-8 shows that for certain periods of time, a process with continuous improvement performs better than one that depends only on innovation. Of course, once an innovation takes place, the immense improvement in the quality measure initially outperforms the small improvements that occur on a gradual basis. This can be useful in

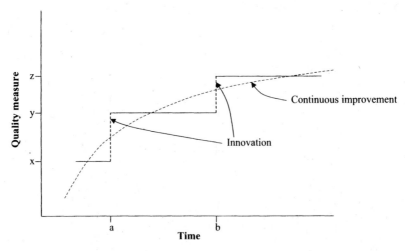

FIGURE 3-8 Impact of innovation and continuous improvement.

gaining market share, but it is also a high-risk strategy because innovations are rare. A company must carefully assess how risk averse it is. If its aversion to risk is high, continuous improvement is its best strategy. A process that is guaranteed to improve gradually is always a wise investment.

One way to promote continuous improvement is through innovative adaptation of the best practices in the industry. To improve its operations, an organization can incorporate information on the companies perceived to be the leaders in the field. Depending on the relative position of the company with respect to the industry leader, gains will be incremental or dramatic. Incorporating such adaptations on an ongoing basis provides a framework for continuous improvement.

Benchmarking

As discussed earlier, the practice of identifying best practices in industry and thereby setting goals to emulate them is known as **benchmarking**. Companies cannot afford to stagnate; this guarantees a loss of market share to the competition. Continuous improvement is a mandate for survival, and such fast-paced improvement is facilitated by benchmarking. This practice enables an organization to accelerate its rate of improvement. While innovation allows an organization to "leapfrog" its competitors, it does not occur frequently and thus cannot be counted on. Benchmarking, on the other hand, is doable. To adopt the best, adapt it innovatively, and thus reap improvements is a strategy for success.

Specific steps for benchmarking vary from company to company, but the fundamental approach is the same. One company's benchmarking may not work at another organization because of different operating concerns. Successful benchmarking reflects the culture of the organization, works within the existing infrastructure, and is harmonious with the leadership philosophy. Motorola, Inc., winner of the Malcolm Baldrige Award for 1988, uses a five-step benchmarking model: (1) Decide what to benchmark; (2) select companies to benchmark; (3) obtain data and collect information; (4) analyze data and form action plans; and (5) recalibrate and start the process again.

AT&T, which has two Baldrige winners among its operating units, uses a nine-step model: (1) Decide what to benchmark; (2) develop a benchmarking plan; (3) select a method to collect data; (4) collect data; (5) select companies to benchmark; (6) collect data during a site visit; (7) compare processes, identify gaps, and make recommendations; (8) implement recommendations; and (9) recalibrate benchmarks.

A primary advantage of the benchmarking practice is that it promotes a thorough understanding of the company's own processes—the company's current profile is well understood. Intensive studies of existing practices often lead to identification of non-value-added activities and plans for process improvement. Second, benchmarking enables comparisons of performance measures in different dimensions, each with the best practices for that particular measure. It is not merely a comparison of the organization with a selected company, but a comparison with several companies that are the best for the measure chosen. Some common performance measures are return on assets, cycle time, percentage of on-time delivery, percentage of damaged goods, proportion of defects, and time spent on administrative functions. The spider chart shown in Figure 3-9 is used to compare multiple performance measures and gaps between the host company and industry benchmark practices. Six performance measures are being considered here. The scales are standardized: say, between 0 and 1, 0 being at the center and 1 at the outer circumference, which represents the most desired value. Best practices for each performance measure are indicated, along with

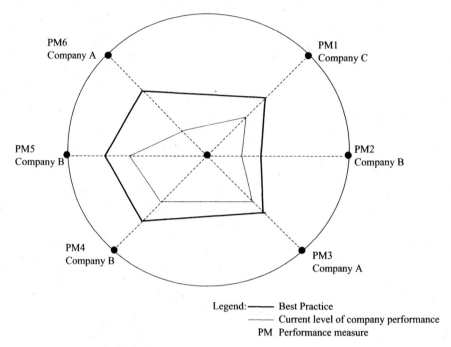

FIGURE 3-9   Spider chart for gap analysis.

the companies that achieve them. The current performance level of the company performing the benchmarking is also indicated in the figure. The difference between the company's level and that of the best practice for that performance measure is identified as the gap. The analysis that focuses on methods and processes to reduce this gap and thereby improve the company's competitive position is known as **gap analysis**.

Another advantage of benchmarking is its focus on performance measures and processes, not on the product. Thus, benchmarking is not restricted to the confines of the industry in which the company resides. It extends beyond these boundaries and identifies organizations in other industries that are superior with respect to the measure chosen. It is usually difficult to obtain data from direct competitors. However, companies outside the industry are more likely to share such information. It then becomes the task of management to find ways to adapt those best practices innovatively within their own environment.

In the United States, one of the pioneers of benchmarking is Xerox Corporation. It embarked on this process because its market share eroded rapidly in the late 1970s to Japanese competition. Engineers from Xerox took competitors' products apart and looked at them component by component. When they found a better design, they sought ways to adapt it to their own products or, even better, to improve on it. Similarly, managers from Xerox began studying the best management practices in the market; this included companies both within and outside the industry. As Xerox explored ways to improve its warehousing operations, it found a benchmark outside its own industry: L. L. Bean, Inc., the outdoor sporting goods retailer.

L. L. Bean has a reputation of high customer satisfaction; the attributes that support this reputation are its ability to fill customer orders quickly and efficiently with minimal errors and to deliver undamaged merchandise. The backbone behind this successful operation is an effective management system aided by state-of-the-art operations planning that addresses

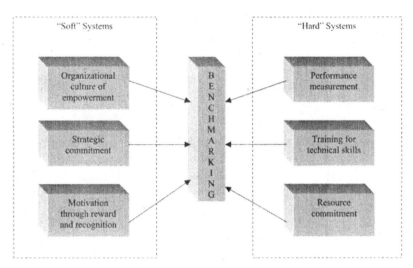

FIGURE 3-10   Role of benchmarking in implementing best practices.

warehouse layout, workflow design, and scheduling. Furthermore, the operations side of the process is backed by an organizational culture of empowerment, management commitment through effective education and training, and a motivational reward system of incentive bonuses.

Figure 3-10 demonstrates how benchmarking brings the "soft" and "hard" systems together. Benchmarking is not merely identification of the best practices. Rather, it seeks to determine how such practices can be adapted to the organization. The real value of benchmarking is accomplished only when the company has integrated the identified best practices successfully into its operation. To be successful in this task, soft and hard systems must mesh. The emerging organizational culture should empower employees to make decisions based on the new practice.

For benchmarking to succeed, management must demonstrate its strategic commitment to continuous improvement and must also motivate employees through an adequate reward and recognition system that promotes learning and innovative adaptation. When dealing with hard systems, resources must be made available to allow release time from other activities, access to information on best practices, and installation of new information systems to manage the information acquired. Technical skills, required for benchmarking such as flowcharting and process mapping, should be provided to team members through training sessions. The team must also identify performance measures for which the benchmarking will take place. Examples of such measures are return on investment, profitability, cycle time, and defect rate.

Several factors influence the adoption of benchmarking; **change management** is one of them. Figure 3-11 illustrates factors that influence benchmarking and the subsequent outcomes that derive from it. In the current environment of global competition, change is a given. Rather than react haphazardly to change, benchmarking provides an effective way to manage it. Benchmarking provides a road map for adapting best practices, a major component of change management. These are process-oriented changes. In addition, benchmarking facilitates cultural changes in an organization. These deal with overcoming resistance to change. This is a people-oriented approach, the objective being to demonstrate that change is not a threat but an opportunity.

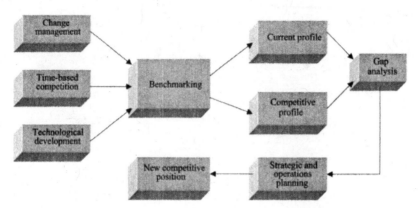

FIGURE 3-11   Influences on benchmarking and its outcomes.

The ability to reduce process time and create a model of quick response is important to all organizations. The concept of **time-based competition** is linked to reductions in **cycle time**, which can be defined as the interval between the beginning and ending of a process, which may consist of a sequence of activities. From the customer's point of view, cycle time is the elapsed time between placing an order and having it fulfilled satisfactorially. Reducing cycle time is strongly correlated with performance measures such as cost, market share, and customer satisfaction. Detailed flowcharting of the process can identify bottlenecks, decision loops, and non-value-added activities. Reducing decision and inspection points, creating electronic media systems for dynamic flow of information, standardizing procedures and reporting forms, and consolidating purchases are examples of tactics that reduce cycle time. Motorola, Inc., for example, reduced its corporate auditing process over a three-year period from an average of seven weeks to five days.

Technological development is another impetus for benchmarking. Consider the micro-electronics industry. Its development pace is so rapid that a company has no choice but to benchmark. Falling behind the competition in this industry means going out of business. In this situation, benchmarking is critical to survival.

Quality Auditing

The effectiveness of management control programs may be examined through a practice known as quality auditing. One reason that management control programs are implemented is to prevent problems. Despite such control, however, problems can and do occur, so, **quality audits** are undertaken to identify problems.

In any quality audit, three parties are involved. The party that requests the audit is known as the *client*, the party that conducts the audit is the *auditor*, and the party being audited is the *auditee*. Auditors can be of two types, internal or external. An *internal auditor* is an employee of the auditee. *External auditors* are not members of the auditee's organization. An external auditor may be a single individual or a member of an independent auditing organization.

Quality audits fulfill two major purposes. The first purpose, performed in the **suitability quality audit,** deals with an in-depth evaluation of the quality program against a reference standard, usually predetermined by the client. Reference standards are set by several organizations, including the American National Standards Institute/American Society for Quality (ANSI/ASQ), International Organization for Standardization (ISO), and British

Standards Institute (BSI). Some ISO standards are discussed later in this chapter. The entire organization may be audited, or specific processes, products, or services may be audited. The second purpose, performed in the **conformity quality audit,** deals with a thorough evaluation of the operations and activities within the quality system and the degree to which they conform to the quality policies and procedures defined.

Quality audits may be categorized as one of three types. The most extensive and inclusive type is the **system audit**. This entails an evaluation of the quality program documentation (including policies, procedures, operating instructions, defined accountabilities, and responsibilities to achieve the quality function) using a reference standard. It also includes an evaluation of the activities and operations that are implemented to accomplish the quality objectives desired. Such audits therefore explore conformance of quality management standards and their implementation to specified norms. They encompass the evaluation of the phases of planning, implementation, evaluation, and comparison. An example of a system audit is a pre-award survey, which typically evaluates the ability of a potential vendor to provide a desired level of product or service.

A second type of quality audit (not as extensive as the system audit) is the **process audit,** which is an in-depth evaluation of one or more processes in the organization. All relevant elements of the identified process are examined and compared to specified standards. Because a process audit takes less time to conduct than a system audit, it is more focused and less costly. If management has already identified a process that needs to be evaluated and improved, the process audit is an effective means of verifying compliance and suggesting places for improvement. A process audit can also be triggered by unexpected output from a process. For industries that use continuous manufacturing processes, such as chemical industries, a process audit is the audit of choice.

The third type of quality audit is the **product audit,** which is an assessment of a final product or service on its ability to meet or exceed customer expectations. This audit may involve conducting periodic tests on the product or obtaining information from the customer on a particular service. The objective of a product audit is to determine the effectiveness of the management control system. Such an audit is separate from decisions on product acceptance or rejection and is therefore not part of the inspection system used for such processes. Customer or consumer input plays a major role in the decision to undertake a product audit. For a company producing a variety of products, a relative comparison of product performance that indicates poor performers could be used as a guideline for a product audit.

Audit quality is heavily influenced by the independence and objectivity of the auditor. For the audit to be effective, the auditor must be independent of the activities being examined. Thus, whether the auditor is internal or external may have an influence on audit quality. Consider the assessment of an organization's quality documentation. It is quite difficult for an internal auditor to be sufficiently independent to perform this evaluation effectively. For such suitability audits, external auditors are preferable. System audits are also normally conducted by external auditors. Process audits can be internal or external, as can product audits. An example of an internal product audit is a *dock audit*, where the product is examined prior to shipment. Product audits conducted at the end of a process line are also usually internal audits. Product audits conducted at the customer site are typically external audits.

Vendor audits are external. They are performed by representatives of the company that is seeking the vendor's services. Knowledge of product and part specifications, contractual obligations and their secrecy, and purchase agreements often necessitate a second-party audit where the client company sends personnel from its own staff to perform the audit.

Conformity quality audits may be carried out by internal or external auditors as long as the individuals are not directly involved in the activities being audited.

Methods for conducting a quality audit are of two types. One approach is to conduct an evaluation of all quality system activities at a particular location or operation within an organization, know as a *location-oriented quality audit*. This audit examines the actions and interactions of the elements in the quality program at that location and may be used to interpret differences between locations. The second approach is to examine and evaluate activities relating to a particular element or function within a quality program at all locations where it applies before moving on to the next function in the program. This is known as a *function-oriented quality audit*. Successive visits to each location are necessary to complete the latter audit. It is helpful in evaluating the overall effectiveness of the quality program and also useful in tracing the continuity of a particular function through the locations where it is applicable.

The utility of a quality audit is derived only when remedial actions in deficient areas, exposed by the quality audit, are undertaken by company management. A quality audit does not necessarily prescribe actions for improvement; it typically identifies areas that do not conform to prescribed standards and therefore need attention. If several areas are deficient, a company may prioritize those that require immediate attention. Only on implementation of the remedial actions will a company improve its competitive position. Tools that help identify critical areas, find root causes to problems, and propose solutions include cause-and-effect diagrams, flowcharts, and Pareto charts; these are discussed later in the chapter.

Vendor Selection and Certification Programs

As discussed in Chapter 2, the modern trend is to establish long-term relationships with vendors. In an organization's pursuit of continuous improvement, the purchaser (customer) and vendor must be integrated in a quality system that serves the strategic missions of both companies. The vendor must be informed of the purchaser's strategies for market-share improvement, advance product information (including changes in design), and delivery requirements. The purchaser, on the other hand, should have access to information on the vendor's processes and be advised of their unique capabilities.

Cultivating a partnership between purchaser and vendor has several advantages. First, it is a win–win situation for both. To meet unique customer requirements, a purchaser can then redesign products or components collaboratively with the vendor. The vendor, who makes those particular components, has intimate knowledge of the components and the necessary processes that will produce the desired improvements. The purchaser is thus able to design its own product in a cost-effective manner and can be confident that the design will be feasible to implement. Alternatively, the purchaser may give the performance specification to the vendor and entrust them with design, manufacture, and testing. The purchaser thereby reduces design and development costs, lowers internal costs, and gains access to proprietary technology through its vendor, technology that would be expensive to develop internally. Through such partnerships, the purchaser is able to focus on its areas of expertise, thereby maintaining its competitive edge. Vendors gain from such partnerships by taking ownership of the product or component from design to manufacture; they can meet specifications more effectively because of their involvement in the entire process. They also gain an expanded insight into product and purchaser requirements through linkage with the purchaser; this helps them better meet those requirements. This, in turn, strengthens the vendor's relationship with the purchaser.

Vendor Rating and Selection

Maintaining data on the continual performance of vendors requires an evaluation scheme. **Vendor rating** based on established performance measures facilitates this process. There are several advantages in monitoring vendor ratings. Since the quality of the output product is a function of the quality of the incoming raw material or components procured through vendors, it makes sense to establish long-term relationships with vendors that consistently meet or exceed performance requirements. Analyzing the historical performance of vendors enables the company to select vendors that deliver their goods on time. Vendor rating goes beyond reporting on the historical performance of the vendor. It ensures a disciplined material control program. Rating vendors also helps reduce quality costs by optimizing the cost of material purchased.

Measures of vendor performance, which comprise the rating scheme, address the three major categories of quality, cost, and delivery. Under quality, some common measures are percent defective as expressed by defects in parts per million, process capability, product stoppages due to poor quality of vendor components, number of customer complaints, and average level of customer satisfaction. The category of cost includes such measures as scrap and rework cost, return cost, incoming-inspection cost, life-cycle costs, and warranty costs. The vendor's maintenance of delivery schedules is important to the purchaser in order to meet customer-defined schedules. Some measures in this category are percent of on-time deliveries, percent of late deliveries, percent of early deliveries, percent of underorder quantity, and percent of overorder quantity.

Which measures should be used are influenced by the type of product or service, the customer's expectations, and the level of quality systems that exists in the vendor's organization. For example, the Federal Express Corporation, winner of the 1990 Malcolm Baldrige National Quality Award in the service category, is *the* name in fast and reliable delivery. FedEx tracks its performance with such measures as late delivery, invoice adjustment needed, damaged packages, missing proof of delivery on invoices, lost packages, and missed pickups. For incoming material inspection, defectives per shipment, inspection costs, and cost of returning shipment are suitable measures. For vendors with statistical process control systems in place, measuring process capability is also useful. Customer satisfaction indices can be used with those vendors that have extensive companywide quality systems in place.

Vendor performance measures are prioritized according to their importance to the purchaser. Thus, a weighting scheme similar to that described in the house of quality (Figure 3-5) is often used. Let's consider a purchaser that uses rework and scrap cost, price, percent of on-time delivery, and percent of underorder quantity as its key performance measures. Table 3-6 shows these performance measures and the relative weight assigned to each one. This company feels that rework and scrap costs are most important, with a weight of 40. Note that price is not the sole determinant; in fact, it received the lowest weighting.

Table 3-6 shows the evaluation of vendors, A, B, and C. For each performance measure, the vendors are rated on a scale 1 to 5, with 1 representing the least desirable performance. A weighted score is obtained by adding the products of the weight and the assigned rating for each performance measure (weighted rating). The weighted scores are then ranked, with 1 denoting the most desirable vendor. From Table 3-6 we can see that vendor B, with the highest weighted score of 350, is the most desirable.

Vendor evaluation in quality programs is quite comprehensive. Even the vendor's culture is subject to evaluation as the purchaser seeks to verify the existence of a quality program.

TABLE 3-6   Prioritizing Vendor Performance Measures Using a Weighting Scheme

Performance Measure	Weight	Vendor A		Vendor B		Vendor C	
		Rating	Weighted Rating	Rating	Weighted Rating	Rating	Weighted Rating
Price	10	4	40	2	20	3	30
Rework and scrap cost	40	2	80	4	160	3	120
Percent of on-time delivery	30	1	30	3	90	2	60
Percent of underorder quantity	20	2	40	4	80	5	100
Weighted score			190		350		310
Rank			3		1		2

Commitment to customer satisfaction as demonstrated by appropriate actions is another attribute the purchaser will examine closely. The purchaser will measure the vendor's financial stability; the purchaser obviously prefers vendors that are going to continue to exist so the purchaser will not be visited with the problems that follow from liquidity or bankruptcy. The vendor's technical expertise relating to product and process design is another key concern as vendor and purchaser work together to solve problems and to promote continuous improvement.

Vendor Certification   **Vendor certification** occurs when the vendor has reached the stage at which it consistently meets or exceeds the purchaser's expectations. Consequently, there is no need for the purchaser to perform routine inspections of the vendor's product. Certification motivates vendors to improve their processes and, consequently, their products and services. A vendor must also demonstrate a thorough understanding of the strategic quality goals of the customer such that its own strategic goals are in harmony with those of the customer. Improving key processes through joint efforts strengthens the relationship between purchaser and vendor. The purchaser should therefore assess the vendor's capabilities on a continuous basis and provide adequate feedback.

A vendor goes through several levels of acceptance before being identified as a long-term partner. Typically, these levels are an approved vendor, a preferred vendor, and finally, a certified vendor: that is, a "partner" in the quality process. To move from one level to the next, the quality of the vendor's product or service must improve. The certification process usually transpires in the following manner. First, the process is documented; this defines the roles and responsibilities of personnel of both organizations. Performance measures, described previously, are chosen, and measurement methods are documented. An orientation meeting occurs at this step.

The next step is to gain a commitment from the vendor. The vendor and purchaser establish an environment of mutual respect. This is important because they must share vital and sometimes sensitive information in order to improve process and product quality. A quality system survey of the vendor is undertaken. In the event that the vendor is certified or registered by a third party, the purchaser may forego its own survey and focus instead on obtaining valid performance measurements. At this point, the purchaser sets acceptable performance standards on quality, cost, and delivery and then identifies those vendors that meet these standards. These are the **approved vendors**.

Following this step, the purchaser decides on what requirements it will use to define its **preferred vendors**. Obviously, these requirements will be more stringent than for approved vendors. For example, the purchaser may give the top 20% of its approved vendors preferred vendor status. Preferred vendors may be required to have a process control mechanism in place that demonstrates its focus on problem prevention (as opposed to problem detection).

At the next level of quality, the **certified vendor**, the criteria entail not only quality, costs, and delivery measures but also technical support, management attitude, and organizational quality culture. The value system for the certified vendor must be harmonious with that of the purchaser. An analysis of the performance levels of various attributes is undertaken, and vendors that meet the stipulated criteria are certified. Finally, a process is established to ensure vendor conformance on an ongoing basis. Normally, such reviews are conducted annually.

3M Company, as part of its vendor management process, uses five categories to address increasing levels of demonstrated quality competence to evaluate its vendors. The first category is the *new vendor*. Their performance capabilities are unknown initially. Interim specifications would be provided to them on an experimental basis. The next category is the *approved vendor*, where agreed-upon specifications are used and a self-survey is performed by the vendor. To qualify at this level, vendors need to have a minimum performance rating of 90% and must also maintain a rating of no less than 88%. Following this is the *qualified vendor*. To enter at this level, the vendor must demonstrate a minimum performance rating of 95% and must maintain a rating of at least 93%. Furthermore, the vendor must show that it meets ISO 9001 standards, or be approved by the U.S. Food and Drug Administration, or pass a quality system survey conducted by 3M. The next category is the *preferred vendor*. To enter this category, the vendor must demonstrate a minimum performance rating of 98% and must maintain a rating of at least 96%. The preferred vendor demonstrates continuous improvement in the process and constantly meets 3M standards. Minimal or no incoming inspection is performed. The highest level of achievement is the *strategic vendor* category. These are typically high-volume, critical-item, or equipment vendors that have entered into strategic partnerships with the company. They share their own strategic plans and cost data, make available their plants and processes for study by representatives from 3M, and are open to joint ventures, where they pursue design and process innovations with 3M. The strategic vendor has a long-term relationship with the company.

Other certification criteria are based on accepted norms set by various agencies. The *International Organization for Standardization* (ISO) is an organization that has prepared a set of standards: ISO 9001, *Quality Management Systems: Requirements*. Certification through this standard sends a message to the purchaser that the vendor has a documented quality system in place. To harmonize with the ISO 9000 standards, an international automotive quality standard, ISO/TS 16949, was developed. In the United States, the big three automakers, DaimlerChrysler, Ford, and General Motors, have been standardizing their requirements for suppliers and now subscribe to the QS 9000, *Quality System Requirements* standards, derived from the ISO 9000 standards. QS 9000, developed through the Automotive Industry Action Group (AIAG), has been instrumental in eliminating multiple audits of suppliers and requirements (often conflicting) from customers. Over 13,000 first-tier suppliers to the big three automobile companies were required to adopt QS 9000 standards. These first-tier suppliers, in turn, created a ripple effect for second-tier and others in the supply chain to move toward adoption of QS 9000 standards.

Another form of certification is attained through meeting the criteria set in the Malcolm Baldrige National Quality Award, which is administered by the U.S. Department of Commerce's National Institute of Standards and Technology (NIST 2007). If vendors can demonstrate that the various criteria set out in the award are met without necessarily applying for or winning the award, that by itself is an accomplishment that purchasers look at carefully. Details of the Malcolm Baldrige Award are described later in this chapter.

3-5   TOOLS FOR CONTINUOUS QUALITY IMPROVEMENT

To make rational decisions using data obtained on a product, process, service, or from a consumer, organizations use certain graphical and analytical tools. We explore some of these tools.

Pareto Diagrams

Pareto diagrams are important tools in the quality improvement process. Alfredo Pareto, an Italian economist (1848–1923), found that wealth is concentrated in the hands of a few people. This observation led him to formulate the *Pareto principle*, that the majority of wealth is held by a disproportionately small segment of the population. In manufacturing or service organizations, for example, problem areas or defect types follow a similar distribution. Of all the problems that occur, only a few are quite frequent; the others seldom occur. These two problem areas are labeled the *vital few* and the *trivial many*. The Pareto principle also lends support to the *80/20 rule*, which states that 80% of problems (nonconformities or defects) are created by 20% of causes. **Pareto diagrams** help prioritize problems by arranging them in decreasing order of importance. In an environment of limited resources, these diagrams help companies decide on the order in which they should address problems.

Example 3-1   We demonstrate the use of Minitab software to construct a Pareto chart. In Minitab, variables are input as columns in a worksheet. Thus, using the data shown in Table 3-7 on customer dissatisfaction in airlines, two columns, one labeled "Reasons" (column C1), and the other labeled "Count" (column C2) are input in the Minitab worksheet. The following point-and-click commands are executed: **Stat > Quality Tools > Pareto Chart**. Select **Chart defects table**. In **Labels in,** enter the name or column number (in this example, C1). In **Frequencies in**, enter the name or column number (in this case, C2) that contains the count data. Other options exist for axis labels and graph titles, Click **OK**.

Figure 3-12 shows a Pareto diagram of reasons for airline customer dissatisfaction. Delays in arrival is the major reason, as indicated by 40% of customers. Thus, this is the problem that the airlines should address first.

TABLE 3-7   **Customer Dissatisfaction in Airlines**

Reasons	Count
Lost baggage	15
Delay in arrival	40
Quality of meals	20
Attitude of attendant	25

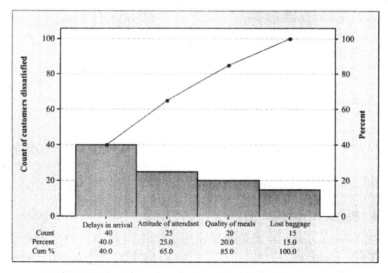

FIGURE 3-12   Pareto diagram for dissatisfied airline customers.

Flowcharts

Flowcharts, which show the sequence of events in a process, are used for manufacturing and service operations. They are often used to diagram operational procedures to simplify a system, as they can identify bottlenecks, redundant steps, and non-value-added activities. A realistic flowchart can be constructed by using the knowledge of the personnel who are directly involved in the particular process. Valuable process information is usually gained through the construction of flowcharts. Figure 3-13 shows a flowchart for patients reporting to the emergency department in a hospital. The chart identifies where delays can occur: for

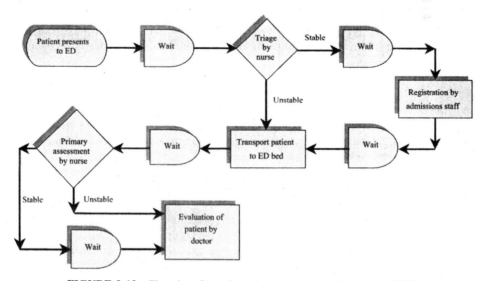

FIGURE 3-13   Flowchart for patients in an emergency department (ED).

example, in several steps that involve waiting. A more detailed flowchart would allow pinpointing of key problem areas that contribute to lengthening waiting time.

Further, certain procedures could be modified or process operations could be combined to reduce waiting time. A detailed version of the flowchart is the *process map*, which identifies the following for each operation in a process: process inputs (e.g., material, equipment, personnel, measurement gage), process outputs (these could be the final results of the product or service), and process or product parameters (classified into the categories of controllable, procedural, or noise). Noise parameters are uncontrollable and could represent the in-flow rate of patients or the absenteeism of employees. Through discussion and data analysis, some of the parameters could be classified as critical. It will then be imperative to monitor the critical parameters to maintain or improve the process.

Cause-and-Effect Diagrams

Cause-and-effect diagrams were developed by Kaoru Ishikawa in 1943 and thus are often called **Ishikawa diagrams**. They are also known as **fishbone diagrams** because of their appearance (in the plotted form). Basically, **cause-and-effect diagrams** are used to identify and systematically list various causes that can be attributed to a problem (or an effect) (Ishikawa 1976). These diagrams thus help determine which of several causes has the greatest effect. A cause-and-effect diagram can aid in identifying the reasons why a process goes out of control. Alternatively, if a process is stable, these diagrams can help management decide which causes to investigate for process improvement. There are three main applications of cause-and-effect diagrams: cause enumeration, dispersion analysis, and process analysis.

Cause enumeration is usually developed through a brainstorming session in which all possible types of causes (however remote they may be) are listed to show their influence on the problems (or effect) in question. In **dispersion analysis**, each major cause is analyzed thoroughly by investigating the subcauses and their impact on the quality characteristic (or effect) in question. This process is repeated for each major cause in a prioritized order. The cause-and-effect diagram helps us analyze the reasons for any variability or dispersion. When cause-and-effect diagrams are constructed for **process analysis,** the emphasis is on listing the causes in the sequence in which the operations are actually conducted. This process is similar to creating a flow diagram, except that a cause-and-effect diagram lists in detail the causes that influence the quality characteristic of interest at each step of a process.

Example 3-2 One of the quality characteristics of interest in automobile tires is the bore size, which should be within certain specifications. In a cause-and-effect diagram, the final bore size is the effect. Some of the main causes that influence the bore size are the incoming material, mixing process, tubing operation, splicing, press operation, operator, and measuring equipment. For each main cause, subcauses are identified and listed. For the raw material category, the incoming quality is affected by such subcauses as vendor selection process (e.g., is the vendor certified?), the content of scrap tire in the raw material, the density, and the ash content.

Using Minitab, for each **Branch** or main cause, create a column and enter the subcauses in the worksheet. Then, execute the following: **Stat > Quality Tools > Cause-and-Effect**. Under **Causes**, enter the name or column number of the main causes. The **Label** for each branch may be entered to match the column names. In **Effect**, input the brief problem description. Click **OK**. Figure 3-14 shows the completed cause-and-effect diagram.

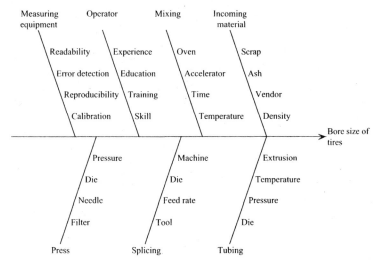

FIGURE 3-14   Cause-and-effect diagram for the bore size of tires.

Scatterplots

The simplest form of a **scatterplot** consists of plotting bivariate data to depict the relationship between two variables. When we analyze processes, the relationship between a controllable variable and a desired quality characteristic is frequently of importance. Knowing this relationship may help us decide how to set a controllable variable to achieve a desired level for the output characteristic. Scatterplots are often used as follow-ups to a cause-and-effect analysis.

Example 3-3   Suppose that we are interested in determining the relationship between the depth of cut in a milling operation and the amount of tool wear. We take 40 observations from the process such that the depth of cut (in millimeters) is varied over a range of values and the corresponding amount of tool wear (also in millimeters) over 40 operation cycles is noted. The data values are shown in Table 3-8.

Using Minitab, choose the commands **Graph > Scatterplot**. Select **Simple** and click **OK**. Under **Y**, enter the column number or name, in this case, "Tool wear." Under **X**, enter the column number or name, in this case, "Depth of cut." Click **OK**.

The resulting scatterplot is shown in Figure 3-15. It gives us an idea of the relationship that exists between depth of cut and amount of tool wear. In this case the relationship is generally nonlinear. For depth-of-cut values of less than 3.0 mm, the tool wear rate seems to be constant, whereas with increases in depth of cut, tool wear starts increasing at an increasing rate. For depth-of-cut values above 4.5 mm, tool wear appears to increase drastically. This information will help us determine the depth of cut to use to minimize downtime due to tool changes.

Multivariable Charts

In most manufacturing or service operations, there are usually several variables or attributes that affect product or service quality. Since realistic problems usually have more than two variables, **multivariable charts** are useful means of displaying collective information.

TABLE 3-8   Data on Depth of Cut and Tool Wear

Observation	Depth of Cut (mm)	Tool Wear (mm)	Observation	Depth of Cut (mm)	Tool Wear (mm)
1	2.1	0.035	21	5.6	0.073
2	4.2	0.041	22	4.7	0.064
3	1.5	0.031	23	1.9	0.030
4	1.8	0.027	24	2.4	0.029
5	2.3	0.033	25	3.2	0.039
6	3.8	0.045	26	3.4	0.038
7	2.6	0.038	27	2.8	0.040
8	4.3	0.047	28	2.2	0.031
9	3.4	0.040	29	2.0	0.033
10	4.5	0.058	30	2.9	0.035
11	2.6	0.039	31	3.0	0.032
12	5.2	0.056	32	3.6	0.038
13	4.1	0.048	33	1.9	0.032
14	3.0	0.037	34	5.1	0.052
15	2.2	0.028	35	4.7	0.050
16	4.6	0.057	36	5.2	0.058
17	4.8	0.060	37	4.1	0.048
18	5.3	0.068	38	4.3	0.049
19	3.9	0.048	39	3.8	0.042
20	3.5	0.036	40	3.6	0.045

Several types of multivariate charts are available (Blazek et al. 1987). One of these is known as a **radial plot**, or **star**, for which the variables of interest correspond to different rays emanating from a star. The length of each ray represents the magnitude of the variable.

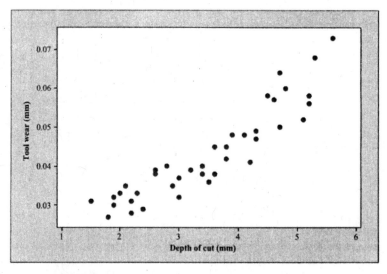

FIGURE 3-15   Scatterplot of tool wear versus depth of cut.

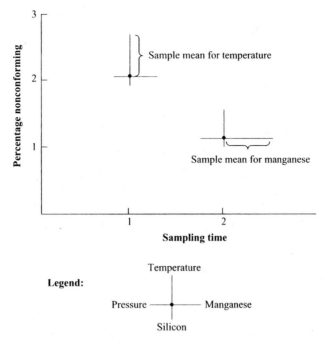

FIGURE 3-16   Radial plot of multiple variables.

Example 3-4   Suppose that the controllable variables in a process are temperature, pressure, manganese content, and silicon content. Figure 3-16 shows radial plots, or stars, for two samples of size 10 taken an hour apart. The sample means for the respective variables are calculated. These are represented by the length of the rays. A relative measure of quality performance is used to locate the center of a star vertically (in this case, the percentage of nonconforming product), while the horizontal axis represents the two sampling times.

Several process characteristics can be observed from Figure 3-16. First, from time 1 to time 2, an improvement in the process performance is seen, as indicated by a decline in the percentage nonconforming. Next, we can examine what changes in the controllable variables led to this improvement. We see that a decrease in temperature, an increase in both pressure and manganese content, and a basically constant level of silicon caused this reduction in the percentage nonconforming.

Other forms of multivariable plots (such as standardized stars, glyphs, trees, faces, and weathervanes) are conceptually similar to radial plots. For details on these forms, refer to Gnanadesikan (1977).

Matrix and Three-Dimensional Plots

Investigating quality improvement in products and processes often involves data that deal with more than two variables. With the exception of multivariable charts, the graphical methods discussed so far deal with only one or two variables. The **matrix plot** is a graphical option for situations with more than two variables. This plot depicts two-variable relationships between a number of variables all in one plot. As a two-dimensional matrix of separate

TABLE 3-9   Data on Temperature, Pressure, and Seal Strength for Plastic Packages

Observation	Temperature	Pressure	Seal Strength	Observation	Temperature	Pressure	Seal Strength
1	180	80	8.5	16	220	40	11.5
2	190	60	9.5	17	250	30	10.8
3	160	80	8.0	18	180	70	9.3
4	200	40	10.5	19	190	75	9.6
5	210	45	10.3	20	200	65	9.9
6	190	50	9.0	21	210	55	10.1
7	220	50	11.4	22	230	50	11.3
8	240	35	10.2	23	200	40	10.8
9	220	50	11.0	24	240	40	10.9
10	210	40	10.6	25	250	35	10.8
11	190	60	8.8	26	230	45	11.5
12	200	70	9.8	27	220	40	11.3
13	230	50	10.4	28	180	70	9.6
14	240	45	10.0	29	210	60	10.1
15	240	30	11.2	30	220	55	11.1

plots, it enables us to conceptualize relationships among the variables. The Minitab software can produce matrix plots.

Example 3-5   Consider the data shown in Table 3-9 on temperature, pressure, and seal strength of plastic packages. Since temperature and pressure are process variables, we want to investigate their impact on seal strength, a product characteristic.

Using Minitab, the data are entered for the three variables in a worksheet. Next, choose **Graph** > **Matrix Plot** and **Matrix of Plots Simple**. Click **OK**. The resulting matrix plot is shown in Figure 3-17. Observe that seal strength tends to increase linearly with temperature

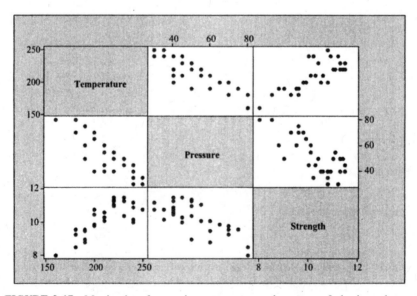

FIGURE 3-17   Matrix plot of strength, temperature, and pressure of plastic packages.

up to a certain point, which is about 210°C. Beyond 210°C, seal strength tends to decrease. The relationship between seal strength and pressure decreases with pressure. Also, the existing process conditions exhibit a relationship between temperature and pressure that decreases with pressure. Such graphical aids provide us with some insight on the relationship between the variables, taken two at a time.

Three-dimensional scatterplots depict the joint relationship of a dependent variable with two independent variables. While matrix plots demonstrate two-variable relationships, they do not show the joint effect of more than one variable on a third variable. Since interactions do occur between variables, a three-dimensional scatterplot is useful in identifying optimal process parameters based on a desired level of an output characteristic.

Example 3-6 Let's reconsider the plastic package data shown in Table 3-9. Suppose that we want to identify the joint relationship of the process variables, temperature and pressure, on seal strength of packages, an output characteristic. Using Minitab, we choose **Graph > 3D Surface Plot > Surface**. Type in the variable names, say strength in **Z**, pressure in **Y**, and temperature in **X**, then click **OK**. Figure 3-18 shows the resulting three-dimensional surface plot of strength versus temperature and pressure. This surface plot helps us identify optimal process parameter values that will maximize a variable. For example, a temperature around 230°C and a pressure around 40 kg/cm^2 appear to be desirable process parameter values for maximizing seal strength.

Failure Mode and Effects Criticality Analysis

Failure mode and effects criticality analysis (FMECA) is a disciplined procedure for systematic evaluation of the impact of potential failures and thereby determining a priority of possible actions that will reduce the occurrence of such failures. It can be applied at the system level, at the design level for a product or service, at the process level for manufacturing or services, or at the functional level of a component or subsystem level.

In products involving safety issues, say the braking mechanism in automobiles, FMECA assists in a thorough analysis of what the various failure modes could be, their impact and

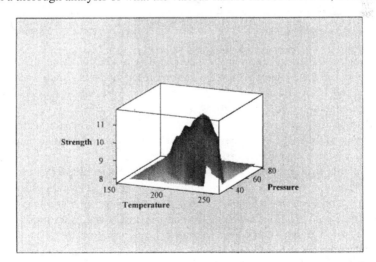

FIGURE 3-18   Three-dimensional surface plot of strength versus temperature and pressure of plastic packages.

effect on the customer, the severity of the failure, the possible causes that may lead to such a failure, the chance of occurrence of such failures, existing controls, and the chance of detection of such failures. Based on the information specified above, a *risk priority number* (RPN) is calculated, which indicates a relative priority scheme to address the various failures. The risk priority number is the product of the severity, occurrence, and detection ratings. The larger the RPN, the higher the priority. Consequently, recommended actions are proposed for each failure mode. Based on the selected action, the ratings on severity, occurrence, and detection are revised. The severity ratings typically do not change since a chosen action normally influences only the occurrence and/or detection of the failure. Only through fundamental design changes can the severity be reduced. Associated with the action selected, a rating that measures the risk associated with taking that action is incorporated. This rating on risk is a measure of the degree of successful implementation. Finally, a weighted risk priority number, which is the product of the revised ratings on severity, occurrence, and detection and the risk rating, is computed. This number provides management with a priority in the subsequent failure-related problems to address.

Several benefits may accrue from using failure modes and effects criticality analysis. First, by addressing *all* potential failures even before a product is sold or service rendered, there exits the ability to improve quality and reliability. A FMECA study may identify some fundamental *design changes* that must be addressed. This *creates* a better product in the first place rather than subsequent changes in the product and/or process. Once a better design is achieved, processes to create such a design can be emphasized. All of this leads to a reduction in development time of products and, consequently, costs. Since an FMECA involves a team effort, it leads to a thorough analysis and thereby identification of all possible failures. Finally, customer satisfaction is improved, with fewer failures being experienced by the customer.

It is important to decide on the level at which FMECA will be used since the degree of detailed analysis will be influenced by this selection. At the system level, usually undertaken prior to the introduction of either a product or service, the analysis may identify the general areas of focus for failure reduction. For a product, for example, this could be suppliers providing components, parts manufactured or assembled by the organization, or the information system that links all the units. A design FMECA is used to analyze product or service designs prior to production or operation. Similarly, a process FMECA could be used to analyze the processing/assembly of a product or the performance of a service. Thus, a hierarchy exists in FMECA use.

After selection of the level and scope of the FMECA, a block diagram that depicts the units/operations and their interrelationships is constructed, and the unit or operation to be studied is outlined. Let us illustrate use of FMECA through an example. Consider an original equipment manufacturer (OEM), with one supplier, who assembles to-order computers, as shown in Figure 3-19. There is flow of information and goods taking place in this chain. We restrict our focus to the OEM, where failure constitutes not meeting customer requirements regarding order quantity, quality, and delivery date.

Now, functional requirements are defined based on the selected level and scope. Table 3-10 lists these requirements based on customer's order quantity, quality, and delivery date. Through group brainstorming, potential failures for each functional requirement are listed. There could be more than one failure mode for each function. Here, for example, a failure in not meeting order quantity could occur due to the supplier and/or the OEM, as shown in Table 3-10. The impact or effects of failures are then listed. In the example, it leads to customer dissatisfaction. Also, for failures in order quantity or delivery date, another effect could be the creation of back orders, if permissible.

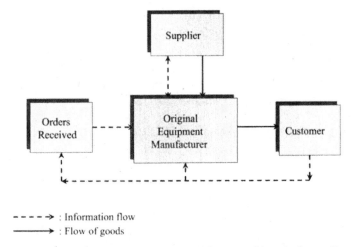

- - - - → : Information flow
———→ : Flow of goods

FIGURE 3-19   Original equipment manufacturer with a single supplier.

The next step involves rating the severity of the failure. Severity ratings are a measure of the impact of such failures on the customer. Such a rating is typically on a discrete scale from 1 (no effect) to 10 (hazardous effect). The Automotive Industry Action Group (AIAG) has some guidelines for severity ratings (El-Haik and Roy 2005), which other industries have adapted correspondingly. Table 3-11 shows rating scores on severity, occurrence, and detection, each of which is on a discrete scale of 1 to 10. AIAG has guidelines on occurrence and detection ratings as well, with appropriate modifications for process functions (Ehrlich 2002). For the example we indicate a severity rating of 6 on failure to meet order quantity or delivery date, while a rating of 7 is assigned to order quality, indicating that it has more impact on customer dissatisfaction, as shown in Table 3-10.

Causes of each failure are then listed, which will lead to suggestion of remedial actions. The next rating relates to the occurrence of failures; larger the rating the more likely the possible happening. The guidelines in Table 3-11 could be used to select the rated value. Here, we deem that a failure in not meeting order quantity or delivery date at the supplier to be more likely (rating of 7 in occurrence) relative to that of the OEM (rating of 4). Further, in not meeting specified quality, we believe that this is less likely to happen at the supplier (supplier rating is 5) and even more remote at the OEM (rating of 3). Existing controls, to detect failures, are studied. Finally, the chance of existing controls detecting failures is indicated by a rating score. In this example, there is a moderately high chance (rating of 4) of detecting lack of capacity at the supplier's location through available capacity/inventory reports, whereas detecting the same at the OEM through capacity reports has a very high chance (rating of 2). A similar situation exists for detecting lack of quality, with the OEM having a high chance of detection (rating of 3) through matching of customer orders with product bar codes relative to that of the supplier (rating of 4) through process control and capability analysis. Finally, in Table 3-10, a risk priority number (RPN) is calculated for each failure mode and listed. Larger RPN values indicate higher priority to the corresponding failure modes. Here, we would address capacity issues at the supplier (RPN of 168), and not meeting due dates due to the supplier (RPN of 168) first, followed by lack of quality at the supplier (RPN of 140).

Continuing on with the FMECA analyses, the next step involves listing specific recommended actions for addressing each failure mode. Table 3-12 presents this phase

TABLE 3-10 Failure Mode and Effects Criticality Analysis

Functional Requirement	Failure Mode	Failure Effects	Severity	Causes	Occurrence	Controls	Detection	Risk Priority Number
Meet customer order quantity	Not meet specified order quantity due to supplier	Dissatisfied customer; back order (if acceptable)	6	Lack of capacity at supplier	7	Available capacity/inventory level reports	4	168
	Not meet specified order quantity due to OEM	Dissatisfied customer; back order (if acceptable)	6	Lack of capacity at OEM	4	Available capacity reports	2	48
Meet customer order quality	Not meet specified order quality at supplier	Dissatisfied customer	7	Lack of process quality control at supplier	5	Process control; capability analysis	4	140
	Not meet specified order quality at OEM	Dissatisfied customer	7	Lack of process quality control at OEM	3	Incoming inspection; matching of customer orders with product bar code	3	63
Meet customer delivery date	Not meet specified due date due to supplier	Dissatisfied customer; back order (if acceptable)	6	Lack of capacity at supplier	7	Available capacity/inventory level reports	4	168
	Not meet specified due date due to OEM	Dissatisfied customer; back order (if acceptable)	6	Lack of capacity at OEM	4	Available capacity reports	2	48

TABLE 3-11 Rating Scores on Severity, Occurrence, and Detection in FMECA

Rating Score	Severity Criteria	Occurrence Criteria	Detection Criteria
10	Hazardous without warning	Very high; $\geq$50% in processes; $\geq$10% in automobile industry	Almost impossible; no known controls available
9	Hazardous with warning	Very high; 1 in 3 in processes; 5% in automobile industry	Very remote chance of detection
8	Very high; customer dissatisfied; major disruption to production line in automobile industry	High; 1 in 8 in processes; 2% in automobile industry	Remote chance that current controls will detect
7	High; customer dissatisfied; minor disruption to production line in automobile industry	High; 1 in 20 in processes; 1% in automobile industry	Very low chance of detection
6	Moderate; customer experiences discomfort	Moderate; 1 in 80 in processes; 0.5% in automobile industry	Low chance of detection
5	Low; customer experiences some dissatisfaction	Moderate; 1 in 400 in processes; 0.2% in automobile industry	Moderate chance of detection
4	Very low; defect noticed by some customers	Moderate; 1 in 2000 in processes; 0.1% in automobile industry	Moderately high change of detection
3	Minor; defect noticed by average customers	Low; 1 in 15,000 in processes; 0.05% in automobile industry	High chance of detection
2	Very minor; defect noticed by discriminating customers	Very low; 1 in 150,000 in processes; 0.01% in automobile industry	Very high chance of detection
1	None; no effect	Remote; <1 in 1,500,000 in processes; $\leq$ 0.001% in automobile industry	Almost certain detection

Source: Adapted from B.H. Ehrlich, *Transactional Six Sigma and Lean Servicing*, St. Lucie Press, 2002; B. El-Haik and D.M. Roy, *Service Design for Six Sigma*, Wiley, New York, 2005.

TABLE 3-12 Impact of Recommended Actions in FMECA Analysis

Functional Requirement	Recommended Actions	Actions Taken	Revised			Revised RPN	Risk	Weighted Risk Priority Number
			Severity	Occurrence	Detection			
Meet customer order quantity	Increase capacity through overtime at supplier	Overtime at supplier	6	4	4	96	3	288
	Flexibility to add suppliers or through overtime at OEM	Overtime at OEM	6	2	2	24	1	24
Meet customer order quality	Identify critical to quality characteristics (CTQ) through Pareto analysis at supplier	Monitor CTQ characteristics at supplier	7	3	3	63	3	189
	Pareto analysis of CTQ characteristics at OEM and appropriate remedial action	Monitor CTQ characteristics at OEM	7	2	2	28	2	56
Meet customer delivery date	Increase capacity through overtime at supplier	Overtime at supplier	6	4	4	96	3	288
	Flexibility to add suppliers or through overtime at OEM	Overtime at OEM	6	2	2	24	1	24
	Reduce downtime at OEM		6	2	2	24	4	96

of the analysis. The objectives of the recommended actions are to reduce the severity and/ or occurrence of the failure modes and to increase their detection through appropriate controls such as evaluation techniques or detection equipment. Whereas severity can be reduced only through a design change, occurrence may be reduced through design or process improvements. Of the possible recommended actions, Table 3-12 lists the action taken for each failure mode and the revised ratings on severity, occurrence, and detection. In this example, with no design changes made, the severity ratings are the same as before. However, the occurrence has been lowered by the corresponding action taken. Also, for certain failure modes (e.g., lack of quality), through monitoring critical to quality (CTQ) characteristics, the chances of detection are improved (lower rating compared to those in Table 3-10). An RPN is calculated, which could then be used for further follow-up actions. Here, it seems that failure to meet order quantity due to lack of capacity at the supplier and the OEM is still a priority issue.

Associated with each action, a rating on a scale of 1 to 5 is used to indicate the risk of taking that action. Here, risk refers to the chance of implementing the action successfully, with a 1 indicating the smallest risk. Using this concept, observe from Table 3-12 that it is easier to add overtime at the OEM (risk rating of 1) compared to that at the supplier (risk rating of 3), since the OEM has more control over its operations. Similarly, for the OEM it is easier to add overtime than it is to reduce downtime (risk rating of 4). Finally, the last step involves multiplying the revised RPN by the risk factor to obtain a weighted risk priority number. The larger this number, the higher the priority associated with the failure mode and the corresponding remedial action. Management could use this as a means of choosing areas to address.

3-6 INTERNATIONAL STANDARDS ISO 9000 AND OTHER DERIVATIVES

Quality philosophies have revolutionized the way that business is conducted. It can be argued that without quality programs the global economy would not exist because quality programs have been so effective in driving down costs and increasing competitiveness. Total quality systems are no longer an option—they are required. Companies without quality programs are at risk. The emphasis on customer satisfaction and continuous quality improvement has necessitated a system of standards and guidelines that support the quality philosophy. To address this need, the International Organization for Standardization (ISO) developed a set of standards, **ISO 9000, 9001**, and **9004**.

The *ISO 9000 standards* referred to as quality management standards (QMSs) were revised in 2000. This series consists of three primary standards: ISO 9000, *Quality Management Systems: Fundamentals and Vocabulary*; ISO 9001, *Quality Management Systems: Requirements*; and ISO 9004, *Quality Management Systems: Guidelines for Performance Improvements*. The American National Standards Institute and the American Society for Quality have brought their standards in line with the ISO standards and are accordingly labeled ANSI/ISO/ASQ Q9000, Q9001, and Q9004, respectively, (ASQ 2000a-c) ISO 9001 is now a more generic standard applicable to manufacturing and service industries, which have the option of omitting requirements that do not apply to them specifically. Further, organizations may be certified and registered only to ISO 9001. ISO 9004 presents comprehensive quality management guidelines that could be used by companies to improve their existing quality systems; these are not subject to audit, and organizations do not register to ISO 9004. A fourth standard in the revision by ISO consists of ISO 19011, *Guidelines on Quality and Environmental Auditing*.

Features of ISO 9000 and ANSI/ISO/ASQ Q9000

There are eight key management principles based on which the revisions of ISO 9000 and the associated ANSI/ISO/ASQ Q9000 have been incorporated. They reflect the philosophy and principles of total quality management, as discussed earlier in the chapter.

With the primary focus on an organization being to meet or exceed customer needs, a desirable shift in the standards has been towards addressing *customer* needs. This blends with our discussion of the philosophies of quality management that equally emphasize this aspect. *Senior management* must set the direction for the vision and mission of the company, with input from *all levels* in order to obtain the necessary buy-in of all people. As stated in one of the points of Deming's System of Profound Knowledge, optimization of the *system* (which may consist of *suppliers* and internal and external customers) is a desirable approach. Further, emphasis on improving the *process* based on *observed data* and *information* derived through analyses of the data, as advocated by Deming, is found to exist in the current revision of the standards. Finally, the principle of *continuous improvement*, advocated in the philosophy of total quality management, is also embraced in the standards.

ISO 9001 and ISO 9004 are a consistent pair. They are designed for use together but may be used independently, with their structures being similar. Two fundamental themes, *customer-related processes* and the concept of *continual improvement*, are visible in the new revision of the standards.

The standards have evolved to a focus on developing and managing effective processes from documenting procedures. An emphasis on the role of top management is viewed along with a data-driven process of identifying measurable objectives and measuring performance against them. Concepts of quality improvement discussed under the Shewhart (Deming) cycle of plan–do–check–act are integrated in the standards.

Other Industry Standards

Various industries are adopting to standards, similar to ISO 9000, but modified to meet their specific needs. A few of these standards are listed:

1. *QS 9000.* In the United States, the *automotive* industry comprised of the big three companies—DaimlerChrysler, Ford, and General Motors adopted to the ANSI/ISO/ASQ QS 9000, *Quality Management Systems: Requirements* standards, thereby eliminating the conflicting requirements for suppliers. Previously, each company had its own requirements for suppliers. An international automotive quality standard has also been developed as ISO/TS 16949, *Quality Systems: Automotive Suppliers*, that would provide for a single third-party registration acceptable to the big three and European automakers.

2. *AS 9100.* The *aerospace* industry, following a process similar to that used in the automotive industry, has developed the standard AS 9100, *Quality Systems: Aerospace-Model for Quality Assurance in Design, Development, Production, Installation, and Servicing.* These standards incorporate the features of ISO 9001 as well as the Federal Aviation Administration (FAA) Aircraft Certification System Evaluation Program, and Boeing's massive D1-9000 variant of ISO 9000. Companies such as Boeing, Rolls-Royce Allison, and Pratt & Whitney use AS 9100 as the basic quality management system for their suppliers.

3. *TL 9000.* This is the standard developed in the *telecommunications service* industry to seek continuous improvement in quality and reliability. The Quality Excellence

for Suppliers of Telecommunications (QuEST) Leadership Forum, formed by leading telecommunications service providers such as BellSouth, Bell Atlantic, Pacific Bell, and Southwestern Bell was instrumental in the creation of this standard. The membership now includes all regional Bell operating companies (RBOCs), AT&T, GTE, Bell Canada, and telecommunications suppliers such as Fujitsu Network Communications, Lucent Technologies, Motorola, and Nortel Networks. The globalization of the telecommunications industry has created a need for service providers and suppliers to implement common quality system requirements. The purpose of the standard is to effectively and efficiently manage hardware, software, and services by this industry. Through the adoption of such a standard, the intent is also to create cost- and performance-based metrics to evaluate efforts in the quality improvement area.

4. *Anticipated developments.* Occupational safety and health is an area that certainly deserves attention for an international standard. The designation of ISO 1800 has been reserved for this sector.

3-7 MALCOLM BALDRIGE NATIONAL QUALITY AWARD[*]

In the United States, the strategic importance of quality control and improvement has been formally recognized through the **Malcolm Baldrige National Quality Award,** which was created by Public Law 100-107 and signed into effect on August 20, 1987. The findings of Public Law 100-107 revealed that poor quality has cost companies as much as 20% of sales revenues nationally and that an improved quality of goods and services goes hand in hand with improved productivity, lower costs, and increased profitability. Furthermore, improved management understanding of the factory floor, worker involvement in quality, and greater emphasis on statistical process control can lead to dramatic improvements in the cost and quality of products and services. Quality improvement programs must be management-led and customer-oriented. All of these findings are consistent with the quality philosophies discussed in Chapter 2.

The U.S. government felt that a national quality award would motivate American companies to improve quality and productivity. The idea was that national recognition of achievements in quality would provide an example to others and would establish guidelines and criteria that could be used by business, industrial, educational, governmental, health care, and other organizations to evaluate their own quality improvement efforts. The objectives of the award are to foster competitiveness among U.S. companies through improvement of organizational performance practices, to facilitate communication and sharing of best practices, and to serve as a working tool for understanding and managing performance. Pursuing the criteria, regardless of whether a company applies for the award, is of major benefit, as it stimulates a companywide quality effort.

Award Eligibility Criteria and Categories

Responsibility for the award is assigned to the National Institute of Standards and Technology (NIST), an agency of the U.S. Department of Commerce. The American Society for Quality assists in administering the award program under contract to NIST.

*Adapted from "2007 Criteria for Performance Excellence—Malcolm Baldrige National Quality Award," U.S. Department of Commerce, National Institute of Standards and Technology, Gaithersburg, MD, www.nist.gov.

Awards are made annually to U.S. companies that excel in quality management and quality achievement; the first award was given in 1988.

Three business eligibility categories have been established for the award: manufacturing, service, and small business. Any for-profit business and subunits headquartered in the United States or its territories, including U.S. subunits of foreign companies, may apply. Not eligible in the business category are local, state, and federal government agencies, trade associations, and nonprofit organizations. Up to two awards may be given in each category each year. The eligibility under each of these categories is defined as follows:

- *Manufacturing:* companies or some subunits that produce and sell manufactured products or manufacturing processes, and producers of agricultural, mining, or construction products
- *Service:* companies or some subunits that sell services
- *Small business:* companies or some subunits engaged in manufacturing and/or the provision of services that have 500 or fewer employees

Nonprofit public, private, and government organizations are also eligible in a separate category. Such organizations may include local, state, and federal government agencies, trade associations, charitable organizations, social-service agencies, credit unions, and professional societies.

Two other award categories exist: *education* and *health care*:

- *Education.* Eligible education organizations include elementary and secondary schools and school districts, colleges, universities and university systems, schools or colleges within universities, professional schools, community colleges, and technical schools. Separate criteria exist for this category.
- *Health care.* Eligible organizations include hospitals, HMOs, long-term care facilities, health care practitioner offices, home health agencies, and dialysis centers. Separate criteria exist for this category.

For-profit education or health care organizations may apply under the service or small business category, as appropriate using the respective criteria or under the health care or education categories, using their respective criteria.

Criteria for Evaluation

Applying for the award is a two-step process. First, potential applicants must establish their eligibility in one of the award categories and may nominate a senior member of their staff to serve on the board of examiners. Second, an application form and a report must be completed. These forms are reviewed in a three-stage process. In the first stage, a review is conducted by at least six members of a board of examiners. At the conclusion of the first stage, the panel determines which applications should be referred for consensus review in the second stage. After the second stage, the panel determines which applications should receive the site visits. The third stage involves an on-site verification of the application report. The site-visit review team develops a report for the panel of judges, which develops a set of recommendations for the National Institute of Standards and Technology. Awards are traditionally presented by the President of the United States.

TABLE 3-13   The 2007 Malcolm Baldrige Award Criteria Categories

2007 Categories		Point Values
1. **Leadership**		120
1.1 Senior leadership	70	
1.2 Governance and social responsibilities	50	
2. **Strategic planning**		85
2.1 Strategic development	40	
2.2 Strategy deployment	45	
3. **Customer and market focus**		85
3.1 Customer and market knowledge	40	
3.2 Customer relationships and satisfaction	45	
4. **Measurement, analysis, and knowledge management**		90
4.1 Measurement, analysis, and review of organizational performance	45	
4.2 Management of information, information technology, and knowledge	45	
5. **Workforce focus**		85
5.1 Workforce engagement	45	
5.2 Workforce environment	40	
6. **Process management**		85
6.1 Work system design	35	
6.2 Work process management and improvementv	50	
7. **Results**		450
7.1 Product and service outcomes	100	
7.2 Customer-focused outcomes	70	
7.3 Financial and market outcomes	70	
7.4 Workforce-focused outcomes	70	
7.5 Process effectiveness outcomes	70	
7.6 Leadership outcomes	70	
Total points		1000

Source: NIST, "2007 Criteria for Performance Excellence—Malcolm Baldrige National Quality Award," U.S. Department of Commerce, National Institute of Standards and Technology, Gaithersburg, MD, www.nist.gov.

The award criteria are built on the following core values[*]: Visionary leadership, customer-driven excellence, organizational and personal learning, valuing employees and partners, agility, focus on the future, managing for innovation, management by fact, social responsibility, focus on results and creating value, and systems perspective.

Seven categories incorporate the core values and concepts. They determine the framework for evaluation. Table 3-13 shows the award criteria categories, subcategories, and point values of the *2007 Criteria for Performance Excellence* in the manufacturing, service, and small business categories. Briefly, the categories may be described as follows:

1. *Leadership.* The leadership category examines *how* the organization's senior leaders guide and sustain the organization. Also examined are the organization's governance and *how* the organization addresses its ethical, legal, and community responsibilities.

*Adapted from NIST, "2007 Baldrige Criteria for Performance Excellence," U.S. Department of Commerce, National Institute of Standards and Technology, Gaithersburg, MD, www.nist.gov.

2. *Strategic planning.* This category examines *how* the organization develops strategic objectives and action plans. Also examined are *how* the chosen strategic objectives and action plans are deployed and changed if the circumstances require, and *how* progress is measured.

3. *Customer and market focus.* The manner in which the organization determines the requirements, needs, expectations, and preferences of customers and markets is addressed in this category. Also of importance is *how* the organization builds relationships with customers and determines the *key* factors that lead to customer acquisition, satisfaction, loyalty and retention, business expansion, and sustainability.

4. *Measurement, analysis, and knowledge management.* This category examines *how* the organization selects, gathers, analyzes, manages, and improves its data, information, and knowledge assets. The manner in which the organization reviews its performance is also examined.

5. *Workforce focus.* This category examines *how* an organization engages, manages, and develops its workforce to utilize their full potential. Building a workforce environment conducive to high performance is imperative.

6. *Process management.* Key aspects of process management, including *key* product, service, and organizational processes for creating *customer* and organizational value, are examined. How is new technology and the need for agility designed into the processes to achieve sustainability?

7. *Results.* This category examines the organization's performance and improvement in all *key* areas: product and service outcomes, customer satisfaction, financial and marketplace outcomes, human resource outcomes, process-effectiveness outcomes, and leadership and social responsibility. Performance levels are examined relative to those of competitors and other organizations, providing similar products and services.

SUMMARY

In the chapter we examined the philosophy of total quality management and the role management plays in accomplishing desired organizational goals and objectives. A company's vision describes what it wants to be; the vision molds quality policy. This policy, along with the support and commitment of top management, defines the quality culture that prevails in an organization. Since meeting and exceeding customer needs are fundamental criteria for the existence and growth of any company, the steps of product design and development, process analysis, and production scheduling have to be integrated into the quality system.

The fundamental role played by top management cannot be overemphasized. Based on a company's strategic plans, the concept of using a balanced scorecard that links financial and other dimensions, such as learning and growth and customers, is a means for charting performance. Exposure to the techniques of failure mode and effects criticality analysis enables adequate product or process design.

The planning tool of quality function deployment is used in an interdisciplinary team effort to accomplish the desired customer requirements. Benchmarking enables a company to understand its relative performance with respect to industry performance measures and thus helps the company improve its competitive position. Adaptation of best practices to the organization's environment also ensures continuous improvement. Vendor quality audits, selection, and certification programs are important because final product quality is influenced by the quality of raw material and components. Since quality decisions are dependent on the collected data and information on products, processes, and customer satisfaction, simple tools for quality improvement that make use of such data have been presented. These include Pareto analysis, flowcharts, cause-and-effect diagrams, and various scatterplots.

Finally, some international standards on quality assurance practices have been depicted. Furthermore, the criteria for the Malcolm Baldrige National Quality Award bestowed annually in the United States have been presented.

CASE STUDY: ROBERT WOOD JOHNSON UNIVERSITY HOSPITAL HAMILTON*

Robert Wood Johnson University Hospital Hamilton (RWJ Hamilton) is a private, not-for-profit acute-care hospital serving more than 350,000 residents in Hamilton Township, New Jersey, on a 68-acre campus. Founded in 1940, the 340,000-square-foot hospital includes services for medical, surgical, obstetric, cardiac, orthopedic, emergency, and intensive care, as well as a comprehensive outpatient cancer center, an Occupational Health Center, and eight child-care centers. Three outdoor areas, designated Grounds For Healing™, offer environments that aid healing through art and nature. And the recently opened 86,000-square-foot Center for Health and Wellness features extensive community health education, fitness and physical therapy facilities, a healthy cooking kitchen, a healthy café, and a day spa.

RWJ Hamilton's more than 1730 employees and 650 medical staff members provide quality care to 14,000 admitted patients and manage more than 50,000 patient visits a year to its emergency department, generating revenues of over $170 million. The hospital is part of the Robert Wood Johnson Health System and Network and is affiliated with the University of Medicine and Dentistry of New Jersey–Robert Wood Johnson Medical School. It is the recipient of the 2004 Malcolm Baldrige National Quality Award in the health care category.

Vision: Growth Through Improvement

RWJ Hamilton, New Jersey's fastest-growing hospital from 1999 to 2003, has attained a market leadership position across the service lines. Its market share in cardiology grew from 20% to nearly 30%; in surgery from 17% to 30%; and in oncology from 13% to above 30%. Its closest competitors' market shares have remained the same or declined in each of these areas. RWJ Hamilton's emergency department volume doubled in that time, making it the volume leader in the area.

Fueling this growth is RWJ Hamilton's continual drive to improve the quality of its health care services and environment—or, as the organization's vision statement phrases it, "to passionately pursue the health and well-being of our patients, employees, and the community." A 2002 Gallup Community Survey ranked the hospital first among all local competitors in customer satisfaction and loyalty, an indication of why the hospital's occupancy rate grew from 70% to 90%, more than 25 percentage points higher than its nearest competitor.

Excellence Through Service

From the patient's bedside to the hospital's boardroom, RWJ Hamilton's mandate is to provide ever-improving service to three customer groups: patients, employees, and

*Adapted from NIST, "Malcolm Baldrige National Quality Award, 2004 Award Recipient Profile," U.S. Department of Commerce, National Institute of Standards and Technology, Gaithersburg, MD, www.nist.gov.

community. The hospital continuously studies the changing market to determine what these customers need today and anticipate what they will need years down the road. It studies trends in the industry, patient preferences, and physician referrals. It gathers and analyzes market research, demographic data, and information on competitors. It continually solicits opinions and recommendations from employees and pays close attention to the results of customer and community surveys. All information is captured in a database called "Voice of the Customer" and is used to design service improvements and set ever-higher Excellence Through Service goals.

Also with a focus on Excellence Through Service, the leadership team works within a system that links all management functions—from planning and implementing policies, new technologies, and new facilities through ongoing cycles of evaluation and improvement—with unhindered communications at all levels. Each executive management team member, including the CEO, holds daily briefings that are designed to share key information with the staff and to answer questions. As a result, over the past four years, employee satisfaction with hospital leadership has improved to almost 100%.

Serving Patients

At RWJ Hamilton, the patient is the central focus of the organization. Multidisciplinary health care teams design service objectives and expected outcomes based on patient needs. Patient care plans are evaluated daily and take into account each patient's language preferences, cultural needs, lifestyle, and quality-of-life issues.

As part of this focus, RWJ Hamilton utilizes its 5-Star Service Standards, which include commitments to customers and co-workers, courtesy and etiquette, and safety awareness, to recruit, train, and evaluate employees. Between 2002 and 2003, the number of training hours for full-time employees was increased from 38 hours to 58 hours, ensuring continued improvement of skills.

RWJ Hamilton's commitment to excellence mandates innovation: in services, processes, and technology. In 1998 it launched the 15/30 program, guaranteeing that every emergency patient sees a nurse within 15 minutes and a physician within 30 minutes or their emergency department charge is waived. And, in a program called Walk in My Shoes, employees work in departments other than their own, creating opportunities for cross-training, sharing information and best practices, and gaining fresh insights. An integrated information technology system enables RWJ Hamilton to acquire and internally share information essential for day-to-day operations and decision making, improving key processes and strategic planning. The hospital's steadily expanding IT capabilities, which RWJ Hamilton calls its "IT Innovation Journey," has also made the hospital a pioneer in technologies that are the future of medical care, such as fully digital radiology, bar coding of medications and blood and radiology products, electronic patient medical records and computerized physician order-entry systems.

This patient-centered focus has led to growing satisfaction with the emergency department—up from 85% in 2001 to 90% in 2004, exceeding national benchmarks. Also, patient satisfaction with the nursing staff has improved from 70% in 1999 to 90% in 2004. In two Gallup Community surveys, RWJ Hamilton's nursing staff has been recognized as the best among local competitors.

RWJ Hamilton scores just as high in clinical results as it does in customer satisfaction results. Through partnerships with physicians and steady improvement of processes, facilities, technology, and training, RWJ Hamilton has reduced its rates of mortality, hospital-acquired infections, and medications errors to among the lowest in the nation.

Physician partners are involved in developing evidence-based clinical guidelines and implementing standards to promote the highest level of care. In an evaluation by the Joint Commission on Accreditation of Healthcare Organizations, RWJ Hamilton ranks among the top 10% of hospitals in the effectiveness of its aspirin and beta-blocker treatment for patients who have suffered a heart attack or congestive heart failure.

Serving Employees and Community

RWJ Hamilton values its employees and promotes their health and safety. Health screening is part of the hiring process and repeated annually. Employees with direct contract with patients receive additional testing and training in safety procedures and the use of personal protection equipment. Over the past four to five years, employee satisfaction has risen in a number of key areas. Between 1999 and 2003, satisfaction with benefits rose from 30% to above 90%; satisfaction with participation in decisions grew from about 40% to 90%; satisfaction with employee recognition from 70% to 97%. Retention rates for employees tipped 96% in 2003; retention of registered nurses reached 98%.

RWJ Hamilton demonstrates deep community concern and involvement. In an average month, the hospital provides more than 900 community residents with free health screenings. Guided by the Medical Advisory Panel, the RWJ Hamilton Center for Health and Wellness provides health education to more than 100,000 people annually. Between 1999 and 2003, hospital donations to community organizations increased from $80,000 to almost $140,000, and Charity Care Dollars increased from approximately $5 million to almost $23 million.

Highlights

- RWJ Hamilton's 15/30 program guarantees that patients coming into the emergency department will see a nurse within 15 minutes and a physician within 30 minutes.
- RWJ Hamilton utilizes its 5-Star Service Standards, which include commitment to customers and co-workers, courtesy and etiquette, and safety awareness, to recruit, train, and evaluate employees.
- RWJ Hamilton, New Jersey's fastest-growing hospital from 1999 to 2003, has attained a market leadership position in an extremely competitive environment.
- Patient satisfaction with the emergency department has improved from 85% in 2001 to 90% in 2004, exceeding the national benchmark.

Questions for Discussion

1. What are some of the unique characteristics of the quality culture at RWJ Hamilton?
2. Who are the customers, and how does RWJ Hamilton ensure customer satisfaction? What procedures are used to address the dynamic needs of customers?
3. If a balanced scorecard approach were to be taken at RWJ Hamilton, describe some of the measures that would be used.
4. What processes does RWJ Hamilton maintain to stay abreast of recent developments? Discuss the emphasis it places on the philosophy of continuous improvement.
5. Discuss the social obligations served by RWJ Hamilton. Does this promote advancing satisfaction levels?
6. Technology is part and parcel of process improvement. Discuss the methods that RWJ Hamilton has used to adopt technology and the subsequent results.

7. Select a particular process in a hospital. Develop a flowchart of the existing system and comment on ways to improve the process.

KEY TERMS

ANSI/ISO/ASQ Q9000; Q9001; Q9004	conformity quality audit
approved vendor	process audit
balanced scorecard	product audit
benchmarking	suitability quality audit
cause-and-effect diagram	system audit
certified vendor	quality function deployment
change management	quality policy
cycle time	risk priority number
empowerment	scatter diagrams
failure mode and effects criticality analysis	scatterplot
flowchart	six-sigma quality
gap analysis	define
house of quality	measure
ISO 9000; 9001, 9004	analyze
Malcolm Baldrige National Quality Award	improve
matrix plot	control
mission statement	three-dimensional scatterplot
multivariable charts	time-based competition
Pareto diagram	TL 9000
performance standards	vendor certification
preferred vendor	vendor rating
QS 9000	vision
quality audit	

EXERCISES

3-1   Describe the total quality management philosophy. Choose a company and discuss how its quality culture fits this theme.

3-2   What are the advantages of creating a long-term partnership with vendors?

3-3   Compare and contrast a company vision, mission, and quality policy. Discuss these concepts in the context of a hospital of your choice.

3-4   Describe Motorola's concept of six sigma quality and explain the level of non-conforming product that could be expected from such a process.

3-5   What are the advantages of using quality function deployment? What are some key ingredients that are necessary for its success?

3-6   Select an organization of your choice in the following categories. Identify the organization's strategy. Based on these strategies, perform a balanced scorecard analysis by indicating possible diagnostic and strategic measures in each of the areas of learning and growth, internal processes, customers, and financial status.
(a) Information technology services

    (b) Health care
    (c) Semiconductor manufacturing
    (d) Pharmaceutical

3-7   Consider the airline transportation industry. Develop a house of quality showing customer requirements and technical descriptors.

3-8   Consider a logistics company transporting goods on a global basis. Identify possible vision and mission statements and company strategies. Conduct a balanced scorecard analysis and indicate suggested diagnostic and strategic measures in each of the areas of learning and growth, internal processes, customers, and financial.

3-9   Consider the logistics company in Exercise 3-8. Conduct a quality function deployment analysis where the objective is to minimize delays in promised delivery dates.

3-10   Describe the steps of benchmarking relative to a company that develops microchips. What is the role of top management in this process?

3-11   What are the various types of quality audits? Discuss each and identify the context in which they are used.

3-12   A financial institution is considering outsourcing its information technology–related services. What are some criteria that the institution should consider? Propose a scheme to select a vendor.

3-13   The area of nanotechnology is of much importance in many phases of our lives—one particular area being development of drugs for Alzheimer's disease. Discuss the role of benchmarking, innovation, and time-based competition in this context.

3-14   In a large city, the mass-transit system, currently operated by the city, needs to be overhauled with projected demand expected to increase substantially in the future. The city government is considering possible outsourcing.
    (a) Discuss the mission and objectives of such a system.
    (b) What are some criteria to be used for selecting a vendor?
    (c) For a private vendor, through a balanced scorecard analysis, propose possible diagnostic and strategic measures.

3-15   What is the purpose of vendor certification? Describe typical phases of certification.

3-16   Discuss the role of national and international standards in certifying vendors.

3-17   The postal system has undertaken a quality improvement project to reduce the number of lost packages. Construct a cause-and-effect diagram and discuss possible measures that should be taken.

3-18   The safe operation of an automobile is dependent on several subsystems (e.g., engine, transmission, braking mechanism). Construct a cause-and-effect diagram for automobile accidents. Conduct a failure mode and effects criticality analysis and comment on areas of emphasis for prevention of accidents.

3-19   Consider Exercise 3-18 on the prevention of automobile accidents. However, in this exercise, consider the driver of the automobile. Construct a cause-and-effect diagram for accidents influenced by the driver. Conduct a failure model and effects criticality analysis considering issues related to the driver, assuming that the automobile is in fine condition.

3-20   You are asked to make a presentation to senior management outlining the demand for a product. Describe the data you would collect and the tools you would use to organize your presentation.

3-21   Consider a visit to your local physician's office for a routine procedure. Develop a flowchart for the process. What methods could be implemented to improve your satisfaction and reduce waiting time?

3-22   What are some reasons for failure of total quality management in organizations? Discuss.

3-23   A product goes through 20 independent operations. For each operation, the first-pass yield is 95%. What is the rolled throughput yield for the process?

3-24   Consider Exercise 3-23. Suppose, through a quality improvement effort, that the first-pass yield of each operation is improved to 98%. What is the percentage improvement in rolled throughput yield?

3-25   Consider Exercise 3-24. Through consolidation of activities, the number of operations has now been reduced to 10, with the first-pass yield of each operation being 98%. What is the percentage improvement in rolled-throughout yield relative to that in Exercise 3-24?

3-26   Discuss the role of established standards and third-party auditors in quality auditing. What is the role of ISO 9000 standards in this context?

3-27   In a printing company, data from the previous month show the following types of errors, with the unit cost (in dollars) of rectifying each error, in Table 3-14.
(a) Construct a Pareto chart and discuss the results.
(b) If management has a monthly allocation of $18,000, which areas should they tackle?

TABLE 3-14

Error Categories	Frequency	Unit Costs
Typographical	4000	0.20
Proofreading	3500	0.50
Paper tension	80	50.00
Paper misalignment	100	30.00
Inadequate binding	120	100.00

3-28   An insurance company is interested in determining whether life insurance coverage is influenced linearly by disposable income. A randomly chosen sample of size 20 produced the data shown in Table 3-15. Construct a scatterplot. What conclusions can you draw?

3-29   Use a flowchart to develop an advertising campaign for a new product that you will present to top management.

3-30   Is accomplishing registration to ISO 9001 standards similar to undergoing an audit process? What are the differences?

3-31   Discuss the emerging role of ISO 9000 standards in the global economy.

TABLE 3-15

Disposable Income ($ thousands)	Life Insurance Coverage ($ thousands)	Disposable Income ($ thousands)	Life Insurance Coverage ($ thousands)
45	60	65	80
40	58	60	90
65	100	45	50
50	50	40	50
70	120	55	70
75	140	55	60
70	100	60	80
40	50	75	120
50	70	45	50
45	60	65	70

3-32 What is the role of the Malcolm Baldrige National Quality Award? How is it different from ISO 9000 standards? Explain the obligations of the Malcolm Baldrige Award winners.

3-33 In a chemical process, the parameters of temperature, pressure, proportion of catalyst, and pH value of the mixture influence the acceptability of the batch. The data from 20 observations are shown in Table 3-16.

(a) Construct a multivariable chart. What inferences can you make regarding the desirable values of the process parameters?

TABLE 3-16

Observation	Temperature (°C)	Pressure (kg/cm^2)	Proportion of Catalyst	Acidity (pH)	Proportion Nonconforming
1	300	100	0.03	20	0.080
2	350	90	0.04	20	0.070
3	400	80	0.05	15	0.040
4	500	70	0.06	25	0.060
5	550	60	0.04	10	0.070
6	500	50	0.06	15	0.050
7	450	40	0.05	15	0.055
8	450	30	0.04	20	0.060
9	350	40	0.04	15	0.054
10	400	40	0.04	15	0.052
11	550	40	0.05	10	0.035
12	350	90	0.04	20	0.070
13	500	40	0.06	10	0.030
14	350	80	0.04	15	0.070
15	300	80	0.03	20	0.060
16	550	30	0.05	10	0.030
17	400	80	0.03	20	0.065
18	500	40	0.05	15	0.035
19	350	90	0.03	20	0.065
20	500	30	0.06	10	0.040

(b) Construct a matrix plot and make inferences on desirable process parameter levels.

(c) Construct contour plots of the proportion nonconforming by selecting two of the process parameters at a time and comment.

REFERENCES

ASQ (2000a). *Quality Management Systems: Fundamentals and Vocabulary*, ANSI/ISO/ASQ Q9000, Milwaukee, WI: American Society for Quality.

——— (2000b). *Quality Management Systems: Requirements*, ANSI/ISO/ASQ Q9001, Milwaukee, WI: American Society for Quality.

——— (2000c). *Quality Management Systems-Guidelines for Performance Improvement*, ANSI/ISO/ASQ Q9004, Milwaukee, WI: American Society for Quality.

Blazek, L. W., B. Novic, and D. M. Scott (1987). "Displaying Multivariate Data Using Polyplots," *Journal of Quality Technology*, 19(2): 69–74.

Buzanis, C. H. (1993). "Hyatt Hotels and Resorts: Achieving Quality Through Employee and Guest Feedback Mechanisms. In *Managing Quality in America's Most Admired Companies*, J. W. Spechler, (Ed.) San Francisco, CA: Berrett-Kochler.

Ehrlich, B. H. (2002). *Transactional Six Sigma and Lean Servicing*. Boca Raton, FL: St. Lucie Press.

El-Haik, B., and D. M. Roy (2005). *Service Design for Six Sigma*. Hoboken, NJ: Wiley.

Gnanadesikan, R. (1977). *Methods of Statistical Data Analysis of Multivariate Observations*. New York: John Wiley.

Ishikawa, K. (1976). *Guide to Quality Control*. Asian Productivity Organization, Nordica International Limited, Hong Kong.

ISO (2000a). *Guidelines on Quality and Environmental Auditing*, ISO 19011. Geneva, Switzerland: International Organization for Standardization.

——— (2000b). *Quality Management Systems: Fundamentals and Vocabulary*, ISO 9000, Geneva, Switzerland: International Organization for Standardization.

——— (2000c). *Quality Management Systems: Requirements*, ISO 9001 Geneva, Switzerland: International Organization for Standardization.

——— (2000d). *Quality Management Systems: Guidelines for Performance Improvement*, ISO 9004, Geneva, Switzerland: International Organization for Standardization.

Kaplan, R. S., and D. P. Norton (1996). *The Balanced Scorecard*, Boston MA: Harvard Business School Press.

Mills, C. A. (1989). *The Quality Audit*, Milwaukee, WI: American Society Quality Control.

Minitab, Inc. (2007). Release 15. State College, PA: Minitab.

NIST (2007). 2007 *Criteria for Performance Excellence—Malcolm Baldrige National Quality Award*. Gaithersburg, MD: U.S. Department of Commerce National Institute of Standards and Technology.

PART II

STATISTICAL FOUNDATIONS AND METHODS OF QUALITY IMPROVEMENT

4

FUNDAMENTALS OF STATISTICAL CONCEPTS AND TECHNIQUES IN QUALITY CONTROL AND IMPROVEMENT

4-1 INTRODUCTION AND CHAPTER OBJECTIVES

In this chapter we build a foundation for the statistical concepts and techniques used in quality control and improvement. Statistics is a subtle science, and it plays an important role in quality programs. Only a clear understanding of statistics will enable you to apply it properly. They are often misused, but a sound knowledge of statistical principles will help you formulate correct procedures in different situations and will help you interpret the results properly. When we analyze a process, we often find it necessary to study its characteristics individually. Breaking the process down allows us to determine whether some identifiable cause has forced a deviation from the expected norm and whether a remedial action needs to be taken. Thus, our objective in this chapter is to review different statistical concepts and techniques along two major themes. The first deals with *descriptive statistics*, those that are used to describe products or processes and their characteristic features, based on collected data. The second theme is focused on *inferential statistics*, whereby conclusions on product or process parameters are made through statistical analysis of data. Such inferences, for example, may be used to determine if there has been a significant improvement in the quality level of a process, as measured by the proportion of nonconforming product.

Fundamentals of Quality Control and Improvement, Third Edition, By Amitava Mitra
Copyright © 2008 John Wiley & Sons, Inc.

4-2 POPULATION AND SAMPLE

A **population** is the set of all items that possess a certain characteristic of interest.

Example 4-1 Suppose that our objective is to determine the average weight of cans of brand A soup processed by our company for the month of July. The population in this case is the set of all cans of brand A soup that are output in the month of July (say, 50,000). Other brands of soup made during this time are not of interest, only the population of brand A soup cans.

A **sample** is a subset of a population. Realistically, in many manufacturing or service industries, it is not feasible to obtain data on every element in the population. Measurement, storage, and retrieval of large volumes of data are impractical, and the costs of obtaining such information are high. Thus, we usually obtain data from only a portion of the population—a sample.

Example 4-2 Consider our brand A soup. To save ourselves the cost and effort of weighing 50,000 cans, we randomly select a sample of 500 cans of brand A soup from the July output.

4-3 PARAMETER AND STATISTIC

A **parameter** is a characteristic of a population, something that describes it.

Example 4-3 For our soup example, we will be looking at the parameter average weight of all 50,000 cans processed in the month of July.

A **statistic** is a characteristic of a sample. It is used to make inferences on the population parameters that are typically unknown.

Example 4-4 Our statistic then is the average weight of a sample of 500 cans chosen from the July output. Suppose that this value is 300 g; this would then be an *estimate* of the average weight of all 50,000 cans. A statistic is sometimes called an *estimator*.

4-4 PROBABILITY

Our discussion of the concepts of probability is intentionally brief. For an in-depth look at probability, see the references at the end of the chapter. The **probability** of an event describes the chance of occurrence of that event. A probability function is bounded between 0 and 1, with 0 representing the definite nonoccurrence of the event and 1 representing the certain occurrence of the event.

The set of all outcomes of an experiment is known as the **sample space** S.

Relative Frequency Definition of Probability

If each event in the sample space is equally likely to happen, the probability of an event A is given by

$$P(A) = \frac{n_A}{N} \tag{4-1}$$

where $P(A)$ = probability of event A, n_A = number of occurrences of event A, and N = size of the sample space.

This definition is associated with the relative frequency concept of probability. It is applicable to situations where historical data on the outcome of interest are available. The probability associated with the sample space is 1 [i.e., $P(S) = 1$].

Example 4-5   A company makes plastic storage bags for the food industry. Out of the hourly production of 2000 500-g bags, 40 were found to be nonconforming. If the inspector chooses a bag randomly from the hour's production, what is the probability of it being nonconforming?

Solution   We define event A as getting a bag that is nonconforming. The sample space S consists of 2000 bags (i, e., $N = 2000$). The number of occurrences of event $A(n_A)$ is 40. Thus, if the inspector is equally likely to choose any one of the 2000 bags,

$$P(A) = \frac{40}{2000} = 0.02$$

Simple and Compound Events

Simple events cannot be broken into other events. They represent the most elementary form of the outcomes possible in an experiment. **Compound events** are made up of two or more simple events.

Example 4-6   Suppose that an inspector is sampling transistors from an assembly line and identifying them as acceptable or not. Suppose the inspector chooses two transistors. What are the simple events? Give an example of a compound event. Find the probability of finding at least one acceptable transistor.

Solution   Consider the following outcomes:

A_1: event that the first transistor is acceptable
D_1: event that the first transistor is unacceptable
A_2: event that the second transistor is acceptable
D_2: event that the second transistor is unacceptable
Four simple events make up the sample space S:

$$S = \{A_1A_2, A_1D_2, D_1A_2, D_1D_2\}$$

These events may be described as follows:

$E_1 = \{A_1A_2\}$: event that the first and second transistors are acceptable
$E_2 = \{A_1D_2\}$: event that the first transistor is acceptable and the second one is not
$E_3 = \{D_1A_2\}$: event that the first transistor is unacceptable and the second one is
                   acceptable
$E_4 = \{D_1D_2\}$: event that both transistors are unacceptable

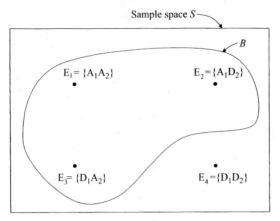

FIGURE 4-1   Venn diagram.

Compound event B is the event that at least one of the transistors is acceptable. In this case, event B consists of the following three simple events: $B = \{E_1, E_2, E_3\}$. Assuming that each of the simple events is equally likely to happen, $P(B) = P(E_1) + P(E_2) + P(E_3) = \frac{1}{4} + \frac{1}{4} + \frac{1}{4} = \frac{3}{4}$. Figure 4-1 shows a Venn diagram, which is a graphical representation of the sample space and its associated events.

Complementary Events

The *complement* of an event A implies the occurrence of everything but A. If we define A^c to be the complement of A, then

$$P(A^c) = 1 - P(A) \qquad (4\text{-}2)$$

Figure 4-2 shows the probability of the complement of an event by means of a Venn diagram. Continuing with Example 4-6, suppose that we want to find the probability of the

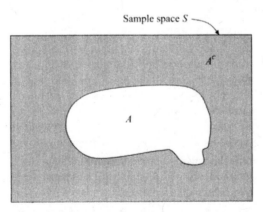

FIGURE 4-2   An event and its complement. *Note*: The shaded area represents $P(A^c)$.

event that both transistors are unacceptable. Note that this is the complement of event B, which was defined as at least one of the transistors being acceptable. So

$$P(B^c) = 1 - P(B) = 1 - \frac{3}{4} = \frac{1}{4}$$

Additive Law

The *additive law* of probability defines the probability of the union of two or more events happening. If we have two events A and B, the union of these two implies that A happens *or* B happens *or* both happen. Figure 4-3 shows the union of two events, A and B. The hatched area in the sample space represents the probability of the union of the two events. The **additive law** is as follows:

$$\begin{aligned} P(A \cup B) &= P(A \text{ or } B \text{ or both}) \\ &= P(A) + P(B) - P(A \cap B) \end{aligned} \qquad (4\text{-}3)$$

Note that $P(A \cap B)$ represents the probability of the intersection of events A and B: that is, the occurrence of both A and B. The logic behind the additive law can easily be seen from the Venn diagram in Figure 4-3, where $P(A)$ represents the area within the boundary-defining event A. Similarly, $P(B)$ represents the area within the boundary-defining event B. The overlap (crosshatched) between areas A and B represents the probability of the intersection, $P(A \cap B)$. When $P(A)$ is added to $P(B)$, this intersection is included twice, so eq. (4-3) adds $P(A)$ to $P(B)$ and subtracts $P(A \cap B)$ once.

Multiplicative Law

The **multiplicative law** of probability defines the probability of the intersection of two or more events. Intersection of a group of events means that all the events in that group occur. In general, for two events A and B,

$$\begin{aligned} P(A \cap B) &= P(A \text{ and } B) = P(A)P(B \mid A) \\ &= P(B)P(A \mid B) \end{aligned} \qquad (4\text{-}4)$$

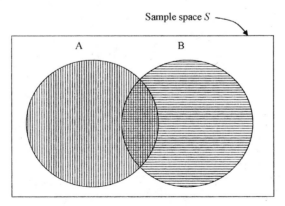

FIGURE 4-3   Union of two events.

The term $P(B|A)$ represents the conditional probability of B given that event A has happened (i.e., the probability that B will occur if A has). Similarly, $P(A|B)$ represents the conditional probability of A given that event B has happened. Of the two forms given by eq. (4-4), the problem will dictate which version to use.

Independence and Mutually Exclusive Events

Two events A and B are said to be **independent** if the outcome of one has no influence on the outcome of the other. If A and B are independent, then $P(B|A) = P(B)$; that is, the conditional probability of B given that A has happened equals the unconditional probability of B. Similarly, $P(A|B) = P(A)$ if A and B are independent. From eq. (4-4), it can be seen that if A and B are independent, the general multiplicative law reduces to

$$P(A \cap B) = P(A \text{ and } B) = P(A)P(B) \quad \text{if } A \text{ and } B \text{ are independent} \qquad (4\text{-}5)$$

Two events A and B are said to be **mutually exclusive** if they cannot happen simultaneously. The intersection of two mutually exclusive events is the null set, and the probability of their intersection is zero. Notationally, $P(A \cap B) = 0$ if A and B are mutually exclusive. Figure 4-4 shows a Venn diagram for two mutually exclusive events. Note that when A and B are mutually exclusive events, the probability of their union is simply the sum of their individual probabilities. In other words, the additive law takes on the following special form:

$$P(A \cup B) = P(A) + P(B) \qquad \text{if } A \text{ and } B \text{ are mutually exclusive}$$

If events A and B are mutually exclusive, what can we say about their dependence or independence? Obviously, if A happens, B cannot happen, and vice versa. Therefore, if A and B are mutually exclusive, they are dependent. If A and B are independent, the additive rule from eq. (4-3) becomes

$$P(A \text{ or } B \text{ or both}) = P(A) + P(B) - P(A)P(B) \qquad (4\text{-}6)$$

Example 4-7

(a) In the production of metal plates for an assembly, it is known from past experience that 5% of the plates do not meet the length requirement. Also, from historical

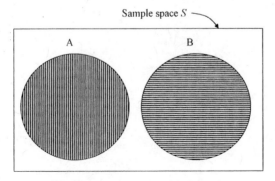

FIGURE 4-4   Mutually exclusive events.

records, 3% of the plates do not meet the width requirement. Assume that there are no dependencies between the processes that make the length and those that trim the width. What is the probability of producing a plate that meets both the length and width requirements?

Solution   Let A be the outcome that the plate meets the length requirement and B be the outcome that the plate meets the width requirement. From the problem statement, $P(A^c) = 0.05$ and $P(B^c) = 0.03$. Then

$$P(A) = 1 - P(A^c) = 1 - 0.05 = 0.95$$
$$P(B) = 1 - P(B^c) = 1 - 0.03 = 0.97$$

Using the special case of the multiplicative law for independent events, we have

$$P(\text{meeting both length } and \text{ width requirements}) = P(A \cap B)$$
$$= P(A)P(B) \text{ (since } A \text{ and } B \text{ are independent events)}$$
$$= (0.95)(0.97) = 0.9215$$

(b) What proportion of the parts will not meet at least one of the requirements?

Solution   The required probability $= P(A^c \text{ or } B^c \text{ or both})$. Using the additive law, we get

$$P(A^c \text{ or } B^c \text{ or both}) = P(A^c) + P(B^c) - P(A^c \cap B^c)$$
$$= 0.05 + 0.03 - (0.03)(0.05) = 0.0785$$

Therefore, 7.85% of the parts will have at least one characteristic (length, width, or both) not meeting the requirements.

(c) What proportion of parts will meet neither length nor width requirements?

Solution   We want to find $P(A^c \cap B^c)$.

$$P(A^c \cap B^c) = P(A^c)P(B^c) = (0.05)(0.03) = 0.0015$$

(d) Suppose the operations that produce the length and the width are not independent. If the length does not satisfy the requirement, it causes an improper positioning of the part during the width trimming and thereby increases the chances of nonconforming width. From experience, it is estimated that if the length does not conform to the requirement, the chance of producing nonconforming widths is 60%. Find the proportion of parts that will neither conform to the length nor the width requirements.

Solution   The probability of interest is $P(A^c \cap B^c)$. The problem states that $P(B^c \mid A^c) = 0.60$. Using the general from of the multiplicative law

$$P(A^c \cap B^c) = P(A^c)P(B^c \mid A^c)$$
$$= (0.05)(0.06) = 0.03$$

So 3% of the parts will meet neither the length nor the width requirements. Notice that this value is different from the answer to part (c), where the events were assumed to be independent.

(e) In part (a), are events A and B mutually exclusive?

Solution   We have found $P(A) = 0.95$, $P(B) = 0.97$, and $P(A \cap B) = 0.9215$. If A and B were mutually exclusive, $P(A \cap B)$ would have to be zero. However, this is not the case, since $P(A \cap B) = 0.9215$. So A and B are *not* mutually exclusive.

(f) Describe two events in this example setting that are mutually exclusive.

Solution   Events A and A^c are mutually exclusive, to name one instance, and $P(A \cap A^c) = 0$, because A and A^c cannot happen simultaneously. This means that it is not possible to produce a part that both meets and does not meet the length requirement.

4-5 DESCRIPTIVE STATISTICS: DESCRIBING PRODUCT OR PROCESS CHARACTERISTICS

Statistics is the science that deals with the collection, classification, analysis, and making of inferences from data or information. Statistics is subdivided into two categories: *descriptive statistics* and *inferential statistics*.

Descriptive statistics describes the characteristics of a product or process using information collected on it. Suppose that we have recorded service times for 500 customers in a fast-food restaurant. We can plot this as a frequency histogram where the horizontal axis represents a range of service time values and the vertical axis denotes the number of service times observed in each time range, which would give us some idea of the process condition. The average service time for 500 customers could also tell us something about the process.

Inferential statistics draws conclusions on unknown product or process parameters based on information contained in a sample. Let's say that we want to test the validity of a claim that the average service time in the fast-food restaurant is no more than 3 minutes (min). Suppose we find that the sample average service time (based on a sample of 500 people) is 3.5 min. We then need to determine whether this observed average of 3.5 min is significantly greater than the claimed mean of 3 min.

Such procedures fall under the heading of inferential statistics. They help us draw conclusions about the conditions of a process. They also help us determine whether a process has improved by comparing conditions before and after changes. For example, suppose that the management of the fast-food restaurant is interested in reducing the average time to serve a customer. They decide to add two people to their service staff. Once this change is implemented, they sample 500 customers and find that the average service time is 2.8 min. The question then is whether this decrease is a statistically significant decrease or whether it is due to random variation inherent to sampling. Procedures that address such problems are discussed later.

Data Collection

To control or improve a process, we need information, or data. Data can be collected in several ways. One of the most common methods is throught *direct observation*. Here, a measurement of the quality characteristic is taken by an observer (or automatically by an instrument); for instance, measurements on the depth of tread in automobile tires taken by an inspector are direct observations. On the other hand, data collected on the performance of a particular brand of hair dryer through questionnaires mailed to consumers are *indirect observations*. In this case, the data reported by the consumers have not been observed by the exprimenter, who has no control over the data collection process. Thus, the data may be flawed because errors can arise from a respondent's incorrect interpretation of a question, an error in estimating the satisfactory performance period, or an inconsistent degree of precision among responders' answers.

Data on quality characteristics are described by a **random variable** and are categorized as *continuous* or *discrete*.

Continuous Variable   A variable that can assume any value on a continuous scale within a range is said to be **continuous**. Examples of continuous variables are the hub length of lawn mower tires, the viscosity of a certain resin, the specific gravity of a toner used in photocopying machines, the thickness of a metal plate, and the time to admit a patient to a hospital. Such variables are measurable and have associated numerical values.

Discrete Variable   Variables that can assume a finite or countably infinite number of values are said to be **discrete**. These variables are counts of an event. The number of defective rivets in an assembly is a discrete random variable. Other examples include the number of paint blemishes in an automobile, the number of operating capacitors in an electrical instrument, and the number of satisfied customers in an automobile repair shop.

Counting events usually costs less than measuring the corresponding continuous variables. The discrete variable is merely classified as being, say, unacceptable or not; this can be done through a go/no-go gage, which is faster and cheaper than finding exact measurements. However, the reduced collection cost may be offset by the lack of detailed information in the data.

Sometimes, continuous characteristics are viewed as discrete to allow easier data collection and reduced inspection costs. For example, the hub diameter in a tire is actually a continuous random variable, but rather than precisely measuring the hub diameter numerically, a go/no-go gage is used to quickly identify the characteristic as either acceptable or not. Hence, the *acceptability* of the hub diameter is a discrete random variable. In this case, the goal is not to know the exact hub diameter but rather to know whether it is within certain acceptable limits.

Accuracy and Precision   The **accuracy** of a data set or a measuring instrument refers to the degree of uniformity of the observations around a desired value such that, on average, the target value is realized. Let's assume that the target thickness of a metal plate is 5.25 mm. Figure 4-5a shows observations spread on either side of the target value in almost equal

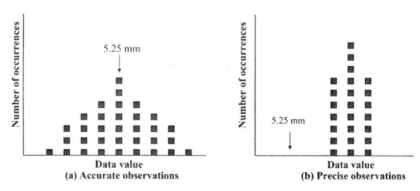

FIGURE 4-5   Accuracy and precision of observations.

proportions; these observations are said to be accurate. Even though individual observations may be quite different from the target value, a data set is considered accurate if the average of a large number of observations is close to the target.

For measuring instruments, accuracy is dependent on **calibration**. If a measuring device is properly calibrated the average output value given by the device for a particular quality characteristic should, after repeated use, equal the true input value.

The **precision** of a data set or a measuring instrument refers to the degree of variability of the observations. Observations may be off the target value but still considered precise, as shown in Figure 4-5b. A sophisticated measuring instrument should show very little variation in output values if a constant value is used multiple times as input. Similarly, sophisticated equipment in a process should be able to produce an output characteristic with as little variability as possible. The precision of the data is influenced by the precision of the measuring instrument. For example, the thickness of a metal plate may be 12.5 mm when measured by callipers; however, a micrometer may yield a value of 12.52 mm, while an optical sensor may give a measurement of 12.523 mm.

Having both accuracy and precision is desirable. In equipment or measuring instruments, accuracy can usually be altered by changing the setting of a certain adjustment. However, precision is an inherent function of the equipment itself and cannot be improved by changing a setting.

Measurement Scales

Four scales of measurement are used to classify data: the nominal, ordinal, interval, and ratio scales. Notice that each scale builds on the previous scale.

1. *Nominal scale.* The scale of measurement is **nominal** when the data variables are simply labels used to identify an attribute of the sample element. Labels can be "conforming and nonconforming" or "critical, major, and minor." Numerical values, even though assigned, are not involved.

2. *Ordinal scale.* The scale of measurement is **ordinal** when the data have the properties of nominal data (i,e., labels) and the data rank or order the observations. Suppose that customers at a clothing store are asked to rate the quality of the store's service. The customers rate the quality according to these responses: 1 (outstanding), 2 (good), 3 (average), 4 (fair), 5 (poor). There are ordinal data. Note that a rating of 1 does not necessarily imply that the service is twice as good as a rating of 2. However, we can say that a rating of 1 is preferable to a rating of 2, and so on.

3. *Interval scale.* The scale of measurement is **interval** when the data have the properties of ordinal data *and* a fixed unit of measure describes the interval between observations. Suppose that we are interested in the temperature of a furnace used in steel smelting. Four readings taken during a 2-hour interval are 2050, 2100, 2150, and 2200 °F. Obviously, these data values ranked (like ordinal data) in ascending order of temperature, indicating the coolest temperature, the next coolest, and so on. Furthermore, the differences between the ranked values can then be compared. Here the interval between the data values 2050 and 2100 represents a 50 °F increase in temperature, as do the intervals between the remaining ranked values.

4. *Ratio scale.* The scale of measurement is **ratio** when the data have the properties of interval data *and* a natural zero exists for the measurement scale. Both the order of, and difference between, observations can be compared and there exists a natural zero

for the measurement scale. Suppose that the weights of four castings are 2.0, 2.1, 2.3, and 2.5 kg. The order (ordinal) of, and difference (interval) in, the weights can be compared. Thus, the increase in weight from 2 to 2.1 is 0.1 kg, which is the same as the increase from 2.3 to 2.4 kg. Also, when we compare the weights of 2.0 and 2.4 kg, we find a meaningful ratio. A casting weighing 2.4 kg is 20% heavier than one weighing 2.0 kg. There is also a natural zero for the scale—0 kg implies no weight.

Measures of Central Tendency

In statistical quality control, data collected need to be described so that analysts can objectively evaluate the process or product characteristic. In this section we describe some of the common numerical measures used to derive summary information from observed values. Measures of central tendency tell us something about the location of the observations and the value about which they cluster and thus help us decide whether the settings of process variables should be changed.

Mean   The **mean** is the simple average of the observations in a data set. In quality control, the mean is one of the most commonly used measures. It is used to determine whether, on average, the process is operating around a desirable target value. The **sample mean**, or average (denoted by $\bar{X}$), is found by adding all observations in a sample and dividing by the number of observations (n) in that sample. If the ith observation is denoted by X_i, the sample mean is calculated as

$$\bar{X} = \frac{\sum_{i=1}^{n} X_i}{n} \tag{4-7}$$

The **population mean** (μ) is found by adding all the data values in the population and dividing by the size of the population (N). It is calculated as

$$\mu = \frac{\sum_{i=1}^{N} X_i}{N} \tag{4-8}$$

The population mean is sometimes denoted as $E(X)$, the expected value of the random variable X. It is also called the *mean of the probability distribution* of X. Probability distributions are discussed later in the chapter.

Example 4-8   A random sample of five observations of the waiting time of customers in a bank is taken. The times (in minutes) are 3, 2, 4, 1, and 2. The sample average ($\bar{X}$), or mean waiting time, is

$$\bar{X} = \frac{3 + 2 + 4 + 1 + 2}{5} = \frac{12}{5} = 2.4 \text{ minutes}$$

The bank can use this information to determine whether the waiting time needs to be improved by increasing the number of tellers.

Median   The **median** is the value in the middle, when the observations are ranked. If there are an even number of observations, the simple average of the two middle numbers is chosen as the median. The median has the property that 50% of the values are less than or equal to it.

Example 4-9

(a) A random sample of 10 observations of piston ring diameters (in millimeters) yields the following values: 52.3, 51.9, 52.6, 52.4, 52.4, 52.1, 52.3, 52.0, 52.5 and 52.5. We first rank the observations:

$$51.9 \quad 52.0 \quad 52.1 \quad 52.3 \quad 52.3$$
$$52.4 \quad 52.4 \quad 52.5 \quad 52.5 \quad 52.6$$

The observations in the middle are 52.3 and 52.4. The median is (52.3 + 52.4)/2 or 52.35.

The median is less influenced by the extreme values in the data set; thus, it is said to be more "robust" than the mean.

(b) A department store is interested in expanding its facilities and wants to do a preliminary analysis of the number of customers it serves. Five weeks are chosen at random, and the number of customers served during those weeks were as follows:

$$3000, \quad 3500, \quad 500, \quad 3300, \quad 3800$$

The median number of customers is 3300, while the mean is 2820. On further investigation of the week with 500 customers, it is found that a major university, whose students frequently shop at the store, was closed for spring break. In this case, the median (3300) is a better measure of central tendency than the mean (2820) because it gives a better idea of the variable of interest. In fact, had the data value been 100 instead of 500, the median would still be 3300, although the mean would decrease further, another demonstration of the robustness of the median.

Outliers (values that are very large or very small compared to the majority of the data points) can have a significant influence on the mean, which is pulled toward the outliers. Figure 4-6 demonstrates the effect of outliers.

Mode   The **mode** is the value that occurs most frequently in the data set. It denotes a "typical" value from the process.

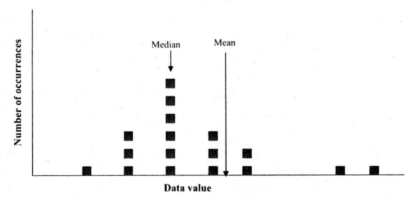

FIGURE 4-6   Effect of outliers on a mean.

Example 4-10   A hardware store wants to determine what size of circular saws it should stock. From past sales data, a random sample of 30 shows the following sizes (in millimeters):

80	120	100	100	150	120	80	150	120	80
120	100	120	120	150	80	120	100	120	80
100	120	120	150	120	100	120	120	100	100

Note that the mode has the highest frequency. In this case, the mode is 120 (13 is the largest number of occurrences). So, the manager may decide to stock more size 120 saws. A data set can have more than one mode, in which case it is said to be *multimodal*.

Trimmed Mean   The **trimmed mean** is a robust estimator of the central tendency of a set of observations. It is obtained by calculating the mean of the observations that remain after a proportion of the high and low values have been deleted. The $\alpha\%$ trimmed mean, denoted by $T(\alpha)$, is the average of the observations that remain after trimming (or deleting) $\alpha\%$ of the high observations and $\alpha\%$ of the low observations. This is a suitable measure when it is believed that existing outliers do not represent usual process characteristics. Thus, analysts will sometimes trim extreme observations caused by a faulty measurement process to obtain a better estimate of the population's central tendency.

Example 4-11   The time taken for car tune-ups (in minutes) is observed for 20 randomly selected cars. The data values are as follows:

15	10	12	20	16	18	30	14	16	15
18	40	20	19	17	15	22	20	19	22

To find the 5% trimmed mean [i.e., $T(0.05)$], first rank the data in increasing order:

10	12	14	15	15	15	16	16	17	18
18	19	19	20	20	20	22	22	30	40

The number of observations (high and low) to be deleted on each side is $(20)(0.05) = 1$. Delete the lowest observation (10) and the highest one (40). The trimmed mean (of the remaining 18 observations) is 18.222, which is obviously more robust than the mean (18.9). For example, if the largest observation of 40 had been 60, the 5% trimmed mean would still be 18.222. However, the untrimmed mean would jump to 19.9.

Measures of Dispersion

An important function of quality control and improvement is to analyze and reduce the variability of a process. The numerical measures of location we have described give us indications of the central tendency, or middle, of a data set. They do not tell us much about the variability of the observations. Consequently, sound analysis requires an understanding of measures of dispersion, which provide information on the variability, or scatter, of the observations around a given value (usually, the mean).

Range   A widely used measure of dispersion in quality control is the **range**, which is the difference between the largest and smallest values in a data set. Notationally, the range R is defined as

$$R = X_L - X_S \tag{4-9}$$

where X_L is the largest observation and X_S is the smallest observation.

Example 4-12   The following 10 observations of the time to receive baggage after landing are randomly taken in an airport. The data values (in minutes) are as follows:

$$15, \quad 12, \quad 20, \quad 13, \quad 22, \quad 18, \quad 19, \quad 21, \quad 17, \quad 20$$

The range $R = 22 - 12 = 10$ minutes. This value gives us an idea of the variability in the observations. Management can now decide whether this spread is acceptable.

Variance   The **variance** measures the fluctuation of the observations around the mean. The larger the value, the greater the fluctuation. The population variance σ^2 is given by

$$\sigma^2 = \frac{\sum\limits_{i=1}^{N}(X_i - \mu)^2}{N} \tag{4-10}$$

where μ is the population mean and N represents the size of the population. The sample variance s^2 is given by

$$s^2 = \frac{\sum\limits_{i=1}^{n}(X_i - \bar{X})^2}{n-1} \tag{4-11}$$

where $\bar{X}$ is the sample mean and n is the number of observations in the sample. In most applications, the sample variance is calculated rather than the population variance because calculation of the latter is possible only when every value in the population is known.

A modified version of eq. (4-11) for calculating the sample variance is

$$s^2 = \frac{\sum\limits_{i=1}^{n}X_i^2 - \left(\sum\limits_{i=1}^{n}X_i\right)^2 \Big/ n}{n-1} \tag{4-12}$$

This version is sometimes easier to use. It involves accumulating the sum of the observations and the sum of squares of the observations as data values become available. These two equations are algebraically equivalent.

Note that in calculating the sample variance, the denominator is $n - 1$, whereas for the population variance the denominator is N. Thus, eq. (4-10) can be interpreted as the average of the squared deviations of the observations from the mean, and eq. (4-11) can be interpreted similarly, except for the difference in the denominator, where $(n - 1)$ is used instead of n. This difference can be explained as follows. First, a population variance σ^2 is a parameter, whereas a sample variance s^2 is an estimator, or a statistic. The value of s^2 can therefore change from sample to sample, whereas σ^2 should be constant.

One desirable property of s^2 is that even though it may not equal σ^2 for every sample, on average s^2 does equal σ^2. This is known as the *property of unbiasedness*, where the mean or

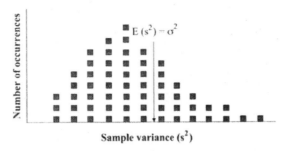

FIGURE 4-7 Sampling distribution of an unbiased sample variance.

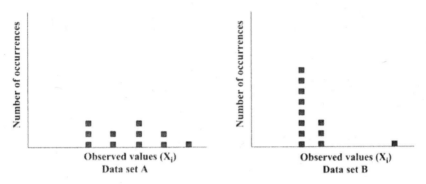

FIGURE 4-8 Variablity of two data sets with the same range.

expected value of the estimator equals the corresponding parameter. If a denominator of $n - 1$ is used to calculate the sample variance, it can be shown that the sample variance is an unbiased estimator of the population variance. On the other hand, if a denominator of n is used, on average the sample variance underestimates the population variance.

Figure 4-7 denotes the sampling distribution (i.e., the relative frequency) of s^2 calculated using eq. (4-11) or (4-12) over repeated samples. Suppose that the value of σ^2 is as shown. If the average value of s^2 is calculated over repeated samples, it will equal the population variance. Technically, the expected value of s^2 will equal σ^2 [i.e., $E(s^2) = \sigma^2$]. If a denominator of n is used in eq. (4-11) or (4-12), $E(s^2)$ will be less than σ^2.

Unlike the range, which uses only the extreme values of the data set, the sample variance incorporates every observation in the sample. Two data sets with the same range can have different variability. As Figure 4-8 shows, data sets A and B have the same range; however, their degree of variability is quite different. The sample variances will thus indicate different degrees of fluctuation around the mean. The units of variance are the square of the units of measurement for the individual values. For example, if the observations are in millimeters, the units of variance are square millimeters.

Standard Deviation Like the variance, the **standard deviation** measures the variability of the observations around the mean. It is equal to the positive square root of the variance. A standard deviation has the same units as the observations and is thus easier to interpret. It is probably the most widely used measure of dispersion in quality control. Using eq. (4-10), the population standard deviation is given by

$$\sigma = \sqrt{\frac{\sum\limits_{i=1}^{N}(X_i - \mu)^2}{N}} \tag{4-13}$$

Similarly, the sample standard deviation s is found using eq. (4-11) or (4-12) as

$$s = \sqrt{\frac{\sum\limits_{i=1}^{n}(X_i - \bar{X})^2}{n-1}} \tag{4-14}$$

$$= \sqrt{\frac{\sum\limits_{i=1}^{n}X_i^2 - \left(\sum\limits_{i=1}^{n}X_i\right)^2 \big/ n}{n-1}} \tag{4-15}$$

As with the variance, the data set with the largest standard deviation will be identified as having the most variability about its average. If the probability distribution of the random variable is known—a normal distribution, say—the proportion of observations within a certain number of standard deviations of the mean can be obtained. Techniques for obtaining such information are discussed in Section 4-6.

Example 4-13   A random sample of 10 observations of the output voltage of transformers is taken. The values (in volts, V) are as follows:

$$9.2, \quad 8.9, \quad 8.7, \quad 9.5, \quad 9.0, \quad 9.3, \quad 9.4, \quad 9.5, \quad 9.0, \quad 9.1$$

Using eq. (4-7), the sample mean $\bar{X}$ is 9.16 V.

Table 4-1 shows the calculations. From the table $\sum(X_i - \bar{X})^2 = 0.644$. The sample variance is given by

$$s^2 = \frac{\sum(X_i - \bar{X})^2}{n-1} = \frac{0.644}{9} = 0.0716\,\text{V}^2$$

TABLE 4-1   Calculation of Sample Variance Using Eq. (4-11) or (4-12)

X_i	X_i^2	Deviation from Mean, $X_i - \bar{X}$	Squared Deviation, $(X_i - \bar{X})^2$
9.2	84.64	0.04	0.0016
8.9	79.21	−0.26	0.0676
8.7	75.69	−0.46	0.2116
9.5	90.25	0.34	0.1156
9.0	81.00	−0.16	0.0256
9.3	86.49	0.14	0.0196
9.4	88.36	0.24	0.0576
9.5	90.25	0.34	0.1156
9.0	81.00	−0.16	0.0256
9.1	81.81	−0.06	0.0036
$\sum X_i = 91.60$	$\sum X_i^2 = 839.70$	$\sum(X_i - \bar{X}) = 0$	$\sum(X_i - \bar{X})^2 = 0.644$

The sample standard deviation given by eq. (4-14) is

$$s = \sqrt{0.0716} = 0.2675 \text{ V}$$

Next, using eq. (4-12) (the calculations are shown in Table 4-1), the sample variance is given by

$$s^2 = \frac{\sum X_i^2 - \left(\sum X_i\right)^2/n}{n-1}$$

$$= \frac{839.70 - (91.60)^2/10}{9}$$

$$= 0.0716 \text{ V}^2$$

The sample standard deviation s is 0.2675 V, as before.

Interquartile Range   The lower quartile, Q_1, is the value such that one-fourth of the observations fall below it and three-fourths fall above it. The middle quartile is the median—half the observations fall below it and half above it. The third quartile, Q_3, is the value such that three-fourths of the observations fall below it and one-fourth above it.

The **interquartile range** (IQR) is the difference between the third quartile and the first quartile. Thus,

$$\text{IOR} = Q_3 - Q_1 \tag{4-16}$$

Note from Figure 4-9 that IQR contains 50% of the observations. The larger the IQR value, the greater the spread of the data. To find the IQR, the data are ranked in ascending order. Q_1 is located at rank $0.25(n+1)$, where n is the number of data points in the sample. Q_3 is located at rank $0.75(n+1)$.

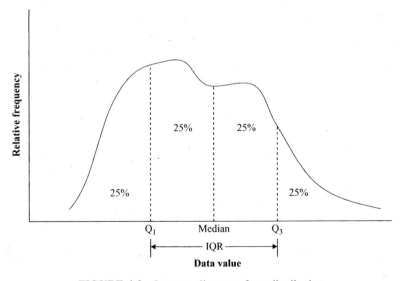

FIGURE 4-9   Interquartile range for a distribution.

Example 4-14   A random sample of 20 observations on the welding time (in minutes) of an operation gives the following values:

$$
\begin{array}{cccccccccc}
2.2 & 2.5 & 1.8 & 2.0 & 2.1 & 1.7 & 1.9 & 2.6 & 1.8 & 2.3 \\
2.0 & 2.1 & 2.6 & 1.9 & 2.0 & 1.8 & 1.7 & 2.2 & 2.4 & 2.2
\end{array}
$$

First let's find the locations of Q_1 and Q_3:

$$
\text{location of } Q_1 = 0.25(n+1) = (0.25)(21) = 5.25
$$
$$
\text{location of } Q_3 = 0.75(n+1) = (0.75)(21) = 15.75
$$

Now let's rank the data values:

Q_1's location $= 5.25$ Q_3's location $= 15.75$

Rank	1	2	3	4	5	6	7	8	9	10	11	12	13	14	15	16	17	18	19	20
Data Value	1.7	1.7	1.8	1.8	1.8	1.9	1.9	2.0	2.0	2.0	2.1	2.1	2.2	2.2	2.2	2.3	2.4	2.5	2.6	2.6

$$Q_1 = 1.825 \qquad\qquad\qquad Q_3 = 2.275$$

Thus, linear interpolation yields a Q_1 of 1.825 and a Q_3 of 2.275. The interquartile range is then

$$
\begin{aligned}
\text{IQR} &= Q_3 - Q_1 \\
&= 2.275 - 1.825 = 0.45 \text{ minute}
\end{aligned}
$$

Measures of Skewness and Kurtosis

In addition to central tendency and dispersion, two other measures are used to describe data sets: the skewness coefficient and the kurtosis coefficient.

Skewness Coefficient   The **skewness coefficient** describes the asymmetry of the data set about the mean. The skewness coefficient is calculated as follows:

$$
\gamma_1 = \frac{n}{(n-1)(n-2)} \frac{\sum_{i=1}^{n} (X_i - \bar{X})^3}{s^3} \tag{4-17}
$$

In Figure 4-10, part a is a negatively skewed distribution (skewed to the left), part b is positively skewed (skewed to the right), and part c is symmetric about the mean. The skewness coefficient is zero for a symmetric distribution, because (as shown in part c) the mean and the median are equal. For a positively skewed distribution, the mean is greater than the median because a few values are large compared to the others; the skewness coefficient will be a positive number. If a distribution is negatively skewed, the mean is less than the median because the outliers are very small compared to the other values, and the skewness coefficient will be negative. The skewness coefficient indicates the degree to which a distribution deviates from symmetry. It is used for data sets that are unimodal (that is, have one mode) and have a sample size of at least 100. The larger the magnitude of the skewness coefficient, the stronger the case for rejecting the notion that the distribution is symmetric.

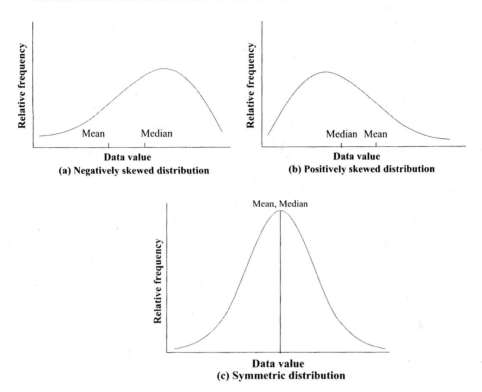

FIGURE 4-10   Symmetric and skewed distributions.

Kurtosis Coefficient   **Kurtosis** is a measure of the peakedness of the data set. It is also viewed as a measure of the "heaviness" of the tails of a distribution. The kurtosis coefficient is given by

$$\gamma_2 = \frac{n(n+1)}{(n-1)(n-2)(n-3)} \frac{\sum_{i=1}^{n}(X_i - \bar{X})^4}{s^4} - \frac{3(n-1)^2}{(n-2)(n-3)} \qquad (4\text{-}18)$$

The **kurtosis coefficient** is a relative measure. For normal distributions (discussed in depth later), the kurtosis coefficient is zero. Figure 4-11 shows a normal distribution (*mesokurtic*), a distribution that is more peaked than the normal (*leptokurtic*), and one that is less peaked than the normal (*platykurtic*). For a leptokurtic distribution, the kurtosis coefficient is greater than zero. The more pronounced the peakedness, the larger the value of the kurtosis coefficient. For platykurtic distributions, the kurtosis coefficient is less than zero. The kurtosis coefficient should only be used to make inferences on a data set when the sample size is at least 100 and the distribution is unimodal.

Example 4-15   A sample of 50 coils to be used in an electrical circuit is randomly selected, and the resistance of each is measured (in ohms). From the data in Table 4-2, the sample mean $\bar{X}$ is found to be

$$\bar{X} = \frac{1505.5}{50} = 30.11$$

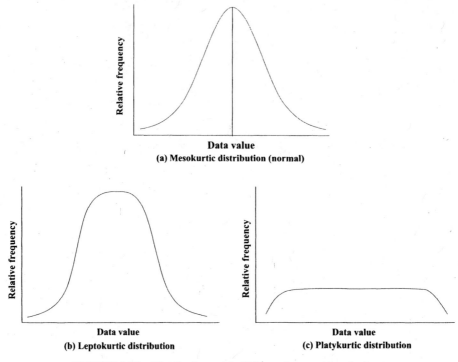

FIGURE 4-11   Distributions with different degrees of peakedness.

TABLE 4-2   Resistance of Coil Data

Observation, i	Resistance, X_i	Observation, i	Resistance, X_i	Observation, i	Resistance, X_i
1	35.1	18	25.8	35	31.4
2	35.4	19	26.4	36	28.5
3	36.3	20	25.6	37	28.4
4	38.8	21	33.1	38	27.6
5	39.0	22	33.6	39	27.6
6	22.5	23	32.3	40	28.2
7	23.7	24	32.6	41	30.8
8	25.0	25	32.2	42	30.6
9	25.3	26	27.5	43	30.4
10	25.0	27	26.5	44	30.5
11	34.7	28	26.9	45	30.5
12	34.2	29	26.7	46	28.5
13	34.4	30	27.2	47	30.2
14	34.7	31	31.8	48	30.1
15	34.3	32	32.1	49	30.0
16	26.4	33	31.5	50	28.9
17	25.5	34	31.2		

Using the data, the following values are obtained:

$$\sum (X_i - \bar{X})^2 = 727.900 \quad \sum (X_i - \bar{X})^3 = 736.321 \quad \sum (X_i - \bar{X})^4 = 26151.892$$

The sample standard deviation is calculated as

$$s = \sqrt{\frac{727.900}{49}} = 3.854$$

The skewness coefficient is

$$\gamma_1 = \frac{50}{(49)(48)} \frac{736.321}{(3.854)^3} = 0.273$$

The positive skewness coefficient indicates that the distribution of the data points is slightly skewed to the right. The kurtosis coefficient is

$$\gamma_2 = \frac{(50)(51)}{(49)(48)(47)} \frac{26,151.892}{(3.854)^4} - \frac{(3)(49)^2}{(48)(47)} = -0.459$$

This implies that the given distribution is less peaked than a normal distribution.

Example 4-16 For the coil resistance data shown in Table 4-2, all of the previous descriptive statistics can be obtained using Minitab. The commands to be used are **Stat > Basic Statistics > Display Descriptive Statistics**. In the **Variables** box, enter the column number or name of the variable (in this case, **Resistance**). By clicking on the **Statistics** button in the dialog box, you may choose the statistics to be displayed. Click **OK**. The output is shown in Figure 4-12.

The mean and the standard deviation of the resistance of coils are 30.11 and 3.854, respectively, and the median is 30.3. The minimum and maximum values are 22.5 and 39.0, respectively, and the interquartile range (IQR) is 5.875. Skewness and kurtosis coefficients are 0.27 and −0.46, matching the previously computed values.

Measures of Association

Measures of association indicate how two or more variables are related to each other. For instance, as one variable increases, how does it influence another variable? Small values of the measures of association indicate a nonexistent or weak relationship between the variables, and large values indicate a strong relationship.

```
Variable     N     Mean  SE Mean  StDev  Variance  Minimum     Q1  Median
Resistance  50   30.110    0.545  3.854    14.855   22.500  26.850  30.300

Variable     Q3  Maximum   Range    IQR  Skewness  Kurtosis
Resistance  32.725  39.000  16.500  5.875      0.27     -0.46
```

FIGURE 4-12 Descriptive statistics for coil resistance using Minitab.

Correlation Coefficient   A **correlation coefficient** is a measure of the strength of the linear relationship between two variables in bivariate data. If two variables are denoted by X and Y, the correlation coefficient r of a sample of observations is found from

$$r = \frac{\sum_{i=1}^{n}(X_i - \bar{X})(Y_i - \bar{Y})}{\sqrt{\sum_{i=1}^{n}(X_i - \bar{X})^2}\sqrt{\sum_{i=1}^{n}(Y_i - \bar{Y})^2}} \tag{4-19}$$

where X_i and Y_i denote the coordinates of the ith observation, $\bar{X}$ is the sample mean of the X_i-values, $\bar{Y}$ is the sample mean of the Y_i-values, and n is the sample size. An alternative version for calculating the sample correlation coefficient is

$$r = \frac{\sum X_i Y_i - (\sum X_i)(\sum Y_i)/n}{\sqrt{[\sum X_i^2 - (\sum X_i)^2/n][\sum Y_i^2 - (\sum Y_i)^2/n]}} \tag{4-20}$$

The sample correlation coefficient r is always between -1 and 1. An r-value of 1 denotes a perfect positive linear relationship between X and Y. This means that as X increases, Y increases linearly and that as X decreases, Y decreases linearly. Similarly, an r-value of -1 indicates a perfect negative linear relationship between X and Y. If the value of r is zero, the two variables X and Y are uncorrelated, which implies that if X increases, we cannot really say how Y would change. A value of r that is close to zero thus indicates that the relationship between the variables is weak.

Figure 4-13 shows plots of bivariate data with different degrees of strength of the linear relationship. Figure 4-13a shows a perfect positive linear relationship between X and Y. As X increases, Y very definitely increases linearly, and vice versa. In Figure 4-13c, X and Y are positively correlated (say, with a correlation coefficient of 0.8) but not perfectly related. Here, on the whole, as X increases, Y tends to increase, and vice versa. Similar analogies can be drawn for Figure 4-13b and d, where X and Y are negatively correlated. In Figure 4-13e, note that it is not evident what happens to Y as X increases or decreases. No general trend can be established from the plot, and X and Y are either uncorrelated or very weakly correlated. Statistical tests are available for testing the significance of the sample correlation coefficient and for determining if the population correlation coefficient is significantly different from zero (Neter et al. 2005), as discussed later in the chapter.

Example 4-17   Consider the data shown in Table 4-3 on the depth of cut and tool wear (in millimeters) in a milling operation. To find the strength of linear relationship between these two variables, we need to compute the correlation coefficient.

Using Minitab, choose **Stat > Basic Statistics > Correlation**. In **Variables**, enter the column number or names of the two variables, Depth and Tool Wear. Click **OK**. The output from Minitab shows the correlation coefficient as 0.915, indicating a strong positive linear relationship between depth of cut and tool wear. Further, Minitab tests the null hypothesis (H_0: $\rho = 0$) versus the alternative hypothesis (H_a: $\rho \neq 0$), where ρ denotes the population correlation coefficient and reports a probability value (***p*-value**) of 0.000. As we discuss in Section 4-7, if the p-value is small relative to a chosen level of significance, we reject the null

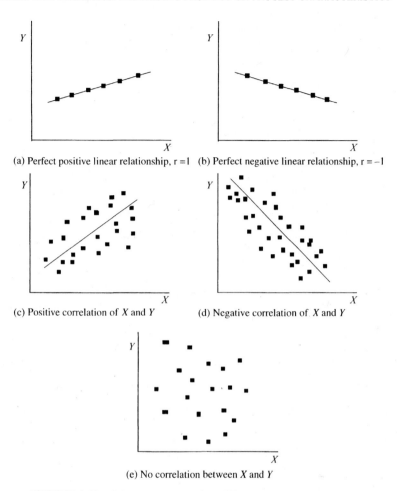

FIGURE 4-13   Scatterplots indicating different degrees of correlation.

hypothesis. So, here, for a chosen level of significance (α) of 0.05 (say), we reject the null hypothesis.

Example 4-18   Now let's use the milling operation data in Table 4-3 to obtain summary descriptive statistics. The data are entered in two columns for the variables depth of cut and tool wear. From the Minitab Windows menu, choose **Stat** > **Basic Statistics** > **Graphical Summary**. Now, select the variable you want to describe—say, depth of cut.

A sample Minitab output of the graphical summary is shown in Figure 4-14. The mean and standard deviation of depth of cut are 3.527 and 1.138, indicating location and dispersion measures, respectively. The first and third quartiles are 2.45 and 4.45, respectively, yielding an interquartile range of 2.0, within which 50% of the observations are contained. Skewness and kurtosis coefficients are also shown. The skewness coefficient is -0.034, indicating that the distribution is close to being symmetrical, although slightly negatively skewed. A kurtosis value of -1.085 indicates that the distribution is less peaked than the normal distribution, which would have a value of zero. The graphical

TABLE 4-3   Milling Operation Data

Observation, i	Depth of Cut, X_i	Tool Wear, Y_i	Observation, i	Depth of Cut, X_i	Tool Wear, Y_i
1	2.1	0.035	21	5.6	0.073
2	4.2	0.041	22	4.7	0.064
3	1.5	0.031	23	1.9	0.030
4	1.8	0.027	24	2.4	0.029
5	2.3	0.033	25	3.2	0.039
6	3.8	0.045	26	3.4	0.038
7	2.6	0.038	27	3.8	0.040
8	4.3	0.047	28	2.2	0.031
9	3.4	0.040	29	2.0	0.033
10	4.5	0.058	30	2.9	0.035
11	2.6	0.039	31	3.0	0.032
12	5.2	0.056	32	3.6	0.038
13	4.1	0.048	33	1.9	0.032
14	3.0	0.037	34	5.1	0.052
15	2.2	0.028	35	4.7	0.050
16	4.6	0.057	36	5.2	0.058
17	4.8	0.060	37	4.1	0.048
18	5.3	0.068	38	4.3	0.049
19	3.9	0.048	39	3.8	0.042
20	3.5	0.036	40	3.6	0.045

summary yields four plots. The first is a frequency distribution with the normal curve superimposed on it. The confidence intervals shown are discussed in Section 4-7, as is the p-value and the hypothesis testing associated with it. The box plot shown in Figure 4-14 is discussed in Chapter 5.

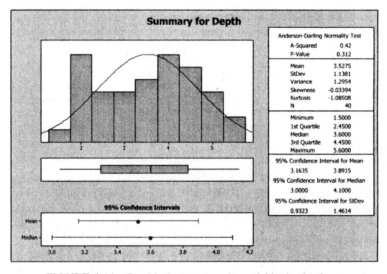

FIGURE 4-14   Graphical summary of a variable depth of cut.

4-6 PROBABILITY DISTRIBUTIONS

Sample data can be described with frequency histograms or variations thereof (such as relative frequency or cumulative frequency, which we discuss later). Data values in a population are described by a probability distribution. As noted previously, random variables may be discrete or continuous. For discrete random variables, a probability distribution shows the values that the random variable can assume and their corresponding probabilities. Some examples of discrete random variables are the number of defects in an assembly, the number of customers served over a period of time, and the number of acceptable compressors.

Continuous random variables can take on an infinite number of values, so the probability distribution is usually expressed as a mathematical function of the random variable. This function can be used to find the probability that the random variable will be between certain bounds. Almost all variables for which numerical measurements can be obtained are continuous in nature: for example, the length of a pin, the diameter of a bolt, the tensile strength of a cable, or the specific gravity of a liquid.

For a discrete random variable X, which takes on the values x_1, x_2, and so on, a **probability distribution function** $p(x)$ has the following properties:

1. $p(x_i) \geq 0$ for all i, where $p(x_i) = P(X = x_i), i = 1, 2, \ldots$

2. $\sum_{\text{all } i} p(x_i) = 1$

When X is a continuous random variable, the **probability density function** is represented by $f(x)$, which has the following properties:

1. $f(x) \geq 0$ for all x, where $P(a \leq x \leq b) = \int_a^b f(x)\, dx$

2. $\int_{-\infty}^{\infty} f(x)dx = 1$

Note the similarity of these two properties to those for discrete random variables.

Example 4-19 Let X denote a random variable that represents the number of defective solders in a printed circuit board. The probability distribution of the discrete random variable X may be given by

x	0	1	2	3
$p(x)$	0.3	0.4	0.2	0.1

This table gives the values taken on by random variable and their corresponding probabilities. For instance, $P(X = 1) = 0.4$; that is, there is a 40% chance of finding one defective solder. A graph of the probability distribution of this discrete random variable is shown in Figure 4-15.

Example 4-20 Consider a continuous random variable X representing the time taken to assemble a part. The variable X is known to be between 0 and 2 minutes, and its probability

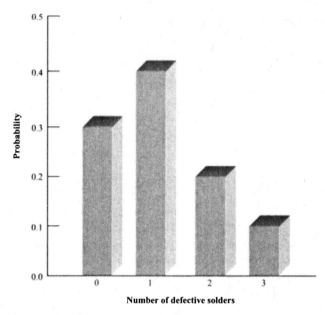

FIGURE 4-15   Probability distribution of a discrete random variable.

density function (pdf), $f(x)$, is given by

$$f(x) = \frac{x}{2}, \quad 0 < x \le 2$$

The graph of this probability density function is shown in Figure 4-16. Note that

$$\int_0^2 f(x)dx = \int_0^2 \frac{x}{2}dx = 1$$

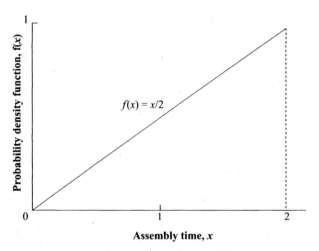

FIGURE 4-16   Probability density function, $0 < x \le 2$.

The probability that X is between 1 and 2 is

$$P(1 \leq X \leq 2) = \int_1^2 \frac{x}{2} dx = \frac{3}{4}$$

Cumulative Distribution Function

The **cumulative distribution function (cdf)** is usually denoted by $F(x)$ and represents the probability of the random variable X taking on a value less than or equal to x, that is

$$F(x) = P(X \leq x)$$

For a discrete random variable,

$$F(x) = \sum_{\text{all } i} p(x_i) \quad \text{for} \quad x_i \leq x \tag{4-21}$$

If X is a continuous random variable,

$$F(x) = \int_{-\infty}^{x} f(t) dt \tag{4-22}$$

Note that $F(x)$ is a nondecreasing function of x such that

$$\lim_{x \to \infty} F(x) = 1 \quad \text{and} \quad \lim_{x \to -\infty} F(x) = 0$$

Expected Value

The **expected value** or mean of a distribution is given by

$$\mu = E(X) = \sum_{\text{all } i} x_i p(x_i) \quad \text{if } X \text{ is discrete} \tag{4-23}$$

and

$$\mu = E(X) = \int_{-\infty}^{\infty} x f(x) dx \quad \text{if } X \text{ is continuous} \tag{4-24}$$

The variance of a random variable X is given by

$$\begin{aligned} \text{Var}(X) &= E[(X - \mu)^2] \\ &= E(X^2) - [E(X)]^2 \end{aligned} \tag{4-25}$$

Example 4-21 For the probability distribution of Example 4-19, regarding the defective solders, the mean μ or expected value $E(X)$ is given by

$$\begin{aligned} \mu = E(X) &= \sum_{\text{all } i} x_i p(x_i) \\ &= (0)(0.3) + (1)(0.4) + (2)(0.2) + (3)(0.1) = 1.1 \end{aligned}$$

The variance of X is

$$\sigma^2 = \text{Var}(X) = E(X^2) - [E(X)]^2$$

First, $E(X^2)$ is calculated as follows:

$$E(X^2) = \sum_{\text{all } i} x_i^2 p(x_i)$$
$$= (0)^2(0.3) + (1)^2(0.4) + (2)^2(0.2) + (3)^2(0.1) = 2.1$$

So

$$\text{Var}(X) = 2.1 - (1.1)^2 = 0.89$$

Hence, the standard deviation of X is $\sigma = \sqrt{0.89} = 0.943$.

Example 4-22   For the probability distribution function in Example 4-20, regarding a part's assembly time, the mean μ, or expected value $E(X)$, is given by

$$E(X) = \int_{-\infty}^{\infty} xf(x)dx = \int_{0}^{2} x\frac{x}{2}dx$$
$$= \frac{2^3}{6} = 1.333 \text{ minutes}$$

Thus, the mean assembly time for this part is 1.333 minutes.

Discrete Distributions

The discrete class of probability distributions deals with those random variables that can take on a finite or countably infinite number of values. Several **discrete distributions** have applications in quality control, three of which are discussed in this section (Montgomery, 2004).

Hypergeometric Distribution   A **hypergeometric distribution** is useful in sampling from a finite population (or lot) without replacement (i.e., without placing the sample elements back in the population) when the items or outcomes can be categorized into one of two groups (usually called *success* and *failure*). If we consider finding a nonconforming item a success, the probability distribution of the number of nonconforming items (x) in the sample is given by

$$p(x) = \frac{\binom{D}{x}\binom{N-D}{n-x}}{\binom{N}{n}}, \qquad x = 0, 1, 2, \ldots, \min(n, D) \qquad (4\text{-}26)$$

where $D =$ number of nonconforming items in the population, $N =$ size of the population, $n =$ size of the sample, $x =$ number of nonconforming items in the sample, and $\binom{D}{x} =$ combination of D items taken x at a time, $D!/(x!(D-x)!)$.

The factorial of a positive integer x is written as $x! = x(x-1)(x-2)\ldots3\cdot2\cdot1$, and $0!$ is defined to be 1. The mean (or expected value) of a hypergeometric distribution is given by

$$\mu = E(X) = \frac{nD}{N} \tag{4-27}$$

The variance of a hypergeometric random variable is given by

$$\sigma^2 = \text{Var}(X) = \frac{nD}{N}\left(1 - \frac{D}{N}\right)\frac{N-n}{N-1} \tag{4-28}$$

Example 4-23   A lot of 20 chips contains 5 nonconforming ones. If an inspector randomly samples 4 items, find the probability of 3 nonconforming chips.

Solution   In this problem, $N = 20$, $D = 5$, $n = 4$, and $x = 3$.

$$P(X = 3) = \frac{\binom{5}{3}\binom{15}{1}}{\binom{20}{4}} = 0.031$$

Using Minitab, click on **Calc > Probability Distributions > Hypergeometric**. The option exists to select one of the following three: **Probability**, which represents the individual probability of the variable taking on a chosen value (x) [i.e., $P(X=x)$]; **Cumulative probability**, which represents $P(X \leq x)$; and **Inverse cumulative probability**, which represents the value a such that $P(X \leq a) = b$, where b is the specified cumulative probability.

Here, we select **Probability**. Input the **Population size** as 20, **Event count in population** as 5, and **Sample size** as 4. Select **Input constant** and input the value of 3. Click **OK**. Minitab outputs $P(X = 3) = 0.03096$.

Binomial Distribution   Consider a series of independent trials where each trial results in one of two outcomes. These outcomes are labeled as either a success or a failure. The probability p of success on any trial is assumed to be constant. Let X denote the number of successes if n such trials are conducted. Then the probability of x successes is given by

$$p(x) = \binom{n}{x}p^x(1-p)^{n-x}, \qquad x = 0, 1, 2, \ldots, n \tag{4-29}$$

and X is said to have a *binomial distribution*. The mean of the binomial random variable is given by

$$\mu = E(X) = np \tag{4-30}$$

and the variance is expressed as

$$\sigma^2 = \text{Var}(X) = np(1-p) \tag{4-31}$$

A **binomial distribution** is a distribution using the two parameters n and p. If the values of these parameters are known, all information associated with the binomial distribution can be determined. Such a distribution is applicable to sampling without replacement from a population (or lot) that is large compared to the sample, or to sampling with replacement from a finite population. It is also used for situations in which items are selected from an ongoing process (i.e., the population size is very large). Tables of cumulative binomial probabilities are shown in Appendix A-1.

Example 4-24   A manufacturing process is estimated to produce 5% nonconforming items. If a random sample of five items is chosen, find the probability of getting two nonconforming items.

   Solution   Here, $n = 5, p = 0.05$ (if success is defined as getting a nonconforming item), and $x = 2$.

$$P(X = 2) = \binom{5}{2}(0.05)^2(0.95)^3 = 0.021$$

This probability may be checked using Appendix A-1.

$$
\begin{aligned}
P(X = 2) &= P(X \leq 2) - P(X \leq 1) \\
&= 0.999 - 0.977 = 0.022
\end{aligned}
$$

The discrepancy between the two values is due to rounding the values of Appendix A-1 to three decimal places. Using Appendix A-1, the complete probability distribution of X, the number of nonconforming items, may be obtained:

x	0	1	2	3	4	5
$p(x)$	0.774	0.203	0.022	0.001	0.000	0.000

The expected number of nonconforming items in the sample is

$$\mu = E(X) = (5)(0.05) = 0.25 \text{ item}$$

while the variance is

$$\sigma^2 = (5)(0.05)(0.95) = 0.2375 \text{ item}^2$$

   To generate the probability distribution of X using Minitab, first create a worksheet by using a column to represent the values of the random variable X (i.e., 0, 1, 2, 3, 4 and 5). Click on **Calc > Probability Distributions > Binominal**. Select **Probability**, and input the **Number of trials** as 5 and **Event probability** as 0.05. Select **Input column**, and input the column number or name of the variable X. Click **OK**. Minitab outputs the probability distribution of X, which matches the values calculated previously.

   The major differences between binomial and hypergeometric distributions are as follows: The trials are independent in a binomial distribution, whereas they are not in a hypergeometric one; the probability of success on any trial remains constant in a binomial distribution but not so in a hypergeometric one. A hypergeometric distribution approaches a binomial distribution as $N \to \infty$ and D/N remains constant.

   The proportion of nonconforming items in sample is frequently used in statistical quality control. This may be expressed as

$$\hat{p} = \frac{x}{n}$$

where X has a binomial distribution with parameters n and p, and x denotes an observed value of X. The probability distribution of $\hat{p}$ is obtained using

$$P(\hat{p} \le a) = P\left(\frac{x}{n} \le a\right) = P(x \le na)$$

$$= \sum_{x=0}^{\lceil na \rceil} \binom{n}{x} p^x (1-p)^{n-x} \tag{4-32}$$

where $\lceil na \rceil$ is the largest integer less than or equal to na. It can be shown that the mean of $\hat{p}$ is p and that the variance of $\hat{p}$ is given by

$$\text{Var}(\hat{p}) = \frac{p(1-p)}{n}$$

Poisson Distribution   A **Poisson distribution** is used to model the number of events that happen within a product unit (number of defective rivets in an airplane wing), space or volume (blemishes per 200 square meters of fabric), or time period (machine breakdowns per month). It is assumed that the events happen randomly and independently.

The Poisson random variable is denoted by X. An observed value of X is represented by x. The probability distribution (or mass) function of the number of events (x) is given by

$$p(x) = \frac{e^{-\lambda}\lambda^x}{x!} \qquad x = 0, 1, 2, \ldots \tag{4-33}$$

where λ is the mean or average number of events that happen over the product, volume, or time period specified. The symbol e represents the base of natural logarithms, which is equal to about 2.7183. The Poisson distribution has one parameter, λ. The mean and the variance of Poisson distribution are equal and are given by

$$\mu = \sigma^2 = \lambda \tag{4-34}$$

The Poisson distribution is sometimes used as an approximation to the binomial distribution when n is large ($n \to \infty$) and p is small ($p \to 0$), such that $np = \lambda$ is constant. That is, a Poisson distribution can be used when all of the following hold:

1. The number of possible occurrences of defects or nonconformities per unit is large.
2. The probability or chance of a defect or nonconformity happening is small ($p \to 0$).
3. The average number of defects or nonconformities per unit is constant.

Appendix A-2 lists cumulative Poisson probabilities for various values of λ

Example 4-25   It is estimated that the average number of surface defects in 20 m^2 of paper produced by a process is 3. What is the probability of finding no more than 2 defects in 40 m^2 of paper through random selection?

Solution   Here, one unit is $40 \, m^2$ of paper. So, λ is 6 because the average number of surface defects per $40 \, m^2$ is 6. The probability is

$$P(X \leq 2) = P(X = 0) + P(X = 1) + P(X = 2)$$
$$= \frac{e^{-6}(6^0)}{0!} + \frac{e^{-6}(6^1)}{1!} + \frac{e^{-6}(6^2)}{2!} = 0.062$$

To calculate this probability using Minitab, click on **Calc > Probability Distributions Poisson**. Select **Cumulative probability**, and input the value of the **Mean** as 6. Select **Input constant** and input the value as 2. Click **OK**. Minitab outputs $P(X \leq 2) = 0.0619688$.

Appendix A-2 also gives this probability as 0.062. The mean and variance of the distribution are both equal to 6. Using Appendix A-2, the probability distribution is as follows:

x	0	1	2	3	4	5	6	7	8
$p(x)$	0.002	0.015	0.045	0.089	0.134	0.161	0.160	0.138	0.103

x	9	10	11	12	13	14	15	16
$p(x)$	0.069	0.041	0.023	0.011	0.005	0.003	0.000	0.000

Continuous Distributions

Continuous random variables may assume an infinite number of values over a finite or infinite range. The probability distribution of a continuous random variable X is often called the *probability density function $f(x)$*. The total area under the probability density function is 1.

Normal Distribution   The most widely used distribution in the theory of statistical quality control is the **normal distribution**. The probability density function of a normal random variable is given by

$$f(x) = \frac{1}{\sqrt{2\pi}\sigma} \exp\left[\frac{-(x - \mu)^2}{2\sigma^2}\right], \quad -\infty < x < \infty \qquad (4\text{-}35)$$

where μ is the population mean, and σ is the population standard deviation.

The two parameters of a normal distribution are the mean and the variance (or standard deviation). Note that the variance σ^2 is the square of the standard deviation. Figure 4-17

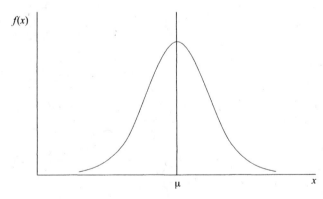

FIGURE 4-17   Normal distribution.

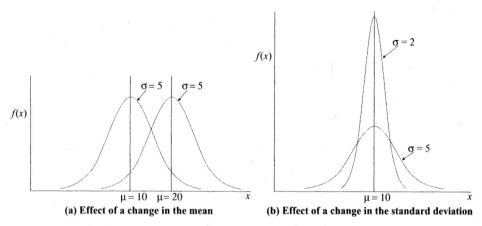

FIGURE 4-18 Effects of the parameters μ and σ^2 on the normal distribution.

shows the normal probability density function. The effect of the parameters μ and σ^2 on the shape of the probability density function is shown in Figure 4-18a and b. A change in the mean μ causes a change in the location of the distribution. As the mean increases, the distribution shifts to the right, and as the mean decreases, the distribution shifts to the left. As the variance σ^2 (or standard deviation) increases, the spread about the mean increases. A normal distribution is symmetric about the mean; that is the mean, median, and mode are equal.

The standard deviation is very important in a normal distribution. The proportion of population values that fall in range $\mu \pm \sigma$ is 68.26%. Similarly, 95.44% of the total area is within $\mu \pm 2\sigma$, and 99.74% of the area is between $\mu \pm 3\sigma$.

Finding the area under a normal curve requires integrating eq. (4-35) within the prescribed limits of the random variable, a fairly involved task. Fortunately, already constructed tables enable us to find this area. Note that because the shape of the density function changes with each possible combination of μ and σ^2, it is impossible to tabulate areas for each conceivable normal distribution. Nevertheless, the area within certain limits for any normal distribution can be found by looking up tabulated areas for a **standard normal distribution**. The standardized normal random variable Z is given by

$$Z = \frac{X - \mu}{\sigma} \qquad (4\text{-}36)$$

The z-value, or standardized value, is the number of standard deviations that a raw, or observed, value x is from the mean. The z-value can be positive or negative. If the z-value is positive, the raw value is to the right of the mean, whereas negative z-values indicate points to the left of the mean. At the mean, the z-value is 0. The distribution of the standardized normal random variable has a mean of 0 and a variance of 1. It is represented as an $N(0, 1)$ variable, where the first parameter represents the mean and the second the variance, and its density function is given by

$$f(z) = \frac{1}{\sqrt{2\pi}} e^{-z^2/2}, \qquad -\infty < z < \infty \qquad (4\text{-}37)$$

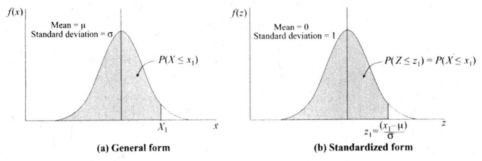

FIGURE 4-19 Normal distributions.

The cumulative distribution function of Z is

$$\Phi(z) = F(z) = \int_{-\infty}^{z} f(t)dt \qquad (4\text{-}38)$$

Figure 4-19a and b show the standard normal distribution and its relationship to the raw variable X.

Appendix A-3 gives values for the cumulative distribution function of Z. The normal distribution has the property that the area between certain limits a and b for a variable X is the same as the area between the standardized values for a and b under the standard normal distribution. Thus, we need only one set of tables—those for the standard normal distribution function—to calculate the area between certain limits for any normal distribution.

Example 4-26 The length of a machined part is known to have a normal distribution with a mean of 100 mm and a standard deviation of 2 mm.

(a) What proportion of the parts will be above 103.3 mm?

Solution Let X denote the length of the part. The parameter values for the normal distribution are $\mu = 100$ and $\sigma = 2$. The probability required is shown in Figure 4-20a. The standardized value of 103.3 corresponds to

$$z_1 = \frac{x_1 - \mu}{\sigma} = \frac{103.3 - 100}{2} = 1.65$$

Thus, $P(X > 103.3) = P(Z > 1.65)$. From Appendix A-3, $P(Z \le 1.65) = 0.9505$, which also equals $P(X \le 103.3)$. So

$$P(Z > 1.65) = 1 - P(Z \le 1.65)$$
$$= 1 - 0.9505 = 0.0495$$

The desired probability $P(X > 103.3)$ is 0.0495, or 4.95%.

To obtain the desired probability using Minitab, click on **Calc > Probability Distributions > Normal**. Select **Cumulative Probability**, and input the value of the **Mean** to be 100 and that of the **Standard deviation** to be 2. Select **Input constant** and input the value of

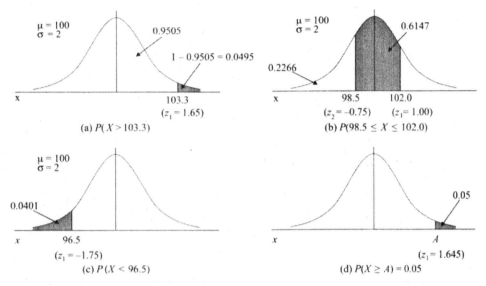

FIGURE 4-20   Calculation of normal probabilities.

103.3. Click **OK**. Minitab outputs $P(X \leq 103.3) = 0.950529$. Hence, $P(X > 103.3) = 0.049471$.

(**b**) What proportion of the output will be between 98.5 and 102.0 mm?

Solution   We wish to find $P(98.5 \leq X \leq 102.0)$, which is shown in Figure 4-20b. The standardized values are computed as

$$z_1 = \frac{102.0 - 100}{2} = 1.00$$

$$z_2 = \frac{98.5 - 100}{2} = -0.75$$

From Appendix A-3, we have $P(Z \leq 1.00) = 0.8413$ and $P(Z \leq -0.75) = 0.2266$. The required probability equals $0.8413 - 0.2266 = 0.6147$. Thus, 61.47% of the output is expected to be between 98.5 and 102.0 mm

(**c**) What proportion of the parts will be shorter than 96.5 mm?

Solution   We want $P(X < 96.5)$, which is equivalent to $P(X \leq 96.5)$, since for a continuous random variable the probability that the variable equals a particular value is zero. The standardized value is

$$z_1 = \frac{96.5 - 100}{2} = -1.75$$

The required proportion is shown in Figure 4-20c. Using Appendix A-3, $P(Z \leq -1.75)$ 0.0401. Thus, 4.01% of the parts will have a length less than 96.5 mm.

(**d**) It is important that not many of the parts exceed the desired length. If a manager stipulates that no more than 5% of the parts should be oversized, what specification limit should be recommended?

Solution   Let the specification limit be A. From the problem information, $P(X \geq A) = 0.05$. To find A, we first find the standardized value at the point where the raw value is A. Here, the approach will be the reverse of what was done for the previous three parts of this example. That is, we are given an area, and we want to find the z-value. Here, $P(X \leq A) = 1 - 0.05 = 0.95$. We look for an area of 0.95 in Appendix A-3 and find that the linearly interpolated z-value is 1.645. Finally, we unstandardize this value to determine the limit A.

$$1.645 = \frac{x_1 - 100}{2}$$

$$x_1 = 103.29 \text{ mm}$$

Thus, A should be set at 103.29 mm to achieve the desired stipulation.

To solve this using Minitab, click on **Calc** > **Probability Distributions** > **Normal**. Select **Inverse cumulative probability**, and input the value of the **Mean** as 100 and **Standard deviation** as 2. **Select Input constant**, and input the value 0.95. Click **OK**. Minitab outputs the value of X as 103.290, for which $P(X \leq x) = 0.95$.

Exponential Distribution   The **exponential distribtuion** is used in reliability analysis to describe the time to the failure of a component or system. Its probability density function is given by

$$f(x) = \lambda e^{-\lambda x}, \quad x \geq 0 \tag{4-39}$$

where λ denotes the failure rate. Figure 4-21 shows the density function. An exponential distribution represents a constant failure rate and is used to model failures that happen randomly and independently. If we consider the typical life cycle of a product, its useful life occurs after the debugging phase and before the wearout phase. During its useful life, the failure rate is fairly constant, and failures happen randomly and independently. An exponential distribution, which has these properties, is therefore appropriate for modeling failures in the useful phase of a product. The mean and the variance of an exponential random variable are given by

$$\mu = \frac{1}{\lambda}, \quad \sigma^2 = \frac{1}{\lambda^2} \tag{4-40}$$

Thus, the mean and the standard deviation are equal for an exponential random variable.

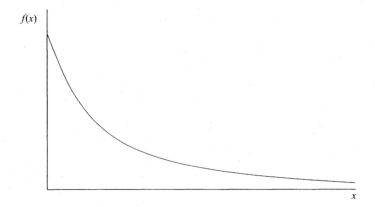

FIGURE 4-21   Exponential density function.

The exponential cumulative distribution function is obtained as follows:

$$F(x) = P(X \leq x)$$
$$= \int_0^x \lambda e^{-\lambda t} dt = 1 - e^{-\lambda x} \tag{4-41}$$

An exponential distribution has the property of being *memoryless*. This means that the probability of a component's exceeding $(s + t)$ time units, given that it has lasted t time units, is the same as the probability of the life exceeding s time units. Mathematically, this property may be represented as

$$P(X > s + t \,|\, X > t) = P(X > s) \qquad \text{for all } s \text{ and } t \geq 0 \tag{4-42}$$

Example 4-27   It is known that a battery for a video game has an average life of 500 hours (h). The failures of batteries are known to be random and independent and may be described by an exponential distribution.

(a) Find the probability that a battery will last at least 600 hours.

Solution   Since the average life, or mean life, of a battery is given to be 500 hours, the failure rate is $\lambda = 1/500$.
If the life of a battery is denoted by X, we wish to find $P(X > 600)$.

$$P(X > 600) = 1 - P(X \leq 600) = 1 - [1 - e^{-(1/500)(600)}] = e^{-1.2} = 0.301$$

(b) Find the probability of a battery failing within 200 hours.

Solution

$$P(X \leq 200) = 1 - e^{-(1/500)(200)} = 1 - e^{-0.4} = 0.330$$

(c) Find the probability of a battery lasting between 300 and 600 hours.

Solution

$$P(300 \leq X \leq 600) = F(600) - F(300) = e^{-(1/500)(300)} - e^{-(1/500)(600)}$$
$$= e^{-0.6} - e^{-1.2} = 0.248$$

To find this probability using Minitab, first create a worksheet by using a column for the variable **X**, and input the values of 300 and 600. Click on **Calc > Probability Distributions Exponential**. Select **Cumulative Probability**, and input the value of **Scale**, which is equal to the mean when the threshold is 0, as 500. Select **Input column**, and input the column number or name of the variable X. Click **OK**. Minitab outputs $P(X \leq 300) = 0.451188$ and $P(X \leq 600) = 0.698806$. The desired probability is 0.247618.

(d) Find the standard deviation of the life of a battery.

Solution

$$\sigma = 1/\lambda = 500 \text{ hours}$$

(e) If it is known that a battery has lasted 300 hours, what is the probability that it will last at least 500 hours?

Solution

$$P(X > 500\,|\,X > 300) = P(X > 200) = 1 - P(X \le 200)$$
$$= 1 - [1 - e^{-(1/500)200}]$$
$$= e^{-0.4} = 0.670 \ .$$

Weibull Distribution   A Weibull random variable is typically used in reliability analysis to describe the time to failure of mechanical and electrical components. It is a three-parameter distribution (Banks 1989; Henley and Kumamoto 1991). A Weibull probability density function is given by

$$f(x) = \frac{\beta}{\alpha}\left(\frac{x - \gamma}{\alpha}\right)^{\beta - 1} \exp\left[-\left(\frac{x - \gamma}{\alpha}\right)^{\beta}\right], \qquad x \ge \gamma \qquad (4\text{-}43)$$

The parameters, are a **location parameter** $\gamma(-\infty < \gamma < \infty)$, a **scale parameter** $\alpha(\alpha > 0)$, and a **shape parameter** $\beta(\beta > 0)$.

Figure 4-22 shows the probability density functions for $\gamma = 0$, $\alpha = 1$, and several values of β. The **Weibull distribution** as a general distribution is important because it can be used to model a variety of situations. The shape varies depending on the parameter values. For certain parameter combinations, it approaches a normal distribution. If $\gamma = 0$ and $\beta = 1$, a Weibull distribution reduces to an exponential distribution. The mean and the variance of the Weibull distribution are

$$\mu = E(X) = \gamma + \alpha\Gamma\left(\frac{1}{\beta} + 1\right) \qquad (4\text{-}44)$$

$$\sigma^2 = \text{Var}(X) = \alpha^2\left\{\Gamma\left(\frac{2}{\beta} + 1\right) - \left[\Gamma\left(\frac{1}{\beta} + 1\right)\right]^2\right\} \qquad (4\text{-}45)$$

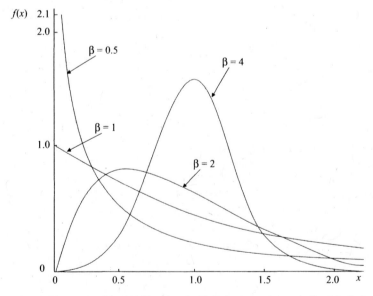

FIGURE 4-22   Weibull probability density functions ($\gamma = 0$, $\alpha = 1$, $\beta = 0.5$, 1, 2, 4).

where $\Gamma(t)$ represents the **gamma function**, given by

$$\Gamma(t) = \int_0^\infty e^{-x} x^{t-1} dx$$

If u is an integer such that $u \geq 1$, then $\Gamma(u) = (u-1)!$, Note that $u! = u(u-1)(u-2)\cdots 1$ and $0! = 1$.

The cumulative distribution function of a Weibull random variable is given by

$$F(x) = 1 - \exp\left[-\left(\frac{x-\gamma}{\alpha}\right)^\beta\right], \qquad x \geq \gamma \qquad (4\text{-}46)$$

Example 4-28 The time to failure for a cathode ray tube can be modeled using a Weibull distribution with parameters $\gamma = 0$, $\beta = \dfrac{1}{3}$ and $\alpha = 200$ hours.

(a) Find the mean time to failure and its standard deviation.

Solution The mean time to failure is given by

$$\begin{aligned}
\mu &= E(X) \\
&= 0 + 200\Gamma(3+1) \\
&= 200\Gamma(4) = 1200 \text{ hours}
\end{aligned}$$

The variance is given by

$$\begin{aligned}
\sigma^2 &= (200)^2\{\Gamma(6+1) - [\Gamma(3+1)]^2\} \\
&= (200)^2\{\Gamma(7) - [\Gamma(4)]^2\} = 2736 \times 10^4
\end{aligned}$$

The standard deviation is $\sigma = 5230.679$ hours.

(b) What is the probability of a tube operating for at least 800 hours?

Solution

$$\begin{aligned}
P(X > 800) &= 1 - P(X \leq 800) \\
&= 1 - \left\{1 - \exp[-(800/200)^{1/3}]\right\} \\
&= \exp[-(4)^{1/3}] = \exp[-1.587] \\
&= 0.204
\end{aligned}$$

To find this probability using Minitab, click on **Calc > Probability Distributions > Weibull**. Select **Cumulative probability**, and input **Shape parameter** as 0.3333, **Scale parameter** as 200, and **Threshold parameter** (for location) as 0. Select **Input constant** and input the value 800. Click **OK**. Minitab outputs $P(X \leq 800) = 0.795529$. The desired probability is thus 0.204471.

Gamma Distribution Another important distribution with applications in reliability analysis is the gamma distribution. Its probability density function is given by

$$f(x) = \frac{\lambda^k}{\Gamma(k)} x^{k-1} e^{-\lambda x}, \qquad x \geq 0 \qquad (4\text{-}47)$$

where k is a **shape** parameter, $k > 0$; and λ is a **scale** parameter, $\lambda > 0$. The mean and the variance of the gamma distribution are

$$\mu = \frac{k}{\lambda} \tag{4-48}$$

$$\sigma^2 = \frac{k}{\lambda} \tag{4-49}$$

Gamma distributions may take on a variety of shapes, similar to the Weibull, based on the choice of the parameters k and λ. If $k = 1$, a special case of the gamma distribution defaults to the exponential distribution. The gamma distribution may be viewed as the sum of k independent and identically distributed exponential distributions, each with parameter λ, the constant failure rate. Thus, it could model the time to the kth failure in a system, where items fail randomly and independently. Minitab may be used to calculate probabilities using the gamma distribution.

Lognormal Distribution   A random variable X has a **lognormal distribution** if $\ln(x)$ has a normal distribution with mean μ and variance σ^2, where ln represents the natural logarithm. Its probability density function is given by

$$f(x) = \frac{1}{\sqrt{2\pi}\,\sigma x} \, \exp\left[-\frac{\left(\ln(x) - \mu\right)^2}{2\sigma^2}\right], \qquad x > 0 \tag{4-50}$$

The mean and variance of the lognormal distribution are

$$E(X) = \exp\left(\mu + \frac{\sigma^2}{2}\right) \tag{4-51}$$

$$\mathrm{Var}(X) = \exp(2\mu + \sigma^2)\left[\exp(\sigma^2) - 1\right] \tag{4-52}$$

The cumulative distribution function is expressed as

$$F(x) = \Phi\left[\frac{\ln(x) - \mu}{\sigma}\right], \quad x > 0 \tag{4-53}$$

where $\Phi(\cdot)$ represents the cdf of the standard normal.

Certain quality characteristics such as tensile strength or compressive strength are modeled by the lognormal distribution. It is used to model failure distributions due to accumulated damage, such as crack propagation or wear. These characteristics typically have a distribution that is positively skewed. It can be shown that the skewness increases rapidly with σ^2, independent of μ. The location and shape parameters are given by e^μ and σ, respectively.

4-7   INFERENTIAL STATISTICS: DRAWING CONCLUSIONS ON PRODUCT AND PROCESS QUALITY

In this section we examine statistical procedures that are used to make inferences about a population (a process or product characteristic), on the basis of sample data. As mentioned previously, analysts use statistics to draw conclusions about a process based on limited information. The two main procedures of inferential statistics are estimation (point and interval) and hypothesis testing.

Usually, the parameters of a process, such as average furnace temperature, average component length, and average waiting time prior to service, are unknown, so these values must be estimated, or claims as to these parameter values must be tested for verification. For a more thorough treatment of estimation and hypothesis testing, see Duncan (1986) or Mendenhall et al. (1993).

Sampling Distributions

An *estimator*, or *statistic* (which is a characteristic of a sample), is used to make inferences as to the corresponding parameter. For example, an estimator of sample mean is used to draw conclusions on the population mean. Similarly, a sample variance is an estimator of the population variance. Studying the behavior of these estimators through repeated sampling allows us to draw conclusions about the corresponding parameters. The behavior of an estimator in repeated sampling is known as the **sampling distribution** of the estimator, which is expressed as the probability distribution of the statistic. Sampling distributions are discussed in greater detail in the section on interval estimation.

The sample mean is one of the most widely used estimators in quality control because analysts frequently need to estimate the population mean. It is therefore of interest to know the sampling distribution of the sample mean; this is described by the **central limit theorem**:

Suppose that we have a population with mean μ and standard deviation σ. If random samples of size n are selected from this population, the following holds if the sample size is large:

1. The sampling distribution of the sample mean will be approximately normal.
2. The mean of the sampling distribution of the sample mean $(\mu_{\bar{x}})$ will be equal to the population mean, μ.
3. The standard deviation of the sample mean is given by $\sigma_{\bar{x}} = \sigma/\sqrt{n}$, known as the *standard error*.

The degree to which a sampling distribution of a sample mean approximates a normal distribution becomes greater as the sample size n becomes larger. Figure 4-23 shows a

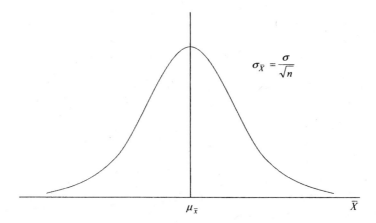

FIGURE 4-23   Sampling distribution of a sample mean.

sampling distribution of a sample mean. A sample size should be 30 or more to allow a close approximation of a normal distribution. However, it has been shown that if a population distribution is symmetric and unimodal, sample sizes as small as 4 or 5 yield sample means that are approximately normally distributed. In the case of a population distribution already being normal, samples of any size (even $n = 1$) will lead to sample means that are normally distributed. Note that the variability of the sample means, as measured by the standard deviation, decreases as the sample size increases.

Example 4-29   The tuft bind strength of a synthetic material used to make carpets is known to have a mean of 50 kg and a standard deviation of 10 kg. If a sample of size 40 is randomly selected, what is the probability that the sample mean will be less than 52.5 kg?

Solution   Using the central limit theorem, the sampling distribution of the sample mean will be approximately normal with a mean $\mu_{\bar{X}}$ of 50 kg and a standard deviation of

$$\sigma_{\bar{X}} = \frac{10}{\sqrt{40}} = 1.581 \text{ kg}$$

We want to find $P(\bar{X} < 52.5)$ and first find the standardized value:

$$z_1 = \frac{\bar{X} - \mu_{\bar{X}}}{\sigma_{\bar{X}}}$$
$$= \frac{52.5 - 50}{1.581} = 1.58$$

Then $P(\bar{X} < 52.5) = P(Z < 1.58) = 0.9429$ using Appendix A-3.

Estimation of Product and Process Parameters

One branch of statistical inference uses sample data to estimate unknown population parameters. There are two types of estimation: *point estimation* and *interval estimation*. In **point estimation**, a single numerical value is obtained as an estimate of the population parameter. In **interval estimation**, a range or interval is determined such that there is some desired level of probability that the true parameter value is contained within it. Interval estimates are also called *confidence intervals*.

Point Estimation   A point estimate consists of a single numerical value that is used to make an inference about an unknown product or process parameter. Suppose that we wish to estimate the mean diameter of all piston rings produced in a certain month. We randomly select 100 piston rings and compute the sample mean diameter, which is 50 mm. The value of 50 mm is thus a point estimate of the mean diameter of all piston rings produced that month. A common convention for denoting an estimator is to use "^" above the corresponding parameter. For example, an estimator of the population mean μ is $\hat{\mu}$, which is the sample mean $\bar{X}$. An estimator of the population variance σ^2 is $\hat{\sigma}^2$, usually noted as the sample variance s^2.

Desirable Properties of Estimators   Two desirable properties of estimators are worth noting here. A point estimator is said to be **unbiased** if the expected value, or mean, of its sampling distribution is equal to the parameter being estimated. A point estimator is said to have a *minimum variance* if its variance is smaller than that of any other point estimator for the parameter under consideration.

The point estimators $\bar{X}$ and s^2 are unbiased estimators of the parameters μ and σ^2, respectively. We know that $E(\bar{X}) = \mu$ and $E(s^2) = \sigma^2$. In fact, using a denominator of $n - 1$ in the computation of s^2 in eq. (4-11) or (4-12) makes s^2 unbiased. The central limit theorem supports the idea that the sample mean is unbiased. Also note from the central limit theorem that the variance of the sample mean $\bar{X}$ is inversely proportional to the square root of the sample size.

Interval Estimation   Interval estimation consists of finding an interval defined by two end-points–say, L and U—such that the probability of the parameter θ being contained in the interval is some value $1 - \alpha$. That is,

$$P(L \leq \theta \leq U) = 1 - \alpha \tag{4-54}$$

This expression represents a two-sided **confidence interval**, with L representing the lower confidence limit and U the upper confidence limit. If a large number of such confidence intervals were constructed from independent samples, then $100(1 - \alpha)\%$ of these intervals would be expected to contain the true parameter value of θ. (Methods for using sample data to construct such intervals are discussed in the next subsections.)

Suppose that a 90% confidence interval for the mean piston ring diameter in millimeters is desired. One sample yields an interval of $(48.5, 51.5)$—that is, $L = 48.5$ mm and $U = 51.5$ mm. Then, if 100 such intervals were constructed (one each from 100 samples), we would expect 90 of them to contain the population mean piston ring diameter. Figure 4-24 shows this concept. The quantity $(1 - \alpha)$ is called the *level of confidence* or the *confidence coefficient*.

Confidence intervals can also be one-sided. An interval of the type

$$L \leq \theta, \qquad \text{such that } P(L \leq \theta) = 1 - \alpha$$

is a one-sided lower $100(1 - \alpha)\%$ confidence interval for θ. On the other hand, an interval of the type

$$\theta \leq U, \qquad \text{such that } P(\theta \leq U) = 1 - \alpha$$

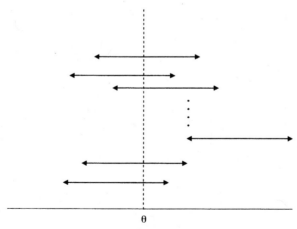

θ

Confidence level = 90%
(90 of the 100 confidence intervals enclose θ)

FIGURE 4-24   Interpreting confidence intervals.

is an upper $100(1-\alpha)\%$ confidence interval for θ. The context of a situation will influence the type of confidence interval to be selected. For example, when the concern is the breaking strength of steel cables, the customer may prefer a one-sided lower confidence interval. Since the exact expression for the confidence intervals is determined by the estimator, we discuss the estimation of several types of parameters next.

Confidence Interval for the Mean

1. *Variance known.* Suppose that we want to estimate the mean μ of a product when the population variance σ^2 is known. A random sample of size n is chosen, and the sample mean $\bar{X}$ is calculated. From the central limit theorem, we know that the sampling distribution of the point estimator $\bar{X}$ is approximately normal with mean μ and variance σ^2/n. A $100(1-\alpha)\%$ two-sided confidence interval for μ is given by

$$\bar{X} - z_{\alpha/2}\frac{\sigma}{\sqrt{n}} \leq \mu \leq \bar{X} + z_{\alpha/2}\frac{\sigma}{\sqrt{n}} \tag{4-55}$$

The value of $z_{\alpha/2}$ is the standard normal variate such that the right tail area of the standardized normal distribution is $\alpha/2$. Equation (4-55) represents an approximate $100(1-\alpha)\%$ confidence interval for any distribution of a random variable X. However, if X is normally distributed, then eq. (4-55) becomes an exact $100(1-\alpha)\%$ confidence interval.

Example 4-30   The output voltage of a power source is known to have a standard deviation of 10 V. Fifty readings are randomly selected, yielding an average of 118 V. Find a 95% confidence interval for the population mean voltage.

Solution   For this example, $n=50$, $\sigma=10$, $\bar{X}=118$, and $1-\alpha=0.95$. From Appendix A-3, we have $z_{.025}=1.96$. Hence, a 95% confidence interval for the population mean voltage μ is

$$118 - \frac{(1.96)(10)}{\sqrt{50}} \leq \mu \leq 118 + \frac{(1.96)(10)}{\sqrt{50}}$$

or

$$115.228 \leq \mu \leq 120.772$$

Hence, there is a 95% chance that the population mean voltage falls within this range.

2. *Variance unknown.* Suppose we have a random variable X that is normally distributed with unknown mean μ and unknown variance σ^2. A random sample of size n is selected, and the sample mean $\bar{X}$ and sample variance s^2 are computed. It is known that the sampling distribution of the quantity $(\bar{X} - \mu)/(s/\sqrt{n})$ is what is known as a **t-distribution** with $(n-1)$ degrees of freedom; that is,

$$\frac{\bar{X} - \mu}{s/\sqrt{n}} \sim t_{n-1} \tag{4-56}$$

where the symbol "$\sim$" stands for "is distributed as." The shape of a t-distribution is similar to that of the standard normal distribution and is shown in Figure 4-25. As the sample size n increases, the t-distribution approaches the standard normal distribution. The number of

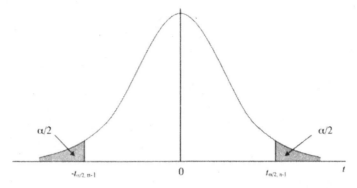

FIGURE 4-25   A t-distribution.

degrees of freedom of t, in this case $(n - 1)$, is the same as the denominator used to calculate s^2 in eq. (4-11) or (4-12). The number of **degrees of freedom** represents the fact that if we are given the sample mean $\bar{X}$ of n observations, then $(n - 1)$ of the observations are free to be any value. Once these $(n - 1)$ values are found, there is only one value for the nth observation that will yield a sample mean of $\bar{X}$. Hence, one observation is "fixed" and $(n - 1)$ are "free".

The values of t corresponding to particular right-hand-tail areas and numbers of degrees of freedom are given in Appendix A-4. For a right-tail area of 0.025 and 10 degrees of freedom, the t-value is 2.228. As the number of degrees of freedom increases for a given right-tail area, the t-value decreases. When the number of degrees of freedom is large (say, greater than 120), notice that the t-value given in Appendix A-4 is equal to the corresponding z-value given in Appendix A-3. A $100(1 - \alpha)\%$ two-sided confidence interval for the population mean μ is given by

$$\bar{X} - t_{\alpha/2,n-1}\frac{s}{\sqrt{n}} \leq \mu \leq \bar{X} + t_{\alpha/2,n-1}\frac{s}{\sqrt{n}} \tag{4-57}$$

where $t_{\alpha/2, n-1}$ represents the axis point of the t-distribution where the right-tail area is $\alpha/2$ and the number of degrees of freedom is $(n - 1)$.

Example 4-31   A new process has been developed that transforms ordinary iron into a kind of superiron called metallic glass. This new product is stronger than steel alloys and is much more corrosion-resistant than steel. However, it has a tendency to become brittle at high temperatures. It is desired to estimate the mean temperature at which it becomes brittle. A random sample of 20 pieces of metallic glass is selected. The temperature at which brittleness is first detected is recorded for each piece. The summary results give a sample mean $\bar{X}$ of 600 °C and a sample standard deviation s of 15 °C. Find a 90% confidence interval for the mean temperature at which metallic glass becomes brittle.

Solution   We have $n = 20$, $\bar{X} = 600$, and $s = 15$. Using the t-distribution tables in Appendix A-4, $t_{.05,19} = 1.729$. A 90% confidence interval for μ is

$$600 - (1.729)\frac{15}{\sqrt{20}} \leq \mu \leq 600 + (1.729)\frac{15}{\sqrt{20}}$$

or

$$594.201 \leq \mu \leq 605.799$$

Example 4-32   Consider the milling operation data on depth of cut shown in Table 4-3. Figure 4-14 shows the Minitab output on descriptive statistics for this variable. Note the 95% confidence interval for the mean μ, which is

$$3.1635 \leq \mu \leq 3.8915$$

Confidence Interval for the Difference Between Two Means

1. *Variances known.*   Suppose that we have a random variable X_1 from a first population with mean μ_1 and variance σ_1^2; X_2 represents a random variable from a second population with mean μ_2 and variance σ_2^2. Assume that μ_1 and μ_2 are unknown and σ_1^2 and σ_2^2 are known. Suppose that a sample of size n_1 is selected from the first population and an independent sample of size n_2 is selected from the second population.

Let the sample means be denoted by $\bar{X}_1$ and $\bar{X}_2$. A $100(1 - \alpha)\%$ two-sided confidence interval for the difference between the two means is given by

$$(\bar{X}_1 - \bar{X}_2) - z_{\alpha/2}\sqrt{\frac{\sigma_1^2}{n_1} + \frac{\sigma_2^2}{n_2}} \leq \mu_1 - \mu_2 \leq (\bar{X}_1 - \bar{X}_2) + z_{\alpha/2}\sqrt{\frac{\sigma_1^2}{n_1} + \frac{\sigma_2^2}{n_2}} \quad (4\text{-}58)$$

2. *Variances unknown.*   Let's consider two cases here. The first is the situation where the unknown variances are equal (or are assumed to be equal)–that is, $\sigma_1^2 = \sigma_2^2$. Suppose that the random variable X_1 is from a normal distribution with mean μ_1 and variance σ_1^2 [i.e., $X_1 \sim N(\mu_1, \sigma_1^2)$] and the random variable X_2 is from $N(\mu_2, \sigma_2^2)$. Using the same notation as before, a $100(1 - \alpha)\%$ confidence interval for the difference in the population means $(\mu_1 - \mu_2)$ is

$$(\bar{X}_1 - \bar{X}_2) - t_{\alpha/2, n_1+n_2-2} s_p \sqrt{\frac{1}{n_1} + \frac{1}{n_2}} \leq \mu_1 - \mu_2 \leq (\bar{X}_1 - \bar{X}_2) + t_{\alpha/2, n_1+n_2-2} s_p \sqrt{\frac{1}{n_1} + \frac{1}{n_2}} \quad (4\text{-}59)$$

where a pooled estimate of the common variance, obtained by combining the information on the two sample variances, is given by

$$s_p^2 = \frac{(n_1 - 1)s_1^2 + (n_2 - 1)s_2^2}{n_1 + n_2 - 2} \quad (4\text{-}60)$$

The validity of assuming that the population variances are equal ($\sigma_1^2 = \sigma_2^2$) can be tested using a statistical test, which we discuss later.

In the second case, the population variances are not equal; that is, $\sigma_1^2 \neq \sigma_2^2$ (a situation known as the *Behrens-Fisher problem*). A $100(1 - \alpha)\%$ two-sided confidence interval is

$$(\bar{X}_1 - \bar{X}_2) - t_{\alpha/2, v}\sqrt{\frac{s_1^2}{n_1} + \frac{s_2^2}{n_2}} \leq \mu_1 - \mu_2 \leq (\bar{X}_1 - \bar{X}_2) + t_{\alpha/2, v}\sqrt{\frac{s_1^2}{n_1} + \frac{s_2^2}{n_2}} \quad (4\text{-}61)$$

where the number of degrees of freedom of t is denoted by v, which is given by

$$v = \frac{(s_1^2/n_1 + s_2^2/n_2)^2}{(s_1^2/n_1)^2/(n_1 - 1) + (s_2^2/n_2)^2/(n_2 - 1)} \quad (4\text{-}62)$$

3. *Paired samples.*   The previous cases have assumed the two samples to be indepen-dent. However, in certain situations, observations in two samples may be paired, to reduce the impact of extraneous factors that are not of interest. In this case the samples are not independent. Consider, for example, the effect of a drug to reduce blood cholesterol levels in diabetic patients. By choosing the same patient, before and after administration of the drug, to observe cholesterol levels, the effect of the patient-related factors on cholesterol levels may be reduced. Thus, the precision of the experiment could be improved.

Let X_{1i} and X_{2i}, $i = 1,2,\ldots, n$ denote the n-paired observations with their differences being $d_i = X_{1i} - X_{2i}$, $i = 1, 2,\ldots, n$. If the mean and standard deviation of the differences are represented by $\bar{d}$ and s_d, respectively, the confidence interval for $\mu_d = \mu_1 - \mu_2$, is similar to that of a single population mean with unknown variance and is given by

$$\bar{d} - t_{\alpha/2,n-1}\frac{s_d}{\sqrt{n}} \leq \mu_d \leq \bar{d} + t_{\alpha/2,n-1}\frac{s_d}{\sqrt{n}} \tag{4-63}$$

The assumption made is that the differences are normally distributed.

Example 4-33   Two operators perform the same machining operation. Their supervisor wants to estimate the difference in the mean machining times between them. No assumption can be made as to whether the variabilities of machining time are the same for both operators. It can be assumed, however, that the distribution of machining times is normal for each operator. A random sample of 10 from the first operator gives an average machining time of 4.2 minutes with a standard deviation of 0.5 minutes. A random sample of 6 from the second operator yields an average machining time of 5.1 minutes with a standard deviation of 0.8 minutes. Find a 95% confidence interval for the difference in the mean machining times between the two operators.

Solution   We have $n_1 = 10, \bar{X}_1 = 4.2, s_1 = 0.5$, and $n_2 = 6, \bar{X}_2 = 5.1, s_2 = 0.8$. Since the assumption of equal variances cannot be made, eq. (4-61) must be used. From eq. (4-62), the number of degrees of freedom of t is

$$v = \frac{(0.25/10 + 0.64/6)^2}{(0.25/10)^2/9 + (0.64/6)^2/5} = 7.393$$

As an approximation, using 7 degrees of freedom rather than the calculated value of 7.393, Appendix A-4 gives $t_{.025,7} = 2.365$. A 95% confidence interval for the difference in the mean machining times is

$$(4.2 - 5.1) - 2.365\sqrt{\frac{0.25}{10} + \frac{0.64}{6}} \leq (\mu_1 - \mu_2) \leq (4.2 - 5.1) + 2.365\sqrt{\frac{0.25}{10} + \frac{0.64}{6}}$$

or

$$-1.758 \leq (\mu_1 - \mu_2) \leq -0.042$$

Confidence Interval for a Proportion   Now let's consider the parameter p, the proportion of successes in a binomial distribution. In statistical quality control, this parameter corresponds to the proportion of nonconforming items in a process or in a large lot, or the proportion of customers that are satisfied with a product or service. A point estimator of p

is $\hat{p}$, the sample proportion of nonconforming items, which is found from $\hat{p} = x/n$, where x denotes the number of nonconforming items and n the number of trials or items sampled. When n is large, a $100(1 - \alpha)\%$ two-sided confidence interval for p is given by

$$\hat{p} - z_{\alpha/2}\sqrt{\frac{\hat{p}(1-\hat{p})}{n}} \leq p \leq \hat{p} + z_{\alpha/2}\sqrt{\frac{\hat{p}(1-\hat{p})}{n}} \tag{4-64}$$

For small n, the binomial tables should be used to determine the confidence limits for p. When n is large and p is small ($np < 5$), the Poisson approximation to the binomial can be used. If n is large and p is neither too small nor too large [$np \geq 5, n(1-p) \geq 5$], the normal distribution serves as a good approximation to the binomial.

Confidence Interval for the Difference Between Two Binomial Proportions   Suppose that a sample of size n_1 is selected from a binomial population with parameter p_1 and a sample of size n_2 is selected from a binomial population with parameter p_2. For large sample sizes of n_1 and n_2, a $100(1 - \alpha)\%$ confidence interval for $p_1 - p_2$ is

$$(\hat{p}_1 - \hat{p}_2) - z_{\alpha/2}\sqrt{\frac{\hat{p}_1(1-\hat{p}_1)}{n_1} + \frac{\hat{p}_2(1-\hat{p}_2)}{n_2}} \leq p_1 - p_2$$
$$\leq (\hat{p}_1 - \hat{p}_2) + z_{\alpha/2}\sqrt{\frac{\hat{p}_1(1-\hat{p}_1)}{n_1} + \frac{\hat{p}_2(1-\hat{p}_2)}{n_2}} \tag{4-65}$$

Example 4-34   Two operators perform the same operation of applying a plastic coating to Plexiglas. We want to estimate the difference in the proportion of nonconforming parts produced by the two operators. A random sample of 100 parts from the first operator shows that 6 are nonconforming. A random sample of 200 parts from the second operator shows that 8 are nonconforming. Find a 90% confidence interval for the difference in the proportion of nonconforming parts produced by the two operators.

Solution   We have $n_1 = 100$, x_1 (number of nonconforming parts produced by the first operator) $= 6$, $n_2 = 200$, $x_2 = 8$, $(1 - \alpha) = 0.90$. From Appendix A-3, using linear interpolation, $z_{.05} = 1.645$ (for the right-tail area of 0.05). So, $\hat{p}_1 = x_1/n_1 = 6/100 = 0.06$, and $\hat{p}_2 = x_2/n_2 = 8/200 = 0.04$. A 90% confidence interval for the difference in the proportion of nonconforming parts is

$$(0.06 - 0.04) - 1.645\sqrt{\frac{(0.06)(0.94)}{100} + \frac{(0.04)(0.96)}{200}} \leq p_1 - p_2$$
$$\leq (0.06 - 0.04) + 1.645\sqrt{\frac{(0.06)(0.94)}{100} + \frac{(0.04)(0.96)}{200}}$$

or

$$-0.025 \leq p_1 - p_2 \leq 0.065$$

Confidence Interval for the Variance   Consider a random variable X from a normal distribution with mean μ and variance σ^2 (both unknown). An estimator of σ^2 is the

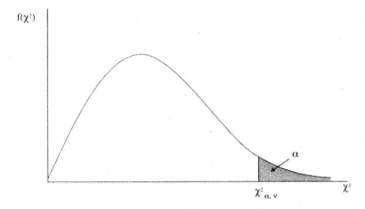

$f(\chi^2)$

$\chi^2_{\alpha, \nu}$

α

χ^2

FIGURE 4-26   Chi-squared distribution.

sample variance s^2. We know that the sampling distribution of $(n - 1)s^2/\sigma^2$ is a **chi-squared** (χ^2) **distribution** with $n - 1$ degrees of freedom. Notationally,

$$\frac{(n-1)s^2}{\sigma^2} = \chi^2_{n-1} \qquad (4\text{-}66)$$

A chi-squared distribution is skewed to the right as shown in Figure 4-26. It is dependent on the number of degrees of freedom ν. Appendix A-5 shows the values of χ^2 corresponding to the right-tail area α for various numbers of degrees of freedom ν. A $100(1 - \alpha)\%$ two-sided confidence interval for the population variance σ^2 is given by

$$\frac{(n-1)s^2}{\chi^2_{\alpha/2,n-1}} \le \sigma^2 \le \frac{(n-1)s^2}{\chi^2_{1-\alpha/2,n-1}} \qquad (4\text{-}67)$$

where $\chi^2_{\alpha/2,n-1}$ denotes the axis point of the chi-squared distribution with $n - 1$ degrees of freedom and a right-tail area of $\alpha/2$.

Example 4-35   The time to process customer orders is known to be normally distributed. A random sample of 20 orders is selected. The average processing time $\bar{X}$ is found to be 3.5 days with a standard deviation s of 0.5 day. Find a 90% confidence interval for the variance σ^2 of the order processing times.

Solution   We have $n = 20$, $\bar{X} = 3.5$, and $s = 0.5$. From Appendix A-5, $\chi^2_{0.05,19} = 30.14$ and $\chi^2_{0.95,19} = 10.12$. A 90% confidence interval for σ^2 is

$$\frac{(19)(0.5)^2}{30.14} \le \sigma^2 \le \frac{(19)(0.5)^2}{10.12}$$

or

$$0.158 \le \sigma^2 \le 0.469$$

Example 4-36   Consider the milling operation data on depth of cut shown in Table 4-3. Figure 4-14 shows the Minitab output on descriptive statistics for this variable with a 95% confidence interval for the standard deviation σ as

$$0.9323 \le \sigma \le 1.4614$$

So the 95% confidence interval for the variance σ^2 is

$$0.869 \leq \sigma^2 \leq 2.136$$

Confidence Interval for the Ratio of Two Variances   Suppose that we have a random variable X_1, from a normal distribution with mean μ_1 and variance σ_1^2, and a random variable X_2, from a normal distribution with mean μ_2 and variance σ_2^2. A random sample of size n_1 is chosen from the first population, yielding a sample variance s_1^2, and a random sample of size n_2 selected from the second population yields a sample variance s_2^2. We know that the ratio of these statistics, that is, the sample variances divided by the population variance, is an **F-distribution** with $(n_1 - 1)$ degrees of freedom in the numerator and $(n_2 - 1)$ in the denominator (Kendall et al. 1998): that is,

$$\frac{s_1^2/\sigma_1^2}{s_2^2/\sigma_2^2} \sim F_{n_1 - 1, n_2 - 1} \tag{4-68}$$

An F-distribution is skewed to the right, as shown in Figure 4-27. It is dependent on both the numerator and denominator degrees of freedom. Appendix A-6 shows the axis points of the F-distribution corresponding to a specified right-tail area α and various numbers of degrees of freedom of the numerator and denominator (v_1 and v_2), respectively. A $100(1 - \alpha)\%$ two-sided confidence interval for σ_1^2/σ_2^2 is given by

$$\frac{s_1^2}{s_2^2} \frac{1}{F_{\alpha/2, v_1, v_2}} \leq \frac{\sigma_1^2}{\sigma_2^2} \leq \frac{s_1^2}{s_2^2} \frac{1}{F_{1 - \alpha/2, v_1, v_2}}$$

The lower-tail F-value, $F_{1 - \alpha/2, v_1, v_2}$, can be obtained from the upper-tail F-value using the following relation:

$$F_{1 - \alpha/2, v_1, v_2} = \frac{1}{F_{\alpha/2, v_2, v_1}} \tag{4-69}$$

Using eq. (4-69) yields a $100(1 - \alpha)\%$ two-sided confidence interval for σ_1^2/σ_2^2 of

$$\frac{s_1^2}{s_2^2} \left(\frac{1}{F_{\alpha/2, v_1, v_2}} \right) \leq \frac{\sigma_1^2}{\sigma_2^2} \leq \frac{s_1^2}{s_2^2} F_{\alpha/2, v_2, v_1} \tag{4-70}$$

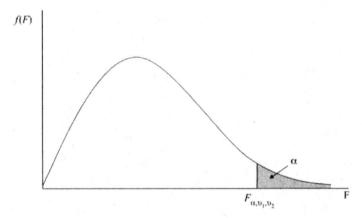

FIGURE 4-27   *F*-distribution.

Example 4-37 The chassis assembly time for a television set is observed for two operators. A random sample of 10 assemblies from the first operator gives an average assembly time of 22 minutes with a standard deviation of 3.5 minutes. A random sample of 8 assemblies from the second operator gives an average assembly time of 20.4 minutes with a standard deviation of 2.2 minutes. Find a 95% confidence interval for the ratio of the variances of the operators' assembly times.

Solution For this problem, $n_1 = 10$, $\bar{X}_1 = 22$, $s_1 = 3.5$, $n_2 = 8$, $\bar{X}_2 = 20.4$, and $s_2 = 2.2$. From Appendix A-6,

$$F_{0.025,9,7} = 4.82 \quad \text{and} \quad F_{0.025,7,9} = 4.20$$

Hence, a 95% confidence interval for the ratio of the variances of the assembly times is

$$\frac{(3.5)^2}{(2.2)^2}\left(\frac{1}{4.82}\right) \leq \frac{\sigma_1^2}{\sigma_2^2} \leq \frac{(3.5)^2}{(2.2)^2}(4.20)$$

or

$$0.525 \leq \frac{\sigma_1^2}{\sigma_2^2} \leq 10.630$$

Table 4-4 lists the formulas for the various confidence intervals and the assumptions required for each.

Hypothesis Testing

Concepts Determining whether claims on product or process parameters are valid is the aim of **hypothesis testing**. Hypothesis tests are based on sample data. A sample statistic used to test hypotheses is known as a **test statistic**. For example, the sample mean length could be a test statistic. Usually, rather than using a point estimate (like the sample mean, which is an estimator of the population mean), a standardized quantity based on the point estimate is found and used as the test statistic. For instance, either the normalized or standardized value of the sample mean could be used as the test statistic, depending on whether or not the population standard deviation is known.

If the population standard deviation is known, the normalized value of the sample mean is the *z-statistic*, given by

$$z = \frac{\bar{x} - \mu}{\sigma/\sqrt{n}}$$

If the population standard deviation is unknown, the standardized value of the sample mean is the *t-statistic*, given by

$$t = \frac{\bar{x} - \mu}{s/\sqrt{n}}$$

Now, how do we test a hypothesis? Suppose that the mean length of a part is expected to be 30 mm. We are interested in determining whether, for the month of March, the mean length differs from 30 mm. That is, we need to test this hypothesis. In any hypothesis-testing problem, there are two hypotheses: the **null hypothesis** H_o and the **alternative hypothesis** H_a. The null hypothesis represents the status quo, or the circumstance being tested (which is not rejected unless proven incorrect). The alternative hypothesis represents what we wish to

TABLE 4-4 Summary of Formulas for Confidence Intervals

Equation Number	Parameter	Assumptions	Two-Sided Confidence Interval
(4-55)	μ	σ^2 *known; n large*	$\bar{X} \pm z_{\alpha/2} \dfrac{\sigma}{\sqrt{n}}$
(4-57)	μ	σ^2 *unknown;* $X \sim N(\mu, \sigma^2)$	$\bar{X} \pm t_{\alpha/2, n-1} \dfrac{s}{\sqrt{n}}$
(4-58)	$\mu_1 - \mu_2$	σ_1^2, σ_2^2 *known;* n_1, n_2 *large*	$(\bar{X}_1 - \bar{X}_2) \pm z_{\alpha/2} \sqrt{\dfrac{\sigma_1^2}{n_1} + \dfrac{\sigma_2^2}{n_2}}$
(4-59), (4-60)	$\mu_1 - \mu_2$	σ_1^2, σ_2^2 *unknown* $X_1 \sim N(\mu_1, \sigma_1^2)$ $X_2 \sim N(\mu_2, \sigma_2^2)$ $\sigma_1^2 = \sigma_2^2$	$(\bar{X}_1 - \bar{X}_2) \pm t_{\alpha/2, n_1+n_2-2}\, s_p \sqrt{\dfrac{1}{n_1} + \dfrac{1}{n_2}}$, where $s_p^2 = \dfrac{(n_1-1)s_1^2 + (n_2-1)s_2^2}{n_1 + n_2 - 2}$
(4-61), (4-62)	$\mu_1 - \mu_2$	σ_1^2, σ_2^2 *unknown* $X_1 \sim N(\mu_1, \sigma_1^2)$ $X_2 \sim N(\mu_2, \sigma_2^2)$ $\sigma_1^2 \neq \sigma_2^2$	$(\bar{X}_1 - \bar{X}_2) \pm t_{\alpha/2, \nu} \sqrt{\dfrac{s_1^2}{n_1} + \dfrac{s_2^2}{n_2}}$ where $\nu = \dfrac{(s_1^2/n_1 + s_2^2/n_2)^2}{(s_1^2/n_1)^2/(n_1-1) + (s_2^2/n_2)^2/(n_2-1)}$
(4-63)	$\mu_d = \mu_1 - \mu_2$	Paired samples; $d_i = X_{1i} - X_{2i} \sim N$	$\bar{d} \pm t_{\alpha/2, n-1} \dfrac{s_d}{\sqrt{n}}$
(4-64)	p	$X \sim binomial(n, p);\ n$ *large*	$\hat{p} \pm z_{\alpha/2} \sqrt{\dfrac{\hat{p}(1-\hat{p})}{n}}$
(4-65)	$p_1 - p_2$	$X_1 \sim binomial\,(n_1, p_1);\ X_2 \sim binomial$ $(n_2, p_2);\ n_1, n_2$ *large*	$(\hat{p}_1 - \hat{p}_2) \pm z_{\alpha/2} \sqrt{\dfrac{\hat{p}_1(1-\hat{p}_1)}{n_1} + \dfrac{\hat{p}_2(1-\hat{p}_2)}{n_2}}$
(4-67)	σ^2	$X \sim N(\mu, \sigma^2)$	$\left[\dfrac{(n-1)s^2}{\chi^2_{\alpha/2, n-1}},\ \dfrac{(n-1)s^2}{\chi^2_{1-\alpha/2, n-1}} \right]$
(4-70)	$\dfrac{\sigma_1^2}{\sigma_2^2}$	$X_1 \sim N(\mu_1, \sigma_1^2);\ X_2 \sim N(\mu_2, \sigma_2^2)$	$\left[\dfrac{s_1^2}{s_2^2} \dfrac{1}{F_{\alpha/2, \nu_1, \nu_2}},\ \dfrac{s_1^2}{s_2^2} F_{\alpha/2, \nu_2, \nu_1} \right]$

prove or establish. It is formulated to contradict the null hypothesis. For the situation we have just described, the hypotheses are

$$H_0: \mu = 30$$
$$H_a: \mu \neq 30$$

where μ represents the mean length of the part. This is a **two-tailed test**; that is, the alternative hypothesis is designed to detect departures of a parameter from a specified value in both directions. On the other hand, if we were interested in determining whether the average length *exceeds* 30 mm, the hypotheses would be

$$H_0: \mu \leq 30$$
$$H_a: \mu > 30$$

This is a **one-tailed test**; that is, the alternative hypothesis detects departures of a parameter from a specified value in only one direction. If our objective were to find whether the average part length is less than 30 mm, the two hypotheses would be

$$H_0: \mu \geq 30$$
$$H_a: \mu < 30$$

This is also a one-tailed test.

In hypothesis testing, the null hypothesis is assumed to be true unless proven otherwise. Hence, if we wish to establish the validity of a certain claim, that claim must be formulated as the alternative hypothesis. If there is statistically significant evidence contradictory to the null hypothesis, the null hypothesis is rejected; otherwise, it is not rejected. Defining what is statistically significant will, of course, depend on what the decision maker deems tolerable. Say we wish to prove that the mean length is less than 30: that is,

$$H_0: \mu \geq 30$$
$$H_a: \mu < 30$$

We'll assume that the population standard deviation σ is 2 mm. We take a sample of size 36 and find the sample mean length to be 25 mm. Is this difference statistically significant?

Detailed expressions for test statistics will be given later, in accordance with the parameters for which hypothesis tests are being performed.

Figure 4-28 shows the sampling distribution of the sample mean $\bar{X}$ under the assumption that $\mu = 30$. According to the central limit theorem, the distribution of $\bar{X}$ will be approximately

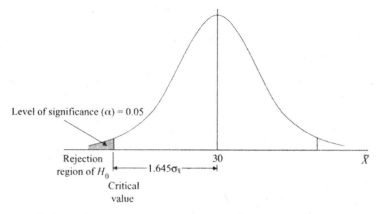

FIGURE 4-28 Sampling distribution of $\bar{X}$ assuming that $\mu = 30$.

normal for large sample sizes. The important question for our scenario is whether the sample mean length of 25 mm is significantly less statistically than the specified value of 30 mm. Can we reject the null hypothesis?

To determine this, we need a cutoff point beyond which the null hypothesis will be rejected. That is, how small must the sample mean be for us to conclude that the mean length is less than 30 mm? There is a critical value, in this case on the left tail, such that if the sample mean (or test statistic) falls below it, we will reject the null hypothesis. This value defines the **rejection region** of the null hypothesis. If the test statistic does not fall in the rejection region, we do not have significant evidence to conclude that the population mean is less than 30, so we will not reject the null hypothesis.

But how is the precise location of the critical value—and hence the rejection region—selected? How small must the sample mean be to be considered significantly less than 30? The answer to this question is influenced by the choice of the **level of significance** of the test. The rejection region is chosen such that if the null hypothesis is true, the probability of the test statistic falling in that region is small (say, 0.01 or 0.05); this probability is known as the *level of significance* and is denoted by α. Hence, the choice of α will dictate the rejection region.

Suppose that for a suitable choice of α (say, 0.05), the **critical value** is found to be $1.645\sigma_{\bar{x}}$ below the population mean of 30. (Details as to how to arrive at an expression for the critical value are given later.) The rejection region is then the shaded portion under the curve where the sample mean is at a distance more than $1.645\sigma_{\bar{x}}$ from the population mean, as shown in Figure 4-28. For our scenario, then,

$$\sigma_{\bar{x}} = \frac{\sigma}{\sqrt{n}} = \frac{2}{\sqrt{36}} = 0.333 \text{ mm}$$

So the critical value is 0.548 [(1.645)(0.333)] units below 30, and the rejection region is $\bar{X} < 29.452$ mm. If a smaller value of α were chosen, the rejection region would shift farther to the left.

For a given α, once the rejection region is selected, a framework for decision making in hypothesis testing is defined. Only if the test statistic falls in the rejection region will the null hypothesis be rejected. In our example, the rejection region was found to be $\bar{X} < 29.452$ mm, which is equivalent to $Z < -1.645$. The sample mean observed is 25 mm. The appropriate decision then is to reject the null hypothesis; that is, 25 mm is significantly less.

Errors in Hypothesis Testing   There are two types of errors in hypothesis testing: type I and type II. In a **type I error**, the null hypothesis is rejected when it is actually true. The probability of a type I error is indicated by α, the level of significance of the test. Thus, $\alpha = P$ (type I error) $= P$(rejecting $H_0 \mid H_0$ is true). For example, in testing (H_0: $\mu \geq 30$) against (H_a: $\mu < 30$), suppose that a random sample of 36 parts yields a sample average length of 28 mm when the true mean length of all parts is really 30 mm. If our rejection region is $\bar{X} < 29.542$, we must reject the null hypothesis. The magnitude of such an error can be controlled by selecting an acceptable level of α.

In a **type II error**, the null hypothesis is not rejected even though it is false. The probability of a type II error is denoted by β. Thus, $\beta = P$(type II error) $= P$(not rejecting $H_0 \mid H_0$ is false). For example, let's test (H_0: $\mu \geq 30$) against (H_a: $\mu < 30$) with a rejection region of $\bar{X} < 29.452$. Now, suppose that the true population mean length of all parts is 28 mm and a sample of 36 parts yields a sample mean of 29.8 mm. In this case, we do not reject the null hypothesis (because 29.8 does not lie in the region $\bar{X} < 29.452$). This is a type II error.

Calculating the probability of a type II error requires information about the population parameter (or at least an assumption about it). In such instances, we predict the probability of a type II error based on the actual or assumed parameter value; this prediction serves as a measure of the goodness of the testing procedure and the acceptability of the chosen rejection region. The values of α and β are inversely related. If all other problem parameters remain the same, β will decrease as α increases, and vice versa. Increasing the sample size can reduce both α and β.

The **power** of a test is the complement of β and is defined as

$$\text{power} = 1 - \beta = P(\text{rejecting } H_0 | H_0 \text{ is false})$$

The power is the probability of correctly rejecting a null hypothesis that is false. Obviously, tests with high powers are the most desirable.

Steps in Hypothesis Testing   In hypothesis testing, different formulas are used with different parameters (such as the population mean or difference between two population means). For each situation, the appropriate test statistic is based on an estimator of the population parameter, and the rejection region is found accordingly. The following steps summarize the *hypothesis-testing procedure*:

Step 1: Formulate the null and alternative hypotheses.

Step 2: Determine the test statistic.

Step 3: Determine the rejection region of the null hypothesis based on a chosen level of significance α.

Step 4: Make the decision. If the test statistic lies in the rejection region, reject the null hypothesis. Otherwise, do not reject the null hypothesis.

In an alternative procedure, the rejection region is not specifically found, although a chosen level of significance α is given. Upon determining the test statistic, the probability of obtaining that value (or an even more extreme value) for the test statistic, assuming that the null hypothesis is true, is computed. This is known as the *probability value* or the ***p*-value** associated with the test statistic. This *p*-value, also known as the **observed level of significance**, is then compared to α, the chosen level of significance. If the *p*-value is smaller than α, the null hypothesis is rejected.

Let's reconsider the mean part length example (see Figure 4-28). Suppose that the observed sample mean $(\bar{X})$ is 25, for a sample of size 36. The standard deviation of the sample mean, $\sigma_{\bar{x}}$, is $2/\sqrt{36} = 0.333$ mm (assuming a population standard deviation of 2). The observed sample mean of 25 is 5 less than the population mean of 30, which corresponds to $(25-30)/0.333 = -15.015$ standard deviations away from the population mean. The probability of observing a sample mean of 25 or less represents the *p*-value and is found using the standard normal table (Appendix A-3):

$$p - \text{value} = P(\bar{X} \leq 25)$$
$$= P(Z \leq -15.015) \simeq 0.0000$$

Therefore, if the chosen level of significance α is 0.05, the *p*-value is essentially zero (which is less than α), so we reject H_0. This means that if the null hypothesis is true, the chance of observing an average of 25 or something even more extreme is highly unlikely. Therefore, since we observed a sample mean of 25, we would be inclined to conclude that the null hypothesis must not be true, and we would therefore reject it.

Hypothesis Testing of the Mean

1. *Variance known.* For this situation, we assume that the sample size is large (allowing the central limit theorem to hold) or that the population distribution is normal. The appropriate test statistics and rejection regions are as follows:

Hypotheses: $H_0: \mu = \mu_0$ $H_0: \mu \leq \mu_0$ $H_0: \mu \geq \mu_0$
$H_a: \mu \neq \mu_0$ $H_a: \mu > \mu_0$ $H_a: \mu < \mu_0$

Rejection region: $|z_0| > z_{\alpha/2}$ $z_0 > z_\alpha$ $z_0 < -z_\alpha$

Test statistic: $z_0 = \dfrac{\bar{X} - \mu_0}{\sigma/\sqrt{n}}$ (4-71)

The steps for testing the hypothesis are the same four steps that were described previously. Let α denote the chosen level of significance and z_α denote the axis point of the standard normal distribution such that the right-tail area is α.

2. *Variance unknown.* In this situation we assume that the population distribution is normal. If the sample size is large ($n \geq 30$), slight departures from normality do not strongly influence the test. The notation refers to t-distributions as described in the section on interval estimation.

Hypotheses: $H_0: \mu = \mu_0$ $H_0: \mu \leq \mu_0$ $H_0: \mu \geq \mu_0$
$H_a: \mu \neq \mu_0$ $H_a: \mu > \mu_0$ $H_a: \mu < \mu_0$

Rejection region: $|t_0| > t_{\alpha/2, n-1}$ $t_0 > t_{\alpha, n-1}$ $t_0 < -t_{\alpha, n-1}$

Test statistic: $t_0 = \dfrac{\bar{X} - \mu_0}{s/\sqrt{n}}$ (4-72)

Example 4-38   In Example 4-31, the mean temperature at which metallic glass becomes brittle was of interest. Now suppose we would like to determine whether this mean temperature exceeds 595 °C. A random sample of 20 is taken, yielding a sample mean $\bar{X}$ of 600 °C and a sample standard deviation s of 15 °C. Use a level of significance α of 0.05.

Solution   The hypotheses are

$$H_0: \mu \leq 595$$
$$H_a: \mu > 595$$

The test statistic is

$$t_0 = \frac{600 - 595}{15/\sqrt{20}} = 1.491$$

From Appendix A-4, $t_{0.05, 19} = 1.729$. The rejection region is therefore $t_0 > 1.729$. since the test statistic t_0 does not lie in the rejection region, we do not reject the null hypothesis.

Thus, even though the sample mean is $600\,^\circ\text{C}$, the 5% level of significance does not allow us to conclude that there is statistically significant evidence that the mean temperature exceeds $595\,^\circ\text{C}$.

Hypothesis Testing of the Correlation Coefficient   Assume that we have bivariate normal data on two variables X and Y and a sample of n observations, (X_i, Y_i), $i = 1, 2, \ldots, n$. Let the population correlation coefficient between X and Y be denoted by ρ, for which an estimator is the sample correlation coefficient r.

Hypotheses:
$$
\begin{array}{ccc}
H_0: \rho = 0 & H_0: \rho \leq 0 & H_0: \rho \geq 0 \\
H_a: \rho \neq 0 & H_a: \rho > 0 & H_a: \rho < 0
\end{array}
$$

Rejection region: $|t_0| > t_{\alpha/2, n-2}$ $\quad t_0 > t_{\alpha, n-2}$ $\quad t_0 < -t_{\alpha, n-2}$

Test statistic:
$$
t_0 = \frac{r\sqrt{n-2}}{\sqrt{1-r^2}} \tag{4-73}
$$

Hypothesis Testing for the Difference Between Two Means

1. *Variances known.* In this situation, we assume that the sample sizes are large enough for the central limit theorem to hold. However, if the population distribution is normal, the test statistic as shown will be valid for any sample size.

Hypotheses:
$$
\begin{array}{ccc}
H_0: \mu_1 - \mu_2 = \mu_0 & H_0: \mu_1 - \mu_2 \leq \mu_0 & H_0: \mu_1 - \mu_2 \geq \mu_0 \\
H_a: \mu_1 - \mu_2 \neq \mu_0 & H_a: \mu_1 - \mu_2 > \mu_0 & H_a: \mu_1 - \mu_2 < \mu_0
\end{array}
$$

Rejection region: $|z_0| > z_{\alpha/2}$ $\quad\quad z_0 > z_\alpha$ $\quad\quad z_0 < -z_\alpha$

Test statistic:
$$
z_0 = \frac{(\bar{X}_1 - \bar{X}_2) - \mu_0}{\sqrt{\sigma_1^2/n_1 + \sigma_2^2/n_2}} \tag{4-74}
$$

Example 4-39   The owner of a local logging operation wants to examine the average unloading time of logs. Two methods are used for unloading. A random sample of size 40 for the first method gives an average unloading time $\bar{X}_1$ of 20.5 minutes. A random sample of size 50 for the second method yields an average unloading time $\bar{X}_2$ of 17.6 minutes. We know that the variance of the unloading times using the first method is 3, while that for the second method is 4. At a significance level α of 0.05, can we conclude that there is a difference in the mean unloading times for the two methods?

Solution   The hypotheses are

$$
\begin{array}{c}
H_0: \mu_1 - \mu_2 = 0 \\
H_a: \mu_1 - \mu_2 \neq 0
\end{array}
$$

The test statistic is

$$
z_0 = \frac{(20.5 - 17.6) - 0}{\sqrt{3/40 + 4/50}} = 7.366
$$

From Appendix A-3, $z_{0.025} = 1.96$. The critical values are ± 1.96, and the rejection region is $|z_0| > 1.96$. Since the test statistic z_0 lies in the rejection region, we reject the null hypothesis and conclude that there is a difference in the mean unloading times for the two methods.

2. *Variances unknown.* Here we assume that each population is normally distributed. If we assume that the population variances, though unknown, are equal (i.e., $\sigma_1^2 = \sigma_2^2$), we get the following:

Hypotheses: $H_0: \mu_1 - \mu_2 = \mu_0$ $H_0: \mu_1 - \mu_2 \leq \mu_0$ $H_0: \mu_1 - \mu_2 \geq \mu_0$
$H_a: \mu_1 - \mu_2 \neq \mu_0$ $H_a: \mu_1 - \mu_2 > \mu_0$ $H_a: \mu_1 - \mu_2 < \mu_0$

Rejection region: $|t_0| > t_{\alpha/2, n_1 + n_2 - 2}$ $t_0 > t_{\alpha, n_1 + n_2 - 2}$ $t_0 < -t_{\alpha, n_1 + n_2 - 2}$

Test statistic:

$$t_0 = \frac{(\bar{X}_1 - \bar{X}_2) - \mu_0}{s_p \sqrt{1/n_1 + 1/n_2}} \tag{4-75}$$

Note: s_p^2 is given by eq. (4-60).

If the population variances cannot be assumed to be equal ($\sigma_1^2 \neq \sigma_2^2$), we have the following:

Hypotheses: $H_0: \mu_1 - \mu_2 = \mu_0$ $H_0: \mu_1 - \mu_2 \leq \mu_0$ $H_0: \mu_1 - \mu_2 \geq \mu_0$
$H_a: \mu_1 - \mu_2 \neq \mu_0$ $H_a: \mu_1 - \mu_2 > \mu_0$ $H_a: \mu_1 - \mu_2 < \mu_0$

Rejection region: $|t_0| > t_{\alpha/2, \nu}$ $t_0 > t_{\alpha, \nu}$ $t_0 < -t_{\alpha, \nu}$

Test statistic:

$$t_0 = \frac{(\bar{X}_1 - \bar{X}_2) - \mu_0}{\sqrt{s_1^2/n_1 + s_2^2/n_2}} \tag{4-76}$$

Note: ν is given eq. (4-62)

Example 4-40 A large corporation is interested in determining whether the average days of sick leave taken annually is more for night-shift employees than for day-shift employees. It is assumed that the distribution of the days of sick leave is normal for both shifts and that the variances of sick leave taken are equal for both shifts. A random sample of 12 employees from the night shift yields an average sick leave $\bar{X}_1$ of 16.4 days with a standard deviation s_1 of 2.2 days. A random sample of 15 employees from the day shift yields an average sick leave $\bar{X}_2$ of 12.3 days with a standard deviation s_2 of 3.5 days. At a level of significance α of 0.05, can we conclude that the average sick leave for the night shift exceeds that in the day shift?

Solution   The hypotheses are

$$H_0: \ \mu_1 - \mu_2 \leq 0$$
$$H_a: \ \mu_1 - \mu_2 > 0$$

The pooled estimate of the variance, s_p^2, from eq. (4-60) is

$$s_p^2 = \frac{(11)(2.2)^2 + (14)(3.5)^2}{25} = 8.990$$

So $s_p = \sqrt{8.990} = 2.998$. The test statistic is

$$t_0 = \frac{(16.4 - 12.3) - 0}{2.998\sqrt{1/12 + 1/15}} = 3.531$$

From Appendix A-4, $t_{0.05,25} = 1.708$. Since the test statistic t_0 exceeds 1.708 and falls in the rejection region, we reject the null hypothesis and conclude that the average sick leave for the night shift exceeds that for the day shift.

Let us now use Minitab to test the hypotheses. Click on **Stat > Basic Statistics > 2-Sample t**. Since we are not given the individual observations in each sample, select **Summarized data**. Assuming the **First** sample corresponds to the night shift and the **Second** corresponds to the day shift, input the values of **Sample size, Mean**, and **Standard deviation** for both samples. Check the box **Assume equal variances**, and click on **Options**. Input for **Confidence level** the value 0.95, **Test difference** the value 0, and select for **Alternative hypothesis greater than**. Click **OK**.

Figure 4-29 shows the output from Minitab, with the point estimate of the difference in the means as 4.1. The test statistic is the t-value of 3.53 with 25 degrees of freedom, and the pooled standard deviation is 2.9983. Using the concept of p-value, shown as 0.001, we reject the null hypothesis since the p-value $< \alpha = 0.05$.

3. *Paired samples*. Here, the samples from the two populations are paired and so are not independent. The differences between the paired observations are assumed to be normally distributed.

Hypotheses:	H_0: $\mu_d = \mu_1 - \mu_2 = \mu_0$	H_0: $\mu_d \leq \mu_0$	H_0: $\mu_d \geq \mu_0$
	H_a: $\mu_d = \mu_1 - \mu_2 \neq \mu_0$	H_a: $\mu_d > \mu_0$	H_a: $\mu_d < \mu_0$
Rejection region:	$|t_0| > t_{\alpha/2, n-1}$	$t_0 > t_{\alpha, n-1}$	$t_0 < -t_{\alpha, n-1}$

Test statistic: $t_0 = \dfrac{\bar{d} - \mu_0}{s_d/\sqrt{n}}$ (4-77)

Hypothesis Testing for a Proportion   The assumption here is that the number of trials n in a binomial experiment is large, that $np \geq 5$, and that $n(1-p) \geq 5$. This allows the distribution of the sample proportion of successes ($\hat{p}$) to approximate a normal distribution.

```
Difference = mu (1) - mu (2)
Estimate for difference:  4.10000
95% lower bound for difference:  2.11647
T-Test of difference = 0 (vs >): T-Value = 3.53  P-Value = 0.001  DF = 25
Both use Pooled StDev = 2.9983
```

FIGURE 4-29   Hypothesis testing on difference of means using Minitab.

Hypotheses:	$H_0: p = p_0$	$H_0: p \leq p_0$	$H_0: p \geq p_0$
	$H_a: p \neq p_0$	$H_a: p > p_0$	$H_a: p < p_0$

Rejection region:	$\lvert z_0 \rvert > z_{\alpha/2}$	$z_0 > z_\alpha$	$z_0 < -z_\alpha$

Test statistic:
$$z_0 = \frac{\hat{p} - p_0}{\sqrt{p_0(1 - p_0)/n}} \tag{4-78}$$

Example 4-41   The timeliness with which due dates are met is an important factor in maintaining customer satisfaction. A medium-sized organization wants to test whether the proportion of times that it does not meet due dates is less than 6%. Based on a random sample of 100 customer orders, they found that they missed the due date five times. What is your conclusion? Test at a level of significance α of 0.05.

Solution   The hypotheses are

$$H_0: p \geq 0.06$$
$$H_a: p < 0.06$$

The test statistic is

$$z_0 = \frac{0.05 - 0.06}{\sqrt{(0.06)(0.94)/100}} = -0.421$$

From Appendix A-3, $z_{0.05} = 1.645$. Since the test statistic z_0 is not less than -1.645, it does not lie in the rejection region. Hence, we do not reject the null hypothesis. At the 5% level of significance, we cannot conclude that the proportion of due dates missed is less than 6%.

Hypothesis Testing for the Difference Between Two Binomial Proportions   Here we assume that the sample sizes are large enough to allow a normal distribution for the difference between the sample proportions. Also, we consider the case for the null hypothesis, where the difference between the two proportions is zero. For a treatment of other cases, such as the hypothesized difference between two proportions being 3%, where the null hypothesis is given by $H_0: p_1 - p_2 = 0.03$, consult Duncan (1986) and Mendenhall et al.(1993).

Hypotheses:	$H_0: p_1 - p_2 = 0$	$H_0: p_1 - p_2 \leq 0$	$H_0: p_1 - p_2 \geq 0$
	$H_a: p_1 - p_2 \neq 0$	$H_a: p_1 - p_2 > 0$	$H_a: p_1 - p_2 < 0$

Rejection region:	$\lvert z_0 \rvert > z_{\alpha/2}$	$z_0 > z_\alpha$	$z_0 < -z_\alpha$

Test statistic:
$$z_0 = \frac{\hat{p}_1 - \hat{p}_2}{\sqrt{\hat{p}(1 - \hat{p})(1/n_1 + 1/n_2)}} \tag{4-79}$$

Note: $\hat{p} = (n_1\hat{p}_1 + n_2\hat{p}_2)/(n_1 + n_2)$ is the pooled estimate of the proportion of nonconforming items.

Example 4-42 A company is interested in determining whether the proportion of non-conforming items is differenet for two of its vendors. A random sample of 100 items from the first vendor revealed 4 nonconforming items. A random sample of 200 items from the second vendor showed 10 nonconforming items. What is your conclusion? Test at a level of significance α of 0.05.

Solution The hypotheses are

$$H_0: p_1 - p_2 = 0$$
$$H_a: p_1 - p_2 \neq 0$$

The pooled estimate of the proportion of nonconforming items is

$$\hat{p} = \frac{(100)(0.04) + (200)(0.05)}{300} = 0.047$$

The test statistic is

$$z_0 = \frac{0.04 - 0.05}{\sqrt{(0.047)(0.953)(1/100 + 1/200)}} = -0.386$$

From Appendix A-3, $z_{0.025} = 1.96$. Since the test statistic z_0 does not lie in the rejection region, we do not reject the null hypothesis. We cannot conclude that the proportion of nonconforming items between the two vendors differs.

Hypothesis Testing for the Variance Assume here that the population distribution is normal.

Hypotheses:
$$H_0: \sigma^2 = \sigma_0^2 \qquad H_0: \sigma^2 \leq \sigma_0^2 \qquad H_0: \sigma^2 \geq \sigma_0^2$$
$$H_a: \sigma^2 = \sigma_0^2 \qquad H_a: \sigma^2 > \sigma_0^2 \qquad H_a: \sigma^2 < \sigma_0^2$$

Rejection region:
$$\chi_0^2 > \chi_{\alpha/2, n-1}^2 \qquad \chi_0^2 > \chi_{\alpha, n-1}^2 \qquad \chi_0^2 < \chi_{1-\alpha, n-1}^2$$
$$\text{or} \quad \chi_0^2 < \chi_{1-\alpha/2, n-1}^2$$

Test statistic:
$$\chi_0^2 = \frac{(n-1)s^2}{\sigma_0^2} \tag{4-80}$$

Example 4-43 The variability of the time to be admitted in a health care facility is of concern. A random sample of 15 patients shows a mean time to admission $\bar{X}$ of 2.2 hours with a standard deviation s of 0.2 hours. Can we conclude that the variance of time to admission is less is than 0.06? Use a level of significance α of 0.01.

Solution The hypotheses are

$$H_0: \sigma^2 \geq 0.06$$
$$H_a: \sigma^2 < 0.06$$

The test statistic is

$$\chi_0^2 = \frac{(14)(0.2)^2}{0.06} = 9.333$$

From Appendix A-5, $\chi^2_{0.99,14} = 4.66$. The test statistic value of 9.333 is not less than 4.66 and so does not lie in the rejection region. Hence, we do not reject the null hypothesis. At the 1% level of significance, we cannot conclude that the variance of time to admission is less than 0.06.

Hypothesis Testing for the Ratio of Two Variances   We assume that both populations are normally distributed.

Hypotheses:

$$H_0: \sigma_1^2 = \sigma_2^2 \qquad H_0: \sigma_1^2 \leq \sigma_2^2 \qquad H_0: \sigma_1^2 \geq \sigma_2^2$$

$$H_a: \sigma^2 \neq \sigma_2^2 \qquad H_a: \sigma_1^2 > \sigma_2^2 \qquad H_a: \sigma_1^2 < \sigma_2^2$$

Rejection region:

$$F_0 > F_{\alpha/2,v_1,v_2} \text{ or } \quad F_0 > F_{\alpha,v_1,v_2} \qquad F_0 < F_{1-\alpha,v_1,v_2}$$

$$F_0 < F_{1-\alpha/2,v_1,v_2}$$

Test statistic: $\qquad F_0 = \dfrac{s_1^2}{s_2^2}$ $\qquad\qquad\qquad$ (4-81)

Example 4-44   The variabilities of the service times of two bank tellers are of interest. Their supervisor wants to determine whether the variance of service time for the first teller is greater than that for the second. A random sample of 8 observations from the first teller yields a sample average $\bar{X}_1$ of 3.4 minutes with a standard deviation s_1 of 1.8 minutes. A random sample of 10 observations from the second teller yields a sample average $\bar{X}_2$ of 2.5 minutes with a standard deviation of 0.9 minutes. Can we conclude that the variance of the service time is greater for the first teller than for the second? Use a level of significance α of 0.05.

Solution   The hypotheses are

$$H_0: \sigma_1^2 \leq \sigma_2^2$$
$$H_a: \sigma_1^2 > \sigma_2^2$$

The test statistic is

$$F = \frac{s_1^2}{s_2^2} = \frac{(1.8)^2}{(0.9)^2} = 4.00$$

From Appendix A-6, $F_{0.05,7,9} = 3.29$. The test statistic lies in the rejection region, so we reject the null hypothesis.

SUMMARY

In this chapter we presented the statistical foundations necessary for quality control and improvement. The procedures for summarizing data that describe product or process characteristics have been discussed. A review of common discrete and continuous probability distributions with applications in quality control has been included. We also presented inferential statistics that can be used for drawing conclusions as to product and process quality. In particular, the topics of estimation and hypothesis testing have been emphasized. The variety of statistical procedures presented are meant to serve as an overview. Technical details have been purposely limited.

APPENDIX: APPROXIMATIONS TO SOME PROBABILITY DISTRIBUTIONS

In some situations, if tables for the needed probability distributions are not available for the parameter values in question or if the calculations using the formula become tedious and prone to error due to round-offs, approximations for the probability distribution under consideration can be considered.

Binomial Approximation of the Hypergeometric

When the ratio of sample size to population size is small—that is n/N is small (≤ 0.1, as a rule of thumb)—the binomial distribution serves as a good approximation to the hypergeometric distribution. The parameter values to be used for the binomial distribution would be the same value of n as in the hypergeometric distribution, and $p = D/N$.

Example 4A-1 Consider a lot of 100 parts, of which 6 are nonconforming. If a sample of 4 parts is selected, what is the probability of obtaining 2 nonconforming items? If a binomial approximation is used, what is the required probability?

Solution Using the hypergeometric distribution, $N = 100$, $D = 6$, $n = 4$, and $x = 2$, we have

$$P(X = 2) = \frac{\binom{6}{2}\binom{94}{2}}{\binom{100}{4}} = 0.017$$

Note that $n/N = 0.04$, which is less than 0.1. Using the binomial distribution as an approximation to the hypergeometric, with $p = 6/100 = 0.06$, yields

$$P(X = 2) = \binom{4}{2}(0.06)^2(0.94)^2 = 0.019$$

Poisson Approximation to the Binomial

In a binomial distribution, if n is large and p is small ($p < 0.1$) such that np is constant, the Poisson distribution serves as a good approximation to the binomial. The parameter λ in the Poisson distribution is used as np. The larger the value of n and the smaller value of p, the better the approximation. As a rule of thumb, when $np < 5$, the approximation is acceptable.

Example 4A-2 A process is known to have a nonconformance rate of 0.02. If a random sample of 100 items is selected, what is the probability of finding 3 nonconforming items?

Solution Using the binomial distribution, $n = 100$, $p = 0.02$, and $x = 3$, we have

$$P(X = 3) = \binom{100}{3}(0.02)^3(0.98)^{97} = 0.182$$

Next, we use the Poisson distribution as an approximation to the binomial. Using $\lambda = (100)(0.02) = 2$ yields

$$P(X = 3) = \frac{e^{-2}(2)^3}{3!} = 0.180$$

Normal Approximation to the Binomial

In a binomial distribution, if n is large and p is close to 0.5, the normal distribution may be used to approximate the binomial. Usually, if p is neither too large nor too small ($0.1 \leq p \leq 0.9$), the normal approximation is acceptable when $np \geq 5$. A continuity correction factor is used by finding an appropriate z-value because the binomial distribution is discrete, whereas the normal distribution is continuous.

Example 4A-3   A process is known to produce about 6% nonconforming items. If a random sample of 200 items is chosen, what is the probability of finding between 6 and 8 nonconforming items?

Solution   If we decide to use the binomial distribution with $n = 200$, $p = 0.06$, we have

$$P(6 \leq X \leq 8) = P(X = 6) + P(X = 7) + P(X = 8)$$
$$= \binom{200}{6}(0.06)^6(0.94)^{194} + \binom{200}{7}(0.06)^7(0.94)^{193} + \binom{200}{8}(0.06)^8(0.94)^{192}$$
$$= 0.0235 + 0.0416 + 0.0641 = 0.1292$$

Using the normal distribution to approximate a binomial probability, the mean is given by $np = (200)(0.06) = 12$, and the variance is given by $np(1 - p) = (200)(0.06)(1 - 0.06) = 11.28$. The required probability is

$$P(6 \leq X \leq 8) \simeq P(5.5 \leq X \leq 8.5)$$

The 0.5 adjustment is often known as the *continuity correction factor*. The binomial random variable is discrete. When using a continuous random variable such as the normal to approximate it, this adjustment makes the approximation better. In other words, $P(X \leq 8)$ is written as $P(X \leq 8.5)$, while $P(X \geq 6)$ is written as $P(X \geq 5.5)$:

$$\text{requited probability} = P\left(\frac{5.5 - 12}{\sqrt{11.28}} \leq z \leq \frac{8.5 - 12}{\sqrt{11.28}}\right)$$
$$= P(-1.94 \leq z \leq -1.04)$$
$$= 0.1492 - 0.0262 = 0.1230$$

Normal Approximation to the Poisson

If the mean λ of a Poisson distribution is large ($\lambda \geq 10$), a normal distribution may be used to approximate it. The parameters of this normal distribution are the mean μ which is set equal to λ and the variance σ^2, which is also set equal to λ. As in Example 4A-3, the continuity correction factor of 0.5 may be used since a discrete random variable is approximated by a continuous one.

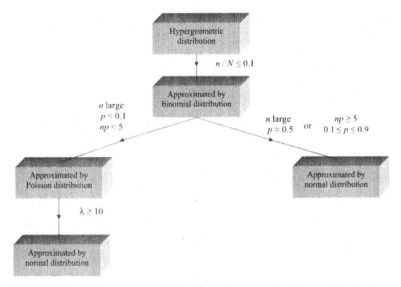

FIGURE 4A-1 Necessary conditions for using approximations to distributions.

Example 4A-4 The number of small businesses that fail each year is known to have a Poisson distribution with a mean of 16. Find the probability that in a given year there will be no more than 18 small business failures.

Solution Using the Poisson distribution ($\lambda = 16$), the required probability is $P(X \leq 18)$. From the table in Appendix A-2, this probability is 0.742. When using the normal distribution as an approximation, the mean and variance are set equal to λ. The 0.5 continuity correction factor is used in a similar manner as in Example 4A-3.

$$P(X \leq 18) \simeq P(X \leq 18.5)$$

$$= P\left(z \leq \frac{18.5 - 16}{\sqrt{16}} \right)$$

$$= P(z \leq 0.625) = 0.7340 \text{ (using linear interpolation)}$$

Figure 4A-1 summarizes the conditions under which the various approximations to these common distributions may be used.

KEY TERMS

accuracy	chi-squared distribution
additive law	confidence interval
alternative hypothesis	continuous variable
association, measure of	correlation coefficient
binomial distribution	critical value
calibration	cumulative distribution function
central limit theorem	data collection
central tendency, measures of	degrees of freedom

descriptive statistics
discrete variable
dispersion, measures of
distribution
   continuous
   discrete
estimation
   interval
   point
events
   complementary
   compound
   independent
   mutually exclusive
   simple
expected value
exponential distribution
F-distribution
gamma distribution
gamma function
hypergeometric distribution
hypothesis testing
   one-tailed
   two-tailed
inferential statistics
interquartile range
interval estimation
kurtosis
kurtosis coefficient
level of significance
lognormal distribution
mean
   population
   sample
measurement scales
   interval
   nominal
   ordinal
   ratio
median

misspecification
mode
multiplicative law
normal distribution
   standard normal
null hypothesis
outlier
p-value
paired samples
parameter
   scale
   shape
point estimation
Poisson distribution
population
power
precision
probability
probability density function
probability distribution function
random variable
range
rejection region
sample
sample space
sampling distribution
skewness coefficient
standard deviation
statistic
statistics
   descriptive
   inferential
t-distribution
test statistic
trimmed mean
type I error
type II error
unbiased
variance
Weibull distribution

EXERCISES

Discussion Questions

4-1   The development of new drugs has to undergo stringent regulations resulting in long lead times and high research and development costs. Suppose that you are in charge of the federal agency that grants drug approvals and you wish to be convinced that a new drug is better than an existing one in terms of increasing the average life of patients.

(a) What are appropriate null and alternative hypotheses?

(b) Explain what type I and II errors are in this context, and discuss their relative seriousness.

(c) How might you reduce the chances of these errors?

(d) What assumptions do you need to make to test the hypothesis in part (a)? How would you test the assumptions? Note that distributional assumptions will be tested in Chapter 5.

4-2   Customers arrive at a department store, randomly and independently.

(a) What is an appropriate distribution for modeling the number of customers that arrive in a 2-hour period?

(b) Under what situations might the stated assumptions not hold?

(c) What information would you need to collect to estimate the probability distribution?

(d) Suppose that a new location is being contemplated for the store. Explain how you would estimate the probability distribution.

4-3   Find an expression for the probability of the union of three events that are mutually independent of each other.

4-4   Explain the difference between accuracy and precision of measurements. How do you control for accuracy? What can you do about precision?

4-5   Explain the different types of measurement scales, and give examples in the following situations:

(a) Gallons of water to put out a fire by the fire department

(b) Response time of an ambulance

(c) Test score in an examination

(d) Customer product preference expressed on a scale of 1 to 5

(e) Categorizing employees in groups based on county of residence

(f) Performance appraisal of employees through placement in categories

4-6   Distinguish between the use of the mean, median, and mode in quality control applications. When do you prefer to use the trimmed mean?

4-7   State the null and alternative hypotheses in the following situations by defining the parameters used. Also, state any assumptions that you need to make to conduct the test:

(a) The Postal Service wishes to prove that the mean delivery time for packages is less than 5 days.

(b) A financial institution believes that it has an average loan processing time of less than 10 days.

(c) A marketing firm believes that the average contract for a customer exceeds $50,000.

(d) A Web-order company wishes to test if it has improved its efficiency of operations by reducing its average response time.

(e) A manufacturer of consumer durables believes that over 70% of its customers are satisfied with the product.

4-8   Distinguish between a hypergeometric and a binomial random variable.

4-9   A 95% confidence interval for the mean thickness of a part in millimeters is (10.2, 12.9). Interpret this interval.

4-10   Refer to Exercise 4-7. For each situation, define a type I and a type II error in the appropriate context. Consider the costs of such errors and discuss the implications.

4-11   Consider the price of homes in a large metropolitan area. What kind of distribution would you expect in terms of skewness and kurtosis? As an index, would the mean or median price be representative? What would the interquartile range indicate?

4-12   With increased air travel, the training of air traffic controllers is vital. However, of those who enter the program, several drop out, for a variety of reasons. In the past, 70% have completed the program. A new program is being developed at significant cost.
   (a) To test the effectiveness of the new program, what set of hypotheses would you test?
   (b) Discuss type I and II errors and their relative consequences.
   (c) Are there other factors to be considered that might improve the outcome? What data would you collect and analyze to validate your answer?

4-13   For each of the following areas, define appropriate quality characteristic(s) and parameters and indicate the hypotheses that you would test:
   (a) Effectiveness of a hospital in satisfying patients, employees, and shareholders (note that different hypotheses may apply for each category)
   (b) Efficiency of a health care facility
   (c) Effectiveness of a call center
   (d) New product development time and costs in the semiconductor industry

Problems

4-14   Based on historical data, it is estimated that 12% of new products will obtain a profitable market share. However, if two products are newly introduced in the same year, there is only a 5% chance of both products becoming profitable. A company is planning to market two new products, 1 and 2, this coming year. What is the probability of
   (a) only product 1 becoming profitable?
   (b) only product 2 becoming profitable?
   (c) at least one of the products becoming profitable?
   (d) neither product becoming profitable?
   (e) either product 1 or product 2 (but not both) becoming profitable?
   (f) product 2 becoming profitable if product 1 is found to be profitable?

4-15   Two types of defects are observed in the production of integrated circuit boards. It is estimated that 6% of the boards have solder defects and 3% have some surface-finish defects. The occurrences of the two types of defects are assumed to be independent of each other. If a circuit board is randomly selected from the output, find the probabilities for the following situations:
   (a) Either a solder defect or a surface-finish defect or both is found.
   (b) Only a solder defect is found.

(c) Both types of defects are found.

(d) The board is free of defects.

(e) If a surface-finish defect is found, what are the chances of also finding a solder defect?

4-16 The settling of unwanted material in a mold is causing some defects in the output. Based on recent data, it is estimated that 5% of the output has one or more defects. In spot checking some parts, an inspector randomly selects two parts. Find the probabilities that:

(a) the first part is defect-free.

(b) the second part is defect-free.

(c) both parts are defect-free.

(d) one of the parts is acceptable.

(e) at least one part is acceptable.

4-17 The following times (in minutes) to process hot-rolled steel are observed for a random sample of size 10 as follows:

$$5.4 \quad 6.2 \quad 7.9 \quad 4.8 \quad 7.5$$
$$6.2 \quad 5.5 \quad 4.5 \quad 7.2 \quad 6.2$$

(a) Find the mean, median, and mode of the processing times. Interpret the differences between them.

(b) Compute the range, variance, and standard deviation of the processing times, and interpret them.

4-18 A pharmaceutical company making antibiotics has to abide by certain standards set by the Food and Drug Administration. The company performs some testing on the strength of the antibiotic. In a case of 25 bottles, 4 bottles are selected for testing. If the case actually contains 5 understrength bottles, what is the probability that the sample chosen will contain no understrength bottles? Exactly 1 understrength bottle? How many understrength bottles would be expected in the sample? What is the standard deviation of understrength bottles in the sample?

4-19 A company involved in making solar panels estimates that 3% of its product is nonconforming. If a random sample of 5 items is selected from the production output, what is the probability that none are nonconforming? That 2 are nonconforming? The cost of rectifying a nonconforming panel is estimated to be $5. For a shipment of 1000 panels, what is the expected cost of rectification?

4-20 A university has purchased a service contract for its computers and pays $20 annually for each computer. Maintenance records show that 8% of the computers require some sort of servicing during the year. Furthermore, it is estimated that the average expenses for each repair, had the university not been covered by the service contract, would be about $200. If the university currently has 20 computers, would you advise buying the service contract? Based on expected costs, for what annual premium per computer will the university be indifferent to purchasing the service contract? What is the probability of the university spending no more than $500 annually on repairs if it does not buy the service contract?

4-21   The probability of an electronic sensor malfunctioning is known to be 0.10. A random sample of 12 sensors is chosen. Find the probability that:
(a) at least 3 will malfunction.
(b) no more than 5 will malfunction.
(c) at least 1 but no more than 5 will malfunction.
(d) What is the expected number of sensors that will malfunction?
(e) What is the standard deviation of the number of sensors that will malfunction?

4-22   A process is known to produce 5% nonconforming items. A sample of 40 items is selected from the process.
(a) What is the distribution of the nonconforming items in the sample?
(b) Find the probability of obtaining no more than 3 nonconforming items in the sample.
(c) Using the Poisson distribution as an approximation to the binomial, calculate the probability of the event in part (b).
(d) Compare the answers to parts (b) and (c). What are your observations?

4-23   The guidance system design of a satellite places several components in parallel. The system will function as long as at least one of the components is operational. In a particular satellite, 4 such components are placed in parallel. If the probability of a component operating successfully is 0.9, what is the probability of the system functioning? What is the probability of the system failing? Assume that the components operate independently of each other.

4-24   In a lot of 200 electrical fuses, 20 are known to be nonconforming. A sample of 10 fuses is selected.
(a) What is the probability distribution of the number of nonconforming fuses in the sample? What are its mean and standard deviation?
(b) Using the binomial distribution as an approximation to the hypergeometric, find the probability of getting 2 nonconforming fuses. What is the probability of getting at most 2 nonconforming fuses?

4-25   A local hospital estimates that the number of patients admitted daily to the emergency room has a Poisson probability distribution with a mean of 4.0. What is the probability that on a given day:
(a) only 2 patients will be admitted?
(b) at most 6 patients will be admitted?
(c) no one will be admitted?
(d) What is the standard deviation of the number of patients admitted?
(e) For each patient admitted, the expected daily operational expenses to the hospital are $800. If the hospital wants to be 94.9% sure of meeting daily expenses, how much money should it retain for operational expenses daily?

4-26   In an auto body shop, it is known that the average number of paint blemishes per car is 3. If 2 cars are randomly chosen for inspection, what is the probability that:
(a) the first car has no more than 2 blemishes?
(b) each of the cars has no more than 2 blemishes?
(c) the total number of blemishes in both of the cars combined is no more than 2?

4-27 The number of bank failures per year among those insured by the Federal Deposit Insurance Company has a mean of 7.0. The failures occur independently. What is the probability that:
(a) there will be at least 4 failures in the coming year?
(b) there will be between 2 and 8 failures, inclusive, in the coming year?
(c) during the next two years there will be at most 8 failures?

4-28 The outside diameter of a part used in a gear assembly is known to be normally distributed with a mean of 40 mm and standard deviation of 2.5 mm. The specifications on the diameter are (36, 45), which means that part diameters between 36 and 45 mm are considered acceptable. The unit cost of rework is $0.20, while the unit cost of scrap is $0.50. If the daily production rate is 2000, what is the total daily cost of rework and scrap?

4-29 The breaking strength of a cable is known to be normally distributed with a mean of 4000 kg and a standard deviation of 25 kg. The manufacturer prefers that at least 95% of its product meet a strength requirement of 4050 kg. Is this requirement being met? If not, by changing the process parameter, what should the process mean target value be?

4-30 The specifications for the thickness of nonferrous washers are 1.0 ± 0.04 mm. From process data, the distribution of the washer thickness is estimated to be normal with a mean of 0.98 mm and a standard deviation of 0.02 mm. The unit cost of rework is $0.10, and the unit cost of scrap is $0.15. For a daily production of 10,000 items:
(a) What proportion of the washers is conforming? What is the total daily cost of rework and scrap?
(b) In its study of constant improvement, the manufacturer changes the mean setting of the machine to 1.0 mm. If the standard deviation is the same as before, what is the total daily cost of rework and scrap?
(c) The manufacturer is trying to improve the process and reduces its standard deviation to 0.015 mm. If the process mean is maintained at 1.0 mm, what is the percent decrease in the total daily cost of rework and scrap compared to that of part (a)?

4-31 A company has been able to restrict the use of electrical power through energy conservation measures. The monthly use is known to be normal with a mean of 60,000 kWh (kilowatt-hour) and a standard deviation of 400 kWh.
(a) What is the probability that the monthly consumption will be less than 59,100 kWh?
(b) What is the probability that the monthly consumption will be between 59,000 and 60,300 kWh?
(c) The capacity of the utility that supplies this company is 61,000 kWh. What is the probability that demand will not exceed supply by more than 100 kWh?

4-32 A component is known to have an exponential time-to-failure distribution with a mean life of 10,000 hours.
(a) What is the probability of the component lasting at least 8000 hours?
(b) If the component is in operation at 9000 hours, what is the probability that it will last another 6000 hours?
(c) Two such components are put in parallel, so that the system will be in operation if at least one of the components is operational. What is the probability of the

system being operational for 12,000 hours? Assume that the components operate independently.

4-33   The time to repair an equipment is known to be exponentially distributed with a mean of 45 min.
   (a) What is the probability of the machine being repaired within half an hour?
   (b) If the machine breaks down at 3 P.M. and a repairman is available immediately, what is the probability of the machine being available for production by the start of the next day? Assume that the repairman is available until 5 P.M.
   (c) What is the standard deviation of the repair time?

4-34   A limousine service catering to a large metropolitan area has found that the time for a trip (from dispatch to return) is exponentially distributed with a mean of 30 minutes.
   (a) What is the probability that a trip will take more than an hour?
   (b) If a limousine has already been gone for 45 minutes, what is the probability that it will return within the next 20 minutes?
   (c) If two limousines have just been dispatched, what is the probability that both will not return within the next 45 minutes? Assume that the trips are independent of each other.

4-35   The time to failure of an electronic component can be described by a Weibull distribution with $\gamma = 0$, $\beta = 0.25$, and $\alpha = 800$ hours.
   (a) Find the mean time to failure.
   (b) Find the standard deviation of the time to failure.
   (c) What is the probability of the component lasting at least 1500 hours?

4-36   The time to failure of a mechanical component under friction may be modeled by a Weibull distribution with $\gamma = 20$ days, $\beta = 0.2$, and $\alpha = 35$ days.
   (a) What proportion of these components will fail within 30 days?
   (b) What is the expected life of the component?
   (c) What is the probability of a component lasting between 40 and 50 days?

4-37   The diameter of bearings is known to have a mean of 35 mm with a standard deviation of 0.5 mm. A random sample of 36 bearings is selected. What is the probability that the average diameter of these selected bearings will be between 34.95 and 35.18 mm?

4-38   Refer to Exercise 4-37. Suppose that the machine is considered to be out of statistical control if the average diameter of a sample of 36 bearings is less than 34.75 mm or greater than 35.25 mm.
   (a) If the true mean diameter of all bearings produced is 35 mm, what is the probability of the test indicating that the machine is out of control?
   (b) Suppose that the setting of the machine is accidentally changed such that the mean diameter of all bearings produced is 35.05 mm. What is the probability of the test indicating that the machine is in statistical control?

4-39   Vendor quality control is an integral part of a total quality system. A soft drink bottling company requires its vendors to produce bottles with an internal pressure strength of at least 300 kg/cm². A vendor claims that its bottles have a mean strength of 310 kg/cm² with a standard deviation of 5 kg/cm². As part of a vendor surveillance program, the bottling company samples 50 bottles from the production and finds the average strength to be 308.6 kg/cm².

    (a) What are the chances of getting that sample average that was observed, or even less, if the assertion by the vendor is correct?

    (b) If the standard deviation of the strength of the vendor's bottles is $8 \, kg/cm^2$, with the mean (as claimed) of $310 \, kg/cm^2$, what are the chances of seeing what the bottling company observed (or an even smaller sample average)?

4-40   An electronic switch has a constant failure rate of 10^{-3} per hour.

    (a) What is the expected life of the switch?

    (b) What is the standard deviation of the life of the switch?

    (c) Find the probability that the switch will last at least 1200 hours.

    (d) What is the probability that the switch will last between 1200 and 1400 hours?

4-41   Refer to the electronic switch in Exercise 4-40. In order to improve the reliability of a system, three such additional switches are used on a standby basis. This means that only when a switch fails, another is activated, and so on. The system operates as long as at least one switch is operational.

    (a) What is mean life of the system?

    (b) What is the standard deviation of the system life?

    (c) What is the probability that the system will operate for at least 5000 hours?

    (d) What is the minimum number of additional switches, on a standby basis, needed if it is desirable for the system to operate at least 3000 hours with a probability of 40%?

4-42   Reinforced concrete beams are used in bridges. However, cracks develop in these beams and it has an accumulation effect over time. The time to failure of such bridges, in days, is modeled by a lognormal distribution, where the mean of the natural logarithm of the failure time is 7.6, with a standard deviation of 2.

    (a) Find the mean life of beams.

    (b) What is the standard deviation of the life of beams?

    (c) Find the probability of a beam lasting more than 4000 days.

4-43   The mean time to assemble a product as found from a sample of size 40 is 10.4 minutes. The standard deviation of the assembly times is known to be 1.2 minutes.

    (a) Find a two-sided 90% confidence interval for the mean assembly time, and interpret it.

    (b) Find a two-sided 99% confidence interval for the mean assembly time, and interpret it.

    (c) What assumptions are needed to answer parts (a) and (b)?

    (d) The manager in charge of the assembly line believes that the mean assembly time is less than 10.8 min. Can he make this conclusion at a significance level α of 0.05?

4-44   A company that dumps its industrial waste into a river has to meet certain restrictions. One particular constraint involves the minimum amount of dissolved oxygen that is needed to support aquatic life. A random sample of 10 specimens taken from a given location gives the following results of dissolved oxygen (in parts per million, ppm):

$$9.0 \quad 8.6 \quad 9.2 \quad 8.4 \quad 8.1$$

$$9.5 \quad 9.3 \quad 8.5 \quad 9.0 \quad 9.4$$

(a) Find a two-sided 95% confidence interval for the mean dissolved oxygen, and interpret it.

(b) What assumptions do you need to make to answer part (a)?

(c) Suppose that the environmental standards stipulate a minimum of 9.5 ppm of average dissolved oxygen. Is the company violating the standard? Test at a level of significance α of 0.05.

4-45 The Occupational Safety and Health Administration (OSHA) mandates certain regulations that have to be adopted by corporations. Prior to the implementation of the OSHA program, a company found that for a sample of 40 randomly selected months, the mean employee time lost due to job-related accidents was 45 hours. After implementation of the OSHA program, for a random sample of 45 months, the mean employee time lost due to job-related accidents was 39 hours. It can be assumed that the variability of time lost due to accidents is about the same before and after implementation of the OSHA program (with a standard deviation being 3.5 hours).

(a) Find a 90% confidence interval for the difference in the mean time lost due to accidents.

(b) Test the hypothesis that implementation of the OSHA program has reduced the mean employee lost time. Use a level of significance of 0.10.

4-46 Refer to Exercise 4-45. Suppose that the standard deviations of the values of lost time due to accidents before and after use of the OSHA program are unknown but are assumed to be equal. The first sample of size 40 gave a mean of 45 hours with a standard deviation of 3.8 hours. Similarly, the second sample of size 45, taken after the implementation of the OSHA program, yielded a mean of 39 hours with a standard deviation of 3.5 hours.

(a) Find a 95% confidence interval for the difference in the mean time lost due to job-related accidents.

(b) What assumptions are needed to answer part (a)?

(c) Can you conclude that the mean employee time lost due to accidents has decreased due to the OSHA program? Use a level of significance of 0.05.

(d) How would you test the assumption of equality of the standard deviations of time lost due to job-related accidents before and after implementation of the OSHA program? Use a level of significance of 0.05.

4-47 A company is experimenting with synthetic fibers as a substitute for natural fibers. The quality characteristic of interest is the breaking strength. A random sample of 8 natural fibers yields an average breaking strength of 540 kg with a standard deviation of 55 kg. A random sample of 10 synthetic fibers gives a mean breaking strength of 610 kg with a standard deviation of 22 kg.

(a) Can you conclude that the variances of the breaking strengths of natural and synthetic fibers are different? Use a level of significance α of 0.05. What assumptions are necessary to perform this test?

(b) Based on the conclusions in part (a), test to determine if the mean breaking strength for synthetic fibers exceeds that for natural fibers. Use a significance level α of 0.10.

(c) Find a two-sided 95% confidence interval for the ratio of the variances of the breaking strengths of natural and synthetic fibers.

(d) Find a two-sided 90% confidence interval for the difference in the mean breaking strength for synthetic and natural fibers.

4-48   Consider the data in Exercise 4-17 on the time (in minutes) to process hot-rolled steel for a sample of size 10.
    (a) Find a 98% confidence interval for the mean time to process hot-rolled steel. What assumptions do you have to make to solve this problem?
    (b) Find a 95% confidence interval for the variance.
    (c) Test the hypothesis that the process variability, as measured by the variance, exceeds 0.80. Use $\alpha = 0.05$.

4-49   Price deregulation in the airline industry has promoted competition and a variety of fare structures. Prior to deciding on a price change, a particular airline is interested in obtaining an estimate of the proportion of the market that it presently captures for a certain city. A random sample of 300 passengers indicates that 80 used that airline.
    (a) Find a point estimate of the proportion of the market that uses this particular airline.
    (b) Find a 95% confidence interval for the proportion that uses this airline.
    (c) Can the airline conclude that its market share is more than 25%? Use a level of significance of 0.01.

4-50   An advertising agency is judged by the increase in the proportion of people who buy a particular product after the advertising campaign is conducted. In a random sample of 200 people prior to the campaign, 40 said that they prefer the product in question. After the advertising campaign, out of a random sample of 300, 80 say they prefer the product.
    (a) Find a 90% confidence interval for the difference in the proportion of people who prefer the stipulated product before and after the advertising campaign.
    (b) Can you conclude that the advertising campaign has been successful in increasing the proportion of people who prefer the product? Use a level of significance of 0.10.

4-51   Two machines used in the same operation are to be compared. A random sample of 80 parts from the first machine yields 6 nonconforming ones. A random sample of 120 parts from the second machine shows 14 nonconforming ones.
    (a) Can we conclude that there is a difference in the output of the machines? Use a level of significance of 0.10.
    (b) Find a 95% confidence interval for difference in the proportion of nonconforming parts between the two machines.

4-52   The precision of equipment and instruments is measured by the variability of their operation under repeated conditions. The output from an automatic lathe producing the diameter (in millimeters) of a part gave the following readings for a random sample of size 10:

$$10.3 \quad 9.7 \quad 9.6 \quad 9.5 \quad 9.9$$
$$10.2 \quad 9.8 \quad 10.1 \quad 10.2 \quad 9.8$$

    (a) Find a 90% confidence interval for the variance of the diameters.
    (b) Find a 90% confidence interval for the standard deviation of the diameters.
    (c) Test the null hypothesis that variance of the diameters does not exceed 0.05 mm. Use a significance level of 0.10.

(d) Does the mean setting for the machine need adjustment? Is it significantly different from 9.5 mm? Test at a significance level of 0.10.

4-53   A company is investigating two potential vendors on the timeliness of their deliveries. A random sample of size 10 from the first vendor produced an average delay time of 4.5 days with a standard deviation of 2.3 days. A random sample of size 12 from the second vendor yielded an average delay time of 3.4 days with a standard deviation of 6.2 days.
   (a) Find a 90% confidence interval for the ratio of the variances of the delay times for the two vendors.
   (b) What assumptions are needed to solve part (a)?
   (c) Can we conclude that the first vendor has a smaller variability regarding delay times than the second? Use a significance level of 0.05.
   (d) Which vendor would you select, and why?

4-54   Twenty five patients of a certain diagnosis-related group were randomly selected, and their systolic blood pressure, blood glucose level, and total cholesterol level were measured. Upon administration of a certain drug, after 6 months the same characteristics were measured for the selected patients. The data are shown in Table 4-5.

TABLE 4-5

	Before			After		
Patient Number	Blood Pressure	Blood Glucose	Total Cholesterol	Blood Pressure	Blood Glucose	Total Cholesterol
1	145	186	240	138	183	233
2	162	142	235	143	150	246
3	128	122	203	125	119	218
4	116	124	222	118	126	230
5	130	121	219	121	132	215
6	132	116	205	134	108	183
7	110	105	195	112	102	192
8	125	119	216	122	107	204
9	139	115	226	125	105	215
10	142	132	231	130	133	225
11	154	152	255	140	150	233
12	124	120	235	125	122	222
13	114	118	212	112	113	214
14	136	131	238	122	126	230
15	150	220	255	144	180	250
16	133	135	232	126	130	224
17	129	119	220	123	109	231
18	108	106	234	114	103	238
19	112	117	194	111	108	204
20	146	122	225	130	117	220
21	153	204	256	132	196	242
22	145	182	248	134	175	240
23	126	140	229	120	135	206
24	138	180	240	124	172	231
25	129	135	218	120	133	204

(a) Find the mean, standard deviation, skewness coefficient, kurtosis coefficient, and interquartile range of systolic blood pressure before the drug was administered and comment on the values.

(b) Find the mean, standard deviation, skewness coefficient, kurtosis coefficient, and interquartile range of systolic blood pressure after the drug was administered and comment on the values.

(c) Can we conclude that the mean systolic blood pressure, before the drug was administered, exceeds 125? Use $\alpha = 0.05$ What is the p-value?

(d) Can we conclude that the drug was effective in reducing mean systolic blood pressure? Use $\alpha = 0.05$ What is the p-value? Explain this p-value.

(e) Did the drug have a significant impact on reducing the average cholesterol level? Use $\alpha = 0.05$. What is the p-value?

(f) Construct descriptive statistics for blood glucose level before the drug was administered and comment.

(g) What is the correlation between blood glucose levels before and after administration of the drug? Test whether the correlation differs from zero. Use $\alpha = 0.05$.

(h) Construct a 98% confidence interval for the variance of systolic blood pressure after administration of the drug.

4-55 Management is interested in increasing the efficiency of processing purchase orders. The time to process purchase orders, in days, was observed for 31 randomly selected customers and is shown in Table 4-6. Following a detailed study of the process, certain recommendations were adopted. Twenty-six customers were randomly selected and their purchase order processing times are shown after the process improvement changes.

(a) Find the mean, standard deviation, skewness coefficient, kurtosis coefficient, and interquartile range of the processing time prior to process changes and comment on the values.

(b) Find the mean, standard deviation, skewness coefficient, kurtosis coefficient, and interquartile range of the processing time after the process changes and comment on the values.

(c) Find a 95% confidence interval for the mean processing time prior to process changes.

(d) Can we conclude that the mean processing time, before the process changes, is less than 10.5 days? Use $\alpha = 0.02$. What is the p-value? If $\alpha = 0.10$, what is your decision? What does this imply?

(e) Can we conclude that the mean processing time, after the process changes, is less than 8.5 days? Use $\alpha = 0.05$. What is the p-value? Explain.

(f) Is there a difference in the variabilities of processing times before and after process changes? Use $\alpha = 0.05$.

(g) Can we conclude that the process changes have been effective in reducing the mean processing time? Use $\alpha = 0.05$. What assumptions do you need to make for conducting the test?

4-56 An insurance company wants to estimate the premium to be charged for a $200,000 homeowner's policy that covers fire, theft, vandalism, and natural calamities. Flood and earthquakes are not covered. The company has estimated from historical data that a total loss may happen with a probability of 0.0005, a 50% loss with a probability of 0.001, and a 25% loss with a probability of 0.01. Ignoring all other

TABLE 4-6

Before Change		After Change	
Customer	Processing Time	Customer	Processing Time
1	9.7	1	7.6
2	11.2	2	8.3
3	10.5	3	7.9
4	8.3	4	8.6
5	9.2	5	8.3
6	8.8	6	6.2
7	10.4	7	9.1
8	11.6	8	8.8
9	10.1	9	8.4
10	9.6	10	7.5
11	8.8	11	9.4
12	12.3	12	6.8
13	10.9	13	8.1
14	11.1	14	9.7
15	8.5	15	7.4
16	11.6	16	9.2
17	12.3	17	8.3
18	10.2	18	8.5
19	10.7	19	6.8
20	11.3	20	7.5
21	9.1	21	7.2
22	9.9	22	9.2
23	10.5	23	8.1
24	11.4	24	7.3
25	8.3	25	7.7
26	8.7	26	8.1
27	9.4		
28	10.3		
29	11.4		
30	8.8		
31	10.2		

losses, what premium should the company charge to make an average net profit of 1.5% of the policy's face value?

4-57 An insertion machine in printed circuit board manufacturing has an insertion rate of 5000 parts per hour. From historical data, it is known that the error rate is 300 parts per million parts inserted. The errors occur randomly and independently.

(a) What is the probability of observing no more than 5 errors in 2 hours of operation?

(b) What is the probability of an error-free insertion (rolled throughput yield) in 2 hours of operation? In 3 hours of operation?

(c) Suppose that the desired probability of an error-free insertion in 2 hours of operation is 0.001. What must be the hourly error rate of the insertion machine to accomplish this?

4-58   A wave soldering process is used in printed circuit boards. It is known that the error rate is 200 per million solders, where errors occur randomly and independently. A given board requires 5000 solders.

(a) What is the probability of 3 or more errors in a circuit board?

(b) Find the probability of a board having no solder errors.

(c) If a new design requires 2000 solders per board, what is the probability of an error-free board?

(d) The cost of rectification of a defective solder is $0.05, and the monthly production rate is 1 million boards. What is the expected cost reduction per month?

REFERENCES

Banks, J. (1989). *Principles of Quality Control*. New York: Wiley.

Duncan, A.J. (1986). *Quality Control and Industrial Statistics*, 5th ed. Homewood, IL: Richard D. Irwin.

Henley, E.J., and H. Kumamoto (1991). *Probabilistic Risk Assessment: Reliability Engineering, Design, and Analysis*, reprint ed. Piscataway, NJ: Institute of Electrical and Electronics Engineers.

Kendall, M.G., A. Stuart, J.K. Ord, S.F. Arnold, and A. O'Hagan (1998). *Kendall's Advanced Theory of Statistics*, Vol. 1, London: Arnold Publishers.

Mendenhall, W., J.E. Reinmuth, and R. Beaver (1993). *Statistics for Management and Economics*, 7th ed. Belmont, CA: Duxbury Press.

Minitab, Inc. (2007). *Release 15*. State College, PA: Minitab.

Montgomery, D.C. (2004). *Introduction to Statistical Quality Control*, 5th ed. Hoboken NJ: Wiley.

Neter, J., M.H. Kutner, C.J. Nachtsheim, and W. Wasserman (2005). *Applied Linear Statistical Models*, 5th ed. Homewood, IL: Richard D. Irwin.

5

DATA ANALYSES AND SAMPLING

5-1   INTRODUCTION AND CHAPTER OBJECTIVES

In this chapter we continue to expand on the various descriptive and inferential statistical procedures described in Chapter 4. Our objective is to analyze empirical data graphically since they provide comprehensive information and are a viable tool for analysis of product and process data. The information they provide on existing product or process characteristics helps us determine whether these characteristics are close to the desired norm. A second objective is to test for distributional assumptions. Recall that in Chapter 4, for testing hypothesis on various parameters such as the population mean or variance, the assumption of normality was made. We present a method for testing the validity of such an assumption. Further, we discuss some transformations to achieve normality for variables that are non-normal. A third objective involves analyzing qualitative data. Such information is typically frequency-type data obtained from product surveys. Finally, we include a discussion of various sampling techniques. The issue of determination of sample size is of paramount importance in quality. Based on the degree of acceptable risks, expressions are presented for the required sample size.

5-2   EMPIRICAL DISTRIBUTION PLOTS

Histograms

Distribution plots are applicable to quantitative data. In such instances, the quality characteristic values are obtained on a measurable scale. Seldom do we get an idea of process

TABLE 5-1 Inside Diameter (in mm) of Metal Sleeves

Sample	Observations X (Five per Sample)				
1	50.05	50.03	50.02	50.00	49.94
2	49.96	49.99	50.03	50.01	49.98
3	50.01	50.01	50.01	50.00	49.92
4	49.95	49.97	50.02	50.10	50.02
5	50.00	50.01	50.00	50.00	50.09
6	50.02	50.05	49.97	50.02	50.09
7	50.01	49.99	49.96	49.99	50.00
8	50.02	50.00	50.04	50.02	50.00
9	50.06	49.93	49.99	49.99	49.95
10	49.96	49.93	50.08	49.92	50.03
11	50.01	49.96	49.98	50.00	50.02
12	50.04	49.94	50.00	50.03	49.92
13	49.97	49.90	49.98	50.01	49.95
14	50.00	50.01	49.95	49.97	49.94
15	49.97	49.98	50.03	50.08	49.96
16	49.98	50.00	49.97	49.96	49.97
17	50.03	50.04	50.03	50.01	50.01
18	49.98	49.98	49.99	50.05	50.00
19	50.07	50.00	50.02	49.99	49.93
20	49.99	50.06	49.95	49.99	50.02

characteristics just by looking at the individual data values gathered from the process. Such data are often voluminous. Frequency distributions and histograms summarize such information and present it in a format that allows us to draw conclusions regarding the process condition.

A **frequency distribution** is a rearrangement of raw data in ascending or descending order of magnitude, such that the quality characteristic is subdivided into classes and the number of occurrences in each class is presented.

Table 5-1 shows the inside diameter (in millimeters) of metal sleeves produced in a machine shop for 100 randomly selected parts. Twenty samples, each of size 5, were taken. Simply looking at the data in Table 5-1 provides little insight about the process. Even though we know that there is variability in the sleeve diameters, we can hardly identify a pattern in the data (what is the degree of variability?) or comment about the central tendency of the process (about which value are most of the observations concentrated?).

A *histogram* is a graphical display of data such that the characteristic is subdivided into classes, or cells. In a **frequency histogram,** the vertical axis usually represents the number of observations in each class. An alternative representation of the vertical axis could be the percentage.

Example 5-1 For the data in Table 5-1, let us use Minitab to construct a histogram. Choose **Graph** > **Histogram** > **Simple**. Click **OK**. Under **Graph variables**, input the column number or name of the variable, in this case Diameter. Click **OK**.

Minitab produces the frequency histogram shown in Figure 5-1. The histogram provides us with a sense of the distribution of the 100 values, where classes of equal width have been created. The midpoints of the classes are 49.00, 49.92, $\cdots$, and so on. A majority of the values

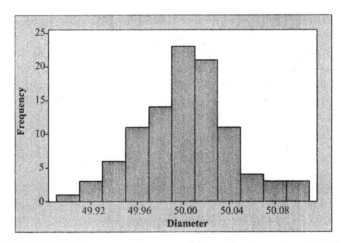

FIGURE 5-1   Frequency histogram of sleeve diameters using Minitab.

are clustered between 49.96 and 50.04. The shape of the distribution resembles a bell-shaped distribution. We will, however, demonstrate a test for normality later in the chapter.

Stem-and-Leaf Plots

Stem-and-leaf plots are another graphical approach to plotting observations and interpreting process characteristics. With frequency histograms, the identities of the individual observations are lost in the process of plotting. In the stem-and-leaf plot, however, individual numerical values are retained. Let's construct a stem-and-leaf plot using the metal sleeves data from Table 5-1.

Example 5-2   Using Minitab, click on **Graph > Stem-and-leaf**. Under **Graph variables**, enter the column number or name of the variable, in this case Diameter. Click **OK**. The output from Minitab is shown in Figure 5-2.

Each data value is split into two parts, the stem and the leaf. For example, the data value 49.90 is displayed with the stem part as 499 and the leaf part as 0. Notice that the decimal

```
Stem-and-leaf of Diameter   N  = 100
Leaf Unit = 0.010


    1     499  0
    7     499  222333
   15     499  44455555
   28     499  6666667777777
   44     499  888888999999999
  (25)    500  0000000000000011111111111
   31     500  22222222223333333
   14     500  444455
    8     500  667
    5     500  8899
    1     501  0
```

FIGURE 5-2   Stem-and-leaf plot for the inside diameter of metal sleeves.

point is implicit to the left of the rightmost digit in the stem. The digit in the leaf portion represents hundredths. The leftmost column in Figure 5-2 represents the cumulative frequency, from the corresponding end, depending on the location of the stem relative to the median. Thus, the value "7" in the second row of the stem "499" tells us that there are seven values ≤ 49.93, while the value "5" in the next to last row of the stem "500" indicates there are five values ≥ 50.08. The value in parentheses, "(25)", indicates the median class, and only for this class the number "25" indicates that there are 25 values in this median class (i.e., between 50.00 and 50.01).

Box Plots

Box plots graphically depict data and also display summary measures (Chambers 1977; Chambers et al. 1983). A **box plot** shows the central tendency and dispersion of a data set and indicates the **skewness** (deviation from symmetry) and **kurtosis** (measure of tail length). The plot also shows outliers.

There are several features of a box plot. The box is bounded by Q_1 and Q_3, the first and third quartiles, respectively, with a line drawn at the median. Hence, the length of the box represents the interquartile range (IQR). Depending on the location of the median relative to the edges of the box, inferences are drawn on the symmetry of the distribution. For a symmetric distribution, the median would be located midway between the edges of the box: that is, at a value $(Q_1 + Q_3)/2$. If the median is closer to Q_3, the data distribution is negatively skewed. Conversely, if the median is closer to Q_1, the distribution is positively skewed. Two lines, known as **whiskers,** are drawn outward from the box. One line extends from the top edge of the box at Q_3 to the maximum data value that is less than or equal to $Q_3 + 1.5(IQR)$. Similarly, a line from the bottom edge of the box at Q_1 extends downward to the minimum value that is greater than or equal to $Q_1 - 1.5(IQR)$. The endpoints of the whiskers are known as the *upper and lower adjacent values*. The length of the whiskers indicates the **tail lengths**. Values that fall outside the adjacent values are candidates for consideration as **outliers**. They are plotted as asterisks (*).

Example 5-3   A private company by the name of Quick Dock operates an unloading facility for supertankers in the port of Singapore. A random sample of size 30 that shows the unloading times (in hours) was collected. To improve the efficiency of unloading, a process study was conducted. Using the changes recommended, the process was modified and a subsequent random sample of size 30 was collected. Both sets of observations are shown in Table 5-2.

   (a) Construct a box plot for unloading times prior to process changes and comment on the process.

Solution   Using Minitab, first create a worksheet with two variables, unloading times before and after process changes. Click on **Graph** > **Boxplot**. Select **One Y – Simple**. In **Variables**, enter the column number or variable name, Unloading-Before in this case. Click **OK**. Figure 5-3 shows the box plot.

Several insights can be gained from the box plot. The box itself, extending from 4.0 to 20.475, contains 50% of the observations. The median, indicated by the line within the box, is at 10.1 and is closer to the bottom edge (Q_1) of the box. So, the distribution is positively skewed. The length of the top whisker is much longer than that of the bottom whisker, indicating a much longer righttail. There are no outliers.

TABLE 5-2   Unloading Times (hours) of Supertankers

	Before Changes				After Changes		
Sample Number	Time	Sample Number	Time	Sample Number	Time	Sample Number	Time
1	9.4	16	1.2	1	2.7	16	4.6
2	1.5	17	2.6	2	14.4	17	0.9
3	14.6	18	7.5	3	3.2	18	2.5
4	10.8	19	19.2	4	17.0	19	1.2
5	18.2	20	1.8	5	1.0	20	14.9
6	19.8	21	34.2	6	12.7	21	1.4
7	23.5	22	18.1	7	4.6	22	7.3
8	5.2	23	7.5	8	6.8	23	4.6
9	9.3	24	3.4	9	22.7	24	6.6
10	9.1	25	12.4	10	2.3	25	4.9
11	13.3	26	30.8	11	8.3	26	2.0
12	26.1	27	5.3	12	2.6	27	4.9
13	2.3	28	39.5	13	7.7	28	9.6
14	30.7	29	4.2	14	21.4	29	1.1
15	22.5	30	2.5	15	1.7	30	1.2

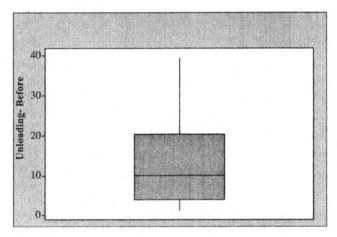

FIGURE 5-3   Box plot of unloading times before process changes.

(b) Construct side-by-side box plots for unloading times before and after process changes. Comment on process improvements.

Solution   Click on **Graph > Boxplot**. Select **Multiple Y's – Simple**. In **Variables**, enter the column numbers or variable names of both, Unloading-Before and Unloading-After, in this case. Click **OK**. Figure 5-4 shows this box plot.

The box plot of the unloading times, after process changes, is quite different from that of unloading times before the process changes. Note that the box extends from 1.925 to 8.625, a much more compact box than before, containing 50% of the observations. The new median is at 4.6, quite a bit below the previous median. The new unloading time distribution is also

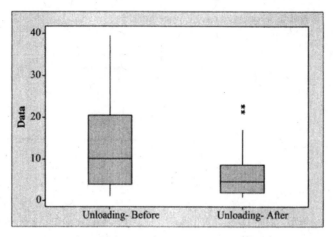

FIGURE 5-4   Box plot of unloading times before and after process changes.

positively skewed, however, the median and variability are much lower. A longer right tail is indicated, as before, but with a shorter tail length. There are two extreme values or outliers on the right tail indicated by the *. These could represent unusually long unloading times, for which, perhaps, special causes could be identified to improve the process further.

Variations of the Basic Box Plot

One variation of the basic form of the box plot is the **notched box plot**. A notch, the width of which corresponds to the length of the confidence interval for the median, is constructed on the box around the median. Assuming that the data values are normally distributed, the standard deviation of the median is given by Kendall et al. (1998) as

$$s_m = \frac{1.25(\text{IQR})}{1.35\sqrt{n}} \tag{5-1}$$

The notch around the median M should start at values of

$$M \pm Cs_m \tag{5-2}$$

where C is a constant representing the level of confidence. For a level of confidence of 95%, $C = 1.96$ can be used. For further details, consult McGill et al. (1978).

Notched box plots are used to determine whether there are significant differences between the medians of the quality characteristic of two plots. If there is no overlap between the notches of two box plots, we can conclude that there is a significant difference between the two medians. This may indicate that the actions taken have changed the process parameter conditions significantly.

5-3   RANDOMNESS OF A SEQUENCE

In this section we present a technique to determine if the sequence of observations collected in a sample, usually as a function of time, is random in nature. If the sequence is nonrandom, it may indicate the existence of special causes. A cause-and-effect analysis may then indicate possible process changes to undertake. On the other hand, if the sequence is deemed to be random, it may imply the presence of common causes, which lead to variability in the characteristic observed. In this case, systemic changes are necessary to create process improvement.

Run Chart

A **run chart** is a plot of the quality characteristic as a function of the order (or usually time) in which the observations are collected. They provide an idea of the clustering of the data or whether the data are from a mixture of, say, two populations. These inferences are based on the number of runs about the median.

A run (about the median) is defined as one or more consecutive data points on the same side (of the median). When counting runs, points that fall exactly on the reference line (median) are ignored. If the pattern is **random**, the actual number of runs should be close to the number of runs expected, based on the assumption of randomness of the pattern. So, when the number of runs observed is much greater than the number of runs expected, it implies the possibility of the data being from a mixture pattern (say, two populations), causing frequent fluctuations about the median. Similarly, when the number of runs observed is much less than the number of runs expected, it may indicate the possibility of clustering of the data (a nonrandom behavior). We test these situations through the procedure of hypothesis testing and use the *p*-**value** approach, described in Chapter 4.

Another pair of situations involves testing for the presence of a trend or oscillation, both examples of nonrandomness. These tests are conducted by using the number of runs up or down. A trend is an unusually long series of consecutive increases or decreases in the data. In counting the run length, we ignore points that repeat the preceding value. A pattern of oscillation is indicated if the number of runs up or down observed is much greater than the number of runs expected. Similarly, a trend in the data is inferred when the number of runs up or down observed is much less than the number of runs expected. A long run length about the median may also indicate a shift in the data.

Example 5-4   The hemoglobin A1C value is a measure of blood glucose level over a period of about three months. Table 5-3 shows hemoglobin A1C values for a diabetic patient taken every three months. Construct a run chart and comment on whether the process shows

TABLE 5-3   Hemoglobin A1C Values for a Diabetic Patient

Observation	A1C	Observation	A1C	Observation	A1C
1	8.3	11	7.9	21	6.7
2	6.5	12	8.4	22	6.6
3	7.2	13	8.2	23	7.6
4	7.0	14	8.6	24	7.4
5	8.1	15	8.8	25	7.8
6	6.8	16	7.8	26	8.0
7	7.1	17	7.4	27	7.7
8	6.6	18	7.2	28	7.6
9	7.3	19	7.1	29	7.8
10	7.5	20	6.8	30	7.5

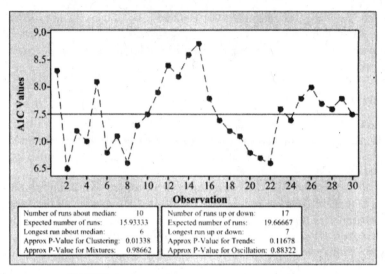

FIGURE 5-5   Run chart of hemoglobin A1C values.

common or special causes of variation. Has there been a significant trend? Test at the 5% level of significance.

Solution   Using Minitab, first create a worksheet with the variable, A1C Values, consisting of 30 observations. Click on **Stat** > **Quality Tools** > **Run Chart**. Since the A1C Values are entered as a single column in the worksheet, select **Single column**, and enter the column number or name. Under **Subgroup size**, enter the value 1. Click **OK**.

Figure 5-5 shows the run chart of hemoglobin A1C values. Note that the number of runs about the median is 10, while the number of runs expected, if the data are from a distribution of random sequence, is 15.933. Also, the number of runs up or down observed is 17, while the number of runs expected, under the null hypothesis of a random sequence, is 19.667. Let us now test for the presence of clustering or mixtures. Note that the p-value for clustering is $0.01338 < \alpha = 0.05$. Hence, we reject the null hypothesis of a random sequence and conclude that there is significant clustering. The number of runs about the median observed is significantly less than that expected, under the assumption of a random sequence. We find that the p-value for testing the presence of a mixture is 0.98662 (the complement of 0.01338), indicating, obviously, that we cannot conclude the presence of a mixture pattern. Hence, a special cause of variation exists due to clustering of the observations. As a follow-up measure, one could investigate if there were certain traits or patient habits that led to clustering.

In testing for the presence of significant trend or oscillation, we consider the Minitab output in Figure 5-5. The p-value for testing trends is $0.11678 > \alpha = 0.05$, and similarly, the p-value for testing oscillations is 0.88322 (the complement of 0.116678). Hence, no significant trend or oscillations can be concluded. From the plot, observe that the longest run down has a length of 7, which might seem to indicate that there had been a significant declining trend. However, in actually testing a hypothesis on the presence of a trend, it was not found to be significant.

5-4   VALIDATING DISTRIBUTIONAL ASSUMPTIONS

In many statistical techniques, the population from which the sample data is drawn is assumed to have a certain specified distribution so that certain inferences can be made about the quality characteristic. Statistical procedures known as **goodness-of-fit tests** are used for testing the information from the empirical cumulative distribution function (cdf) obtained from the sample versus the theoretical cumulative distribution function, based on the hypothesized distribution. Moreover, parameters of the hypothesized distribution may be specified or estimated from the data. The test statistic could be a function of the difference between the frequency observed and that expected, as determined on the basis of the distribution that is hypothesized. Goodness-of-fit tests may include chi-squared tests (Duncan 1986), Kolmogorov–Smirnov tests (Massey 1951), or the Anderson–Darling test (Stephens 1974), among others. Graphical methods such as **probability plotting** may be used along with such tests.

Probability Plotting

In probability plotting, the sample observations are ranked in ascending order from smallest to largest. Thus, the observations $x_1, x_2, \cdots, x_n$ are ordered as $x_{(1)}, x_{(2)}, \cdots, x_{(n)}$, where $x_{(1)}$ denotes the smallest observation and so on. The **empirical cdf** of the ith ranked observation, $x_{(i)}$, is given by

$$F_i = \frac{i - 0.5}{n}. \tag{5-3}$$

The **theoretical cdf**, based on the hypothesized distribution, at $x_{(i)}$, is given by $G(x_{(i)})$, where $G(\cdot)$ is calculated using specified parameters or estimates from the sample. A probability plot displays the plot of $x_{(i)}$, on the horizontal axis, versus F_i and $G(x_{(i)})$ on the vertical axis. The **vertical axis** is scaled such that if the data is from the hypothesized distribution, the plot of $x_{(i)}$ versus $G(x_{(i)})$ will be a **straight line**. Thus, departures of $F(\cdot)$ from $G(\cdot)$ are visually easy to detect. The closer the plotted values of $F(\cdot)$ are to the fitted line, $G(\cdot)$, the stronger the support for the null hypothesis. A test statistic is calculated where large deviations between $F(\cdot)$ and $G(\cdot)$ lead to large values of the test statistic, or alternatively, small **p-values**. Hence, if the observed p-value is less than α, the chosen level of significance, the null hypothesis representing the hypothesized distribution is rejected.

In a **normal probability plot**, suppose that we are testing the null hypothesis that the data are from a normal distribution, with mean and standard deviation not specified. Then $G(x_{(i)})$ will be $\Phi((x_{(i)} - \bar{x})/s)$, $\Phi(\cdot)$ represents the cdf of a standard normal distribution and $\bar{x}$ and s are sample estimates of the mean and standard deviation, respectively.

Minitab allows probability plotting based on a variety of distributions, such as normal, lognormal, exponential, gamma, Weibull, logistic, or loglogistic. A common test statistic used for such tests is the **Anderson–Darling statistic,** which measures the area between the fitted line (based on the hypothesized distribution) and the empirical cdf. This statistic is a squared distance that is weighted more heavily in the tails of the distribution. Smaller values of this statistic lead to the nonrejection of the null hypothesis and we confirm validity of the observations being likely from the hypothesized distribution.

Example 5-5   A sample of 50 coils to be used in an electrical circuit is selected randomly and their resistances measured in ohms (Ω). The data values are shown in Table 4-2. A normal probability plot is constructed using Minitab. First, a worksheet consisting of the data values on coil resistance is created. Choose **Graph > Probability Plot**. Select **Single** and click **OK**. In **Graph variable**, input the name of the variable, say Resistance, and click **OK**. Click on **Distribution** and in the drop-down menu, select **Normal**. Click **OK**. The resulting normal probability plot is shown in Figure 5-6, where a straight line is fitted through the points

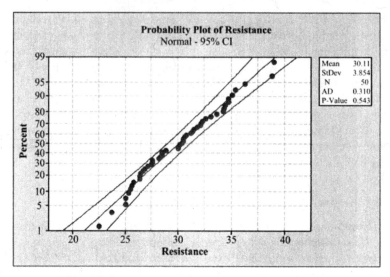

FIGURE 5-6   Normal probability plot of coil resistance.

plotted. Observe that a majority of the points are in close proximity to the straight line. A few points on the extremes deviate from the line.

Minitab also displays 95% confidence intervals for the fitted distribution (as hypothesized in the null hypothesis) on a pointwise basis. Usually, points may fall outside the confidence intervals near the tails. In the lower half of the plot, points to the right of the confidence band indicate that there are fewer data in the left tail relative to what is expected based on the hypothesized distribution. Conversely, in the upper half, points to the right of the confidence band indicate that there are more data in the right tail relative to what is expected. Here, the points plotted seem fairly close to the fitted line, supporting the null hypothesis that the data are from a normal distribution. Further, the Anderson–Darling test statistic is shown to be 0.310, with a p-value of 0.543. With this large p-value, for any reasonable level of significance (α), we do not reject the null hypothesis of normality.

With the validation of the assumption of normality, we can make several inferences from Figure 5-6. First, we can estimate the population mean by reading the value of the 50th percentile off the fitted straight line; it appears to be approximately 30.0 Ω. Second, we can estimate the population standard deviation as the difference between the 84th and 50th percentile data values. This is because the 84th percentile of a normal distribution is approximately 1 standard deviation away from the mean (the 50th percentile). From Figure 5-6, the population standard deviation is estimated to be (34.5 − 30.0) $\Omega = 4.5\,\Omega$. Third, we can estimate the proportion of a process output that does not meet certain specification limits. For example, if the lower specification limit of the resistance of cables is 25 Ω, from the figure we estimate that approximately 10% of the output is less than the limit specified.

Example 5-6   Consider the data in Table 5-2 on the unloading times of supertankers after changes have been made in improving the process. Can you conclude that the distribution of unloading times is normal? Use $\alpha = 0.05$. If the distribution is deemed not normal, can you conclude if it is exponential? Estimate the parameter of the exponential distribution.

Solution   Using Minitab, select **Stat > Quality Tools > Individual Distribution Iden-tification**. Select **Single column**, and input the column number or name of the variable,

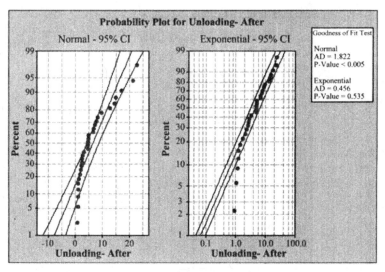

FIGURE 5-7   Normal and exponential probability plots of unloading times.

Unloading-After. Minitab provides the option of checking against hypothesized distributions such as normal, exponential, lognormal, gamma, Weibull, logistic, or extreme values (smallest or largest). Here, rather than validate with all distributions, we select two distributions: **Normal** and **Exponential**. Click **OK**.

Figure 5-7 shows the Minitab output with probability plots of unloading times after process changes using the normal and the exponential as hypothesized distributions. Note that for the normal probability plot, the values plotted are systematically away from the fitted straight line. The value of the Anderson–Darling statistic is 1.822 with a p-value < 0.005. Hence, we reject the null hypothesis of normality of the distribution of unloading times. When considering the exponential probability plot, the majority of the observations are within a narrow band around the fitted line, with the exception of three observations that are unusually small. The Anderson–Darling statistic is 0.456 with a p-value of 0.535. So, for $\alpha = 0.05$, we do not reject the null hypothesis of the distribution of unloading times being exponential. Also, a separate part of the output from Minitab gives an estimate of the scale parameter for the exponential distribution to be 6.56, which is also the estimate of the mean.

5-5   TRANSFORMATIONS TO ACHIEVE NORMALITY

Inferences on population parameters typically require distributional assumptions on the quality characteristic. Recall, from Chapter 4 that confidence intervals and hypothesis testing on parameters such as the population mean (with unknown standard deviation), difference in the means of two populations (with unknown standard deviations), population variance, or the ratio of two population variances require the assumption of normality of the quality characteristic. Hence, for proper use of these inferential methods, if the characteristic is inherently not distributed normally, we investigate procedures to transform the original variable such that the transformed variable satisfies the assumption of normality.

TABLE 5-4   Guidelines on Common Transformations

Data Characteristics	Type of Transformation
Right-skewed data; nonconstant variance across different values; standard deviation proportional to the square of the mean	$Y_T = \dfrac{1}{Y}$
Right-skewed data; quite concentrated	$Y_T = \dfrac{1}{\sqrt{Y}}$
Right-skewed data with nonnegative values; standard deviation proportional to the mean	$Y_T = \ln(Y)$
Poisson data; discrete count data; standard deviation proportional to square root of mean	$Y_T = \sqrt{Y}$
Binomial data, consisting of proportion of successes (p)	$Y_T = \arcsin\left(\sqrt{p}\right)$ or $Y_T = \ln\dfrac{p}{1-p}$

Some Common Transformations

Based on the shape of the distribution and characteristics associated with the data, guidelines on some commonly used transformations are shown in Table 5-4, where Y represents the original variable, and Y_T, the transformed variable.

Power Transformations

These transformations are of the type

$$Y_T = \begin{cases} (Y^q), & q > 0 \\ -(Y^q), & q < 0 \\ \ln(Y), & q = 0 \end{cases} \qquad (5\text{-}4)$$

The impact of such transformations is on changing the distributional shape. Values of the power coefficient, $q > 1$, shift weight to the upper tail of the distribution and reduce negative skewness. The higher the power, stronger the effect. Figure 5-8 demonstrates the effect of q on a negatively skewed variable. Similarly, values of $q < 1$ pull in the upper tail and reduce positive skewness. The lower the power, the stronger the effect. To preserve the order of the data values, a minus sign is added to the transformed variable after raising to powers less than zero. Figure 5-9 shows the effect of q on a positively skewed variable.

Johnson Transformation

Johnson (1949) developed three families of distributions of a variable, Y, that are easily transformed to a standard normal distribution. These are labeled S_B, S_L, and S_U, where the subscripts $B, L,$ and U refer to the variable being bounded, bounded from below or lognormal, and unbounded, respectively. Table 5-5 shows the Johnson family of distributions along with the transformation function and conditions on the parameters.

To fit a nonnormal data set using a Johnson distribution, algorithms have been developed that consider almost all potential transformation functions from the Johnson system, with the parameters being estimated from the data set (Chou et al., 1998). One approach uses a set of four sample percentiles to choose the family and estimate the unknown parameters. For some constant $s > 1$, let four symmetric standard normal deviates be denoted by $-sz, -z, z,$ and sz

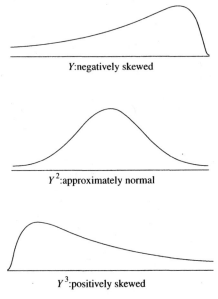

Y:negatively skewed

Y^2:approximately normal

Y^3:positively skewed

FIGURE 5-8 Impact of q on a negatively skewed variable.

for any constant $z > 0$. Let the cumulative distribution functions (cdf) at these points be denoted by q_1, q_2, q_3, and q_4, where, for example, $q_1 = \Phi(-sz)$, where $\Phi(\cdot)$ represents the cdf of a standard normal variate. Further, let Y_i denote the q_ith quantile of the distribution of Y, $i = 1, 2, 3, 4$. The **quantile ratio** is defined as

$$QR = \frac{(Y_4 - Y_3)(Y_2 - Y_1)}{(Y_3 - Y_2)^2} \tag{5-5}$$

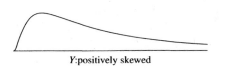

Y:positively skewed

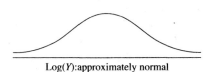

$\text{Log}(Y)$:approximately normal

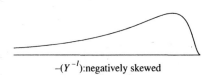

$-(Y^{-1})$:negatively skewed

FIGURE 5-9 Impact of q on a positively skewed variable.

TABLE 5-5   Transformation for Johnson System of Distributions

Johnson Family	Transformation Function	Parameter Conditions
S_B	$z = \gamma + \eta \ln \dfrac{Y - \varepsilon}{\lambda + \varepsilon - Y}$	$\eta, \lambda \geq 0,\ -\infty < \gamma < \infty$ $-\infty < \varepsilon < \infty,\ \varepsilon < Y < \varepsilon + \lambda$
S_L	$z = \gamma + \eta \ln(Y - \varepsilon)$	$\eta > 0,\ -\infty < \gamma < \infty,$ $-\infty < \varepsilon < \infty,\ Y > \varepsilon$
S_U	$z = \gamma + \eta \sinh^{-1}\left(\dfrac{Y - \varepsilon}{\lambda}\right),$ where $\sinh^{-1}(Y) = \ln(Y + \sqrt{1 + Y^2})$	$\eta, \lambda \geq 0,\ -\infty < \gamma < \infty$ $-\infty < \varepsilon < \infty,\ -\infty < Y < \infty$

In practice, QR is estimated by $\widehat{QR}$ based on estimates $\widehat{Y}_i, i = 1, 2, 3, 4$. Slifker and Shapiro (1980) have shown that for $s = 3$, QR can be used to discriminate among the three families of Johnson distributions as follows:

- Y has an S_B distribution, yields QR < 1.
- Y has an S_L distribution, yields QR $= 1$.
- Y has an S_U distribution, yields QR > 1.

Minitab considers all potential transformations, with the parameters being estimated, for a given value of z. It transforms the data and checks for normality using the Anderson–Darling test and the corresponding p-value. The routine selects the transformation function that yields the largest p-value that is greater than a default value (0.10). Selecting the transformation with the largest p-value ensures that the transformation selected provides the "most conformance" to normality. If the p-value is not found to be greater than the default value, no transformation is appropriate.

Example 5-7   Consider the data on the unloading time of supertankers, before process changes, in Table 5-2. Select a suitable transformation in case the distribution does not pass the test for normality using an $\alpha = 0.05$. Check for normality of the transformed variable. Explore using Johnson transformation and comment.

Solution   Using Minitab, we first display a graphical summary of the unloading times before process changes. Select **Stat > Basic Statistics > Graphical Summary**. In the window for **Variables**, input the column number or name of the variable, Unloading-Before. Click **OK**. Figure 5-10 shows the graphical summary. The distribution of unloading times is somewhat skewed to the right (skewness coefficient 0.802). Using the Anderson–Darling normality test, we reject the null hypothesis of normality (p-value $= 0.023 < \alpha = 0.05$).

With the distribution being right-skewed and consisting of nonnegative values, we consider the natural logarithm transform. Use **Calc > Calculator**. Indicate a **column number** to store the transformed variable. Under **Expression**, type in Loge(Unloading-Before). Alternatively, one may use the drop-down menu of stored functions that are available. Click **OK**. Now, construct a normal probability plot of the transformed variable, using commands described previously. Figure 5-11 shows the normal probability plot of ln(Unloading-Before). Note that the p-value associated with the Anderson–Darling test for normality is $0.175 > \alpha = 0.05$. So we do not reject the null hypothesis of normality for this transformed variable.

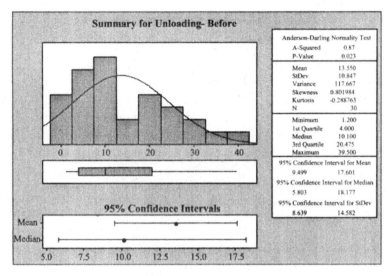

FIGURE 5-10   Graphical summary of unloading times before process changes.

Let us now use the Johnson transformation on the original unloading times before process changes and explore. Click on **Stat** > **Quality Tools** > **Johnson Transformation**. With the data being arranged as a **Single column**, select this, and input the column number or name of the variable, Unloading-Before. For **Store transformed data in**, input the column number. Click **OK**. Figure 5-12 shows the output of using Johnson transformation. The transformed data passes the Anderson–Darling normality test with flying colors (p-value $= 0.991$). The identified distribution is of type S_B, bounded, with the equation of transformation function shown. The output shows normal probability plots for the original and the transformed variables.

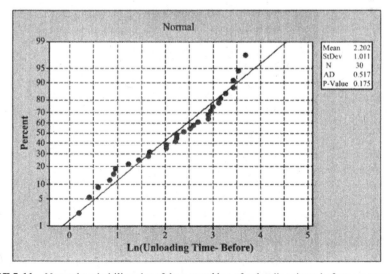

FIGURE 5-11   Normal probability plot of the natural log of unloading times before process changes.

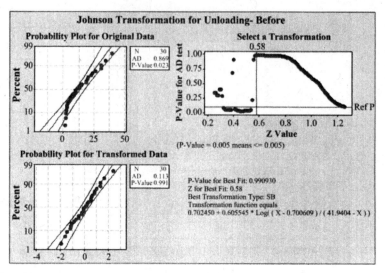

FIGURE 5-12   Johnson transformation of unloading times before process changes.

5-6   ANALYSIS OF COUNT DATA

In many quality improvement projects, the type of data available is of count or frequency type rather than a physical measurement of a variable. An example could be the number of minor, major, or fatal accidents in a chemical company. Here, we could be interested in testing a hypothesis on the proportion of accidents of each type. A second application involves the categorization of the count data by two or more categories. For example, we could have data on the number of minor, major, or fatal accidents categorized by the size of the company: say, small, medium, or large. The objective is to test if the classification variables, type of accident and size of company, are independent. Such a procedure is known as **contingency tables** analysis.

Hypothesis Test on Cell Probabilities

Consider an extension of the binomial experiment, where each independent trial may result in one of k classes or cells. When $k = 2$, we have the binomial situation. Let p_i, $i = 1, 2, \cdots, k$ denote the hypothesized probability of the outcome being in cell i. If the number of outcomes observed in cell i is denoted by n_i, $i = 1, 2, \cdots, k$, we have

$$n_1 + n_2 + \cdots + n_k = n \quad \text{and} \quad p_1 + p_2 + \cdots + p_k = 1$$

where n represents the total number of trials. This is known as a **multinomial experiment**. The expected frequency in cell i is given by

$$E(n_i) = np_i \tag{5-6}$$

A test statistic is a function of the squared difference between the frequencies observed and expected for each cell and is expressed as

$$X^2 = \sum_{i=1}^{k} \frac{[n_i - E(n_i)]^2}{E(n_i)} = \sum_{i=1}^{k} \frac{(n_i - np_i)^2}{np_i} \tag{5-7}$$

It is known that when n is large, X^2 has a chi-squared distribution with $(k - 1)$ degrees of freedom. Thus, critical values may be found from Appendix A-5 for a chosen level of significance (α). Large values of X^2 will lead to rejection of the null hypothesis, which makes an assertion about the values of the cell probabilities. As a rule of thumb, for the chi-squared approximation to hold the expected frequency for each cell should be at least 5.

Example 5-8   A marketing company is attempting to determine consumer preference among three brands of a product. Each customer is asked to indicate his or her preference from among the three brands presented (A, B, and C). Following is the summarized frequency response from 300 consumers:
Is there a brand preference? Test using $\alpha = 0.05$.

Brand	A	B	C
Frequency	70	140	90

Solution   The hypotheses are:

$$H_0 : p_A = p_B = p_C = 1/3 \quad \text{(i.e., no brand preference)}$$

$$H_a : \text{At least one } p_i \text{ is different from } 1/3$$

The test statistic is calculated as

$$X^2 = \frac{(70 - 100)^2}{100} + \frac{(140 - 100)^2}{100} + \frac{(90 - 100)^2}{100} = 26$$

From the chi-squared tables, $\chi^2_{0.05,2} = 5.99$. Since the test statistic exceeds 5.99, we reject the null hypothesis and conclude that consumers have a brand preference for the product.

Contingency Tables

In **contingency tables** analysis, we test if the classification variables are independent of each other, based on count data for each cell. Further, for categorical variables, measures of association may also be found. For simplicity, we present our discussion for two-way tables, even though the concepts extend to the more general situation.
Let us define the following notation:

i: row number, $i = 1, 2, \cdots, r$
j: column number, $j = 1, 2, \cdots, c$
n_{ij}: observed frequency in cell (i, j)
$E(n_{ij})$: expected frequency in cell (i, j)
r_i: row total for row i, $i = 1, 2, \cdots, r$
c_j: column total for column j, $j = 1, 2, \cdots, c$
n: total number of observations $= \sum_{i=1}^{r} \sum_{j=1}^{c} n_{ij}$

Under the null hypothesis of independence of the classification variables, the expected frequencies are given by

$$E(n_{ij}) = \frac{r_i c_j}{n} \tag{5-8}$$

TABLE 5-6   Customer Satisfaction Data in the Hospitality Industry

Degree of Satisfaction Rating	Location					
	1	2	3	4	5	6
1	20	15	20	15	10	30
2	30	25	30	10	25	20
3	40	40	25	25	40	25
4	30	60	40	35	30	20
5	40	30	50	45	25	30

The test statistic is derived as a function of the squared difference between the observed and expected frequencies for each cell and is given by

$$X^2 = \sum_{i=1}^{r} \sum_{j=1}^{c} \frac{[n_{ij} - E(n_{ij})]^2}{E(n_{ij})} \tag{5-9}$$

The test statistic, given by eq. (5-9), has approximately a chi-squared distribution with degrees of freedom of $(r-1)(c-1)$. As a rule of thumb, for the chi-squared approximation to hold, the expected frequency should be at least 5 for each cell.

Example 5-9   A company in the hospitality industry owns hotels in six geographically dispersed locations. Normally, on completion of a stay, the customer is asked to complete a satisfaction survey, where the responses are on an ordinal scale (say, 1 to 5) with the following implications: 1, poor; 2, below average; 3, above average; 4, good; 5, excellent. Table 5-6 shows the responses tallied for each cell. Can we conclude that the degree of customer satisfaction is independent of the location of the facility? Use $\alpha = 0.05$.

Solution   Using Minitab, first create a worksheet with the following variables: **Rating** (stands for the rating score labels), **Location** (representing the six locations), and **Frequency** (indicating the count observed for the corresponding cell). Click on **Stat > Tables > Cross Tabulation** and **Chi-Square**. Under **Categorical variables, For rows,** input Rating; **For Columns,** input Location; for **Frequencies are in,** input Frequency. Click on **Chi-Square** button, and choose to **Display Chi-square analysis** and **Expected cell counts**. Click **OK**. The output from Minitab is shown in Figure 5-13. Note that for each cell, the observed and expected counts are displayed. The test statistic value is 61.515 with 20 degrees of freedom and a p-value of 0.000. Thus, we reject the null hypothesis of independence of location versus degree of customer satisfaction. From the analysis we conclude that degree of customer satisfaction is influenced by the location of the facility. A follow-up analysis may involve determining the facilities that do not meet customer expectations and identifying remedial measures.

Measures of Association

In contingency tables, measures of association between the categorical variables are based on the values of the test statistic, X^2, given by eq. (5-9). When $X^2 = 0$, there is exact agreement between the cell frequencies observed and expected under the null hypothesis of independence. As the values of X^2 increase, they imply dependence between the categorical variables.

```
Rows: Rating    Columns: Location

              1         2         3         4         5         6      All

1            20        15        20        15        10        30      110
          20.00     21.25     20.63     16.25     16.25     15.63   110.00

2            30        25        30        10        25        20      140
          25.45     27.05     26.25     20.68     20.68     19.89   140.00

3            40        40        25        25        40        25      195
          35.45     37.67     36.56     28.81     28.81     27.70   195.00

4            30        60        40        35        30        20      215
          39.09     41.53     40.31     31.76     31.76     30.54   215.00

5            40        30        50        45        25        30      220
          40.00     42.50     41.25     32.50     32.50     31.25   220.00

All         160       170       165       130       130       125      880
         160.00    170.00    165.00    130.00    130.00    125.00   880.00

Cell Contents:          Count
                        Expected count


Pearson Chi-Square = 61.515, DF = 20, P-Value = 0.000
```

FIGURE 5-13   Contingency table analysis of customer data in the hospitality industry.

One measure is the **mean-squared contingency,** given by

$$\Phi^2 = \frac{X^2}{n} \tag{5-10}$$

The maximum value of X^2 is $n(q-1)$, where $q = \min(r, c)$. So the maximum value of Φ is $\sqrt{q-1}$. For a 2×2 table, Φ lies between 0 and 1.

Another measure is **Cramer's** V, given by

$$V = \sqrt{\frac{X^2}{n(q-1)}} \tag{5-11}$$

This measure lies between 0 and 1. If the calculated test statistic, X^2, is significant, as determined from the chi-squared test, each of the measures Φ and V will be considered as significantly different from zero.

5-7   CONCEPTS IN SAMPLING

In quality control it is often not feasible, due to a lack of time and resources, to obtain data regarding a certain quality characteristic for each item in a population. Furthermore, sample data provide adequate information about a product or process characteristic at a fraction of the cost. It is therefore important to know how samples are selected and the properties of various sampling procedures.

A **sampling design** is a description of the procedure by which the observations in a sample are to be chosen. It does not necessarily deal with the measuring instrument to be used. For example, a sampling design might specify choosing every tenth item produced.

In the context of sampling, an **element** is an object (or group of objects) for which data or information are gathered. A **sampling unit** is an individual element or a collection of nonoverlapping elements from a population. A **sampling frame** is a list of all sampling units. For example, if our interest is confined to a set of parts produced in the month of July (sampling element), the sampling unit could be an individual part, while the sampling frame would be a list of the part numbers of all the items produced.

Sampling Designs and Schemes

A major objective of any sampling design or scheme is to select the sample in such a way as to accurately portray the population from which it is drawn. After all, a sample is supposed to be representative of the population.

Sampling in general has certain advantages. If the measurement requires destroying the item being measured (destructive testing), we cannot afford to obtain data from each item in the population. Also, in measurements involving manual methods or high production rates, inspector fatigue may result, which would yield inaccurate data.

Errors in Sampling   There are three sources of errors in sample surveys. The first source is **random variation**. The inherent nature of sampling variability causes such errors to occur. The more sophisticated the measuring instrument, the lower the random variation.

Misspecification of the population is a second source of error. This type of error occurs in public opinion polling, in obtaining responses regarding consumer satisfaction with a product, in listing a sampling frame incorrectly, and so on.

The third source of error deals with **nonresponses** (usually, in sample surveys). This category also includes situations where a measurement is not feasible due to an inoperative measuring instrument, a shortage of people responsible for taking the measurement, or other such reasons.

Simple Random Sample   One of the most widely used sampling designs in quality control is the simple random sample. Suppose that we have a finite population of N items from which a sample of n items is to be selected. If samples are chosen such that each possible sample of size n has an equal chance of being selected, the sampling process is said to be random, and the sample obtained is known as a **simple random sample**.

Random-number tables (or computer-generated uniform random numbers) can be used to draw a simple random sample. For example, if there are 1000 elements in the population, the three-digit numbers 000 through 999 are used to identify each element. To start, a random number is selected; one element corresponds to this number. This selection continues until the desired sample of size n is chosen. If a random number that has already been used comes up, it is ignored and another one is chosen.

In estimating the population mean (μ) by the sample mean ($\bar{X}$), the conclusions from the central limit theorem, discussed in Chapter 4, are used to describe the **precision** of the estimator. The precision is the inverse of the sampling variance and is estimated by

$$\hat{\sigma}_{\bar{x}}^2 = \frac{s^2}{n} \left(\frac{N-n}{N} \right) \tag{5-12}$$

where s^2 represents the sample variance. The term $(N-n)/N$ is known as the *finite population correction factor*. If N is very large compared to n, the expression for $\hat{\sigma}_{\bar{x}}^2$ is s^2/n.

Stratified Random Sample   Sometimes the population from which samples are selected is heterogeneous. For instance, consider the output from two operators who are known to differ greatly in their performance. Rather than randomly selecting a sample from their combined output, a random sample is selected from the output of each operator. In this way both operators are fairly represented, so we can determine whether there are significant differences between them. Another application may deal with estimation of the average response time of emergency services, such as ambulances in a city. Given the traffic density in different parts of the city, location of the ambulance centers, and/or location of hospitals, it could be meaningful to divide the city into certain segments or strata. Then, random samples could be chosen from each stratum for the purpose of estimation. Thus, a **stratified random sample** is obtained by separating the elements of the population into nonoverlapping distinct groups (called *strata*) and then selecting a simple random sample from each stratum.

When the variance of the observations within a stratum is smaller than the overall population variance, stratified sampling may yield a variance of the estimator that is smaller than that found from a simple random sample of the same size. Another advantage of stratified sampling is that the cost of collecting data could be reduced. Through proper selection of strata such that elements within a stratum are homogeneous, smaller samples could be selected from strata where unit costs of sampling are higher. Finally, through stratified sampling, it is possible to obtained parameter estimates for each stratum, which is of importance in heterogeneous populations.

We consider a common scheme of **proportional allocation** where the sample size is partitioned among the strata in the same proportion as the size of the strata to the population. Let N_i, $i = 1, 2, \cdots, k$ represent the size of the k strata in the population of size N, where $N = \sum_{i=1}^{k} N_i$. If the sample size is represented by n, the size from each strata using proportional allocation is given by

$$n_i = n\frac{N_i}{N}, \quad i = 1, 2, \cdots, k \tag{5-13}$$

Estimates of the mean and variance of each stratum are

$$\bar{x}_i = \frac{\sum_{j=1}^{n_i} x_{ij}}{n_i} \tag{5-14}$$

$$s_i^2 = \frac{\sum_{j=1}^{n_i} (x_{ij} - \bar{x}_i)^2}{n_i - 1}, \quad i = 1, 2, \cdots, k$$

Further, the estimate of the population mean and the variance of the estimator, for a stratified proportional sample, are given by

$$\bar{x}_{st} = \frac{1}{N} \sum_{i=1}^{k} N_i \bar{x}_i$$

$$\text{Var}(\bar{x}_{st}) = \frac{1}{N^2} \sum_{i=1}^{k} N_i^2 \left(\frac{N_i - n_i}{N_i} \right) \frac{s_i^2}{n_i}, \quad i = 1, 2, \cdots, k \tag{5-15}$$

Cluster Sample   In the event that a sampling frame is not available, or if obtaining samples from all segments of the population is not feasible for geographic reasons, a **cluster sample** can be used. Here, the population is first divided into groups of elements, or clusters. Clusters are randomly selected, and a census of data is obtained (i.e., all of the elements within the chosen clusters are examined). If a company has plants throughout the southeastern United

States, it may not be feasible to sample from each plant. Clusters are then defined (say, one for each plant), and some of the clusters are then randomly chosen (say, three of the five clusters). A census of data is then obtained for each selected cluster. Cluster sampling is less costly, relative to a simple or stratified random sampling, if the cost of obtaining a frame with a listing of all population elements is high. Also, if the cost of obtaining observations increases with the distance separating the elements (i.e., travel cost), cluster sampling may be preferred.

Sampling error may be reduced by choosing many small clusters rather than a few large clusters, since we reduce the probability of excluding certain groups of elements. Let M and m denote the number of clusters in the population and the number of sampled clusters, respectively, with n_i representing the size of the ith cluster. Let t_i represent the total of the measurements in cluster i, (i.e., $t_i = \sum_j x_{ij}$). The estimate of the population mean and the variance of the estimator, using cluster sampling, is given by

$$\bar{x}_{\text{cl}} = \frac{\sum_{i=1}^{m} t_i}{\sum_{i=1}^{m} n_i}$$

(5-16)

$$\text{Var}(\bar{x}_{\text{cl}}) = \left(\frac{M-m}{Mm\bar{n}^2}\right) \frac{\sum_{i=1}^{m}(t_i - \bar{x}_{\text{cl}}n_i)^2}{m-1}$$

where $\bar{n} = \sum_{i=1}^{m} n_i/m$.

Sample Size Determination

The size of a sample has a direct impact on the reliability of the information provided by the data. The larger the sample size, the more valuable the data. It is usually of interest to know the minimum sample size that can be used to estimate an unknown product or process parameter accurately, under certain acceptable levels of tolerance. Here, we focus on **simple random samples**.

Bound on the Error of Estimation and Associated Confidence Level

Estimating the Population Mean   Suppose that the mean of a product or process characteristic is of interest (e.g., the mean time to process a transaction in a marketing firm). There is a $(1 - \alpha)$ probability that the difference between the estimated value and the actual value will be no greater than some number B. Figure 5-14 shows the principle behind selecting an appropriate sample size for estimating the population mean. The quantity B, sometimes referred to as the *tolerable error bound*, is given by

$$B = z_{\alpha/2}\sigma_{\bar{x}} = z_{\alpha/2}\frac{\sigma}{\sqrt{n}}$$

(5-17)

or

$$n = \frac{z_{\alpha/2}^2\sigma^2}{B^2}$$

(5-18)

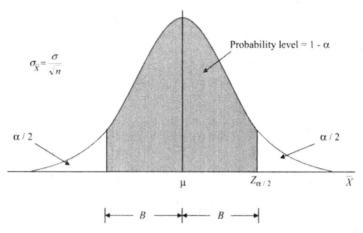

FIGURE 5-14   Sample size determination for estimating the population mean μ.

Example 5-10   An analyst wishes to estimate the average bore size of a large casting. Based on historical data, it is estimated that the standard deviation of the bore size is 4.2 mm. If it is desired to estimate with a probability of 0.95 the average bore size to within 0.8 mm, find the appropriate sample size.

Solution   We have $\hat{\sigma} = 4.2$, $B = 0.8$, and $z_{0.025} = 1.96$. Thus,

$$\text{sample size } n = \frac{(1.96)^2(4.2)^2}{(0.8)^2} = 105.88 \simeq 106$$

Estimating the Population Proportion   Consider a binomial population where the objective is to estimate the proportion of "successes" (p). Examples might include estimating the proportion of nonconforming stamped parts in a press shop or the proportion of unsatisfied customers in a restaurant. Here again, we must select a tolerable error bound B such that that estimate will have a probability of $(1 - \alpha)$ of being within B units of the parameter value. We use the concept of the sampling distribution of the sample proportion of successes ($\hat{p}$), which is approximately normal for large sample sizes. The equation for determining the sample size n is given by

$$B = z_{\alpha/2}\sigma_{\hat{p}} = z_{\alpha/2}\sqrt{\frac{p(1-p)}{n}} \tag{5-19}$$

or

$$n = \frac{z_{\alpha/2}^2 p(1-p)}{B^2} \tag{5-20}$$

Since the true parameter value p is not known, there are a couple of ways in which eq. 5-20 can be modified. First, if a historical estimate of p is available (say, $\hat{p}$), it can be used in place of p. Second, if no prior information is available, a conservative estimate of n can be calculated by using $p = 0.5$. Using $p = 0.5$ maximizes the value of $p(1-p)$ and hence produces a conservative estimate.

Example 5-11   In the production of rubber tubes, the tube stock first has to be cut into a piece of a specified length. This piece is then formed into a circular shape and joined using pressure and the correct temperature. The operator training and such process parameters as temperature, pressure, and die size influence the production of conforming tubes. We want to estimate with a probability of 0.90 the proportion of nonconforming tubes to within 4%. How large a sample should be chosen if no prior information is available on the process?

Solution   Using the preceding notation, $B = 0.04$. From Appendix A-3, $z_{0.05} = 1.645$. Since no information on p, the proportion of nonconforming tubes, is available, use $p = 0.5$. Hence,

$$n = \frac{(1.645)^2 (0.5)(0.5)}{(0.04)^2} = 422.8 \simeq 423$$

If a prior estimate of p had been available, it would have reduced the required sample size.

Estimating the Difference of Two Population Means

Suppose that our interest is to estimate the difference in the mean project completion time, in a consulting firm, before (μ_1) and after (μ_2) some process improvement changes. The estimator is the difference in the sample means, $\bar{x}_1 - \bar{x}_2$. We have

$$B = z_{\alpha/2} \sqrt{\frac{\sigma_1^2}{n_1} + \frac{\sigma_2^2}{n_2}}$$

where σ_1^2 and σ_2^2 represent the population variances, respectively, before and after changes, and n_1 and n_2 represent the corresponding sample sizes. Under the assumption of equal sample sizes ($n_1 = n_2 = n$), we have

$$n = \frac{z_{\alpha/2}^2 (\sigma_1^2 + \sigma_2^2)}{B^2} \tag{5-21}$$

Estimating the Difference of Two Population Proportions

A company wishes to estimate the difference in the proportion of satisfied customers between its two branch offices. If the proportion of successes in each population is defined as p_1 and p_2, respectively, then for large samples, we have

$$B = z_{\alpha/2} \sqrt{\frac{p_1(1 - p_1)}{n_1} + \frac{p_2(1 - p_2)}{n_2}}$$

where n_1 and n_2 represent the corresponding sample sizes from each population. Under the assumption of equal sample sizes ($n_1 = n_2 = n$), we have

$$n = \frac{z_{\alpha/2}^2 [p_1(1 - p_1) + p_2(1 - p_2)]}{B^2} \tag{5-22}$$

When historical estimates of p_1 and p_2 are available, those estimates, $\widehat{p}_1$ and $\widehat{p}_2$, are used in eq. (5-22). Alternatively, with no prior information, using $p_1 = 0.5$ and $p_2 = 0.5$ will provide a conservative estimate.

Controlling the Type I Error, Type II Error, and Associated Parameter Shift

In the previous discussion on calculating sample sizes, we were interested in controlling the error bound (B) and the associated confidence level $(1 - \alpha)$. Recall that in the hypothesis testing context, the type I error (rejecting a null that is true) is represented by α. Now suppose that we are also interested in controlling the probability of a type II error (not rejecting the null when it is false), β, associated with a shift (δ) in the parameter value. How large a sample must be chosen to meet these tolerable conditions? Sometimes, instead of β, the power of the test $(1 - \beta)$ may be indicated.

Let us demonstrate the derivation concept in the context of estimating the population mean (μ): say, the mean delivery time of packages by a company. Due to an anticipated increase in volume, we would like to determine if the mean delivery time exceeds a hypothesized value (μ_0). If the mean does increase by an amount δ, we would like to detect such with a power given by $(1 - \beta)$. The hypothesis being tested are: $H_0: \mu \leq \mu_o; H_a: \mu > \mu_0$. The critical value and the associated type I and II errors, considering the sampling distribution of $\bar{X}$ when σ is known, is shown in Figure 5-15.

Using the distribution under H_0, when $\mu = \mu_0$, the critical value (CV) is obtained as

$$CV = \mu_0 + z_{1-\alpha} \frac{\sigma}{\sqrt{n}} \tag{5-23}$$

where $z_{1-\alpha}$ is the $(1 - \alpha)$th quantile of the standard normal distribution. Under H_a, when $\mu = \mu_1 = \mu_0 + \delta$, the critical value is given by

$$CV = \mu_1 - z_{1-\beta} \frac{\sigma}{\sqrt{n}} \tag{5-24}$$

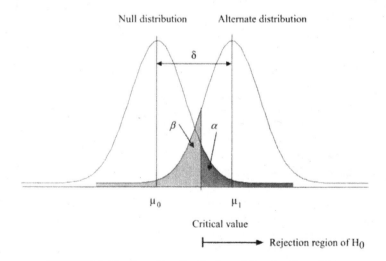

FIGURE 5-15   Sampling distribution of $\bar{X}$ under H_0 and H_a.

Equating eqs. (5-23) and (5-24), we obtain the sample size as

$$n = \frac{(z_{1-\alpha} + z_{1-\beta})^2 \sigma^2}{\delta^2} \tag{5-25}$$

When we have a **two-tailed hypothesis** test of the type H_0: $\mu = \mu_0$ versus H_a: $\mu \neq \mu_0$, a slight modification of eq. 5-23 is necessary. In this case, the rejection region is on both tails of the null distribution, with the area on each tail being $\alpha/2$. So, in computing the necessary sample size, replace $z_{1-\alpha}$ with $z_{1-\alpha/2}$ in eq. (5-25).

A similar approach may be conducted for determining the sample size when the parameter of interest is the difference in the means of two populations, proportion of successes in a binomial population, or difference in the proportion of successes of two binomial populations.

Minitab has an attractive feature in determination of sample sizes under a variety of parameter testing situations in its **Power and Sample Size** option. Information is input on the given level of the type I error (α), specification of the form of H_a (not equal to, less than, or greater than), and an estimate of the population standard deviation (σ). Minitab then outputs either the sample size, power, or the degree of difference that one wishes to detect (δ) when any two of the three quantities are specified.

Example 5-12   A company is interested in determining whether a product advertising campaign that has been used in one geographical region has had a positive impact. It wishes to detect an increase in average sales per customer of $50 with a probability of 0.98. The tolerable level of a type I error is 0.05. Assuming that the estimate of the standard deviation of sales per customer is $70 in both the region where the campaign was used and elsewhere, how large a sample must be chosen from each region? Assume equal sample sizes.

Solution   The hypotheses are H_0: $\mu_1 - \mu_2 \leq 0$; H_a: $\mu_1 - \mu_2 > 0$, where μ_1 and μ_2 represent mean sales per customer in the region where advertising campaign was used and not used, respectively. Also, $\hat{\sigma}_1 = \hat{\sigma}_2 = 70$, with $\alpha = 0.05$ and power $= 0.98$ when δ (difference in mean sales) $= 50$. Using Minitab, click on **Stat > Power and Sample Size > 2- Sample t**. Under **Difference**, input 50; under **Power Values**, input 0.98; under **Standard deviation**, input 70. Click on **Options**. For **Alternative Hypothesis**, select greater than and input **significance level** as 0.05. Click **OK**. The output from Minitab indicates that the sample size for each group should be 55.

SUMMARY

In this chapter we present some analytical and graphical tools for describing populations and making inferences on their parameters. Using observed values of the quality characteristic, empirical distribution plots such as frequency histograms, stem-and-leaf plots, or box plots are methods for displaying distributional features. Such plots indicate the location and shape of the empirical distribution, whether they are skewed, and the degree of variability. They can also be used to identify any outliers.

A test for indicating the randomness of a sequence of observations, usually in time, is discussed through a run chart. Since many parametric tests require assumptions on the distribution of the quality characteristic, typically normality, methods for testing such assumptions are presented. To justify the use of certain statistical inferential procedures when the distribution of the characteristic is nonnormal, certain transformations that are useful in achieving normality are exhibited. The use of power transformations and a family of transformations known as the Johnson family are discussed.

In many quality improvement analyses, the data that are collected may not be quantitative. For example, in consumer surveys, product preference may be observed in terms of frequency or count data. Methods of analysis of count data are presented. When classification of the observations is through two or more categorical variables, contingency table analysis is used to test independence of the classification variables. Measures of association between such categorized variables are indicated.

An important issue in product/process analysis involves determining how the sampling procedure is to be conducted. In this context, methods such as simple random sampling, stratified sampling, and cluster sampling are addressed. A follow-up issue is the size of the sample to select, based on certain tolerable levels of risk. For the case of simple random samples, expressions are derived for the appropriate sample size in various inferential contexts. Thus, the chapter provides a good exposure to how much data should be collected, how to collect it, and how to analyze it to make inference on product and process characteristics.

KEY TERMS

Anderson–Darling test
association measures
box plot
   notched
chi-squared test
clustering
contingency tables
count data
cumulative distribution function
distribution
   empirical
error bound
exponential probability plot
frequency
   distribution
   histogram
interquartile range
kurtosis
misspecification
mixture pattern
multinomial experiment
nonresponse
normal probability plot
oscillation
outlier
p-value
power
proportional allocation

quantile ratio
quartile
random sequence
run chart
sample
   cluster
   simple random
   stratified
sampling
   designs
   distribution
   element
   errors in
   frame
   unit
sample size
   determination of
skewness
stem-and-leaf plot
tail length
trend
transformations
   Johnson
   power
type I error
type II error
whisker

EXERCISES

Discussion Questions

5-1 Explain some specific parametric tests that require the distributional assumption of normality. What do you do if the assumption is not satisfied?

5-2 A county wishes to estimate the average income of people, where it is known that income levels vary quite a bit among the residents in sections of the county. What type of sampling scheme should be used?

5-3 Data from a survey of customers are on an ordinal scale (1 to 7) regarding satisfaction with the services provided in a bank by three tellers. If we wish to determine if the degree of customer satisfaction is independent of the teller, what statistical test should be used?

5-4 A financial institution is contemplating the offering of three different types of savings accounts. They select 100 customers randomly and obtain a response on which account each person would select. What statistical test would you use to determine if customers have a preference for any account?

5-5 Explain the relationship between a type I error, power, degree of difference that one wishes to detect in a parameter value, and sample size. How can a type I error be reduced and the power be increased, for a given difference in the parameter?

5-6 Explain type I and type II errors in the context of sampling from customers' accounts to identify billing errors in a large retail store. What are the associated costs of these two types of errors?

5-7 A fast-food restaurant in an urban area experiences a higher traffic rate during the early morning and lunch hours. In order to conduct a customer satisfaction survey, discuss possible sampling procedures.

Problems

5-8 A random sample of 50 observations on the waiting time (in seconds) of customers before speaking to a representative at a call center is as follows:

33.2	29.4	36.5	38.1	30.0
29.1	32.2	29.5	36.0	31.5
34.5	33.6	27.4	30.4	28.4
32.6	30.4	31.8	29.8	34.6
30.7	31.9	32.3	28.2	27.5
34.9	32.8	27.7	28.4	28.8
30.2	26.8	27.8	30.5	28.5
31.8	29.2	28.6	27.5	28.5
30.8	31.8	29.1	26.9	34.2
33.5	27.4	28.5	34.8	30.5

(a) Construct a histogram and comment on the process.

(b) What assumptions are necessary to test if the mean waiting time is less than 32 seconds?

(c) Make an appropriate transformation to satisfy the assumption stated in part (b) and validate it. Use $\alpha = 0.05$.

(d) Test to determine if the mean waiting time is less than 32 seconds. Use $\alpha = 0.05$.

(e) Find a 90% confidence interval for the variance of waiting time.

5-9   Using the call waiting time of customers data in Exercise 5-8, construct a stem-and-leaf plot. Construct a box plot and comment on the distribution. Are there any outliers? Construct a 95% confidence interval for the median.

5-10   The pH values of a dye for 30 samples taken consecutively over time are listed row-wise:

20.3	15.5	18.2	18.0	20.5	22.8
21.6	21.0	22.5	23.8	23.9	24.2
23.6	24.9	27.4	25.5	20.9	25.8
24.6	25.5	27.3	26.4	26.8	27.5
26.4	26.8	27.2	27.1	27.4	27.8

(a) Can we conclude that the sequence of pH values over time is nonrandom? Use $\alpha = 0.05$.

(b) Assume that dyes are produced in batches, with 10 random samples taken from each of three batches. Explain possible special causes in the process.

(c) What type of special cause (clustering, mixture, trend, oscillation) do you suspect?

(d) What type of sampling scheme was used in this process, and is it appropriate?

5-11   Percentage investment in new-product development is monitored for a pharmaceutical company, by quarter. The observations are listed row-wise:

32.9	31.5	34.3	36.8	35.0
29.4	33.2	37.8	35.0	32.7
28.5	30.4	32.6	31.5	30.6
35.8	36.4	34.2	35.0	33.5
31.8	32.5	28.4	33.8	35.1
30.2	33.0	34.6	32.4	32.0

(a) Can we conclude that the sequence of investment percentages is random? Use $\alpha = 0.05$.

(b) Is there clustering? Is there a trend? Use $\alpha = 0.05$.

(c) Can you conclude that the distribution of investment percentages is normal? Use $\alpha = 0.01$.

(d) Find a 98% confidence interval for the mean percentage investment.

(e) Find a 98% confidence interval for the standard deviation of the percentage investment.

(f) Can we conclude that the mean percentage investment in new products exceeds 31%? Use $\alpha = 0.01$. What is the p-value? Explain.

5-12   An automobile manufacturing company is attempting to cope with rising health care costs of its employees. It offers three types of plans. Based on a random sample of 200 employees selected last year, Table 5-7 shows the number that selected each plan. The same table also shows, for a random sample of 150 employees selected in the current year, the number that selected each plan. Can we conclude that employee preferences for the health care plans have changed from last year? Use $\alpha = 0.05$. What is the p-value?

TABLE 5-7

Year	Plan 1	Plan 2	Plan 3
Last year	50	40	110
Current year	15	45	90

5-13   The impact of three different advertising techniques is being studied by a marketing firm. Sales, in thousand dollars, categorized in four groups are shown for each advertising technique for 200 randomly selected customers, exposed to each technique, in Table 5-8. Can we conclude that the advertising technique has an impact on sales? Use $\alpha = 0.10$. What is the p-value?

TABLE 5-8

Sales (thousands)	Technique 1	Technique 2	Technique 3
0–99	60	40	75
100–199	85	30	70
200–299	20	70	30
300–399	35	60	25

5-14   A survey asked customers to rate their overall satisfaction with a service company as well as the speed in responding to their requests. For both questions, a rating scale of 1 to 5 was used with the following guidelines: 1, poor; 2, below average; 3, above average; 4, good; and 5, excellent. Table 5-9 shows the tabulated frequencies based on responses to both questions. Does response speed influence overall satisfaction? Use $\alpha = 0.01$. Find Cramer's index of association.

TABLE 5-9

Response Speed Rating	Overall Satisfaction Rating				
	1	2	3	4	5
1	40	3	0	0	0
2	1	30	8	6	5
3	2	5	49	15	45
4	0	6	33	45	60
5	0	12	15	50	70

5-15   The Small Business Administration (SBA) wishes to estimate the mean annual sales of companies that employ fewer than 20 persons. Historical data suggest that the standard deviation of annual sales of such companies is about $5500.

(a) If the SBA wants to estimate mean annual sales to within \$1000 with a probability of 0.90, how large a sample should be chosen?

(b) Suppose that the SBA has a tolerable type I error rate of 10%. It wishes to detect a difference in mean annual sales of \$500 with a probability of 0.95. How large a sample should be chosen?

5-16 A company's quality manager wants to estimate, with a probability of 0.90, the copper content in a mixture to within 4%.

(a) How many samples must be selected if no prior information is available on the proportion of copper in the mixture?

(b) Suppose that the manager has made some changes in the process and wishes to determine if the copper content exceeds 15%. The manager would like to detect increases of 2% or more, with a probability of 0.98. How large a sample should be chosen for a chosen level of significance of 2%?

5-17 The personnel manager wants to determine if there is a difference in the average time lost, due to absenteeism, between two plants. From historical data, the estimated standard deviations of lost time are 200 and 250 minutes, respectively, for plants 1 and 2.

(a) Assuming that equal sample sizes will be selected from each plant, what should the sample size be if the bound on the error of estimation is 40 minutes with a probability of 0.90?

(b) If the unit sampling costs are less in plant 1 compared to plant 2, such that a sample that is twice as large could be selected from plant 1, what are the respective sample sizes?

(c) Suppose that it is desired to detect a difference of 30 minutes in the lost time with a probability of 0.80. Assume that there is no significant difference in the standard deviations of lost time. For a chosen level of significance of 0.10, what should the sample size be, assuming samples of equal size from both plants?

5-18 By incorporating certain process improvement methods, a company believes that it has reduced the proportion of nonconforming product and wishes to test this belief.

(a) If the operations manager selects an error bound for the difference in the proportion nonconforming before and after implementation of process changes as 4%, with a probability of 0.90, what sample size should be selected? Assume that samples of equal size will be chosen for before and after implementation of process changes.

(b) What measures could you take to reduce the sample size?

(c) Suppose that the hypothesized proportion nonconforming is 8% prior to process changes. We wish to detect reductions of 5% or more with a probability of 0.80. For a chosen level of significance of 10%, what should the sample sizes be? Assume samples of equal size before and after process changes.

5-19 The production of nonconforming items is of critical concern because it increases costs and reduces productivity. Identifying the causes behind the production of unacceptable items and taking remedial action are steps in the right direction. To begin, we decide to estimate the proportion of nonconforming items in a process to within 3% with a probability of 0.98.

(a) Previous information suggests that the percentage nonconforming is approximately 4%. How large a sample must be chosen?

(b) If no prior information is available on the proportion nonconforming, how large a sample must be chosen?

5-20   Management is exploring the possibility of adding more agents at the check-in counter of an airline to reduce the waiting time of customers. Based on available information, the standard deviation of the waiting times is approximately 5.2 minutes. If management wants to estimate the mean waiting time to within 2 minutes, with a probability of 0.95, how many samples must be selected?

5-21   It is desired to estimate the average consumer spending on durable goods, annually, in a defined geographical region. Spending patterns are influenced by disposable income. Based on county tax records, the population has been segmented into three groups: (1) annual income $< \$50,000$; (2) between \$50,000 and \$100,000; (3) over \$100,000. The number of people in each category has been identified as 2000, 7000, and 1000, respectively. Sampling costs limit the total sample size to be 200.

(a) Using a stratified proportional sample, what should the sample sizes be from each group?

(b) Using the sample sizes found in part (a), the following sample means and standard deviations (in thousands of dollars) were found: $\bar{x}_1 = 3.5$, $s_1 = 1.2$; $\bar{x}_2 = 7.6$, $s_2 = 2.8$; $\bar{x}_3 = 15.1$, $s_3 = 6.4$. Estimate the mean and standard deviation of the amount spent on durables.

(c) Find a bound on the error of estimation using a 95% level of confidence.

(d) Suppose that you had selected a simple random of the same total size from the population and that the results indicated the following sample mean and standard deviation: $\bar{x} = 6.8$, $s = 5.6$. Find a bound on the error of estimation using a 95% level of confidence.

(e) What is the percentage improvement in precision by using a stratified proportional sample relative to a simple random sample?

5-22   Consider Exercise 5-21 and suppose that a cluster sample is chosen, where the randomly groups selected are (2) and (3). The total of all observations from group 2 is \$7,100,000, while that from group 3 is \$15,200,000.

(a) Estimate the mean spending and its standard error.

(b) Find a bound on the error of estimation using a 95% level of confidence.

(c) What are the disadvantages of using cluster samples in this situation?

REFERENCES

Chambers, J. M. (1977). *Computational Methods for Data Analysis*. New York: Wiley.

Chambers, J. M., W. S. Cleveland, B. Kleiner, and P. A. Tukey (1983). *Graphical Methods for Data Analysis*. Belmont, CA: Wadsworth.

Chou, Y., A. M. Polansky, and R. L. Mason (1998). "Transforming Nonnormal Data to Normality in Statistical Process Control", *Journal of Quality Technology*, 22: 223–229.

Duncan, A. J. (1986). *Quality Control and Industrial Statistics*, 5th ed. Homewood, IL: Richard D. Irwin.

Johnson, N. L. (1949). "Systems of Frequency Curves Generated by Methods of Translation", *Biometrika*, 36: 149–176.

Kendall, M. G., A. Stuart, J. K. Ord, S. F. Arnold, and A. O'Hagan (1998). *Kendall's Advanced Theory of Statistics*, Vol. 1, London: Arnold Publishers.

Massey, F. J., Jr. (1951). "The Kolmogorov–Smirnov Test of Goodness of Fit", "*Journal of the American Statistical Association*", 46: 68–78.

McGill, R., J. W. Tukey, and W. A. Larsen (1978). "Variations of Box Plots", *The American Statistician*, 32 (1): 12–16.

Minitab, Inc. (2007). Release 15. State College, PA: Minitab.

Slifker, J. F., and S. S. Shapiro (1980). "The Johnson System: Selection and Parameter Estimation", *Technometrics*, 22: 239–246.

Stephens, M. A. (1974). "EDI Statistics for Goodness of Fit and Some Comparisons", *Journal of the American Statistical Association*, 69: 730–737.

PART III

STATISTICAL PROCESS CONTROL

6

STATISTICAL PROCESS CONTROL USING CONTROL CHARTS

6-1   INTRODUCTION AND CHAPTER OBJECTIVES

We have discussed at length the importance of satisfying the customer by improving the product or service. A process capable of meeting or exceeding customer requirements is a key part of this endeavor. Part III of the book deals with the topic of process control and improvement. It provides the necessary background for understanding statistical process control through control charts. In this chapter we build the foundation for using control charts. The objectives of this chapter are to introduce the principles on which control charts are based. The basic features of the control charts, along with the possible inferential errors and how they may be reduced, are presented. Various types of out-of-control patterns are also discussed. In Chapter 7 we examine control charts for variables, and in Chapter 8 we discuss control charts for attributes. Process capability analysis is covered in Chapter 9.

A **control chart** is a graphical tool for monitoring the activity of an ongoing process. Control charts are sometimes referred to as **Shewhart control charts**, because Walter A. Shewhart first proposed their general theory. The values of the quality characteristic are plotted along the vertical axis, and the horizontal axis represents the samples, or subgroups (in order of time), from which the quality characteristic is found. Samples of a certain size (say, 4 or 5 observations) are selected, and the quality characteristic (say, average length) is calculated based on the number of observations in the sample. These characteristics are then plotted in the order in which the samples were taken. Figure 6-1 shows a typical control chart.

Examples of quality characteristics include average length, average diameter, average tensile strength, average resistance, and average service time. These characteristics are *variables*, and numerical values can be obtained for each. The term *attribute* applies to such

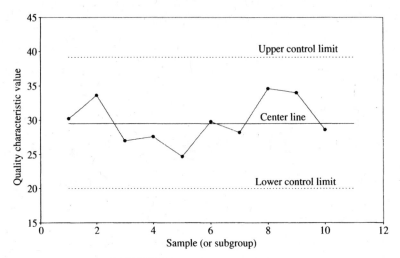

FIGURE 6-1 Typical control chart.

quality characteristics as the proportion of nonconforming items, the number of noncon-formities in a unit, and the number of demerits per unit.

Three lines are indicated on the control chart. The **centerline**, which typically represents the average value of the characteristic being plotted, is an indication of where the process is centered. Two limits, the **upper control limit** and the **lower control limit**, are used to make decisions regarding the process. If the points plot within the control limits and do not exhibit any identifiable pattern, the process is said to be *in statistical control*. If a point plots outside the control limits or if an identifiable nonrandom pattern exists (such as 12 out of 14 successive points plotting above the centerline), the process is said to be *out of statistical control*. Details are given in Section 6-5 on the rules for identifying out-of-control conditions.

Several benefits can be realized by using control charts. Such charts indicate the following:

1. *When to take corrective action.* A control chart indicates when something may be wrong so that corrective action can be taken.

2. *Type of remedial action necessary.* The pattern of the plot on a control chart diagnoses possible causes and hence indicates possible remedial actions.

3. *When to leave a process alone.* Variation is part of any process. A control chart shows when an exhibited variability is normal and inherent such that no corrective action is necessary. As explained in Chapter 2, inappropriate overcontrol through frequent adjustments only increases process variability.

4. *Process capability.* If the control chart shows a process to be in statistical control, we can estimate the capability of the process and hence its ability to meet customer requirements. This helps product and process design.

5. *Possible means of quality improvement.* The control chart provides a baseline for instituting and measuring quality improvement. Control charts also provide useful information regarding actions to take for quality improvement.

6-2 CAUSES OF VARIATION

Variability is a part of any process, no matter how sophisticated, so management and employees must understand it. Several factors over which we have some control, such as methods, equipment, people, materials, and policies, influence variability. Environmental factors also contribute to variability. The causes of variation can be subdivided into two groups: common causes and special causes. Control of a process is achieved through the elimination of special causes. Improvement of a process is accomplished through the reduction of common causes.

Special Causes

Variability caused by **special** or *assignable* **causes** is something that is not inherent in the process. That is, it is not part of the process as designed and does not affect all items. Special causes can be the use of a wrong tool, an improper raw material, or an incorrect procedure. If an observation falls outside the control limits or a nonrandom pattern is exhibited, special causes are assumed to exist, and the process is said to be out of control. One objective of a control chart is to detect the presence of special causes as soon as possible to allow appropriate corrective action. Once the special causes are eliminated through **remedial actions**, the process is again brought to a state of statistical control.

Deming believed that 15% of all problems are due to special causes. Actions on the part of both management and employees will reduce the occurrence of such causes.

Common Causes

Variability due to **common** or *chance* **causes** is something inherent to a process. It exists as long as the process is not changed and is referred to as the natural variation in a process. It is an inherent part of the process *design* and effects all items. This variation is the effect of many small causes and cannot be totally eliminated. When this variation is random, we have what is known as a stable system of common causes. A process operating under a stable system of common causes is said to be in **statistical control**. Examples include inherent variation in incoming raw material from a qualified vendor, the vibration of machines, and fluctuations in working conditions.

Management alone is responsible for common causes. Deming believed that about 85% of all problems are due to common causes and hence can be solved only by action on the part of management. In a control chart, if quality characteristic values are within control limits and no nonrandom pattern is visible, it is assumed that a system of common causes exists and that the process is in a state of statistical control.

6-3 STATISTICAL BASIS FOR CONTROL CHARTS

Basic Principles

A control chart has a centerline and lower and upper control limits. The centerline is usually found in accordance with the data in the samples. It is an indication of the mean of a process and is usually found by taking the average of the values in the sample. However, the centerline can also be a desirable target or standard value.

Normal distributions play an important role in the use of control charts (Duncan 1986). The values of the statistic plotted on a control chart (e.g., average diameter) are assumed to have an approximately normal distribution. For large sample sizes or for small sample sizes with a population distribution that is unimodal and close to symmetric, the central limit theorem states that if the plotted statistic is a sample average, it will tend to have a normal distribution. Thus, even if the parent population is not normally distributed, control charts for averages and related statistics are based on normal distributions.

The control limits are two lines, one above and one below the centerline, that aid in the decision-making process. These limits are chosen so that the probability of the sample points falling between them is almost 1 (usually about 99.7% for 3σ limits) if the process is in statistical control. As discussed previously, if a system is operating under a stable system of common causes, it is assumed to be in statistical control. Typical control limits are placed at 3 standard deviations away from the mean of the statistic being plotted. Normal distribution theory states that a sample statistic will fall within the limits 99.74% of the time if the process is in control. If a point falls outside the control limits, there is a reason to believe that a special cause exists in the system. We must then try to identify the special cause and take corrective action to bring the process back to control.

The most common basis for deciding whether a process is out of control is the presence of a sample statistic outside the control limits. Other rules exist for determining out-of-control process conditions and are discussed in Section 6-5. These rules focus on nonrandom or systematic behavior of a process as evidenced by a nonrandom plot pattern. For example, if seven successive points plot above the centerline but within the upper control limit, there is a reason to believe that something might be wrong with the process. If the process were in control, the chances of this happening would be extremely small. Such a pattern might suggest that the process mean has shifted upward. Hence, appropriate actions would need to be identified in order to lower the process mean.

A control chart is a means of **online process control**. Data values are collected for a process, and the appropriate sample statistics (such as sample mean, sample range, or sample standard deviation) based on the quality characteristic of interest (such as diameter, length, strength, or response time) are obtained. These sample statistics are then plotted on a control chart. If they fall within the control limits and do not exhibit any systematic or nonrandom pattern, the process is judged to be in statistical control. If the control limits are calculated from current data, the chart tells us whether the process is presently in control. If the control limits were calculated from previous data based on a process that was in control, the chart can be used to determine whether the current process has drifted out of control.

Control charts are important management control tools. If management has some target value in mind for the process mean (say, average part strength), a control chart can be constructed with that target value as the centerline. Sample statistics, when plotted on the control chart, will show how close the actual process output comes to the desired standard. If the deviation is unsatisfactory, management will have to come up with remedial actions.

Control charts help management set realistic goals. For example, suppose the output of a process shows that the average part strength is 3000 kg, with a standard deviation of 100 kg. If management has a target average strength of at least 3500 kg, the control chart will indicate that such a goal is unrealistic and may not be feasible for the existing process. Major changes in the system and process, possibly only through action on the part of management, will be needed to create a process that will meet the desired goal.

If a process is under statistical control, control chart information can estimate such process parameters as the mean, standard deviation, and the proportion of nonconforming items (also known as *fallout*). These estimates can then be used to determine the capability of the process. **Process capability** refers to the ability of the process to produce within desirable specifications. Conclusions drawn from studies on process capability have a tremendous influence on major management decisions such as whether to make or buy, how to direct capital expenditures for machinery, how to select and control vendors, and how to implement process improvements to reduce variability. Process capability is discussed in Chapter 9.

For variables, the value of a quality characteristic is measurable numerically. Control charts for variables are constructed to show measures of central tendency as well as dispersion. Variable control charts display such information as sample mean, sample range, sample standard deviation, cumulative sum, individual values, and moving average. Control charts for variables are described in Chapter 7. Attributes, on the other hand, indicate the presence or absence of a condition. Typical attribute charts deal with the fraction of nonconforming items, the number of nonconforming items, the total number of nonconformities, the number of nonconformities per unit, or the number of demerits per unit. Control charts for attributes are described in Chapter 8.

There are several issues pertinent to the construction of a control chart: the number of items in a sample, the frequency with which data are sampled, how to minimize errors in making inferences, the analysis and interpretation of the plot patterns, and rules for determining out-of-control conditions. We discuss these issues in the following sections.

Selection of Control Limits

Let θ represent a quality characteristic of interest and $\hat{\theta}$ represent an estimate of θ. For example, if θ is the mean diameter of parts produced by a process, $\hat{\theta}$ would be the sample mean diameter of a set of parts chosen from the process. Let $E(\hat{\theta})$ represent the mean, or expected value, and let $\sigma(\hat{\theta})$ be the standard deviation of the estimator $\hat{\theta}$.

The centerline and **control limits** for this arrangement are given by

$$
\begin{aligned}
\text{CL} &= E(\hat{\theta}) \\
\text{UCL} &= E(\hat{\theta}) + k\sigma(\hat{\theta}) \\
\text{LCL} &= E(\hat{\theta}) - k\sigma(\hat{\theta})
\end{aligned}
\tag{6-1}
$$

where k represents the number of standard deviations of the sample statistic that the control limits are placed from the centerline. Typically, the value of k is chosen to be 3 (hence the name 3σ *limits*). If the sample statistic is assumed to have an approximately normal distribution, a value of $k = 3$ implies that there is a probability of only 0.0026 of a sample statistic falling outside the control limits if the process is in control.

Sometimes, the selection of k in eq. (6-1) is based on a desired probability of the sample statistic falling outside the control limits when the process is in control. Such limits are known as **probability limits**. For example, if we want the probability that the sample statistic will fall outside the control limits to be 0.002, Appendix A-3 gives $k = 3.09$ (assuming that the sample statistic is normally distributed). The probabilities of the sample statistic falling above the upper control limit and below the lower control limit are each equal to 0.001. Using this principle, the value of k and hence the control limits can be found for any desired probability.

The choice of k is influenced by error considerations also. As discussed in the next section, two types of errors (I and II) can be made in making inferences from control charts. The choice of a value k is influenced by how significant we consider the impact of such errors to be.

Example 6-1 A semiautomatic turret lathe machines the thickness of a part that is subsequently used in an assembly. The process mean is known to be 30 mm with a standard

TABLE 6-1   Average Part Thickness Values

Sample	Average Part Thickness, $\bar{x}$ (mm)	Sample	Average Part Thickness, $\bar{x}$ (mm)	Sample	Average Part Thickness, $\bar{x}$ (mm)
1	31.56	6	31.45	11	30.20
2	29.50	7	29.70	12	29.10
3	30.50	8	31.48	13	30.85
4	30.72	9	29.52	14	31.55
5	28.92	10	28.30	15	29.43

deviation of 1.5 mm. Construct a control chart for the average thickness using 3σ limits if samples of size 5 are randomly selected from the process. Table 6-1 shows the average thickness of 15 samples selected from the process. Plot these on a control chart, and make inferences.

Solution   The centerline is

$$CL = 30 \, mm$$

The standard deviation of the sample mean $\bar{X}$ is given by

$$\sigma_{\bar{X}} = \frac{\sigma}{\sqrt{n}} = \frac{1.5}{\sqrt{5}} = 0.671 \, mm$$

Assuming a normal distribution of the sample mean thickness, the value of k in eq. (6-1) is selected as 3. The control limits are calculated as follows:

$$UCL = 30 + (3)(0.671) = 32.013$$
$$LCL = 30 - (3)(0.671) = 27.987$$

The centerline and control limits are shown in Figure 6-2. The sample means for the 15 samples shown in Table 6-1 are plotted on this control chart. Figure 6-2 shows that all of the sample means are within the control limits. Also, the pattern of the plot does not exhibit any nonrandom behavior. Thus, we conclude that the process is in control.

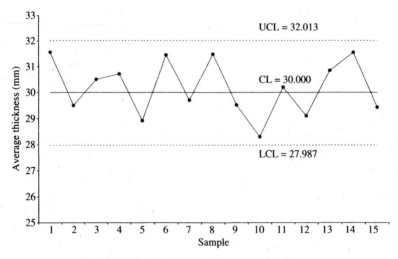

FIGURE 6-2   Control chart for average thickness.

Errors in Making Inferences from Control Charts

Making inferences from a control chart is analogous to testing a hypothesis. Suppose that we are interested in testing the null hypothesis that the average diameter θ of a part from a particular process is 25 mm. This situation is represented by the null hypothesis H_0: $\theta = 25$; the alternative hypothesis is H_a: $\theta \neq 25$. The rejection region of the null hypothesis is thus two-tailed. The control limits are the critical points that separate the rejection and acceptance regions. If a sample value (sample average diameter, in this case) falls above the upper control limit or below the lower control limit, we reject the null hypothesis. In such a case, we conclude that the process mean differs from 25 mm and the process is therefore out of control. Types I and II errors can occur when making inferences from control charts.

Type I Errors   **Type I errors** result from inferring that a process is out of control when it is actually in control. The probability of a type I error is denoted by α. Suppose that a process is in control. If a point on the control chart falls outside the control limits, we assume that the process is out of control. However, since the control limits are a finite distance (usually, 3 standard deviations) from the mean, there is a small chance (about 0.0026) of a sample statistic falling outside the control limits. In such instances, inferring that the process is out of control is a wrong conclusion. Figure 6-3 shows the probability of making a type I error in control charts. It is the sum of the two tail areas outside the control limits.

Type II Errors   **Type II errors** result from inferring that a process is in control when it is really out of control. If no observations fall outside the control limits, we conclude that the process is in control. Suppose, however, that a process is actually out of control. Perhaps the process mean has changed (say, an operator has inadvertently changed a depth of cut or the quality of raw materials has decreased). Or, the process could go out of control because the process variability has changed (due to the presence of a new operator). Under such circumstances, a sample statistic could fall within the control limits, yet the process would be out of control—this is a type II error.

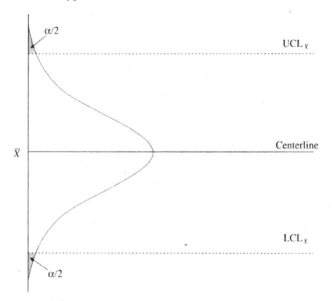

FIGURE 6-3   Type I error in control charts.

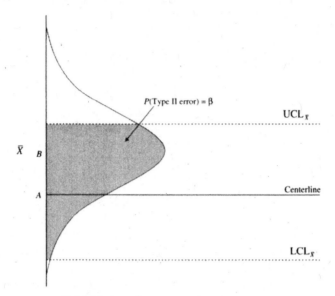

FIGURE 6-4   Type II error in control charts.

Let's consider Figure 6-4, which depicts a process going out of control due to a change in the process mean from A to B. For this situation, the correct conclusion is that the process is out of control. However, there is a strong possibility of the sample statistic falling within the control limits (as indicated by the shaded area), in which case we would conclude that the process is in control and thus make a type II error.

Example 6-2   A control chart is to be constructed for the average breaking strength of nylon fibers. Samples of size 5 are randomly chosen from the process. The process mean and standard deviation are estimated to be 120 kg and 8 kg, respectively.

(a) If the control limits are placed 3 standard deviations from the process mean, what is the probability of a type I error?

Solution   From the problem statement, $\hat{\mu} = 120$ and $\hat{\sigma} = 8$. The centerline for the control chart is at 120 kg. The control limits are

$$\text{UCL}_{\bar{X}} = 120 + 3\left(\frac{8}{\sqrt{5}}\right) = 130.733 \qquad \text{LCL}_{\bar{X}} = 120 - 3\left(\frac{8}{\sqrt{5}}\right) = 109.267$$

These limits are shown in Figure 6-5a.

Since the control limits are 3 standard deviations from the mean, the standardized normal value at the upper control limit is

$$Z = \frac{\bar{X} - \mu}{\sigma_{\bar{X}}}$$
$$= \frac{130.733 - 120}{8/\sqrt{5}} = 3.00$$

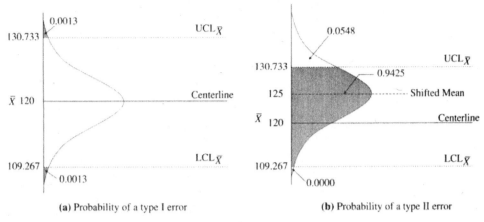

(a) Probability of a type I error   (b) Probability of a type II error

FIGURE 6-5   Control charts for Example 6-2.

Similarly, the Z-value at the lower control limit is -3.00. For these Z-values in the standard normal table in Appendix A-3, each tail area is found to be 0.0013. The probability of a type I error, as shown by the shaded tail areas in Figure 6-5a, is therefore 0.0026.

  (b) If the process mean shifts to 125 kg, what is the probability of concluding that the process is in control and hence making a type II error on the first sample plotted after the shift?

Solution   The process mean shifts to 125 kg. Assuming that the process standard deviation is the same as before, the distribution of the sample means is shown in Figure 6-5b. The probability of concluding that the process is in control is equivalent to finding the area between the control limits under the distribution shown in Figure 6-5b. We find the standardized normal value at the upper control limit as

$$Z_1 = \frac{130.733 - 125}{8/\sqrt{5}} = 1.60$$

From the standard normal table in Appendix A-3, the tail area above the upper control limit is 0.0548. The standardized normal value at the lower control limit is

$$Z_2 = \frac{109.267 - 125}{8/\sqrt{5}} = -4.40$$

From Appendix A-3, the tail area below the lower control limit is approximately 0.0000. The area between the control limits is $1 - (0.0548 + 0.0000) = 0.9452$. Hence, the probability of concluding that the process is in control and making a type II error is 0.9452, or 94.52%. This implies that for a shift of this magnitude, there is a pretty good chance of not detecting it in the first sample drawn after the shift.

  (c) What is the probability of detecting the shift by the second sample plotted after the shift if the samples are chosen independently?

Solution   The probability of detecting the shift by the second sample is P (detecting shift on sample 1) $+ P$(not detecting shift in sample 1 and detecting shift in sample 2). This first probability was found in part (b) to be 0.0548. The second probability, using eqs. (4-2) and (4-5), is found to be $(1 - 0.0548)(0.0548) = 0.0518$, assuming independence of the two

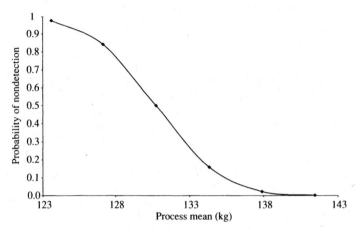

FIGURE 6-6 Operating characteristic curve for a control chart.

samples. The total probability is $0.0548 + 0.0518 = 0.1066$. Thus, there is a 10.66% chance of detecting a shift in the process by the second sample.

Operating Characteristic Curve   An **operating characteristic** (OC) **curve** is a measure of goodness of a control chart's ability to detect changes in process parameters. Specifically, it is a plot of the probability of the type II error versus the shift of a process parameter value from its in-control value. OC curves enable us to determine the chances of not detecting a shift of a certain magnitude in a process parameter on a control chart.

A typical OC curve is shown in Figure 6-6. The shape of an OC curve is similar to an inverted S. For small shifts in the process mean, the probability of nondetection is high. As the change in the process mean increases, the probability of nondetection decreases; that is, it becomes more likely that we will detect the shift. For large changes, the probability of nondetection is very close to zero. The ability of a control chart to detect changes quickly is indicated by the steepness of the OC curve and the quickness with which the probability of nondetection approaches zero. Calculations for constructing an operating characteristic curve are identical to those for finding the probability of a type II error.

Example 6-3   Refer to the data in Example 6-2 involving the control chart for the average breaking strength of nylon fibers. Samples of size 5 are randomly chosen from a process whose mean and standard deviation are estimated to be 120 kg and 8 kg, respectively. Construct the operating characteristic curve for increases in the process mean from 120 kg.

Solution   A sample calculation for the probability of not detecting the shift when the process mean increases to 125 kg is given in Example 6-2. This same procedure is used to calculate the probabilities of nondetection for several values of the process mean. Table 6-2 displays some sample calculations. The vertical axis of the operating characteristic curve in Figure 6-6 represents the probabilities of nondetection given in Table 6-2 (these values are also the probabilities of a type II error). The graph shows that for changes in the process mean exceeding 15 kg, the probability of nondetection is fairly small (less than 10%), while shifts of 5 kg or less have a high probability (over 85%) of nondetection.

TABLE 6-2 Probabilities for OC Curve

Process Mean	Z-Value at UCL, Z_1	Area Above UCL	Z-Value at LCL, Z_2	Area Below LCL	Probability of Nondetection, β
123.578	2.00	0.0228	−4.00	0.0000	0.9772
127.156	1.00	0.1587	−5.00	0.0000	0.8413
130.733	0.00	0.5000	−6.00	0.0000	0.5000
134.311	−1.00	0.8413	−7.00	0.0000	0.1587
137.888	−2.00	0.9772	−8.00	0.0000	0.0228
141.466	−3.00	0.9987	−9.00	0.0000	0.0013

Effect of Control Limits on Errors in Inference Making

The choice of the control limits influences the likelihood of the occurrence of type I and type II errors. As the control limits are placed farther apart, the probability of a type I error decreases (refer to Figure 6-3). For control limits placed 3 standard deviations from the centerline, the probability of a type I error is about 0.0026. For control limits placed 2.5 standard deviations from the centerline, Appendix A-3 gives the probability of a type I error as 0.0124. On the other hand, for control limits placed 4 standard deviations from the mean, the probability of a type I error is negligible. If a process is in control, the chance of a sample statistic falling outside the control limits decreases as the control limits expand. Hence, the probability of making a type I error decreases, too. The control limits could be placed sufficiently far apart, say 4 or 5 standard deviations on each side of the centerline, to reduce the probability of a type I error, but doing so affects the probability of making a type II error.

Moving the control limits has the opposite effect on the probability of type II error. As the control limits are placed farther apart, the probability of a type II error increases (refer to Figure 6-4). Ideally, to reduce the probability of a type I error, we would tend to have the control limits placed closer to each other. But this, of course, has the detrimental effect of increasing the probability of a type I error. Thus, the two types of errors are inversely related to each other as the control limits change. As the probability of a type I error decreases, the probability of a type II error increases.

If all other process parameters are held fixed, the probability of a type II error will decrease with an increase in sample size. As n increases, the standard deviation of the sampling distribution of the sample mean decreases. Thus, the control limits will be drawn closer, and the probability of a type II error will be reduced. Figure 6-7 demonstrates this effect. The new sample size is larger than the old sample size. The sampling distribution of the new sample mean has a reduced variance, so the new control limits are closer to each other. As can be seen from the figure, the probability of a type II error is smaller for the larger sample.

Because of the inverse relationship between type I and type II errors, a judicious choice of control limits is desirable. In the majority of uses, the control limits are placed at 3 standard deviations from the centerline, thereby restricting the probability of a type I error to 0.0026. The reasoning behind this choice of limits is that the chart user does not want to look unnecessarily for special causes in a process when there are none. By placing the control limits at 3 standard deviations, the probability of a false alarm is small, and minimal resources will be spent on locating nonexistent problems with the process. However, the probability of a type II error may be large for small shifts in the process mean.

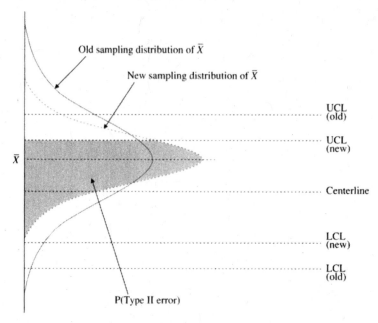

FIGURE 6-7   Effect of an increased sample size on the probability of a type II error.

If it is more important to detect small changes in the process than to avoid spending time looking for nonexistent problems, it may be desirable to place the control limits closer (at, say, 2 or 2.5 standard deviations). For sophisticated processes, it is often crucial to detect small changes as soon as possible, because the impact on downstream activities is enormous if they are not detected right away. In this case, tighter control limits are preferable, even if this means incurring some costs for unnecessary investigation of problems when the process is in control.

Warning Limits

Warning limits are usually placed at 2 standard deviations from the centerline. When a sample statistic falls outside the warning limits but within the control limits, the process is not considered to be out of control, but users are now alerted that the process may be going out of control. For a normally distributed sample statistic, Appendix A-3 gives the probability of it falling in the band between the warning limit and the control limit to be 0.043 (i.e., there is about a 4.3% chance of this happening). Thus, a sample statistic outside the warning limits is reason to be wary. If two out of three successive sample statistics fall within the warning/control limit on a given side, the process may indeed be out of control, because the probability of this happening in an in-control process is very small (0.0027, obtained from $2 \times 3 \times 0.0215 \times 0.0215 \times 0.9785$).

Effect of Sample Size on Control Limits

The **sample size** usually has an influence on the standard deviation of the sample statistic being plotted on the control chart. For example, consider a control chart for the sample mean

$\bar{X}$. The standard deviation of $\bar{X}$ is given by

$$\sigma_{\bar{X}} = \frac{\sigma}{\sqrt{n}}$$

where σ represents the process standard deviation and n is the sample size. We see that the standard deviation of $\bar{X}$ is inversely related to the square root of the sample size. Since the control limits are placed a certain number of standard deviations (say, 3) from the centerline, an increase in the sample size causes the control limits to be drawn closer. Similarly, decreasing the sample size causes the limits to expand. Increasing the sample size provides more information, intuitively speaking, and causes the sample statistics to have less variability. A lower variability reduces the frequency with which errors occur in making inferences.

Average Run Length

An alternative measure of the performance of a control chart, in addition to the OC curve, is the **average run length**(ARL). This denotes the number of samples, on average, required to detect an out-of-control signal. Suppose that the rule used to detect an out-of-control condition is a point plotting outside the control limits. Let P_d denote the probability of an observation plotting outside the control limits. Then the run length is 1 with a probability of P_d, 2 with a probability of $(1 - P_d)P_d$, 3 with a probability of $(1 - P_d)^2 P_d$, and so on. The average run length is given by

$$\text{ARL} = \sum_{j=1}^{\infty} j(1 - P_d)^{j-1} P_d$$

$$= P_d \sum_{j=1}^{\infty} j(1 - P_d)^{j-1} \tag{6-2}$$

The infinite series inside the summation is obtained from $1/[1 - (1 - P_d)]^2$. Hence, we have

$$\text{ARL} = \frac{P_d}{[1-(1-P_d)]^2} = \frac{1}{P_d} \tag{6-3}$$

For a process in control, P_d is equal to α, the probability of a type I error. Thus, for 3σ control charts with the selected rule for the detection of an out-of-control condition, ARL is $1/0.0026 \simeq 385$. This indicates that an observation will plot outside the control limits every 385 samples, on average. For a process in control, we prefer the ARL to be large because an observation plotting outside the control limits represents a false alarm.

For an out-of-control process, it is desirable for the ARL to be small because we want to detect the out-of-control condition as soon as possible. Let's consider a control chart for the process mean and suppose that a change takes place in this parameter. In this situation, $P_d = 1 - \beta$, where β is the probability of a type II error. So, $\text{ARL} = 1/(1 - \beta)$.

We have computed β, the probability of a type II error, for a control chart on changes in the process mean (see Figure 6-4). Because it is straightforward to develop a general expression for β in terms of the shift in the process mean (expressed in units of the process standard deviation, σ), we can also construct ARL curves for the control chart. Figure 6-8 shows ARL curves for sample sizes of 1, 2, 3, 4, 5, 7, 9, and 16 for a control chart for the mean where the shifts in the mean are shown in units of σ. Note that if we wish to detect a shift of 1.0σ in the process mean, using a sample of size 5, the average number of samples required will be about 4. In case this ARL is not suitable and we wish to reduce it, the sample size could be increased to 7,

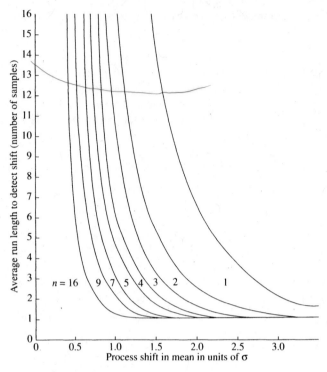

FIGURE 6-8   ARL curves for control charts for the mean.

whereupon the ARL is reduced to approximately 3. When we need to express ARL in terms of the expected number of individual units sampled, I, the expression is

$$I = n(\text{ARL}) \tag{6-4}$$

where n denotes the sample size.

Example 6-4   Let's reconsider Example 6-3 on the average breaking strength of nylon fibers. Our sample size is 5. Table 6-2 shows calculations for β, the probability of nondetection, for different values of the process mean. Table 6-3 displays the values of ARL for each change in the process mean. The change in the process mean, from the in-control value of 120, is shown in multiples of the process standard deviation, σ.

TABLE 6-3   Computation of ARL for Changes in Process Mean

Process Mean	Shift in Process Mean in Units of $\sigma_{\bar{x}}$	Shift in Process Mean in Units of σ	P_d	ARL
123.578	1	0.4472	0.0228	43.86
127.156	2	0.8945	0.1587	6.30
130.733	3	1.3416	0.5000	2.00
134.311	4	1.7889	0.8413	1.19
137.888	5	2.2360	0.9772	1.02
141.466	6	2.6832	0.9987	1.00

From Table 6-3 we find that for shifts in the process mean of 2.5 or more standard deviations, the control chart is quite effective because the ARL is slightly above 1. This indicates that, on average, the out-of-control condition will be detected on the first sample drawn after the shift takes place. For a shift in the process mean of 1.34σ, the ARL is 2, while for a smaller shift in the process mean of 0.89σ, the ARL is above 6. These values of ARL represent a measure of the strength of the control chart in its ability to detect process changes quickly. For a small shift in the process mean of about 0.45σ, about 44 samples, on average, will be required to detect the shift.

6-4   SELECTION OF RATIONAL SAMPLES

Shewhart described the fundamental criteria for the selection of rational subgroups, or rational samples, the term we use in this book. The premise is that a **rational sample** is chosen in such a manner that the variation within it is considered to be due only to common causes. So, samples are selected such that if special causes are present, they will occur between the samples. Therefore, the differences *between* samples will be maximized, and differences *within* samples will be minimized.

In most cases, the sampling is done by time order. Let's consider a job shop with several machines. Samples are collected at random times from each machine. Control charts for the average value of the characteristic for each machine are plotted separately. If two operators are producing output, samples are formed from the output of each operator, and a separate control chart is plotted for each operator. If output between two shifts differs, the two outputs should not be mixed in the sampling process. Rather, samples should first be selected from shift 1 and a control chart constructed to determine the stability of that shift's output. Next, rational samples are selected from shift 2 and a control chart constructed for this output.

Selection of the sample observations is done by the **instant-of-time method** (Besterfield 2003). Observations are selected at approximately the same time for the population under consideration. This method provides a time frame for each sample, which makes the identification of problems simpler. The instant-of-time method minimizes variability within a sample and maximizes variability between samples if special causes are present.

Sample Size

Selecting sample size—the number of items in each sample—is a necessity in using control charts. The degree of shift in the process parameter expected to take place will influence the choice of sample size. As noted in the discussion of operating characteristic curves, large shifts in a process parameter (say, the process mean) can be detected by smaller sample sizes than those needed to detect smaller shifts. Having an idea of the degree of shift we wish to detect enables us to select an appropriate sample size. If we can tolerate smaller changes in the process parameters, a small sample size might suffice. Alternatively, if it is important to detect slight changes in process parameters, a larger sample size will be needed.

Frequency of Sampling

The **sampling frequency** must be decided prior to the construction of control charts. Choosing large samples very frequently is the sampling scheme that provides the most information. However, this is not always feasible because of resource constraints. Other options include choosing small sample sizes at frequent intervals or choosing large sample sizes at infrequent intervals. In practice, the former is usually adopted.

Other factors also influence the frequency of sampling and the sample size. The type of inspection needed to obtain the measurement—that is, destructive or nondestructive—can be a factor. The current state of the process (in control or out of control) is another factor. If the process is stable, we might get by with sampling at infrequent intervals. However, for processes that indicate greater variability, we would need to sample more frequently.

The cost of sampling and inspection per unit is another area of concern. The choice of sample size is influenced by the loss incurred due to a nonconforming item being passed on to the consumer. These intangible costs are sometimes hard to identify and quantify. Because larger sample sizes detect shifts in process parameters sooner than smaller sample sizes, they can be the most cost-effective choice.

6-5   ANALYSIS OF PATTERNS IN CONTROL CHARTS

One of the main objectives of using control charts is to determine when a process is out of control so that necessary actions may be taken. Criteria other than a plotted point falling outside the control limits are also used to determine whether a process is out of control. We discuss some **rules for out-of-control processes** next.

Later, we examine some typical control chart patterns and the reasons for their occurrence. As mentioned previously, plot patterns often indicate whether or not a process is in control; a systematic or nonrandom pattern suggests an out-of-control process. Analyzing these patterns is more difficult than plotting the chart. Identifying the causes of nonrandom patterns requires knowledge of the process, equipment, and operating conditions as well as of their impact on the characteristic of interest.

Some Rules for Identifying an Out-of-Control Process

Rule 1   A process is assumed to be out of control if a single point plots outside the control limits.

This is the most commonly used rule. If the control limits are placed at 3 standard deviations from the mean of the quality characteristic being plotted (assuming a normal distribution), the probability of a point falling outside these limits if the process is in control is very small (about 0.0026). Figure 6-9 depicts this situation.

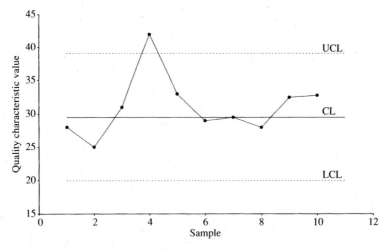

FIGURE 6-9   Out-of-control patterns: Rule 1.

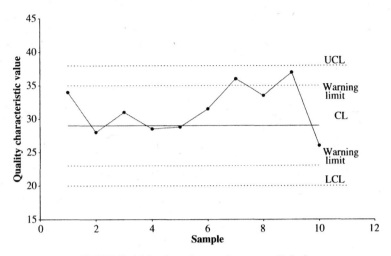

FIGURE 6-10   Out-of-control patterns: Rule 2.

Rule 2   A process is assumed to be out of control if two out of three consecutive points fall outside the 2σ warning limits on the same side of the centerline.

As noted in Section 6-3, warning limits at 2 standard deviations of the quality characteristic from the centerline can be constructed. These are known as 2σ limits. If the process is in control, the chance of two out of three points falling outside the warning limits is small. In Figure 6-10, observe that samples 7 and 9 fall above the upper 2σ limit. We can infer that this process has gone out of control, so special causes should be investigated.

Rule 3   A process is assumed to be out of control if four out of five consecutive points fall beyond the 1σ limit on the same side of the centerline.

If the control limits are first determined, the standard deviation can be calculated. Note that the distance between the centerline and the upper control limit is 3 standard deviations (assuming 3σ limits). Dividing this distance by 3 gives the standard deviation of the characteristic being plotted. Adding and subtracting this standard deviation from the centerline value gives the 1σ limits. Consider Figure 6-11, for which samples 4, 5, 6, and

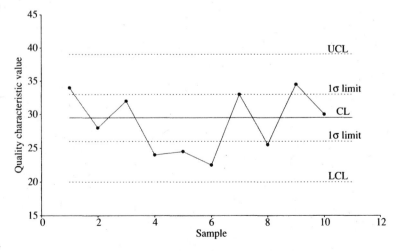

FIGURE 6-11   Out-of-control patterns: Rule 3.

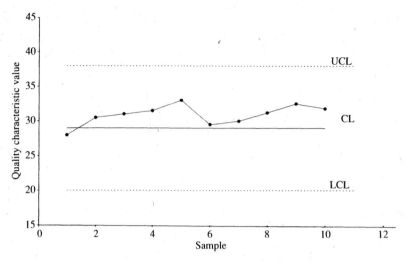

FIGURE 6-12   Out-of-control patterns: Rule 4.

8 plot below the lower 1σ limit. Based on Rule 3, this process would be considered out of control.

Rule 4   A process is assumed to be out of control if nine or more consecutive points fall to one side of the centerline.

For a process in control, a roughly equal number of points should be above or below the centerline, with no systematic pattern visible. The condition stated in Rule 4 is highly unlikely if a process is in control. For instance, if nine or more consecutive points plot above the centerline on an $\bar{X}$-chart, an upward shift in the process mean may have occurred. In Figure 6-12, samples 2, 3, 4, 5, 6, 7, 8, 9, and 10 plot above the centerline. The process is assumed to be out of control.

Rule 5   A process is assumed to be out of control if there is a run of six or more consecutive points steadily increasing or decreasing.

A run is a sequence of like observations. Thus, if three successive points increase in magnitude, we would have a run of three points. In Figure 6-13, samples 2 to 8 show a continual increase; so this process would be deemed out of control.

Interpretation of Plots

The five rules for determining out-of-control conditions are not all used simultaneously. Rule 1 is used routinely along with a couple of the other rules (say, Rules 2 and 3). The reason for not using all of them simultaneously is that doing so increases the chance of a type I error. In other words, the probability of a false alarm increases as more rules are used to determine an out-of-control state. Even though the probability of the stated condition occurring is rather small for any one rule with an in-control process, the **overall type I error rate**, based on the number of rules that are used, may not be small.

Suppose that the number of independent rules used for out-of-control criteria is k. Let α_i be the probability of a type I error of rule i. Then, the overall probability of a type I

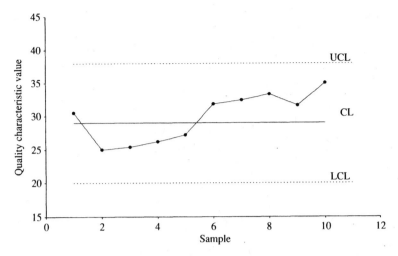

FIGURE 6-13   Out-of-control patterns: Rule 5.

error is

$$\alpha = 1 - \prod_{i=1}^{k}(1 - \alpha_i) \tag{6-5}$$

Suppose that four independent rules are being used to determine whether a process is out of control. Let the probability of a type I error for each rule be given by $\alpha_1 = 0.005$, $\alpha_2 = 0.02$, $\alpha_3 = 0.03$, and $\alpha_4 = 0.05$. The overall false alarm rate, or the probability of a type I error, would be

$$\alpha = 1 - (0.995)(0.98)(0.97)(0.95) = 0.101$$

If several more rules were used simultaneously, the probability of type I error would become too large to be acceptable. Note that the relationship in eq. (6-5) is derived under the assumption that the rules are independent. The rules, however, are not independent, so eq. (6-5) is only an approximation to the probability of a type I error. For more information, see Walker et al. (1991).

Using many rules for determining out-of-control conditions complicates the decision process and sabotages the purpose of using control limits. One of the major advantages of control charts is that they are easy to construct, interpret, and use.

In addition to the rules that we've been discussing, there are many other nonrandom patterns that a control chart user has to interpret judiciously. It is possible for the process to be out of control yet for none of the five rules to be applicable. This is where experience, judgment, and interpretive skills come into play. Consider, for example, Figure 6-14. None of the five rules for out-of-control conditions apply even though the pattern is clearly nonrandom. The systematic nature of the plot and the somewhat cyclic behavior are important clues. This pattern probably means that special causes are present and the process is out of control. You should always keep an eye out for nonrandom patterns when you examine control charts.

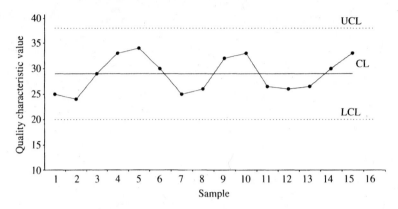

FIGURE 6-14   Nonrandom pattern in a control chart.

Determination of Causes of Out-of-Control Points

The task of the control chart user does not end with the identification of out-of-control points. In fact, the difficult part begins when out-of-control points have been determined. Now we must pinpoint the causes associated with these points—not always a trivial task. This requires a thorough knowledge of the process and the sensitivity of the output quality characteristic to the process parameters.

Determination of cause is usually a collective effort, with people from product design, process design, tooling, production, purchasing, and vendor control involved. A cause-and-effect chart is often an appropriate tool here. Once special causes have been identified, appropriate remedial actions need to be proposed. Typical control chart patterns for out-of-control processes, along with their possible causes, are discussed in Chapter 7. These include a sudden shift in the pattern level, a gradual shift in the pattern level, a cyclic pattern, or a mixture pattern, among others.

6-6   MAINTENANCE OF CONTROL CHARTS

Although the construction of control charts is an important step in statistical process control, it should be emphasized that quality control and improvement are an ongoing process. Therefore, implementation and **control chart maintenance** are a vital link in the quality system. When observations plotted on control charts are found to be out of control, the centerline and control limits need to be revised; this, of course, will eliminate these out-of-control points. There are some exceptions to the elimination of out-of-control points, however, especially for points below the lower control limit. These are discussed in Chapters 7 and 8. Once computation of the revised centerline and control limits is completed, these lines are drawn on charts where future observations are to be plotted. The process of revising the centerline and control limits is ongoing.

Proper placement of the control charts on the shop floor is important. Each person who is associated with a particular quality characteristic should have easy access to the chart. When a statistical process control system is first implemented, the chart is usually placed in a conspicuous place where operators can look at it. The involvement of everyone from the operator to the manager to the chief executive officer is essential to the success of the program. The control charts should be attended to by everyone involved. Proper maintenance of these charts on a regular ongoing basis helps ensure the success of the quality systems approach. If a particular quality characteristic becomes insignificant, its control chart can be replaced by others that are relevant. Products, processes, vendors, and equipment change with time. Similarly, the different control charts that are kept should be chosen to reflect important characteristics of the current environment.

SUMMARY

This chapter has introduced the basic concepts of control charts for statistical process control. The benefits that can be derived from using control charts have been discussed. This chapter covers the statistical background for the use of control charts, the selection of the control limits, and the manner in which inferences can be drawn from the charts. The two types of errors that can be encountered in making inferences from control charts are discussed. Guidelines for the proper selection of sample size and rules for determining out-of-control conditions have been explored. Several control chart patterns have been studied with a focus on identifying possible special causes. Since this chapter is intended solely to explain the fundamentals of control charts, such technical details as formulas for various types of control charts have intentionally been omitted here. We discuss these in Chapters 7 and 8, and you may also find them in Banks (1989), Montgomery (2004), and Wadsworth et al. (2001).

KEY TERMS

average run length

centerline

common cause

control chart

control chart maintenance

control limits

   lower control limit

   upper control limit

instant-of-time method

online process control

operating characteristic curve

probability limits

process capability

rational samples

remedial actions

rules for out-of-control processes

sample size

sampling frequency

Shewhart control charts

special cause

statistical control

type I error

   overall rate

type II error

warning limits

EXERCISES

Discussion Questions

6-1   What are the benefits of using control charts?

6-2   Explain the difference between common causes and special causes. Give examples of each.

6-3   Explain the rationale behind placing the control limits at 3 standard deviations from the mean.

6-4 Define and explain type I and type II errors in the context of control charts. Are they related?
How does the choice of control limits influence these two errors?

6-5 What are warning limits, and what purpose do they serve?

6-6 What is the utility of the operating characteristic curve? How can the discriminatory power of the curve be improved?

6-7 Describe the role of the average run length (ARL) in the selection of control chart parameters. Explain how ARL influences sample size.

6-8 Discuss the relationship between ARL and type I and II errors.

6-9 How are rational samples selected? Explain the importance of this in the total quality systems approach.

6-10 State and explain each rule for determining out-of-control points.

6-11 What are some reasons for a process to go out of control due to a sudden shift in the level?

6-12 Explain some causes that would make the control chart pattern follow a gradually increasing trend.

Problems

6-13 What is meant by an overall type I error rate? If Rules 1, 2, and 3 of this chapter are used simultaneously, assuming independence, what is the probability of an overall type I error if 3σ control limits are used?

6-14 The diameter of cotter pins produced by an automatic machine is a characteristic of interest. Based on historical data, the process average diameter is 15 mm with a process standard deviation of 0.8 mm. If samples of size 4 are randomly selected from the process:
(a) Find the 1σ and 2σ control limits.
(b) Find the 3σ control limits for the average diameter.
(c) What is the probability of a false alarm?
(d) If the process mean shifts to 14.5 mm, what is the probability of not detecting this shift on the first sample plotted after the shift? What is the ARL?
(e) What is the probability of failing to detect the shift by the second sample plotted after the shift?
(f) Construct the OC curve for this control chart.
(g) Construct the ARL curve for this control chart.

6-15 The length of industrial filters is a quality characteristic of interest. Thirty samples, each of size 5, are chosen from the process. The data yields an average length of 110 mm, with the process standard deviation estimated to be 4 mm.
(a) Find the warning limits for a control chart for the average length.
(b) Find the 3σ control limits. What is the probability of a type I error?
(c) If the process mean shifts to 112 mm, what are the chances of detecting this shift by the third sample drawn after the shift?

(d) What is the chance of detecting the shift for the first time on the second sample point drawn after the shift?

(e) What is the ARL for a shift in the process mean to 112 mm? How many samples, on average, would it take to detect a change in the process mean to 116 mm?

6-16 The tensile strength of nonferrous pipes is of importance. Samples of size 5 are selected from the process output, and their tensile strength values are found. After 30 such samples, the process mean strength is estimated to be 3000 kg with a standard deviation of 50 kg.

(a) Find the 1σ and 2σ control limits. For the 1σ limits, what is the probability of concluding that the process is out of control when it is really in control?

(b) Find the 3σ limits.

(c) If Rules 1 and 2 are used simultaneously to detect out-of-control conditions, assuming independence, what is the overall probability of a type I error if 3σ control limits are used?

6-17 Suppose 3σ control limits are constructed for the average temperature in a furnace. Sample of size 4 were selected with the average temperature being 5000°C and a standard deviation of 50°C.

(a) Find the 3σ control charts.

(b) Suppose that Rules 2 and 3 are used simultaneously to determine out-of-control conditions. What is the overall probability of a type I error assuming independence of the rules?

(c) Approximately how many samples, on average, will be analyzed before detecting a change when Rules 2 and 3 are used simultaneously?

(d) If the process average temperature drops to 4960°C, what is the probability of failing to detect this change by the third sample point drawn after the change?

(e) What is the probability of the shift being detected within the first two samples?

6-18 A manager is contemplating using Rules 1 and 4 for determining out-of-control conditions. Suppose that the manager constructs 3σ limits.

(a) What is the overall type I error probability assuming independence of the rules?

(b) On average, how many samples will be analyzed before detecting a change in the process mean? Assume that the process mean is now at 110 mm (having moved from 105 mm) and that the process standard deviation is 6 mm. Samples of size 4 are selected from the process.

6-19 The time to deliver packaged containers by a logistics company is found from samples of size 4. The mean and standard deviation of delivery times is estimated to be 140 hours and 6 hours, respectively.

(a) Find the 2σ and 3σ control limits for the average delivery time.

(b) Explain a type I and type II error specifically in this context.

(c) Suppose that Rules 1 and 3 are used simultaneously to detect out-of-control conditions. Assuming independence of the rules, what is the overall probability of a type I error for 3σ control limits?

(d) If the mean delivery time shifts to 145 hours, what is the probability of not detecting this by the second sample after the shift?

(e) What is the ARL? Explain.

6-20   A health care facility is monitoring daily expenditures for a certain diagnosis-related group (DRG). Individual observations are selected. After 50 samples, the average and standard deviation of daily expenditures (in hundreds of dollars) are estimated to be 15 and 2, respectively.
   (a) Find the 3σ control limits.
   (b) Suppose that Rules 1 and 2 are used simultaneously for the detection of out-of-control conditions. Assuming independence of the rules, what is the overall probability of a type I error? Explain the meaning of a type I error in this context.
   (c) Suppose that the average daily expenditures for the same DRG increases to $1750. What is the chance of detecting this shift by the second sample drawn after the shift?
   (d) What is the ARL?

REFERENCES

ASQ(1993). ANSI/ISO/ASQ. *Statistics—Vocabulary and Symbols–Statistical Quality Control*, A3534-2. Milwaukee, Wi: American Society for Quality.

Banks, J. (1989). *Principles of Quality Control*. New York: Wiley.

Besterfield, D. H. (2003). *Quality Control*. 7th ed. Upper Saddle River, N J: Prentice Hall.

Duncan, A. J. (1986). *Quality Control and Industrial Statistics*. 5th ed. Homewood, IL: Richard Irwin.

Montgomery, D. C. (2004). *Introduction to Statistical Quality Control*. 5th ed. Hoboken, NJ: Wiley.

Wadsworth, H. M., K. S. , Stephens, and A. B. Godfrey, (2001). *Modern Methods for Quality Control and Improvement,* 2nd ed. New York: Wiley.

Walker, E., J. W. Philpot, and J. Clement (1991). "False Signal Rates for the Shewhart Control Chart with Supplementary Runs Tests," *Journal of Quality Technology,* 23 (3):247–252.

7

CONTROL CHARTS FOR VARIABLES

7-1 INTRODUCTION AND CHAPTER OBJECTIVES

In Chapter 6 we introduced the fundamentals of control charts. In this chapter we look at the details of **control charts for variables**—quality characteristics that are measurable on a numerical scale. Examples of **variables** include length, thickness, diameter, breaking strength, temperature, acidity, viscosity, order processing time, and waiting time for service. We must be able to control the mean value of a quality characteristic as well as its variability. The **mean** gives an indication of the central tendency of a process, and the **variability** provides an idea of the process dispersion. Therefore, we need information about both these statistics to keep a process in control.

Let's consider Figure 7-1. A change in the **process mean** of a quality characteristic (say, length of a part) is shown in Figure 7-1a, where the **mean shifts** from μ_o to μ_1. It is, of course, important that this change be detected because if the **specification limits** are as shown in Figure 7-1a, a change in the process mean would change the proportion of parts that do not meet specifications. Figure 7-1b shows a change in the dispersion of the process; the **process standard deviation** has changed from σ_0 to σ_1, with the process mean remaining stationary at μ_0. Note that the proportion of the output that does not meet specifications has increased. Control charts aid in detecting such changes in process parameters.

Variables provide more information than attributes. **Attributes** deal with qualitative information such as whether an item is nonconforming or what the number of nonconformities in an item is. Thus, attributes do not show the degree to which a quality characteristic

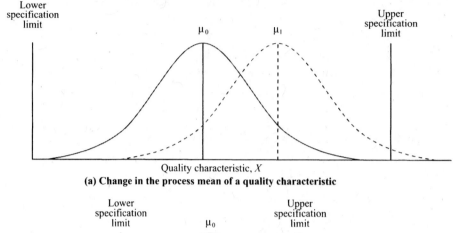

(a) Change in the process mean of a quality characteristic

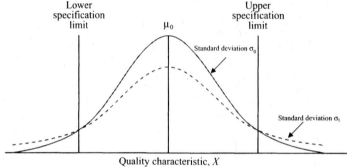

(b) Change in the dispersion of a quality characteristic

FIGURE 7-1   Changes in the mean and dispersion of a process.

is nonconforming. For instance, if the specifications on the length of a part are 40 ± 0.5 mm and a part has length 40.6 mm, attribute information would indicate as nonconforming both this part *and* a part of length 42 mm. The degree to which these two lengths deviate from the specifications is lost in attribute information. This is not so with variables, however, because the numerical value of the quality characteristic (length, in this case) is used in the control chart.

The cost of obtaining variable data is usually higher than for attributes because attribute data are collected by means such as go/no-go gages, which are easier to use and therefore less costly. The total cost of data collection is the sum of two components: the fixed cost and the variable unit cost. Fixed costs include the cost of the inspection equipment; variable unit costs include the cost of inspecting units. The more units inspected, the higher the variable cost, whereas the fixed cost is unaffected. As the use of automated devices for measuring quality characteristic values spreads, the difference in the variable unit cost between variables and attributes may not be much. However, the fixed costs, such as investment costs, may increase.

7-2   SELECTION OF CHARACTERISTICS FOR INVESTIGATION

In small organizations as well as in large ones, many possible product and process quality characteristics exist. A single component usually has several quality characteristics, such as

TABLE 7-1 Pareto Analysis of Defects for Assembly Data

Defect Code	Defect	Frequency	Percentage
1	Outside diameter of hub	30	8.82
2	Depth of keyway	20	5.88
3	Hub length	60	17.65
4	Inside diameter of hub	90	26.47
5	Width of keyway	30	8.82
6	Thickness of flange	40	11.77
7	Depth of slot	50	14.71
8	Hardness (measured by Brinell hardness number)	20	5.88

length, width, height, surface finish, and elasticity. In fact, the number of quality character-istics that affect a product is usually quite large. Now multiply such a number by even a small number of products and the total number of characteristics quickly increases to an unman-ageable value. It is normally not feasible to maintain a control chart for each possible variable.

Balancing feasibility and completeness of information is an ongoing task. Accomplishing it involves selecting a few vital quality characteristics from the many candidates. Selecting which quality characteristics to maintain control charts on requires giving higher priority to those that cause more nonconforming items and that increase costs. The goal is to select the "vital few" from among the "trivial many." This is where **Pareto analysis** comes in because it clarifies which are the "important" quality characteristics.

When nonconformities occur because of different defects, the frequency of each defect can be tallied. Table 7-1 shows the Pareto analyses for various defects in an assembly. Alterna-tively, the cost of producing the nonconformity could be collected. Table 7-1 shows that the three most important defects are the inside hub diameter, the hub length, and the slot depth.

Using the percentages given in Table 7-1, we can construct a Pareto diagram like the one shown in Figure 7-2. The defects are thus shown in a nonincreasing order of occurrence. From the figure we can see that if we have only enough resources to construct three variable charts, we will choose inside hub diameter (code 4), hub length (code 3), and slot depth (code 7).

Once quality characteristics for which control charts are to be maintained have been identified, a scheme for obtaining the data should be set up. Quite often, it is desirable

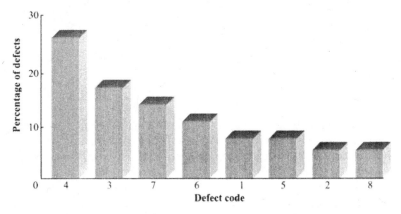

FIGURE 7-2   Pareto diagram for assembly data.

to measure process characteristics that have a causal relationship to product quality characteristics. Process characteristics are typically controlled directly through control charts. In the assembly example of Table 7-1, we might decide to monitor process variables (cutting speed, depth of cut, and coolant temperature) that have an impact on hub diameter, hub length, and slot depth. Monitoring process variables through control charts implicitly controls product characteristics.

7-3 PRELIMINARY DECISIONS

Certain decisions must be made before we can construct control charts. Several of these were discussed in detail in Chapter 6.

Selection of Rational Samples

The manner in which we sample the process deserves our careful attention. The sampling method should maximize differences between samples and minimize differences within samples. This means that separate control charts may have to be kept for different operators, machines, or vendors.

Lots from which sample are chosen should be homogeneous. As mentioned in Chapter 6, if our objective is to determine shifts in process parameters, samples should be made up of items produced at nearly the same time. This gives us a time reference and will be helpful if we need to determine special causes. Alternatively, if we are interested in the nonconformance of items produced since the previous sample was selected, samples should be chosen from items produced since that time.

Sample Size

Sample sizes are normally between 4 and 10, and it is quite common in industry to have sample sizes of 4 or 5. The larger the sample size, the better the chance of detecting small shifts. Other factors, such as cost of inspection or cost of shipping a nonconforming item to the customer, also influence the choice of sample size.

Frequency of Sampling

The sampling frequency depends on the cost of obtaining information compared to the cost of not detecting a nonconforming item. As processes are brought into control, the frequency of sampling is likely to diminish.

Choice of Measuring Instruments

The accuracy of the measuring instrument directly influences the quality of the data collected. Measuring instruments should be calibrated and tested for dependability under controlled conditions. Low-quality data lead to erroneous conclusions. The characteristic being controlled and the desired degree of measurement precision both have an impact on the choice of measuring instrument. In measuring dimensions such as length, height, or thickness, something as simple as a set of calipers or a micrometer may be acceptable. On the other hand, measuring the thickness of silicon wafers may require complex optical sensory equipment.

Design of Data Recording Forms

Recording forms should be designed in accordance with the control chart to be used. Common features for data recording forms include the sample number, the date and time when the sample was selected, and the raw values of the observations. A column for comments about the process is also useful.

7-4 CONTROL CHARTS FOR THE MEAN AND RANGE

Development of the Charts

Step 1: Using a preselected sampling scheme and sample size, record on the appropriate forms, measurements of the quality characteristic selected.

Step 2: For each sample, calculate the sample mean and range using the following formulas:

$$\overline{X} = \frac{\sum_{i=1}^{n} X_i}{n} \tag{7-1}$$

$$R = X_{\max} - X_{\min} \tag{7-2}$$

where X_i represents the ith observation, n is the sample size, $X_{\max}$ is the largest observation, and $X_{\min}$ is the smallest observation.

Step 3: Obtain and draw the centerline and the **trial control limits** for each chart. For the $\overline{X}$-chart, the centerline $\overline{\overline{X}}$ is given by

$$\overline{\overline{X}} = \frac{\sum_{i=1}^{g} \overline{X}_i}{g} \tag{7-3}$$

where g represents the number of samples. For the R-chart, the centerline $\overline{R}$ is found from

$$\overline{R} = \frac{\sum_{i=1}^{g} R_i}{g} \tag{7-4}$$

Conceptually, the **3σ control limits** for the $\overline{X}$-chart are

$$\overline{\overline{X}} \pm 3\sigma_{\overline{X}} \tag{7-5}$$

Rather than compute $\sigma_{\overline{x}}$ from the raw data, we can use the relation between the process standard deviation σ (or the standard deviation of the individual items) and the mean of the ranges $(\overline{R})$. Multiplying factors used to calculate the centerline and control limits are given in Appendix A-7. When sampling from a population that is normally distributed, the distribution of the statistic $W = R/\sigma$ (known as the relative range) is dependent on the sample size n. The mean of W is represented by d_2 and is tabulated in Appendix A-7. Thus, an estimate of the process standard deviation is

$$\hat{\sigma} = \frac{\overline{R}}{d_2} \tag{7-6}$$

The control limits for an $\overline{X}$-**chart** are therefore estimated as

$$(UCL_{\overline{X}}, LCL_{\overline{X}}) = \overline{\overline{X}} \pm \frac{3\hat{\sigma}}{\sqrt{n}}$$

$$= \overline{\overline{X}} \pm \frac{3\overline{R}}{\sqrt{n}d_2}$$

$$(UCL_{\overline{X}}, LCL_{\overline{X}}) = \overline{\overline{X}} \pm A_2\overline{R} \qquad (7\text{-}7)$$

where $A_2 = 3/\sqrt{n}\,d_2$ and is tabulated in Appendix A-7. Equation (7-7) is the working equation for determining the $\overline{X}$-chart control limits, given $\overline{R}$.

The control limits for the R-chart are conceptually given by

$$(UCL_R, \ LCL_R) = \overline{R} \pm 3\sigma_R \qquad (7\text{-}8)$$

Since $R = \sigma W$, we have $\sigma_R = \sigma\sigma_w$. In Appendix A-7, σ_w is tabulated as d_3. Using eq. (7-6), we get

$$\hat{\sigma}_R = \frac{\overline{R}}{d_2}d_3$$

The control limits for the R-**chart** are estimated as

$$UCL_R = \overline{R} + 3d_3\frac{\overline{R}}{d_2} = D_4\overline{R}$$

$$LCL_R = \overline{R} - 3d_3\frac{\overline{R}}{d_2} = D_3\overline{R}$$

$$(7\text{-}9)$$

where

$$D_4 = 1 + \frac{3d_3}{d_2} \quad \text{and} \quad D_3 = \max\left(0, 1 - \frac{3d_3}{d_2}\right)$$

Equation (7-9) is the working equation for calculating the control limits for the R-chart. Values of D_4 and D_3 are tabulated in Appendix A-7.

Step 4: Plot the values of the range on the control chart for range, with the centerline and the control limits drawn. Determine whether the points are in statistical control. If not, investigate the special causes associated with the **out-of-control** points (see the **rules** for this in Chapter 6) and take appropriate remedial action to eliminate special causes.

Typically, only some of the rules are used simultaneously. The most commonly used criterion for determining an out-of-control situation is the presence of a point outside the control limits.

An R-chart is usually analyzed before an $\overline{X}$-chart to determine out-of-control situations. An R-chart reflects process variability, which should be brought in to control first. As shown by eq. (7-7), the control limits for an $\overline{X}$-chart involve the process variability and hence $\overline{R}$. Therefore, if an R-chart shows an out-of-control situation, the limits on the $\overline{X}$-chart may not be meaningful.

Let's consider Figure 7-3. On the R-chart, sample 12 plots above the upper control limit and so is out of control. The $\overline{X}$-chart, however, does not show the process to be out of control. Suppose that the special cause is identified as a problem with a new vendor, who supplies raw materials and components. The task is to eliminate the cause, perhaps by choosing a new vendor or requiring evidence of statistical process control at the vendor's plant.

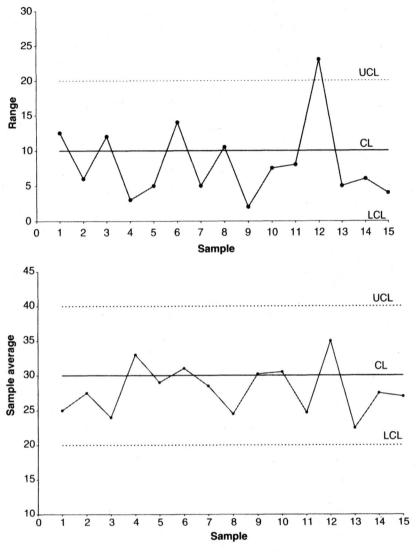

FIGURE 7-3   Plot of sample values on $\overline{X}$-and R-charts.

Step 5: Delete the out-of-control point(s) for which **remedial actions** have been taken to remove special causes (in this case, sample 12) and use the remaining samples (here they are samples 1 to 11 and 13 to 15) to determine the revised centerline and control limits for the $\overline{X}$- and R-charts.

These limits are known as the **revised control limits.** The cycle of obtaining information, determining the trial limits, finding out-of-control points, identifying and correcting special causes, and determining revised control limits then continues. The revised control limits will serve as trial control limits for the immediate future until the limits are revised again. This ongoing process is a critical component of continuous improvement.

A point of interest regarding the revision of R-charts concerns observations that plot below the lower control limit when the lower control limit is greater than zero. Such points that fall below LCL_R are, statistically speaking, out of control; however, they are also desirable because they indicate unusually small variability within the sample which is, after all, one of our main objectives. It is most likely that such small variability is due to special causes.

If the user is convinced that the small variability does indeed represent the operating state of the process during that time, an effort should be made to identify the causes. If such conditions can be created consistently, process variability will be reduced. The process should be set to match those favorable conditions, and the observations should be retained for calculating the revised centerline and the revised control limits for the R-chart.

Step 6: Implement the control charts.

The $\overline{X}$- and R-charts should be implemented for future observations, using the revised centerline and control limits. The charts should be displayed in a conspicuous place where they will be visible to operators, supervisors, and managers. Statistical process control will be effective only if everyone is committed to it—from the operator to the chief executive officer.

Example 7-1 Consider a process by which coils are manufactured. Samples of size 5 are randomly selected from the process, and the resistance values (in ohms) of the coils are measured. The data values are given in Table 7-2, as are the sample mean $\overline{X}$ and the range R. First, the sum of the ranges is found and then the centerline $\overline{R}$. We have

$$\overline{R} = \frac{\sum\limits_{i=1}^{g} R_i}{g} = \frac{87}{25} = 3.48$$

For a sample of size 5, Appendix A-7 gives $D_4 = 2.114$ and $D_3 = 0$. The trial control limits for the R-chart are calculated as follows:

$$UCL_R = D_4\overline{R} = (2.114)(3.48) = 7.357$$
$$LCL_R = D_3\overline{R} = (0)(3.48) = 0$$

The centerline on the $\overline{X}$-chart is obtained as follows:

$$\overline{\overline{X}} = \frac{\sum\limits_{i=1}^{g} \overline{X}_i}{g} = \frac{521.00}{25} = 20.840$$

Appendix A-7, for $n = 5$, gives $A_2 = 0.577$. Hence, the trial control limits on the $\overline{X}$-charts are

$$UCL_{\overline{X}} = \overline{\overline{X}} + A_2\overline{R} = 20.84 + (0.577)(3.48) = 22.848$$
$$LCL_{\overline{X}} = \overline{\overline{X}} - A_2\overline{R} = 20.84 - (0.577)(3.48) = 18.832$$

We can use Minitab to construct trial $\overline{X}$- and R-charts for the data in Table 7-2. **Choose Stat > Control Charts > Variables Charts for subgroups > X bar-R.** Indicate whether the subgroups are arranged in a single column or in rows, input, in this case column numbers C1 to C5, since in the worksheet for this example a subgroup is entered as a row across five columns, Click on **X bar-R chart options**, select **Estimate**, and under **Method for estimating standard deviation**, select **Rbar.** Click **OK.** Figure 7-4 shows the Minitab $\overline{X}$- and R-charts with 3σ limits. Observe that sample 3 is above the upper control limit on the R-chart and samples 22 and 23 are below and above the $\overline{X}$ -chart control limit, respectively. When the special causes for these three samples were investigated, operators found that the large value

TABLE 7-2   Coil Resistance Data

Sample	Observation (Ω)	$\overline{X}$	R	Comments
1	20, 22, 21, 23, 22	21.60	3	
2	19, 18, 22, 20, 20	19.80	4	
3	25, 18, 20, 17, 22	20.40	8	New vendor
4	20, 21, 22, 21, 21	21.00	2	
5	19, 24, 23, 22, 20	21.60	5	
6	22, 20, 18, 18, 19	19.40	4	
7	18, 20, 19, 18, 20	19.00	2	
8	20, 18, 23, 20, 21	20.40	5	
9	21, 20, 24, 23, 22	22.00	4	
10	21, 19, 20, 20, 20	20.00	2	
11	20, 20, 23, 22, 20	21.00	3	
12	22, 21, 20, 22, 23	21.60	3	
13	19, 22, 19, 18, 19	19.40	4	
14	20, 21, 22, 21, 22	21.20	2	
15	20, 24, 24, 23, 23	22.80	4	
16	21, 20, 24, 20, 21	21.20	4	
17	20, 18, 18, 20, 20	19.20	2	
18	20, 24, 22, 23, 23	22.40	4	
19	20, 19, 23, 20, 19	20.20	4	
20	22, 21, 21, 24, 22	22.00	3	
21	23, 22, 22, 20, 22	21.80	3	
22	21, 18, 18, 17, 19	18.60	4	High temperature
23	21, 24, 24, 23, 23	23.00	3	Wrong die
24	20, 22, 21, 21, 20	20.80	2	
25	19, 20, 21, 21, 22	20.60	3	
		Sum = 521.00	Sum = 87	

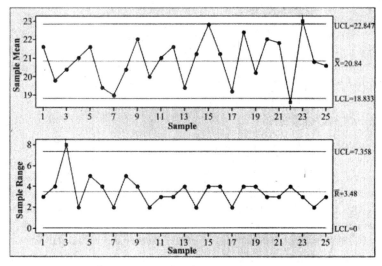

FIGURE 7-4 $\overline{X}$- and R-charts for data on coil resistance using Minitab.

for the range in sample 3 was due to the quality of raw materials and components purchased from a new vendor. Management decided to require the new vendor to provide documentation showing that adequate control measures are being implemented at the vendor's plant and that subsequent deliveries of raw materials and components will conform to standards.

When the special causes for samples 22 and 23 were examined, operators found that the oven temperature was too high for sample 22 and the wrong die was used for sample 23. Remedial actions were taken to rectify these situations.

With samples 3, 22, and 23 deleted, the revised centerline on the R-chart is

$$\bar{R} = \frac{72}{22} = 3.273$$

The revised control limits on the R-chart are

$$\text{UCL}_R = D_4\bar{R} = (2.114)(3.273) = 6.919$$
$$\text{LCL}_R = D_3\bar{R} = (0)(3.273) = 0$$

The revised centerline on the $\bar{X}$-chart is

$$\bar{\bar{X}} = \frac{459}{22} = 20.864$$

The revised control limits on the $\bar{X}$-chart are

$$\text{UCL}_{\bar{X}} = \bar{\bar{X}} + A_2\bar{R} = 20.864 + (0.577)(3.273) = 22.753$$
$$\text{LCL}_{\bar{X}} = \bar{\bar{X}} - A_2\bar{R} = 20.864 - (0.577)(3.273) = 18.975$$

Note that sample 15 falls slightly above the upper control limit on the $\bar{X}$-chart. On further investigation, no special causes could be identified for this sample. So, the revised limits will be used for future observations until a subsequent revision takes place.

Variable Sample Size

So far, our sample size has been assumed to be constant. A change in the sample size has an impact on the control limits for the $\bar{X}$- and R-charts. It can be seen from eqs. (7-7) and (7-9) that an increase in the sample size n reduces the width of the control limits. For an $\bar{X}$-chart, the width of the control limits from the centerline is inversely proportional to the square root of the sample size. Appendix A-7 shows the pattern in which the values of the control chart factors A_2, D_4, and D_3 decrease with an increase in sample size.

Standardized Control Charts

When the sample size varies, the control limits on an $\bar{X}$- and an R-chart will change, as discussed previously. With fluctuating control limits, the rules for identifying out-of-control conditions we discussed in Chapter 6 become difficult to apply—that is, except for Rule 1 (which assumes a process to be out of control when an observation plots outside the control limits). One way to overcome this drawback is to use a standardized control chart. When we standardize a statistic, we subtract its mean from its value and divide this value by its standard deviation. The standardized values then represent the deviation from the mean in units of standard deviation. They are dimensionless and have a mean of zero. The control limits, on a standardized chart are at ± 3 and are therefore constant. It's easier to interpret shifts in the process from a standardized chart than from a chart with fluctuating control limits.

Let the sample size for sample i be denoted by n_i, and let $\bar{X}_i$ and s_i denote its average and standard deviation, respectively. The mean of the sample averages is found as

$$\bar{\bar{X}} = \frac{\sum_{i=1}^{g} n_i \bar{X}_i}{\sum_{i=1}^{g} n_i} \tag{7-10}$$

An estimate of the process standard deviation, $\hat{\sigma}$, is the square root of the weighted average of the sample variances, where the weights are 1 less the corresponding sample sizes. So,

$$\hat{\sigma} = \sqrt{\frac{\sum_{i=1}^{g} (n_i - 1) s_i^2}{\sum_{i=1}^{g} (n_i - 1)}} \tag{7-11}$$

Now, for sample i, the standardized value for the mean, Z_i, is obtained from

$$Z_i = \frac{\bar{X}_i - \bar{\bar{X}}}{\hat{\sigma} / \sqrt{n_i}} \tag{7-12}$$

where $\bar{\bar{X}}$ and $\hat{\sigma}$ are given by eqs. (7-10) and (7-11), respectively. A plot of the Z_i values on a control chart, with the centerline at 0, the upper control limit at 3, and the lower control limit at -3, represents a standardized control chart for the mean.

To standardize the range chart, the range R_i for sample i is first divided by the estimate of the process standard deviation, $\hat{\sigma}$, given by eq. (7-11), to obtain

$$r_i = R_i / \hat{\sigma} \tag{7-13}$$

The values of r_i are then standardized by subtracting its mean d_2 and dividing by its standard deviation d_3 (Nelson 1989). The factors d_2 and d_3 are tabulated for various sample sizes in Appendix A-7. So, the standardized value for the range, k_i, is given by

$$k_i = \frac{r_i - d_2}{d_3} \tag{7-14}$$

These values of k_i are plotted on a control chart with a centerline at 0, and upper and lower control limits at 3 and -3, respectively.

Control Limits for a Given Target or Standard

Management sometimes wants to specify values for the process mean and standard deviation. These values may represent goals or desirable standard or **target values.** Control charts based on these target values help determine whether the existing process is capable of meeting the desirable standards. Furthermore, they also help management set realistic goals for the existing process.

Let $\bar{X}_0$ and σ_0 represent the target values of the process mean and standard deviation, respectively. The centerline and control limits based on these standard values for the $\bar{X}$-chart are given by

$$\begin{aligned} \text{CL}_{\bar{X}} &= \bar{X}_0 \\ \text{UCL}_{\bar{X}} &= \bar{X}_0 + 3 \frac{\sigma_0}{\sqrt{n}} \\ \text{LCL}_{\bar{X}} &= \bar{X}_0 - 3 \frac{\sigma_0}{\sqrt{n}} \end{aligned} \tag{7-15}$$

Let $A = 3/\sqrt{n}$. Values for A are tabulated in Appendix A-7. Equation (7-15) may be rewritten as

$$
\begin{aligned}
\mathrm{CL}_{\bar{X}} &= \bar{X}_0 \\
\mathrm{UCL}_{\bar{X}} &= \bar{X}_0 + A\sigma_0 \\
\mathrm{LCL}_{\bar{X}} &= \bar{X}_0 - A\sigma_0
\end{aligned}
\tag{7-16}
$$

For the R-chart, the centerline is found as follows, Since $\hat{\sigma} = \bar{R}/d_2$, we have

$$
\mathrm{CL}_R = d_2\sigma_0
\tag{7-17}
$$

where d_2 is tabulated in Appendix A-7. The control limits are

$$
\begin{aligned}
\mathrm{UCL}_R &= \bar{R} + 3\sigma_R = d_2\sigma_0 + 3d_3\sigma_0 \\
&= (d_2 + 3d_3)\sigma_0 = D_2\sigma_0
\end{aligned}
\tag{7-18}
$$

where $D_2 = d_2 + 3d_3$ (Appendix A-7) and $\sigma_R = d_3\sigma$.

Similarly,

$$
\begin{aligned}
\mathrm{LCL}_R &= \bar{R} - 3\sigma_R = d_2\sigma_0 - 3d_3\sigma_0 \\
&= (d_2 - 3d_3)\sigma_0 = D_1\sigma_0
\end{aligned}
\tag{7-19}
$$

where $D_1 = d_2 - 3d_3$ (Appendix A-7).

We must be cautious when we interpret control charts based on target or standard values. Sample observations can fall outside the control limits even though no special causes are present in the process. This is because these desirable standards may not be consistent with the process conditions. Thus, we could waste time and resources looking for special causes that do not exist.

On an $\bar{X}$-chart, plotted points can fall outside the control limits because a target process mean is specified as too high or too low compared to the existing process mean. Usually, it is easier to meet a desirable target value for the process mean than it is for the process variability. For example, adjusting the mean diameter or length of a part can often be accomplished by simply changing controllable process parameters. However, correcting for R-chart points that plot above the upper control limit is generally much more difficult.

An R-chart based on target values can also indicate excessive process variability without special causes present in the system. Therefore, meeting the target value σ_0 may involve drastic changes in the process. Such an R-chart may be implying that the existing process is not capable of meeting the desired standard. This information enables management to set realistic goals.

Example 7-2   Refer to the coil resistance data in Example 7-1. Let's suppose that the target values for the average resistance and standard deviation are 21.0 and 1.0 Ω, respectively. The sample size is 5. The centerline and the control limits for the $\bar{X}$-chart are as follows:

$$
\begin{aligned}
\mathrm{CL}_{\bar{X}} &= \bar{X}_0 = 21.0 \\
\mathrm{UCL}_{\bar{X}} &= \bar{X}_0 + A\sigma_0 = 21.0 + (1.342)(1.0) = 22.342 \\
\mathrm{LCL}_{\bar{X}} &= \bar{X}_0 - A\sigma_0 = 21.0 - (1.342)(1.0) = 19.658
\end{aligned}
$$

The centerline and control limits for the R-chart are

$$
\begin{aligned}
\mathrm{CL}_R &= d_2\sigma_0 = (2.326)(1.0) = 2.326 \\
\mathrm{UCL}_R &= D_2\sigma_0 = (4.918)(1.0) = 4.918 \\
\mathrm{LCL}_R &= D_1\sigma_0 = (0)(1.0) = 0
\end{aligned}
$$

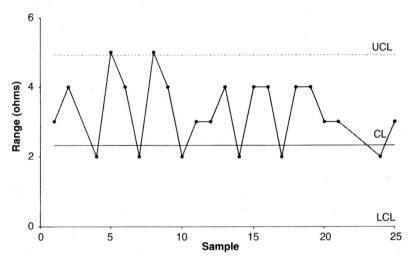

FIGURE 7-5   *R*-chart based on a standard value.

Figure 7-5 shows the control chart for the range based on the standard value. Since the control charts were revised in Example 7-1, we plot the 22 in-control samples and exclude samples 3, 22, and 23 because we are assuming that remedial actions have eliminated those causes. Now we can see how close the in-control process comes to meeting the stipulated target values.

The process seems to be out of control with respect to the given standard. Samples 5 and 8 are above the upper control limit, and a majority of the points lie above the centerline. Only six of the points plot below the centerline. Figure 7-5 thus reveals that the process is not capable of meeting company guidelines. The target standard deviation σ_0 is 1.0. The estimated process standard deviation from Example 7-1 (calculated after the process was brought to control) is

$$\hat{\sigma} = \frac{\bar{R}}{d_2} = \frac{3.50}{2.326} = 1.505$$

This estimate exceeds the target value of 1.0. Management must look at common causes to reduce the process variability if the standard is to met. This may require major changes in the methods of operation, the incoming material, or the equipment. Process control will not be sufficient to achieve the desired target.

The $\bar{X}$-chart based on the standard value is shown in Figure 7-6. Several points fall outside the control limits—four points below and two points above. In Example 7-1, the revised centerline for the $\bar{X}$-chart was found to be 20.864. Our target centerline is now 21.0. Adjusting controllable process parameters could possibly shift the average level up to 21.0. However, the fact that there are points outside both the upper and lower control limits signifies that process variability is the issue here.

Common causes must be examined: That is, reducing variability will only be achieved through process improvement. Figure 7-6 indicates that the target standard deviation of 1.0 is not realistic for the current process. Unless management makes major changes in the process, the target value will not be met. Actions on the part of the operators alone are unlikely to cause the necessary reduction in process variability.

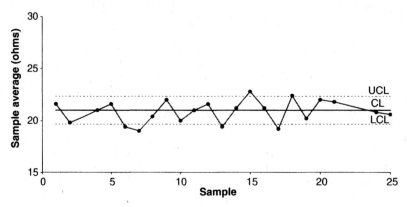

FIGURE 7-6 $\overline{X}$-chart based on a standard value.

Interpretation and Inferences from the Charts

The difficult part of analysis is determining and interpreting the special causes and selecting remedial actions. Effective use of control charts requires operators who are familiar with not only the statistical foundations of control charts but also the process itself. They must thoroughly understand how the different controllable parameters influence the dependent variable of interest. The quality assurance manager or analyst should work closely with the product design engineer and the process designer or analyst to come up with optimal policies.

In Chapter 6 we discussed five rules for determining out-of-control conditions. The presence of a point falling outside the 3σ limits is the most widely used of those rules. Determinations can also be made by interpreting typical plot patterns. Once the special cause is determined, this information plus a knowledge of the plot can lead to appropriate remedial actions.

Often, when the R-chart is brought to control, many special causes for the-$\overline{X}$-chart are eliminated as well. The $\overline{X}$-chart monitors the centering of the process because $\overline{X}$ is a measure of the center. Thus, a jump on the $\overline{X}$-chart means that the process average has jumped and an increasing trend indicates the process center is gradually increasing. Process centering usually takes place through adjustments in machine settings or such controllable parameters as proper tool, proper depth of cut, or proper feed. On the other hand, reducing process variability to allow an R-chart to exhibit control is a difficult task that is accomplished through quality improvement.

Once a process is in statistical control, its capability can be estimated by calculating the process standard deviation. This measure can then be used to determine how the process performs with respect to some stated specification limits. The **proportion** of **nonconforming** items can be estimated. Depending on the characteristic being considered, some of the output may be reworked, while some may become scrap. Given the unit cost of rework and scrap, an estimate of the total cost of rework and scrap can be obtained. **Process capability** measures are discussed in more detail in Chapter 9. From an R-chart that exhibits control, the process standard deviation can be estimated as

$$\hat{\sigma} = \frac{\overline{R}}{d_2}$$

where $\overline{R}$ is the centerline and d_2 is a factor tabulated in Appendix A-7. If the distribution of the quality characteristic can be assumed to be normal, then given some specification limits, the standard normal table can be used to determine the proportion of output that is nonconforming.

Example 7-3   Refer to the coil resistance data in Example 7-1. Suppose that the specifications are $21 \pm 3\,\Omega$.

(a) Determine the proportion of the output that is nonconforming, assuming that coil resistance is normally distributed.

Solution   From the revised R-chart, we found the centerline to be $\bar{R} = 3.50$. The estimated process standard deviation is

$$\hat{\sigma} = \frac{\bar{R}}{d_2} = \frac{3.50}{2.236} = 1.505$$

The revised centerline on the $\bar{X}$-chart is $\bar{\bar{X}} = 20.864$, which we use as an estimate of the process mean. Figure 7-7 shows the proportion of the output that is nonconforming. The standardized normal value at the lower specification limit (LSL) is found as

$$z_1 = \frac{18 - 20.864}{1.505} = -1.90$$

The standardized normal value at the upper specification limit (USL) is

$$z_2 = \frac{24 - 20.864}{1.505} = 2.08$$

From Appendix A-3 we find that the proportion of the product below the LSL is 0.0287, and the proportion above the USL is 0.0188. Thus, the total proportion of nonconforming output is 0.0475.

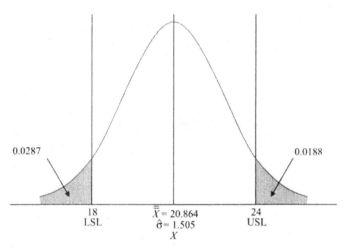

FIGURE 7-7   Proportion of nonconforming output.

(b) If the daily production rate is 10,000 coils and if coils with a resistance less than the LSL cannot be used for the desired purpose, what is the loss to the manufacturer if the unit cost of scrap is 50 cents?

Solution   The daily cost of scrap is

$$(10,000)(0.0287)(\$0.50) = \$143.50$$

Control Chart Patterns and Corrective Actions

A nonrandom identifiable pattern in the plot of a control chart might provide sufficient reason to look for **special causes** in the system. **Common causes** of variation are inherent to a system; a system operating under only common causes is said to be in a state of statistical control. Special causes, however, could be due to periodic and persistent disturbances that affect the process intermittently. The objective is to identify the special causes and take appropriate remedial action.

Western Electric Company engineers have identified 15 typical patterns in control charts. Your ability to recognize these patterns will enable you to determine *when* action needs to be taken and *what* action to take (AT&T, 1984). We discuss nine of these patterns here.

Natural Patterns   A natural pattern is one in which no identifiable arrangement of the plotted points exists. No points fall outside the control limits, the majority of the points are near the centerline, and few points are close to the control limits (Gitlow et al. 1989). Natural patterns are indicative of a process that is in control; that is, they demonstrate the presence of a stable system of common causes. A natural pattern is shown in Figure 7-8.

Sudden Shifts in the Level   Many causes can bring about a sudden change (or jump) in pattern level on an $\overline{X}$- or R-chart. Figure 7-9 shows a sudden shift on an $\overline{X}$-chart. Such jumps occur because of changes—intentional or otherwise—in such process settings as temperature, pressure, or depth of cut. A sudden change in the average service level, for

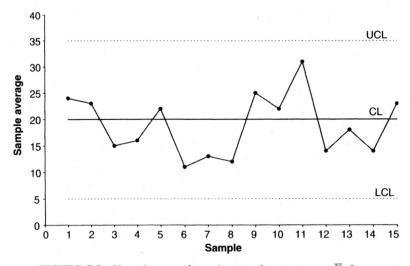

FIGURE 7-8   Natural pattern for an in-control process on an $\overline{X}$-chart.

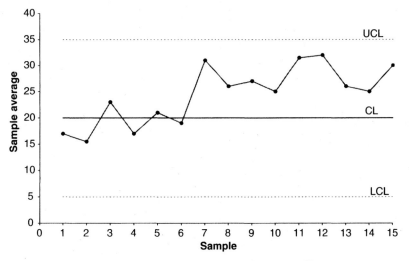

FIGURE 7-9   Sudden shift in pattern level on an $\overline{X}$-chart.

example, could be a change in customer waiting time at a bank because the number of tellers changed. New operators, new equipment, new measuring instruments, new vendors, and new methods of processing are other reasons for sudden shifts on $\overline{X}$- and R-charts.

Gradual Shifts in the Level   Gradual shifts in level occur when a process parameter changes gradually over a period of time. Afterward, the process stabilizes. An $\overline{X}$-chart might exhibit such a shift because the incoming quality of raw materials or components changed over time, the maintenance program changed, or the style of supervision changed. An R-chart might exhibit such a shift because of a new operator, a decrease in worker skill due to fatigue or monotony, or a gradual improvement in the incoming quality of raw materials because a vendor has implemented a statistical process control system. Figure 7-10 shows an $\overline{X}$-chart exhibiting a gradual shift in the level.

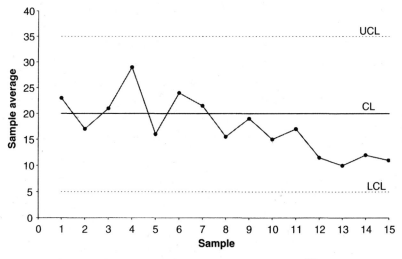

FIGURE 7-10   Gradual shift in pattern level on an $\overline{X}$-chart.

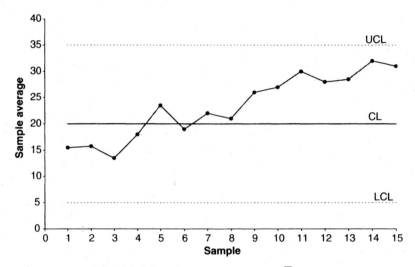

FIGURE 7-11   Trending pattern on an $\overline{X}$-chart.

Trending Pattern   Trends differ from gradual shifts in level in that trends do not stabilize or settle down. Trends represent changes that steadily increase or decrease. An $\overline{X}$-chart may exhibit a trend because of tool wear, die wear, gradual deterioration of equipment, buildup of debris in jigs and fixtures, or gradual change in temperature. An R-chart may exhibit a trend because of a gradual improvement in operator skill resulting from on-the-job training, or a decrease in operator skill due to fatigue. Figure 7-11 shows a trending pattern on an $\overline{X}$-chart.

Cyclic Patterns   Cyclic patterns are characterized by a repetitive periodic behavior in the system. Cycles of low and high points will appear on the control chart. An $\overline{X}$-chart may exhibit cyclic behavior because of a rotation of operators, periodic changes in temperature and humidity (such as a cold-morning startup), periodicity in the mechanical or chemical properties of the material, or seasonal variation of incoming components. An R-chart may exhibit cyclic patterns because of operator fatigue and subsequent energization following breaks, a difference between shifts, or periodic maintenance of equipment. Figure 7-12

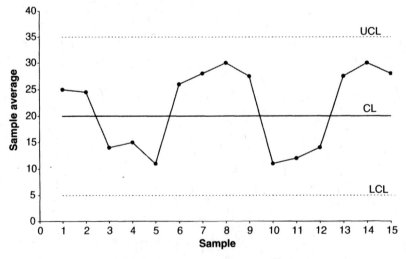

FIGURE 7-12   Cyclic pattern on an $\overline{X}$-chart.

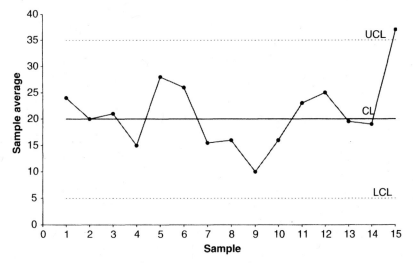

FIGURE 7-13   Freak pattern on an $\overline{X}$-chart.

shows a cyclic pattern for an $\overline{X}$-chart. If samples are taken too infrequently, only the high or the low points will be represented, and the graph will not exhibit a cyclic pattern. If control chart users suspect cyclic behavior, they should take samples frequently to investigate the possibility of a cyclic pattern.

Wild Patterns   Wild patterns are divided into two categories: freaks and bunches (or groups). Control chart points exhibiting either of these two properties are, statistically speaking, significantly different from the other points. Special causes are generally associated with these points.

   Freaks are caused by external disturbances that influence one or more samples. Figure 7-13 shows a control chart exhibiting a freak pattern. Freaks are plotted points too small or too large with respect to the control limits. Such points usually fall outside the control limits and are easily distinguishable from the other points on the chart. It is often not difficult to identify special causes for freaks. You should make sure, however, that there is no measurement or recording error associated with the freak point. Some special causes of freaks include sudden, very short-lived power failures; the use of a new tool for a brief test period; and the failure of a component.

   Bunches, or **groups**, are clusters of several observations that are decidedly different from other points on the plot. Figure 7-14 shows a control chart pattern exhibiting bunching behavior. Possible special causes of such behavior include the use of a new vendor for a short period time, use of a different machine for a brief time period, and new operator used for a short period.

Mixture Patterns (or the Effect of Two or More Populations)   A mixture pattern is caused by the presence of two or more populations in the sample and is characterized by points that fall near the control limits, with an absence of points near the centerline, A mixture pattern can occur when one set of values is too high and another set too low because of differences in the incoming quality of material from two vendors. A remedial action would be to have a separate control chart for each vendor. Figure 7-15 shows a mixture pattern. On an $\overline{X}$-chart, a mixture pattern can also result from over control. If an operator chooses to adjust the machine

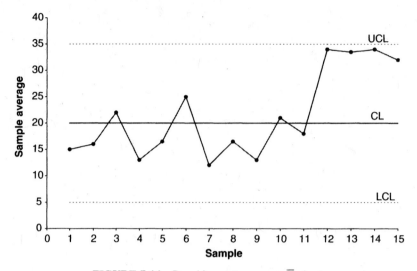

FIGURE 7-14   Bunching pattern on an $\overline{X}$-chart.

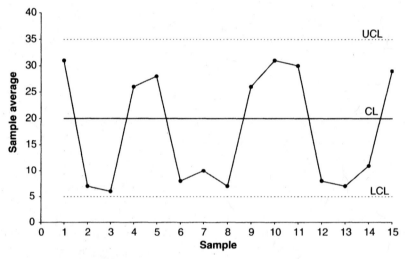

FIGURE 7-15   Mixture pattern on an $\overline{X}$-chart.

or process *every* time a point plots near a control limit, the result will be a pattern of large swings. Mixture patterns can also occur on both $\overline{X}$- and R-charts because of two or more machines being represented on the same control chart. Other examples include two or more operators being represented on the same chart, differences in two or more pieces of testing or measuring equipment, and differences in production methods of two or more lines.

Stratification Patterns   A stratification pattern is another possible result when two or more population distributions of the same quality characteristic are present. In this case, the output is combined, or mixed (say, from two shifts), and samples are selected from the mixed output. In this pattern, the majority of the points are very close to the centerline, with very few points near the control limits, Thus, the plot can be misinterpreted as indicating unusually good control. A stratification pattern is shown in Figure 7-16. Such a plot could have resulted from

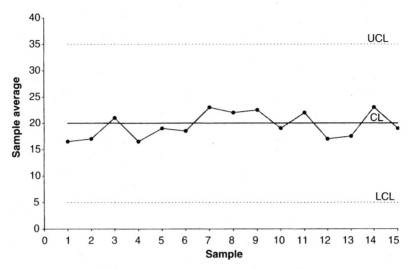

FIGURE 7-16   Stratifications pattern on an $\bar{X}$-chart.

plotting data for samples composed of the combined output of two shifts, each different in its performance. It is possible for the sample average (which is really the average of parts chosen from both shifts) to fluctuate very little, resulting in a stratification pattern in the plot. Remedial measures in such situations involve having separate control charts for each shift. The method of choosing rational samples should be carefully analyzed so that component distributions are not mixed when samples are selected.

Interaction Patterns   An interaction pattern occurs when the level of one variable affects the behavior of other variables associated with the quality characteristic of interest. Furthermore, the combined effect of two or more variables on the output quality characteristic may be different from the individual effect of each variable. An interaction pattern can be detected by changing the scheme for rational sampling. Suppose that in a chemical process the temperature and pressure are two important controllable variables that affect the output quality characteristic of interest. A low pressure and a high temperature may produce a very desirable effect on the output characteristic, whereas a low pressure by itself may not have that effect. An effective sampling method would involve controlling the temperature at several high values and then determining the effect of pressure on the output characteristic for each temperature value Samples composed of random combinations of temperature and pressure may fail to identify the interactive effect of those variables on the output characteristic. The control chart in Figure 7-17 shows interaction between variables. In the first plot, the temperature was maintained at level A; in the second plot, it was held at level B. Note that the average level and variability of the output characteristic change for the two temperature levels. Also, if the R-chart shows the sample ranges to be small, information regarding the interaction could be used to establish desirable process parameter settings.

Control Charts for Other Variables   The control chart patterns described in this section also occur in control charts besides $\bar{X}$- and R-charts. When found in other types of control charts, these patterns may indicate different causes than those we discussed in this section, but similar reasoning can be used to determine them. Furthermore, both the preliminary considerations and the steps for constructing control charts described earlier also apply to other control charts.

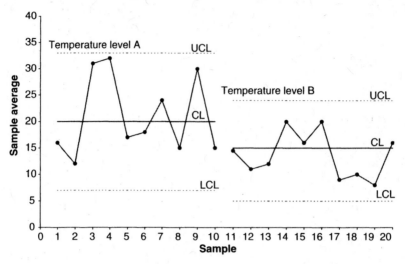

FIGURE 7-17   Interaction pattern between variables on an $\overline{X}$-chart.

7-5   CONTROL CHARTS FOR THE MEAN AND STANDARD DEVIATION

Although an *R*-chart is easy to construct and use, a standard deviation chart (**s-chart**) is preferable for larger sample sizes (equal to or greater than 10, usually). As mentioned in Chapter 4, the range accounts for only the maximum and minimum sample values and consequently is less effective for large samples. The sample standard deviation serves as a better measure of process variability in these circumstances. The **sample standard deviation** is given by

$$s = \sqrt{\frac{\sum_{i=1}^{n} (X_i - \overline{X})^2}{n-1}} \tag{7-20}$$

$$= \sqrt{\frac{\sum_{i=1}^{n} X_i^2 - \left(\sum_{i=1}^{n} X_i\right)^2 / n}{n-1}} \tag{7-21}$$

If the population distribution of a quality characteristic is normal with a **population standard deviation** denoted by σ, the mean and standard deviation of the sample standard deviation are given by

$$E(s) = c_4 \sigma \tag{7-22}$$

$$\sigma_s = \sigma \sqrt{1 - c_4^2} \tag{7-23}$$

respectively, where c_4 is a factor that depends on the sample size and is given by

$$c_4 = \left[\frac{2}{n-1}\right]^{1/2} \frac{\Gamma(n/2)}{\Gamma[(n-1)/2]} \tag{7-24}$$

Values of c_4 are tabulated in Appendix A-7 [see also Montgomery (2004) and Wadsworth et al. (2001).]

No Given Standards

The centerline of a standard deviation chart is

$$\text{CL}_s = \bar{s} = \frac{\sum_{i=1}^{g} s_i}{g} \tag{7-25}$$

where g is the number of samples and s_i is the standard deviation of the ith sample. The upper control limit is

$$\text{UCL}_s = \bar{s} + 3\sigma_s = \bar{s} + 3\sigma\sqrt{1 - c_4^2}$$

In accordance with eq. (7-22), an estimate of the population standard deviation σ is

$$\hat{\sigma} = \frac{\bar{s}}{c_4} \tag{7-26}$$

Substituting this estimate of $\hat{\sigma}$ in the preceding expression yields

$$\text{UCL}_s = \bar{s} + \frac{3\bar{s}\sqrt{(1 - c_4^2)}}{c_4} = B_4\bar{s}$$

where $B_4 = 1 + 3\left(\sqrt{1 - c_4^2}/c_4\right)$ and is tabulated in Appendix A-7. Similarly,

$$\text{LCL}_s = \bar{s} - \frac{3\bar{s}\sqrt{(1 - c_4^2)}}{c_4} = B_3\bar{s}$$

where $B_3 = \max\left[0, 1 - 3\left(\sqrt{1 - c_4^2}/c_4\right)\right]$ and is also tabulated in Appendix A-7. Thus, the 3σ control limits are

$$\begin{aligned} \text{UCL}_s &= B_4\bar{s} \\ \text{LCL}_s &= B_3\bar{s} \end{aligned} \tag{7-27}$$

The centerline of the chart for the mean $\bar{X}$ is given by

$$\text{CL}_{\bar{X}} = \bar{\bar{X}} = \frac{\sum_{i=1}^{g} \bar{X}_i}{g} \tag{7-28}$$

The control limits on the $\bar{X}$-chart are

$$\bar{\bar{X}} \pm 3\sigma_{\bar{X}} = \bar{\bar{X}} \pm \frac{3\sigma}{\sqrt{n}}$$

Using eq. (7-26) to obtain $\hat{\sigma}$, we find the control limits to be

$$\text{UCL}_{\bar{X}} = \bar{\bar{X}} + \frac{3\bar{s}}{c_4\sqrt{n}} = \bar{\bar{X}} + A_3\bar{s}$$

$$\text{LCL}_{\bar{X}} = \bar{\bar{X}} - \frac{3\bar{s}}{c_4\sqrt{n}} = \bar{\bar{X}} - A_3\bar{s} \tag{7-29}$$

where $A_3 = 3/(c_4\sqrt{n})$ and is tabulated in Appendix A-7.

The process of constructing trial control limits, determining special causes associated with out-of-control points, taking remedial actions, and finding the revised control limits is similar to that explained in the section on $\bar{X}$- and R-charts. The s-chart is constructed first. Only if it is in control should the $\bar{X}$-chart be developed, because the standard deviation of $\bar{X}$ is dependent on $\bar{s}$. If the s-chart is not in control, any estimate of the standard deviation of $\bar{X}$ will be unreliable, which will in turn create unreliable control limits for $\bar{X}$.

Given Standard

If a target standard deviation is specified as σ_0, the centerline of the s-chart is found by using eq. (7-22) as

$$CL_s = c_4\sigma_0 \qquad (7\text{-}30)$$

The upper control limit for the s-chart is found by using eq. (7-23) as

$$
\begin{aligned}
UCL_s \quad &= c_4\sigma_0 + 3\sigma_s = c_4\sigma_0 + 3\sigma_0\sqrt{1 - c_4^2} \\
&= \left(c_4 + 3\sqrt{1 - c_4^2}\right)\sigma_0 = B_6\sigma_0
\end{aligned}
$$

where $B_6 = c_4 + 3\sqrt{1 - c_4^2}$ and is tabulated in Appendix A-7. Similarly, the lower control limit for the s-chart is

$$LCL_s = \left(c_4 - 3\sqrt{1 - c_4^2}\right)\sigma_0 = B_5\sigma_0$$

where $B_5 = \max\left[0, c_4 - 3\sqrt{1 - c_4^2}\right]$ and is tabulated in Appendix A-7. Thus, the control limits for the s-chart are

$$
\begin{aligned}
UCL_s &= B_6\sigma_0 \\
LCL_s &= B_5\sigma_0
\end{aligned}
\qquad (7\text{-}31)
$$

If a target value for the mean is specified as $\bar{X}_0$, the centerline is given by

$$CL_{\bar{X}} = \bar{X}_0 \qquad (7\text{-}32)$$

Equations for the control limits will be the same as those given by eq. (7-16) in the section on $\bar{X}$- and R-charts:

$$
\begin{aligned}
UCL_{\bar{X}} &= \bar{X}_0 + A\sigma_0 \\
LCL_{\bar{X}} &= \bar{X}_0 - A\sigma_0
\end{aligned}
\qquad (7\text{-}33)
$$

where $A = 3/\sqrt{n}$ and is tabulated in Appendix A-7.

Example 7-4 The thickness of the magnetic coating on audio tapes is an important characteristic. Random samples of size 4 are selected, and the thickness is measured using an optical instrument. Table 7-3 shows the mean $\bar{X}$ and standard deviation s for 20 samples. The specifications are 38 ± 4.5 micrometers (μm). If a coating thickness is less than the specifications call for, that tape can be used for a different purpose by running it through another coating operation.

TABLE 7-3   Data for Magnetic Coating Thickness (μm)

Sample	Sample Mean, $\bar{X}$	Sample Standard Deviation, s	Sample	Sample Mean, $\bar{X}$	Sample Standard Deviation, s
1	36.4	4.6	11	36.7	5.3
2	35.8	3.7	12	35.2	3.5
3	37.3	5.2	13	38.8	4.7
4	33.9	4.3	14	39.0	5.6
5	37.8	4.4	15	35.5	5.0
6	36.1	3.9	16	37.1	4.1
7	38.6	5.0	17	38.3	5.6
8	39.4	6.1	18	39.2	4.8
9	34.4	4.1	19	36.8	4.7
10	39.5	5.8	20	37.7	5.4

(a) Find the trial control limits for an $\bar{X}$- and an s-chart.

Solution   The standard deviation chart must first be constructed. The centerline of the s-chart is

$$CL_s = \bar{s} = \frac{\sum_{i=1}^{20} s_i}{20} = \frac{95.80}{20} = 4.790$$

The control limits for the s-chart are

$$UCL_s = B_4\bar{s} = (2.266)(4.790) = 10.854$$
$$LCL_s = B_3\bar{s} = (0)(4.790) = 0$$

Figure 7-18 shows this standard deviation control chart. None of the points fall outside the control limits, and the process seems to be in a state of control, so the $\bar{X}$-chart is

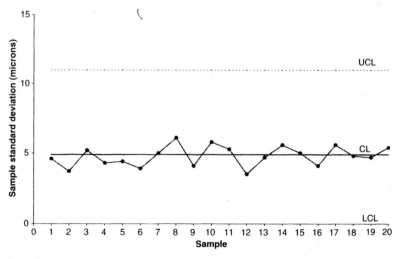

FIGURE 7-18 s-Chart for magnetic coating thickness.

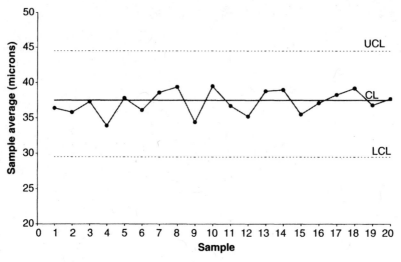

FIGURE 7-19 $\bar{X}$-chart for magnetic coating thickness.

constructed next. The centerline of the $\bar{X}$-chart is

$$CL_{\bar{x}} = \bar{\bar{X}} = \frac{\sum\limits_{i=1}^{20} \bar{X}_i}{20} = \frac{743.5}{20} = 37.175$$

The control limits for the $\bar{X}$-chart are

$$UCL_{\bar{x}} = \bar{\bar{X}} + A_3\bar{s} = 37.175 + (1.628)(4.790) = 44.973$$
$$LCL_{\bar{x}} = \bar{\bar{X}} - A_3\bar{s} = 37.175 - (1.628)(4.790) = 29.377$$

Figure 7-19 depicts the $\bar{X}$-chart. All the points are within the control limits, and no unusual nonrandom pattern is visible on the plot.

(b) Assuming special causes for the out-of-control points, determine the revised control limits.

Solution In this case, the revised control limits will be the same as the trial control limits because we believe that no special causes are present in the system.

(c) Assuming the thickness of the coating to be normally distributed, what proportion of the product will not meet specifications?

Solution The process standard deviation may be estimated as

$$\hat{\sigma} = \frac{\bar{s}}{c_4} = \frac{4.790}{0.9213} = 5.199$$

To find the proportion of the output that does not meet specifications, the standard normal values at the upper and lower specification limits (USL and LSL) must be found. At the lower specification limit we get

$$z_1 = \frac{33.5 - 37.175}{5.199} = -0.71$$

The area below the LSL, found by using the standard normal table in Appendix A-3, is 0.2389. Similarly, the standard normal value at the upper specification limit is

$$z_2 = \frac{42.5 - 37.175}{5.199} = 1.02$$

From Appendix A-3, the area above the USL is 0.1539. Hence, the proportion of product not meeting specifications is $0.2389 + 0.1539 = 0.3928$.

(**d**) Comment on the ability of the process to produce items that meet specifications.

Solution   A proportion of 39.28% of product not meeting specifications is quite high. On the other hand, we found the process to be in control. This example teaches an important lesson. It is possible for a process to be in control and still not produce conforming items. In such cases, management must look for the prevailing common causes and come up with ideas for process improvement. The existing process is not capable of meeting the stated specifications.

(**e**) If the process average shifts to 37.8 μm, what proportion of the product will be acceptable?

Solution   If the process average shifts to 37.8 μm, the standard normal values must be recalculated. At the LSL,

$$z_1 = \frac{33.5 - 37.8}{5.199} = -0.83$$

From the standard normal table in Appendix A-3, the area below the LSL is 0.2033. The standard normal value at the USL is

$$z_2 = \frac{42.5 - 37.8}{5.199} = 0.90$$

The area above the USL is 0.1841. So, the proportion nonconforming is $0.2033 + 0.1841 = 0.3874$. Although this change in the process average does reduce the proportion nonconforming, 38.74% nonconforming is still quite significant.

If output that falls below the LSL can be salvaged at a lower expense than that for output above the USL, the company could consider adjusting the process mean in the downward direction to reduce the proportion above the USL. Being aware of the unit costs associated with salvaging output outside the specification limits will enable the company to choose a target value for the process mean. Keep in mind, though, that this approach does not solve the basic problem. The underlying problem concerns the process variability. To make the process more capable, we must find ways of reducing the process standard deviation. This cannot come through process control, because the process is currently in a state of statistical control. It must come through process improvement, some analytical tools of which are discussed at length in Chapter 5.

7-6 CONTROL CHARTS FOR INDIVIDUAL UNITS

For some situations in which the rate of production is low, it is not feasible for a sample size to be greater than 1. Additionally, if the testing process is destructive and the cost of the item is expensive, the sample size might be chosen to be 1. Furthermore, if every manufactured unit from a process is inspected, the sample size is essentially 1. Service applications in marketing and accounting often have a sample size of 1.

In a control chart for individual units—for which the value of the quality characteristic is represented by X—the variability of the process is estimated from the **moving range** (MR), found from two successive observations. The moving range of two observations is simply the result of subtracting the lesser value. Moving ranges are correlated because they use common-rather than independent values in their calculations, That is, the moving range of observations 1 and 2 correlates with the moving range of observations 2 and 3. Because they are correlated, the pattern of the MR-chart must be interpreted carefully. Neither can we assume, as we have in previous control charts, that X-values in a chart for individuals will be normally distributed. So we must first check the distribution of the individual values. To do this, we might conduct an initial analysis using frequency histograms to identify the shape of the distribution, its skewness, and its kurtosis. Alternatively, we could conduct a test for normality. This information will tell us whether we can make the assumption of a normal distribution when we establish the control limits.

No Given Standards

An estimate of the process standard deviation is given by

$$\hat{\sigma} = \frac{\overline{\text{MR}}}{d_2}$$

where $\overline{\text{MR}}$ is the average of the moving ranges of successive observations, Note that if we have a total of g individual observations, there will be $g-1$ moving ranges. The centerline and control limits of the MR-chart are

$$\begin{aligned} \text{CL}_{\text{MR}} &= \overline{\text{MR}} \\ \text{UCL}_{\text{MR}} &= D_4\overline{\text{MR}} \\ \text{LCL}_{\text{MR}} &= D_3\overline{\text{MR}} \end{aligned} \tag{7-34}$$

For $n = 2$, $D_4 = 3.267$, and $D_3 = 0$, the control limits become

$$\begin{aligned} \text{UCL}_{\text{MR}} &= 3.267\overline{\text{MR}} \\ \text{LCL}_{\text{MR}} &= 0 \end{aligned}$$

The centerline of the X-chart is

$$\text{CL}_X = \bar{X} \tag{7-35}$$

The control limits of the X-chart are

$$\begin{aligned} \text{UCL}_X &= \bar{X} + 3\frac{\overline{\text{MR}}}{d_2} \\ \text{LCL}_X &= \bar{X} - 3\frac{\overline{\text{MR}}}{d_2} \end{aligned} \tag{7-36}$$

where (for $n = 2$) Appendix A-7 gives $d_2 = 1.128$.

Given Standard

The preceding derivation is based on the assumption that no standard values are given for either the mean or the process standard deviation. If standard values are specified as $\bar{X}_0$ and

σ_0, respectively, the centerline and control limits of the X-chart are

$$\begin{aligned} \mathrm{CL}_X &= \bar{X}_0 \\ \mathrm{UCL}_X &= \bar{X}_0 + 3\sigma_0 \\ \mathrm{LCL}_X &= \bar{X}_0 - 3\sigma_0 \end{aligned} \qquad (7\text{-}37)$$

Assuming $n = 2$, the MR-chart for standard values has the following centerline and control limits:

$$\begin{aligned} \mathrm{CL}_{\mathrm{MR}} &= d_2\sigma_0 = (1.128)\sigma_0 \\ \mathrm{UCL}_{\mathrm{MR}} &= D_4 d_2\sigma_0 = (3.267)(1.128)\sigma_0 = (3.685)\sigma_0 \\ \mathrm{LCL}_{\mathrm{MR}} &= D_3 d_2\sigma_0 = 0 \end{aligned} \qquad (7\text{-}38)$$

One advantage of an X-chart is the ease with which it can be understood. It can also be used to judge capability of a process by plotting the upper and lower specification limits on the chart itself. However, it has several disadvantages compared to an $\bar{X}$-chart. An X-chart is not as sensitive to changes in the process parameters. It typically requires more samples to detect parametric changes of the same magnitude. The main disadvantage of an X-chart, though, is that the control limits can become distorted if the individual items don't fit a normal distribution.

Example 7-5   Table 7-4 shows the Brinell hardness numbers of 20 individual steel fasteners and the moving ranges. The testing process dents the parts so that they cannot be used for their intended purpose. Construct the X-chart and MR-chart based on two successive observations. Specification limits are 32 ± 7.

Solution   Note that there are 19 moving-range values for 20 observations. The average of the moving ranges is

$$\overline{\mathrm{MR}} = \frac{\sum \mathrm{MR}_i}{19} = \frac{96}{19} = 5.053$$

which is also the centerline of the MR-chart. The control limits for the MR-chart are

$$\begin{aligned} \mathrm{UCL}_{\mathrm{MR}} &= D_4 \overline{\mathrm{MR}} = (3.267)5.053 = 16.508 \\ \mathrm{LCL}_{\mathrm{MR}} &= D_3 \overline{\mathrm{MR}} = (0)5.053 = 0 \end{aligned}$$

TABLE 7-4   Brinell Hardness Data for Individual Fasteners

Sample	Brinell Hardness	Moving Range	Sample	Brinell Hardness	Moving Range
1	36.3	–	11	29.4	1.1
2	28.6	7.7	12	35.2	5.8
3	32.5	3.9	13	37.7	2.5
4	38.7	6.2	14	27.5	10.2
5	35.4	3.3	15	28.4	0.9
6	27.3	8.1	16	33.6	5.2
7	37.2	9.9	17	28.5	5.1
8	36.4	0.8	18	36.2	7.7
9	38.3	1.9	19	32.7	3.5
10	30.5	7.8	20	28.3	4.4

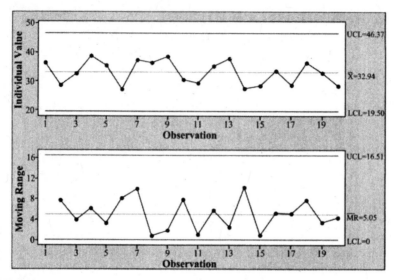

FIGURE 7-20 Control charts for individual values (*X*-chart) and moving range (MR-chart) for hardness of steel fasteners.

We can use Minitab to construct control charts for individual values and moving ranges for the steel fastener hardness data in Table 7-4. Click on **Stat** > **Control Charts** > **Variables Charts for Individuals** > **I – MR.** In **Variables,** input the column number or variable name, Brinell hardness in this case. Click **OK.** Figure 7-20 shows the trial control charts. No points plot outside the control limits on the MR-chart. Since the MR-chart exhibits control, we can construct the *X*-chart for individual data values. The centerline of the *X*-chart is

$$\bar{X} = \frac{\sum X_i}{20} = \frac{658.7}{20} = 32.935$$

The control limits for the *X*-chart are given by

$$UCL_X = \bar{X} + \frac{3\overline{MR}}{d_2} = 32.935 + \frac{3(5.053)}{1.128} = 46.374$$

$$LCL_X = \bar{X} - \frac{3\overline{MR}}{d_2} = 32.935 - \frac{3(5.053)}{1.128} = 19.496$$

The *X*-chart is also shown in Figure 7-20. No out-of-control points are visible. Comparing the individual values with the specification limits, we find no values outside the specification limits. Thus, the observed nonconformance rate is zero and the process is capable.

7-7 CONTROL CHARTS FOR SHORT PRODUCTION RUNS

Organizations, both manufacturing and service, are faced with short production runs for several reasons. Product specialization and being responsive to customer needs are two important reasons. Consider a company that assembles computers based on customer orders. There is no guarantee that the next 50 orders will be for a computer with the same hardware and software features.

$\bar{X}$- and R-Charts for Short Production Runs

Where different parts may be produced in the short run, one approach is to use the deviation from the nominal value as the modified observations. The nominal value may vary from part to part. So, the deviation of the observed value O_i from the nominal value N is given by

$$X_i = O_i - N, \qquad i = 1, 2, \ldots, n \tag{7-39}$$

The procedure for the construction of the $\bar{X}$- and R-charts is the same as before using the modified observations, X_i. Different parts are plotted on the same control chart so as to have the minimum information (usually, at least 20 samples) required to construct the charts, even though for each part there are not enough samples to justify construction of a control chart.

Several assumptions are made in this approach. First, it is assumed that the process standard deviation is approximately the same for all the parts. Second, what happens when a nominal value is not specified (which is especially true for characteristics that have one-sided specifications, such as breaking strength)? In such a situation, the process average based on historical data may have to be used.

Z-MR Chart

When individuals' data are obtained on the quality characteristic, an approach is to construct a standardized control chart for individuals (Z-chart) and a moving-range (MR) chart. The standardized value is given by

$$Z = \frac{\text{individual value} - \text{process mean}}{\text{process standard deviation}} \tag{7-40}$$

The moving range is calculated from the standardized values using a length of size 2. Depending on how each group (part or product) is defined, the process standard deviation for group i is estimated by

$$\hat{\sigma}_i = \frac{\overline{MR}_i}{d_2} \tag{7-41}$$

where $\overline{MR}_i$ represents the average moving range for group i, and d_2 is a control chart factor found from Appendix A-7. Minitab provides several options for selecting computation of the process mean and process standard deviation. For each group (part or product), the mean of the observations in that group could be used as an estimate of the process mean for that group. Alternatively, historical values of estimates may be specified as an option.

In estimating the process standard deviation for each group (part or product), Minitab provides options for defining groups as follows: by runs; by parts, where all observations on the same part are combined in one group; constant (combine all observations for all parts in one group); and relative to size (transform the original data by taking the natural logarithm and then combine all into one group).

The relative-to-size option assumes that variability increases with the magnitude of the quality characteristic. The natural logarithm transformation stabilizes the variance. A common estimate ($\hat{\sigma}$) of the process standard deviation is obtained from the transformed data. The *constant* option that pools all data assumes that the variability associated with all groups is the same, implying that product or part type or characteristic size has no influence. This option must be used only if there is enough information to justify the

TABLE 7-5   Data on Short Production Runs

Run Number	Part Number	Quality Characteristic	Run Number	Part Number	Quality Characteristic
1	A	30	5	B	44
	A	25		B	41
	A	28		B	45
2	B	42	6	D	35
	B	40		D	32
3	A	31		D	33
	A	29	7	B	43
4	C	54		B	45
	C	56		B	40
	C	53		B	42

assumption. It produces a single estimate ($\hat{\sigma}$) of the common process standard deviation. The option of pooling by parts assumes that all runs of a particular part have the same variability. It produces an estimate ($\hat{\sigma}_i$) of the process standard deviation for each part group. Finally, the option of pooling by runs assumes that part variability may change from run to run. It produces an estimate of the process standard deviation for each run, independently.

Example 7-6   Data on short production runs on the diameters of four parts (A, B, C, and D) are shown in Table 7-5. It is believed that the processes for manufacturing the four parts have different variabilities. Since parts are manufactured based on demand, they are not necessarily produced in the same run. Construct an appropriate control chart and comment on the process.

Solution   A worksheet is created in Minitab using the data provided. Click on **Stat > Control Charts > Variables Charts for Individuals > Z − MR.** In **Variables,** enter the

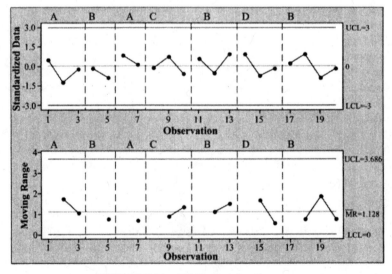

FIGURE 7-21   Z-MR chart for diameter.

column number or name of the variable, Diameter, in this case. In **Part indicator,** enter the column number or name, in this case the Part number. Click on **Z-MR Options,** select **Estimate.** Under **How to define groups of observations,** select **By parts.** Select **Average moving range** to estimate the standard deviation of each group, with **Length of moving range** as 2. Click **OK.** Figure 7-21 shows the Z-MR chart for diameter of parts. All of the points on Z-MR charts are within the control limits with no unusual patterns. Note that the upper and lower control limits on the Z-chart are at 3 and -3, respectively, with the centerline at 0.

7-8 OTHER CONTROL CHARTS

In previous sections we have examined commonly used control charts. Now we look at several other control charts. These charts are specific to certain situations. Procedures for constructing $\bar{X}$- and R-charts and interpreting their patterns apply to these charts as well, so they are not repeated here.

Cumulative Sum Control Chart for the Process Mean

In Shewhart control charts such as the $\bar{X}$- and R-charts, a plotted point represents information corresponding to that observation only. It does not use information from previous observations. On the other hand, a **cumulative sum chart**, usually called a **cusum chart**, uses information from all of the prior samples by displaying the cumulative sum of the deviation of the sample values (e.g., the sample mean) from a specified target value.

The cumulative sum at sample number m is given by

$$S_m = \sum_{i=1}^{m}(\bar{X}_i - \mu_0) \tag{7-42}$$

where $\bar{X}_i$ is the sample mean for sample i and μ_0 is the target mean of the process.

Cusum charts are more effective than Shewhart control charts in detecting relatively small shifts in the process mean (of magnitude $0.5\sigma_{\bar{X}}$ to about $2\sigma_{\bar{X}}$). A cusum chart uses information from previous samples, so the effect of a small shift is more pronounced. For situations in which the sample size n is 1 (say, when each part is measured automatically by a machine), the cusum chart is better suited than a Shewhart control chart to determining shifts in the process mean. Because of the magnified effect of small changes, process shifts are easily found by locating the point where the slope of plotted cusum pattern changes.

There are some disadvantages to using cusum charts, however. First, because the cusum chart is designed to detect small changes in the process mean, it can be slow to detect large changes in the process parameters. Because a decision criterion is designed to do well under a specific situation does not mean that it will perform equally well under different situations. Details on modifying the decision process for a cusum chart to detect large shifts can be found in Hawkins (1981, 1993), Lucas (1976, 1982), and Woodall and Adams (1993). Second, the cusum chart is not an effective tool in analyzing the historical performance of a process to see whether it is in control or to bring it in control. Thus, these charts art typically used for well-established processes that have a history of being stable.

Recall that for Shewhart control charts, the individual points are assumed to be uncorrelated. Cumulative values are, however, related. That is, S_{i-1} and S_i are related because $S_i = S_{i-1} + (\bar{X}_i - \mu_0)$. It is therefore possible for a cusum chart to exhibit runs or other patterns as a result of this relationship. The rules for describing out-of-control conditions based on the plot patterns of Shewhart charts may therefore not be applicable to cusum charts. Finally, training workers to use and maintain cusum charts may be more costly than for Shewhart charts.

Cumulative sum charts can model the proportion of nonconforming items, the number of nonconformities, the individual values, the sample range, the sample standard deviation, or the sample mean. In this section we focus on their ability to detect shifts in the process mean.

Suppose that the target value of a process mean when the process is in control is denoted by μ_0. If the process mean shifts upward to a higher value μ_1. an upward drift will be observed in the value of the cusum S_m given by eq. (7-42) because the old lower value μ_0 is still used in the equation even though the X-values are now higher. Similarly, if the process mean shifts to a lower value μ_2, a downward trend will be observed in S_m. The task is to determine whether the trend in S_m is significant so that we can conclude that a change has taken place in the process mean.

In the situation where individual observations ($n = 1$) are collected from a process to monitor the process mean, eq. (7-42) becomes

$$
\begin{aligned}
S_m &= \sum_{i=1}^{m} (X_i - \mu_0) \\
&= (X_m - \mu_0) + S_{m-1}
\end{aligned}
\tag{7-43}
$$

where $S_0 = 0$.

Tabular Method

Let us first consider the case of individual observations (X_i) being drawn from a process with mean μ_0 and standard deviation σ. When the process is in control, we assume that $X_i \sim N(\mu_0, \sigma)$. In the tabular cusum method, deviations above μ_0 are accumulated with a statistic S^+, and deviations below μ_0 are accumulated with a statistic S^-. These two statistics, S^+ and S^-, are labeled one-sided upper and lower cusums, respectively, and are given by

$$
S_m^+ = \max[0, X_i - (\mu_0 + K) + S_{m-1}^+]
\tag{7-44}
$$

$$
S_m^- = \max[0, (\mu_0 - K) - X_i + S_{m-1}^-]
\tag{7-45}
$$

where $S_0^+ = S_0^- = 0$.

The parameter K in eqs. (7-44) and (7-45) is called the *allowable slack* in the process and is usually chosen as halfway between the target value μ_0, and the shifted value μ_1, that we are interested in detecting. Expressing the shift (δ) in standard deviation units, we have $\delta = |\mu_1 - \mu_0|/\sigma$, leading to

$$
K = \frac{\delta}{2}\sigma = \frac{|\mu_1 - \mu_0|}{2}
\tag{7-46}
$$

Thus, examining eqs. (7-44) and (7-45), we find that S_m^+ and S_m^- accumulate deviations from the target value μ_0 that are greater than K. Both are reset to zero upon becoming negative. In practice, $K = k\sigma\delta$, where k is in units of standard deviation. In eq. (7-46), $k = 0.5$.

A second parameter in the decision-making process using cusums is the *decision interval* H, to determine out-of-control conditions. As before, we set $H = h\sigma$, where h is in standard deviation units. When the value of S_m^+ or S_m^- plots beyond H, the process will be considered to be out of control. When $k = 1/2$, a reasonable value of h is 5 (in standard deviation units), which ensures a small average run length for shifts of the magnitude of 1 standard deviation that we wish to detect (Hawkins 1993). It can be shown that for a small value of β, the probability of a type II error, the decision interval is given by

$$H = \frac{-\sigma^2 \ln(\alpha)}{\mu_1 - \mu_0} \tag{7-47}$$

Thus, if sample averages are used to construct cusums in the above procedures, σ^2 will be replaced by σ^2/n in eq. (7-47), assuming samples of size n.

To determine when the shift in the process mean was most likely to have occurred, we will monitor two counters, N^+ and N^-. The counter N^+ notes the number of consecutive periods that $S_m{}^+$ is above 0, whereas N^- tracks the number of consecutive periods that $S_m{}^-$ is above zero. When an out-of-control condition is detected, one can count backward from this point to the time period when the cusum was above zero to find the first period in which the process probably shifted. An estimate of the new process mean may be obtained from

$$\hat{\mu} = \left\{ \mu_0 + K + \frac{S_m^+}{N^+} \qquad \text{if } S_m^+ > H \right. \tag{7-48}$$

or from

$$\hat{\mu} = \left\{ \mu_0 - K - \frac{S_m^-}{N^-} \qquad \text{if } S_m^- > H \right. \tag{7-49}$$

Example 7-7   In the preparation of a drug, the percentage of calcium is a characteristic we want to control. Random samples of size 5 are selected, and the average percentage of calcium is found. The data values from 15 samples are shown in Table 7-6. From historical data, the standard deviation of the percentage of calcium is estimated as 0.2%. The target value for the average percentage of calcium content is 26.5%. We decide to detect shifts in the average percentage of calcium content of 0.1%.

Solution   Here $\mu_0 = 26.5$, $\mu_1 = 26.6$, $\hat{\sigma} = 0.2$, and $n = 5$. So, since sample averages are being monitored, $\sigma_{\bar{x}} = 0.2/\sqrt{5} = 0.089$. Now $K = (26.6 - 26.5)/2 = 0.05$, while $k = 0.05/0.089 = 0.5618$. Assuming that $h = 5$, the decision interval $H = 5(0.089) = 0.445$. Table 7-7 shows the values computed for $S_m{}^+$, N^+, $S_m{}^-$, and N^-. If we wish to detect an upward drift of the process mean, using $S_m{}^+$, we find that the first time when $S_m{}^+ > H = 0.445$ is at sample

TABLE 7-6   Average Percentage of Calcium

Sample	Average Percentage of Calcium, $\overline{X}$	Sample	Average Percentage of Calcium, $\overline{X}$
1	25.5	9	26.4
2	26.0	10	26.3
3	26.6	11	26.9
4	26.8	12	27.8
5	27.5	13	26.2
6	25.9	14	26.8
7	27.0	15	26.6
8	25.4		

TABLE 7-7   Tabular Cumulative Sums for Average Calcium Percentage

Sample Number	$\bar{X}_m$	$\bar{X}_m - 26.55$	S_m^+	N^+	$26.45 - \bar{X}_m$	S_m^-	N^-
1	25.5	−1.05	0	0	0.95	0.95	1
2	26.0	−0.55	0	0	0.45	1.40	2
3	26.6	0.05	0.05	1	−0.15	1.25	3
4	26.8	0.25	0.30	2	−0.35	0.90	4
5	27.5	0.95	1.25	3	−1.05	0	0
6	25.9	−0.65	0.60	4	0.55	0.55	1
7	27.0	0.45	1.05	5	−0.55	0	0
8	25.4	−1.15	0	0	1.05	1.05	1
9	26.4	−0.15	0	0	0.05	1.10	2
10	26.3	−0.25	0	0	0.15	1.25	3
11	26.9	0.35	0.35	1	−0.45	0.80	4
12	27.8	1.25	1.60	2	−1.35	0	0
13	26.2	−0.35	1.25	3	0.25	1.25	1
14	26.8	0.25	1.50	4	−0.35	0	0
15	26.6	0.05	1.55	5	−0.15	0	0

number 5. Note that for $m = 5$, $N^+ = 3$. We estimate that the upward shift in the process average occurs between samples 2 and 3. An estimate of the new process mean is

$$\hat{\mu} = 26.5 + 0.05 + \frac{1.25}{3} = 26.967$$

On the contrary, if we were interested in detecting downward drifts in the process mean using the results of Table 7-7, it would have been detected on the first sample itself. Note that $S_1^- = 0.95 > H = 0.445$.

A similar decision may be arrived at using Minitab. Click on **Stat > Control Charts > Time-Weighted Charts > CUSUM**. In the worksheet, all observations were entered in one column. Input the column number or name, in this case Calcium. Input **Subgroup size as 1** and **Target** as 26.5. Click on **Cusum Options**. Under **Parameters,** input for **Standard deviation** the value 0.089. Click on **Plan/Type,** under **Type of CUSUM** select One-sided. Under **CUSUM Plan,** input for **h** the value 5.0, and for **k**, the value 0.5618, Click **OK**. Figure 7-22 shows the one-sided cumulative sum charts using Minitab. Conclusions drawn from this figure are similar to those made earlier.

V-Mask Method

In the V-mask approach, a template known as a V-mask, proposed by Barnard (1959), is used to determine a change in the process mean through the plotting of cumulative sums. Figure 7-23 shows a V-mask, which has two parameters, the lead distance d and the angle θ of each decision line with respect to the horizontal. The V-mask is positioned such that point P coincides with the last plotted value of the cumulative sum and line OP is parallel to the horizontal axis. If the values plotted previously are within the two arms of the V-mask—that is, between the upper decision line and the lower decision line—the process is judged to be in control. If any value of the cusum lies outside the arms of the V-mask, the process is considered to be out of control.

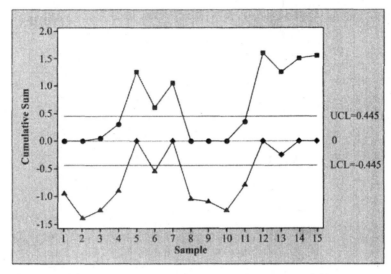

FIGURE 7-22   One-sided cumulative sum charts for average calcium percentage.

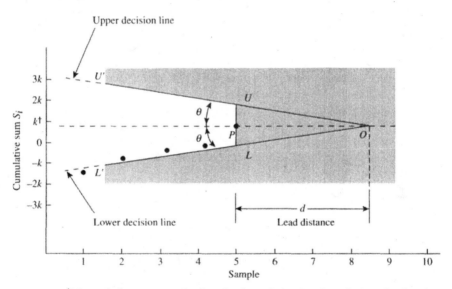

†k is a scale factor representing the ratio of a vertical scale unit to a horizontal scale unit.

FIGURE 7-23   V-mask for making decisions with cumulative sum charts.

In Figure 7-23, notice that a strong upward shift in the process mean is visible for sample 5. This shift makes sense given the fact that the cusum value for sample 1 is below the lower decision line, indicating an out-of-control situation. Similarly, the presence of a plotted value above the upper decision line indicates a downward drift in the process mean.

Determination of V-Mask Parameters   The two parameters of a V-mask, d and θ, are determined based on the levels of risk that the decision maker is willing to tolerate. These risks are the type I and type II errors described in Chapter 6. The probability of **type I error**,

α, is the risk of concluding that a process is out of control when it is really in control. The probability of a **type II error**, β, is the risk of failing to detect a change in the process parameter and concluding that the process is in control when it is really out of control. Let $\Delta \bar{X}$ denote the amount of shift in the process mean that we want to be able to detect, and let $\sigma_{\bar{X}}$ denote the standard deviation of $\bar{X}$. Next, consider the equation

$$\delta = \frac{\Delta \bar{X}}{\sigma_{\bar{X}}} \tag{7-50}$$

where δ represents the degree of shift in the process mean, relative to the standard deviation of the mean, that we wish to detect. Then, the **lead distance** for the V-mask is given by

$$d = \frac{2}{\delta^2} \ln \frac{1 - \beta}{\alpha} \tag{7-51}$$

If the probability of a type II error, β, is selected to be small, then eq. (7-51) reduces to

$$d = -\frac{2}{\delta^2} \ln(\alpha) \tag{7-52}$$

The **angle of decision line** with respect to the horizontal is obtained from

$$\theta = \tan^{-1} \frac{\Delta \bar{X}}{2k} \tag{7-53}$$

where k is a scale factor representing the ratio of a vertical-scale unit to a horizontal-scale unit on the plot. The value of k should be between $\sigma_{\bar{X}}$ and $2\sigma_{\bar{X}}$, with a preferred value of $2\sigma_{\bar{X}}$.

One measure of a control chart's performance is the average run length (ARL). (We discussed ARL in Chapter 6.) This value represents the average number of points that must be plotted before an out-of-control condition is indicated. For a Shewhart control chart, if p represents the probability that a single point will fall outside the control limits, the average run length is given by

$$\text{ARL} = \frac{1}{p} \tag{7-54}$$

For 3σ limits on a Shewhart $\bar{X}$-chart, the value of p is about 0.0026 when the process is in control. Hence, the ARL for an $\bar{X}$-chart exhibiting control is

$$\text{ARL} = \frac{1}{0.0026} = 385$$

The implication of this is that, on average, if the process is in control, every 385th sample statistic will indicate an out-of-control state. The ARL is usually larger for a cusum chart than for a Shewhart chart. For example, for a cusum chart with comparable risks, the ARL is around 500. Thus, if the process is in control, on average, every 500th sample statistic will indicate an out-of-control situation, so there will be fewer false alarms.

Example 7-8 Refer to Example 7-7 and the data on the average percentage of calcium from 15 samples shown in Table 7-6. Construct a cumulative sum chart and make a decision using the V-mask method.

Solution As shown in Example 7-7, $\sigma_{\bar{X}} = 0.089$. Now, the deviation of each sample mean, $\bar{X}_i$ from the target mean $\mu_0 = 26.5$ is found, following which the cumulative sum S_i is

TABLE 7-8 Cumulative Sum of Data for Calcium Content

Sample, i	Deviation of Sample Mean from Target, $\overline{X}_i - \mu_0$	Cumulative Sum, S_i	Sample, i	Deviation of Sample Mean from Target, $\overline{X}_i - \mu_0$	Cumulative Sum, S_i
1	−1.0	−1.0	9	−0.1	−1.4
2	−0.5	−1.5	10	−0.2	−1.6
3	0.1	−1.4	11	0.4	−1.2
4	0.3	−1.1	12	1.3	0.1
5	1.0	−0.1	13	−0.3	−0.2
6	−0.6	−0.7	14	0.3	0.1
7	0.5	−0.2	15	0.1	0.2
8	−1.1	−1.3			

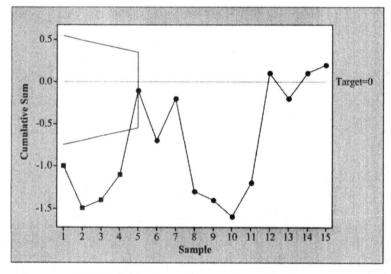

FIGURE 7-24 Cumulative sum chart using V-mask.

obtained. These values are shown in Table 7-8. Using Minitab, click on **Stat > Control Charts > Time-Weighted Charts > CUSUM.** Input **Subgroup size** as **1,** and **Target** as 26.5. Click on **Cusum Options.** Under **Parameters,** input for **Standard deviation** the value 0.089. Click on **Plan/Type,** under **Type of CUSUM,** select Two-sided. Under **CUSUM Plan,** input for **h** the value of 5, and for **k,** the value 0.5618. Click **OK.** Figure 7-24 shows the cumulative sum chart using a V-mask. When the V-mask is centered on sample 5, note that samples 1, 2, 3, and 4 fall below the decision line. We conclude that the process is out of control.

Designing a Cumulative Sum Chart for a Specified ARL The average run length can be used as a design criteria for control charts. If a process is in control, the ARL should be long, whereas if the process is out of control, the ARL should be short. Recall that δ is the degree of shift in the process mean, relative to the **standard deviation of the sample mean,** that we are interested in detecting; that is, $\delta = \Delta \overline{X} / \sigma_{\overline{X}}$. Let $L(\delta)$ denote the desired ARL when a shift in the process mean is on the order of δ. An ARL curve is a plot of δ versus its corresponding average run length, $L(\delta)$. For a process in control, when $\delta = 0$, a large value of $L(0)$ is

TABLE 7-9   Selection of Cumulative Sum Control Charts Based on Specified ARL

δ = Deviation from Target Value (standard deviations)		$L(0)$ = Expected Run Length When Process Is in Control					
		50	100	200	300	400	500
0.25	$(k/\sigma_{\bar{x}})\tan\theta$	0.125			0.195		0.248
	d	47.6			46.2		37.4
	$L(0.25)$	28.3			74.0		94.0
0.50	$(k/\sigma_{\bar{x}})\tan\theta$	0.25	0.28	0.29	0.28	0.28	0.27
	d	17.5	18.2	21.4	24.7	27.3	29.6
	$L(0.5)$	15.8	19.0	24.0	26.7	29.0	30.0
0.75	$(k/\sigma_{\bar{x}})\tan\theta$	0.375	0.375	0.375	0.375	0.375	0.375
	d	9.2	11.3	13.8	15.0	16.2	16.8
	$L(0.75)$	8.9	11.0	13.4	14.5	15.7	16.5
1.0	$(k/\sigma_{\bar{x}})\tan\theta$	0.50	0.50	0.50	0.50	0.50	0.50
	d	5.7	6.9	8.2	9.0	9.6	10.0
	$L(1.0)$	6.1	7.4	8.7	9.4	10.0	10.5
1.5	$(k/\sigma_{\bar{x}})\tan\theta$	0.75	0.75	0.75	0.75	0.75	0.75
	d	2.7	3.3	3.9	4.3	4.5	4.7
	$L(1.5)$	3.4	4.0	4.6	5.0	5.2	5.4
2.0	$(k/\sigma_{\bar{x}})\tan\theta$	1.0	1.0	1.0	1.0	1.0	1.0
	d	1.5	1.9	2.2	2.4	2.5	2.7
	$L(2.0)$	2.26	2.63	2.96	3.15	3.3	3.4

Source: A. H. Bowker and G.J. Lieberman, *Engineering Statistics*, 2nd ed.,© 1972. p. 498. Reprinted by permission of Prentice Hall, Upper Saddle River, NJ.

desirable, For a specified value of δ, we may have a desirable value of $L(\delta)$. Thus, two points on the ARL curve, $[0, L(0)]$ and $[\delta, L(\delta)]$, are specified. The goal is to find the cusum chart parameters d and θ that will satisfy these desirable goals.

Bowker and Lieberman (1972) provide a table (see Table 7-9) for selecting the V-mask parameters d and θ when the objective is to minimize $L(\delta)$ for a given δ. It is assumed that the decision maker has a specified value of L (0) in mind. Table 7-9 gives values for $(k/\sigma_{\bar{x}})\tan\theta$ and d, and the minimum value of $L(\delta)$ for a specified δ. We use this table in Example 7-9.

Example 7-9   Suppose that for a process in control, we want an ARL of 400. We also decide to detect shifts in the process mean of magnitude $1.5\sigma_{\bar{x}}$—that is $\Delta\bar{X} = 1.5\sigma_{\bar{x}}$—which means that $\delta = 1.5$, Find the parameters of a V-mask for this process.

Solution   From Tables 7-9 for $L(0) = 400$ and $\delta = 1.5$, we have

$$(k/\sigma_{\bar{x}})\tan\theta = 0.75$$
$$d = 4.5$$
$$L(1.5) = 5.2$$

If k, the ratio of the vertical scale to the horizontal scale, is selected to be $2\sigma_{\bar{x}}$, we have

$$2\tan\theta = 0.75 \quad \text{or} \quad \tan\theta = 0.375$$

The angle of the V-mask is

$$\theta = \tan^{-1}(0.375) = 20.556°$$

If we feel that 5.2 is too large a value of $L(1.5)$, we need to reduce the average number of plotted points it takes to first detect a shift in the process mean of magnitude $1.5\sigma_{\bar{x}}$. Currently, it takes about 5.2 points, on average, to detect a shift of this magnitude. Assume that we prefer $L(1.5)$ to be less than 5.0. From Table 7-9, for $\delta = 1.5$ and $L(1.5) < 5.0$, we could choose $L(1.5) = 4.6$, which corresponds to $(k/\sigma_{\bar{x}})\tan\theta = 0.75$, and $d = 3.9$. If k is chosen to be $2\sigma_{\bar{x}}$, we get

$$\tan\theta = \frac{0.75}{2} = 0.375$$

Hence, $\theta = 20.556°$ (the same value as before) and $d = 3.9$ (a reduced value). For $L(1.5) = 4.6$, which is less than 5.0, we have increased the sensitivity of the cusum chart to detect changes of magnitude $1.5\sigma_{\bar{x}}$, but in doing so, we have reduced the ARL for $\delta = 0$ [i.e., $L(0)$] to 200 from the previous value of 400. So now every 200th observation, on average, will be plotted as an out-of-control point when the process is actually in control.

Cumulative Sum for Monitoring Process Variability

Cusum charts may also be used to monitor process variability as discussed by Hawkins (1981). Assuming that $X_i \sim N(\mu_0, \sigma)$, the standardized value Y_i is obtained first as $Y_i = (X_i - \mu_0)/\sigma$. A new standardized quantity (Hawkins 1993) is constructed as follows:

$$v_i = \frac{\sqrt{|y_i|} - 0.822}{0.349} \tag{7-55}$$

where it is suggested that the v_i are sensitive to both variance and mean changes. For an in-control process, v_i is distributed approximately $N(0, 1)$. Two one-sided standardized cusums are constructed as follows, to detect scale changes:

$$S_m^+ = \max[0, v_m - k + S_{m-1}^+] \tag{7-56}$$

$$S_m^- = \max[0, -k - v_m + S_{m-1}^-] \tag{7-57}$$

where $S_0^+ = S_0^- = 0$. The values of h and k are selected using guidelines similar to those discussed in the section on cusum for the process mean. When the process standard deviation increases, the values of S_m^+ in eq. (7-56) will increase. When S_m^+ exceeds h, we will detect an out-of-control condition. Similarly, if the process standard deviation decreases, values of S_m^- will increase.

Moving-Average Control Chart

As mentioned previously, standard *Shewhart control charts* are quite *insensitive* to *small shifts*, and cumulative sum charts are one way to alleviate this problem. A control chart using the moving-average method is another. Such charts are effective for detecting shifts of small magnitude in the process mean. Moving-average control charts can also be used in situations for which the *sample size is 1*, such as when product characteristics are *measured automatically* or when the time to produce a unit is long. It should be noted that by their very nature, *moving-average values are correlated*.

Suppose that samples of size n are collected from the process. Let the first t sample means be denoted by $\bar{X}_1, \bar{X}_2, \bar{X}_3, \ldots, \bar{X}_t$. (One sample is taken for each time step.) The moving average of width w (i.e., w samples) at time step t is given by

$$M_t = \frac{\bar{X}_t + \bar{X}_{t-1} + \cdots + \bar{X}_{t-w+1}}{w} \qquad (7\text{-}58)$$

At any time step t, the moving average is updated by dropping the oldest mean and adding the newest mean. The variance of each sample mean is

$$\text{Var}(\bar{X}_t) = \frac{\sigma^2}{n}$$

where σ^2 is the population variance of the individual values. The variance of M_t is

$$
\begin{aligned}
\text{Var}(M_t) &= \frac{1}{w^2} \sum_{i=t-w+1}^{t} \text{Var}(\bar{X}_i) \\
&= \frac{1}{w^2} \sum_{i=t-w+1}^{t} \frac{\sigma^2}{n} \qquad (7\text{-}59) \\
&= \frac{\sigma^2}{nw}
\end{aligned}
$$

The centerline and control limits for the moving-average chart are given by

$$
\begin{aligned}
\text{CL} &= \bar{\bar{X}} \\
\text{UCL} &= \bar{\bar{X}} + 3\frac{\sigma}{\sqrt{nw}} \qquad (7\text{-}60) \\
\text{LCL} &= \bar{\bar{X}} - 3\frac{\sigma}{\sqrt{nw}}
\end{aligned}
$$

From eq. (7-60), we can see that as w increases, the width of the control limits decreases. So, *to detect shifts of smaller magnitudes, larger values of w should be chosen.*
For the startup period (when $t < w$), the moving average is given by

$$M_t = \frac{\sum_{i=1}^{t} \bar{X}_i}{t}, \qquad t = 1, 2, \ldots, w-1 \qquad (7\text{-}61)$$

The control limits for this startup period are

$$
\begin{aligned}
\text{UCL} &= \bar{\bar{X}} + \frac{3\sigma}{\sqrt{nt}}, \qquad t = 1, 2, \quad \ldots \quad , w-1 \\
\text{LCL} &= \bar{\bar{X}} - \frac{3\sigma}{\sqrt{nt}}, \qquad t = 1, 2, \quad \ldots \quad , w-1
\end{aligned} \qquad (7\text{-}62)
$$

Since these control limits change at each sample point during this startup period, an alternative procedure would be to use the ordinary $\bar{X}$-chart for $t < w$ and use the moving-average chart for $t \geq w$.

Example 7-10 The amount of a coloring pigment in polypropylene plastic, produced in batches, is a variable of interest. For 20 random samples of size 5, the average amount of

TABLE 7-10   Data and Results for a Moving-Average Control Chart (kg)

Sample	Sample Average, $\overline{X}_t$	Moving Average, M_t	Control Limits for M_t LCL	Control Limits for M_t UCL
1	25.0	25.0	24.929	25.391
2	25.4	25.2	24.997	25.323
3	25.2	25.2	25.027	25.293
4	25.0	25.15	25.045	25.275
5	25.2	25.16	25.057	25.263
6	24.9	25.12	25.066	25.254
7	25.0	25.12	25.066	25.254
8	25.4	25.12	25.066	25.254
9	24.9	25.07	25.066	25.254
10	25.2	25.10	25.066	25.254
11	25.0	25.07	25.066	25.254
12	25.7	25.20	25.066	25.254
13	25.0	25.20	25.066	25.254
14	25.1	25.15	25.066	25.254
15	25.0	25.17	25.066	25.254
16	24.9	25.12	25.066	25.254
17	25.0	25.12	25.066	25.254
18	25.1	25.02	25.066	25.254
19	25.4	25.08	25.066	25.254
20	25.8	25.20	25.066	25.254

pigment (in kilograms) is shown in Table 7-10. Construct a moving-average control chart of width 6. The process has up to this point been in control with an average range $\overline{R}$ of 0.40 kg.

Solution   Table 7-10 shows the values computed for the moving average M_t based on a width w of 6. For values of $t < 6$, the moving average is calculated using eq. (7-61). For $t \geq 6$, M_t is calculated using eq. (7-58). Also shown in Table 7-10 are the lower and upper control limits for the moving-average chart. To find these limits, eq. (7-62) is used for $t < 6$, and eq. (7-60) is used for $t \geq 6$. The mean of the sample averages is

$$\overline{\overline{X}} = \frac{\sum\limits_{t=1}^{20} \overline{X}_t}{20} = \frac{503.2}{20} = 25.16$$

Since $\overline{R} = 0.40$, an estimate of the process standard deviation is

$$\hat{\sigma} = \frac{\overline{R}}{d_2} = \frac{0.40}{2.326} = 0.172$$

To calculate the control limits, consider sample 3:

$$\text{UCL} = \overline{\overline{X}} + 3\frac{\sigma}{\sqrt{nt}} = 25.16 + 3\frac{0.172}{\sqrt{(5)(3)}} = 25.293$$

$$\text{LCL} = \overline{\overline{X}} - 3\frac{\sigma}{\sqrt{nt}} = 25.16 - 3\frac{0.172}{\sqrt{(5)(3)}} = 25.027$$

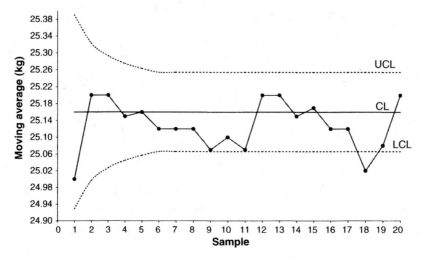

FIGURE 7-25   Moving average control chart.

For samples 6 to 20, the control limits stay the same; the LCL is 25.066 kg and the UCL is 25.254 kg. Figure 7-25 shows a plot of the moving averages and control limits. The moving average for sample 18 plots below the lower control limit, indicating that the process mean has drifted downward. Special causes should be investigated for this out-of-control condition, and appropriate corrective action should be taken.

Exponentially Weighted Moving-Average or Geometric Moving-Average Control Chart

The preceding discussion showed that a moving-average chart can be used as an alternative to an ordinary $\overline{X}$-chart to detect small changes in process parameters. The moving-average method is basically a weighted-average scheme. For sample t, the sample means $\overline{X}_t, \overline{X}_{t-1}, \ldots, \overline{X}_{t-w+1}$ are each weighted by $1/w$ [see eq. (7-58)], while the sample means for time steps less than $(t-w+1)$ are weighted by zero. Along similar lines, a chart can be constructed based on varying weights for the prior observations. More weight can be assigned to the most recent observation, with the weights decreasing for less recent observations. A geometric moving-average control chart, also known as an exponentially weighted moving-average (EWMA) chart, is based on this premise. One of the advantages of a geometric moving-average chart over a moving-average chart is that the former is more effective in detecting small changes in process parameters. The geometric moving average at time step t is given by

$$G_t = r\overline{X}_t + (1-r)G_{t-1} \tag{7-63}$$

where r is a weighting constant ($0 < r \le 1$) and G_0 is $\overline{\overline{X}}$. By using eq. (7-63) repeatedly, we get

$$\begin{aligned} G_t &= r\overline{X}_t + r(1-r)\overline{X}_{t-1} + r(1-r)^2 G_{t-2} \\ &= r\overline{X}_t + r(1-r)\overline{X}_{t-1} + r(1-r)^2 \overline{X}_{t-2} + \ldots + (1-r)^t G_0 \end{aligned} \tag{7-64}$$

Equation (7-64) shows that the weight associated with the ith mean from $t(\overline{X}_{t-i})$ is $r(1-r)^i$. The weights decrease geometrically as the sample mean becomes less recent. The sum of all the weights is 1. Consider, for example, the case for which $r = 0.3$. This implies that in calculating G_t, the most recent sample mean $(\overline{X}_t)$ has a weight of 0.3, the next most recent observation $(\overline{X}_{t-1})$ has a weight of $(0.3)(1-0.3) = 0.21$, the next observation $(\overline{X}_{t-2})$ has a weight of $0.3(1-0.3)^2 = 0.147$, and so on. Here, G_0 has a weight of $(1-0.3)^t$. Since these weights appear to decrease exponentially, eq. (7-64) describes what is known as the exponentially weighted moving-average model.

If the sample means $\overline{X}_1, \overline{X}_2, \overline{X}_3, ..., \overline{X}_{t-1}$ are assumed to be independent of each other and if the population standard deviation is σ, the variance of G_t, is given by

$$\text{Var}(G_t) = \left(\frac{\sigma^2}{n}\right)\left(\frac{r}{2-r}\right)\left[1 - (1-r)^{2t}\right] \tag{7-65}$$

For large values of t, the standard deviation of G_t, is

$$\sigma_G = \sqrt{\text{Var}(G_t)} = \sqrt{\frac{\sigma^2}{n}\left(\frac{r}{2-r}\right)}$$

The upper and lower control limits are

$$\text{UCL} = \overline{\overline{X}} + 3\sigma\sqrt{\frac{r}{(2-r)n}}$$
$$\text{LCL} = \overline{\overline{X}} - 3\sigma\sqrt{\frac{r}{(2-r)n}} \tag{7-66}$$

For small values of t, the control limits are found using eq. (7-65) to be

$$\text{UCL} = \overline{\overline{X}} + 3\sigma\sqrt{\frac{r}{n(2-r)}\left[1 - (1-r)^{2t}\right]}$$
$$\text{LCL} = \overline{\overline{X}} - 3\sigma\sqrt{\frac{r}{n(2-r)}\left[1 - (1-r)^{2t}\right]} \tag{7-67}$$

A geometric moving-average control chart is based on a concept similar to that of a moving-average chart. By choosing an adequate set of weights, however, where recent sample means are more heavily weighted, the ability to detect small changes in process parameters is increased. If the **weighting factor** r is selected as

$$r = \frac{2}{w+1} \tag{7-68}$$

where w is the **moving-average span**, the moving-average method and the geometric moving-average method are equivalent. There are guidelines for choosing the value of r. If our goal is to detect small shifts in the process parameters as soon as possible, we use a small value of r (say, 0.1). If we use $r = 1$, the geometric moving-average chart reduces to the standard Shewhart chart for the mean.

Example 7-11 Refer to Example 7-10 regarding the amount of a coloring pigment in polypropylene plastic. Table 7-11 gives the sample averages for 20 samples of size 5. Construct a geometric moving-average control chart using a weighting factor r of 0.2.

TABLE 7-11 Data and Results for a Geometric Moving-Average Control Chart (kg)

Sample	Sample Average, $\bar{X}_t$	Geometric Moving Average, G_t	Control Limits for Geometric Average	
			LCL	UCL
1	25.0	25.128	25.114	25.206
2	25.4	25.182	25.101	25.219
3	25.2	25.186	25.094	25.226
4	25.0	25.149	25.090	25.230
5	25.2	25.159	25.087	25.233
6	24.9	25.107	25.086	25.234
7	25.0	25.086	25.085	25.235
8	25.4	25.149	25.084	25.236
9	24.9	25.099	25.084	25.236
10	25.2	25.119	25.084	25.236
11	25.0	25.095	25.083	25.237
12	25.7	25.216	25.083	25.237
13	25.0	25.173	25.083	25.237
14	25.1	25.158	25.083	25.237
15	25.0	25.127	25.083	25.237
16	24.9	25.081	25.083	25.237
17	25.0	25.065	25.083	25.237
18	25.1	25.072	25.083	25.237
19	25.4	25.138	25.083	25.237
20	25.8	25.270	25.083	25.237

Solution For the data in Table 7-11 the mean of the sample averages is

$$\bar{\bar{X}} = \frac{503.2}{20} = 25.160$$

Since $\bar{R}$ is given as 0.40 in Example 7-10, the estimated process standard deviation is

$$\hat{\sigma} = \frac{\bar{R}}{d_2} = \frac{0.40}{2.326} = 0.172$$

The geometric moving average for sample 1 using eq. (7-63) is (for $G_0 = \bar{\bar{X}}$)

$$G_1 = r\bar{X}_1 + (1-r)G_0$$
$$(0.2)(25.0) + (1-0.2)(25.16) = 25.128$$

The remaining geometric moving averages are calculated similarly. These values are shown in Table 7-11. The control limits for sample 1 are calculated using eq. (7-67):

$$\text{UCL} = 25.160 + (3)(0.172)\sqrt{\frac{0.2}{(5)(2-0.2)}\left[1-(1-0.2)^2\right]}$$

$$= 25.206$$

$$\text{LCL} = 25.160 + (3)(0.172)\sqrt{\frac{0.2}{(5)(2-0.2)}\left[1-(1-0.2)^2\right]}$$

$$= 25.114$$

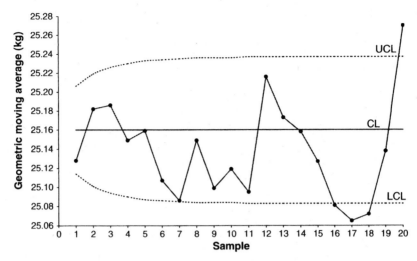

FIGURE 7-26   Geometric moving-average control chart for amount of coloring pigment.

Similar computations are performed for the remaining samples. For large values of t (say, $t = 15$), the control limits are found by using eq. (7-66):

$$UCL = 25.160 + (3)(0.172)\sqrt{\frac{0.2}{(2-0.2)(5)}}$$
$$= 25.237$$
$$LCL = 25.160 - (3)(0.172)\sqrt{\frac{0.2}{(2-0.2)(5)}}$$
$$= 25.083$$

Figure 7-26 shows a plot of this geometric moving-average control chart. Notice that samples 16,17, and 18 are below the lower control limit and that sample 20 plots above the upper control limit. The special causes for these points should be investigated in order to take remedial action. Note that in the moving-average chart in Figure 7-27, sample 18 plotted below the lower control limit, but samples 16, 17, and 20 were within the control limits. The geometric moving-average chart (Figure 7-26), which is a little more sensitive to small shifts in the process parameters than the moving-average chart, identifies these additional points as being out of control.

Trend Chart (Regression Control Chart)

In certain circumstances, such as those involving tool wear or die wear, the characteristic of interest (say, the average outside diameter of a part) is expected to increase or decrease gradually. Hence, rather than being horizontal, the centerline will slope upward or downward. The control limits for the sample average will be parallel to the centerline. Under such conditions, the allowable initial and final values of the process mean will be determined by the specification limits. Furthermore, it is assumed that specification-limit range is wider than the range of the inherent process variability. This implies that the process can be allowed, for a certain period of time, to drift upward or downward as long as the output remains within the specifications.

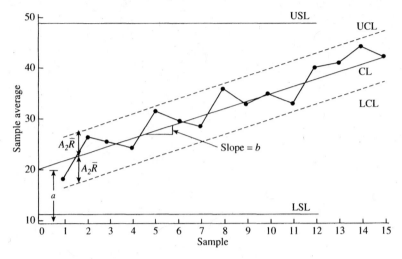

FIGURE 7-27   Trend chart.

The ordinate-intercept and slope of the centerline are influenced by the plot pattern. If the **principle of least squares** (which minimizes the sum of the squared deviations of the observed values from the fitted values) is used to find the best-fitting straight line through the plotted points, the equation of this fitted centerline is of the form

$$C = a + bi \qquad (7\text{-}69)$$

where C = fitted value of the sample average for sample i

$\quad a$ = intersection point of the fitted centerline with the vertical axis

$\quad b$ = slope of the fitted centerline

$\quad i$ = sample number

Estimates for the parameters of the fitted centerline are given by

$$a = \frac{(\Sigma \bar{X})(\Sigma i^2) - (\Sigma \bar{X} i)(\Sigma i)}{g\Sigma i^2 - (\Sigma i)^2}$$

$$b = \frac{g(\Sigma \bar{X} i) - (\Sigma \bar{X})(\Sigma i)}{g\Sigma i^2 - (\Sigma i)^2} \qquad (7\text{-}70)$$

where $\bar{X}$ is the sample average for sample i and g is the number of samples.

The upper and the lower control limits are drawn parallel to the fitted centerline and have slope b. Using the same principle discussed in Section 7-4 for $\bar{X}$- and R-charts, the control limits are placed at $3\sigma_{\bar{X}}$ from the centerline. If the factors from Appendix A-7 are used, the control limits are $A_2\bar{R}$ from the centerline. In other words, the control limits are

$$\text{UCL} = (a + A_2\bar{R}) + bi$$

$$\text{LCL} = (a - A_2\bar{R}) + bi \qquad (7\text{-}71)$$

Careful attention should be given to setting the equipment so that virtually all of the parts produced conform to specifications. This consideration determines when to take remedial action for the drift (e.g., by changing the tool) and when the control chart will have to start a new cycle. Typically, the initial and final positions of the centerline are set 3σ from the appropriate specification limits.

Let's consider a situation where the outside diameter of a part is increasing due to the wear of a cutting tool. When a new tool replaces the old one, the centerline will be set initially at 3σ above the lower specification limit. As the tool wears out, the tool would be replaced when the centerline is 3σ below the upper specification limit. This recommendation is based on the assumption that the distribution of the quality characteristic is normal. If the distribution is indeed normal, only about 0.13% of the output will fall outside the specification limit for this centerline arrangement. Figure 7-27 is an example of a trend chart.

Example 7-12 Over the course of machining the diameter of steel hubs, tool wear is gradual. To determine the optimal point to change tools, samples of size 4 are randomly selected, and the mean and range of the hub diameters are found. Table 7-12 shows the sample mean $\overline{X}$ and range R for 25 such samples. Find the centerline and control limits of the trend chart for the sample average. If the specification limits state that the hub diameter must be between 34 and 78 mm, when should the tool be changed?

Solution The calculations for determining the vertical axis intercept and the slope of the centerline are found using eq. (7-70):

$$a = \frac{(1386.4)(5525) - (19,471.6)(325)}{(25)(5525) - (325)^2} = 40.972$$

$$b = \frac{(25)(19,471.6) - (1386.4)(325)}{(25)(5525) - (325)^2} = 1.114$$

The centerline is therefore given by $C = 40.972 + 1.114i$.
From the data, $\overline{R} = 11.84$. From Appendix A-7, $A_2 = 0.729$. The control limits are

$$\begin{aligned}
\text{UCL} &= (a + A_2\overline{R}) + bi \\
&= [40.972 + (0.729)(11.84)] + 1.114i \\
&= 49.6034 + 1.114i
\end{aligned}$$

$$\begin{aligned}
\text{LCL} &= (a - A_2\overline{R}) + bi \\
&= [40.972 - (0.729)(11.84)] + 1.114i \\
&= 32.341 + 1.114i
\end{aligned}$$

The tool should be changed when the centerline reaches a value 3σ below the upper specification limit. An estimate of the process standard deviation σ is given by

TABLE 7-12 Sample Mean and Range for Tool Wear Data (mm)

Sample	Average, $\overline{X}$	Range, R	Sample	Average, $\overline{X}$	Range, R	Sample	Average, $\overline{X}$	Range, R
1	36.2	8.0	10	52.6	7.8	19	64.2	13.5
2	42.4	11.8	11	50.4	11.3	20	61.4	9.4
3	38.6	6.2	12	59.5	15.1	21	66.7	16.6
4	45.5	14.3	13	60.5	11.7	22	63.2	12.2
5	53.1	16.2	14	53.8	8.8	23	62.1	10.5
6	46.7	9.5	15	54.5	12.8	24	64.5	12.6
7	55.4	10.2	16	61.2	14.5	25	69.6	14.7
8	42.8	12.0	17	60.4	12.0			
9	57.3	13.9	18	63.8	10.4			

$\overline{R}/d_2 = 11.84/2.059 = 5.750$ mm. Thus, the maximum value that the centerline should be allowed to reach is $78.0 - (3)(5.750) = 60.75$ mm. Once the centerline reaches this limit, the tool should be changed. Alternatively, if the rate of tool wear per unit of production were given, we could estimate the number of units that would be produced between tool changes.

Modified Control Chart

In our discussions of all the control charts except the trend chart, we have assumed that the process spread (6σ) is close to and hopefully less than the difference between the specification limits. That is, we hope that $6\sigma < (USL - LSL)$. Now we assume that the natural process spread of 6σ is *significantly* less than the difference between the specification limits: that is, the process capability ratio $= (USL - LSL)/6\sigma \gg 1$. Figure 7-28 depicts this situation.

So far, the specification limits have not been placed on $\overline{X}$-charts. One reason for this is that the specification limits correspond to the conformance of individual items. If the distribution of individual items is plotted, it makes sense to show the specification limits on the X-chart, but a control chart for the mean $\overline{X}$ deals with averages, not individual values. Therefore, plotting the specification limits on an $\overline{X}$-chart is not appropriate. For a modified control chart, however, the specification limits are shown.

Our objective here is to determine bounds on the process mean such that the proportion of nonconforming items does not exceed a desirable value δ. The focus is not on detecting the statistical state of control, because a process can drift out of control and still produce parts that conform to specifications. In fact, we assume that the process variability is in a state of **statistical control**. An estimate of the process standard deviation σ is obtained from either the mean ($\overline{R}$) of the R-chart or from the mean ($\overline{s}$) of the s-chart. Furthermore, we assume that the distribution of the individual values is normal and that a change in the process mean can be accomplished without much difficulty. Our aim in constructing a modified control chart is to determine whether the process mean μ is within certain acceptable limits such that the proportion of nonconforming items does not exceed the chosen value δ.

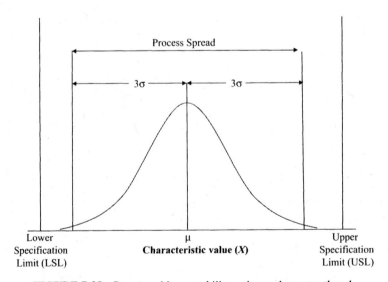

FIGURE 7-28   Process with a capability ratio much greater than 1.

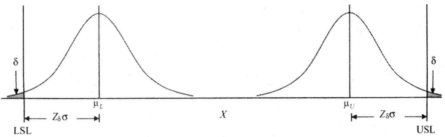

(a) Distribution of individual values at two means where standard deviation of X is σ.

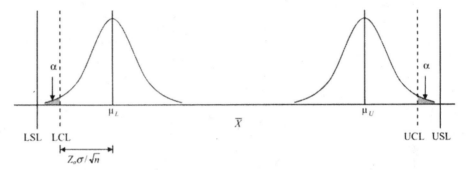

(b) Distribution of sample means where standard deviation of $\overline{X}$ is $\sigma/\sqrt{n}$.

FIGURE 7-29   Determination of modified control chart limits.

Let's consider Figure 7-29a, which shows a distribution of individual values at two different means: one the lowest allowable mean (μ_L) and the other the highest allowable mean (μ_U). Suppose that the process standard deviation is σ. If the distribution of individual values is normal, let z_δ denote the standard normal value corresponding to a tail area of δ. By the definition of a standard normal value, z_δ represents the number of standard deviations that LSL is from μ_L and that USL is from μ_U. So the distance between LSL and μ_L is $z_\delta\sigma$, which is also the same as the distance between USL and μ_U. The bounds within which the process mean should be contained so that the fraction nonconforming does not exceed δ are $\mu_L \leq \mu \leq \mu_U$. From Figure 7-29a,

$$\mu_L = \text{LSL} + z_\delta\sigma$$
$$\mu_U = \text{USL} - z_\delta\sigma \tag{7-72}$$

Suppose that a type I error probability of α is chosen. The control limits are placed such that the probability of a type I error is α, as shown in Figure 7-29b. The control limits are placed at each end to show that the sampling distribution of the sample mean can vary over the entire range. Figure 7-29b shows the distribution of the sample means. Given the sampling distribution of $\overline{X}$, with standard deviation $\sigma_{\overline{X}} = \sigma/\sqrt{n}$, the upper and lower control limits as shown in Figure 7-29b are

$$\text{UCL} = \mu_U + \frac{z_\alpha\sigma}{\sqrt{n}}$$

$$\text{LCL} = \mu_L - \frac{z_\alpha\sigma}{\sqrt{n}} \tag{7-73}$$

By substituting for μ_L and μ_U, following equations are obtained:

$$UCL = USL - \left(z_\delta - \frac{z_\alpha}{\sqrt{n}} \right) \sigma$$

$$LCL = LSL + \left(z_\delta - \frac{z_\alpha}{\sqrt{n}} \right) \sigma$$

(7-74)

If the process standard deviation σ is to be estimated from an R-chart, then $\bar{R}/d_2$ is substituted for σ in eq. (7-74). Alternatively, if σ is to be estimated from an s-chart, $\bar{s}/c_4$ is used in place of σ in eq. (7-74).

Example 7-13   The nitrogen content of a certain fertilizer mix is a characteristic of interest. Random samples of size 5 are selected, and the percentage of nitrogen content is found for each. The sample mean $\bar{X}$ and range R are shown for 20 samples in Table 7-13. The specification limits allow values from 12 to 33%. A process nonconformance rate of more than 1% is unacceptable. We want to construct 3σ modified control limits: that is, limits with a type I error rate α of 0.0026.

Solution   First, we determine whether the variability of the process is in control. Using the data in Table 7-13,

$$\bar{R} = \frac{39.8}{20} = 1.99$$

The centerline of the R-chart is 1.99. The control limits for the R-chart are

$$UCL_R = D_4 \bar{R} = (2.114)(1.99) = 4.207$$
$$LCL_R = D_3 \bar{R} = (0)(1.99) = 0$$

By checking the range values, we find that the variability is in control. Next, we calculate the modified control limits. For $\alpha = 0.0026$, the standard normal tables show that z_α is 2.79. For $\delta = 0.01$, we find that $z_\delta = 2.33$. An estimate of the process standard deviation σ is

$$\hat{\sigma} = \frac{\bar{R}}{d_2} = \frac{1.99}{2.326} = 0.856$$

TABLE 7-13   Sample Average and Range Values for the Percentage of Nitrogen Content

Sample	Sample Average, $\bar{X}$	Range, R	Sample	Sample Average, $\bar{X}$	Range, R
1	14.8	2.2	11	25.0	2.1
2	15.2	1.6	12	16.4	1.8
3	16.7	1.8	13	18.6	1.5
4	15.5	2.0	14	23.9	2.3
5	18.4	1.8	15	17.2	2.1
6	17.6	1.9	16	16.8	1.6
7	21.4	2.2	17	21.1	2.0
8	20.5	2.3	18	19.5	2.2
9	22.8	2.5	19	18.3	1.8
10	16.9	1.8	20	20.2	2.3

The modified control limits are

$$UCL = USL - \left(z_\delta - \frac{z_\alpha}{\sqrt{n}}\right)\sigma$$

$$= 33 - \left(2.33 - \frac{2.79}{\sqrt{5}}\right)(0.856) = 32.074$$

$$LCL = LSL + \left(z_\delta - \frac{z_\alpha}{\sqrt{n}}\right)\sigma$$

$$= 12 + \left(2.33 - \frac{2.79}{\sqrt{5}}\right)(0.856) = 12.926$$

Thus, the sample average values shown in Table 7-13 are within the control limits. All values are well below the upper control limit and are closer to the lower control limit. The closest value to the lower control limit is 14.8%. Therefore, the process is currently able to meet the desired standards.

Acceptance Control Chart

In the preceding discussion we outlined a procedure for obtaining the modified control limits given the sample size n, the proportion nonconforming δ, and the acceptable level of probability of type I error, α. In this section we discuss a procedure to calculate the control chart limits when the sample size is known and when we have a specified level of proportion nonconforming (γ) that we desire to detect with a probability of $(1 - \beta)$. Such a control chart is known as an *acceptance control chart*. The same assumptions are made here as for modified control charts. That is, we assume that the inherent process spread (6σ) is much less than the difference between the specification limits, the process variability is in control, and the distribution of the individual values is normal.

Figure 7-30a shows the distribution of the individual values and the borderline locations of the process mean so that the proportion nonconforming does not exceed the desirable level of γ. From Figure 7-30a we have

$$\begin{aligned}
\mu_U &= USL - z_\gamma\sigma \\
\mu_L &= LSL + z_\gamma\sigma
\end{aligned} \tag{7-75}$$

Figure 7-30b shows the distribution of the sample mean and the bounds within which the process mean must lie for the probability of detecting a nonconformance proportion of γ to be $(1 - \beta)$. From Figure 7-30b we have

$$\begin{aligned}
UCL &= \mu_U - z_\beta\frac{\sigma}{\sqrt{n}} \\
LCL &= \mu_L + z_\beta\frac{\sigma}{\sqrt{n}}
\end{aligned} \tag{7-76}$$

Substituting from eq. (7-75), we get

$$\begin{aligned}
UCL &= USL - \left(z_\gamma + \frac{z_\beta}{\sqrt{n}}\right)\sigma \\
LCL &= LSL + \left(z_\gamma + \frac{z_\beta}{\sqrt{n}}\right)\sigma
\end{aligned} \tag{7-77}$$

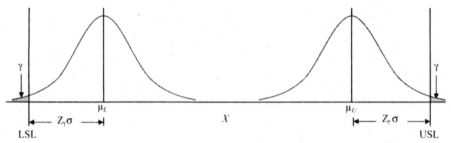

(a) Distribution of X where standard deviation of X is σ.

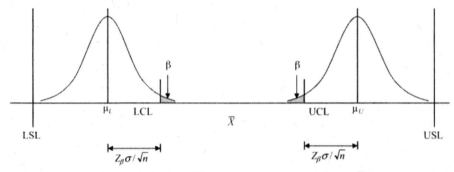

(b) Distribution of $\bar{X}$ where standard deviation of $\bar{X}$ is $\sigma/\sqrt{n}$.

FIGURE 7-30 Determination of acceptance control chart limits.

If an R-chart is used to control the process variability, σ is estimated by $\bar{R}/d_2$. If an s-chart is used, σ is estimated by $\bar{s}/c_4$. These estimates are then used in eq. (7-77).

Example 7-14 Refer to Example 7-13 and the nitrogen content data. Suppose that the proportion of nonconforming product is 3% and we want to detect an out-of-control condition with a probability of 0.95. Find the acceptance control chart limits.

Solution For this problem, $\gamma = 0.03$, $1 - \beta = 0.95$, LSL = 12%, USL = 33%. From the standard normal tables, $z_{0.03} \simeq 1.88$ and $z_{0.05} = 1.645$. From Example 7-13, $\bar{R} = 1.99$ and $n = 5$. We have

$$\hat{\sigma} = \frac{\bar{R}}{d_2} = \frac{1.99}{2.326} = 0.856$$

The acceptance control chart limits are

$$\text{UCL} = \text{USL} - \left(z_\gamma + \frac{z_\beta}{\sqrt{n}} \right)\sigma$$

$$= 33 - \left(1.88 + \frac{1.645}{\sqrt{5}} \right)(0.856) = 30.761$$

$$\text{LCL} = \text{LSL} + \left(z_\gamma + \frac{z_\beta}{\sqrt{n}} \right)\sigma$$

$$= 12 + \left(1.88 + \frac{1.645}{\sqrt{5}} \right)(0.856) = 14.239$$

None of the sample averages are outside these control limits.

TABLE 7-14   Applications of Control Charts for Variables in the Service Sector

Quality Characteristic	Control Chart
Response time for correspondence in a consulting firm or financial institution	$\overline{X}$ and $R(n < 10)$; $\overline{X}$ and $s(n \geq 10)$
Waiting time in a restaurant	$\overline{X}$ and $R(n < 10)$; $\overline{X}$ and $s(n \geq 10)$
Processing time of claims in an insurance company	$\overline{X}$ and $R(n < 10)$; $\overline{X}$ and $s(n \geq 10)$
Waiting time to have cable installed	$\overline{X}$ and $R(n < 10)$; I and MR $(n = 1)$
Turnaround time in a laboratory test in a hospital	I and MR $(n = 1)$
Use of electrical power or gas on a monthly basis	Moving average; EWMA; Trend Chart
Admission time into an intensive care unit	I and MR $(n = 1)$
Blood pressure or cholesterol ratings of a patient over time	I and MR $(n = 1)$

The principles of modified control charts and acceptance control charts can be combined to determine an acceptable sample size n for chosen levels of δ, α, γ, and β. By equating the expressions for the UCL in eqs. (7-74) and (7-77), we have

$$\text{USL} - \left(z_\delta - \frac{z_\alpha}{\sqrt{n}}\right)\sigma = \text{USL} - \left(z_\gamma + \frac{z_\beta}{\sqrt{n}}\right)\sigma$$

which yields

$$n = \left(\frac{z_\alpha + z_\beta}{z_\delta - z_\gamma}\right)^2 \qquad (7\text{-}78)$$

Some examples of variables control chart applications in the service sector are shown in Table 7-14. Note that the size of the subgroup in the data collected will determine, in many instances, the use of the X-MR chart (for individuals' data), $\overline{X}$ and R (when the subgroup size is small, usually less than 10), and $\overline{X}$ and s (when the subgroup size is large, usually equal to or greater than 10).

7-9   MULTIVARIATE CONTROL CHARTS

The control charts mentioned thus far have dealt with controlling one characteristic. However, in real-world situations, we often deal with two or more variables simultaneously. For instance, we may want to simultaneously control both the length and the inside diameter of a pipe. In other words, both the length and the inside diameter must be acceptable for the pipe to be usable. Controlling both characteristics separately may not yield a product in which both variables are acceptable.

Controlling Several Related Quality Characteristics

Suppose we have two quality characteristics that must both be in control for the process to be in control. If control charts for the averages of these two characteristics are kept independently, the result is a rectangular **control region** on a two-dimensional plot. The boundaries of this region are basically the upper and lower control limits of the two quality

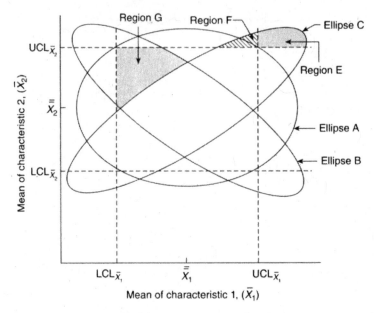

FIGURE 7-31 Elliptical control region.

characteristics and are calculated using eq. (7-5). If the bivariate observation of sample means $(\overline{X}_1, \overline{X}_2)$ plots within the control limits, the process would seem to be in control.

Such rectangular boundaries, however, can often be incorrect. An actual control region for two characteristics is elliptic in nature (see Figure 7-31). The equation of a statistic that incorporates two characteristics is an ellipse, as we will see in eq. (7-80). If the two characteristics are independent of each other, the major and minor axes of the ellipse are parallel to the respective plot axes (see ellipse A in Figure 7-31). If the pair of sample means $(\overline{X}_1, \overline{X}_2)$ falls within the boundary of the ellipse, the process is said to be in control. If two characteristics are negatively correlated, the shape of the control ellipse will be similar to that of ellipse B. If the two variables are positively correlated, the control ellipse will be similar to that of ellipse C.

Figure 7-31 shows that if the variables are positively correlated and we use the rectangular region erroneously as the control region, we draw various incorrect conclusions. For instance, if $(\overline{X}_1, \overline{X}_2)$ falls in region E or region F, the process is in control even though the point falls outside the rectangular region. A point in region G, on the other hand, is within the rectangular region, but the process is nonetheless out of control.

The degree of correlation between the variables influences the magnitude of the errors encountered in making inferences. If a separate $\overline{X}$-chart is constructed for each characteristic based on a type I error probability of α and a rectangular control region is used, then for independent variables the probability of a type I error for the joint control procedure is

$$\alpha' = 1 - (1 - \alpha)^p \tag{7-79}$$

where p represents the number of jointly controlled variables. The probability of all p sample means plotting within the rectangular region is $(1 - \alpha)^p$.

Moderate or large values of p have a major impact on the errors associated with inference making. Suppose that individual control chart limits are constructed using a type I error probability of 0.0026. If we have four independent characteristics (i.e., $p = 4$), the **overall type I error** probability (α') for the joint control procedure is

$$\alpha' = 1 - (0.9974)^4 = 0.0104$$

If the variables are not independent, the magnitude of the type I error will be difficult to obtain. In practice, a control ellipse should be chosen so that the probability of the sample means being plotted within the elliptical region when the process is in control is $(1 - \alpha)$, where α is the desired overall probability of a type I error.

Hotelling's T^2 Control Chart and Its Variations

Suppose that we have two quality characteristics, X_1 and X_2, distributed jointly according to a bivariate normal distribution. Assume that the target mean values of the characteristics are represented by $\overline{\overline{X}}_1$ and $\overline{\overline{X}}_2$, respectively. Let the sample means be $\overline{X}_1$ and $\overline{X}_2$, with sample variances s_1^2 and s_2^2, and the covariance between the two variables be represented by s_{12} for a sample of size n. Under these conditions, the statistic

$$T^2 = \frac{n}{s_1^2 s_2^2 - s_{12}^2} \left[s_2^2 \left(\overline{X}_1 - \overline{\overline{X}}_1 \right)^2 + s_1^2 \left(\overline{X}_2 - \overline{\overline{X}}_2 \right)^2 - 2 s_{12} \left(\overline{X}_1 - \overline{\overline{X}}_1 \right) \left(\overline{X}_2 - \overline{\overline{X}}_2 \right) \right] \qquad (7\text{-}80)$$

is distributed according to Hotelling's T^2-distribution with 2 and $(n - 1)$ degrees of freedom (Hotelling 1947). The 2 in this case comes from the two characteristics being considered, and the $(n - 1)$ is the degrees of freedom associated with the sample variance. If the calculated value of T^2 given by eq. (7-80) exceeds $T_{\alpha,2,n-1}^2$ the point on the T^2- distribution such that the proportion to the right is α, then at least one of the characteristics is out of control.

This procedure can be shown graphically. Equation (7-80) represents the control ellipses shown in Figure 7-31. If the variables are independent, the covariance between them is zero (i.e., $s_{12} = 0$), the control ellipse is A, and the joint control region is represented by the area within the control ellipse A. If a plot of the bivariate means $(\overline{X}_1, \overline{X}_2)$ falls within this control region, we can assume a state of statistical control. If the two variables are positively correlated, then $s_{12} > 0$, and the control ellipse is similar to ellipse C. If the variables are negatively correlated, then $s_{12} < 0$, and the control ellipse will be similar to ellipse B.

Hotelling's control ellipse procedure has several disadvantages. First, the time sequence of the plotted points $(\overline{X}_1, \overline{X}_2)$ is lost. This implies that we cannot check for runs in the plotted pattern as with control charts. Second, the construction of the control ellipse becomes quite difficult for more than two characteristics. To overcome these disadvantages, the values of T^2 given by eq. (7-80) are plotted on a control chart on a sample-by-sample basis to preserve the time order in which the data values are obtained. Such a control chart has an upper control limit of $T_{\alpha,p,n-1}^2$, where p represents the number of characteristics. Patterns of nonrandom runs can be investigated in such plots.

Values of Hotelling's T^2 percentile points can be obtained from the percentile points of the **F-distribution** given in Appendix A-6 by using the relation

$$T_{\alpha,p,n-1}^2 = p \left(\frac{n-1}{n-p} \right) F_{\alpha,p,n-p} \qquad (7\text{-}81)$$

where $F_{\alpha,p,n-p}$ represents the point on the F-distribution such that the proportion to the right is α, with p degrees of freedom in the numerator and $(n - p)$ degrees of freedom in the denominator.

If more than two characteristics are being considered, the value of T^2 given by eq. (7-80) for a sample can be generalized as

$$T^2 = n\left(\overline{X} - \overline{\overline{X}}\right)' S^{-1}\left(\overline{X} - \overline{\overline{X}}\right) \tag{7-82}$$

where $\overline{X}$ represents the vector of sample means of p characteristics for a sample of size n, $\overline{\overline{X}}$ represents the vector of target values for each characteristic, and S denotes the variance–covariance matrix of the p quality characteristics.

In practice, $\overline{\overline{X}}$ and S are usually estimated from sample information when the process is in control. Under such conditions, the upper control limit of the T^2-chart given by eq. (7-81) can be modified to take the following form (Alt 1982):

$$\text{UCL} = \left(\frac{mnp - mp - np + p}{mn - m - p + 1}\right) F_{\alpha,p,mn-m-p+1} \tag{7-83}$$

where m represents the number of samples, each of size n, used to estimate $\overline{\overline{X}}$ and S. The value of T^2 for each of the m samples is calculated using eq. (7-82) and is then compared to the UCL given by eq. (7-83). If the value of T^2 for the jth sample (i.e., T_j^2) is above the UCL, it is treated as an out-of-control point, and investigative action is begun.

To calculate T_j^2, the vector of the sample means is given by

$$\overline{X}_j = \begin{bmatrix} \overline{X}_{1j} \\ \overline{X}_{2j} \\ \vdots \\ \overline{X}_{pj} \end{bmatrix}, \qquad j = 1, 2, \ldots, m$$

where $\overline{X}_{ij}$ represents the sample mean of the ith characteristic for the jth sample and is found from

$$\overline{X}_{ij} = \frac{\sum\limits_{k=1}^{n} X_{ijk}}{n}, \qquad \begin{cases} i = 1, 2, \ldots, p \\ j = 1, 2, \ldots, m \end{cases} \tag{7-84}$$

where X_{ijk} represents the value of the kth observation of the ith characteristic in the jth sample.

The sample variances for the ith characteristic in the jth sample are given by

$$s_{ij}^2 = \frac{1}{n-1} \sum\limits_{k=1}^{n} \left(X_{ijk} - \overline{X}_{ij}\right)^2, \qquad \begin{cases} i = 1, 2, \ldots, p \\ j = 1, 2, \ldots, m \end{cases} \tag{7-85}$$

The covariance between characteristics i and h in the jth sample is calculated from

$$s_{ihj} = \frac{1}{n-1} \sum\limits_{k=1}^{n} \left(X_{ijk} - \overline{X}_{ij}\right)\left(X_{hjk} - \overline{X}_{hj}\right), \qquad \begin{cases} j = 1, 2, \ldots, m \\ i \neq h \end{cases} \tag{7-86}$$

The vector $\overline{\overline{X}}$ of target means of each characteristic for m samples is estimated as

$$\overline{\overline{X}}_i = \frac{\sum\limits_{j=1}^{m} \overline{X}_{ij}}{m}, \qquad i = 1, 2, \ldots, p \tag{7-87}$$

The elements of the variance–covariance matrix S in eq. (7-82) are estimated from the following average for m samples:

$$s_i^2 = \frac{\sum\limits_{j=1}^{m} s_{ij}^2}{m}, \qquad i = 1, 2, \ldots, p \tag{7-88}$$

and

$$s_{ih} = \frac{\sum_{j=1}^{m} s_{ihj}}{m}, \qquad i \neq h \tag{7-89}$$

Finally, the vector $\overline{\overline{X}}$ is estimated using the elements $\{\overline{\overline{X}}_i\}$, and the matrix S is estimated as follows (only the upper diagonal part is shown because the matrix is symmetric):

$$S = \begin{bmatrix} s_1^2 & s_{12} & s_{13} & \cdots & s_{1p} \\ & s_2^2 & s_{23} & \cdots & s_{2p} \\ & & & & \vdots \\ & & & & s_p^2 \end{bmatrix} \tag{7-90}$$

The use of eq. (7-82) requires inverting this matrix.

A Hotelling's control chart is constructed using the upper control limit given by eq. (7-83) and the plotted values of T^2 for each sample given by eq. (7-82), where the vector $\overline{\overline{X}}$ and the matrix S are found using the preceding procedure. A sample value of T^2 above the upper control limit indicates an out-of-control situation. How do we determine which quality characteristic caused the out-of-control state?

Even with only two characteristics ($p = 2$), the situation can be complex. If the two quality characteristics are highly positively correlated, we expect the averages for each characteristic in the sample to maintain the same relationship relative to the process average $\overline{\overline{X}}$. For example, in the jth sample, if $\overline{X}_{1j} > \overline{\overline{X}}_1$, we could expect $\overline{X}_{2j} > \overline{\overline{X}}_2$. Similarly, if $\overline{X}_{1j} < \overline{\overline{X}}_1$, we would expect $\overline{X}_{2j} < \overline{\overline{X}}_2$, which would confirm that the sample averages for each characteristic move in the same direction relative to their means.

If the two characteristics are highly positively correlated and $\overline{X}_{1j} > \overline{\overline{X}}_1$, we would not expect that $\overline{X}_{2j} < \overline{\overline{X}}_2$. However, should this occur, this sample may show up as an out-of-control point in Hotelling's T^2 procedure, thereby indicating that the bivariate process is out of control. This same inference can be made using individual 3σ control limit charts constructed for each characteristic if $\overline{X}_{1j}$ exceeds $\overline{\overline{X}}_1 + 3\sigma_{\overline{X}_1}$ or $\overline{X}_{2j}$ exceeds $\overline{\overline{X}}_2 + 3\sigma_{\overline{X}_2}$. However, individual quality characteristic means can plot within the control limits on separate control charts even though the T^2 plots above the UCL on the joint control chart. Using joint control charts for characteristics that need to be considered simultaneously is thus advantageous. However, note that an individual chart for a quality characteristic can sometimes indicate an out-of-control condition when the joint control chart does not.

In general, larger sample sizes are needed to detect process changes with positively correlated characteristics than with negatively correlated characteristics. Furthermore, for highly positively correlated characteristics, larger sample sizes are needed to detect large positive shifts in the process means than to detect small positive shifts.

Generally speaking, if an out-of-control condition is detected by a Hotelling's control chart, individual control intervals are calculated for each characteristic for that sample. If the probability of a type I error for a joint control procedure is α, then for sample j, the individual control interval for the ith quality characteristic is

$$\overline{\overline{X}}_i \pm t_{\alpha/2p, m(n-1)} s_i \sqrt{\frac{m-1}{mn}}, \qquad i = 1, 2, \ldots, p \tag{7-91}$$

where $\overline{\overline{X}}_i$ and s_i^2 are given by eqs. (7-87) and (7-88), respectively. If $\overline{X}_{ij}$ falls outside this interval, the corresponding characteristic should be investigated for a lack of control. If special causes are detected, the sample that contains information relating to all the characteristics should be deleted when the upper control limit is recomputed.

Usage and Interpretations

The multivariate T^2 control chart is normally used in two phases. In phase 1 the charts are used to establish control. Here, identified out-of-control points are deleted, assuming that adequate remedial actions will be taken, and the procedure is repeated until all retained observations are in control. The formula given in eq. (7-83) for the UCL is used in phase 1.

Phase 2 is used for monitoring future observations from the process. The observations retained at the end of phase 1 (denoting the number by m) are used to calculate the control limits for phase 2. The upper control limit for the T^2 control chart in phase 2 is given by

$$\text{UCL} = \frac{p(m+1)(n-1)}{mn-m-p+1} F_{\alpha,p,mn-m-p+1} \tag{7-92}$$

As described previously, even though the T^2 control chart is useful in detecting shifts in the process mean vector, it does not identify which specific variables(s) are responsible. One approach, in this context, is the T^2 decomposition method. The concept is to determine the individual contributions of each of the p variables, or combinations thereof, to the overall T^2 statistic. These individual contributions or partial T^2 statistics are found as follows:

$$D_i = T^2 - T^2_{1,2,\ldots,i-1,i+1,\ldots p}, \qquad i = 1,2,\ldots p \tag{7-93}$$

where $T^2_{1,2,\ldots,i-1,i+1,\ldots p}$ denotes the T^2 statistic when the ith variable is left out. Large values of D_i will indicate a significant impact of variable i for the particular observation under investigation.

Individual Observations with Unknown Process Parameters

The situation considered previously dealt with subgroups of data, where the sample size (n) for each subgroup exceeds 1. In this section we consider individual observations and assume that the process parameters, mean vector or the elements of the variance–covariance matrix, are unknown. As before, we use the two-phase approach, where in phase 1 we use the preliminary data to retain observations in control.

The value of T^2, when individual observations are obtained, is given by

$$T^2 = (X - \overline{X})' S^{-1} (X - \overline{X}) \tag{7-94}$$

In eq. (7-94), the process mean vector is estimated from the observations by $\overline{X}$, while the process variance–covariance matrix is estimated, using the data, by S. The upper control limit in this situation is given by

$$\text{UCL} = \frac{(m-1)^2}{m} B\left(\alpha, \frac{p}{2}, \frac{m-p-1}{2}\right) \tag{7-95}$$

where $B(\alpha, p/2, (m-p-1)/2)$ denotes the upper αth quantile of the beta distribution with parameters $p/2$ and $(m-p-1)/2$.

If an observation vector has a value of T^2, given by eq. (7-94), that exceeds the value of UCL, given by eq. (7-95), it is deleted from the preliminary data set. Revised estimates of the process mean vector and variance–covariance matrix elements are found using the remaining observations and the process is repeated until no further observations are deleted. We now proceed to phase 2 to monitor future observations. The estimates $\overline{X}$ and S obtained at the end of phase 1 are used to calculate T^2, using eq. (7-94), for new observations. Assuming that the number of observations retained at the end of phase 1 is given by m, the upper control limit for phase 2 is obtained as

$$\text{UCL} = \frac{p(m+1)(m-1)}{m(m-p)} F_{\alpha,p,m-p} \tag{7-96}$$

Hence, values of T^2 for new observations will be compared with the UCL given by eq. (7-96) to determine out-of-control conditions.

Generalized Variance Chart

The multivariate control charts discussed previously dealt with monitoring the process mean vector. Here, we introduce a procedure to develop a multivariate dispersion chart to monitor process variability based on the sample variance–covariance matrix S. A measure of the sample generalized variance is given by $|S|$, the determinant of the sample variance–covariance matrix.

Denoting the mean and variance of $|S|$ by $E(|S|)$ and $V(|S|)$, respectively, and using the property that most of the probability distribution of $|S|$ is contained in the interval $E|S| \pm 3\sqrt{V(|S|)}$, expressions for the parameters of the control chart for $|S|$ may be obtained. It is known that

$$E(|S|) = b_1|\Sigma| \tag{7-97}$$

$$V(|S|) = b_2|\Sigma|^2 \tag{7-98}$$

where Σ represents the process variance–covariance matrix, and

$$b_1 = \frac{\prod_{i=1}^{p}(n-i)}{(n-1)^p} \tag{7-99}$$

$$b_2 = \frac{\prod_{i=1}^{p}(n-i)\left[\prod_{j=1}^{p}(n-j+2) - \prod_{j=1}^{p}(n-j)\right]}{(n-1)^{2p}} \tag{7-100}$$

Since Σ is usually unknown, it is estimated based on sample information. From eq. (7-97), an unbiased estimator of $|\Sigma|$ is $|S|/b_1$. Using eqs. (7-97) and (7-98), the centerline and control limits for the $|S|$ chart are given by

$$\begin{aligned}
\text{CL} &= b_1|\Sigma| \\
\text{UCL} &= |\Sigma|\left(b_1 + 3b_2^{1/2}\right) \\
\text{LCL} &= |\Sigma|\left(b_1 - 3b_2^{1/2}\right)
\end{aligned} \tag{7-101}$$

When a target value for Σ, say Σ_0, is specified, $|\Sigma|$ is replaced by $|\Sigma_0|$ in eq. (7-101). Alternatively, the sample estimate of $|\Sigma|$ given by $|S|/b_1$ will be used to compute the centerline and control limits in eq. (7-101). In the event that LCL from eq. (7-101) is computed to be less than zero, it is converted to zero.

For a given sample j, $|S_j|$, the determinant of the variance–covariance matrix for sample j, is computed and plotted on the generalized variance chart. If the plotted value of $|S_j|$ is outside the control limits, we flag the process and look for special causes.

Even though the generalized sample variance chart is useful, as it aggregates the variability of several variables into one index, it has to be used with caution. This is because many different S_j matrices may give the same value of $|S_j|$, while the variance structure could be quite different. Hence, a univariate range (R) chart or standard deviation (s) chart may help us understand the variables that contribute to make the combined impact on the generalized variance to be significant.

Example 7-15 The single-strand break factor (a measure of the breaking strength) and weight of textile fibers (hanks per pound) are both of interest in keeping a fabric-production process in control. Table 7-15 shows data for the two characteristics for 20 samples of size 4. For instance, the bivariate observations for sample 1 are (80, 19), (82, 22), (78, 20), and (85, 20).

To construct the Hotelling's T^2-chart, we need to calculate the sample means, sample variances and covariances, and T^2- values; these are shown in Table 7-16. We use $p = 2$ for the number of characteristics, $n = 4$ for the sample size, and $m = 20$ for the number of samples. The calculations proceed as follows.

The sample means of each characteristic are found for each sample by using eq. (7-84). For sample 1, the mean of the break-factor readings is

$$\overline{X}_{11} = \frac{80 + 82 + 78 + 85}{4} = 81.25$$

Similarly, the mean fiber weight for sample 1 is 20.25 hanks per pound. This procedure is repeated for each sample, yielding the results shown in Table 7-16.

TABLE 7-15 Bivariate Data for the Fabric-Production Example

Sample	Single-Strand Break Factor, $i = 1$				Weight of Textile Fibers, $i = 2$			
1	80	82	78	85	19	22	20	20
2	75	78	84	81	24	21	18	21
3	83	86	84	87	19	24	21	22
4	79	84	80	83	18	20	17	16
5	82	81	78	86	23	21	18	22
6	86	84	85	87	21	20	23	21
7	84	88	82	85	19	23	19	22
8	76	84	78	82	22	17	19	18
9	85	88	85	87	18	16	20	16
10	80	78	81	83	18	19	20	18
11	86	84	85	86	23	20	24	22
12	81	81	83	82	22	21	23	21
13	81	86	82	79	16	18	20	19
14	75	78	82	80	22	21	23	22
15	77	84	78	85	22	19	21	18
16	86	82	84	84	19	23	18	22
17	84	85	78	79	17	22	18	19
18	82	86	79	83	20	19	23	21
19	79	88	85	83	21	23	20	18
20	80	84	82	85	18	22	19	20

TABLE 7-16   Fabric-Product Statistics for Construction of Hotelling's T^2-Chart

Sample, j	Sample Means		Sample Variances		Sample Covariance s_{12j}	Hotelling's T_j^2	Generalized Variance, $|S_j|$
	Break Factor, $\overline{X}_{1j}$	Weight, $\overline{X}_{2j}$	s_{1j}^2	s_{2j}^2			
1	81.25	20.25	8.92	1.58	0.92	0.78	13.28
2	79.50	21.00	15.00	6.00	−9.00	5.25	9.00
3	85.00	21.50	3.33	4.33	3.00	5.98	5.44
4	81.50	17.75	5.67	2.92	1.17	7.95	15.17
5	81.75	21.00	10.92	4.67	5.33	1.03	22.50
6	85.50	21.25	1.67	1.58	0.17	6.72	2.61
7	84.75	20.75	6.25	4.25	4.58	3.36	5.55
8	80.00	19.00	13.33	4.67	−7.33	5.27	8.44
9	86.25	17.50	2.25	3.67	−2.50	15.25	2.00
10	80.50	18.75	4.33	0.92	−0.50	4.86	3.72
11	85.25	22.25	0.92	2.92	0.92	10.08	1.83
12	81.75	21.75	0.92	0.92	0.58	3.17	0.50
13	82.00	18.25	8.67	2.92	−0.33	4.74	25.17
14	78.75	22.00	8.92	0.67	1.33	10.66	4.17
15	81.00	20.00	16.67	3.33	−7.33	1.21	1.78
16	84.00	20.50	2.67	5.67	−2.67	1.45	8.00
17	81.50	19.00	12.33	4.67	3.00	2.31	48.55
18	82.50	20.75	8.33	2.92	−4.50	0.41	4.05
19	83.75	20.50	14.25	4.33	3.17	1.06	51.72
20	82.75	19.75	4.92	2.92	2.92	0.25	5.83
Means	$\overline{\overline{X}}_1 = 82.46$	$\overline{\overline{X}}_2 = 20.17$	$s_1^2 = 7.51$	$s_2^2 = 3.29$	$s_{12} = -0.35$		

Next, the sample variance of each characteristic is found for each sample [eq. (7-85)]. For the single-strand break factor ($i = 1$), the sample variance of sample 1 is

$$s_{11}^2 = \frac{1}{3}[(80 - 81.25)^2 + (82 - 81.25)^2 + (78 - 81.25)^2 + (85 - 81.25)^2] = 8.92$$

The covariances between the two characteristics are then calculated for each sample using eq. (7-86). For sample 1, we use

$$s_{121} = \frac{1}{3}[(80 - 81.25)(19 - 20.25) + (82 - 81.25)(22 - 20.25)$$
$$+ (78 - 81.25)(20 - 20.25) + (85 - 81.25)(20 - 20.25)]$$
$$= 0.917$$

Next, estimates of the target mean of each characteristic are found [eq. (7-87)]. Estimates of the elements of the variance–covariance matrix S are found in accordance with eqs. (7-88) and (7-89). The values of T^2 are given by eq. (7-80).

The upper control limit for the T^2-chart for an overall type I error probability of 0.0054 is found using eq. (7-83):

$$\text{UCL} = \left[\frac{(20)(4)(2) - (20)(2) - (4)(2) + 2}{(20)(4) - 20 - 2 + 1} \right] F_{0.0054, 2, (20)(4) - 20 - 2 + 1}$$

$$= 1.932 F_{0.0054, 2, 59} = (1.932)(6.406) = 12.376$$

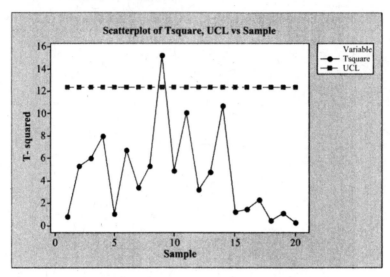

FIGURE 7-32 T^2 control chart for bivariate fabric-production data.

The value of $F_{0.0054,2,59}$ may be approximated from Appendix A-6 using linear interpolation.

The multivariate T^2-chart for these two characteristics in Figure 7-32 shows sample 9 to be out of control. From Table 7-16, $T^2 = 15.25$ for this sample, which exceeds the upper control limit of 12.376. The next step is to investigate *which* characteristic is causing this condition.

The mean of the ranges ($\bar{R}$) of the single-strand break factors for the 20 samples is 5.75, while that for the weight of the fibers is 3.95. When individual 3σ limits are constructed ($\alpha/2p = 0.0054/4 = 0.0013$, which yields a z-value of 3.00), the control limits are found as follows.

For the single-strand break factor, the upper control limit is

$$\begin{aligned} \text{UCL} &= \bar{\bar{X}}_1 + A_2\bar{R} \\ &= 82.46 + (0.729)(5.75) \\ &= 86.652 \end{aligned}$$

For the fiber weight, the upper control limit is

$$\begin{aligned} \text{UCL} &= \bar{\bar{X}}_2 + A_2\bar{R} \\ &= 20.17 + (0.729)(3.95) \\ &= 23.050 \end{aligned}$$

Note that for sample 9, the single-strand break factor does not exceed the UCL of its separate control chart. The fiber weight sample mean for sample 9 is also less than the UCL of its own control chart. Hence, in this case, the joint control chart indicates an out-of-control condition that individual control charts do not detect.

Our next step should be to determine *what* caused the single-strand break factor or fiber weight to go out of control. Once the necessary corrective action has been taken, the Hotelling T^2 control limits should be revised by deleting sample 9 for both characteristics.

If we had decided to use the individual control limits for a joint control procedure as given by eq. (7-91), the limits would be as follows. For the single-strand break factor,

$$\overline{\overline{X}}_1 \pm t_{0.0054/(2)(2),20(3)} \, s_1 \sqrt{\frac{19}{(20)(4)}} = 82.46 \pm t_{0.0013,60}(2.740)\sqrt{0.2375}$$

$$= 82.46 \pm (3.189)(1.3353)$$
$$= (78.202, 86.718)$$

The sample mean of sample 9 for this characteristic does not exceed the upper control limit of 86.718 and so would not be considered out of control. The control limits for the fiber weight are found in a similar manner:

$$\overline{\overline{X}}_2 \pm t_{0.0013,60} \, s_2 \sqrt{\frac{19}{(20)(4)}} = 20.17 \pm (3.189)(\sqrt{3.29})(0.4873)$$

$$= (17.351, 22.989)$$

For sample 9, the mean fiber weight is within these calculated limits and so would also be considered in control.

Thus, based on the joint control procedure, this process has been identified as being out of control. Remedial actions based on the special causes for sample 9 should be identified and implemented. The control limits for Hotelling's T^2-chart should be revised by deleting the observations for this sample.

This example illustrates that a joint control chart can indicate an out-of-control process even though the individual charts do not. It is thus imperative to control the quality characteristics jointly. However, when the T^2-chart shows an out-of-control process *and* an individual chart plots points outside its control limits, we may have a possible cause, one that we can monitor more closely and take remedial action on.

Example 7-16   Consider Example 7-15 and the data on single-strand break factor and weight of textile fibers as given in Table 7-15. Construct a generalized variance chart and comment on the process.

Solution   The calculations in Table 7-16 yield our sample estimate for the variance–covariance matrix as

$$S = \begin{bmatrix} 7.51 & -0.35 \\ -0.35 & 3.29 \end{bmatrix}$$

The determinant of S is obtained as $|S| = 24.585$. Computations for the constants produce

$$b_1 = \frac{(3)(2)}{3^2} = 0.667$$

$$b_2 = \frac{(3)(2)[5(4) - 3(2)]}{3^{2(2)}} = 1.037$$

Hence, the centerline and control limits are

$$CL = 24.585$$

$$UCL = \frac{24.585}{0.667}\left[0.667 + (3)(1.037)^{1/2}\right] = 137.189$$

$$LCL = \frac{24.585}{0.667}\left[0.667 - (3)(1.037)^{1/2}\right] = -88.019 \rightarrow 0$$

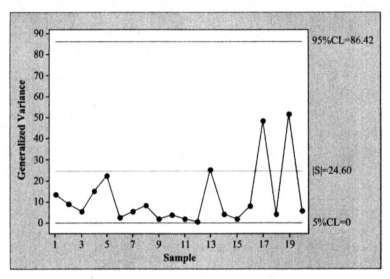

FIGURE 7-33   Generalized variance chart.

The values of $|S_j|$ for each sample are shown in Table 7-16. Note than none of the generalized sample variances fall outside the control limits.

Minitab may be used to construct the generalized sample variance plot. A worksheet is first created using the data in Table 7-15 and putting it in a format compatible to Minitab. Select **Stat > Control Charts > Multivariate Charts.** Choose **Generalized Variance.** Under **Variables,** input the column number or names of the two variables Break Factor and Weight. Under **Subgroup sizes,** input the column number or name of the variable that indicates the subgroup number that each observation belongs to. If desirable, under **Gen Var Options**, may select **Confidence Limits** and identify the desired confidence level (say, 95%). Click **OK.**

Figure 7-33 shows the generalized variance chart. Note that all of the plotted values of $|S_j|$ are well below the 95% confidence limit. As regards the impact on variability of the combined effect of break factor and weight of fibers, we do not discern out-of-control conditions.

SUMMARY

This chapter has introduced different types of control charts that can be used with quality characteristics that are variables (i.e., they are measurable on a numerical scale). Details as to the construction, analysis, and interpretation of each chart have been presented. Guidelines were provided for the appropriate settings in which each control chart may be used. The rationale behind each type of control chart has been discussed. A set of general considerations that deserve attention prior to the construction of a control chart was given. Statistical process control by means of control charts for variables is the backbone of many processes. Procedures to construct and maintain these control charts were discussed at length.

General guidelines are presented for selection of the type of variables control chart to construct based on the nature of the data collected. When the subgroup size is 1, a control chart for individuals and moving range (I – MR) is used. For small subgroups ($n < 10$), charts for the mean and range ($\overline{X}$ and R) are used; for larger subgroups ($n \geq 10$), charts for the mean and standard deviation ($\overline{X}$ and s) are appropriate. When it is of interest to detect small deviations of a process from a state of control, the cumulative sum chart is an option. We also discussed multivariate control charts where more than one product or process variables are of interest. A T^2-chart for controlling the process mean vector and a generalized variance chart for monitoring the process variability are presented.

KEY TERMS

attribute

average run length

causes

  common

  special

centerline

control chart patterns

  bunches/groups

  cyclic

  freaks

  gradual shift in level

  interaction

  mixture or the effect of two or more populations

  natural

  stratification

  sudden shift in level

  trend

  wild

control charts for variables

  acceptance control chart

  cumulative sum chart (cusum chart)

  generalized variance chart

  geometric moving-average chart

  Hotelling's T^2 control chart

  modified control chart

  moving-average chart

  R-chart

  short production runs

  s-chart

  standardized

  trend chart (regression control chart)

  X-chart

  $\bar{X}$-chart

  Z-MR chart

control limits

  lower

  revised limits

  trial limits

  upper

F-distribution

geometric moving average

  exponentially weighted moving average

  weighting factor

moving average

  span

moving range

multivariate control charts

  control region

out of control

  rules

Pareto analysis

principle of least squares

process

  mean

  shift in mean

  standard deviation

process capability

proportion nonconforming

range

remedial actions

sample

  frequency of

  rational selection of

  size

specification limits

standard deviation

  population

  sample

  sample mean

standardized control chart

statistical control

target value

3σ control limits

type I error

  overall rate

type II error

V-mask

  lead distance

  angle of decision line

variable

EXERCISES

Discussion Questions

7-1   What are the advantages and disadvantages of using variables rather than attributes in control charts?

7-2   Describe the use of the Pareto concept in the selection of characteristics for control charts.

7-3   Discuss the preliminary decisions that must be made before you construct control chart. What concepts should be followed when selecting rational samples?

7-4   Discuss specific characteristics that could be monitored through variables control charts, the form of data to collect, and the appropriate control chart in the following situations:

   (a) Waiting time to check in baggage at an airport counter
   (b) Product assembly time in a hardware company
   (c) Time to develop a proposal based on customer solicitation
   (d) Emission level of carbon monoxide for a certain model of automobile
   (e) Detection of a change in average response time to customer queries, when the degree of change is small
   (f) Changes in blood pressure of a patient over a period of time
   (g) When to change a tool in a machining operation when the tool wears out with use and it is desired that product nonconformance level not exceed a stated level
   (h) Acceptance of products manufactured in batches, where batch means of the characteristic selected are determined, with the ability to detect a set proportion nonconformance with a desired level of probability

7-5   What are some considerations in the interpretation of control charts based on standard values? Is it possible for a process to be in control when its control chart is based on observations from the process but to be out of control when the control chart is based on a specified standard? Explain.

7-6   A start-up company, promoting the development of new products, can afford only a few observations from each product. Thus, a critical quality characteristic is selected for monitoring from each product. What type of control chart would be suitable in this context? What assumptions are necessary?

7-7   Patient progress in a health care facility for a certain diagnosis-related group have a few vital characteristics (systolic blood pressure, diastolic blood pressure, total cholesterol, weight) monitored over time. The characteristics, however, are not independent of each other. Target values for each characteristic are specified. What is an appropriate control chart in this context?

7-8   Explain the difference in interpretation between an observation falling below the lower control limit on an $\bar{X}$-chart and one falling below the lower control limit on an R-chart. Discuss the impact of each on the revision of control charts in the context of response time to fire alarms.

7-9   A new hire has been made in a management consulting firm and data are monitored on response time to customer queries. Discuss what the patterns on an $\overline{X}$- and R-chart might look like as learning on the job takes place.

7-10   A financial institution wants to improve proposal preparation time for its clients. Discuss the actions to be taken in reducing the average preparation time and the variability of preparation times.

7-11   Control charts are maintained on individual values on patient recovery time for a certain diagnosis-related group. What precautions should be taken in using such charts and what are the assumptions?

7-12   Explain the concept of process capability and when it should be estimated. What is its impact on nonconformance? Discuss in the context of project completion time of the construction of an office building.

7-13   What are the advantages and disadvantages of cumulative sum charts compared to Shewhart control charts?

7-14   What are the conditions under which a moving-average control chart is preferable? Compare the moving-average chart with the geometric moving-average chart.

7-15   Discuss the appropriate settings for using a trend control chart.

7-16   Discuss the appropriate setting for using a modified control chart and an acceptance control chart. Compare and contrast the two charts.

7-17   What is the motivation behind constructing multivariate control charts? What advantages do they have over control charts for individual characteristics?

7-18   Lung congestion may occur in illness among infants. However, it is not easily verifiable without radiography. To monitor an ill infant to predict whether lung opacity will occur on a radiograph, data are kept on age, respiration rate, heart rate, temperature, and pulse oximetry. Target values for each variable are identified. What control chart should you use in this context?

Problems

7-19   A soft drink bottling company is interested in controlling its filling operation. Random samples of size 4 are selected and the fill weight is recorded. Table 7-17 shows the data for 24 samples. The specifications on fill weight are 350 ± 5 grams (g). Daily production rate is 20,000 bottles.
(a) Find the trial control limits for the $\overline{X}$- and R-charts.
(b) Assuming special causes for out-of-control points, find the revised control limits.
(c) Assuming the distribution of fill weights to be normal, how many bottles are nonconforming daily?
(d) If the cost of rectifying an underfilled bottle is $0.08 and the lost revenue of an overfilled bottle is $0.03, what is monthly revenue lost on average?
(e) If the process average shifts to 342 g, what is the probability of detecting it on the next sample drawn after the shift?
(f) What proportion of the output is nonconforming at the level of process average indicated in part (e)?

TABLE 7-17

Sample	Observations (g)				Sample	Observations (g)			
1	352	348	350	351	13	352	350	351	348
2	351	352	351	350	14	356	351	349	352
3	351	346	342	350	15	353	348	351	350
4	349	353	352	352	16	353	354	350	352
5	351	350	351	351	17	351	348	347	348
6	353	351	346	346	18	353	352	346	352
7	348	344	350	347	19	346	348	347	349
8	350	349	351	346	20	351	348	347	346
9	344	345	346	349	21	348	352	351	352
10	349	350	352	352	22	356	351	350	350
11	353	352	354	356	23	352	348	347	349
12	348	353	346	351	24	348	353	351	352

7-20   A major automobile company is interested in reducing the time that customers have to wait while having their car serviced with one of the dealers. They select four customers randomly each day and find the total time that each customer has to wait (in minutes) while his or her car is serviced. From these four observations, the sample average and range are found. This process is repeated for 25 days. The summary data for these observations are

$$\sum_{i=1}^{25} \overline{X}_i = 1000 \qquad \sum_{i=1}^{25} R_i = 250$$

(a) Find the $\overline{X}$- and R-chart control limits
(b) Assuming that the process is in control and the distribution of waiting time is normal, find the percentage of customers who will not have to wait more than 50 minutes.
(c) Find the 2σ control limits.
(d) The service manager is developing a promotional program and is interested in reducing the average waiting time to 30 minutes by employing more mechanics. If the plan is successful, what proportion of the customers will have to wait more than 40 minutes? More than 50 minutes?

7-21   Flight delays are of concern to passengers. An airline obtained observations on the average and range of delay times of flights (in minutes), each chosen from a sample of size 4, as shown in Table 7-18. Construct appropriate control charts, and comment on the performance level. What are the chances of meeting a goal of no more than a 10-minute delay?

7-22   In a textile company, it is important that the acidity of the solution used to dye fabric be within certain acceptable values. Data values are gathered for a control chart by randomly taking four observations from the solution and determining the average pH value and range. After 25 such samples, the following summary information is obtained.

$$\sum_{i=1}^{25} \overline{X}_i = 195 \qquad \sum_{i=1}^{25} R_i = 10$$

The specifications for the pH value are 7.5 ± 0.5.

TABLE 7-18

Observation	Average Delay	Range	Observation	Average Delay	Range
1	6.5	2.1	14	9.2	3.5
2	11.1	3.8	15	7.8	2.2
3	15.8	4.6	16	10.6	4.1
4	10.9	4.2	17	10.7	4.2
5	11.2	4.0	18	8.8	3.8
6	5.6	3.5	19	9.8	3.6
7	10.4	4.1	20	10.2	3.6
8	9.8	2.0	21	9.0	4.2
9	7.7	3.2	22	8.5	3.3
10	8.6	3.8	23	9.8	4.0
11	10.5	4.2	24	7.7	2.8
12	10.2	3.8	25	10.5	3.2
13	10.5	4.0			

(a) Find the $\bar{X}$ and R-chart control limits.

(b) Find the 1σ and 2σ $\bar{X}$-chart limits.

(c) What fraction of the output is nonconforming (assuming a normal distribution of pH values)?

7-23 The bore size on a component to be used in assembly is a critical dimension. Samples of size 4 are collected and the sample average diameter and range are calculated. After 25 samples, we have

$$\sum_{i=1}^{25} \bar{X}_i = 107.5 \qquad \sum_{i=1}^{25} R_i = 12.5$$

The specifications on the bore size are 4.4 ± 0.2 mm. The unit costs of scrap and rework are $2.40 and $0.75, respectively. The daily production rate is 1200.

(a) Find the $\bar{X}$- and R-chart control limits.

(b) Assuming that the process is in control, estimate its standard deviation.

(c) Find the proportion of scrap and rework.

(d) Find the total daily cost of scrap and rework.

(e) If the process average shifts to 4.5 mm, what is the impact on the proportion of scrap and rework produced?

7-24 The time to be seated at a popular restaurant is of importance. Samples of five randomly selected customers are chosen and their average and range (in minutes) are calculated. After 30 such samples, the summary data values are

$$\sum_{i=1}^{30} \bar{X}_i = 306 \qquad \sum_{i=1}^{30} R_i = 24$$

(a) Find the $\bar{X}$- and R-chart control limits.

(b) Find the 1σ and 2σ $\bar{X}$- chart limits.

(c) The manager has found that customers usually leave if they are informed of an estimated waiting time of over 10.5 minutes. What fraction of customers will this restaurant lose? Assume a normal distribution of waiting times.

TABLE 7-19

Sample	Sample Average, $\overline{X}$	Sample Standard Deviation, (s)	Sample	Sample Average, $\overline{X}$	Sample Standard Deviation, (s)
1	10.19	0.15	11	10.18	0.16
2	9.80	0.12	12	9.85	0.15
3	10.12	0.18	13	9.82	0.06
4	10.54	0.19	14	10.18	0.34
5	9.86	0.14	15	9.96	0.11
6	9.45	0.09	16	9.57	0.09
7	10.06	0.16	17	10.14	0.12
8	10.13	0.18	18	10.08	0.15
9	9.82	0.14	19	9.82	0.09
10	10.17	0.13	20	10.15	0.12

7-25   The thickness of sheet metal (mm) used for making automobile bodies is a characteristic of interest. Random samples of size 4 are taken. The average and standard deviation are calculated for each sample and are shown in Table 7-19 for 20 samples. The specification limits are 9.95 ± 0.3 mm.

(a) Find the control limits for the $\overline{X}$- and s-charts. If there are out-of-control points, assume special causes and revise the limits.

(b) Estimate the process mean and the process standard deviation.

(c) If the thickness of sheet metal exceeds the upper specification limit, it can be reworked. However, if the thickness is less than the lower specification limit, it cannot be used for its intended purpose and must be scrapped for other uses. The cost of rework is \$0.25 per linear foot, and the cost of scrap is \$0.75 per linear foot. The rolling mills are 100 feet in length. The manufacturer has four such mills and runs 80 batches on each mill daily. What is the daily cost of rework? What is the daily cost of scrap?

(d) If the manufacturer has the flexibility to change the process mean, should it be moved to 10.00?

(e) What alternative courses of action should be considered if the product is nonconforming?

7-26   Light bulbs are tested for their luminance, with the intensity of brightness desired to be within a certain range. Random samples of five bulbs are chosen from the output, and the luminance is measured. The sample mean $\overline{X}$ and the standard deviation s are found. After 30 samples, the following summary information is obtained:

$$\sum_{i=1}^{30} \overline{X}_i = 2550 \qquad \sum_{i=1}^{30} s_i = 195$$

The specifications are 90 ± 15 lumens.

(a) Find the control limits for the $\overline{X}$- and s-charts.

(b) Assuming that the process is in control, estimate the process mean and process standard deviation.

(c) Comment on the ability of the process to meet specifications. What proportion of the output is nonconforming?

(d) If the process mean is moved to 90 lumens, what proportion of output will be nonconforming? What suggestions would you make to improve the performance of the process?

7-27 The advertised weight of frozen food packages is 16 (oz) and the specifications are 16 ± 0.3 oz. Random samples are of size 8 are selected from the output and weighted. The sample mean and standard deviation are calculated. Information on 25 such samples yields the following

$$\sum_{i=1}^{25} \overline{X}_i = 398 \qquad \sum_{i=1}^{25} s_i = 3.00$$

(a) Determine the centerlines and control limits for the $\overline{X}$- and s-charts.

(b) Estimate the process mean and standard deviation, assuming that the process is in control.

(c) Find the 1σ and 2σ control limits for each chart.

(d) What proportion of the output is nonconforming? Is the process capable?

(e) What proportion of the output weighs less than the advertised weight?

(f) If the manufacturer is interested in reducing potential complaints and lawsuits from customers who feel that they have been cheated by packages weighing less than what is advertised, what action should the manufacturer take?

7-28 The baking time of painted corrugated sheet metal is of interest. Too much time will cause the paint to flake, and too little time will result in an unacceptable finish. The specifications on baking time are 10 ± 0.2 minutes. Random samples of size 6 are selected and their baking times noted. The sample means and standard deviations are calculated for 20 samples, with the following results:

$$\sum_{i=1}^{20} \overline{X}_i = 199.8 \qquad \sum_{i=1}^{25} s_i = 1.40$$

(a) Calculate the centerline and control limits for the $\overline{X}$- and s-charts.

(b) Estimate the process mean and standard deviation, assuming the process to be in control.

(c) Is the process capable? What proportion of the output is nonconforming?

(d) If the mean of the process can be shifted to 10 minutes, would you recommend such a change?

(e) If the process mean changes to 10.2 minutes, what is the probability of detecting this change on the first sample taken after the shift? Assume that the process variability has not changed.

7-29 The level of dissolved oxygen in water was measured every 2 hours in a river where industrial plants discharge processed waste. Each observation consists of four samples, from which the sample mean and range of the amount of dissolved oxygen in parts per million are calculated. Table 7-20 shows the results of 25 such observations. Discuss the stability of the amount of dissolved oxygen. Revise the control limits, if necessary, assuming special causes for the out-of-control points. Suppose that environmental standards call for a minimum of 4 ppm of dissolved oxygen. Are these standards being achieved? Discuss.

TABLE 7-20

Observation	Average Level of Dissolved Oxygen	Range	Observation	Average Level of Dissolved Oxygen	Range
1	7.4	2.1	14	4.3	2.0
2	8.2	1.8	15	5.8	1.4
3	5.6	1.4	16	5.4	1.2
4	7.2	1.6	17	8.3	1.9
5	7.8	1.9	18	8.0	2.3
6	6.1	1.5	19	6.7	1.5
7	5.5	1.1	20	8.5	1.3
8	6.0	2.7	21	5.7	2.4
9	7.1	2.2	22	8.3	2.1
10	8.3	1.8	23	5.8	1.6
11	6.4	1.2	24	6.8	1.8
12	7.2	2.1	25	5.9	2.1
13	4.2	2.5			

7-30   In a gasoline-blending plant, the quality of the output as indicated by its octane rating is measured for a sample taken from each batch. The observations from 20 such samples are shown in Table 7-21. Construct a chart for the moving range of two successive observations and a chart for individuals.

TABLE 7-21

Sample	Octane Rating	Sample	Octane Rating	Sample	Octane Rating	Sample	Octane Rating
1	89.2	6	87.5	11	85.4	16	90.3
2	86.5	7	92.6	12	91.6	17	85.6
3	88.4	8	87.0	13	87.7	18	90.9
4	91.8	9	89.8	14	85.0	19	82.1
5	90.3	10	92.2	15	91.5	20	85.8

7-31   Automatic machines that fill packages of all-purpose flour to a desired standard weight need to be monitored closely. A random sample of four packages is selected and weighed. The average weight is then computed. Observations from 15 such samples are shown in Table 7-22. The desired weight of packages is 80 oz. Historical information on the machine reveals that the standard deviation of the weights of individual packages is 0.2 oz. Assume an acceptable type I error rate of 0.05. Also assume that it is desired to detect shifts in the process mean of 0.15 oz. Construct a cumulative sum chart and determine whether the machine needs to be adjusted to meet the target weight.

TABLE 7-22

Sample	Average Weight	Sample	Average Weight	Sample	Average Weight	Sample	Average Weight
1	80.2	5	80.1	9	79.7	13	79.8
2	80.0	6	80.4	10	79.5	14	80.4
3	79.6	7	79.5	11	80.3	15	80.2
4	80.3	8	79.4	12	80.5		

7-32   The bending strength of the poles is a consideration to a manufacturer of fiberglass fishing rods. Samples are chosen from the process, and the average bending strength of four samples is found. The target mean bending strength is 30 kg, with a process standard deviation of 0.8 kg. Suppose that it is desired to detect a shift of $\pm 0.75\sigma_{\bar{x}}$ from the target value. If the process mean is not significantly different from the target value, it is also desirable for the average run length to be 300. Find the parameters of an appropriate V-mask. If the manufacturer desires that the average run length not exceed 13 so as to detect shifts of the magnitude indicated, what will be the parameters of a V-mask if there is some flexibility in the allowable ARL when the process is in control?

7-33   The average time (minutes) that a customer has to wait for the arrival of a cab after calling the company has been observed for random samples of size 4. The data for 20 such samples are shown in Table 7-23. Previous analysis gave the upper and lower control limits for an $\bar{X}$-chart when the process was in control as 10.5 and 7.7 minutes respectively. What is your estimate of the standard deviation of the waiting time for a customer? Construct a moving-average control chart using a span of 3. What conclusions can you draw from the chart?

TABLE 7-23

Sample	Average Waiting Time	Sample	Average Waiting Time	Sample	Average Waiting Time	Sample	Average Waiting Time
1	8.4	6	9.4	11	8.8	16	9.9
2	6.5	7	10.2	12	10.0	17	10.2
3	10.8	8	8.1	13	9.5	18	8.3
4	9.7	9	7.4	14	9.6	19	8.6
5	9.0	10	9.6	15	8.3	20	9.9

7-34   Consider Exercise 7-33 which deals with the average wait time for the arrival of a cab. Using the data in Table 7-23, construct a geometric moving-average control chart. Use a weighing factor of 0.10. What conclusions can you draw from the chart? How is it different from the moving-average control chart constructed in Exercise 7-33?

7-35   Consider the data on purchase order processing time for customers shown in Table 4-6. Construct an individuals and a moving-range chart on the data before process improvement changes are made, and comment on process stability.

7-36 Consider the data on the waiting time (seconds) of customers before speaking to a representative at a call center, shown in Exercise 5-8. The data, in sequence, are to be read across the row, before moving to the next row.
(a) Construct an individuals and moving-range chart, and comment on the process.
(b) Construct a moving-average chart, with a window of 3, and comment on the process.
(c) Construct an exponentially weighted moving-average chart, with a weighting factor of 0.2, and comment on the process.
(d) Construct a cumulative sum chart, with a target value of 30 seconds, and comment on the process. Assume a type I error level of 0.05.

7-37 In an injection molding process, the die wears out gradually. To account for this wear, it is suggested that a trend chart be constructed for the outside diameter of the component produced. Samples of size 5 are selected and the sample average $\overline{X}$ and range R are found. The results of 20 such samples are shown in Table 7-24. Construct the centerline and control limits of a trend chart for the sample average. Is the process in control? If the process is out of control, assume special causes, and revise the limits. Suppose that the specification limits are 110 ± 8 mm. At what point should the die be changed?

TABLE 7-24

Sample	Sample Average, $\overline{X}$ (mm)	Sample Range, R (mm)	Sample	Sample Average, $\overline{X}$ (mm)	Sample Range, R (mm)
1	107.6	3.1	11	111.6	2.3
2	104.3	2.6	12	113.3	2.5
3	103.5	2.8	13	109.8	2.4
4	105.7	2.4	14	110.3	2.1
5	104.8	3.2	15	108.6	2.6
6	108.5	2.5	16	112.7	1.8
7	109.7	2.8	17	114.2	2.8
8	105.3	1.7	18	115.5	3.0
9	112.6	2.4	19	112.8	2.7
10	110.5	2.0	20	116.2	2.2

7-38 The percentage of potassium in a compound is expected to be within the specification limits of 18 to 35%. Samples of size 4 are selected, and the mean and range of 25 such samples are shown in Table 7-25. It is desirable for the process nonconformance to be within 1.5%. If the acceptable level of type I error is 0.05, find the modified control limits for the process mean.

7-39 Refer to Example 7-13 and the data for the nitrogen content in a certain fertilizer mix. If it is desired that the proportion nonconforming be within 0.5% and the level of Type I error be limited to 0.025, find the modified control limits for the process mean.

7-40 Refer to Exercise 7-38 and the data for the percentage of potassium content in a compound. Suppose that we wish to detect an out-of-control condition with a probability of 0.90 if the process is producing at a nonconformance rate of 4%. Determine the acceptance control chart limits.

TABLE 7-25

Sample	Sample Average, $\overline{X}$ (%)	Sample Range, R (%)	Sample	Sample Average, $\overline{X}$ (%)	Sample Range, R (%)
1	23.0	1.9	14	23.6	2.0
2	20.0	2.3	15	20.8	1.6
3	24.0	2.2	16	20.2	2.1
4	19.6	1.6	17	19.5	2.3
5	20.5	1.8	18	22.7	2.5
6	22.8	2.4	19	21.2	1.9
7	19.3	2.3	20	22.9	2.2
8	21.6	2.0	21	20.6	2.1
9	20.3	2.1	22	23.5	2.4
10	19.6	1.7	23	21.6	1.8
11	24.2	2.3	24	22.6	2.3
12	21.9	1.8	25	20.5	2.2
13	20.6	1.8			

7-41 Refer to Example 7-14. If the nonconformance production rate is 2% and we wish to detect this with a probability of 0.98, what should be the acceptance control chart limits?

7-42 A component to be used in the assembly of a transmission mechanism is manufactured in a process for which the two quality characteristics of tensile strength (X_1) and diameter (X_2) are of importance. Twenty samples, each of size 4, are obtained from the process. For each component, measurements on the tensile strength and diameter are taken and are shown in Table 7-26. Construct a multivariate Hotelling's T^2 control chart using an overall type I error probability of 0.01.

7-43 It is desired to monitor project completion times by analysts in a consulting company. The magnitude and complexity of the project influence the completion time. It is also believed that the variability in completion time increases with the magnitude of the completion time. Table 7-27 shows recent project completion times (days) along with their complexity. Complexity is indicated by letters A, B, and C, with complexity increasing from A to B and B to C. Construct an appropriate control chart and comment on the process.

7-44 Consider the data on the parameters in a chemical process of temperature, pressure, proportion of catalyst, and pH value of mixture as indicated in Table 3-16.
(a) Construct a Hotelling's T^2-chart and comment on process stability. Which process parameters, if any, would you investigate further?
(b) Analyze process variability through a generalized variance chart.

7-45 Consider the data on 25 patients, of a certain diagnosis-related group, on systolic blood pressure, blood glucose level, and total cholesterol level as shown in Table 4-5. The table shows values on these variables before and after administration of a certain drug. Assume that these variables are not independent of each other. What is an appropriate control chart to use?

TABLE 7-26

Sample	Tensile Strength (1000 kg)				Diameter (cm)			
1	66	70	68	72	16	18	15	20
2	75	60	70	75	17	22	18	19
3	65	70	70	65	20	18	15	18
4	72	70	75	65	19	20	15	17
5	73	74	72	70	21	21	23	19
6	72	74	73	74	21	19	20	18
7	63	62	65	66	22	20	24	22
8	75	84	75	66	22	20	20	22
9	65	69	77	71	18	16	18	18
10	70	68	67	67	18	17	19	18
11	80	75	70	69	24	18	20	22
12	68	65	80	50	20	21	20	22
13	74	80	76	74	19	17	20	21
14	76	74	75	73	20	17	18	18
15	71	70	74	73	18	16	17	18
16	68	67	70	69	18	16	19	20
17	72	76	75	77	22	19	23	20
18	76	74	75	77	19	23	20	21
19	72	74	73	75	20	18	20	19
20	72	68	74	70	21	19	18	20

TABLE 7-27

Project	Complexity	Completion Time	Project	Complexity	Completion Time
1	B	80	14	A	36
2	B	65	15	C	190
3	A	22	16	C	150
4	C	135	17	C	220
5	B	90	18	B	85
6	A	34	19	B	75
7	A	42	20	B	60
8	A	38	21	B	72
9	C	120	22	A	32
10	B	70	23	A	44
11	B	60	24	A	38
12	A	40	25	C	160
13	A	35			

(a) Construct a Hotelling's T^2-chart using data before drug administration and comment on patient stability.

(b) Construct an individuals and a moving-range chart for blood glucose level before drug administration and comment. Are the conclusions from parts (a) and (b) consistent? Explain.

REFERENCES

Alt, F. B. (1982). "Multivariate Quality Control: State of the Art", *American Society for Quality Control Annual Quality Congress Transactions* pp. 886–893.

AT&T (1984). *Statistical Quality Control Handbook*. New York: AT&T.

Barnard, G. A. (1959). "Control Charts and Stochastic Processes", *Journal of the Royal Statistical Society, Series B*, 21, 239 –257.

Bowker, A. H., and G. J. Lieberman (1972). *Engineering Statistics*, 2nd ed. Englewood Cliffs, NJ, Prentice Hall.

Gitlow, H., S. Gitlow, A. Oppenheim, and R. Oppenheim (1989). *Tools and Methods for the Improvement of Quality*. Homewood, IL, Richard D. Irwin.

Hawkins, D. M. (1981). "A Cusum for a Scale Parameter", *Journal of Quality Technology*, 13(4): 228–231.

——(1993). "Cumulative Sum Control Charting: An Underutilized SPC Tool", Quality Engineering, 5 (3): 463–477.

Hotelling, H. (1947). "Multivariate Quality Control". In *Techniques of Statistical Analysis*, C. Eisenhart, M. W. Hastny, and W. A.Wallis, Eds., New York, McGraw-Hill.

Lucas, J. M. (1976). "The Design and Use of V-Mask Control Schemes", *Journal of Quality Technology*, 8(1): 1–12.

——(1982). "Combined Shewhart–Cusum Quality Control Schemes", Journal of Quality Technology, 14(2): 51–59.

Minitab, Inc. (2007). *Release 15*. State College, PA: Minitab.

Montgomery, D. C. (2004). *Statistical Quality Control*, 5th ed. Hoboken NJ: Wiley.

Nelson, L. S. (1989). "Standardization of Control Charts", *Journal of Quality Technology*, 21 (4): 287–289.

Wadsworth, H. M. Jr., K. S. Stephens, and A. B. Godfrey (2001). *Methods for Quality Control and Improvement*, 2nd ed., New York: Wiley.

Woodall, W. H., and B. M. Adams (1993). "The Statistical Design of Cusum Charts", *Quality Engineering*, 5 (4): 559–570.

8

CONTROL CHARTS FOR ATTRIBUTES

8-1 INTRODUCTION AND CHAPTER OBJECTIVES

In Chapter 7 we discussed statistical process control using control charts for variables. In this chapter we examine control charts for attributes. An **attribute** is a quality characteristic for which a numerical value is not specified. It is measured on a nominal scale; that is, it does or does not meet certain guidelines, or it is categorized according to a scheme of labels. For instance, the taste of a certain dish is labeled as acceptable or unacceptable, or is categorized as exceptional, good, fair, or poor. Our objective is to present various types of control charts for attributes.

A quality characteristic that does not meet certain prescribed standards (or specifications) is said to be a **nonconformity** (or defect). For example, if the length of steel bars is expected to be 50 ± 1.0 cm, a length of 51.5 cm is not acceptable. A product with one or more nonconformities, such that it is unable to meet the intended standards and is unable to function as required, is a **nonconforming item** (or **defective**). It is possible for a product to have several nonconformities without being classified as a nonconforming item.

The different types of **control charts** considered in this chapter are grouped into three categories. The first category includes control charts that focus on proportion: the proportion of nonconforming items (p-chart) and the number of nonconforming items (np-chart). These two charts are based on binomial distributions. The second category deals with two charts that focus on the nonconformity itself. The chart for the total number of nonconformities (c-chart) is based on the Poisson distribution. The chart for nonconformities per unit (u-chart) is applicable to situations in which the size of the

Fundamentals of Quality Control and Improvement, Third Edition, By Amitava Mitra
Copyright © 2008 John Wiley & Sons, Inc.

sample unit varies from sample to sample. In the third category, the chart for demerits per unit (*U*-chart) deals with combining nonconformities on a weighted basis, such that the weights are influenced by the severity of each nonconformity. Finally, we include a section on charts for highly conforming processes.

8-2 ADVANTAGES AND DISADVANTAGES OF ATTRIBUTE CHARTS

Advantages

Certain quality characteristics are best measured as attributes. For instance, the taste of a food item is specified as acceptable or not. There are circumstances in which a quality characteristic can be measured as a variable but is instead measured as an attribute because of limited time, money, worker availability, or other resources. Let's consider the inside diameter of a hole. This characteristic could be measured with an inside micrometer, but it may be more convenient and cost-effective to use a go/no-go gage. Of course, the assumption is that attribute information is sufficient; otherwise, the quality characteristic may have to be dealt with as a variable.

In most manufacturing and service operations there are numerous quality characteristics that can be analyzed. If a variable chart (such as an $\bar{X}$- or *R*-chart) is selected, one variable chart is needed for each characteristic. The total number of control charts being constructed and maintained can be overwhelming. A control chart for attributes can provide overall quality information at a fraction of the cost.

Let's consider a simple component for which the three quality characteristics of length *L*, width *W*, and height *H* are important. If variable charts are constructed, we will need three charts. However, we can get by with one attribute chart if an item is simply classified as nonconforming when either the length, width, or height does not conform to specifications. The attribute chart thus summarizes the information for all three characteristics. An attribute chart can also be used to summarize information about several components that make up a product.

Attributes are encountered at all levels of an organization: the company, plant, department, work center, and machine (or operator) level. **Variable charts** are typically used at the lowest level, the machine level. When we do not know what is causing a problem, it is sensible to start at a general level and work to the specific. For example, we know that a high proportion of nonconforming items is being detected at the company level, so we keep an attribute chart at the plant level to determine which plants have high proportions of nonconforming items. Once we have identified these plants, we might use an attribute chart for output at the departmental level to pinpoint problem areas. When the particular work center thought to be responsible for the increase in the production of nonconforming items is identified, we could then focus on the machine or operator level and try to further pinpoint the source of the problem. Attribute charts assist in going from the general to a more focused level. Once the lowest problem level has been identified, a variable chart is then used to determine specific causes for an out-of-control situation.

Disadvantages

Attribute information indicates whether a certain quality characteristic is within specification limits. It does not state the degree to which specifications are met or not met. For example, the specification limits for the diameter of a part are 20 ± 0.1 mm. Two parts, one with diameter 20.2 mm and the other with diameter 22.3 mm, are both classified as nonconforming, but their relative conformance to the specifications is not noted on attribute information.

Mean A: Target value.
Mean B: Variables chart reacts to change in process mean.
Mean C: Attribute chart reacts.

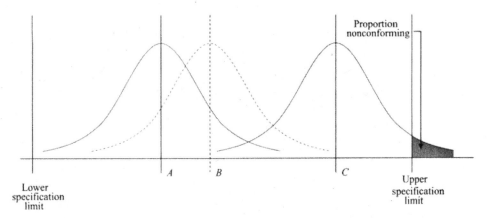

FIGURE 8-1   Forewarning of a lack of process control as indicated by a variable chart.

Variable information, on the other hand, indicates the level of the data values. Variable control charts thus provide more information on the performance of a process. Specific information about the process mean and variability can be obtained. Furthermore, for out-of-control situations, variable plots generally provide more information as to the potential causes and hence make identification of remedial actions easier.

If we can assume that the process is very capable (i.e., its inherent variability is much less than the spread between the specification limits), variable charts can forewarn us when the process is about to go out of control. This, of course, allows us to take corrective action before any nonconforming items are produced. A variable chart can indicate an upcoming out-of-control condition even though items are not yet nonconforming.

Figure 8-1 depicts this situation. Suppose that the target value of the process mean is at *A*, the process is very capable (note the wide spread between the distribution and the specification limits), and the process is in control. The process mean now shifts to *B*. The variable chart indicates an out-of-control condition. Because of the wide spread of the specification limits, no nonconforming items are produced when the process mean is at *B*, even though the variable chart is indicating an out-of-control process. The attribute chart, say, a chart for the proportion of nonconforming items, does not detect a lack of control until the process parameters are sufficiently changed such that some nonconforming items are produced. Only when the process mean shifts to *C* does the attribute chart detect an out-of-control situation. However, were the specifications equal to or tighter than the inherent variability of the process, attribute charts would indicate an out-of-control process in a time frame similar to that for the variable chart.

Attribute charts require larger sample sizes than variable charts to ensure adequate protection against a certain level of process changes. Larger sample sizes can be problematic if the measurements are expensive to obtain or the testing is destructive.

8-3   PRELIMINARY DECISIONS

The choice of sample size for attribute charts is important. It should be large enough to allow nonconformities or nonconforming items to be observed in the sample. For example, if a process has a nonconformance rate of 2.5%, a sample size of 25 is not sufficient because the average number of nonconforming items per sample is only 0.625. Thus, misleading inferences might be made, since no nonconforming items would be observed for many samples. We might erroneously attribute a better nonconformance rate to the process than what actually exists. A sample size of 100 here is sufficient, as the average number of nonconforming items per sample would thus be 2.5.

For situations in which summary measures are required, attribute charts are preferred. Information about the output at the plant level is often best described by proportion-nonconforming charts or charts on the number of nonconformities. These charts are effective for providing information to upper management. On the other hand, variable charts are more meaningful at the operator or supervisor level because they provide specific clues for remedial actions.

What is going to constitute a nonconformity should be properly defined. This definition will depend on the product, its functional use, and customer needs. For example, a scratch mark in a machine vise might not be considered a nonconformity, whereas the same scratch on a television cabinet would. One-, two-, or three-sigma zones and rules pertaining to these zones (discussed in Chapter 6) will not be used because the underlying distribution theory is nonnormal.

8-4 CHART FOR PROPORTION NONCONFORMING: p-CHART

A chart for the proportion of nonconforming items (p-chart) is based on a binomial distribution. For a given sample, the proportion nonconforming is defined as

$$\hat{p} = \frac{x}{n} \tag{8-1}$$

where x is the number of nonconforming items in the sample and n represents the sample size.

For a **binomial distribution** to be strictly valid, the probability of obtaining a nonconforming item must remain constant from item to item. The samples must be identical and are assumed to be independent.

Recall from Chapter 4 that the distribution of the number of nonconforming items in a sample, as given by a binomial distribution, is

$$P(X = x) = \frac{n!}{(n-x)!\,x!} p^x (1-p)^{n-x}, \qquad x = 0, 1, 2, \ldots, n \tag{8-2}$$

where p represents the probability of getting a nonconforming item on each trial or unit selected. Given the mean and standard deviation of X, the mean of the sample proportion nonconforming is

$$E(\hat{p}) = p \tag{8-3}$$

and the variance of $\hat{p}$ is

$$\text{Var}(\hat{p}) = \frac{p(1-p)}{n} \tag{8-4}$$

These measures are used to determine the centerline and control limits for p-charts.

A *p*-chart is one of the most versatile control charts. It is used to control the acceptability of a single quality characteristic (say, the width of a part), a group of quality characteristics of the same type or on the same part (the length, width, or height of a component), or an entire product. Furthermore, a *p*-chart can be used to measure the quality of an operator or machine, a work center, a department, or an entire plant. A *p*-chart provides a fair indication of the general state of the process by depicting the average quality level of the proportion nonconforming. It is thus a good tool for relating information about the average quality level to top management.

For that matter, a *p*-chart can be used as a measure of the performance of top management. For instance, the performance of the chief executive officer of a company can be evaluated by considering the proportion of nonconforming items produced at the company level. Comparing this information with historical values indicates whether improvements for which the CEO can take credit have occurred. Values of the proportion nonconforming can also serve as a benchmark against which to compare future output.

A *p*-chart can provide a source of information for *improving* product quality. It can be used to develop new concepts and ideas. Current values of the proportion nonconforming indicate whether a particular idea has been successful in reducing the proportion nonconforming.

Use of a *p*-chart may, as a secondary objective, identify the circumstances for which $\bar{X}$- and *R*-charts should be used. Variable charts are more sensitive to variations and are useful in diagnosing causes; a *p*-chart is more useful in locating the source of the difficulty. The *p*-charts are based on the normal approximation to the binomial distribution. For approximately large samples, for a process in control, the average run length is approximately 385 for 3σ control charts.

Construction and Interpretation

The general procedures from Chapters 6 and 7 for the construction and interpretation of control charts apply to control charts for attributes as well. They are summarized here.

Step 1: *Select the objective.* Decide on the level at which the *p*-chart will be used (i.e., the plant, the department, or the operator level). Decide to control either a single quality characteristic, multiple characteristics, a single product, or a number of products.

Step 2: *Determine the sample size and the sampling interval.* An appropriate sample size is related to the existing quality level of the process. The sample size must be large enough to allow the opportunity for some nonconforming items to be present on average. The sampling interval (i.e., the time between successive samples) is a function of the production rate and the cost of sampling, among other factors.

A bound for the sample size, *n*, can be obtained using these guidelines. Let $\bar{p}$ represent the process average nonconforming rate. Then, based on the number of nonconforming items that you wish to be represented in a sample—say 5, on average—the bound is expressed as

$$n\bar{p} \geq 5 \quad \text{or} \quad n \geq \frac{5}{\bar{p}} \tag{8-5}$$

Step 3: *Obtain the data, and record on an appropriate form.* Decide on the measuring instruments in advance.

A typical data sheet for a *p*-chart is shown in Table 8-1. The data and time at which the sample is taken, along with the number of items inspected and the number of nonconforming items, are recorded. The proportion nonconforming is found by dividing the

TABLE 8-1   Data for a *p*-Chart

Sample	Date	Time	Number of Items Inspected, n	Number of Nonconforming Items, x	Proportion Nonconforming, $\hat{p}$	Comments
1	10/15	9:00 A.M.	400	12	0.030	
2	10/15	9:30 A.M.	400	10	0.025	
3	10/15	10:00 A.M.	400	14	0.035	New vendor
.	.	.	.	.	.	.
.	.	.	.	.	.	.
.	.	.	.	.	.	.

number of nonconforming items by the sample size. Usually, 25 to 30 samples should be taken prior to performing an analysis.

Step 4: *Calculate the centerline and the trial control limits.* Once they are determined, draw them on the *p*-chart. Plot the values of the proportion nonconforming ($\hat{p}$) for each sample on the chart. Examine the chart to determine whether the process is in control.

The rules for determining out-of-control conditions are the same as those we discussed in Chapters 6 and 7, so are not repeated in this chapter. As usual, the most common criterion for an out-of-control condition is the presence of a point plotting outside the control limits. The means of calculating the centerline and control limits are as follows.

No Standard Specified   When no standard or target value of the proportion non-conforming is specified, it must be estimated from the sample information. Recall that for each sample, the sample proportion nonconforming ($\hat{p}$) is given by eq. (8-1). The average of the individual sample proportion nonconforming is used as the centerline (CL$_p$). That is,

$$CL_p = \bar{p} = \frac{\sum_{i=1}^{g} \hat{p}_i}{g} = \frac{\sum_{i=1}^{g} x_i}{ng} \tag{8-6}$$

where g represents the number of samples. The variance of ($\hat{p}$) is given by eq. (8-4). Since the true value of p is not known, the value of $\bar{p}$ given by eq. (8-6) is used as an estimate. In accordance with the concept of 3σ limits discussed in previous chapters, the control limits are given by

$$UCL_p = \bar{p} + 3\sqrt{\frac{\bar{p}(1-\bar{p})}{n}}$$
$$LCL_p = \bar{p} - 3\sqrt{\frac{\bar{p}(1-\bar{p})}{n}} \tag{8-7}$$

Standard Specified   If the target value of the proportion of nonconforming items is known or specified, the centerline is selected as that target value. In other words, the centerline is given by

$$CL_p = p_0 \tag{8-8}$$

where p_0 represents the **standard** or target value. The control limits in this case are also based on the target value. Thus,

$$
\begin{aligned}
\text{UCL}_p &= p_0 + 3\sqrt{\frac{p_0(1-p_0)}{n}} \\
\text{LCL}_p &= p_0 - 3\sqrt{\frac{p_0(1-p_0)}{n}}
\end{aligned}
\tag{8-9}
$$

If the lower control limit for p turns out to be negative for eq. (8-7) or (8-9), the lower control limit is simply counted as zero because the smallest possible value of the proportion nonconforming is zero.

Step 5: *Calculate the revised control limits.* Analyze the plotted values of $\hat{p}$ and the pattern of the plot for out-of-control conditions. Typically, one or a few of the rules are used concurrently. On detection of an out-of-control condition, identify the special cause, and propose remedial actions. The out-of-control point or points for which remedial actions have been taken are then deleted, and the revised process average $\bar{p}_r$ is calculated from the remaining number of samples.

The investigation of special causes must be conducted in an objective manner. If a special cause cannot be identified or a remedial action implemented for an out-of-control sample point, that sample is not deleted in the calculation of the new average proportion nonconforming.

The revised centerline and control limits are given by

$$
\begin{aligned}
\text{CL}_p &= \bar{p}_r \\
\text{UCL}_p &= \bar{p}_r + 3\sqrt{\frac{\bar{p}_r(1-\bar{p}_r)}{n}} \\
\text{LCL}_p &= \bar{p}_r - 3\sqrt{\frac{\bar{p}_r(1-\bar{p}_r)}{n}}
\end{aligned}
\tag{8-10}
$$

For p-charts based on a standard p_0, the revised limits do not change from those given by eq. (8-9).

Step 6: *Implement the chart.* Use the revised centerline and control limits of the p-chart for future observations as they become available. Revise periodically the chart using guidelines similar to those discussed for variable charts.

A few unique features are associated with implementing a p-chart. First, if the p-chart continually indicates an increase in the value of the average proportion nonconforming, management should investigate the reasons behind this increase rather than constantly revising upward the centerline and control limits. If such an upward movement of the centerline is allowed to persist, it will become more difficult to bring down the level of nonconformance to the previously desirable values. Only if you are sure that the process cannot be maintained at the present average level of nonconformance given the resource constraints should you consider moving the control limits upward.

Possible reasons for an increased level of nonconforming items include a lower incoming quality from vendors or a tightening of specification limits. An increased level of nonconformance can also occur because existing limits are enforced more stringently.

Since the goal is to seek quality improvement continuously, a sustained downward trend in the proportion nonconforming is desirable. Management should revise the average proportion-nonconforming level downward when it is convinced that the proportion nonconforming at the better level can be maintained. Revising these limits provides an incentive not only to maintain this better level but also to seek further improvements.

Example 8-1   Twenty-five samples of size 50 are chosen from a plastic-injection molding machine producing small containers. The number of nonconforming containers for each sample is shown in Table 8-2, as is the proportion nonconforming for each sample, using eq. (8-1). The average proportion nonconforming, using eq. (8-6), is

$$\bar{p} = \frac{90}{1250} = 0.072$$

TABLE 8-2   Data for Nonconforming Containers

Sample	Date	Time	Number of Items Inspected, n	Number of Nonconforming Items, x	Proportion Nonconforming, $\hat{p}$	Comments
1	10/6	8:30	50	4	0.08	
2	10/6	9:30	50	2	0.04	
3	10/6	10:00	50	5	0.10	
4	10/6	10:20	50	3	0.06	
5	10/7	8:40	50	2	0.04	
6	10/7	9:50	50	1	0.02	
7	10/7	10:10	50	3	0.06	
8	10/7	10:50	50	2	0.04	
9	10/8	9:10	50	5	0.10	
10	10/8	9:40	50	4	0.08	
11	10/8	10:40	50	3	0.06	
12	10/8	11:20	50	5	0.10	
13	10/9	8:20	50	5	0.10	
14	10/9	9:10	50	2	0.04	
15	10/9	9:50	50	3	0.06	
16	10/9	10:20	50	2	0.04	
17	10/10	8:40	50	4	0.08	
18	10/10	9:30	50	10	0.20	Drop in
19	10/10	10:10	50	4	0.08	pressure
20	10/10	11:30	50	3	0.06	
21	10/11	8:20	50	2	0.04	
22	10/11	9:10	50	5	0.10	
23	10/11	9:50	50	4	0.08	
24	10/11	10:20	50	3	0.06	
25	10/11	11:30	50	4	0.08	
			1250	90		

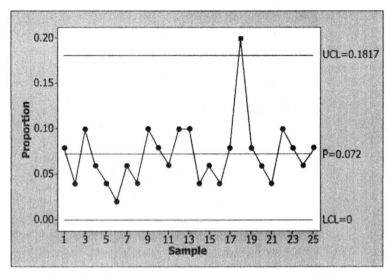

FIGURE 8-2 Proportion-nonconforming chart for containers.

This is the centerline of the *p*-chart. Next, the trial control limits are found using eq. (8-7):

$$CL_p = 0.072$$

$$UCL_p = 0.072 + 3\sqrt{\frac{(0.072)(1 - 0.072)}{50}} = 0.182$$

$$LCL_p = 0.072 - 3\sqrt{\frac{(0.072)(1 - 0.072)}{50}} = -0.038 \rightarrow 0$$

Since the calculated value of the lower control limit is negative, it is converted to zero.

This *p*-chart is shown in Figure 8-2. Using Minitab, a worksheet listing the number inspected and number of nonconforming items for each sample is first created. Select **Stat > Control Charts > Attribute Charts > p**. In the **Variables** window, input the column number or name of the variable representing the number of nonconforming items. Under **Subgroup sizes**, enter the column number or name of the corresponding variable. Click **OK**. Note that the process is in control with the exception of sample 18, which has a proportion nonconforming of 0.20—above the upper control limit. Assume that upon investigating special causes for sample 18, we found that cause to be a drop in pressure inside the mold cavity, as indicated by the comments in Table 8-2. Remedial action is taken to eliminate this special cause, sample 18 is deleted, and the revised centerline and control limits are then found:

$$CL_p = \frac{90 - 10}{1200} = 0.067$$

$$UCL_p = 0.067 + 3\sqrt{\frac{(0.067)(1 - 0.067)}{50}} = 0.173$$

$$LCL_p = 0.067 - 3\sqrt{\frac{(0.067)(1 - 0.067)}{50}} = -0.039 \rightarrow 0$$

The remaining samples are now in control.

Example 8-2   Management has decided to set a standard of 3% for the proportion of nonconforming test tubes produced in a plant. Data collected from 20 samples of size 100 are shown in Table 8-3, as is the proportion of nonconforming test tubes for each sample. The centerline and control limits, based on the specified standard, are found to be

$$CL_p = p_0 = 0.030$$

$$UCL_p = 0.030 + 3\sqrt{\frac{(0.03)(1-0.03)}{100}} = 0.081$$

$$LCL_p = 0.030 - 3\sqrt{\frac{(0.03)(1-0.03)}{100}} = -0.021 \rightarrow 0$$

A p-chart using these limits is shown in Figure 8-3. From the figure we can see that the process is out of control; samples 8 and 11 plot above the upper control limit. The special cause for sample 8 is improper alignment of the die, for which remedial action is taken. For sample 11, no special cause is identified. Hence, the process is viewed to be out of control with respect to the current standard. From Figure 8-3, observe that the general level of the proportion nonconforming does not match the standard value of 3%; the actual proportion nonconforming appears higher. Using the data from Table 8-3, the process average proportion nonconforming is

$$\bar{p} = \frac{84}{2000} = 0.042$$

TABLE 8-3   Data for Nonconforming Test Tubes

Sample	Date	Time	Number of Items Inspected, n	Number of Nonconforming Items, x	Proportion Nonconforming, $\hat{p}$	Comments
1	9/8	8:20	100	4	0.04	
2	9/8	8:45	100	2	0.02	
3	9/8	9:10	100	5	0.05	
4	9/8	9:30	100	3	0.03	
5	9/9	9:00	100	6	0.06	
6	9/9	9:20	100	4	0.04	
7	9/9	9:50	100	3	0.03	
8	9/9	10:20	100	9	0.09	Die not
9	9/10	9:10	100	5	0.05	aligned
10	9/10	9:40	100	6	0.06	
11	9/10	10:20	100	9	0.09	
12	9/10	10:45	100	3	0.03	
13	9/11	8:30	100	3	0.03	
14	9/11	8:50	100	4	0.04	
15	9/11	9:40	100	2	0.02	
16	9/11	10:30	100	5	0.05	
17	9/12	8:40	100	3	0.03	
18	9/12	9:30	100	1	0.01	
19	9/12	9:50	100	4	0.04	
20	9/12	10:40	100	3	0.03	
			2000	84		

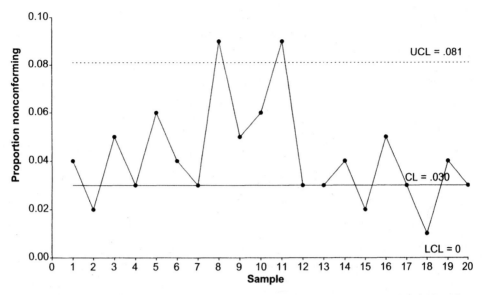

FIGURE 8-3   Proportion-nonconforming chart for test tubes (centerline and control limits based on a standard).

This value exceeds the desired standard of 3%. From Figure 8-3 only three points are below the standard of 3%. This confirms our suspicion that the process mean is greater than the desired standard value. If sample 8 is eliminated following removal of its special cause, the revised process average is

$$\bar{p} = \frac{84 - 9}{1900} = 0.039$$

When the control limits are calculated based on this revised average, we have

$$\text{CL}_p = 0.039$$

$$\text{UCL}_p = 0.039 + 3\sqrt{\frac{(0.039)(1 - 0.039)}{100}} = 0.097$$

$$\text{LCL}_p = 0.039 - 3\sqrt{\frac{(0.039)(1 - 0.039)}{100}} = -0.019 \rightarrow 0$$

If a control chart were constructed using these values, the remaining samples (including sample 11) would indicate a process in control, with the points hovering around 0.039, the calculated average.

With the standard of 3% in mind, we must conclude that the process is currently out of control because sample 11 would plot above the upper control limit value of 0.081, as found using the standard. We have no indication of the special causes for the out-of-control point (sample 11), so we cannot take remedial action to bring the process to control. Furthermore, the process average proportion nonconforming should be reduced; it is too far from the desired value of 3%. Actions to accomplish this task must originate from management and may require major changes in the product, process, or incoming material quality. Operator-assisted measures are not sufficient to bring the process to control.

Another way to view this situation would be to conclude that the present process is not capable of meeting the desired standard of a 3% nonconformance rate. In this case, basic changes that require management input are needed to improve quality. If management decides not to allocate resources to improve the process, they may have to increase the 3% value to a more realistic standard. Otherwise, the control chart will continue to indicate a lack of control, and unnecessary blame may be assigned to operators.

Variable Sample Size

There are many reasons why samples vary in size. In processes for which 100% inspection is conducted to estimate the proportion nonconforming, a change in the rate of production may cause the sample size to change. A lack of available inspection personnel or a change in the unit cost of inspection are other factors that can influence the sample size.

A change in sample size causes the control limits to change, although the centerline remains fixed. As the sample size increases, the control limits become narrower. As stated previously, the sample size is also influenced by the existing average process quality level. For a given process proportion nonconforming, the sample size should be chosen carefully so that there is ample opportunity for nonconforming items to be represented. Thus, changes in the quality level of the process may require a change in the sample size.

Control Limits for Individual Samples   Control limits can be constructed for individual samples. If no standard is given and the sample average proportion nonconforming is $\bar{p}$, the control limits for sample i with size n_i are

$$\text{UCL} = \bar{p} + 3\sqrt{\frac{\bar{p}(1-\bar{p})}{n_i}}$$

$$\text{LCL} = \bar{p} - 3\sqrt{\frac{\bar{p}(1-\bar{p})}{n_i}}$$

(8-11)

Example 8-3   Twenty random samples are selected from a process that makes vinyl tiles. The sample size as well as the number of nonconforming tiles are shown in Table 8-4. First, we construct the control limits for each sample in a proportion-nonconforming chart. The centerline is

$$\text{CL} = \bar{p} = \frac{353}{4860} = 0.0726$$

The control limits for each sample, given by eq. (8-11), are

$$\text{UCL} = 0.0726 + 3\sqrt{\frac{(0.0726)(0.9274)}{n_i}}$$

$$= 0.726 + \frac{0.7784}{\sqrt{n_i}}$$

$$\text{LCL} = 0.0726 - \frac{0.7784}{\sqrt{n_i}}$$

TABLE 8-4   Vinyl Tile Data for Individual Control Limits

Sample, i	Number of Tiles Inspected, n_i	Number of Nonconforming Tiles	Proportion Nonconforming, $\hat{p}_i$	Upper Control Limit Based on n_i	Lower Control Limit Based on n_i
1	200	14	0.070	0.128	0.018
2	180	10	0.056	0.131	0.015
3	200	17	0.085	0.128	0.018
4	120	8	0.067	0.144	0.002
5	300	20	0.067	0.118	0.028
6	250	18	0.072	0.122	0.023
7	400	25	0.062	0.112	0.034
8	180	20	0.111	0.131	0.015
9	210	27	0.129	0.126	0.019
10	380	30	0.079	0.113	0.033
11	190	15	0.079	0.129	0.016
12	380	26	0.068	0.113	0.033
13	200	10	0.050	0.128	0.018
14	210	14	0.067	0.126	0.019
15	390	24	0.061	0.112	0.033
16	120	15	0.125	0.144	0.002
17	190	18	0.095	0.129	0.016
18	380	19	0.050	0.113	0.033
19	200	11	0.055	0.128	0.018
20	180	12	0.067	0.131	0.015
	4860	353			

Table 8-4 shows the sample proportion nonconforming and the control limits for each sample. The p-chart with the individual control limits plotted is shown in Figure 8-4. From this figure we can see that sample 9 is out of control. The proportion-nonconforming value of 0.129 plots above the upper control limit of 0.126 for that sample. Special causes should be investigated for this sample, and remedial actions taken. When the centerline and control limits are revised, this sample is then deleted.

Standardized Control Chart   Another approach to varying sample size is to construct a chart of *normalized* or *standardized* values of the proportion nonconforming. The value of the proportion nonconforming for a sample is expressed as the sample's deviation from the average proportion nonconforming in units of standard deviations. In Chapter 4 we discussed the sampling distribution of a sample proportion nonconforming ($\hat{p}$). It showed that the mean and standard deviation of $\hat{p}$ are given by

$$E(\hat{p}) = p \qquad (8\text{-}12)$$

$$\sigma_{\hat{p}} = \sqrt{\frac{p(1-p)}{n}} \qquad (8\text{-}13)$$

respectively, where p represents the true process nonconformance rate and n is the sample size. In practice, p is usually estimated by $\bar{p}$, the sample average proportion nonconforming. So, in the working versions of eqs. (8-12) and (8-13), p is replaced by $\bar{p}$.

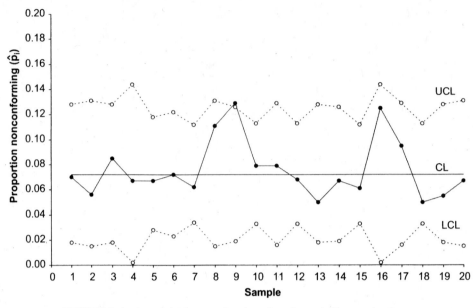

FIGURE 8-4 Proportion-nonconforming chart for variable sample sizes.

The **standardized value** of the proportion nonconforming for the ith sample may be expressed as

$$Z = \frac{\hat{p}_i - \bar{p}}{\sqrt{\bar{p}(1 - \bar{p})/n_i}} \tag{8-14}$$

where n_i is the size of ith sample. One of the advantages of a standardized control chart for the proportion nonconforming is that only one set of control limits needs to be constructed. These limits are placed ± 3 standard deviations from the centerline. Additionally, tests for runs and pattern recognition are difficult to apply to circumstances in which individual control limits change with each sample. They can, however, be applied in the same manner as with other variable charts when a standardized p-chart is constructed. The centerline on a standardized p-chart is at 0, the UCL is at 3, and the LCL is at -3.

Example 8-4 Use the vinyl tile data in Table 8-4 to construct a standardized p-chart.

Solution The standard deviation of the sample proportion nonconforming $(\hat{p}_i)$ for the ith sample is

$$\sigma_{\hat{p}_i} = \sqrt{\frac{\bar{p}(1 - \bar{p})}{n_i}} = \sqrt{\frac{(0.0726)(0.9274)}{n_i}} = \frac{0.2595}{\sqrt{n_i}}$$

For this value of $\sigma_{\hat{p}_i}$, the standardized value of $\hat{p}_i$ is calculated using eq. (8-14) with $\bar{p} = 0.0726$. Table 8-5, shows the standardized Z-values of the proportion nonconforming. The standardized p-chart is shown in Figure 8-5. The control limits are at ± 3, and the centerline is at 0. From Figure 8-5, observe that sample 9 is above the upper control limit, indicating an out-of-control situation. This is in agreement with the conclusions we reached in Example 8-3 using individual control limits.

TABLE 8-5   Standardized *p*-Chart Data for the Vinyl Tile Example

Sample, i	Number of Tiles Inspected, n_i	Number of Nonconforming Tiles	Proportion Nonconforming, $\hat{p}_i$	Standard Deviation, $\sigma_{\hat{p}_i}$	Standardized $\hat{p}_i$, Z-Value
1	200	14	0.070	0.0183	−0.142
2	180	10	0.056	0.0193	−0.858
3	200	17	0.085	0.0183	0.678
4	120	8	0.067	0.0237	−0.236
5	300	20	0.067	0.0150	−0.373
6	250	18	0.072	0.0164	−0.037
7	400	25	0.062	0.0130	−0.815
8	180	20	0.111	0.0193	1.990
9	210	27	0.129	0.0179	3.151
10	380	30	0.079	0.0133	0.481
11	190	15	0.079	0.0188	0.340
12	380	26	0.068	0.0133	−0.346
13	200	10	0.050	0.0183	−1.235
14	210	14	0.067	0.0179	−0.313
15	390	24	0.061	0.0131	−0.885
16	120	15	0.125	0.0237	2.211
17	190	18	0.095	0.0188	1.191
18	380	19	0.050	0.0133	−1.699
19	200	11	0.055	0.0183	−0.962
20	180	12	0.067	0.0193	−0.290

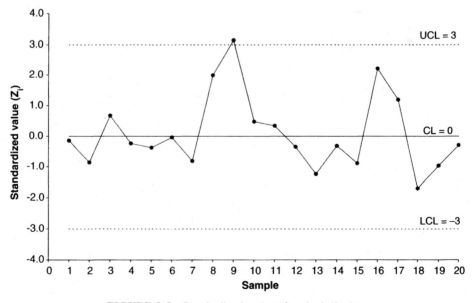

FIGURE 8-5   Standardized *p*-chart for vinyl tile data.

Special Considerations for *p*-Charts

Necessary Assumptions   Recall that the proportion of nonconforming items is based on a binomial distribution. With this distribution, the probability of occurrence of nonconforming items is assumed to be constant for each item, and the items are assumed to be independent of each other with respect to meeting the specifications. The latter assumption may not be valid if products are manufactured in groups. Let's suppose that for a steel manufacturing process, a batch produced in a cupola does not have the correct proportion of an additive. If a sample steel ingot chosen from that batch is found to be nonconforming, other samples from the same batch are also likely to be nonconforming, so the samples are not independent of each other. Furthermore, the likelihood of a sample chosen from the bad batch being nonconforming will be different from that of a sample chosen from a good batch.

Observations Below the Lower Control Limit   The presence of points that plot below the lower control limit in a *p*-chart, even though they indicate an out-of-control situation, are desirable because they also indicate an improvement in the process. Once the special causes for such points have been identified, these points should not be deleted. In fact, the process should be set to the conditions that led to these points in the first place. We must be certain, however, that errors in measurement or recording are not responsible for such low values.

Comparison with a Specified Standard   Management sometimes sets a standard value for the proportion nonconforming (p_0). When control limits are based on a standard, care must be exercised in drawing inferences about the process. If the process yields points that fall above the upper control limit, special causes should be sought. Once they have been found and appropriate actions have been taken, the process performance may again be compared to the standard. If the process is still out of control even though no further special causes can be located, management should question whether the existing process is capable of meeting the desired standards. If resource limitations prevent management from being able to implement process improvement, it may not be advisable to compare the process to the desirable standards. Only if the standards are conceivably attainable should the control limits be based on them.

Impact of Design Specifications   Since a nonconforming product means that its quality characteristics do not meet certain specifications, it is possible for the average proportion nonconforming to be too high even though the process is stable and in control. Only a fundamental change in the design of the product or in the specifications can reduce the proportion nonconforming in this situation. Perhaps tolerances should be loosened; this approach should be taken only if there is no substantial deviation in meeting customer requirements. Constant feedback between marketing, product/process design, and manufacturing will ensure that customer needs are met in the design and manufacture of the product.

Information About Overall Quality Level   A *p*-chart is ideal for aggregating information. For a plant with many product lines, departments, and work centers, a *p*-chart can combine information and provide a measure of the overall product nonconformance rate. It can also be used to evaluate the effectiveness of managers and even the chief executive officer.

8-5 CHART FOR NUMBER OF NONCONFORMING ITEMS: *np*-CHART

As an alternative to calculating the proportion nonconforming, we can count the number of nonconforming items in samples and use the count as the basis for the control chart.

Operating personnel sometimes find it easier to relate to the number nonconforming than to the proportion nonconforming. The assumptions made for the construction of proportion-nonconforming charts apply to number-nonconforming charts as well. The number of nonconforming items in a sample is assumed to be given by a binomial distribution. The same principles apply to number-nonconforming charts also, and constructing an *np*-chart is similar to constructing a *p*-chart.

There is one drawback to the *np*-chart: If the sample size changes, the centerline and control limits change as well. Making inferences in such circumstances is difficult. Thus, an *np*-chart should not be used when the sample size varies.

No Standard Given

The centerline for an *np*-chart is given by

$$\text{CL}_{np} = \frac{\sum_{i=1}^{g} x_i}{g} = n\bar{p} \tag{8-15}$$

where x_i represents the number nonconforming for the *i*th sample, g is the number of samples, n is the sample size, and $\bar{p}$ is the sample average proportion nonconforming. Since the number of nonconforming items is n times the proportion nonconforming, the average and standard deviation of the number nonconforming are n times the corresponding value for the proportion nonconforming. Thus, the standard deviation of the number nonconforming is

$$\sigma_{np} = \sqrt{n\bar{p}(1 - \bar{p})} \tag{8-16}$$

The control limits for an *np*-chart are

$$\begin{aligned} \text{UCL}_{np} &= n\bar{p} + 3\sqrt{n\bar{p}(1 - \bar{p})} \\ \text{LCL}_{np} &= n\bar{p} - 3\sqrt{n\bar{p}(1 - \bar{p})} \end{aligned} \tag{8-17}$$

If the lower control limit calculation yields a negative value, it is converted to zero.

Standard Given

Let's suppose that a specified standard for the number of nonconforming items is np_0. The centerline and control limits are given by

$$\begin{aligned} \text{CL}_{np} &= np_0 \\ \text{UCL}_{np} &= np_0 + 3\sqrt{np_0(1 - p_0)} \\ \text{LCL}_{np} &= np_0 - 3\sqrt{np_0(1 - p_0)} \end{aligned} \tag{8-18}$$

Example 8-5   Data for the number of dissatisfied customers in a department store observed for 25 samples of size 300 are shown in Table 8-6. Construct an *np*-chart for the number of dissatisfied customers.

Solution   The centerline for the *np*-chart is

$$\text{CL}_{np} = \frac{184}{20} = 9.2$$

TABLE 8-6   Number of Dissatisfied Customers

Sample	Number of Dissatisfied Customers	Sample	Number of Dissatisfied Customers
1	10	11	6
2	12	12	19
3	8	13	10
4	9	14	7
5	6	15	8
6	11	16	4
7	13	17	11
8	10	18	10
9	8	19	6
10	9	20	7
			184

The control limits are found using eq. (8-17):

$$UCL_{np} = 9.2 + 3\sqrt{9.2\left(1 - \frac{9.2}{300}\right)} = 18.159$$

$$LCL_{np} = 9.2 - 3\sqrt{9.2\left(1 - \frac{9.2}{300}\right)} = 0.241$$

This *np*-chart is shown in Figure 8-6. Note that sample 12 plots above the upper control limit and so indicates an out-of-control state. Special causes should be investigated, and after

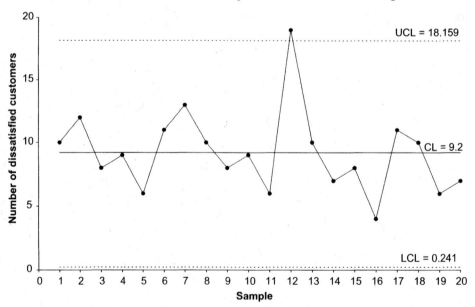

FIGURE 8-6   Number-nonconforming chart for dissatisfied customers.

remedial actions have been taken, limits are revised by deleting sample 12. The revised centerline and control limits are

$$CL_{np} = \frac{184 - 19}{19} = 8.684$$

$$UCL_{np} = 8.684 + 3\sqrt{8.684\left(1 - \frac{8.684}{300}\right)} = 17.396$$

$$LCL_{np} = 8.684 - 3\sqrt{8.684\left(1 - \frac{8.684}{300}\right)} = -0.028 \rightarrow 0$$

These limits should be used for future observations.

8-6   CHART FOR THE NUMBER OF NONCONFORMITIES: c-CHART

A *nonconformity* is defined as a quality characteristic that does not meet some specification. A nonconforming item has one or more nonconformities that make it nonfunctional. It is also possible for a product to have one or more nonconformities and still conform to standards. The p- and np-charts deal with nonconforming items. A c-chart is used to track the total number of nonconformities in samples of constant size. When the sample size varies, a u-chart is used to track the number of nonconformities per unit.

In constructing c- and u-charts, the size of the sample is also referred to as the *area of opportunity*. The area of opportunity may be single or multiple units of a product (e.g., 1 TV set or a collection of 10 TV sets). For items produced on a continuous basis, the area of opportunity could be 100 m^2 of fabric or 50 m^2 of paper. As with p-charts, we must be careful about our choice of area of opportunity. When both the average number of nonconformities per unit and the area of opportunity are small, most observations will show zero non-conformities, with one nonconformity showing up occasionally and two or more non-conformities even less frequently. Such information can be misleading, so it is beneficial to choose a large enough area of opportunity so that we can expect some nonconformities to occur. If the average number of nonconformities per TV set is small (say, about 0.08), it would make sense for the sample size to be 50 sets rather than 10.

The occurrence of nonconformities is assumed to follow a **Poisson distribution**. This distribution is well suited to modeling the number of events that happen over a specified amount of time, space, or volume. Certain assumptions must hold for a Poisson distribution to be used. First, the opportunity for the occurrence of nonconformities must be large, and the average number of nonconformities per unit must be small. An example is the number of flaws in 100 m^2 of fabric. Theoretically, this number could be quite large, but the average number of flaws in 100 m^2 of fabric is not necessarily a large value. The second assumption is that the occurrences of nonconformities must be independent of each other. Suppose that 100 m^2 of fabric is the sample size. A nonconformity in a certain segment of fabric must in no way influence the occurrence of other nonconformities. Third, each sample should have an equal likelihood of the occurrence of nonconformities; that is, the prevailing conditions should be consistent from sample to sample. For instance, if different rivet guns are used to install the rivets in a ship, the opportunity for defects may vary for different guns, so the Poisson distribution would not be strictly applicable.

Because the steps involved in the construction and interpretation of c-charts are similar to those for p-charts, we only point out the differences in the formulas. If x represents the number of nonconformities in the sample unit and c is the mean, then the Poisson distribution yields (as discussed in Chapter 4)

$$p(x) = \frac{e^{-c}c^x}{x!} \tag{8-19}$$

where $p(x)$ represents the probability of observing x nonconformities. Recall that in the Poisson distribution, the mean and the variance are equal.

No Standard Given

The average number of nonconformities per sample unit is found from the sample observations and is denoted by $\bar{c}$. The centerline and control limits are

$$\begin{aligned} \text{CL}_c &= \bar{c} \\ \text{UCL}_c &= \bar{c} + 3\sqrt{\bar{c}} \\ \text{LCL}_c &= \bar{c} - 3\sqrt{\bar{c}} \end{aligned} \tag{8-20}$$

If the lower control limit is found to be less than zero, it is converted to zero.

Standard Given

Let the specified goal for the number of nonconformities per sample unit be c_0. The centerline and control limits are then calculated from

$$\begin{aligned} \text{CL}_c &= c_0 \\ \text{UCL}_c &= c_0 + 3\sqrt{c_0} \\ \text{LCL}_c &= c_0 - 3\sqrt{c_0} \end{aligned} \tag{8-21}$$

Many of the special considerations discussed for p-charts, such as observations below the lower control limit, comparison with a specified standard, impact of design specifications, and information about overall quality level, apply to c-charts as well.

Example 8-6 Samples of fabric from a textile mill, each $100\,\text{m}^2$, are selected, and the number of occurrences of foreign matter are recorded. Data for 25 samples are shown in Table 8-7. Construct a c-chart for the number of nonconformities.

Solution The average number of nonconformities based on the sample information is found as follows. The centerline is given by

$$\bar{c} = \frac{189}{25} = 7.560$$

The control limits are given by

$$\begin{aligned} \text{UCL}_c &= 7.560 + 3\sqrt{7.560} = 15.809 \\ \text{LCL}_c &= 7.560 - 3\sqrt{7.560} = -0.689 \rightarrow 0. \end{aligned}$$

TABLE 8-7   Foreign Matter Data

Sample	Nonconformities	Sample	Nonconformities
1	5	14	11
2	4	15	9
3	7	16	5
4	6	17	7
5	8	18	6
6	5	19	10
7	6	20	8
8	5	21	9
9	16	22	9
10	10	23	7
11	9	24	5
12	7	25	7
13	8		

The *c*-chart is shown in Figure 8-7. We used Minitab to construct this chart. We selected **Stat > Control Charts > Attribute Charts > c**, indicated the name of the variable (in this case, Nonconformities) and clicked **OK**. For sample 9, the number of nonconformities is 16, which exceeds the upper control limit of 15.809. Assuming remedial action has been taken for the special causes, the centerline and control limits are revised as follows (sample 9 is deleted):

$$\bar{c} = \frac{189 - 16}{24} = 7.208$$

$$\text{UCL}_c = 7.208 + 3\sqrt{7.208} = 15.262$$
$$\text{LCL}_c = 7.208 - 3\sqrt{7.208} = -0.846 \rightarrow 0$$

After this revision, the remaining sample points are found to be within the limits.

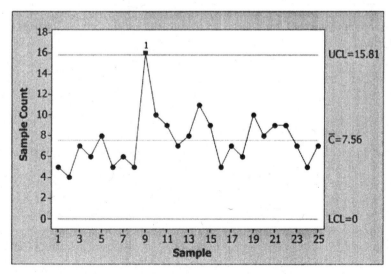

FIGURE 8-7   *c*-Chart for foreign matter data.

Probability Limits

The c-chart is based on the Poisson distribution. Thus, for a chosen level of type I error [i.e., the probability of concluding that a process is out of control when it is in control (a false alarm)], the control limits should be selected using this distribution.

The 3σ limits, shown previously, are not necessarily symmetrical. This means that the probability of an observation falling outside either control limit may not be equal. Appendix A-2 lists cumulative probabilities for the Poisson distribution for a given mean. Suppose that the process mean is c_0 and symmetrical control limits are desired for type I error of α. The control limits should then be selected such that

$$P(X > \text{UCL} \mid c_0) = P(X < \text{LCL} \mid c_0) = \frac{\alpha}{2} \tag{8-22}$$

Since the Poisson distribution is discrete, the upper control limit is found from the smallest integer x^+ such that

$$P(X \leq x^+) \geq 1 - \frac{\alpha}{2} \tag{8-23}$$

Similarly, the lower control limit is found from the largest integer x^- such that

$$P(X \leq x^-) \leq \frac{\alpha}{2} \tag{8-24}$$

Example 8-7   The average number of surface imperfections in painted sheet metal of size $200 \, \text{m}^2$ is 9. Find the probability limits for a type I error of 0.01.

   Solution   From the information given, $c_0 = 9$ and $\alpha = 0.01$. Using Appendix A-2 and incorporating eqs. (8-23) and (8-24), we have

$$P(X \leq 17 \mid c_0 = 9) = 0.995$$
$$P(X \leq 1 \mid c_0 = 9) = 0.001$$

The upper control limit is chosen as 17, and the lower control limit as 1. Note that, strictly speaking, these are not symmetrical probability limits because $P(x \leq 1) = 0.001 < 0.005$, while $P(x > 17) = 0.005$. These are the 0.001 and 0.995 probability limits. The designed type I error is 0.006, which is less than the prescribed level of 0.01.

8-7   CHART FOR NUMBER OF NONCONFORMITIES PER UNIT: u-CHART

A c-chart is used when the sample size is constant. If the area of opportunity changes from one sample to another, the centerline and control limits of a c-chart change as well. For situations in which the sample size varies, a u-chart is used. For companies that inspect all items produced or services rendered for the presence of nonconformities, the output per production run can vary because of fluctuating supplies of labor, machinery, and raw material; consequently, the number inspected per production run changes, thus causing varying sample sizes. When the sample size varies, a u-chart is constructed to monitor the number of nonconformities per unit. Even though the control limits change as the sample size varies, the centerline of a u-chart remains constant, which permits meaningful comparisons between the samples.

Variable Sample Size and No Specified Standard

When the sample size varies, the number of nonconformities per unit for the ith sample is given by

$$u_i = \frac{c_i}{n_i} \tag{8-25}$$

where c_i is the number of nonconformities in the ith sample and n_i is the size of the ith sample. Note that the sample size n_i need not always be an integer. Let's suppose the number of weaving nonconformities is being counted from finished pieces of cloth such that $100\,m^2$ represents 1 unit. If three samples of 250, 100, and $350\,m^2$ are inspected, the corresponding values of the sample size are 2.5, 1, and 3.5 units, respectively.

The average number of nonconformities per unit ($\bar{u}$), which is also the centerline of a u-chart, is given by

$$\bar{u} = \frac{\sum\limits_{i=1}^{g} c_i}{\sum\limits_{i=1}^{g} n_i} \tag{8-26}$$

The control limits are given by

$$\begin{aligned}
\text{UCL}_u &= \bar{u} + 3\sqrt{\frac{\bar{u}}{n_i}} \\
\text{LCL}_u &= \bar{u} - 3\sqrt{\frac{\bar{u}}{n_i}}
\end{aligned} \tag{8-27}$$

It can been seen from eq. (8-27) that the control limits draw closer as the sample size increases. The same behavior is observed for p-charts for variable sample sizes. Thus, the options discussed previously for p-charts with variable sample sizes apply to u-charts as well. Equation (8-27) provides control limits that vary based on each sample size. Note that if $n_i = 1$, all formulas for u-charts equal those for c-charts.

Example 8-8   The number of nonconformities in carpets is determined for 20 samples, but the amount of carpet inspected for each sample varies. Results of the inspection are shown in Table 8-8. Construct a control chart for the number of nonconformities per $100\,m^2$.

TABLE 8-8   Data for Nonconformities in Carpets

Sample, i	Amount Inspected (m^2)	Number of Nonconformities, c_i	Sample, i	Amount Inspected (m^2)	Number of Nonconformities, c_i
1	200	5	11	300	9
2	300	14	12	250	16
3	250	8	13	200	12
4	150	8	14	250	10
5	250	12	15	100	6
6	100	6	16	200	8
7	200	20	17	200	5
8	150	10	18	100	5
9	150	6	19	300	14
10	250	10	20	200	8
					192

TABLE 8-9   Control Limits for Nonconformities per Unit in Carpets

Sample, i	Sample size, n_i	Nonconformities per 100 m^2, u_i	Upper Control Limit	Lower Control Limit
1	2	2.500	9.274	0.092
2	3	4.667	8.431	0.935
3	2.5	3.200	8.789	0.577
4	1.5	5.333	9.984	0
5	2.5	4.800	8.789	0.577
6	1	6.000	11.175	0
7	2	10.000	9.274	0.092
8	1.5	6.667	9.984	0
9	1.5	4.000	9.984	0
10	2.5	4.000	8.789	0.577
11	3	3.000	8.431	0.935
12	2.5	6.400	8.789	0.577
13	2	6.000	9.274	0.092
14	2.5	4.000	8.789	0.577
15	1	6.000	11.175	0
16	2	4.000	9.274	0.092
17	2	2.500	9.274	0.092
18	1	5.000	11.175	0
19	3	4.667	8.431	0.935
20	2	4.000	9.274	0.092
	41			

Solution   With 100 m^2 as a unit, the sample sizes are computed for each sample. Table 8-9 shows the sample sizes and the number of nonconformities per unit for each sample. Equation (8-25) is used to calculate u_i. For example, for the first sample, $u_1 = 5/2 = 2.5$. The centerline $\bar{u}$ is found by using eq. (8-26):

$$\bar{u} = \frac{\sum c_i}{\sum n_i} = \frac{192}{41} = 4.683$$

The control limits are found by using eq. (8-27). Table 8-9 shows the control limits for each sample. For instance, for the first sample, the control limits are found as follows:

$$UCL_u = 4.683 + 3\sqrt{\frac{4.683}{2}} = 9.273$$

$$LCL_u = 4.683 - 3\sqrt{\frac{4.683}{2}} = 0.092$$

The control limits for the other samples are calculated in the same manner. If a lower control limit calculation is a negative value, the limit is converted to zero.

Figure 8-8 shows this control chart for the number of nonconformities per unit. We used Minitab to construct this chart. To do this, we selected **Stat > Control Charts > Attribute Charts > U**, indicated names for the variables that represent the number of nonconformities and the sample size, and then clicked **OK**. We can see that sample 7 plots above the upper control limit. After the special cause has been identified and appropriate corrective action

TABLE 8-10   Attribute Control Chart Applications in the Service Sector

Quality Characteristic	Control Chart
Proportion of income tax returns that have errors	*p*-chart
Proportion of billing errors by a service provider	*p*-chart
Proportion of on-time shipments by a logistics company	*p*-chart
Proportion of payments in Medicare/Medicaid in error	*p*-chart
Proportion of clinical tests performed inaccurately	*p*-chart
Proportion of cases with side effects of medication and treatment	*p*-chart
Number of billing errors per 100 accounts	*c*- or *u*-chart
Number of errors per week in deliveries to patients	*u*-chart
Number of daily customer complaints per occupant in a hotel	*u*-chart

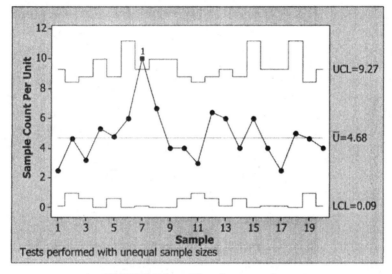

FIGURE 8-8   *u*-Chart for carpet data.

taken, the revised centerline is found by deleting sample 7:

$$\bar{u} = \frac{192 - 20}{39} = 4.410$$

The revised control limits are found using this revised value of $\bar{u}$ in eq. (8-27). The remaining samples are within these revised control limits.

Attribute charts have the potential for numerous applications in the nonmanufacturing sector as well. Table 8-10 shows some examples of application of *p*-, *c*-, and *u*-charts in the service sector.

8-8   CHART FOR DEMERITS PER UNIT: *U*-CHART

The *c*- and *u*-charts treat all types of nonconformities equally, regardless of their degree of severity. Let's suppose that in inspecting computer monitors, we find that one monitor has trouble retaining consistent color and a second monitor has five scratch marks on its surface. Using either a *c*- or *u*-chart, the relative importance of monitor 2's defects, in terms of the number of nonconformities, is five times as great as that of monitor 1. However, the single defect associated with monitor 1 is much more serious than monitor 2's scratch marks. An alternative approach assigns weights to nonconformities according to their relative degree of severity (Besterfield 2003). This quality rating system, which rates demerits per unit and is called the *U*-chart, thus overcomes the deficiency of the *c*- and *u*-charts. These are often helpful in service applications.

Classification of Nonconformities

Several systems classify nonconformities according to their degree of seriousness. A defect that causes severe injury compared to that which may lead to minor problems in the functioning of a product are obviously of different degrees of importance. Using this analogy, defects may be classified into the categories of critical, serious, major, or minor as an example. The definition for each of these categories will be influenced by the product/service and the user who makes a determination of the severity of each type of defect.

Once a classification of defects or nonconformities has been established, demerits per unit are assigned to each class. Control charts are then constructed for demerits per unit. The definitions of the classes are not rigid; users adapt them as they see fit. The classification system mentioned here is but one example. The number of categories and the definitions of each should relate specifically to the problem environment. One organization may have three categories of nonconformities—critical, major, and minor— each with its own definitions. Another organization may define nonconformities as either serious or not serious. The assigned **weights for defects** from each category is user dependent. For example, a weight system of 100, 50, 10, and 1 could be chosen for the categories of very serious, serious, major, and minor, respectively.

Construction of a *U-Chart*

Suppose that we have four categories of nonconformities. In general, the procedure described in this section can be applied to any given number of categories. Let the sample size be n, and let c_1, c_2, c_3, and c_4 denote the total number of nonconformities in a sample for the four categories. Let w_1, w_2, w_3, and w_4 denote the weights assigned to each category. We assume that nonconformities in each category are independent of defects in the other categories. We'll also assume that the occurrence of nonconformities in any category is represented by a Poisson distribution. The applicability of a Poisson distribution to such circumstances is discussed in Section 8-6 (*c*-charts).

For a sample of size n, the total number of demerits is given by

$$D = w_1c_1 + w_2c_2 + w_3c_3 + w_4c_4 \tag{8-28}$$

The demerits per unit for the sample are given by

$$U = \frac{D}{n} = \frac{w_1c_1 + w_2c_2 + w_3c_3 + w_4c_4}{n} \tag{8-29}$$

where the quantity U is a linear combination of independent Poisson random variables. The centerline of the *U*-chart is given by

$$\bar{U} = w_1\bar{u}_1 + w_2\bar{u}_2 + w_3\bar{u}_3 + w_4\bar{u}_4 \tag{8-30}$$

where $\bar{u}_1, \bar{u}_2, \bar{u}_3$, and $\bar{u}_4$ represent the average number of nonconformities per unit in their respective classes. Computation of $\bar{u}_1, \bar{u}_2, \bar{u}_3$, and $\bar{u}_4$ is similar to the calculation of $\bar{u}$ discussed in Section 8-7. If the control limits are based on standard values, those values (say u_{10}, u_{20}, u_{30}, and u_{40}) should be substituted for $\bar{u}$ in each category.

The estimated standard deviation of U is given by

$$\hat{\sigma}_U = \sqrt{\frac{w_1^2 \bar{u}_1 + w_2^2 \bar{u}_2 + w_3^2 \bar{u}_3 + w_4^2 \bar{u}_4}{n}} \tag{8-31}$$

The control limits for the U-chart are given by

$$\begin{aligned} \text{UCL}_U &= \bar{U} + 3\hat{\sigma}_U \\ \text{LCL}_U &= \bar{U} - 3\hat{\sigma}_U \end{aligned} \tag{8-32}$$

If the lower control limit is calculated to be less than zero, it is converted to zero.

Example 8-9 A department store obtains feedback on customer satisfaction regarding a certain product. Twenty random samples, each involving 10 customers, are taken in which customers are asked about the number of serious, major, and minor nonconformities that they have experienced. Clear definitions of each category are provided. The results are shown in Table 8-11. The weights assigned to a serious, major, and minor nonconformity are 50, 10, and 1, respectively. Construct a control chart for the number of demerits per unit.

Solution For each sample, the total number of demerits given by eq. (8-28) is shown in Table 8-11. The table also shows the number of demerits per unit U, given by eq. (8-29). To

TABLE 8-11   Data for Nonconformities in a Department Store Customer Survey

Sample	Serious Nonconformities, c_1	Major Nonconformities, c_2	Minor Nonconformities, c_3	Total Demerits, D	Demerits per Unit, U
1	1	4	2	92	9.2
2	0	3	8	38	3.8
3	0	5	10	60	6.0
4	1	2	5	75	7.5
5	0	6	2	62	6.2
6	0	0	8	8	0.8
7	0	7	5	75	7.5
8	1	1	1	61	6.1
9	1	3	2	82	8.2
10	0	4	12	52	5.2
11	1	5	3	103	10.3
12	2	0	2	102	10.2
13	0	0	9	9	0.9
14	0	6	8	68	6.8
15	1	12	10	180	18.0
16	0	5	7	57	5.7
17	0	1	1	11	1.1
18	1	2	5	75	7.5
19	0	5	6	56	5.6
20	0	3	8	38	3.8
	9	74	114		

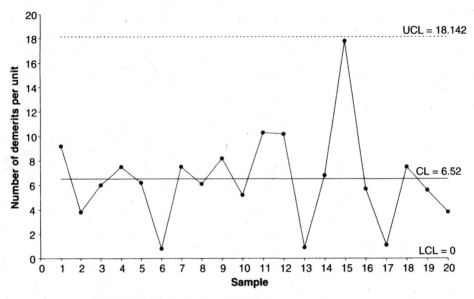

FIGURE 8-9 *U*-chart for department store customer survey.

find the centerline $\bar{U}$, the average number of nonconformities per unit for each defect category is calculated. For "serious" nonconformities

$$\bar{u}_1 = \frac{9}{(20)(10)} = 0.045$$

Similarly,

$$\bar{u}_2 = \frac{74}{200} = 0.37 \quad \text{and} \quad \bar{u}_3 = \frac{114}{200} = 0.57$$

Using eq. (8-30), the centerline of the *U*-chart is

$$\bar{U} = (50)(0.045) + (10)(0.37) + (1)(0.57) = 6.52$$

The estimated standard deviation of *U*, using eq. (8-31), is

$$\hat{\sigma}_U = \sqrt{\frac{(50)^2(0.045) + (10)^2(0.37) + (1)^2(0.57)}{10}} = 3.874$$

Hence, the control limits are

$$\text{UCL}_U = 6.52 + (3)(3.874) = 18.142$$
$$\text{LCL}_U = 6.52 - (3)(3.874) = -5.102 \rightarrow 0$$

Figure 8-9 shows this *U*-chart. Note that all the points plot within the control limits.

8-9 CHARTS FOR HIGHLY CONFORMING PROCESSES

The *p*-chart for proportion nonconforming, discussed previously, was based on the normal distribution as an approximation to the binomial distribution. When *p* is neither too large nor too small and n is large such that $np \geq 5$, the normal distribution serves as an adequate approximation. So, the 3σ limits based on the normal distribution restrict the type I error to about 0.0027. When p is very small, say in the parts-per-million (ppm) range and n is not very large, the normal distribution is not a good approximation to the binomial. Thus, for highly conforming processes, an alternative to the p-chart is necessary. Similarly, for monitoring nonconformities (as in a c- or u-chart) for processes with very low defect rates, an alternative is desirable. Other drawbacks of the traditional p- or u-chart for highly conforming processes include an increased false alarm rate (type I error) and an increased probability of failing to detect a process change (type II error). Further, when the proportion nonconforming is very small, the calculated lower control limit may turn out to be negative, which is then converted to zero. In such cases, an observation cannot fall below the lower control limit. This leads to an inability to detect process improvement, if in fact one does occur. Given the drawbacks of a p-, c-, or u-chart for very good processes, a few alternative approaches are suggested.

Transformation to Normality

Suppose we assume that the occurrence of nonconforming items or nonconformities is modeled by a Poisson distribution with a constant rate of occurrence. It is known that the times between the occurrences of events (in this case, between the occurrence of non-conforming items or nonconformities) are independent and exponentially distributed. The number of conforming items produced between occurrences of nonconforming items will be the variable to be monitored. This variable, which has an exponential distribution, can be transformed into a Weibull distribution, which resembles a normal distribution. Denoting X_i as the number of conforming items produced between successive nonconforming items, it has been shown (Nelson 1994a,b) that the power transformation

$$Y_i = X_i^{1/3.6} = X_i^{0.277} \tag{8-33}$$

yields values of Y that are approximately normally distributed. Hence, an individuals and moving-range chart for the Y-values could be monitored. If we denote the mean of Y by $\bar{Y}$ and the mean of the moving ranges of the Y-values with a window of 2 by $\overline{MR}$, the centerline and control limits on an individuals chart for Y are (for $n = 2$, $d_2 = 1.128$)

$$\begin{aligned} \text{CL} &= \bar{Y} \\ (\text{UCL}, \text{LCL}) &= (\bar{Y} + 2.66\overline{MR}, \ \bar{Y} - 2.66\overline{MR}) \end{aligned} \tag{8-34}$$

Note that values of Y above the upper control limit indicate an improvement in quality, while values of Y below the lower control limit indicate a deterioration in quality.

Use of Exponential Distribution for Continuous Variables

When the process under consideration is continuous, an alternative under the assumption that the occurrence of nonconformities conforms to a Poisson process with mean λ is to monitor the time or number of items required (Q) to observe exactly one nonconformity. It is known that the distribution of Q is exponential with parameter λ (mean $= 1/\lambda$). Hence, the probability of observing Q units in order to observe a defect is

$$F(Q) = 1 - e^{-\lambda Q}, \qquad Q \geq 0 \tag{8-35}$$

Probability limits may be calculated based on a chosen type I error rate, α. Here, because of the exponential distribution not being symmetrical, probability limits rather than 3σ limits for Q are preferred. The centerline and control limits are given as

$$\text{CL} = -\frac{1}{\lambda}\ln\left(\frac{1}{2}\right) = \frac{0.6931}{\lambda} \tag{8-36}$$

$$\text{LCL} = -\frac{1}{\lambda}\ln\left(1 - \frac{\alpha}{2}\right) \tag{8-37}$$

$$\text{UCL} = -\frac{1}{\lambda}\ln\left(\frac{\alpha}{2}\right) \tag{8-38}$$

In monitoring Q, which is either the time or number of items to observing a nonconformity, we can detect process improvement as well as deterioration. When $Q > \text{UCL}$, a likely improvement has taken place. Any time that a nonconformity is detected, the quantity Q is reset to zero, to keep track of the subsequent conforming items prior to a nonconformity being observed. When the process mean is not known, it may be estimated from prior samples as the average of observed Q values.

Use of Geometric Distribution for Discrete Variables

For a highly conforming process, nonconforming items are few and far between. This implies that monitoring the number of nonconforming items will yield most observations with values of zero, which prevents the detection of a change in the nonconformance rate. An alternative is to monitor the number of items until a nonconforming item is found, sometimes referred to as the number of trials up to the first success (X). When a nonconforming item is found, the count starts anew. The discrete variable X, as defined, has a geometric distribution, given by

$$P(X = x) = (1 - p)^{x-1}p, \qquad x = 1, 2, \ldots \tag{8-39}$$

where p represents the probability of success (nonconforming item) on each trial. The mean and variance of X are given by

$$E(X) = \frac{1}{p} \quad \text{and} \quad \text{Var}(X) = \frac{(1-p)}{p^2} \tag{8-40}$$

Probability Limits

Let us assume that the probability of a type I error (α), will be divided equally on both sides of the control limits. The centerline and control limits for the count to a nonconforming item are given by (Xie et al. 2002)

$$\text{CL} = \frac{1}{p}$$
$$\text{UCL} = \frac{\ln(\alpha/2)}{\ln(1-p)} \tag{8-41}$$
$$\text{LCL} = \frac{\ln(1 - \alpha/2)}{\ln(1-p)}$$

These control limits are highly asymmetric, and a logarithmic scale for the vertical axis for monitoring X is recommended. Detection of improvement in the process is usually associated with a value of X falling above the UCL.

A bound can be established for the minimum sample size (n) necessary to detect improvement, if the current level of proportion nonconforming is p. The probability of no nonconforming items in a sample of size n is

$$P(0 \text{ nonconforming items}) = (1 - p)^n \tag{8-42}$$

Assuming that a one-sided limit is being used to detect improvement, with a type I error of α, we have a lower bound for n given by

$$(1 - p)^n < \alpha \quad \text{or} \quad n > \frac{\ln(\alpha)}{\ln(1 - p)} \tag{8-43}$$

Example 8-10 In a microelectronics manufacturing process, the defect sequence and the count of the number of items until a nonconforming item is found are shown in Table 8-12. Based on past data from the process, the process nonconformance rate is believed to be 800 parts per million (ppm). Using a type I error of 0.0027, construct an appropriate control chart and comment if the process is in control.

Solution The centerline and control limits for monitoring the number of items until a nonconforming item is found are calculated using a nonconformance rate of 800 ppm.

$$CL = \frac{1}{0.0008} = 1250$$

$$UCL = \frac{\ln(0.00135)}{\ln(0.9992)} = 8256.26$$

$$LCL = \frac{\ln(0.99865)}{\ln(0.9992)} = 1.69$$

TABLE 8-12 Number of Items Until a Nonconforming Item

Defect Sequence	Number of Items	Defect Sequence	Number of Items	Defect Sequence	Number of Items
1	256	11	919	21	387
2	85	12	851	22	172
3	2013	13	3780	23	1386
4	2080	14	157	24	1927
5	696	15	1112	25	731
6	2548	16	1294	26	884
7	572	17	2078	27	342
8	135	18	555	28	1479
9	656	19	288	29	2640
10	1686	20	2156	30	1845

If a plot is constructed of the observed values (vertical axis on a logarithmic scale), no points are outside the control limits and no discernible patterns are visible. We conclude that the process is in control for an assumed nonconformance rate of 800 ppm.

8-10 OPERATING CHARACTERISTIC CURVES FOR ATTRIBUTE CONTROL CHARTS

An operating characteristic (OC) curve plots the probability of incorrectly concluding that a process is in control as a function of a process parameter. In other words, it is a graph of the probability of a type II error (denoted by β) versus the value of a process parameter.

The choice of process parameter depends on the type of attribute chart under consideration. For a p-chart, the parameter of interest is usually the true process proportion nonconforming (p). An OC curve represents a measure of goodness of a control chart. It can be used to gauge the ability of a chart to detect changes in the process parameter values (Wadsworth et al. 2001). The probability of a change in a process parameter not being detected is related to the probability of a plotted point falling within the control limits. An OC curve is a measure of the sensitivity of a control chart in detecting small changes in process parameters.

The location of the control limits influences the probability of a type I error α, which is the probability of incorrectly concluding that a process is out of control when it is really in control. A type I error is thus a *false alarm*. For a process in control, a measure of goodness of the control chart is a large value of the average run length as discussed in Chapter 6. In this situation, ARL $= 1/\alpha$. So choosing a small value of α and widening the control limits will increase the ARL. For example, if α is 0.05, the ARL is 20. If such a small value of ARL is not acceptable and we reduce α to 0.005 by making the control limits wider, the ARL increases to 200. This implies that with these wider control limits, on average, 1 out of every 200 samples will plot outside the control limits and indicate an out-of-control condition.

For a p-chart, if the process proportion nonconforming is some value p, the probability of a type II error is

$$\beta = P(\hat{p} < \text{UCL}_p \,|\, p) - P(\hat{p} \le \text{LCL}_p \,|\, p)$$
$$= P(X < n\text{UCL} \,|\, p) - P(X \le n\text{LCL} \,|\, p) \tag{8-44}$$

where n is the sample size, $\hat{p}$ is the sample proportion nonconforming, and X represents the number of nonconforming items. Recall that X is a binomial random variable with parameters n and p. The probability values needed for eq. (8-44) are found from the binomial tables in Appendix A-1. Since X has to be an integer, and neither $n\text{UCL}$ nor $n\text{LCL}$ are necessarily integers, an adjustment must be made. Let r_1 and r_2 be defined as follows:

$$r_1 = \lceil n\text{UCL} \rceil$$
$$r_2 = \lfloor n\text{LCL} \rfloor \tag{8-45}$$

where $\lceil n\text{UCL} \rceil$ denotes the largest integer less than or equal to $n\text{UCL}$ and $\lfloor n\text{LCL} \rfloor$ denotes the smallest integer greater than or equal to $n\text{LCL}$. The probability of a type II error may then be expressed as

$$\beta = P(X \le r_1) - P(X \le r_2) \tag{8-46}$$

Example 8-11   Refer to Example 8-1, which deals with the number of nonconforming containers produced by a plastic-injection molding process. The revised control limits for the p-chart are $UCL_p = 0.173$ and $LCL_p = 0$, with the revised centerline at 0.067. The sample size is 50. Construct an OC curve as a function of the process average proportion nonconforming.

Solution   The calculations correspond to a true process proportion nonconforming of 0.10. We assume that a type II error is committed when an observation falls strictly within the control limits. The probability of a type II error found using a binomial distribution is

$$\beta = P[X < (50)(0.173) \,|\, p = 0.10] - P[X \le 50(0) \,|\, p = 0.10]$$
$$= P(X < 8.65 \,|\, p = 0.10) - P(X \le 0 \,|\, p = 0.10)$$
$$= P(X \le 8 \,|\, p = 0.10) - P(X \le 0 \,|\, p = 0.10)$$
$$= \sum_{i=1}^{8} \binom{50}{i} (0.10)^i (0.90)^{50-i} = 0.937$$

If a Poisson approximation to the binomial is used when n is large and p is small (such that np is less than or equal to 5), we have $np = (50)(0.10) = 5$. Next, using the Poisson cumulative probability tables in Appendix A-2, we get

$$\beta = P(X \le 8 \,|\, np = 5) - P(X \le 0 \,|\, np = 5)$$
$$= 0.932 - 0.007 = 0.925$$

We can repeat this approach for other values of p to obtain P (type II error) for each case. Table 8-13 shows the probability of a type II error for several values of p.

A plot of β versus p gives us an OC curve for the p-chart in Example 8-1. This gives an indication of the effectiveness of the p-chart in detecting changes in the process proportion nonconforming. For instance, if the proportion nonconforming changes from the current value of 0.067 to 0.08, there is only a 3.9% probability of detecting this change by the first sample taken after the change. This p-chart is thus not very sensitive to small changes in p. On the other hand, if the value of p changes to 0.15, the chart has a 33.9% chance of detecting this change by the first sample. As the value of p deviates further from its current state, the probability of detection improves. If the value of p changes to 0.28, there is a 93.8% chance of detecting this by the first sample after the change takes place.

This example demonstrates the construction of one segment of the OC curve, the region where p increases. The same procedure can be used to calculate the OC curve for the region where the value of p decreases.

TABLE 8-13   Data for Constructing an OC Curve

Process Proportion Nonconforming, p	$P(X \le 8 \,|\, p)$	$P(X \le 0 \,|\, p)$	P (type II error), β
0.08	0.979	0.018	0.961
0.09	0.960	0.011	0.949
0.10	0.932	0.007	0.925
0.15	0.662	0.001	0.661
0.20	0.333	0.000	0.333
0.28	0.062	0.000	0.062
0.40	0.002	0.000	0.002

TABLE 8-14   Data for the OC Curve for a *c*-Chart for the Foreign Matter Example

c	$P(X \leq 15 \mid c)$	$P(X \leq 0 \mid c)$	P (type II error), β
0.5	1.000	0.607	0.393
1	1.000	0.368	0.632
3	1.000	0.050	0.950
5	1.000	0.007	0.993
7	0.998	0.001	0.997
9	0.978	0.000	0.978
10	0.951	0.000	0.951
12	0.844	0.000	0.844
14	0.669	0.000	0.669
18	0.287	0.000	0.237
20	0.157	0.000	0.157

Operating characteristic curves for the other attribute charts are constructed similarly. Let's consider a chart for the number of nonconformities. If the process average number of nonconformities is c, the probability of a type II error is

$$\beta = P(X < \mathrm{UCL}_c \mid c) - P(X \leq \mathrm{LCL}_c \mid c) \qquad (8\text{-}47)$$

where X represents the number of nonconformities for a process average of c. Incidentally, X is distributed according to a Poisson random variable with mean c. Since the value of X must be an integer and UCL_c and LCL_c need not be integers, we have

$$\beta = P(X \leq \lceil \mathrm{UCL} \rceil \mid c) - P(X \leq \lfloor \mathrm{LCL} \rfloor \mid c) \qquad (8\text{-}48)$$

where $\lceil \mathrm{UCL} \rceil$ represents the largest integer less than or equal to the UCL, and $\lfloor \mathrm{LCL} \rfloor$ represents the smallest integer greater than or equal to the LCL.

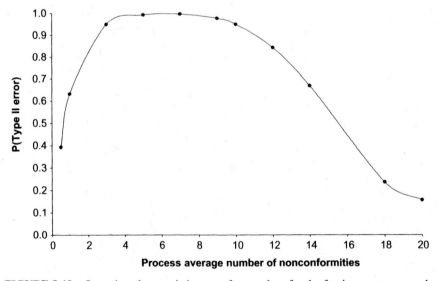

FIGURE 8-10   Operating characteristic curve for a *c*-chart for the foreign matter example.

Example 8-12   Refer to Example 8-6, which deals with the number of occurrences of foreign matter in fabric samples. The revised centerline of the c-chart is 7.208, and the revised control limits are $UCL_c = 15.262$, $LCL_c = 0$. Construct an OC curve for this c-chart.

Solution   We assume that a type II error is committed when an observation falls strictly within the control limits. Table 8-14 gives the probabilities for various values of c. Using eq. (8-48), we get

$$\beta = P(X < 15.262 \,|\, c) - P(X \le 0 \,|\, c)$$
$$= P(X \le 15 \,|\, c) - P(X \le 0 \,|\, c)$$

Using Appendix A-2, we find the probabilities shown in Table 8-14. Figure 8-10 shows the OC curve for this c-chart. The same line of reasoning used with OC curves for p-charts can be used to make inferences from the OC curve in Figure 8-10.

SUMMARY

This chapter has introduced a variety of control charts for attributes. Although attribute charts do not provide as much information as variable charts for the same sample size, they have certain advantages. They are a good tool for summarizing information and for providing data at the aggregate level. Attribute charts are useful when starting a quality control program. They provide guidance as to where variable charts can eventually be used.

Three main categories of attribute charts have been discussed in this chapter. The first deals with products or services that are nonconforming. The p-chart for the proportion nonconforming and the np-chart for the number nonconforming are in this category. The second category involves the number of nonconformities, or defects, and includes the c-chart for the number of nonconformities and the u-chart for the number of nonconformities per unit. For highly conforming processes, a p-chart or c-chart may not be appropriate since the majority of the samples will have no nonconforming items or no nonconformities. In this context, a chart for the number of items until a nonconforming item is found is introduced. The third category involves a weighting scheme for classifications of nonconformities based on their severity. The U-chart for demerits per unit was discussed in this section.

When making inferences from any control chart, there is always the risk of incorrectly declaring a process to be out of control (a type I error) or incorrectly declaring a process to be in control (a type II error). For a process in control, a measure of goodness of a control chart is the probability of a type I error or, implicitly, the average run length, which is the reciprocal of the probability of a type I error. Another measure of goodness of a control chart's performance is its operating characteristic curve, which shows the probability of a type II error as a function of the value of a process parameter such as the process proportion nonconforming or the process average number of nonconformities.

KEY TERMS

attribute	control chart
average sample size	attributes
binomial distribution	demerits per unit (U-chart)

highly conforming processes	defective
number of nonconformities (*c*-chart)	demerits
number of nonconformities	exponential distribution
per unit (*u*-chart)	geometric distribution
number of nonconforming items	nonconformance classification
(*np*-chart)	nonconforming item
proportion nonconforming (*p*-chart)	nonconformity
standardized *p*-chart	operating characteristic curve
variables	Poisson distribution
defect	probability limits
major	standard or target
minor	weights for nonconformities
serious	
very serious	

EXERCISES

Discussion Questions

8-1   Distinguish between a nonconformity and a nonconforming item. Give examples of each in the following contexts:
(a) Financial institution
(b) Hospital
(c) Microelectronics manufacturing
(d) Law firm
(e) Nonprofit organization

8-2   What are the advantages and disadvantages of control charts for attributes over those for variables?

8-3   Discuss the significance of an appropriate sample size for a proportion-nonconforming chart.

8-4   The CEO of a company has been charged with reducing the proportion nonconforming of the product output. Discuss which control charts should be used and where they should be placed.

8-5   How does changing the sample size affect the centerline and the control limits of a *p*-chart?

8-6   What are the advantages and disadvantages of the standardized *p*-chart as compared to the regular proportion-nonconforming chart?

8-7   Discuss the assumptions that must be satisfied to justify using a *p*-chart. How are they different from the assumptions required for a *c*-chart?

8-8   Is it possible for a process to be in control and still not meet some desirable standards for the proportion nonconforming? How would one detect such a condition, and what remedial actions would one take?

8-9   Discuss the role of the customer in influencing the proportion-nonconforming chart. How would the customer be integrated into a total quality systems approach?

8-10   Discuss the impact of the control limits on the average run length and the operating characteristic curve.

8-11   Explain the conditions under which a u-chart would be used instead of a c-chart.

8-12   Explain why a p- or c-chart is not appropriate for highly conforming processes.

8-13   Distinguish between 3σ limits and probability limits. When would you consider constructing probability limits?

8-14   Meeting customer due dates is an important goal. What attribute or variables control charts would you select to monitor? Discuss the underlying assumptions in each case.

8-15   Explain the setting under which a U-chart would be used. How does the U-chart incorporate the user's perception of the relative degree of severity of the different categories of defects?

8-16   Which type of control chart (p-, np-, c-, u-, U-, or charts for highly conforming processes) is most appropriate to monitor the following situations?
    (a) Number of potholes in highways
    (b) Proportion of customers who are satisfied with the operation of the local housing authority
    (c) Satisfaction of customers at a restaurant where customers consider the quality of food and attitude of the server to be more important than the décor
    (d) Number of errors in account transactions at a bank where the number of accounts varies from month to month
    (e) Number of automobile accident claims filed per month at the insurance dealer, assuming a stable number of insured persons
    (f) Responsiveness to customer needs of a local library where customers value the longer hours over short lines to check out books
    (g) Needlestick in patients in a hospital
    (h) Loan defaults in a financial institution in a year
    (i) Number of surgical errors in a health care facility
    (j) Number of thefts in a city over a period of time
    (k) Proportion of people that favor the construction of condominiums in a certain neighborhood
    (l) Number of weld defects in the construction of aircraft
    (m) Number of seeds that germinate from a large pack of seeds sold by a nursery when the number of seeds in a packet may vary
    (n) Performance of the braking mechanism of a certain model of automobile (the car is expected to stop within a certain distance)
    (o) Number of firemen to have on duty to provide an acceptable level of service (each fire usually requires four firemen; the department has data on the number of calls they receive during the specified period of time)
    (p) Performance of data-entry operators (if one or more input errors are made, the file is rendered useless and a revision is necessary)
    (q) Control of the number of typographical errors per page at a document preparation center (data are chosen randomly and obtained daily on the number of errors in 30 pages)

(r) Control of the wiring and transistor defects in an electronic component (wiring defects are considered more serious)

(s) Number of traffic accidents in a city per month

(t) Proportion of successful patent applications by a drug manufacturer

Problems

8-17 Every employee in a check-processing department goes through a four-month training period, after which the employee is responsible for their operation. The work of one employee who has been on the job for eight months is being studied. Table 8-15 shows the number of errors and the number of items sampled over a period of two months. The first 16 samples were each chosen from 400 items, and the remaining 9 samples were each chosen from 300 items. Determine whether the employee's performance can be judged stable. Comment on the capability of the employee.

TABLE 8-15

Sample	Errors	Items Sampled	Sample	Errors	Items Sampled
1	12	400	14	18	400
2	9	400	15	8	400
3	13	400	16	6	400
4	7	400	17	4	300
5	6	400	18	6	300
6	10	400	19	5	300
7	14	400	20	8	300
8	7	400	21	10	300
9	5	400	22	7	300
10	6	400	23	4	300
11	4	400	24	5	300
12	9	400	25	3	300
13	11	400			

8-18 The number of customers who are not satisfied with the service provided in a retail store is found for 20 samples of size 100 and is shown in Table 8-16. Construct a

TABLE 8-16

Sample	Number of Dissatisfied Customers	Sample	Number of Dissatisfied Customers
1	2	11	5
2	5	12	4
3	4	13	2
4	3	14	5
5	4	15	3
6	2	16	12
7	3	17	3
8	2	18	2
9	4	19	5
10	11	20	2

control chart for the proportion of dissatisfied customers. Revise the control limits, assuming special causes for points outside the control limits.

8-19   Refer to Exercise 8-18. Management believes that the dissatisfaction rate is 3%, so establish control limits based on this value. Comment on the ability of the store to meet this standard. If management were to set the standard at 2%, can the store meet this goal? What actions would you recommend?   100 × 20

8-20   Health care facilities must conform to certain standards in submitting bills to Medicare/Medicaid for processing. The number of bills with errors and the number sampled are shown in Table 8-17. Construct an appropriate control chart and comment on the performance of the billing department. Revise the control limits, if necessary, assuming special causes for out-of-control points. Comment on the capability of the department.

TABLE 8-17

Observation	Bills with Errors	Number Sampled	Observation	Bills with Errors	Number Sampled
1	8	400	14	3	300
2	6	400	15	5	300
3	4	400	16	8	300
4	9	400	17	11	500
5	7	400	18	13	500
6	5	400	19	8	500
7	5	300	20	7	500
8	7	300	21	8	500
9	4	300	22	4	500
10	15	300	23	3	500
11	6	300	24	7	500
12	7	300	25	6	500
13	4	300			

8-21   Observations are taken from the output of a company making semiconductors. Table 8-18 shows the sample size and the number of nonconforming semiconductors for each sample. Construct a p-chart by setting up the exact control limits for each sample. Are any samples out of control? If so, assuming special causes, revise the centerline and control limits.

8-22   Refer to Exercise 8-21 and the data shown in Table 8-18. Construct a standardized p-chart and discuss your conclusions.

8-23   The quality of service in a hospital is tracked by determining the proportion of medication errors; this is done by dividing the number of medication errors by 1000 patient-days for each observation. The results of 25 such samples (in percentage of medication errors) are shown in Table 8-19. Construct an appropriate control chart, and comment on the quality level. Is a goal of error-free performance reasonable to expect from this system?

TABLE 8-18

Observation	Items Inspected	Nonconforming Items	Observation	Items Inspected	Nonconforming Items
1	80	3	14	90	4
2	120	6	15	160	5
3	60	4	16	230	3
4	150	5	17	200	12
5	140	8	18	150	8
6	150	10	19	210	6
7	160	7	20	190	4
8	90	6	21	160	9
9	100	5	22	100	8
10	160	12	23	100	12
11	110	8	24	90	7
12	100	5	25	160	10
13	200	14			

TABLE 8-19

Sample	Medication Errors (%)	Sample	Medication Errors (%)
1	2.6	14	2.0
2	1.9	15	4.2
3	2.8	16	2.2
4	2.9	17	1.8
5	2.4	18	2.4
6	1.8	19	2.3
7	2.3	20	1.6
8	2.1	21	1.9
9	1.4	22	2.0
10	1.7	23	2.2
11	2.2	24	2.1
12	2.0	25	2.3
13	1.2		

8-24 A health care facility is interested in monitoring the primary C-section rate. Monthly data on the number of primary C-sections collected over the last two and a half years is shown in Table 8-20.

(a) Is the process in control?

(b) There is pressure to make these data public. Can we conclude that the C-section rates had shifted to a higher level in the last six months relative to the previous six months?

(c) What is your prediction on the C-section rate if no changes are made in current obstetrics practices?

(d) Based on benchmarking with comparable facilities in similar metropolitan areas, is it feasible currently to achieve a C-section rate of 10%?

TABLE 8-20

Month	C-Sections	Deliveries	Month	C-Sections	Deliveries
1	54	350	16	70	449
2	50	384	17	62	366
3	63	415	18	65	405
4	69	422	19	52	386
5	55	395	20	55	392
6	63	412	21	61	408
7	67	407	22	66	442
8	67	415	23	53	426
9	51	377	24	47	385
10	62	404	25	79	413
11	64	377	26	60	388
12	62	382	27	69	411
13	62	425	28	61	378
14	55	410	29	70	392
15	68	426	30	65	420

8-25   Refer to Exercise 8-18 and the data shown in Table 8-16. Construct a control chart for the number of dissatisfied customers. Revise the chart, assuming special causes for points outside the control limits.

8-26   The number of processing errors per 100 purchase orders is monitored by a company with the objective of eliminating such errors totally. Table 8-21 shows samples that were selected randomly from all purchase orders. The company is in the process of testing the effects of a new purchase order form that it has designed. The last five samples were made using the new form. Construct a control chart that the company can use for monitoring the quality characteristic selected. What is the effect of the newly designed purchase order form? Is the company capable of achieving the desired goal?

TABLE 8-21

Sample	Processing Errors	Sample	Processing Errors
1	6	14	3
2	4	15	6
3	2	16	1
4	3	17	5
5	4	18	2
6	7	19	6
7	5	20	4
8	7	21	2
9	11	22	3
10	4	23	2
11	2	24	1
12	5	25	2
13	4		

8-27   The number of dietary errors is found from a random sample of 100 trays chosen on a daily basis in a health care facility. The data for 25 such samples are shown in Table 8-22.
(a) Construct an appropriate control chart and comment on the process.
(b) How many dietary errors do you predict if no changes are made in the process?
(c) Is the system capable of reducing dietary errors to 2, on average, per 100 trays, if no changes are made in the process?

TABLE 8-22

Sample Number	Number of Dietary Errors	Sample Number	Number of Dietary Errors
1	9	14	8
2	6	15	8
3	4	16	7
4	7	17	6
5	5	18	4
6	6	19	12
7	16	20	7
8	8	21	6
9	7	22	8
10	9	23	6
11	3	24	8
12	6	25	5
13	10		

8-28   A building contractor subcontracts to a local merchant a job involving hanging wallpaper. To have an idea of the quality level of the merchant's work, the contractor randomly selects $300\,\text{m}^2$ and counts the number of blemishes. The total number of blemishes for 30 samples is 80. Construct the centerline and control limits for an appropriate chart. Is it reasonable for the contractor to set a goal of an average of 0.5 blemish per $100\,\text{m}^2$?

8-29   The number of imperfections in bond paper produced by a paper mill is observed over a period of several days. Table 8-23 shows the area inspected and the number of imperfections for 25 samples. Construct a control chart for the number of imperfections per square meter. Revise the limits if necessary, assuming special causes for the out-of-control points.

8-30   Refer to Exercise 8-29. If we want to control the number of imperfections per $100\,\text{m}^2$, how would this affect the control chart? What would the control limits be? In terms of decision making, would there be a difference between this problem and Exercise 8-29, depending on which chart is constructed? What conclusions can you draw from this?

8-31   The director of the pharmacy department is interested in benchmarking the level of operations in the unit. The director has defined *medication errors* as being any one of the following: wrong medication; wrong dose; administered to the wrong patient; administered at the wrong time; incorrectly repeating the medication; or omitting the medication. The number of orders filled per day by the pharmacy varies. Table 8-24 shows the number of orders filled and the number of medication errors for 25 days. Construct an appropriate control chart and comment on the stability of the process.

TABLE 8-23

Sample	Area Inspected (m^2)	Imperfections	Sample	Area Inspected (m^2)	Imperfections
1	150	6	14	300	8
2	100	8	15	300	12
3	200	5	16	200	6
4	150	4	17	150	4
5	250	10	18	200	7
6	100	11	19	150	14
7	150	3	20	100	4
8	200	5	21	100	8
9	300	10	22	200	9
10	250	10	23	300	12
11	100	5	24	250	7
12	200	4	25	200	5
13	250	12			

TABLE 8-24

Sample Number	Number of Orders Filled	Number of Medication Errors	Sample Number	Number of Orders Filled	Number of Medication Errors
1	1200	11	14	1200	16
2	1160	10	15	1150	14
3	1210	12	16	1100	23
4	1300	9	17	1160	14
5	1120	10	18	1300	16
6	1150	12	19	1100	10
7	1100	14	20	1180	12
8	1320	12	21	1220	14
9	1240	10	22	1240	13
10	1180	15	23	1120	16
11	1140	4	24	1150	13
12	1120	13	25	1180	12
13	1220	7			

8-32   Nonconformities in automobiles fall into three categories: serious, major, and minor. Twenty-five samples of five automobiles are chosen, and the total number of nonconformities in each category is reported. Table 8-25 shows the results. Assuming a weighing system of 50, 10, and 1 for serious, major, and minor nonconformities, respectively, construct a demerits per unit control chart. Revise the control limits if necessary, assuming special causes for points that are out of control.

8-33   Refer to Exercise 8-32. If the weighing systems were different (i.e., 10, 5, and 1), how would the centerline and the control limits change for the U-chart? Discuss changes, if any, in the inferences made about the process.

8-34   The Joint Commission on Accreditation of Healthcare Organizations (JCAHO) requires an accounting of significant medication errors. Data collected over the last 25 months, shown in Table 8-26, indicate the number of orders filled and the number

TABLE 8-25

Sample	Serious Defects	Major Defects	Minor Defects	Sample	Serious Defects	Major Defects	Minor Defects
1	0	5	8	14	0	7	12
2	0	3	2	15	0	2	8
3	1	0	6	16	0	4	3
4	1	2	1	17	1	0	5
5	0	6	8	18	0	3	2
6	0	3	3	19	0	5	8
7	0	1	10	20	0	2	6
8	1	2	5	21	1	1	4
9	0	4	9	22	0	3	10
10	2	6	6	23	0	2	12
11	1	3	2	24	0	4	7
12	0	5	8	25	0	2	4
13	0	0	9				

TABLE 8-26

Sample	Orders Filled	Significant Errors	Sample	Orders Filled	Significant Errors
1	80	3	14	90	4
2	120	6	15	160	5
3	60	4	16	230	3
4	150	5	17	200	12
5	140	8	18	150	8
6	150	10	19	210	6
7	160	7	20	190	4
8	90	6	21	160	9
9	100	5	22	100	8
10	160	12	23	100	12
11	110	8	24	90	7
12	100	5	25	160	10
13	200	14			

of significant medication errors. Each order is classified either as having or as not having significant medication errors.

(a) Construct an appropriate control chart and comment on the process.

(b) If the process is not in control, assuming special causes and appropriate remedial actions, revise the centerline and control limits.

(c) What is your level of expectation from this process?

(d) What would you do if the goal is to reduce the proportion of significant medication errors to 1%. Is this currently achievable?

8-35   Refer to Exercise 8-18. Construct an OC curve for the p-chart. If the process proportion of dissatisfied customers were to rise to 7%, what is the probability of not detecting this shift on the first sample drawn after the change has taken place? What is the probability of detecting the shift by the third sample?

8-36   Refer to Exercise 8-27. Construct an OC curve for the c-chart. If the process average number of dietary errors per 100 trays increases to 10, what is the probability of detecting this on the first sample drawn after the change?

8-37   Refer to Exercise 8-36. Set up 2σ control limits. What is the probability of detecting a change in the process average number of dietary errors per 100 trays to 8 on the first sample drawn after the change? Explain under what conditions you would prefer to have these 2σ control limits over the traditional 3σ limits.

8-38   The number of heart surgery complications is rare. To monitor the effectiveness of such surgeries, data are recorded on the number of such procedures until a complication occurs. These complications occur independently with a constant probability of occurrence and follow a geometric distribution. Table 8-27 shows such data for a sequence of 25 complications. It is estimated from past data that the complication rate is 0.1%. Construct an appropriate control chart and comment on the process assuming a type I error rate of 0.005. What would the control limits be for a type I error rate of 0.05? What factors would influence your selection of the type I error rate?

TABLE 8-27

Complication Sequence	Number of Procedures	Complication Sequence	Number of Procedures	Complication Sequence	Number of Procedures
1	654	10	1654	18	1794
2	981	11	892	19	1112
3	1508	12	750	20	652
4	436	13	1333	21	1050
5	1202	14	1404	22	1085
6	889	15	909	23	1422
7	1854	16	822	24	688
8	3068	17	1609	25	1095
9	704				

8-39   Consider Exercise 8-38. If you were interested in detecting an improvement in the process using a one-sided limit, what should the minimum sample size be for an α of 0.005? What should it be for an α of 0.05? What conclusions can you draw from these results?

8-40   Consider Exercise 8-38 under the assumption that the complication rate is 0.1%. If you were to construct a p-chart using two-sided 3σ limits, what would the minimum sample size be to detect an improvement in the process?

8-41   Consider Exercise 8-38. Determine the sensitivity of the control limits on the complication rate, using values of 0.2% and 0.5%.

8-42   Consider Exercise 8-38 under the assumption that the complication rate is 0.1% and a type I error of 0.005. If you reduced the upper control limit to half of its previous value, what type of process complication rates, on average, will this new limit be able to detect improvements?

TABLE 8-28

Sample	Number Restrained	Number of Items Not Checked	Sample	Number Restrained	Number of Items Not Checked
1	100	4	14	40	3
2	50	3	15	80	2
3	80	8	16	80	4
4	60	4	17	120	6
5	120	14	18	100	8
6	100	8	19	120	10
7	80	4	20	80	5
8	100	5	21	60	8
9	60	2	22	100	4
10	80	12	23	100	5
11	100	7	24	80	4
12	120	5	25	120	5
13	60	15			

8-43   Consider Exercise 8-38. However, now assume that the interval between complications follows an exponential distribution. Construct an appropriate control chart and comment on the process assuming a type I error rate of 0.005.

8-44   JCAHO has standards pertaining to patient restraint use. A checklist has been developed that is to be used each time a restraint is used. The checklist contains five items, all of which should be checked. Table 8-28 shows data collected for 25 months that indicate the number of patients restrained and the number of items that were not checked prior to restraint use.
   (a) Is the process stable?
   (b) If not, assuming identifiable remedial actions with the special causes, revise the centerline and control limits.
   (c) What is your expectation from this process?

REFERENCES

ASQ (1994). *Quality Systems Terminology, ANSI/ASQ Standard A8402*, Milwaukee, WI: American Society for Quality.

Besterfield, D. H. (2003). *Quality Control*, 7th ed. Upper Saddle Rider, NJ: Prentice Hall.

Minitab, Inc. (2007). Release 15, State College, PA: Minitab.

Montgomery, D. C. (2004). *Introduction to Statistical Quality Control*, 5th ed. Hoboken, NJ: Wiley.

Nelson, L. S. (1994a). "Shewhart Control Charts with Unequal Subgroup Sizes". *Journal of Quality Technology*, 26(1): 64–67.

——— (1994b). "A Control Chart for Parts-per-Million Nonconforming Items". *Journal of Quality Technology*, 26(3): 239–240.

Wadsworth, H. M., K. S. Stephens and A. B. Godfrey (2001). *Modern Methods for Quality Control and Improvement*, 2nd ed. New York: Wiley.

Xie, M., T. N. Goh and V. Kuralmani (2002). *Statistical Models and Control Charts for High Quality Process*. Boston, MA: Kluwer.

9

PROCESS CAPABILITY ANALYSIS

9-1 INTRODUCTION AND CHAPTER OBJECTIVES

In Chapters 6, 7, and 8 we discussed various methods of monitoring a process using control charts. In this chapter we analyze whether a product or service meets the specifications required by the customer. We define measures that indicate the ability of the process to meet specifications; these are, in some sense, measures of process performance. This chapter also deals with determining tolerances on assemblies when those of individual components are known, and vice versa. Note that process capability analysis should be conducted only when a process is in a state of statistical control.

The objectives of this chapter are to present some of the commonly used process capability measures, demonstrate procedures for their computation, interpret them, and discuss any associated assumptions. While many of the measures are interpreted in the context of normality of the distribution of the characteristic, methods are presented for dealing with nonnormality. For discrete variables satisfying the binomial or Poisson distribution, capability measures are also discussed. Since observed values of the quality characteristic are influenced by the measuring instrument, measures of precision of the instrument as well as the impact of various operators who use the instrument are also of interest, and appropriate measures are presented.

Fundamentals of Quality Control and Improvement, Third Edition, By Amitava Mitra
Copyright © 2008 John Wiley & Sons, Inc.

9-2   SPECIFICATION LIMITS AND CONTROL LIMITS

The terms **specifications limits** and **tolerance limits** are often used interchangeably and are defined as the acceptable bounds on quality characteristics. Such bounds could be based on meeting customer needs or functional requirements of the product/service.

Tolerance limits are generally preferred in evaluating manufacturing or service requirements, whereas specification limits are more appropriate for categorizing materials, products, or services in terms of their stated requirements.

For example, one specification for a building crane is a hoist load of 5000 ± 300 kg. To satisfy this criterion, the diameter of the steel cable has to be 4 ± 0.2 cm. This manufacturing requirement for the cable diameter can also be viewed as a tolerance. In general, tolerances are a subset of specifications. Usually, tolerances pertain to physical requirements (such as length, diameter, thickness, etc.), whereas specifications include all requirements.

Tolerance limits can be two-sided (with upper and lower limits) or one-sided with either upper or lower limits. A **lower tolerance limit** defines the lower conformance boundary for an individual unit of manufacturing or service operation; an **upper tolerance limit** defines the upper conformance boundary. For example, a hub diameter requirement of 4 ± 0.1 mm is a two-sided tolerance limit. A cable tensile strength requirement of $(500 - 20)$ kg is a one-sided limit for which the lower tolerance limit is 480 kg.

Specification limits are determined by the needs of the customer. What the customer wants in a product or service is analyzed through market research and incorporated through product or service design. These limits are placed on a product characteristic by designers and engineers to ensure adequate functioning of the product.

As discussed in previous chapters, control limits identify the variation that exists between samples, or subgroups, of measurements. They do not apply to individual units except in the case of control charts for individual measurements. Control limits reflect the variability of the process and have no relationship to specification limits, which are chosen to meet customer needs for the product or service.

To clarify the difference between specification limits and control limits, let's consider a situation in which the process is in control but some of the product does not meet specifications. We will assume that the distribution of the quality characteristic X is normal. We have control charts for the mean $\bar{X}$ and range R, from which we have found the process to be in control. Figure 9-1a shows the $\bar{X}$-chart control limits, which are 3 standard deviations of the sample mean ($3\sigma_{\bar{X}}$) from the process average. From the central limit theorem we know that the standard deviation of the individual items, σ, is related to the standard deviation of the sample mean by $\sigma_{\bar{X}} = \sigma/\sqrt{n}$, where n is the sample size; thus, $\sigma = \sqrt{n}\,\sigma_{\bar{X}}$.

Figure 9-1b shows the distribution of X (assuming a normal distribution) of individual items. This figure indicates the expected degree of variation of individual items if the process is in control. The **process spread** (i.e., the distance expected between the maximum and minimum values of the characteristic, assuming normality) is 6σ. The normality assumption implies that about 99.74% of the individual items will lie in a range of $\pm 3\sigma$ about the mean, a value that we assume constitutes practically all items produced by the process. If we superimpose the upper and lower specification limits (USL and LSL) on Figure 9-1b, we see that some items lie outside the specifications. The control limits are influenced by the process spread but are unrelated to the specification limits.

Because of the conceptual difference between control limits and specification limits, the latter should not be superimposed on control charts for the average $\bar{X}$. Remember that the control limits on an $\bar{X}$-chart are a measure of variability of the sample means. Specification limits represent the acceptable bounds of variability for individual items.

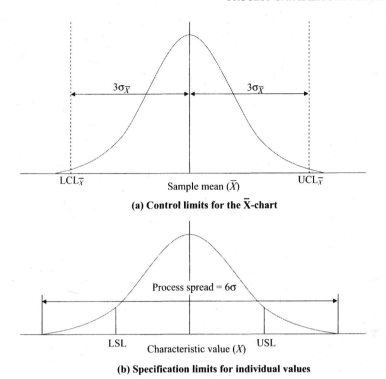

(a) Control limits for the $\overline{X}$-chart

(b) Specification limits for individual values

FIGURE 9-1   Difference between control limits and specification limits.

9-3   PROCESS CAPABILITY ANALYSIS

The determination of process capability begins only after the process has been brought to a state of statistical control. A process is said to be in *statistical control* when the only sources of variation in the system are common causes. Details as to how this state is achieved are discussed in Chapters 6, 7, and 8. Identifying special causes is the first step toward achieving this objective. Taking corrective action to eliminate special causes gets us to a process in statistical control.

Process Capability

Process capability represents the performance of a process in a state of statistical control. It is determined by the total variability that exists because of all common causes present in the system. As we've discussed previously, common causes are inherent to a system—they always exist. Thus, process capability can also be viewed as the variation in the product quality characteristic that remains after all special causes have been removed. The product's performance is then predictable because the special causes are gone. This allows us to determine the ability of the product to meet customer expectations.

A common measure of process capability is given by 6σ, which is also called the *process spread* (see Figure 9-2). We assume that the distance of 6σ encompasses virtually all values of the output quality characteristic. If a normal distribution for the output quality characteristic can be assumed, 99.74% of the distribution will lie within the bounds of $\pm 3\sigma$ on either side of the mean.

Process Capability Analysis   **Process capability analysis** estimates process capability. It involves estimating the process mean and standard deviation of the quality characteristic.

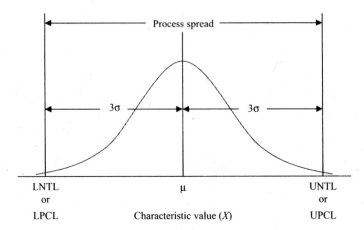

FIGURE 9-2   Natural tolerance limits and process spread.

Additionally, the form of the relative frequency distribution of the characteristic of interest is estimated. If specification limits are known, a process capability analysis will also estimate the proportion of nonconforming product.

Frequently, a process capability study involves observing a quality characteristic of the product. For example, the mean and standard deviation of the bore size of a component are found to be 12 and 0.1 cm, respectively. Since this information usually pertains to the product rather than the process, this analysis should, strictly speaking, be called a *product* analysis study. A true process capability study in this context would involve collecting and analyzing data that relate to process parameters. These parameters could be the depth of cut, the type of tool used, the tool material, the type of jigs or fixtures, or the rate of feed of the tool. The objective is to find the relationship between the process parameters (or process settings) and the product characteristics. Any problems with product characteristics can then be related to process parameters, and remedial actions can be identified on a timely basis.

Benefits of Process Capability Analysis   An ongoing quality improvement program requires continual estimation of the process parameters. Monitoring these parameters will ensure the best performance that the process is capable of achieving. These may include centering the process average on the target value and determining whether the process spread can be reduced by buying new equipment or procuring higher-quality raw materials. Analyzing the flexibility of the specifications by identifying customer needs is another possibility. Here are some benefits of process capability analysis:

1. *Uniformity of output.*   By conducting process capability studies and making necessary adjustments in the process parameters, variability is closely controlled. Any undersirable shape in the distribution of the quality characteristic is evaluated, and feasible changes in the process parameters are made early on.

2. *Maintained or improved quality.*   This is consistent with the goal of quality improvement in an ongoing cycle. Process capability analysis indicates whether new equipment is necessary. As these changes occur, the new capability can be determined.

3. *Product and process design facilitated.*   Information obtained from process capability analysis provides vital feedback for design. This is essential because product

designers must be aware of inherent variation. Designing product tolerances that the process is not capable of achieving makes for longer lead times in the design.

4. *Assistance in vendor selection and control.*   Companies can require their vendors to report process capability information to guide them in choosing vendors. Moreover, for vendors already selected, regular reporting of process capability information is an effective way to control and improve quality at the vendor's premises.

5. *Reduction in total cost.*   This occurs because internal and external failure costs are lowered. By constantly monitoring process parameters, fewer nonconforming items are produced.

9-4 NATURAL TOLERANCE LIMITS

Natural tolerance limits, also known as **process capability limits**, are established or influenced by the process itself. They represent the inherent variation in the quality characteristic of the individual items produced by a process in control. They are estimated based on the population of values or, more typically, from large representative samples.

Assuming a normal distribution of the quality characteristic X, the *upper natural tolerance limit* (UNTL), or the *upper process capability limit* (UPCL), is 3 standard deviations above the process mean; the *lower natural tolerance limit* (LNTL), or the *lower process capability limit* (LPCL), is 3 standard deviations below the process men. Thus, we have

$$
\begin{aligned}
\text{UNTL} &= \mu + 3\sigma \\
\text{LNTL} &= \mu - 3\sigma
\end{aligned}
\tag{9-1}
$$

where μ represents the process mean and σ represents the process standard deviation, which is the standard deviation of the individual items.

Figure 9-2 shows the upper and lower natural tolerance limits, or process capability limits. The assumption of a normal distribution of characteristic X implies that nearly all (approximately 99.74%) the items produced will have a value within the bounds of the natural tolerance limits. Thus, for all practical purposes, these limits indicate the degree of inherent variation that exists in the process.

If the distribution were nonnormal, we would first determine whether it conforms to any well-known distribution. In Chapter 4 we discuss some common distributions. Then using the type of distribution that models the characteristic, the natural tolerance limits would be found such that nearly the entire distribution is contained within those limits.

The population values for the process mean μ and standard deviation σ needed in eq. (9-1) are usually estimated from samples. Hence, the sample mean $\bar{X}$ and the sample standard deviation s are often used to obtain the natural tolerance limits. Estimates of the process mean and standard deviation can also be obtained from control chart information.

Example 9-1   The diameter of a part has to fit an assembly. The specifications for the diameter are 5 ± 0.015 cm. The samples taken from the process in control yield a sample mean $\bar{X}$ of 4.99 cm and a sample standard deviation s of 0.004 cm. Find the natural tolerance limits of the process. Would you consider adjusting the process center?

Solution The upper and lower natural tolerance limits based on the sample estimates are found using eq. (9-1):

$$\text{UNTL} = 4.99 + (3)(0.004) = 5.002$$
$$\text{LNTL} = 4.99 - (3)(0.004) = 4.978$$

Assuming a normal distribution of diameters, the process spread is $(6)(0.004) = 0.024$ cm, which is the difference between the natural tolerance limits. For the current process, we would expect the diameters to lie between 5.002 and 4.978 cm.

The difference between the specification limits is 0.03 cm. If the process were left in its original state, some proportion of the parts would fall below the lower specification limit of 4.985 cm. To calculate this proportion, the standardized normal value at the LSL must first be found as follows:

$$z = \frac{\text{LSL} - \bar{X}}{s} = \frac{4.985 - 4.99}{0.004} = -1.25$$

This Z-value is found in Appendix A-3, the proportion below the LSL is 0.1056. Thus, it would be desirable to adjust the process center to the target value of 5 cm. If this is done, since the process spread is 0.024 cm and the difference between the specification limits is 0.03 cm, virtually all parts would fall between the specification limits, and we would have a capable process.

Statistical Tolerance Limits

Statistical tolerance limits are the limits of an interval that (with a given level of confidence γ) contains at least a specified proportion $(1 - \alpha)$ of the population. These limits are found from sampling information. For example, if we conclude, using a level of confidence of 0.98 and samples of size 10, that 95% of the part lengths fall between 30 and 35 mm, the statistical tolerance limits are 30 and 35 mm. Sample estimates are used to infer population parameters; the limits are influenced by the sample size. As the sample size becomes large, the statistical tolerance limits approach the values found using the population parameters. Estimation of statistical tolerance limits is discussed in Section 9-10. Estimation of these limits is based on a normal distribution and on nonparametric methods.

9-5 SPECIFICATIONS AND PROCESS CAPABILITY

Technically, there might not be any mathematical relationship between the process capability limits (or the natural tolerance limits) and the specification limits. The former are determined by the condition of the process and its inherent variability; the latter are influenced by the needs of the customer. There is, however, a desired relationship between these two sets of limits. The specification limits are preferably outside the natural tolerance limits, in which case most of the units produced will be acceptable. There are three cases that arise regarding this relationship. In our discussion of these cases, we assume a normal distribution of the quality characteristic of interest.

Case I: Process Spread Less Than Specification Spread If the process spread is less than the difference between the specification limits, the process is quite capable. This case, shown in Figure 9-3a, represents the preferred situation. If the process mean μ is at the target value (assumed to be midway between the specification limits), all items produced are well within specifications. In fact, there is some flexibility for the process to go out of

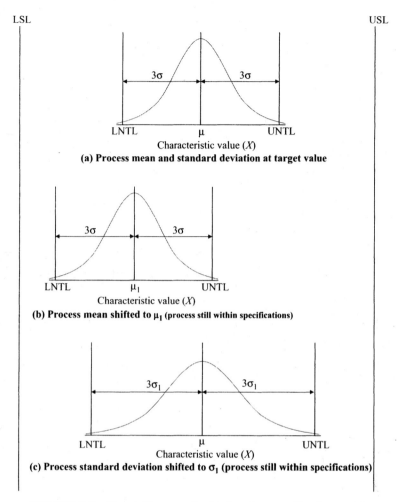

FIGURE 9-3   Case I: Process spread is less than specification spread.

control and still produce items within specifications. For instance, the process mean could shift to μ_1 (Figure 9-3b), or the process standard deviation could increase to σ_1 (Figure 9-3c), yet the items produced would still meet specifications. Of course, if a control chart is kept, any time that an out-of-control signal is observed, action is taken to bring the process back to control.

Case II: Process Spread Equal to Specification Spread   If the process spread is the same as the difference between the specification limits, we have an acceptable or adequate situation in which there is no room for error. If the distribution of the characteristic can be assumed to be normal and the process is in control, virtually all (99.74%) of the items produced will be within specifications. Figure 9-4a shows this situation. However, if such a process goes out of control (say, the process mean shifts from μ to μ_1), a proportion of the product will immediately be nonconforming (below the LSL), as shown in Figure 9-4b. An increase in the standard deviation will also result in a proportion of the product being nonconforming.

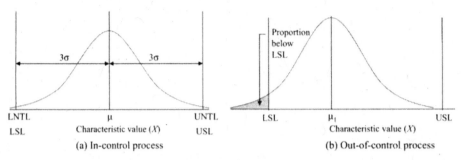

(a) In-control process   (b) Out-of-control process

FIGURE 9-4   Case II: Process spread is equal to specification spread.

Case III: Process Spread Greater Than Specification Spread   An undesirable situation exists when the process spread is greater than the difference between the specification limits. The inherent variability in the process exceeds the specification spread even though the process is in control. Figure 9-5 depicts this situation, for which some proportion of the items produced will not meet specifications. If there is a shift in the mean or an increase in the standard deviation, an increasing proportion of the product will not meet specifications. Such a process is not capable. There are several corrective approaches.

1. The possibility of increasing the specification limits can be explored. Careful consideration must be given to meeting the needs of the customer, however, for these needs determine the specification limits.

2. Measures to reduce the process spread can be investigated. Investing in new equipment, better raw material, or experienced operators are ways to achieve this reduction. The financial aspects of the investment decision are usually dealt with prior to implementing these measures.

3. If it is not economically feasible to reduce the process variability through large capital investments, an alternative may be to shift the process average to achieve a desirable balance in the proportion of scrap and rework. The cost of scrap per unit is usually more than that of rework; producing less scrap and more rework is thus a feasible short-term plan. Of course, in the long run a company must strive for no scrap or rework.

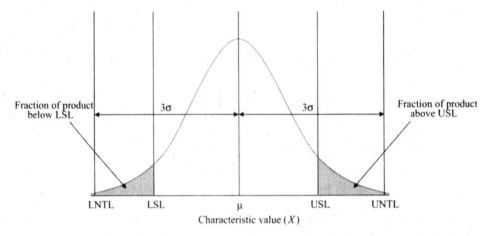

FIGURE 9-5   Case III: Process spread is greater than specification spread.

9-6 PROCESS CAPABILITY INDICES

A process should first be analyzed to verify that it is in control before its capability is estimated. In this section we assume that the process output (i.e., the distribution of the quality characteristic under consideration) is normal. This enables us to estimate the proportion of nonconforming product. The assumption of normality can be validated by empirical plots of histograms, normal probability plots, or statistical tests for goodness of fit such as chi-squared tests or the Kolmogorov-Smirnov test (Cochran 1952; Duncan 1986; Mage 1982; Massey 1951; Nelson 1979; Shapiro 1980).

The process capability index is an easily understood aggregate measure of the goodness of the process performance. The ability to meet specifications is the criterion used for measuring the attractiveness of the process. The capability indices we describe here are nondimensional, which makes them even more versatile and appealing because they do not depend on the specific process parameter units (Kane 1986). The indices incorporate the location and/or the variation in the process.

C_p Index

A common measure for describing the potential of a process to meet specifications is the C_p **index**. It relates the process spread (the difference between the natural tolerance limits) to specification spread, assuming two-sided specification limits. It is given by

$$C_p = \frac{\text{USL} - \text{LSL}}{6\sigma} \tag{9-2}$$

where USL and LSL represent the upper and lower specification limits, respectively, and σ represents the process standard deviation.

When σ is unknown, it is replaced by its estimate, $\hat{\sigma}$. The sample standard deviation s is one estimate of σ. Another estimate of σ can be obtained from the control chart information for the range chart when the process is in control. It is given by $\hat{\sigma} = \bar{R}/d_2$ where $\bar{R}$ is the mean of the ranges and d_2 is a factor for constructing the control chart (see Appendix A-7). If a chart for the standard deviation s is constructed, $\hat{\sigma}$ can be obtained from $\hat{\sigma} = \bar{s}/c_4$, where $\bar{s}$ is the mean of the sample standard deviations and c_4 is a factor for constructing control charts (Appendix A-7).

A process that is centered between the specification limits will produce a minimum proportion of items that fall outside those limits. Note that it is desirable to have ($C_p \geq 1$). When $C_p = 1$, the process spread equals the specification spread, and the process is said to be barely capable. If the process is centered, only 0.26% of the parts will fall outside the specification limits. Such a case was demonstrated in case II in Section 9-5 . Figure 9-4a shows that when $C_p = 1$, the natural tolerance limits coincide with the specification limits. The C_p-value thus represents the process potential.

If the process is not centered, it is possible that even for a process with $C_p > 1$, some proportion of the product will be nonconforming. However, when $C_p > 1$, as shown in Figure 9-3, there is some flexibility; that is, the process can go out of control yet still produce conforming items.

If $C_p < 1$, it implies that the inherent variability in the process, as measured by the process spread 6σ, is greater than the specification spread. For this situation, a process can be in control and still not meet specifications, as described in case III and shown in Figure 9-5.

Equation 9-2 shows that C_p is the ratio of the allowable process spread to the actual process spread. Since C_p does not take into account the location of the process, it is really a measure of *process potential* and not process performance. Other capability indices, such as CPU, CPL, C_{pk}, C_{pm}, and C_{pmk}, measure process performance.

Upper and Lower Capability Indices

Suppose that only a single specification limit is given. Indices can be derived that measure shifts in the process mean μ relative to the process spread. For a given upper specification limit, the **upper capability index** (CPU, or C_p upper) is given by

$$CPU = \frac{USL - \mu}{3\sigma} \tag{9-3}$$

It is desirable to have CPU $\geq$ 1. Note that the denominator of eq. (9-3) is half the process spread. Because nonconformance occurs only if the quality characteristic value exceeds the USL, the farther the USL is from the process mean μ, the less the chance for any nonconforming product. Assuming normality, if CPU = 1, only 0.13% of the product will be above the USL and thus nonconforming.

Similarly, if only the lower specification limit is given, the **lower capability index** (CPL, or C_p lower) is given by

$$CPL = \frac{\mu - LSL}{3\sigma} \tag{9-4}$$

Along the same lines, it is desirable to have CPL $\geq$ 1. For eqs. (9-3) and (9-4), if the process parameters μ and σ are unknown, the sample mean $\bar{X}$ and the sample standard deviation s, respectively, are used as estimates.

The indices CPU and CPL are useful in evaluating the process performance relative to the specification limits. They also aid in determining process parameter settings (such as the process mean μ) or process parameter requirements (such as the process standard deviation σ). Some recommended values for the process capability indices are shown in Table 9-1. A C_p-value of at least 1.33 for a process with two-sided specifications implies that the process standard deviation σ will be no more than one-eighth of the specification spread. Many companies prefer $C_p \geq 1.33$. This ensures an extremely low (0.007%) nonconformance rate. Furthermore, a C_p-value of at least 1.67 implies a process standard deviation of no more than one-tenth of the specification spread. Companies desiring a goal of a "six sigma process" are aiming for a C_p-value of 2.

TABLE 9-1   Recommended Minimum Values for the Process Capability Indices

Process	Two-Sided Specifications	One-Sided Specification
Existing process	1.33	1.25
New process	1.50	1.45
Safety, strength, or critical parameter		
Existing process	1.50	1.45
New process	1.67	1.60

Source: Adapted from D.C. Montgomery, *Introduction to Statistical Quality Control*, 3rd ed., 1996. Reprinted by permission of John Wiley & Sons, Inc.

Example 9-2   The relative humidity in a greenhouse is expected to be between 65 and 85%. Random samples taken over a span of one week yield the following values: 60, 78, 70, 84, 81, 80, 85, 60, 88, 75. Find and interpret the process capability index.

Solution   The specification limits are LSL $= 65\%$ and USL $= 85\%$. Assume that the process is in control. The sample mean and standard deviation are found as 76.1 and 9.905, respectively.

The C_p index is

$$C_p = \frac{\text{USL} - \text{LSL}}{6s} = \frac{85 - 65}{(6)(9.905)} = 0.337$$

This value of C_p, which is less than 1, indicates that the process is not capable of meeting the specifications. Remedial actions that will reduce the process variability must be identified.

Suppose that the only specification is the lower limit of 65%. Let's find the lower capability index. Using eq. (9-4) and replacing μ and σ by their estimates $\bar{X}$ and s, respectively, gives

$$\text{CPL} = \frac{\bar{X} - \text{LSL}}{3s} = \frac{76.1 - 65}{(3)(9.905)} = 0.374$$

The calculated value of CPL, which is less than 1, is undesirable. However, if the process variability cannot be reduced, another option would be to increase the process mean such that it is sufficiently above the LSL; that is, the average humidity level will be significantly above 65%.

To increase the CPL to 1, assuming that the process standard deviation cannot be reduced, the target value of the process mean should be

$$\mu = \text{LSL} + 3s = 65 + (3)(9.905) = 94.715\%$$

C_{pk} Index

We know that process variability is not the only parameter that influences a process's ability to produce a conforming product. The location of the process mean is another parameter that affects process capability (Gunter 1989). Although the C_p index does not incorporate the process location, other indices do.

One index that accounts for this location, the $\boldsymbol{C_{pk}}$ **index,** is used when the process mean is not at the target value, which is assumed to be halfway between the specification limits. The C_{pk} index is given by

$$C_{pk} = \min\left\{ \frac{\text{USL} - \mu}{3\sigma}, \frac{\mu - \text{LSL}}{3\sigma} \right\} \tag{9-5}$$

$$= \min\{\text{CPU}, \text{CPL}\}$$

It can be seen from eq. (9-5) that C_{pk} represents the scaled distance, relative to 3 standard deviations (i.e., half the process spread), between the process mean and the closest specification limit. Desirable values are $C_{pk} \geq 1$. Whereas the C_p index represents the process potential, the C_{pk}-value represents the actual capability of the process with the existing parameter values; it measures process performance.

Figure 9-6 shows the distribution of the quality characteristic X for a process that is not capable ($C_{pk} < 1$). Note that the process spread (6σ) is less than the specification spread. The

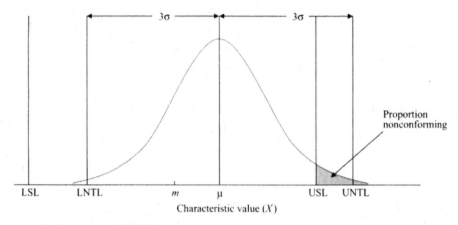

FIGURE 9-6   Process that is not capable ($C_{pk} < 1$).

value of C_p is greater than 1, indicating that the process can potentially meet specifications. However, the process mean μ is shifted to the right and is closer to the USL. The distance between USL and μ is less than 3σ, which is half the process spread. Hence, some proportion of the product will lie above the USL. A corrective measure in this case could be to adjust the process mean downward toward the midpoint m of the specification limits. Virtually all the product will then be conforming.

If management assigns equal significance to values falling above the USL or below the LSL, the optimal setting for the process mean is midway between the specification limits. A measure of the deviation of the process mean from this target value m is given by the *scaled distance k*:

$$k = \frac{|m - \mu|}{(\text{USL} - \text{LSL})/2} \tag{9-6}$$

where $m = (\text{USL} + \text{LSL})/2$. Note that the denominator of eq. (9-6) is half the allowable process spread. An estimate of k is obtained by using the sample mean $\bar{X}$ as an estimate of the process mean μ in eq. (9-6). The relationship between C_p and C_{pk} is given by

$$C_{pk} = C_p(1 - k) \tag{9-7}$$

If $\text{LSL} \leq \mu \leq \text{USL}$, we observe that $0 \leq k \leq 1$. If the process mean is at the target value m, then $k = 0$ and $C_{pk} = C_p$. If the process mean is at the USL or LSL, $k = 1$ and $C_{pk} = 0$.

How do the C_p and C_{pk} indices compare? The C_p-value is a measure of the process potential; it does not change as the process mean changes. A $C_p \geq 1$ is desirable, and a value of $C_p < 1$ indicates that the process is not capable. The C_{pk}-value incorporates both the process mean and the standard deviation to measure actual process performance. If the process is centered (i.e., the process mean is equal to the target value), $C_{pk} = C_p$. A standard for benchmarking processes is a value of $C_{pk} = 1$, in which case practically all the product is conforming. Note that $C_{pk} \leq C_p$—always. When the process mean is outside the specifications, C_{pk} is negative.

Example 9-3   In an electrical circuit, the capacitance of a component should be between 25 and 40 picofarads (pF). A sample of 25 components yields a mean of 30 pF and a standard deviation of 3 pF. Calculate the process capability index C_{pk}, and comment on the process performance. If the process is not capable, what proportion of the product is nonconforming, assuming a normal distribution of the characteristic?

Solution   The C_{pk} index is estimated as

$$C_{pk} = \min\left\{\frac{40-30}{3(3)}, \frac{30-25}{3(3)}\right\}$$
$$= \min\{1.111, 0.556\} = 0.556$$

Because $C_{pk} < 1$, the process is unable to produce only conforming product at its current setting. Corrective actions would be to move the process mean toward the target value of 32.5 pF or, if feasible, to reduce the process variability.

The standardized normal value (Z-value) at the LSL is

$$Z_{LSL} = \frac{25-30}{3} = \frac{-5}{3} = -1.67$$

The standardized value at the USL is

$$Z_{USL} = \frac{40-30}{3} = 3.33$$

After checking the standardized normal distribution in Appendix A-3, we find the area below the LSL to be 0.0475 and that above the USL to be $1 - 0.9996 = 0.0004$ (which is negligible). Thus, 4.75% of the product lies below the LSL. The process performance can be improved if the process mean is shifted to the target value of 32.5 pF. In this case, the Z-value at the LSL would be $Z_{LSL} = (25-32.5)/3 = -2.50$, and that at the USL would be $Z_{USL} = 2.50$. According to Appendix A-3, the proportion of the product below the LSL is 0.0062, with the same proportion above the USL. The total proportion nonconforming would be 0.0124, or 1.24%.

Capability Ratio

A measure of the ability of a process to produce items within specification limits is based on the amount of the specification range, $USL - LSL$, that is used by the process. As the process spread increases, it tends to use more of this specification band. The **capability ratio** (CR) is defined as

$$CR = \frac{6\sigma}{USL - LSL} \tag{9-8}$$

We can see from eq. (9-8) that the capability ratio is the reciprocal of the C_p index: that is, $CR = 1/C_p$. An estimate of CR is obtained by substituting an estimate s for the process standard deviation σ. Often, CR is expressed as a percentage to indicate the percentage of specification range used by the process. A $CR \leq 1$ is desirable. Processes that are centered and have a large value of C_p will obviously use much less of the specification range.

Using the specification range is analogous to the total variation of measured observations from a process. Such variability is identified as the sum of two components, variation in the process and variation in measurement. The latter is usually referred to as *gage variability*. An estimate of gage capability can be obtained through repeated measurements of given parts. Using control charts for the range of measurements that represent the magnitude of measurement error, we can estimate the standard deviation of measurement error. Knowing the total variance of measured observations (by subtracting the variance of measurement error) enables us to estimate the process variance. The capability ratio is a measure of process *potential* because it calculates the percentage of the specification range used under the ideal circumstance of a centered process.

Example 9-4   A process in control has an estimated standard deviation of 3 mm. The specification limits for the corresponding product are 100 ± 7 mm. Estimate the capability ratio of the process and comment on the process potential.

Solution   The capability ratio is estimated as

$$\text{CR} = \frac{6\hat{\sigma}}{\text{USL} - \text{LSL}} = \frac{(6)(3)}{107 - 93} = 1.286$$

So, the percent of the specification range used by the process is 128.6%, which is 28.6% more than is permissible. Even if the process were centered at the target value of 100, which is the most favorable situation, it would still not meet the specifications.

Taguchi Capability Index, C_{pm}

Taguchi (1985, 1986) stressed quality improvement by emphasizing the reduction in variability around a target value, *T*. The proposed **index, C_{pm}**, is defined as

$$C_{pm} = \frac{\text{USL} - \text{LSL}}{6\tau} \tag{9-9}$$

where τ is the standard deviation from the target value and is given by

$$\tau^2 = E\left[(X - T)^2\right] \tag{9-10}$$

Eq. (9-10) can also be expressed as

$$\tau^2 = E\left[(X - T)^2\right] = E\left[(X - \mu)^2\right] + (\mu - T)^2$$
$$= \sigma^2 + (\mu - T)^2 \tag{9-11}$$

where μ and σ^2 represent the process mean and variance, respectively. Therefore, C_{pm} can be expressed as

$$C_{pm} = \frac{\text{USL} - \text{LSL}}{6\sqrt{\sigma^2 + (\mu - T)^2}} = \frac{C_p}{\sqrt{1 + \delta^2}} \tag{9-12}$$

where

$$\delta = \frac{\mu - T}{\sigma} \tag{9-13}$$

represents the deviation of the process mean from the target value in units of standard deviation.

The indices C_{pk} and C_{pm} are regarded as second-generation capability indices, developed from the original C_p index. A third-generation capability index that incorporates the features of C_{pk} and C_{pm} is the C_{pmk} index (Pearn et al. 1992), given by

$$C_{pmk} = \frac{\min[(\text{USL} - \mu), (\mu - \text{LSL})]}{3\sqrt{\sigma^2 + (\mu - T)^2}} \tag{9-14}$$

Note that C_{pmk} takes into consideration process location and variability as well as deviation of the process mean from the target value. It is known that $C_p \geq C_{pk} \geq C_{pmk}$ and $C_p \geq C_{pm} \geq C_{pmk}$. A good review of the various capability indices may be found in Kotz and

Johnson (1993, 2002). A couple of relationships among these indices are as follows:

$$C_{pmk} = \begin{cases} \dfrac{C_{pk}}{\sqrt{1 + [(\mu - T)/\sigma]^2}} & (9\text{-}15) \\[2em] \dfrac{C_{pm}C_{pk}}{C_p} & (9\text{-}16) \end{cases}$$

Confidence Intervals and Hypothesis Testing on Capability Indices

The expressions for the various capability indices involve population parameters (μ or σ or both). In practice, they are replaced by their sample estimates ($\bar{X}$ and s), leading to point estimates $\hat{C}_p, \hat{C}_{pk}, \hat{C}_{pm},$ or $\hat{C}_{pmk}$. Here, we provide expressions for the confidence interval for C_p and C_{pk}, under the *assumption of normality* of the distribution of the quality characteristic. A $100(1-\alpha)\%$ confidence interval for C_p is given by

$$\hat{C}_p \sqrt{\frac{\chi^2_{1-\alpha/2, n-1}}{n-1}} \leq C_p \leq \hat{C}_p \sqrt{\frac{\chi^2_{\alpha/2, n-1}}{n-1}} \tag{9-17}$$

where $\chi^2_{1-\alpha/2, n-1}$ and $\chi^2_{\alpha/2, n-1}$ are the lower and upper $\alpha/2$ percentage points on the chi-square distribution with $(n-1)$ degrees of freedom.

Example 9-5 In an assembly operation in a semiconductor manufacturing company, the lower and upper specification limits are given by 4.8 and 5.2 seconds. A random sample of 25 completion times gave a mean and standard deviation of 5.12 and 0.06 seconds, respectively. Can we conclude that the C_p index for this operation exceeds 1, so as to be considered acceptable by the customer? Test at a significance level of 0.05.

Solution We are given the following: LSL $= 4.8$, USL $= 5.2$, $\bar{X} = 5.12$, $s = 0.06$, $n = 25$, $\alpha = 0.05$. The hypotheses are:

$$H_o : C_p \leq 1.00$$
$$H_a : C_p > 1.00$$

we have $\hat{C}_p = (5.2\text{–}4.8)/(6)(0.06) = 1.111$. Given a one-sided hypothesis test with $\alpha = 0.05$, a 95% lower confidence limit for C_p is obtained as

$$\text{LCL} = \hat{C}_p \sqrt{\frac{\chi^2_{0.95, 24}}{24}} = 1.111 \sqrt{\frac{13.85}{24}} = 0.844$$

Since the hypothesized value of $C_p = 1.00 > \text{LCL} = 0.844$, we do not reject H_0. Hence, we cannot conclude that this is a capable operation.

When the quality characteristic is normally distributed, an approximate $100(1-\alpha)\%$ confidence interval (Kushler and Hurley 1992) on C_{pk} is given by

$$\hat{C}_{pk} \pm Z_{\alpha/2} \sqrt{\frac{1}{9n} + \frac{\hat{C}_{pk}^2}{2(n-1)}} \tag{9-18}$$

where n represents the sample size used to calculate $\hat{C}_{pk}$ and $Z_{\alpha/2}$ represents the standard normal value for a tail area of $\alpha/2$. Hypothesis tests may also be performed on C_{pk} as was demonstrated for C_p.

Comparison of Capability Indices

While C_p measures process potential, C_{pk} measures the actual process yield. When a process is exactly centered between the specification limits, $C_p = C_{pk}$. Under the assumption of normality, the process yield, which represents the proportion of the output that lies within the specification limits, is given by

$$\text{percent yield} = 100\left[\Phi\left(\frac{\text{USL} - \mu}{\sigma}\right) - \Phi\left(\frac{\text{LSL} - \mu}{\sigma}\right)\right] \qquad (9\text{-}19)$$

where Φ denotes the standard normal cumulative distribution function. For a process centered between the specification limits, the yield (Boyles 1991) may be expressed as a function of C_p:

$$\begin{aligned}\text{percent yield} &= 100[\Phi(3C_p) - \Phi(-3C_p)] \\ &= 100[2\Phi(3C_p) - 1]\end{aligned} \qquad (9\text{-}20)$$

Recall that C_{pk} measures actual rather than potential process capability. Using eq. (9-5), we can see that

$$\begin{aligned}\text{USL} &\geq \mu + 3\sigma C_{pk} \\ \text{LSL} &\leq \mu - 3\sigma C_{pk}\end{aligned}$$

with strict equality holding for the specification limit that is closer to μ. The value of C_{pk} equals 1 if and only if the natural tolerance limits do not extend beyond the specification limits, with at least one of the natural tolerance limits coinciding with a specification limit.

Let's consider Figure 9-7, where USL = 62 mm, LSL = 38 mm, and T = 50 mm; three processes, A, B, and C, with different means and standard deviations, are shown.

Table 9-2 shows the values of the process capability indices C_p, C_{pk}, and C_{pm} as well as the percentage yield, under the assumption of normality for each process distribution. Note that C_{pk} is 1 for each process. The values of C_p, however, are 1, 2, and 4 for processes A, B, and C, respectively. Thus, process C has the most potential. For process A, the values

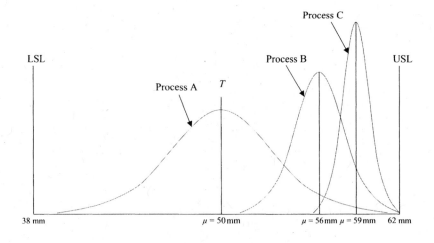

FIGURE 9-7 Effect of process parameters on capability indices.

TABLE 9-2   Process Capability Indices for Processes with Various Parameters

Process	Process Mean and Standard Deviation	δ	C_p	C_{pk}	C_{pm}	Percent Yield
A	$\mu = 50$, $\sigma = 4$	0	1	1	1.000	99.74
B	$\mu = 56$, $\sigma = 2$	3	2	1	0.632	99.87
C	$\mu = 59$, $\sigma = 1$	9	4	1	0.442	99.87

of C_p and C_{pk} are both equal because the process mean is centered between the specification limits.

From Table 9-2 we can see that for processes B and C, whose means deviate from the target by different amounts, the values of C_{pk} are the same. This is because C_{pk} is influenced by both the process mean and the process standard deviation. So, for process C, even though the shift in the process mean from the midrange of the specification limits is greater than for process B, the smaller variability of process C compensates for it.

The percentage yield, shown in Table 9-2, varies between 99.74 and 99.87% for the three processes, which have the same value of C_{pk}. Thus, for a given value of C_{pk}, say 1, the actual process yield has an upper and a lower bound. A continuum of processes, each having the same value of C_{pk}, can be visualized as shown in Figure 9-7 where the actual process yield is contained within two bounds. It can be shown, using eq. (9-20), that for a fixed value of C_{pk}, the bounds on the process yield are given by

$$100\left[2\Phi(3C_{pk}) - 1\right] \leq \text{percent yield} \leq 100\Phi(3C_{pk}) \qquad (9\text{-}21)$$

The lower bound is achieved when the process mean is centered between the specification limits, while the upper bound is achieved as the process mean approaches USL or LSL and the process standard deviation approaches zero. Hence, we can conclude that C_{pk} approximates the actual process yield, as defined by upper and lower bounds.

The Taguchi capability index, C_{pm}, is also shown in Table 9-2 for the three processes in Figure 9-7. Note that for process A, where the process mean coincides with the target value and the midrange specification value, all three indices C_p, C_{pk}, and C_{pm} are equal. As the process mean moves away from the target value, the value of C_{pm} decreases. For process C, whose mean is 9 standard deviations from the target value, C_{pm} is 0.442, while for procees B, with mean 3 standard deviations from the target value, C_{pm} is 0.632. The value of C_{pk} is 1 for all three processes, indicating that it is not a suitable measure of process centering. On the contrary, for a fixed value of μ in the interval (LSL, USL), C_{pk} becomes arbitrarily large as the process standard deviation σ approaches zero. Thus, a large value of C_{pk} does not imply much about the distance between the process mean and the target value.

There are some similarities between C_{pk} and C_{pm}. For a fixed value of the process standard deviation σ, both indices coincide with C_p when the process mean μ equals the target value T. They both decrease as the process mean moves away from the target values. If $\mu \geq$ USL or $\mu \leq$ LSL, $C_{pk} \leq 0$. On the other hand, C_{pm} approaches zero asymptotically as $|\mu - T| \to \infty$. For a fixed μ, as $\sigma \to 0$, C_{pk} increases without bound. However, C_{pm} has an upper bound given by

$$C_{pm} < \frac{\text{USL} - \text{LSL}}{6|\mu - T|} \qquad (9\text{-}22)$$

The right-hand side of eq. (9-22) is the limiting value of C_{pm} as $\sigma \to 0$ and is also equal to the C_p-value of a process with $\sigma = |\mu - T|$. A necessary condition for $C_{pm} \geq 1$ is

$$|\mu - T| < \frac{\text{USL} - \text{LSL}}{6} \tag{9-23}$$

Note that when the target value is at the center of the specification limits, a C_{pm}-value of 1 or more implies that the process mean μ lies within the middle third of the specification band. Similarly, $C_{pm} \geq \frac{4}{3}$ implies that $|\mu - T| < (\text{USL} - \text{LSL})/8$. Therefore, a value of C_{pm} indicates a constraint on the difference between the process mean μ and the target value T. Hence, the value of C_{pm} measures process centering in terms of the variation of the process mean from the target value.

An estimate of C_{pm} can be obtained using an estimate of τ, where τ^2 is given by Eqs. (9-10) or (9-11). Taguchi (1985) proposed the following estimator:

$$\hat{\tau}^2 = \frac{1}{n} \sum_{i=1}^{n} (X_i - T)^2 \tag{9-24}$$

$$= \hat{\sigma}^2 + (\bar{X} - T)^2 \tag{9-25}$$

where $\bar{X}$ is the sample mean and

$$\hat{\sigma}^2 = \frac{(n-1)s^2}{n} \tag{9-26}$$

The sample variance s^2 is given by

$$s^2 = \frac{\sum (X_i - \bar{X})^2}{n-1} \tag{9-27}$$

It can be shown (Boyles 1991) that $\hat{\tau}^2$ is an unbiased estimate of τ^2.

Example 9-6   A process in control has an estimated standard deviation of 2 mm. The product produced by this process has specification limits of 120 ± 8 mm and a target value of 120 mm. Calculate the process capability indices C_p, CPL, CPU, C_{pk}, C_{pm}, and C_{pmk} for the process if the process mean shifts from 118 mm first to 122 mm and then to 124 mm, but the process variability remains the same.

Solution   Figure 9-8 shows the process distributions for various values of the process mean. The calculations for the process capability indices when the process mean is at 118 are as follows:

$$C_p = \frac{128 - 112}{(6)(2)} = 1.333$$

$$\text{CPL} = \frac{118 - 112}{(3)(2)} = 1.000 \quad \text{CPU} = \frac{128 - 118}{(3)(2)} = 1.667$$

$$C_{pk} = \min\{\text{CPU, CPL}\} = 1.000 \quad C_{pm} = \frac{1.333}{\sqrt{1 + (-1)^2}} = 0.943$$

$$C_{pmk} = \frac{6}{3\sqrt{4 + 4}} = 0.707$$

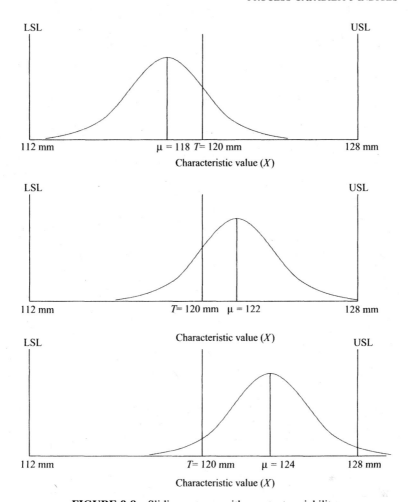

FIGURE 9-8   Sliding process with constant variability.

Table 9-3 lists values for the various process capability indices for process means (in mm) of 118, 122, and 124. Note that the process potential, C_p, remains constant for the different locations of the process mean. The CPL index increases from 1 to 2 as the process mean shifts from 118 to 124. Conversely, the CPU index decreases from 1.667 to 0.667 over this same range. The C_{pk} index is 1.000 when the process mean is at 118 or 122. Note that for these two situations, the process distribution is shifted by the same amount on either side of the midpoint of the specifications of 120. The C_{pm}-value is 0.943 for these two distributions. As the process mean shifts to 124, the C_{pk}-value decreases to 0.667 and the C_{pm}-value decreases from 0.943 to 0.596.

TABLE 9-3   Process Capability Indices for a Sliding Process with Constant Variability

Process Mean	δ	C_p	CPL	CPU	C_{pk}	C_{pm}	C_{pmk}
118	−1	1.333	1.000	1.667	1.000	0.943	0.707
122	1	1.333	1.667	1.000	1.000	0.943	0.707
124	2	1.333	2.000	0.667	0.667	0.596	0.298

The value of C_{pmk} is 0.707 when the process mean is at 118. Similar to the behavior of C_{pk} and C_{pm}, when the mean shifts to 122, the value of C_{pmk}, remains at 0.707, as the mean shift is an equal amount from the target value. As the mean shifts farther away from the target value, C_{pmk} reduces to 0.298 when the mean is at 124.

While both C_{pk} and C_{pm} demonstrate a decreasing trend as the process mean shifts away from the target value, their interpretations are somewhat different. Because C_{pk} is a measure of the actual process yield, when the process mean is at 118 and 122, the C_{pk}-values are the same; this implies that the process yield will be the same under these two situations. On the other hand, C_{pm} measures the deviation of the process mean from the target value. Thus, C_{pm} has the smallest value of 0.596 when the process mean deviates from the target value by 2 standard deviations. The larger the deviation of the mean from the target value, the smaller the value of C_{pm}.

Effect of Measurement Error on Capability Indices

Whenever measurements are involved, the variability of the observations is dependent on the variability of the product characteristic, which is itself dependent on the inherent process variation and measurement variation. Variability in measurements is, of course, a function of the measuring instrument used. In the preceding section we assumed that measurement error had a negligible impact on the observations. Here we study the effect of measurement error on the process potential as it affects the C_p index and the capability ratio CR. A measured value from a process X_m is the sum of the "true" value X plus the measurement error ε. So, we have

$$X_m = X + \varepsilon \tag{9-28}$$

In eq. (9-28), we can only observe X_m; we cannot observe X. However, if the measurement error ε can be estimated, we can obtain an estimate of the true process variability. Assuming that the measurement error is independent of the value being measured and using eq. (9-28), we have

$$\sigma_m^2 = \sigma^2 + \sigma_e^2 \tag{9-29}$$

where σ_m^2 represents the variance of the measured observations, σ^2 is the true process variance, and σ_e^2 is the variance of the measurement errors. Now, when we calculate capability indices, we use σ_m rather than the true process standard deviation σ.

Measurement errors are usually distributed normally. So, $6\sigma_e$ is a good representation of the range of measurement errors. Also, every measuring instrument is typically rated by its manufacturer for a prescribed level of precision. An estimate of the measurement error is obtained through an index known as the **precision-to-tolerance ratio (r)**, given by

$$r = \frac{6\sigma_e}{\text{USL} - \text{LSL}} \tag{9-30}$$

Equation (9-30) represents the percentage of the tolerance range used by the measurement error. Manufacturers of measuring equipment provide information to estimate r. Alternatively, σ_e can be estimated through gage capability studies, where repeated measurements of given parts are obtained. Control charts for the range of measurements, which represent the magnitude of measurement error, can be constructed, and σ_e can be estimated using

$$\hat{\sigma}_e = \frac{\bar{R}}{d_2} \tag{9-31}$$

where $\bar{R}$ is the average of the ranges of the replications and d_2 is a control chart factor, based on the sample size, obtained from Appendix A-7.

Using eq. (9-30), we can rewrite eq. (9-29) as

$$\sigma_m^2 = \sigma^2 + \left(r\frac{USL - LSL}{6} \right)^2$$

The relationship between the observed capability index C_p^* based on measured observations, and the "true" capability index C_p is now defined as follows:

$$C_p^* = \frac{USL - LSL}{6\sigma_m} = \frac{USL - LSL}{6\sqrt{\sigma^2 + \left(r\frac{USL - LSL}{6} \right)^2}}$$

$$= \frac{1}{\sqrt{(1/C_p)^2 + r^2}} \tag{9-32}$$

The true capability index may now be expressed as

$$C_p = \frac{1}{\sqrt{(1/C_p^*)^2 - r^2}} = \frac{1}{\sqrt{(CR^*)^2 - r^2}} \tag{9-33}$$

where CR* represents the capability ratio as calculated from the measured observations .

We can see from eq. (9-32) that when there is no measurement error ($r = 0$), C_p^* equals C_p, the true process capability. The assumption of negligible measurement error supports our contention that the observed capability index C_p^* is a good approximation of the true index C_p. However, even if the process variability shrinks to zero, from eq. (9-32), an upper bound on C_p is given by

$$C_p^* \le \frac{1}{r} \tag{9-34}$$

Gage Repeatability and Reproducibility

Measurement error or variation can be divided into two components, repeatability and reproducibility. **Gage repeatability** represents the inherent variation of the gage or measuring device. It is the variation that is observed when the same operator measures the same part using the same device repeatedly. **Gage reproducibility**, a function of the variability of the operators, is the variation that is observed when different operators measure different parts using the same device. So, two variances constitute gage reproducibility: one between operators and one due to the interaction between operators and parts. The interaction variance represents the variation in the average part sizes measured by each operator. For example, one operator may have more variability when measuring smaller part dimensions, while another operator may exhibit more variability when measuring larger part dimensions.

Figure 9-9 shows the components of the total variability of measured observations. Using eq. (9-29), we express the total variance of measured observations σ_m^2 as

$$\sigma_m^2 = \sigma^2 + \sigma_e^2$$

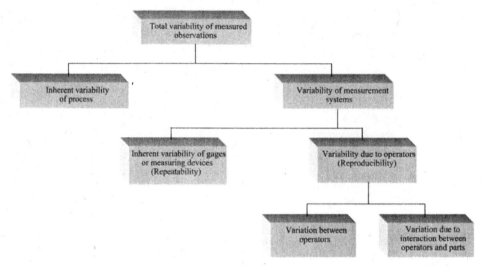

FIGURE 9-9 Components of total variability of measured observations.

where

$$\sigma_e^2 = \sigma_t^2 + \sigma_p^2 \tag{9-35}$$

In eq. (9-35), σ_t^2 and σ_p^2 represent the variance in repeatability and reproducibility of observations, respectively.

Evaluation of Measurement Systems

In any measuremnt or gage system, the concepts of accuracy and precision are important. **Accuracy** is a measure of the location of the distribution of measurements from the "true" or known value. Hence, if a part whose true quality characteristic value is known is measured repeatedly by an inspector, accuracy as measured by **gage bias** is the difference between the observed average measurement and the reference or true value. One particular metric expresses the gage bias as a percentage of the total variation. If the characteristic has a normal distribution, 5.15 standard deviations covers about 99% of the distribution. So

$$\% \text{ bias} = \frac{\text{gage bias}}{5.15\sigma_m}(100) \tag{9-36}$$

Gage *linearity* is a measure of the difference in accuracy values through the expected operating range of the gage. A gage is expected to have the same accuracy for all sizes of the characteristic being measured. To estimate gage linearity, several parts are selected over the operating range. The bias for these parts is found by taking the difference between the average of the observed measurements and the true value. A regression line is then fitted to predicting the bias based on the true values. The slope of the fitted regression line is an estimate of gage linearity, with percent linearity being the slope expressed as a percentage. Another measure of gage linearity is expressed as

$$\text{linearity} = (\text{slope of fitted line})(5.15\,\sigma_m) \tag{9-37}$$

where $5.15\,\sigma_m$ is a measure of the overall total variation in the process, which includes the variability in the process and the gage system.

Gage **stability**, on the other hand, is a measure of accuracy over time when the same device is used. Statistical stability of the measurement system (i.e., special causes in the measurement system must be eliminated) is therefore desirable in order to make inferences as to process capability.

Metrics for Evaluation of Measurement Systems

The measurement system standard deviation is given by

$$\sigma_e = \sqrt{\sigma_t^2 + \sigma_p^2} \qquad (9\text{-}38)$$

while the standard deviation of the measured observations is expressed as

$$\sigma_m = \sqrt{\sigma^2 + \sigma_e^2} \qquad (9\text{-}39)$$

where σ^2 represents the process variance.

- *%R&R.* A common measure is the percent of total variation consumed by the measurement system for repeatability and reproducibility, known as percentage repeatability and reproducibility (R&R):

$$\% R\&R = \frac{\sigma_e}{\sigma_m}(100) \qquad (9\text{-}40)$$

It is desirable that $\% R\&R < 10\%$, while acceptable values are $\% R\&R < 30\%$.

- *Precision to tolerance ratio.* This metric compares the spread in the measurement system to the tolerance spread and is expressed as the *P/T* ratio (r) given by eq. (9-30). It is desirable that $P/T < 10\%$.

- *Percentage of process variation.* Here, the variability of the measurement system is expressed as a percentage of the process variation and is given by

$$\frac{\sigma_e}{\sigma}(100) \qquad (9\text{-}41)$$

   This ratio does not depend on the specification limits.

- *Number of distinct categories.* This represents the number of distinct categories within the process data that the measurement system can discern or distinguish. It is expressed as

$$\frac{\sigma}{\sigma_e}(1.41) \qquad (9\text{-}42)$$

The desirable number of distinct categories is ≥ 4. If the number of distinct categories is < 2, the measurement system is of no value. When the number of distinct categories is 2, the data can be divided into two groups (say, high and low). Along these lines, when the number of categories is 3, the data can be divided into three groups: say, low, middle, and high.

Gage R&R studies usually require balanced designs (equal number of observations per operator and part) and replicates. In a **crossed design**, each part is measured multiple times by each operator, whereas in a **nested design**, each part is measured by only one operator. So, there is no operator-by-part interaction. Measurements that are destructive (i.e., breaking strength of cables) are therefore conceptually of nested design.

Preparation for a Gage Repeatability and Reproducibility Study

An R&R study must be planned outlining the nature of the design (crossed versus nested) to be used. In situations involving destructive testing, a nested design is used. A decision has to be made on the number of inspectors, number of parts, and the number of repeat readings (for nondestructive testing in a crossed design) for each inspector and part combination. When the intention is to estimate reproducibility, at least two operators must be chosen. Usually, from the operating range of the quality characteristic, at least 10 parts should be selected. To estimate repeatability, at least two repeat readings of each part must be taken. Inspection bias is eliminated by taking measurements of the parts selected, in random order.

In an *attribute* gage R&R study, the part quality characteristic is identified only as conforming or not, rather than the degree to which it differs from a set of given specification limits. An example is the use of a go/no-go gage for measuring the diameter of a part. The criterion for the acceptability of an attribute gage is that *all* measurement decisions must agree. This implies that for each part measured, all inspectors must agree over all replications.

In gage R&R studies for variables, Minitab uses the $\bar{X}$ and R method or the analysis of variance (ANOVA) method. The $\bar{X}$ and R method has the drawback of being unable to measure the interaction between operators and parts. In the ANOVA method, the factors, operators (inspectors) and parts, are considered to be random. The variance components are reproducibility (variation due to operator and variation due to operator $\times$ part interaction), repeatability (variation inherent in the gage), and the part-to-part or process variation. The ANOVA table initially includes the main effects of operators, parts, and operator $\times$ part interaction. If the p-value for the operator $\times$ part interaction term is greater than 0.25, a reduced model is fitted to calculate the variance components.

Example 9-7   Measurements on the thickness of support beams are taken at random by three operators. The same beam is measured twice by each operator. Table 9-4 shows the data for 20 such beams. Specifications on thickness are 50 ± 8 mm. Comment on the capability of the measurement system. Use the ANOVA method. What is the precision-to-tolerance ratio? What is the observed process potential, and what is the estimate of the true process potential?

Solution   Using Minitab, we choose **Stat > Quality Tools > Gage Study > Gage R&R study (crossed)**. Enter the column numbers in the data set that represents the part number, operators, and measurement data. Select **Options** and input process tolerance. Figure 9-10 shows the Minitab graphical output, while Figure 9-11 displays the Minitab ANOVA output and other relevant gage R&R information. A two-way ANOVA table with interaction between operators and parts is shown in Figure 9-11. Note that the p-value for operator part interaction is quite small (0.000), so the interaction effect is significant. From the figure it is observed that the variance of measurements due to the gage system $(\hat{\sigma}_e^2)$ is 0.2442. The process (part-to-part) variance $(\hat{\sigma}^2)$ is 15.6808, while the total variance of the measured value $(\hat{\sigma}_m^2)$ is 15.9250. The precision-to-tolerance ratio is 0.1853, or 18.53%, implying that the measurement system uses 18.53% of the tolerance limits. It is desirable for this to be less than or equal to 10%.

The metric % R&R, which compares the standard deviation associated with repeatability and reproducibility of the gage system relative to the total variation, is found to be 12.38% from Figure 9-11. Since % R&R > 10%, although the measurement system is acceptable, attention could be paid to improve it. Since repeatability is the larger proportion of the total gage R&R, alternative measurement devices could be considered.

TABLE 9-4   Data on Beam Thickness (mm)

	Operator 1		Operator 2		Operator 3	
	Measurements		Measurements		Measurements	
Part Number	1	2	1	2	1	2
1	43.4	43.7	43.0	43.1	42.8	43.2
2	51.8	51.7	52.2	52.5	52.3	51.8
3	54.9	54.4	55.0	54.6	54.8	55.2
4	47.2	47.5	46.2	46.5	46.3	46.9
5	46.5	46.8	45.3	45.9	45.2	45.4
6	49.3	50.0	49.0	49.6	49.8	50.5
7	52.6	52.4	52.9	53.2	52.9	52.7
8	55.8	55.5	55.3	54.7	56.0	56.3
9	48.3	48.0	49.6	49.1	47.9	48.4
10	47.8	48.1	48.4	47.3	47.4	48.2
11	53.6	53.8	54.0	53.2	53.7	54.2
12	52.7	52.4	51.8	52.8	53.1	53.6
13	44.1	44.2	44.9	44.4	43.6	43.2
14	46.6	46.9	47.0	46.3	46.2	46.9
15	56.7	56.2	56.9	56.2	57.3	57.6
16	53.8	53.9	53.4	54.3	53.4	54.1
17	48.3	48.6	48.0	48.6	48.2	47.8
18	51.5	51.4	51.9	51.2	51.8	52.3
19	53.8	53.6	53.3	53.9	54.7	53.6
20	47.7	47.9	48.3	47.9	47.5	47.8

In this example, while the variance component between the operators is negligible, the interaction between operators and parts is significant. The number of distinct categories is indicated as 11, where desirable values are ≥4, implying that the measurement system is quite discriminatory. The total variation in the system is quite high, and as indicated in

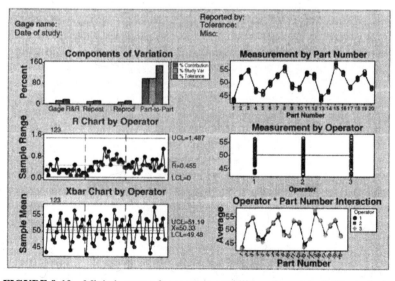

FIGURE 9-10   Minitab output for gage repeatability and reproducibility analysis.

Two-Way ANOVA Table With Interaction

Source	DF	SS	MS	F	P
Part Number	19	1794.37	94.4404	265.691	0.000
Operator	2	0.11	0.0531	0.149	0.862
Part Number * Operator	38	13.51	0.3555	2.674	0.000
Repeatability	60	7.97	0.1329		
Total	119	1815.96			

Source	VarComp	%Contribution (of VarComp)
Total Gage R&R	0.2442	1.53
Repeatability	0.1329	0.83
Reproducibility	0.1113	0.70
Operator	0.0000	0.00
Operator*Part Number	0.1113	0.70
Part-To-Part	15.6808	98.47
Total Variation	15.9250	100.00

Process tolerance = 16

Source	StdDev (SD)	Study Var (6 * SD)	%Study Var (%SV)	%Tolerance (SV/Toler)
Total Gage R&R	0.49415	2.9649	12.38	18.53
Repeatability	0.36458	2.1875	9.14	13.67
Reproducibility	0.33357	2.0014	8.36	12.51
Operator	0.00000	0.0000	0.00	0.00
Operator*Part Number	0.33357	2.0014	8.36	12.51
Part-To-Part	3.95990	23.7594	99.23	148.50
Total Variation	3.99062	23.9437	100.00	149.65

Number of Distinct Categories = 11

FIGURE 9-11 Minitab output using ANOVA method for repeatability and reproducibility study.

Figure 9-11, it is about 150% of the tolerance. Thus, the current process is *not capable* of meeting specifications. The part-to-part or process variation consumes about 148.50% of the tolerance while accounting for about 98.47% of the variance. This is the dominant factor contributing to variability.

Figure 9-10 confirms the results in Figure 9-11. First, observe that on the $\bar{X}$- chart, just about all points fall outside the control limits. This is as expected, because the $\bar{X}$- chart represents the discriminating power of the measuring instrument. The R-chart shows the magnitude of the measurement errors and also is a measure of gage capability. It shows that operator 1 tends to have smaller variability within parts then do operators 2 and 3.

Since the estimated standard deviation of all measured values is 3.9906, the process potential is given by

$$C_p^* = \frac{USL - LSL}{6\sigma_m} = \frac{16}{(6)(3.9906)} = 0.668$$

Thus, the process is not capable. The true process potential, after discounting for the variability in the measurement system, is estimated as

$$C_p = \frac{1}{\sqrt{(1/0.668)^2 - (0.1853)^2}} = 0.673$$

TABLE 9-5   Nonconformance Rates (ppm) for Normally Distributed Processes with One-Sided Specifications for Selected Values of CPL or CPU

Capability Index CPL or CPU	Nonconformance Rate $P(X > USL)$ or $P(X < LSL)$	Capability Index CPL or CPU	Nonconformance Rate $P(X > USL)$ or $P(X < LSL)$
0.50	66,807	1.33	33
0.60	35,930	1.40	13.4
0.70	17,864	1.50	3.4
0.80	8,196	1.60	0.793
0.90	3,467	1.70	0.170
1.00	1,350	1.80	0.033
1.10	483	1.90	0.006
1.20	159	2.00	0.001
1.30	48		

With an estimated value of C_p less than 1, management should focus on methods to reduce the inherent process variability.

C_p Index and the Nonconformance Rate

The process capability index C_p measures process potential and represents the ability of the process to produce a conforming product. Assuming a normal distribution of the quality characteristic, if the process mean is centered between the specification limits, the process fallout or nonconformance rate in parts per million (ppm) can be found for different values of C_p. Similarly, for one-sided specification limits, the nonconformance rate can be found for various values of CPU or CPL.

Table 9-5 shows the nonconformance rate, in parts per million, for selected values of CPU or CPL, for one-sided specifications. As an example, for a process with an upper specification limit, if CPU = 0.5, the nonconformance rate is given by $P[Z > 3(CPU)] = = P[Z > 1.5] = 0.066807 = 66,807$ ppm. Observe that for a CPU or CPL of 1.0, the nonconformance rate is 1350 ppm. For two-sided specification limits, if the process is centered between the specifications, with a C_p index of 1.0, the nonconformance rate would be 2700 ppm. Companies generally desire a C_p of 1.33; for two-sided specification limits, the nonconformance rate for such processes is 66 ppm. Motorola's concept of a process with six-sigma capability, allowing for a drift in the process mean of 1.5 standard deviations, permits only 3.4 ppm outside the specification limit that is closer to the mean.

9-7   PROCESS CAPABILITY ANALYSIS PROCEDURES

As explained earlier in the chapter, process capability analysis may involve estimating process parameters such as the mean and standard deviation, the form of the distribution of the characteristic, or the proportion of the nonconforming product. Several approaches exist for estimating the process standard deviation, some of which we describe here.

Estimating Process Mean and Standard Deviation

The process mean measures the location of the process; the process standard deviation reflects the variability of the process. An idea of the distribution of the quality characteristic is obtained by constructing a frequency distribution. This empirical distribution can be used to validate claims regarding the hypothesized distribution of the characteristic.

Using Individual Observations   Before starting process capability analysis using individual observations, a minimum number of such measurements (usually, at least 50) must be taken. The process mean μ is estimated by the sample mean $\bar{X}$, while the process standard deviation σ is estimated by the sample standard deviation s.

Next, a frequency or relative frequency histogram is constructed. This tells us about the behavior of the characteristic. In many cases, a normal distribution is assumed when we calculate the proportion of nonconforming product; histograms help determine the validity of this assumption. Procedures for testing a hypothesis regarding a specific population distribution (e.g., normal) are known as *goodness-of-fit tests* and can be found in most books on statistics (Duncan 1986).

If the distribution is close to normal, the process spread is $PS = 6s$. An estimate of the process capability limits, or the natural tolerance limits, is given by $\bar{X} \pm 3s$. If the process is centered, a measure of the process capability is obtained using the C_p-value. If the process is not centered, the C_{pk} index is a better measure of process capability.

Using Control Chart Information   Control charts are the major tools for analyzing an existing process and bringing it to control, so it makes sense to use control charts to study process control. A control chart will indicate whether special causes exist in the process and thus signal whether we are in a position to estimate the process capability.

Using Variable Charts   If charts for the mean $\bar{X}$ and range R are used, once the process is in control, the process mean and standard deviation can be estimated as $\hat{\mu} = \bar{\bar{X}}$ and $\hat{\sigma} = \bar{R}/d_2$, respectively, where d_2 is a control chart factor tabulated in Appendix A-7, $\bar{\bar{X}}$ is the center line on the $\bar{X}$-chart, and $\bar{R}$ is the centerline on the R-chart.

If control charts for the mean $\bar{X}$ and standard deviation s are used, then, for a process in control, the mean and standard deviation can be estimated from $\hat{\mu} = \bar{\bar{X}}$ and $\hat{\sigma} = \bar{s}/c_4$, respectively, where c_4 is a control chart factor tabulated in Appendix A-7, $\bar{\bar{X}}$ is the centerline on the $\bar{X}$-chart, and $\bar{s}$ is the centerline on the s-chart.

If control charts for individual values X and the moving range of two consecutive observations are used, once the process is in a state of statistical control, its mean and standard deviation can be estimated from $\hat{\mu} = \bar{X}$ and $\hat{\sigma} = \overline{MR}/d_2$, respectively, where d_2 is a control chart factor tabulated in Appendix A-7 and is found using a sample size of 2, $\bar{X}$ is the centerline on the X-chart, and $\overline{MR}$ is the average of the moving ranges from pairs of consecutive observations.

Example 9-8   Let us consider the data on the inside diameter of metal sleeves, shown in Table 5-1. The specification limits are 50.00 ± 0.05 mm, with a target value of 50. We use Minitab to conduct the capability analysis. First, we constructed control charts for individuals and moving range by setting up the data in a single column, with the rows being merged into a column. Hence, the first 5 observations were from sample 1, the next 5 from sample 2, and so on. From the I-MR charts, applied successively, three observations were outside the control limits. Assessing special causes and appropriate remedial action, after deleting the three observations, all were found to be in control.

Now that the process is in control, we select the commands **Stat > Quality Tools > Capability Analysis > Normal**. Select **Options**, and input a Target value of 50. Input the **Lower spec** and **Upper spec** as 49.5 and 50.5, respectively, and click **OK**.

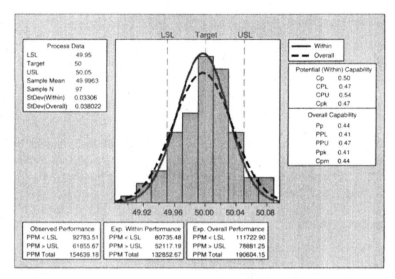

FIGURE 9-12   Process capability anaylsis for sleeve diameter.

Output from Minitab is shown in Figure 9-12. Note that the distribution resembles normality. Further, a test for normality (using the Anderson–Darling test in Minitab) was also performed (p-value $= 0.168$). Being satisfied with the distributional assumption of normality, we investigate output capability indices. All the indices (C_p, C_{pk}) are well below 1, indicating that the current process is not capable.

Two sets of capability indices are shown in Figure 9-12, based on the standard deviation estimate. The *within* standard deviation is based on the variation *within* subgroups. Here, the estimate is based on $\hat{\sigma} = \overline{MR}/d_2$. The *overall* standard deviation is based on the variability of all observations about the grand mean. The capability indices (P_p, P_{pk}) calculated using overall standard deviation are therefore interpreted as long-term measures. Figure 9-12 also provides an estimate of the proportion nonconforming (using within standard deviation and overall standard deviation). Here, in the short term, the expected proportion nonconforming is 13.28%, while in the long term, it is 19.06%.

9-8   CAPABILITY ANALYSIS FOR NONNORMAL DISTRIBUTIONS

In the previous sections we utilized the distributional assumption of normality to calculate proportion nonconforming. When the quality characteristic distribution is inherently nonnormal, a few approaches exist for conducting capability analysis.

Identification of Appropriate Distribution

In this approach, based on knowledge of the process by which the quality characteristic is chosen, a *distributional fitting* is attempted from a specified list of available distributions, such as Weibull, gamma, lognormal, and exponential. Some general families of distributions may also be considered, such as the Johnson family of distributions discussed in Chapter 5. When an acceptable fit is identified through a statistical goodness-of-fit test, that particular distribution, with parameters estimated from the data, is used to obtain measures of capability and proportion nonconforming.

Box-Cox Transformation

This is a form of the power transformation discussed in Chapter 5 to ensure that the transformed variable follows normality. The Box–Cox method estimates a value for λ that minimizes the standard deviation of a standardized transformed variable. It is defined as

$$Y_T = \begin{cases} Y^\lambda, & \lambda \neq 0 \\ \ln(\lambda), & \text{when } \lambda = 0 \end{cases} \qquad (9\text{-}43)$$

Minitab provides a point estimate for λ as well as a 95% confidence interval for λ. If the 95% confidence interval contains some of the commonly interpretable transformations (-2, -1, -0.5, 0.5, 2), those are usually selected. Note that $\lambda = -0.5$ corresponds to the square root of the reciprocal transformation.

Using Attribute Charts

Attribute charts are constructed for selected specifications. If the acceptable bounds for quality characteristics are known, the chart for the proportion nonconforming (p-chart) can be constructed to classify the product as conforming or not conforming. The same holds true for the chart for the number nonconforming (np-chart). Similarly, the control chart for the number of nonconformities (c-chart) can be developed if the definition of a nonconformity (i.e., a quality characteristic that does not meet certain specifications) is known. The same also applies to a chart for the number of nonconformities per unit (u-chart) or the number of demerits per unit (U-chart).

For a p-chart, the measure of process capability is the centerline $\bar{p}$, which represents the average proportion of nonconforming items produced by the process. Similarly, for an np-chart, the measure of process capability is the centerline $n\bar{p}$, the average number of nonconforming items. For the c-chart, the capability measure is $\bar{c}$, the average number of nonconformities; for the u-chart, it is $\bar{u}$, the average number of nonconformities per unit. For the U-chart, it is $\bar{U}$, the average number of demerits per unit. These measures are estimates of the overall process capability. The upper control limits on each of these charts represent an upper bound on the nonconforming items or nonconformities. This is a measure of the "worst quality" that can be expected from the process.

Although the centerlines of the attribute charts provide an aggregate measure of process capability, one drawback is that information from attribute charts does not indicate the degree of product nonconformance. In other words, were the products barely outside the specifications, or were they significantly outside? Was the process mean off the target value of the quality characteristic, or was the process variability too high? This information, which is necessary to formulate remedial actions, is not available from attribute charts. Variable charts, on the other hand, provide measures of the process mean and standard deviation and also present possible reasons for product nonconformance, which leads to corrective actions.

Using a Nonparametric Approach

Here, the concept is to estimate *quantiles* of the distribution of the quality characteristic based on the observations. If $X_{0.99865}$ and $X_{0.00135}$ represent the 99.865th and 0.135th percentiles, respectively, a quantile-based process capability ratio is estimated by

$$C_{pq} = \frac{USL - LSL}{X_{0.99865} - X_{0.00135}} \qquad (9\text{-}44)$$

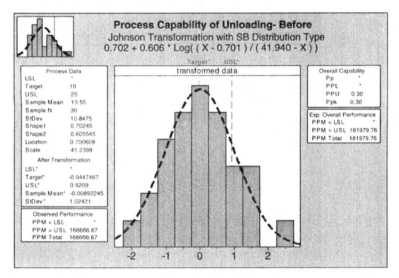

FIGURE 9-13   Process capability analysis of nonnormal data.

Given a set of observations, an empirical estimate of the distribution function and therefore quantiles, can always be obtained. Observe that for a normal distribution, $X_{0.99865} = \mu + 3\sigma$ and $X_{0.00135} = \mu - 3\sigma$.

Example 9-9   Consider the data on the unloading time of supertankers, before process changes, shown initially in Table 5-2. In Example 5-7 it was demonstrated that the distribution of these unloading times did not conform to a normal distribution (p-value $= 0.023$). Suppose that the upper specification limit is 25 hours, with the target value being 10 hours. Perform a capability analysis and comment on the process.

Solution   We use Minitab and execute the commands **Stat** > **Quality Tools** > **Capability Analysis** > **Nonnormal**. After indicating the column in which the data are stored, an option exists to specify from a list of available distributions one to use to determine the degree of fit, or select a Johnson transformation (discussed in Chapter 5). In this example we select the Johnson transformation, input an Upper spec value of 25, and click on **Options**. Here, we input a **Target** value of 10 and click **OK**. Figure 9-13 indicates the Minitab output.

From the figure, observe that the transformed variable is indicated, the distribution of it conforming to normality. A separate test for normality using the Anderson–Darling test confirms this (p-value $= 0.991$). Long-term capability measures of CPU and C_{pk} are 0.30, with the expected nonconformance rate being 18.20%. The current process is not capable. Conducting capability analysis under the assumption of a normal distribution would not be appropriate in this instance. One might verify that the value of C_{pk} in that case would be inflated to 0.35, with the nonconformance rate being 14.77%, which would not be representative of the current process.

9-9   SETTING TOLERANCES ON ASSEMBLIES AND COMPONENTS

Frequently, we need to determine the tolerance or specification limits that should be set on assemblies or subassemblies, given the tolerances of the individual components. In this section we assume that the processes making the assemblies or components are barely

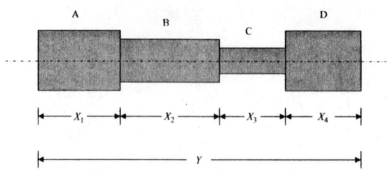

FIGURE 9-14 Assembly of four components.

capable. This means that the spread between the tolerances is the same as the process spread (i.e., six times the process standard deviation). Thus, the process potential as measured by the C_p index is 1, so that virtually all products will be within the specification limits.

Tolerances on Assemblies and Subassemblies

Assemblies and subassemblies are formed by combining two or more **components.** The dimension of interest in an assembly may be the sum or the difference of the individual components. Let's consider an assembly made of the four components shown in Figure 9-14. The components are welded together, and the weld thickness is negligible. The quality characteristic of interest is the length of the assembly, Y. Denoting the length of components A, B, C, and D as X_1, X_2, X_3, and X_4, respectively, the length of the assembly is expressed as

$$Y = X_1 + X_2 + X_3 + X_4 \tag{9-45}$$

Now let's look at another assembly of components A and B, shown in Figure 9-15. Here, the dimension of interest is the exposed length of the longer part, which is given by

$$Y = X_1 - X_2 \tag{9-46}$$

In general, the dimension of interest is expressed as a linear combination of some individual component dimensions, for example,

$$Y = a_1 X_1 + a_2 X_2 + \cdots + a_k X_k$$
$$= \sum_{i=1}^{k} a_i X_i \tag{9-47}$$

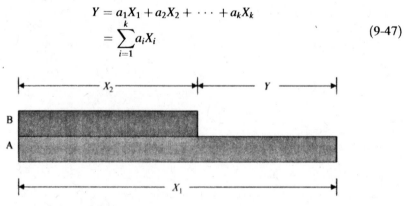

FIGURE 9-15 Assembly of two components.

where k represents the number of different individual dimensions. Note that the component dimensions X_i and hence the assembly dimensions Y are random variables. Suppose that the component dimensions X_i are random variables with mean μ_i and standard deviation σ_i.

Two important properties of the characteristic Y are its mean and variance. If the component dimensions are independent of each other, the mean of Y is given by

$$\mu_Y = \sum_{i=1}^{k} a_i \mu_i \tag{9-48}$$

where μ_i is the mean of the variable X_i and the values of a_i are constants. So, in Figure 9-14 $k = 4$, $a_1 = 1$, $a_2 = 1$, $a_3 = 1$, and $a_4 = 1$. The mean of Y is $\mu_Y = \mu_1 + \mu_2 + \mu_3 + \mu_4$. For the two-component assembly in Figure 9-15, $\mu_Y = \mu_1 - \mu_2$.

In general, the variance of the dimension Y is given by

$$\text{Var}(Y) = \sigma_Y^2 = \sum_{i=1}^{k} a_i^2 \, \text{Var}(X_i) = \sum_{i=1}^{k} a_i^2 \sigma_i^2 \tag{9-49}$$

Equation (9-49) states that the variance of the dimension of interest (Y) is the sum of the weighted variances of the individual dimensions; the weights are the square of the coefficients of the linear combination. We assume that the individual dimensions are independent of each other.

Using eq. (9-49), we get the following for the four-component assembly in Figure 9-14: $\sigma_Y^2 = \sigma_1^2 + \sigma_2^2 + \sigma_3^2 + \sigma_4^2$. For the two-component assembly in Figure 9-15: $\sigma_Y^2 = \sigma_1^2 + \sigma_2^2$.

If we can assume a certain distribution of the component dimension, we can derive the distribution for the assembly dimension. In particular, if each X_i is normally distributed with mean μ_i and standard deviation σ_i and is independent of all other X_i, the distribution of Y will also be normal, with mean given by eq. (9-48) and variance given by eq. (9-49). We can use this property to find the proportion of assemblies whose dimension of interest lies between certain bounds. Assuming a normal distribution of Y, the natural tolerance limits are

$$\mu_Y \pm 3\sigma_Y \tag{9-50}$$

where σ_Y represents the standard deviation of the dimension Y.

Example 9-10   Refer to the four-component assembly shown in Figure 9-14. Suppose that the mean lengths of the four components and their respective tolerances are as shown in the following table:

Component	Mean Length (cm)	Tolerances (cm)
A	2	2 ± 0.3
B	5	5 ± 0.2
C	6	6 ± 0.2
D	7	7 ± 0.1

Assuming a normal distribution for the individual component dimensions, find the natural tolerance limits for the assembly length. The design specifications for assembly length are

20 ± 0.3 cm. What proportion of the assemblies will be nonconforming? Comment on the process capability to make assemblies that meet the design specifications.

Solution   We will assume that the tolerances on the individual components have been set such that they are at the natural tolerance limits. The standard deviations of the component dimensions are estimated assuming a 6σ spread between the tolerances. Our results are as follows:

Component A: $\sigma_1 = (2.3 - 1.7)/6 = 0.100$ cm

Component B: $\sigma_2 = (5.2 - 4.8)/6 = 0.067$ cm

Component C: $\sigma_3 = (6.2 - 5.8)/6 = 0.067$ cm

Component D: $\sigma_4 = (7.1 - 6.9)/6 = 0.033$ cm

The mean assembly length is found using eq. (9-48): $\mu_Y = \mu_1 + \mu_2 + \mu_3 + \mu_4 = 20$ cm. The variance of the assembly length is found using eq. (9-49): $\sigma_Y^2 = (0.1)^2 + (0.067)^2 + (0.067)^2 + (0.033)^2 = 0.020$. The standard deviation of the assembly length is $\sigma_Y = \sqrt{0.020} = 0.142$ cm.

From eq. (9-50), the natural tolerance limits for the assembly length are

$$\mu_Y \pm 3\sigma_Y = 20 \pm (3)(0.142) = 20 \pm 0.426$$
$$= (19.574, 20.426) \text{ cm}$$

Since the individual component lengths are normally distributed, the assembly lengths will also be normally distributed with a mean of 20 cm and a standard deviation of 0.142 cm. Furthermore, virtually all (99.74%) the assemblies will have a length between 19.574 and 20.426 cm.

The design specifications for the assembly length are 20 ± 0.3 cm. The proportion of nonconforming assemblies is now calculated. The standardized normal values at the USL and LSL are

$$Z_{USL} = \frac{20.3 - 20}{0.142} = 2.11 \qquad Z_{LSL} = \frac{19.7 - 20}{0.142} = -2.11$$

From the standard normal tables in Appendix A-3, the proportion above the USL is 0.0174, and that below the LSL is also 0.0174. The total proportion of nonconforming assemblies is 0.0348, or 3.48%.

Tolerance Limits on Individual Components

The tolerances or specification limits on an assembly are often determined from product function and customer needs. The designer determines individual component tolerances that will give the specified assembly tolerances. As mentioned previously, although product design should incorporate customer preferences, the design must also be realistic and achievable. If the process that creates the assembly is not capable of meeting specifications, this must be addressed in the product/process design phase. With this in mind, we now demonstrate how tolerances of components are determined.

Example 9-11   Refer to the four-component assembly shown in Figure 9-14. Suppose that the lengths of the components are normally and independently distributed with the following means:

Component	Mean Length (cm)
A	2
B	5
C	6
D	7

If the specifications for the assembly length are 20 ± 0.3 cm, what should the individual component tolerances be? Assume that the component tolerances are equal and that the specifications are barely equal to the natural tolerance limits, implying a capability ratio of 1.

Solution The length of the assembly, Y, is the sum of the individual component lengths; that is, $Y = X_1 + X_2 + X_3 + X_4$. Because the X-values are normally distributed, Y is normally distributed. Assuming that the specification limits equal the natural tolerance limits, the standard deviation of the assembly length σ_Y, is estimated as $\sigma_Y = (20.3 - 19.7)/6 = 0.100$.

Now, assuming the individual component lengths to be independent, using eq. (9-57), we have $\sigma_Y^2 = \sigma_1^2 + \sigma_2^2 + \sigma_3^2 + \sigma_4^2$, where σ_i^2 represents the variance of the dimension X_i (for $i = 1$, 2, 3, 4). Assuming that $\sigma_1^2 = \sigma_2^2 = \sigma_3^2 = \sigma_4^2$, we get $\sigma_Y^2 = 4\sigma_1^2$, yielding $\sigma_1^2 = 0.0025$. The standard deviation of X_1 is $\sigma_1 = \sqrt{0.0025} = 0.05$ cm. Therefore, $\sigma_2 = \sigma_3 = \sigma_4 = \sigma_1 = 0.05$ cm.

The tolerances on the individual component dimensions are as follows:

Component	Tolerance (cm)
A	$2 \pm (3)(0.05) = (1.85, 2.15)$
B	$5 \pm (3)(0.05) = (4.85, 5.15)$
C	$6 \pm (3)(0.05) = (5.85, 6.15)$
D	$7 \pm (3)(0.05) = (6.85, 7.15)$

In the manufacture of these components, the process capability should be determined and compared to these natural tolerances. Doing so will enable us to evaluate whether the process is capable of making components that meet these desired tolerances.

Tolerance on Mating Parts

Mating parts (e.g., a shaft and a bearing, a pin and a sleeve, a piston and a cylinder) represent a special form of assembly. In such assemblies, the type of fit between the mating parts is classified into three categories.

Clearance Fit   For a **clearance fit**, the size of the hole prior to assembly is always larger than the size of the shaft. In the assembly there is always some room for play between the shaft and the hole. The natural tolerance limits for the hole are outside the natural tolerance limits for the shaft; in no instance will a shaft be larger than the hole. Figure 9-16 shows an assembly in which the shaft diameter X_s is smaller than the hole diameter X_h. The clearance is also shown. Normality is assumed for the distribution of both the shaft and

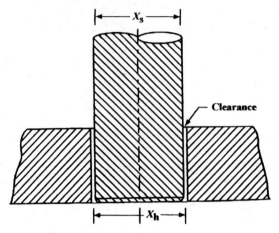

FIGURE 9-16   Clearance fit in an assembly.

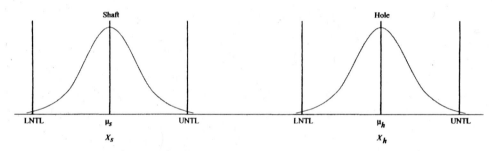

FIGURE 9-17   Distribution of shaft and hole diameters in a clearance fit.

hole diameters; Figure 9-17 indicates their relative positioning. The mean shaft diameter is indicated by μ_s, and the mean hole diameter is denoted by μ_h. Given the natural tolerance limits for each, the shafts will all be smaller than the smallest hole, which will therefore provide a clearance fit. The assembly of a piston within a cylinder in an automobile engine is an example of such a fit.

Interference Fit   For an **interference fit**, the size of the hole before assembly is always smaller than the size of the shaft. The shaft therefore has to be forced into the hole. The range of the natural tolerance limits for the hole is below that for the shaft, and all hole diameters will be smaller than that of the smallest shaft. An example of such a fit is a pin that is forced into a sleeve to stay in place. Such a condition is illustrated by switching the hole and shaft diameter distributions in Figure 9-17.

Transition Fit   For a **transition fit**, there is either a clearance or an interference in the assembly. In some instances, the hole diameter will be larger than the shaft diameter, providing a clearance; in others, the shaft diameter will be greater than the hole diameter, in which case, an interference fit will result. The natural tolerance ranges of the hole and shaft can overlap. Depending on the relative location of the means and the variability of the hole and shaft diameters, an assembly will be either a clearance or an interference fit. Figure 9-18 shows the distributions of the shaft and the hole diameters, indicated by X_s and X_h. Note that

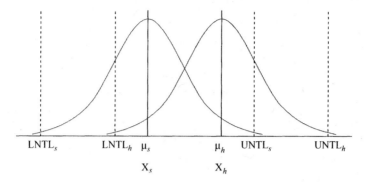

FIGURE 9-18   Distribution of shaft and hole diameters in a transition fit.

there is some overlap between the natural tolerance limits of the shaft (LNTL$_s$, UNTL$_s$) and that of the hole (LNTL$_h$, UNTL$_h$).

Example 9-12   The specifications for the outside diameter of a shaft are 9.0 ± 0.10 cm and those for the inside diameter of a bearing are 9.1 ± 0.13 cm. Assume that it is possible to make each component such that the natural tolerance limits equal the specifications (implying a process capability ratio of 1). Also assume that the parts are produced independently and that the diameters are normally distributed with their means at the respective target values. If a clearance fit is desired between the shaft and the bearing, what proportion of the assemblies will be unacceptable?

Solution   Let X_s denote the shaft outside diameter and X_b the bearing inside diameter. Let d represent the difference between the bearing inside diameter and the shaft outside diameter (i.e., $d = X_b - X_s$). The mean of the shaft diameter, given by μ_s is 9.0 cm; the mean of the bearing diameter, μ_b is 9.1 cm. The mean of the difference between the bearing and the shaft diameters is $\mu_d = \mu_b - \mu_s = 9.1 - 9.0 = 0.1$ cm. The standard deviation of the shaft outside diameter is found from the relation: $6\sigma_s = 9.1 - 8.9 = 0.2$, yielding $\sigma_s = 0.033$ cm. The standard deviation of the bearing inside diameter is found from: $6\sigma_b = 9.23 - 8.97 = 0.26$, yielding $\sigma_b = 0.043$ cm.

Since $d = X_b - X_s$, the variance of the difference between the bearing and shaft diameters is found as: $\text{Var}(d) = \sigma_b^2 + \sigma_s^2 = (0.043)^2 + (0.033)^2 = 0.00294$. The standard deviation of d is $\sigma_d = \sqrt{0.00294} = 0.054$ cm.

Since the bearing and shaft diameters are each independently normally distributed, the distribution of the difference d between the bearing and shaft diameters is also normally distributed with mean $\mu_d = 0.1$ and standard deviation $\sigma_d = 0.054$ cm.

Since a clearance fit is desired, unacceptable assemblies are those for which $d < 0$. To determine the proportion of unacceptable assemblies, the standard normal value is found at $d = 0$:

$$Z = \frac{0 - \mu_d}{\sigma_d} = \frac{0 - 0.10}{0.054} = -1.85$$

Based on the standard normal tables in Appendix A-3, the proportion of nonconforming assemblies is 0.0322. In 3.22% of the assemblies, the shaft outside diameter will exceed the bearing inside diameter and therefore be unacceptable.

Nonlinear Combinations of Random Variables

In some situations, the variable of interest is a nonlinear function of the individual variables $(X_i, i = 1, 2, \ldots, n)$. For example, the volume of a rectangular container is equal to the product of the length, width, and height. In containers used for transporting goods, the volume may be the characteristic of interest. Assume that a derived variable (Y) is given by

$$y = f(x_1, x_2, \ldots, x_n) \tag{9-51}$$

If the nominal or target values of the X_i's are given by μ_i, $i = 1, 2, \ldots, n$, using a Taylor series expansion around these nominal values, we have

$$y = f(\mu_1, \mu_2, \ldots, \mu_n) + \sum_{i=1}^{n} (x_i - \mu_i) \left.\frac{\partial f}{\partial x_i}\right|_{\mu_1, \mu_2, \ldots, \mu_n} + R \tag{9-52}$$

where R represents higher-order terms. If higher-order terms are neglected, the mean and the variance are approximately given by

$$\mu_y \simeq f(\mu_1, \mu_2, \ldots, \mu_n) \tag{9-53}$$

$$\sigma_y^2 \simeq \sum_{i=1}^{n} \left(\left.\frac{\partial f}{\partial x_i}\right|_{\mu_1, \mu_2, \ldots, \mu_n}\right)^2 \sigma_i^2 \tag{9-54}$$

where σ_i^2 represents the variance of X_i.

Example 9-13   A logistics company uses containers for loading and unloading from ships to railroad cars. The specifications on the container volume (V) are $60 \pm 0.5 \, \text{m}^3$. The specifications on the length (L), width (W), and height (H) of containers barely equal their natural tolerance limits and are given (in meters) by L: 4 ± 0.06; W: 3 ± 0.03; H: 5 ± 0.03. Assume that these dimensions are normally and independently distributed with their means at the nominal values. Find the natural tolerance limits on volume. What proportion of the containers will be nonconforming? What is the C_p index for volume?

Solution   The defining relation is given by $V = WLH$. The mean volume (μ_V) is obtained approximately as

$$\mu_V = \mu_W \mu_L \mu_H = (3)(4)(5) = 60$$

The standard deviation of the container dimensions is obtained, under the assumption of the capability ratio being 1. We have $\sigma_W = 0.03/3 = 0.01$, $\sigma_L = 0.06/3 = 0.02$, and $\sigma_H = 0.03/3 = 0.01$. The variance of volume (σ_V^2) is obtained approximately as

$$\sigma_V^2 \simeq (\mu_L \mu_H)^2 \sigma_W^2 + (\mu_W \mu_H)^2 \sigma_L^2 + (\mu_W \mu_L)^2 \sigma_H^2$$
$$= (20)^2 (0.01)^2 + (15)^2 (0.02)^2 + (12)^2 (0.01)^2 = 0.1444$$

This yields $\sigma_V = 0.38$. The natural tolerance limits on container volume are (under the assumption of normality)

$$\mu_V \pm 3\sigma_V \quad \text{or} \quad 60 \pm 1.14 \, \text{m}^3$$

Using the specifications on volume, the proportion of containers that are conforming is found to be

$$\Phi\left(\frac{60.5 - 60}{0.38}\right) - \Phi\left(\frac{59.5 - 60}{0.38}\right)$$
$$= \Phi(1.32) - \Phi(-1.32) = 0.9066 - 0.0934 = 0.8134$$

The percentage of nonconforming containers $= 1 - 0.8134 = 0.1866 = 18.66\%$. Finally, the process capability ratio is obtained as

$$C_p = \frac{60.5 - 59.5}{(6)(0.38)} = 0.439$$

This indicates that the current process is not capable.

9-10   ESTIMATING STATISTICAL TOLERANCE LIMITS OF A PROCESS

The natural tolerance limits define the inherent capability of a process. Consequently, it is of interest to estimate these limits once a process is in statistical control. If control charts are used to bring a process to control, information from these charts can be used to determine the natural tolerance limits of the process. We discussed this approach earlier. In this approach we assume that reliable estimates based on elaborate historical data are available for the process mean and process standard deviation. If, however, data from a small sample are used to generate estimates of the process mean and standard deviation, statistical tolerance limits can be found. These limits are estimates of the natural tolerance limits, and as the sample size increases, the statistical tolerance limits approach the natural tolerance limits.

Let's suppose that a desired level of the process capability index (assuming two-sided specification limits) is 1.33. As discussed previously, for this condition to be met, the actual process spread must be three-fourths of the specification limit spread. In other words, the specification limits must be 4 standard deviations from the process mean, and the natural tolerance limits must be 3 standard deviations from the process mean.

Statistical Tolerance Limits Based on Normal Distribution

Let's suppose that the quality characteristic X is normally distributed with mean μ and standard deviation σ. Statistical tolerance limits that encompass $100(1 - \alpha)\%$ of the product can be constructed as

$$\mu \pm Z_{\alpha/2}\sigma$$

In practice, however, the process mean and standard deviation are usually unknown; they are estimated from sample estimates. So for a sample of size n, the sample mean is $\bar{X}$ and the sample standard deviation is s. Since $\bar{X}$ and s are estimates of μ and σ it is not necessarily true that $\bar{X} \pm Z_{\alpha/2}s$ will include $100(1 - \alpha)\%$ of the distribution. Note that $\bar{X}$ and s are random variables, whereas μ and σ are constants. Different samples will lead to different values of $\bar{X}$

and s, which will in turn lead to different estimates of the limits. Some of these intervals may be very different from $\mu \pm Z_{\alpha/2}\sigma$ and may not include $100(1 - \alpha)\%$ of the product.

Values of a constant k are tabulated such that in a large proportion (γ) of these intervals $\bar{X} \pm ks$, at least $100(1 - \alpha)\%$ of the distribution will be included. This constant k depends on the sample size n, the level of confidence γ, and the percentage of the distribution, $100(1 - \alpha)\%$ that is to be minimally included in the interval. Two-sided statistical tolerance limits are calculated from

$$\bar{X} \pm ks \qquad (9\text{-}55)$$

where k is found from tables in ISO 16269–6(2005): Determination of statistical tolerance intervals.

If one-sided statistical tolerance limits are desired, a one-sided upper statistical tolerance limit is $\bar{X} + ks$, and a one-sided lower statistical tolerance limit is $\bar{X} - ks$.

Note that there is a fundamental difference between confidence intervals and statistical tolerance limits. Confidence intervals are found for a parameter of the process. A 95% confidence interval for the process mean signifies that if a large number of such confidence intervals are constructed, 95% of them will enclose the process mean. Statistical tolerance limits, on the other hand, are designed to contain at least a specified proportion, $100(1 - \alpha)\%$, of the population, with a certain probability γ. As the sample size becomes large, the width of the confidence interval diminishes. In the limiting case, as the sample size n approaches infinity, the width of the confidence interval approaches zero. The statistical tolerance limits, in this case where $n \rightarrow \infty$, approach the corresponding limits for the population. Hence, for $1 - \alpha = 0.95$ and a two-sided confidence interval, the value of k approaches 1.96 as n becomes large. The value of 1.96 is the standardized normal value (Z-value) for $1 - \alpha = 0.95$.

Nonparametric Statistical Tolerance Limits

Nonparametric statistical tolerance limits do not depend on the distribution of the quality characteristic. These limits are valid for any continuous probability distribution. They are based on the largest and smallest observations in the sample.

For two-sided tolerance limits, the number of observations n required to obtain a probability γ that at least $100(1 - \alpha)\%$ of the distribution will lie between the smallest and largest sample observations is

$$n \simeq 0.5 + \left(\frac{2 - \alpha}{\alpha}\right) \frac{\chi^2_{1-\gamma,4}}{4} \qquad (9\text{-}56)$$

Here $\chi^2_{1-\gamma,4}$ is the upper $100(1 - \gamma)$ percentile point of the chi-squared distribution with 4 degrees of freedom; χ^2 is tabulated in Appendix A-5.

If a one-sided nonparametric lower statistical tolerance limit is desired such that there is a probability γ that at least $100(1 - \alpha)\%$ of the population is greater than the smallest sample value, the sample size is given by

$$n = \frac{\ln(1 - \gamma)}{\ln(1 - \alpha)} \qquad (9\text{-}57)$$

Equation (9-57) is also used to determine the sample size required to construct a one-sided nonparametric upper statistical tolerance limit. There is then a probability γ that at least 100 $(1 - \alpha)\%$ of the population will be less than the largest value.

Nonparametric tolerance limits usually require large sample sizes when both the desired probability level and the desired percentage of the population to be included in the limits are high. Available information on the quality characteristic should be used to estimate the form of the distribution. The sample size needed to achieve a comparable degree of confidence and proportion of population inclusion within the tolerance limits for various distributions is usually less than that for nonparametric tolerance limits.

Example 9-14 Compute the sample size for two-sided nonparametric tolerance limits. They should contain 95% of the population with a probability of 0.99. The quality characteristic is the concentration of potassium in a chemical compound in parts per million.

Solution We have $1 - \alpha = 0.95$ and $\gamma = 0.99$. From the chi-squared tables in Appendix A-5, we have $\chi^2_{0.01,4} = 13.28$. The required sample size is

$$n = 0.5 + \frac{2 - 0.05}{0.05}\left(\frac{13.28}{4}\right) = 129.98 \simeq 130$$

Thus, a sample of size 130 is chosen. After ranking the sample values in ascending order, the minimum and maximum values are 28 and 49. The nonparametric tolerance limits are (28,49). We are confident with a probability level of 0.99, that 95% of the population will be contained within these limits.

Example 9-15 Compute the sample size for a one-sided upper nonparametric tolerance limit. It should contain 98% of the distribution with a probability of 0.95. The quality characteristic is the number of grams of fat in 10 kg of processed poultry.

Solution We have $1 - \alpha = 0.98$ and $\gamma = 0.95$. Using eq. (9-57), the required sample size is

$$n = \frac{\ln(1 - \gamma)}{\ln(1 - \alpha)} = \frac{\ln(0.05)}{\ln(0.98)} = 148.28 \simeq 149$$

Thus, a sample of size 149 is chosen. The values in the sample are ranked; the maximum value is 36 g. We are confident with a probability level of 0.95 that 98% of the population will have a fat content of less than 36 g per 10 kg.

SUMMARY

In this chapter we have dealt with an important aspect that concerns all processes; the determination of the inherent capability of the process. Capability analysis should be performed on processes that are in control. This allows us to infer whether the process will be able to produce items that conform to desired specifications. For two-sided limits, the process spread must be less than the specification spread For the process to be considered capable. The process must be centered at the most desirable location to minimize the production of nonconforming items. For one-sided specification limits, the process mean and spread should be such that virtually all items will meet the specifications. Changing the process mean, which requires an adjustment of certain process parameters, is usually an easier task than reducing the process variability. The process spread may be reduced through fundamental changes in the process initiated by management.

Various measures of process capability involving the process mean, process standard deviation, and specification limits have been developed in this chapter. Procedures have been described for determining the proportion of nonconforming product. Procedures for determining process capability have been discussed. We also described the determination of tolerances for assemblies and subassemblies and the setting of component tolerances depending on assembly specifications. Methods for determining statistical tolerance limits have also been discussed.

The procedures described in this chapter help us decide whether an existing process is capable or needs to be changed. Continual process improvement is a goal that should be adopted by every company.

KEY TERMS

assemblies
Box–Cox transformation
capability ratio
clearance fit
components
gage
  bias
  linearity
  R&R
    repeatability
    reproducibility
  stability
interference fit
mating parts
measurement errors
natural tolerance limits
precision-to-tolerance ratio
process capability
process capability analysis

process capability indices
  C_p
  C_{pk}
  C_{pm}
  C_{pmk}
  C_{pq}
  lower capability index
  upper capability index
process capability limits
process spread
specification limits
statistical tolerance limits
  based on normality assumption
  nonparametric approach
tolerance limits
  lower tolerance limit
  upper tolerance limit
transition fit

EXERCISES

Discussion Questions

9-1 Explain the difference between specification limits and control limits. Is there a desired relationship between the two?

9-2 Explain the difference between natural tolerance limits and specification limits. How does a process capability index incorporate both of them? What assumptions are made in constructing the natural tolerance limits?

9-3 What are statistical tolerance limits? Explain how they differ from natural tolerance limits.

9-4   Is it possible for a process to be in control and still produce nonconforming output? Explain. What are some corrective measures under these circumstances?

9-5   What are the advantages of having a process spread that is less than the specification spread? What should be the value of C_p be in this situation? Could C_{pk} be ≤ 1 here?

9-6   Compare the capability indices C_{pk}, C_{pm}, and C_{pmk}, and discuss what they measure in the process. When would you use C_{pq}?

9-7   What condition must exist prior to calculating the process capability? Discuss how process capability can be estimated through control charts.

9-8   Suppose that the time to complete a project is the sum of several independent operations. If the means and standard deviations of the independent operations are known, determine the mean and standard deviation of the project completion time. If the operations are not independent, what effect will this have on the mean and standard deviation of completion time?

9-9   Suppose that the dimension of an assembly has to be within certain tolerances. Discuss how tolerances could be set for the components, given that the difference between two component dimensions comprises this assembly dimension. Assume that the inherent variability of each component is equal.

9-10 Discuss the importance of identifying an appropriate distribution of the quality characteristic in process capability analysis. Address this in the context of waiting time for service in a fast-food restaurant during lunch hour.

9-11 Discuss how the precision of a measurement system affects the process potential in the context of measuring unloading times of supertankers. What bounds exist on the observed process potential?

9-12 Distinguish between gage repeatability and gage reproducibility in the context of measuring unloading times of supertankers.

Problems

9-13 A pharmaceutical company producing vitamin capsules desires a proportion of calcium content between 40 and 55 ppm. A random sample of 20 capsules chosen from the output yields a sample mean calcium content of 44 ppm with a standard deviation of 3 ppm. Find the natural tolerance limits of the process. If the process is in control at the present values of its parameters, what proportion of the output will be nonconforming, assuming a normal distribution of the characteristic?

9-14 For Exercise 9-13, find the C_p index. Comment on the ability of the process to meet specifications. What proportion of the specification range is used up by the process? If it is easier to change the process mean than to change its variability, to what value should the process mean be set to minimize the proportion of nonconforming product?

9-15 The emergency service unit in a hospital has a goal of 3.5 minutes for the waiting time of patients before being treated. A random sample of 20 patients is chosen and the sample average waiting time is found to be 2.3 minutes with a sample standard deviation of 0.5 minutes. Find an appropriate process capability index. Comment on the ability of the emergency service unit to meet the desirable goal, assuming normality. What are some possible actions to consider?

9-16 Refer to Exercise 9-13. Find the process capability index C_{pk}, and comment on process performance. If the target value is 47.5 ppm, find the C_{pm}, and C_{pmk} indices and comment on their values. If the process center is shifted to the midpoint between the specification limits, what proportion of the product will be nonconforming? Has it improved relative to the present setting of the process mean?

9-17 The diameter of a forged part has specifications of 120 ± 5mm. A sample of 25 parts chosen from the process gives a sample mean of 122 mm with a sample standard deviation of 2 mm.
 (a) Find the C_{pk} index for the process, and comment on its value. What is the proportion of nonconforming parts assuming normality? If the target value is 120 mm, find the C_{pm} and C_{pmk} indices and comment on their values. If the process mean is to be set at the target value, how much of a reduction would occur in the proportion nonconforming?
 (b) Parts with a diameter below the lower specification limit cost $1.00 per part to be used in another assembly; those with a diameter above the upper specification limit cost $0.50 per part for rework. If the daily production rate is 30,000 parts, what is the daily total cost of nonconformance if the process is maintained at its current setting? If the process mean is set at the target value, what is the daily total cost of nonconformance?

9-18 The waiting time in minutes before being served in a local post office is observed for 50 randomly chosen customers:
 (a) Test for normality using $\alpha = 0.05$. What inferences can you draw?
 (b) Estimate the mean and standard deviation of the waiting times.
 (c) If the goal of the post office is for the waiting time not to exceed 4 minutes, find the capability indices CPU and C_{pk} and comment on these values. Assuming normality, what proportion of the customers, if any, will have to wait for more than 4 minutes?

2.1	0.5	3.6	1.4	2.0	1.9	2.4	2.7	2.1	1.8
0.8	0.4	4.2	3.5	2.5	4.6	3.8	1.5	4.5	3.9
4.8	2.8	1.9	1.2	3.2	5.5	2.5	3.8	5.0	4.6
1.6	2.5	2.4	1.9	2.0	2.1	2.8	1.6	3.8	4.2
3.5	5.2	3.1	1.6	1.5	3.5	5.2	4.8	3.9	2.6

9-19 A major automobile company is interested in reducing the time that customers have to wait while having their car serviced with one of the dealers. They select four customers randomly each day and find the total time that each of those customers has to wait (in minutes) while having his or her car serviced. Next, from these four observations, the sample average and range are found. This process is repeated for 25 days.

The summary data for these observations are

$$\sum_{i=1}^{25} \bar{X}_i = 1000 \quad \sum_{i=1}^{25} R_i = 250$$

(a) Find the $\bar{X}$- and R-chart control limits.

(b) Assuming that the process is in control and that a desirable value on the upper bound of the waiting time is 50 minutes, calculate a process capability index, and comment on its value.

(c) Assuming a normal distribution of waiting times, find the proportion of customers that will have to wait more than 50 minutes.

(d) To reduce the waiting time of customers, the service manager hires some additional mechanics, which reduces the average waiting time to 35 minutes. What proportion of the customers will still have to wait more than 50 minutes if the variability in service times is the same as before?

9-20   Light bulbs are tested for their luminance, with the intensity of brightness desired to be within a certain range. Random samples of 5 bulbs are chosen from the output and their luminance values measured. The sample mean $\bar{X}$ and standard deviation s are found. After 30 samples, the following summary information is obtained:

$$\sum_{i=1}^{30} \bar{X}_i = 2550 \quad \sum_{i=1}^{30} s_i = 195$$

The specifications are 90 ± 15 lumens.

(a) Find the control limits for the $\bar{X}$- and s-charts.

(b) Assuming that the process in control, estimate the process mean and process standard deviation.

(c) Find the process capability indices C_p and C_{pk} and comment on their values.

(d) If the target value is 90 lumens, find the capability indices C_{pm} and C_{pmk}.

(e) What proportion of the output is nonconforming, assuming a normal distribution of the quality characteristic?

(f) If the process mean is moved to 88 lumens, what proportion of the output is nonconforming? What are your proposals to improve process performance?

9-21   The amount of a preservative added to dairy products should not exceed certain levels of 23 ± 6 mg (set by the Food and Drug Administration). Samples of size 5 of processed cheese produced the values of the average and range shown in Table 9-6.

(a) Construct appropriate control charts and determine stability of the process.

(b) If the process is out of control, assuming remedial actions will be taken, estimate the process mean and standard deviation.

(c) Assuming normality and a target value of 23 mg, determine the indices C_p, C_{pk}, C_{pm} and C_{pmk}.

(d) What proportion of the dairy products meets government standards, assuming normality?

(e) Find a 95% confidence interval for C_{pk}, assuming normality.

(f) Can we conclude that C_{pk} is less than 1? Use $\alpha = 0.05$.

TABLE 9-6

Observation	Average Level of Preservative (mg)	Range	Observation	Average Level of Preservative (mg)	Range
1	22	5	14	22	7
2	26	4	15	20	5
3	26	6	16	24	8
4	24	7	17	25	6
5	22	3	18	23	8
6	21	5	19	20	5
7	29	7	20	22	4
8	25	8	21	22	6
9	22	4	22	23	7
10	25	6	23	24	5
11	25	9	24	22	6
12	22	3	25	25	5
13	21	5			

9-22   Consider the assembly of three components shown in Figure 9-19. The tolerances for these three components are given in Table 9-7. Assume that the tolerances on the components are independent of each other and that the lengths of the components are normally distributed with a capability ratio of 1. What is the tolerance of the gap? Assuming normality, if specifications for the gap are 0.9 ± 0.201 cm, what proportion of the assemblies will not meet specifications? How could the proportion of nonconforming assemblies be reduced?

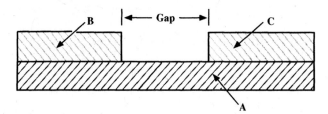

FIGURE 9-19   Assembly of three components.

TABLE 9-7

Component	Mean Length (cm)	Tolerance (cm)
A	10	10 ± 0.5
B	4	4 ± 0.2
C	5	5 ± 0.1

9-23   In Exercise 9-22, suppose that the specifications for the gap are 1.05 ± 0.15 cm. An assembly with a gap exceeding the upper specification limit is scrapped, whereas that with a gap less than the lower specification limit can be reworked to increase the gap dimension. The unit cost of rework is \$0.15 and that for scrap is \$0.40. If the daily production rate is 2000, calculate the daily total cost of scrap and rework. How can this cost be reduced?

9-24   Refer to the four-component assembly shown in Figure 9-14. Assume that the length of each component is normally and independently distributed with the means shown in Table 9-8. The specifications for the assembly length are 35 ± 0.5 cm. Assuming that the natural tolerances (or process spread) for all four components are equal to each other and that the specifications barely match the natural tolerance limits, find the tolerances on the individual components.

TABLE 9-8

Component	Mean Length (cm)
A	3
B	8
C	10
D	14

9-25   Refer to Exercise 9-24. Suppose that the specifications for the assembly length are 35 ± 0.3 cm and that the tolerances of A and C are equal, but those for B and D are each twice as large as that for A. In addition, assume that the specifications are barely equal to the natural tolerance limits. Find the tolerances for each component.

9-26   Consider the two-component assembly shown in Figure 9-15. Suppose that the specifications for the dimension X_2 are 5 ± 0.05 cm and those for X_1 are 12 ± 0.15 cm. Find the specifications for the dimension Y. Assume that the specification limits equal the natural tolerance limits. For what proportion of the assemblies will the dimension Y exceed the value 7.10 cm? Assume that the component dimensions X_1 and X_2 are normally distributed.

9-27   Consider the two-component assembly shown in Figure 9-15. Suppose that the mean lengths are given as $\mu_1 = 14$ cm and $\mu_2 = 8$ cm. Assuming that the specifications for Y are 6 ± 0.2 cm, what are the tolerances for X_1 and X_2? Assume that the variance of X_1 is three times as large as that of X_2.

9-28   Four metal plates, each of thickness of 3 cm, are welded together to form a subassembly. The specifications for the thickness of each plate are 3 ± 0.2 cm. Assuming the weld thickness to be negligible, determine the tolerances for the assembly thickness.

9-29   Consider Figure 9-16, which shows the assembly of a shaft in a bearing. The specifications for the shaft diameter are 6 ± 0.06 cm, and those for the hole diameter are 6.2 ± 0.03 cm.
(a) Find the probability of the assembly having a clearance fit.
(b) What is the probability of the assembly having an interference fit?

9-30   The specifications for a shaft diameter in an assembly are 5 ± 0.03 cm, and those for the hole are 5.25 ± 0.08 cm. If the assembly is to have a clearance of 0.18 ± 0.05 cm, what proportion of the assemblies will be acceptable?

9-31   Refer to Exercise 9-30. If there is too much clearance between the hole and the shaft, a wobble will result. Clearances above 0.05 cm are not desirable and cause a wobble. Find the probability of a wobble.

9-32 In a piston assembly, the specifications for the piston diameter are 12 ± 0.5 cm, and those for the cylinder diameter are 12.10 ± 0.4 cm. Assume that the natural tolerance limits coincide with the specifications. A clearance fit is required for the assembly. What proportion of the assemblies will be nonconforming, assuming a normal distribution of the piston and cylinder diameters? Clearances more than 0.8 cm are undesirable. What proportion of the assemblies will not meet this stipulation?

9-33 A logistics firm has identified four operations, which are to be conducted in succession, for an order to be processed. The tolerances (in hours) are shown in Table 9-9. Assume that the tolerances are independent of each other and that the time in each phase is normally distributed. Further assume that for each operation, the processes are barely capable.
(a) Find the natural tolerance limits for order completion time.
(b) If the company sets a goal of 23.5 hours, what proportion of the orders will fail to satisfy this goal?
(c) Find an appropriate capability index and comment.
(d) Using a methods study, the company has improved operation 3 to a mean time of 7.0 hours. What proportion of the orders will now meet the goal?

TABLE 9-9

Operation	Mean Time (hours)	Tolerance (hours)
1	6	6 ± 0.6
2	4	4 ± 0.6
3	8	8 ± 0.8
4	5	5 ± 0.3

9-34 Measurements on the pH values of a chemical compound are taken at random by two operators. Fifteen samples are randomly chosen, with each operator measuring each sample twice. The data are shown in Table 9-10. Specifications on pH are 6.5 ± 0.05. Comment on the capability of the measurement system. Calculate % gage R&R, precision-to-tolerance ratio, % of process variation, and number of distinct categories and comment on them. What are the observed process potential and the true process potential? Find a 95% confidence interval for C_p.

9-35 Consider the data on call waiting time of customers in a call center (Exercise 5-8). The call center has set a goal of waiting time not to exceed 35 seconds.
(a) Test to see (using $\alpha = 0.05$) if conducting capability analysis using normal distribution is appropriate.
(b) If not, consider a Box–Cox transformation and conduct capability analysis. Report appropriate capability indices and the percentage nonconformance.
(c) Consider conducting capability analysis using a Weibull distribution. Comment on the results.
(d) What are the drawbacks of conducting a capability analysis using the normal distribution in this example?

9-36 A cylindrical piece is used in an assembly in which the weight is to be controlled. The tolerances on diameter and height, on the basis of 5 observations, are 2 ± 0.06 cm and

TABLE 9-10

Sample Number	Operator 1's pH Values		Operator 2's pH Values	
	Measurement 1	Measurement 2	Measurement 1	Measurement 2
1	6.28	6.24	6.25	6.26
2	6.12	6.18	6.06	6.11
3	6.64	6.58	6.61	6.69
4	6.85	6.92	6.78	6.84
5	6.92	6.83	6.85	6.91
6	6.43	6.52	6.38	6.45
7	6.36	6.44	6.35	6.42
8	6.50	6.56	6.46	6.51
9	6.75	6.68	6.69	6.75
10	6.68	6.60	6.65	6.72
11	6.34	6.39	6.28	6.36
12	6.54	6.62	6.48	6.56
13	6.82	6.73	6.76	6.83
14	6.48	6.40	6.49	6.44
15	6.53	6.59	6.47	6.53

6 ± 0.06 cm, respectively. Assume that the dimensions are independent of each other and are each normally distributed with a capability ratio of 1. What are the natural tolerance limits on the volume of the cylinder? If specifications on volume are 18.5 ± 0.6 cm^3, what proportion of the cylinders will be nonconforming, assuming normality? What is the C_p index? If the target volume is 18.5, what is the C_{pm} index? Test a hypothesis to determine if C_p is less than 0.6, using $\alpha = 0.05$.

9-37 In solar cells, the exposed surface area is the characteristic of interest. The tolerances on the length and width of the cells are 4 ± 0.06 cm and 5 ± 0.09 cm, respectively. Assuming these dimensions to be independent of each other and each normally distributed with a capability ratio of 1, what are the natural tolerance limits on surface area? If the design calls for specifications on the surface area to be 20 ± 0.4, what proportion of the cells will be unacceptable assuming normality? What is the C_p index? If we did not know the distribution of surface area, how would a capability index be found?

9-38 The body mass index (BMI) is a measure of obesity and equals a person's weight (in kilograms) divided by the height (in meters) squared. For a certain diagnosis-related group of 20 patients, the following natural tolerances were obtained on weight (60 ± 5 kg) and height (1.7 ± 0.09 m). Assume that the distribution of weight and height are each normal. What are the natural tolerances on BMI for this group of patients? A BMI of 30.0 or more is considered to represent obesity. What proportion of the patients do you expect to be obese, assuming normality? Find the C_p index and establish a 95% lower confidence interval.

9-39 Find the sample size required for two-sided nonparameteric statistical tolerance limits for the viscosity of a grease used as a lubricant. It should contain 99% of the population with a probability of 0.95. How will the interval be found?

9-40   Refer to Exercise 9-39. Find the sample size needed to construct a one-sided lower nonparametric statistical tolerance limit. It should contain 90% of the population with a probability of 0.95. How will the limit be found?

REFERENCES

ASQ(1993). ANSI/ISO/ASQ. A3534-2. *Statistics—Vocabulary and Symbols—Statistical Quality Control*, Milwaukee, WI:American Society for Quality.

Boyles, R. A. (1991). "The Taguchi Capability Index," *Journal of Quality Technology*, 23 (1): 17–26.

Cochran, W. G. (1952). "The χ^2 Test of Goodness of Fit," *Annals of Mathematical Statistics*, 23: 315–345.

Duncan, A. J. (1986). *Quality Control and Industrial Statistics*, 5th ed. Homewood, IL: Richard D. Irwin.

Gunter, B. H. (1989). "The Use and Abuse of C_{pk}," *Quality Progress*, 22 (1): 72–73.

Kane, V. E. (1986). "Process Capability Indices," *Journal of Quality Technology*, 18 (1): 41–52.

Kotz, S., and N. L. Johnson (1993). *Process Capability Indices*. New York: Chapman & Hall.

——— (2002). "Process Capability Indices: A Review, 1992–2000," *Journal of Quality Technology*, 34 (1): 2–53.

Kushler, R. H., and P. Hurley (1992). "Confidence Bounds for Capability Indices," *Journal of Quality Technology*, 24 (4):188–195.

Mage, D. T. (1982). "An Objective Graphical Method for Testing Normal Distributional Assumptions Using Probability Plots," *The American Statistician*, 36 (2): 116–120.

Massey, F. J., Jr. (1951). "The Kolmogorov-Smirnov Test of Goodness of Fit," *Journal of the American Statistical Association*, 46: 68–78.

Minitab Inc. (2007). *Release 15*. State College, PA:Minitab.

Nelson, W. (1979). *How to Analyze Data with Simple Plots*, Milwaukee, WI: American Society for Quality Control.

Pearn, W. L., S. Kotz, and N. L. Johnson (1992). "Distributional and Inferential Properties of Process Capabilities," *Journal of Quality Technology*, 24:216–231.

Shapiro, S. S. (1980). *How to Test for Normality and Other Distributional Assumptions*, Milwaukee, WI: American Society for Quality Control.

Taguchi, G. (1985). "A Tutorial on Quality Control and Assurance: The Taguchi Methods," presented at the 1985 ASA Annual Meeting, Las Vegas, NV.

——— (1986). *Introduction to Quality Engineering*, Tokyo, Japan: Asian Productivity Organization.

PART IV

ACCEPTANCE SAMPLING

10

ACCEPTANCE SAMPLING PLANS FOR ATTRIBUTES AND VARIABLES

10-1   INTRODUCTION AND CHAPTER OBJECTIVES

In this chapter we examine acceptance sampling and procedures for product acceptance or rejection. Acceptance sampling plans where inspection is by attributes are discussed. In these plans, a product item is classified as conforming or not, but the degree of conformance is not specified. Although in certain sampling plans, the terms *defect* and *defective* are used interchangeably with *nonconformity* and *nonconforming* item, we use the latter two terms exclusively in this chapter. We also discuss acceptance plans where inspection is by variables. In these plans, the quality characteristic is expressed as a numerical value.

Acceptance sampling can be performed during inspection of incoming raw materials, components, and assemblies, in various phases of in-process operations, or during final product inspection. It can be used as a form of product inspection between companies and their vendors, between manufacturers and their customers, or between departments or divisions within the same company.

Note that acceptance sampling does not control or improve the quality level of the process. As stressed previously, quality cannot be inspected into a product or service; quality must be designed and built into it. Because of the very nature of sampling, acceptance

Fundamentals of Quality Control and Improvement, Third Edition, By Amitava Mitra
Copyright © 2008 John Wiley & Sons, Inc.

sampling procedures will accept some lots and reject others, even though they are of the same quality. Therefore, methods of process control and improvement are essential; they are the only way to get to maximize quality.

In addition to demonstrating procedures to select an acceptance sampling plan, another objective is to discuss how prior information along with information from the sample selected can be used to make decisions.

10-2   ADVANTAGES AND DISADVANTAGES OF SAMPLING

Keeping in mind that acceptance sampling plans are auditing procedures, let's compare them to 100% inspection (sometimes referred to as *screening*). Sampling is advantageous in that:

1. If inspection is destructive, 100% inspection is not feasible.
2. Sampling is more economical and causes less damage due to handling. If inspection cost is high or if inspection time is long, limited resources may make sampling preferable.
3. Sampling reduces inspection error. In high-quantity, repetitive inspection, such as 100% inspection, inspector fatigue can prevent the identification of all nonconformities or nonconforming units.
4. Sampling provides a strong motivation to improve quality because an entire batch or lot may be rejected.

Sampling plans are disadvantageous in that:

1. There is a risk of rejecting "good" lots or accepting "poor" lots, identified as the producer's risk and consumer's risk, respectively.
2. There is less information about the product compared to that obtained from 100% inspection.
3. The selection and adoption of a sampling plan require more time and effort in planning and documentation.

10-3   PRODUCER AND CONSUMER RISKS

In acceptance sampling, units are randomly chosen from a batch, lot, or process. There are two types of risk inherent in any sampling plan as discussed in the following material:

Producer's Risk: The risk associated with rejecting a "good" lot, due to the inherent nature of random sampling, is defined as a producer's risk. The notion of the quality level of lots that defines acceptable level or "good" product will be influenced by the needs of the customer. *Acceptable quality level* (AQL) is the terminology used to define this level of quality.

Consumer's Risk: The risk associated with accepting a "poor" lot, due to the inherent nature of random sampling, is defined as a consumer's risk. Further, norms of customer requirements will govern the definition of a "poor" lot. *Limiting quality level* (LQL) or *rejectable quality level* (RQL) is the terminology used to defined this level of unacceptable quality. An alternative terminology, when the quality level is expressed in percentage nonconformance, is *lot tolerance percent defective* (LTPD).

Thus, when we state a producer's risk in a sampling plan, we must correspondingly state a desirable level of quality that we prefer to accept. For example, if we state that the producer's risk is 5% for an AQL of 0.02, it means that we consider batches that are 2% nonconforming to be good and prefer to reject such batches no more than 5% of the time. If the consumer's risk is 10% for an LQL of 0.08, this means that batches that are 8% nonconforming are poor and we prefer to accept these batches no more than 10% of the time.

10-4   OPERATING CHARACTERISTIC CURVE

The **operating characteristic (OC) curve** measures the performance of a sampling plan. It plots the probability of accepting the lot versus the proportion nonconforming of the lot. It shows the discriminatory power of the sampling plan. For all sampling plans, we want to accept lots with a low proportion nonconforming most of the time and we do not want to accept batches with a high proportion nonconforming very often. The OC curve indicates the degree to which we achieve this objective.

What is the ideal OC curve? Let's suppose that we have chosen a proportion nonconforming level p_0 such that if a lot has a proportion nonconforming less than or equal to p_0, we consider it to be a good lot and we accept it. On the other hand, if the proportion nonconforming of the lot exceeds p_0, we consider the lot to be poor and we reject it. The ideal OC curve for these circumstances is shown in Figure 10-1. The probability of acceptance of the lot, P_a, is 1 for values of the proportion nonconforming $p \leq p_0$, and 0 for $p > p_0$. This assumes that we have selected a quality level p_0 that reflects the demarcation between acceptable and unacceptable lots. A sampling plan with such an OC curve is totally discriminatory in nature.

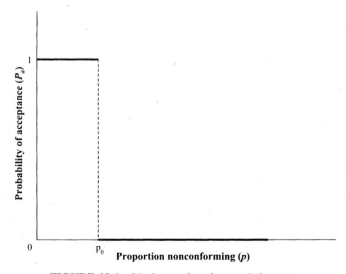

FIGURE 10-1   Ideal operating characteristic curve.

In practice, however, the shape of the OC curve is not ideal. To construct the OC curve for a single sampling plan, let N denote the lot size, n the **sample size**, and c the **acceptance number**. A random sample of size n is chosen from the lot of size N. If the observed number of nonconforming items or nonconformities is less than or equal to c, the lot is accepted. Otherwise, the lot is rejected.

To construct a **type A OC curve**, we assume that the sample is chosen from an isolated lot of finite size. The probability of accepting the lot is calculated based on a hypergeometric distribution. The probability of finding x nonconforming items in the sample is given by

$$P(x) = \frac{\binom{D}{x}\binom{N-D}{n-x}}{\binom{N}{n}} \qquad (10\text{-}1)$$

where D represents the number of nonconforming items in the lot. Since the lot will be accepted if c or fewer nonconforming items are found, the probability of lot acceptance is:

$$P_a = P(x \le c) = \sum_{x=0}^{c} P(x) \qquad (10\text{-}2)$$

where $P(x)$ is given by eq. (10-1).

To construct a **type B OC curve**, we assume that a stream of lots is produced by the process and that the lot size is large (at least 10 times) compared to the sample size. As discussed in Chapter 4, a binomial distribution can be used to find the probability of observing x nonconforming items in a sample of size n. Assuming the lot proportion nonconforming is p, this probability is given by

$$P(x) = \binom{n}{x} p^x (1-p)^{n-x} \qquad (10\text{-}3)$$

The lot acceptance probability is then given by eq. (10-2), where $P(x)$ is given by eq. (10-3). Alternatively, the cumulative binomial probability tables given in Appendix A-1 can be used if the appropriate parameter values n and p are tabulated.

If the lot size is large and the probability of a nonconforming item is small, as explained in Chapter 4, a Poisson distribution can be used as an approximation to the binomial distribution. The probability of x nonconforming items in the sample is found from

$$P(x) = \frac{\lambda^x e^{-\lambda}}{x!} \qquad (10\text{-}4)$$

where $\lambda = np$ represents the average number of nonconforming items in the sample. The probability of lot acceptance, P_a, can then be found from eq. (10-2), where $P(x)$ is given by eq. (10-4). The cumulative Poisson probability distribution tabulated in Appendix A-2 can be used for appropriate values of the parameter λ. In this chapter we use the Poisson distribution to find probabilities associated with lot acceptance whenever appropriate. Because the lot size is usually large compared to the sample size and the values of the proportion nonconforming p we are interested in are small, the Poisson distribution provides a reasonable approximation.

TABLE 10-1   Lot Acceptance Probabilities for Different Values of Proportion Nonconforming for the Sampling Plan $N = 2000$, $n = 50$, $c = 2$

Proportion Nonconforming, p	np	Probability of Lot Acceptance, P_a	Proportion Nonconforming, p	np	Probability of Lot Acceptance, P_a
0.0	0.0	1.000	0.08	4.00	0.238
0.005	0.25	0.997	0.09	4.50	0.174
0.01	0.50	0.986	0.10	5.00	0.125
0.02	1.00	0.920	0.11	5.50	0.088
0.03	1.50	0.809	0.12	6.00	0.062
0.04	2.00	0.677	0.13	6.50	0.043
0.05	2.50	0.544	0.14	7.00	0.030
0.06	3.00	0.423	0.15	7.50	0.020
0.07	3.50	0.321			

Example 10-1 Construct an OC curve for a single sampling plan where the lot size is 2000, the sample size is 50, and the acceptance number is 2.

Solution We are given $N = 2000$, $n = 50$, and $c = 2$. The probability of lot acceptance is equivalent to the probability of obtaining 2 or fewer nonconforming items in the sample. The Poisson probability distribution in Appendix A-2 is used to obtain the lot acceptance probability for different values of the proportion nonconforming p. Let's suppose that p is 0.02 (i.e., the batch is 2% nonconforming). Since $np = (50)(0.02) = 1.0$, the probability P_a of accepting the lot (using Appendix A-2) is 0.920. Table 10-1 shows values of P_a for various values of p. In some instances, the probability values are linearly interpolated from Appendix A-2. A plot of these values, the OC curve, is shown in Figure 10-2.

The discriminating power of the sampling plan $N = 2000$, $n = 50$, $c = 2$ can be seen from the OC curve in Figure 10-2. If a series of batches, each of which is 1% nonconforming, comes in for inspection, then (using this plan) the probability of lot acceptance is 0.986. It means that, on average, about 986 out of 1000 such batches will be accepted by the sampling plan. On the other hand, if batches are 5% nonconforming, only

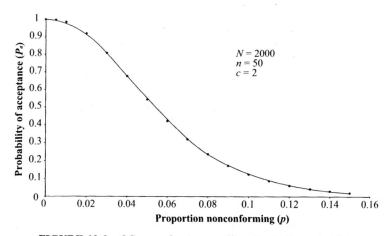

FIGURE 10-2 OC curve for the sampling plan in Example 10-1.

about 544 out of 1000 batches will be accepted. As the lot quality becomes poorer, the probability of lot acceptance decreases, as it should. The steeper the drop in the probability of lot acceptance as lot quality worsens, the higher the discriminatory power of the sampling plan.

Producer and consumer risk can also be demonstrated through the OC curve. Suppose that our numerical definition of good quality (indicated by the AQL) is 0.01 and that of poor quality (indicated by the LQL) is 0.11. From the OC curve in Figure 10-2, the producer's risk α is $1 - 0.986 = 0.014$. We consider batches that are 1% nonconforming to be good. If our sampling plan is used, such batches will be rejected about 1.4% of the time. Batches that are 11% nonconforming, on the other hand, will be accepted 8.8% of the time. The consumer's risk is therefore 8.8%.

Effect of the Sample Size and the Acceptance Number

The parameters n and c of the sampling plan affect the shape of the OC curve. As long as the lot size N is significantly large compared to the sample size n, the lot size does not have an appreciable impact on the shape of OC curve. For fixed values of N and c, as the sample size becomes larger, the slope of the OC curve becomes steeper, implying a greater discriminatory power.

Figure 10-3 shows the OC curves for those sampling plans. Note that, for lots that are 2% nonconforming, the sampling plan $N = 2000$, $n = 50$, $c = 2$ will accept such lots about 92% of the time. However, for the same lots, the sampling plan $N = 2000$, $n = 200$, $c = 2$ will accept them only 23.8% of the time. Changing the sample size from 50 to 200 causes a drop in the acceptance probability of 68.2%.

For fixed values of the lot size N and the sample size n, as the acceptance number decreases, the slope of the OC curve becomes steeper. Figure 10-4 shows the OC curves for four sampling plans. Note that the probability of acceptance decreases for a given lot quality as the acceptance number c decreases.

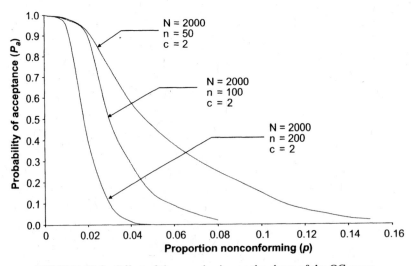

FIGURE 10-3   Effect of the sample size on the shape of the OC curve.

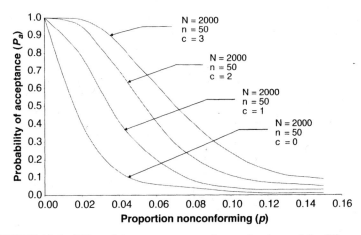

FIGURE 10-4   Effect of the acceptance number on the shape of the OC curve.

One comment needs to be made regarding the case where the acceptance number is zero. The OC curve starts dropping drastically even as the proportion nonconforming deviates slightly from zero. This may not be desirable from a producer's point of view. For example, if lots that are 0.5% nonconforming are considered acceptable, the sampling plan $N = 2000$, $n = 50$, $c = 0$ will reject such lots about 22% of the time, implying a high value of the producer's risk. Sampling plans with $c = 0$ do not have the desirable inverted-S shape of the ideal OC curve. They are, however, stringent and serve a need.

The chosen values of n and c should be such that they match the goals of the user. Given some desirable producer's risk and the associated quality level of a good lot (AQL) and/or a desirable consumer's risk and an associated quality level of a poor lot (LQL), the combination of n and c that produces an OC curve that matches these goals will provide an acceptable sampling plan.

10-5   TYPES OF SAMPLING PLANS

There are, generally speaking, three types of attribute sampling plans: single, double and multiple. In a **single sampling plan**, the information obtained from one sample is used to make a decision to accept or reject a lot. There are two parameters in this sampling plan: the sample size n and the acceptance number c. The plan operates as follows. A random sample of size n is selected from the batch. The number of nonconforming items or nonconformities in the sample is found and compared to the acceptance number c. If the observed number is less than or equal to the acceptance number, the lot is accepted. If more than c nonconforming items or nonconformities are found in the sample, the lot is rejected.

A **double sampling plan** involves making a decision to accept the lot, reject the lot, or take a second sample. If the inference from the first sample is that the lot quality is quite good, the lot is accepted. If the inference is poor lot quality, the lot is rejected. If the first sample gives an inference of neither good nor poor quality, a second sample is taken. Thereafter, based on the combined number of nonconforming items or nonconformities in both samples, a decision is made to accept or reject the lot. The parameters of a double sampling plan are as follows:

n_1:   size of the first sample

c_1:   acceptance number for the first sample

r_1:   rejection number for the first sample

n_2:   size of the second sample

c_2:   acceptance number for the second sample

r_2:   rejection number for the second sample

Let's consider the following double sampling plan where attribute inspection is conducted to find the number of nonconforming items:

$$N = 5000$$
$$n_1 = 40 \qquad n_2 = 60$$
$$c_1 = 1 \qquad c_2 = 5$$
$$r_1 = 4 \qquad r_2 = 6$$

The working procedure for this plan is initially, to select a random sample of 40 items from the lot of size 5000. If 1 or fewer nonconforming items are found, the lot is accepted, but if 4 or more nonconforming items are found, the lot is rejected. If the observed number of nonconforming items is 2 or 3, a second sample of size 60 is selected. If the combined number of nonconforming items from both samples is less than or equal to 5, the lot is accepted; if it is 6 or more, the lot is rejected.

Although double sampling plans are more complicated than single sampling plans, usually fewer items need to be sampled, on average, to make a decision regarding the lot. This is because a demonstration of extremely good or extremely poor batch quality in the first sample causes acceptance or rejection of the lot without the need for a second sample.

Multiple sampling plans are an extension of double sampling plans. Three, four, five, or as many samples as desired may be needed to make a decision regarding the lot. The sampling plan can be terminated at any stage once the acceptance or rejection criteria have been met. The sample sizes in a multiple sampling plan are usually less than those for an equivalent double sampling plan, which in turn are usually less than that for an equivalent single sampling plan.

The ultimate extension of the multiple sampling plan is the sequential sampling plan, which is an item-by-item inspection plan. After each item is inspected, a decision is made to accept the lot, reject the lot, or choose another item for inspection, depending on whether the observed cumulative number of nonconforming items is less than or equal to the acceptance number, greater than or equal to the rejection number, or in between the two.

Advantages and Disadvantages

We can design single, double, or multiple sampling plans that are equivalent in the sense that they have the same probability of lot acceptance for batches of a given quality. Therefore, we need to consider the advantages and disadvantages of these types when we select a sampling plan.

As far as simplicity is concerned, the single sampling plan is the best, followed by double and then multiple sampling plans. Administrative costs for record keeping, training, and inspection are the least for single and the highest for multiple sampling plans.

On average, for equivalent plans, the number of items inspected to make a decision regarding the lot is usually more for a single sampling plan. This is because double and multiple sampling plans use fewer items in their samples, so if the lots are of very good or poor quality, a decision to accept or reject them is made quickly. Inspection costs will therefore be the most for single, and the least for multiple sampling plans.

The information content of the samples is a function of the sample size; the more samples we inspect, the more information we have about the product and consequently the process. Single sampling plans provide the most information, and multiple sampling plans the least.

10-6 EVALUATING SAMPLING PLANS

The OC curve is one measure of the performance of a sampling plan. We also use other measures to evaluate the goodness of a sampling plan. These involve the average quality level of batches leaving the inspection station, the average number of items inspected before making a decision on the lot, and the average amount of inspection per lot if a rejected lot goes through 100% inspection. We discuss single sampling plans here, but the concepts apply to all three plans.

Average Outgoing Quality

Let's first consider the concept of **rectifying inspection** as it applies to lots that are rejected through sampling plans. Usually, such lots go through 100% inspection, known as **screening**, where nonconforming items are replaced with conforming ones. Such a procedure is known as *rectification inspection* because it affects the quality of the product that leaves the inspection station. Nonconforming items found in the sample are also replaced.

The average outgoing quality (AOQ) is the average quality level of a series of batches that leave the inspection station, assuming rectifying inspection, after coming in for inspection at a certain quality level p. The AOQ is not the quality level of a single batch that leaves the inspection station. For instance, a batch with incoming quality level p will leave the inspection station with about the same quality level if accepted by the sampling plan. We assume the sample is a small enough proportion of the lot such that if nonconforming items are found in the sample and replaced with conforming ones, the quality level of the lot is not significantly affected. Similarly, another batch with the same incoming quality p that is rejected by the sampling plan will be screened and so will leave the inspection station with no nonconforming items. (This assumes that screening detects all nonconforming items.) The AOQ measures the average quality level of a large number of such batches of incoming quality p, the proportion nonconforming in the lots, assuming rectification.

Taking N as the lot size, n as the sample size, p as the incoming lot quality, and P_a as the probability of accepting the lot using the given sampling plan, the average, outgoing quality is given by

$$\text{AOQ} = \frac{P_a p (N-n)}{N} \tag{10-5}$$

To understand this equation, note that n items in the sample will have no nonconforming items after they have been inspected. If the lot is rejected by the sampling plan, the $N - n$ items left in the lot go through screening so that no nonconforming items are in the outgoing product. Only if the lot is accepted by the sampling plan will the $N - n$ items left in the lot leave the inspection station with $p(N - n)$ nonconforming items. However, the probability that the batch will be accepted by the sampling plan is P_a. So $P_a p(N - n)$ is the number of nonconforming items per lot expected to leave the inspection station. The average proportion nonconforming is given by eq. (10-5).

The value of AOQ depends on the incoming quality level p of the batches. Thus, an AOQ curve that evaluates the effectiveness of the sampling plan for various levels of incoming quality is usually constructed. Let's, consider the single sampling plan $N = 2000$, $n = 50$, $c = 2$. Suppose that the incoming quality of batches is 2% nonconforming. From the Poisson cumulative distribution tables in Appendix A-2, the probability P_a of accepting the lot using the sampling plan is 0.920. The average outgoing quality is

$$AOQ = \frac{(0.920)(0.02)(2000 - 50)}{2000} = 0.0179$$

Thus, if batches come in as 2% nonconforming, the average outgoing quality is 1.79%.

Example 10-2 Construct the AOQ curve for the sampling plan $N = 2000$, $n = 50$, $c = 2$.

Solution The probability of lot acceptance for various values of the incoming lot quality p is listed in Table 10-1. Using these values of P_a and p, the values of AOQ are calculated for different values of p.

Figure 10-5 shows the AOQ curve for the sampling plan $N = 2000$, $n = 50$, $c = 2$. Note that when the incoming quality is very good, the average outgoing quality is also very good. When the incoming quality is very poor, the average outgoing quality is good because most of the lots are rejected by the sampling plan and go through screening. In between these extremes, the AOQ curve reaches a maximum, AOQL.

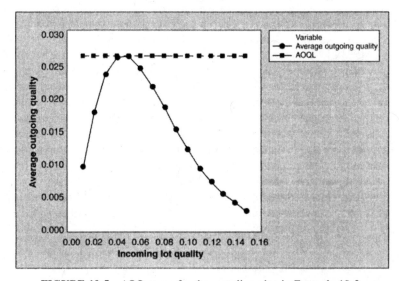

FIGURE 10-5 AOQ curve for the sampling plan in Example 10-2.

Average Outgoing Quality Limit   The *average outgoing quality limit* (AOQL) is the maximum value, or peak, of the AOQ curve. It represents the worst average quality that would leave the inspection station, assuming rectification, regardless of the incoming lot quality. The AOQL value is also a measure of goodness of a sampling plan. Note that the protection offered by the sampling plan, in terms of the AOQL value, does not apply to individual lots. It holds for the average quality of a series of batches.

Consider Example 10-2 and the AOQ curve in Figure 10-5. The AOQL value is approximately 0.0265, or 2.65%. This means that for the sampling plan above, $N = 2000$, $n = 50$, $c = 2$, we have, some protection against the worst quality for a series of batches that leave the inspection program. The average quality level will not be poorer than 2.65% nonconforming. However, it is possible for an individual lot to have an outgoing quality level of more than 2.65% nonconforming. The AOQL value and the shape of the AOQ curve depend on the particular sampling plan. Sampling plans are designed such that their AOQL does not exceed a certain specified value.

Average Total Inspection

If rectifying inspection is conducted for lots rejected by the sampling plan, another evaluation measure is the *average total inspection* (ATI). The ATI represents the average number of items inspected per lot. If a lot has no nonconforming items, it will obviously be accepted by the chosen sampling plan, and only n items (the sample size) will be inspected for a lot. At the other extreme, if the lot has 100% nonconforming items, the number inspected per lot will be N (the lot size) assuming that rejected lots are screened. For a lot quality between these extremes, the average amount inspected per lot will vary between these two values. For single sampling plans, the average total inspection per lot for lots with an incoming quality level p is given by

$$\text{ATI} = n + (1 - P_a)(N - n) \tag{10-6}$$

Here P_a represents the probability of accepting a lot that has an incoming quality level of p. A plot of the average total inspection versus p is an ATI curve. Note that for an individual lot, the amount inspected is either n or N.

For a double sampling plan, the ATI is given by

$$\text{ATI} = n_1(P_{a1}) + (n_1 + n_2)P_{a2} + N(1 - P_{a1} - P_{a2}) \tag{10-7}$$

where P_{a1} represents the probability of accepting the lot on the first sample, and P_{a2} represents the probability of lot acceptance on the second sample. Computation of P_{a1} and P_{a2} will be discussed further in the discussion of double sampling in Section 10-8.

Example 10-3   Construct the ATI curve for the sampling plan where $N = 2000$, $n = 50$, $c = 2$.

Solution   Consider the calculations for a given value of the lot quality p of 0.02. As shown in Table 10-1, the probability of accepting such a lot using the sampling plan is $P_a = 0.920$. The ATI for this value of p is

$$\text{ATI} = 50 + (1 - 0.920)(2000 - 50) = 206$$

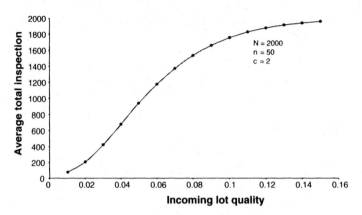

FIGURE 10-6   ATI curve for the sampling plan in Example 10-3.

For other values of p, the ATI is found in the same manner. The ATI curve is plotted in Figure 10-6. Given the unit cost of inspection, the ATI curve can be used to estimate the average inspection cost if the quality level of incoming batches is known.

Average Sample Number

The average number of items inspected for a series of lots with a given incoming lot quality in order to make a decision is known as the *average sample number* (ASN). Assume that inspection is not curtailed for a single sampling plan when making a decision. For example, if 3 nonconforming items are found by the twentieth unit when using a single sampling plan $N = 800$, $n = 60$, $c = 2$, even though a decision can be made after the twentieth unit to reject the lot, inspection continues for all 60 items in the sample. Under this assumption, the average sample number for a single sampling plan is equal to the sample size n.

For a double sampling plan, the ASN is given by

$$\begin{aligned} \text{ASN} &= n_1 P_1 + (n_1 + n_2)(1 - P_1) \\ &= n_1 + n_2(1 - P_1) \end{aligned} \tag{10-8}$$

where P_1 is the probability of making a decision on the first sample. Again it is assumed that there is no curtailment on the first or second sample when making a decision.

There is a valid reason for not curtailing inspection. The estimate of the lot proportion nonconforming is biased if inspection is curtailed. Consider the situation for which the inspection of the first 3 items in a single sampling plan yields all nonconforming items. If the rejection number is 3 and inspection is curtailed on the third item, an estimate of the lot proportion nonconforming would be 1.00 (or 100%), which is quite misleading. Sometimes, for a double sampling plan, inspection is curtailed on the second sample. This leads to a lowering of the average sample number. In cases where inspection costs per unit are high and batches are of neither very good nor very poor quality, curtailing inspection on the second sample will lower the costs associated with decision making.

Note that for a given sample, the number of items inspected before making a decision is either n_1 or $(n_1 + n_2)$. There is a probability P_1 that a decision will be made after inspecting n_1 items; the probability of inspecting $(n_1 + n_2)$ items prior to making a decision is $1 - P_1$. This is the rationale behind the form of eq. (10-8). The probability P_1 can be expressed as

$$P_1 = P(\text{lot accepted on first sample}) + P(\text{lot rejected on first sample})$$
$$= P(x \leq c_1) + P(x \geq r_1)$$

where x represents the number of nonconforming items or nonconformities, c_1 is the acceptance number of the first sample, and r_1 is the rejection number of the first sample.

For a multiple sampling plan, the same concept used for the derivation of the ASN of a double sampling plan can be used. We have

$$\text{ASN} = n_1 P_1 + (n_1 + n_2)P_2 + \ldots + (n_1 + n_2 + \cdots + n_k)P_k \qquad (10\text{-}9)$$

where k represents the number of levels of samples, n_i is the sample size at the ith level, and P_i represents the probability of making a decision at the ith level.

A plot of ASN values as a function of the lot proportion nonconforming p is known as the *ASN curve*.

Example 10-4   For the double sampling plan $N = 3000$, $n_1 = 40$, $c_1 = 1$, $r_1 = 4$, $n_2 = 80$, $c_2 = 3$, $r_2 = 4$, find the average sample number for batches with a proportion nonconforming of 0.02, assuming no curtailment.

Solution   First, calculate P_1, the probability of making a decision after the first sample:

$$P_1 = P(x \leq 1) + P(x \geq 4)$$

where x represents the observed number of nonconforming items. From the cumulative Poisson tables in Appendix A-2, we get.

$$P_1 = P[x \leq 1 \mid n_1 p = (40)(0.02)] + P[x \geq 4 \mid n_1 p = (40)(0.02)]$$
$$= 0.809 + (1 - 0.991) = 0.818$$

The average sample number for batches with a proportion nonconforming of 0.02 is

$$\text{ASN} = n_1 + n_2(1 - P_1)$$
$$= 40 + (80)(1 - 0.818)$$
$$= 54.56$$

This value represents the average number of items inspected prior to making a decision. It suggests that because of the low value of p, most of the batches will be accepted on the first sample. On the basis of the first sample, lots will be accepted about 80.9% of the time, and they will be rejected about 0.9% of the time.

The ASN values are calculated for several values of the proportion nonconforming p in order to construct the ASN curve, which is shown in Figure 10-7. Note that as the proportion nonconforming p increases from zero, the ASN rises until it reaches a peak and then starts falling. For very small and very large values of p, the value of ASN approaches the first sample size n_1, because a decision is made for these lots on the first sample. For lots of intermediate quality, the ASN value is higher because a second sample must be taken more frequently to make a decision.

For single sampling plans, ASN is constant and is represented by a horizontal line. Typically, for equivalent plans, the ASN for a double sampling plan is below that for a single sampling plan. It sometimes happens that the middle segment of the ASN curve for a double sampling plan is above the ASN for a single sampling plan. Management must then use

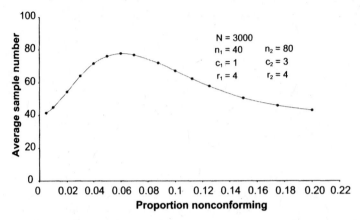

FIGURE 10-7   ASN curve for the double sampling plan in Example 10-4.

historical information to obtain an idea of the process quality. If the quality level p happens to fall in the segment where the ASN is greater for a double sampling plan, the choice of a single sampling plan may be justified to cut down on inspection time and costs. On the other hand, estimates of very high or low levels of process quality may justify the use of a double sampling plan that will yield smaller values of ASN. If curtailed inspection is used on the second sample of a double sampling plan, the ASN curve will be lowered further and will become even more attractive than a single sampling plan.

10-7 BAYES' RULE AND DECISION MAKING BASED ON SAMPLES

Information obtained from samples may be considered sequential in nature. Based on past knowledge (objective or otherwise), one may formulate a **prior distribution** on the unknown population parameter. Using sample information, this prior distribution could be updated to develop a **posterior distribution** of the parameter. Consequently, this posterior distribution becomes the prior as further samples are taken and all of the accumulated information is used to make decisions.

This updating of probabilities, associated with parameters, is accomplished through **Bayes' formula**. Let $\{B_1, B_2, \ldots, B_m\}$ represent a partition of the sample space, where the events $\{B_i, i = 1, 2, \ldots, m\}$ are not directly observable. The random sample results in the occurrence of some event $\{A\}$. Further, the conditional probabilities $\{P(A \mid B_i), i = 1, 2, \ldots, m\}$ are known. Assume that our degree of belief on the occurrence of the events $\{B_i, i = 1, 2, \ldots, m\}$ is represented by some prior probabilities $\{P(B_i), i = 1, 2, \ldots, m\}$. Then, on observance of the event $\{A\}$, the posterior probabilities $\{P(B_i \mid A), i = 1, 2, \ldots, m\}$ are given by

$$P(B_i \mid A) = \frac{P(B_iA)}{P(A)} = \frac{P(A \mid B_i)P(B_i)}{\sum_{i=1}^{m} P(A \mid B_i)P(B_i)}, \qquad i = 1, 2, \ldots, m \qquad (10\text{-}10)$$

Example 10-5   Three vendors, B_1, B_2, and B_3, provide microchips for a semiconductor manufacturer. Based on each vendor's historical data, the probabilities of vendor B_1, B_2, and B_3 producing a defective microchip are 0.01, 0.03, and 0.05, respectively. Since the parts all look alike and have no company identification, it is not possible to identify the vendor they came from. Assume that 30% of the parts are provided by vendor B_1, with the corresponding percentages for vendors B_2 and B_3 being 50% and 20%, respectively. Upon inspection of a part, it is found to be nonconforming. What is the probability that it was produced by vendor B_1? By vendor B_2? By vendor B_3?

Solution   Let the event $\{A\}$ represent observing a defective part. The prior probabilities of the defective part coming from each vendor are $P(B_1) = 0.30$, $P(B_2) = 0.50$, $P(B_3) = 0.20$. It is known that $P(A\,|\,B_1) = 0.01$, $P(A\,|\,B_2) = 0.03$, and $P(A\,|\,B_3) = 0.05$. So the posterior probabilities are obtained as follows:

$$P(B_1\,|\,A) = \frac{(0.01)(0.30)}{(0.01)(0.30) + (0.03)(0.50) + (0.05)(0.20)} = 0.107$$

$$P(B_2\,|\,A) = \frac{(0.03)(0.50)}{(0.01)(0.30) + (0.03)(0.50) + (0.05)(0.20)} = 0.536$$

$$P(B_3\,|\,A) = 1 - (0.107 + 0.536) = 0.357$$

Note that whereas the prior odds of the defective part being from B_1, B_2, or B_3 were 0.3, 0.5, and 0.2, respectively, the posterior odds, after observing the defective part, are quite different. The posterior probability of the vendor being B_2 or B_3 has increased, while that of B_1 has decreased. Thus, the sample information is being used to update the prior probabilities.

The concept of revising prior probabilities through sample information may be further extended in decision making in order to minimize expected opportunity losses. As explained previously, when decisions are made based on sample information, two types of errors may occur: Types I and II errors. In the context of acceptance sampling, a type I error is referred to as the producer's risk, while a type II error is denoted as the consumer's risk. In the framework of hypothesis testing, emphasis is placed on keeping the type I error to a small (or acceptable) level. However, in the premise of using sampling plans to make decisions, it is the cost associated with making a type I or type II error that could influence our decision. Thus, the *joint impact* of the *cost* and the *chance* of each type of error are evaluated. We demonstrate the procedure through an example.

Example 10-6   An electronics company is considering the addition of a new product to its line. It currently has 40,000 customers. Management estimates product development costs to be $1,200,000, which may be assumed to be the overhead costs. Variable manufacturing costs per unit are expected to be $800, along with variable selling expense of $100 per unit. Units will be produced on a demand basis. From a market survey, it is estimated that the new product could be priced at $1500 per unit. Based on previous experience with similar products, market research has come up with estimates of the proportion of existing customers that may purchase the new product and the corresponding probabilities of such happening (Table 10-2). Should the company develop the new product?

Solution   Let us first determine the break-even value of the demand proportion (p) for the new product. With the overhead costs being $1,200,000 and the profit margin/unit being $600, 2000 units is the break-even volume, leading to a break-even demand proportion of

TABLE 10-2   Probability Distribution of Demand for a New Product

Demand Proportion p	Probability
0.02	0.2
0.04	0.4
0.06	0.3
0.08	0.1

$p = 0.05$, since the company has 40,000 customers. So if $p < 0.05$, the company will experience an opportunity loss if the new product is developed. Similarly if $p > 0.05$, an opportunity loss will occur if the new product is not developed. Alternatively, if $p \geq 0.05$ and the company develops the new product, or if $p < 0.05$ and the company does not develop the new product, a correct decision will have been made and there will be no opportunity loss. The *conditional opportunity loss* function (COL) if $p < 0.05$ and the decision is to develop the new product is

$$\text{COL} = \$1,2000,000 - (600p)(40,000) = \$24,000,000(0.05 - p)$$

Similarly, if $p > 0.05$ and the decision is to not develop the new product, we have

$$\text{COL} = \$24,000,000(p - 0.05)$$

The COLs are a measure of the costs associated with errors in decision making.

The conditional opportunity losses for each state of nature, in this case the demand proportion (p) for the new product and action (here to develop or not develop the new product) combination, are shown in Table 10-3. Using these conditional opportunity losses and the prior probability distribution associated with demand for the new product, the expected opportunity losses (EOLs) for each of the two actions are found. Note that the EOLs are found by weighting the COLs by the prior probability estimate of demand. It is observed that the "optimal" decision at this point, without any sample information, is not to develop the new product, since this action has a smaller EOL.

The **expected value of perfect information** (EVP1) is the expected opportunity loss of the optimal action, which in this case is $144,000. If we had perfect information, we would choose the correct action and the opportunity loss would be zero. EVPI is therefore an upper bound on the amount we should be willing to spend to gain additional information before making a decision.

TABLE 10-3   Evaluation of Actions Without Sampling

Demand Proportion, p	Probability, $P(p)$	Action	
		Develop New Product, COL	Do Not Develop New Product, COL
0.02	0.2	720,000	0
0.04	0.4	240,000	0
0.06	0.3	0	240,000
0.08	0.1	0	720,000
		EOL = $240,000	EOL = $144,000

TABLE 10-4   Revised Probability Distribution of Demand for a New Product

Demand Proportion, p	Prior Probability, $P(p)$	$P(x=3\mid p)$	$P(x=3 \text{ and } p)$	Revised $P(p)$
0.02	0.2	0.0065	0.00130	0.0233
0.04	0.4	0.0364	0.01456	0.2609
0.06	0.3	0.0860	0.02580	0.4624
0.08	0.1	0.1414	0.01414	0.2534
			$P(x=3)=0.0558$	

Suppose that it is desired to obtain more information and a random sample of size 20 is selected from the existing customers, where each is asked whether he or she will buy the new product. From the sample results, three stated that they would buy. Let us incorporate this sample information and update the prior probability of demand using Bayes' rule. Table 10-4 shows the posterior or revised probability of demand. Note that the binomial distribution is used to obtain $P(x=3\mid p)$ for the various values of p and $n=20$. We find that the sample results have had a significant impact on the estimates of probability of demand.

Using the revised $P(p)$, as shown in Table 10-4, the expected opportunity losses are calculated for the two actions. Using the sample information it may be verified that

$$\text{EOL(develop new product)} \qquad = \$79,392$$
$$\text{EOL(do not develop new product)} = \$293,424$$

The optimal decision at this phase, incorporating the sample information, is to develop the new product.

10-8   LOT-BY-LOT ATTRIBUTE SAMPLING PLANS

Attribute sampling plans are designed to make a decision regarding items that are submitted for inspection in lots. Batch production is typical of many industries in which inspection is conducted in lots. The objective is to find suitable sample sizes and acceptance numbers of sampling plans that meet certain levels of stipulated risks (such as the producer's risk, consumer's risk, or both).

Single Sampling Plans

Single sampling plans deal with making a decision regarding a lot of size N based on information contained in one sample of size n. The acceptance number c of the sampling plan represents the number of nonconforming items or nonconformities, depending on the circumstances, that cannot be exceeded in the sample in order for the lot to be accepted.

The OC Curve   The OC curve of a single sampling plan has been described in detail. It represents the probability of accepting the lot, P_a, as a function of the lot quality, which is simply the proportion nonconforming p if items are classified only as conforming or not. The effects of the parameters n and c on the shape of the OC curve have also been discussed. A study of these effects enables us to choose appropriate values of n and c, given desirable levels of protection against the producer's and consumer's risks.

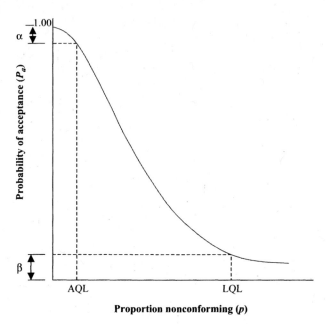

FIGURE 10-8   OC curve showing (AQL, $1-\alpha$) and (LQL, β) for a sampling plan.

Let's suppose that both the acceptable quality level (AQL), which is the measure of a good lot and is associated with the producer's risk α, and the limiting quality level (LQL), which is the measure of a poor lot and is associated with the consumer's risk are specified. The OC curve in Figure 10-8 shows the relationship between these parameters. For a sampling plan specified by n and c, lots with a proportion nonconforming level of AQL that come in for inspection should be accepted $100(1-\alpha)\%$ of the time. Similarly, if the proportion non-conforming of batches coming in for inspection is LQL, they should be accepted $100\beta\%$ of the time. A suitable choice of n and c ensures that good lots will be accepted a large percentage of the time and that bad lots will be accepted infrequently.

Design of Single Sampling Plans   Now we will discuss several approaches for designing single sampling plans. Basically, these approaches involve determining the sample size n and acceptance number c of the plan. The criteria selected influences the parameters of the plan. Sometimes, more than one plan will satisfy the criteria. What follows is a series of procedures for determining the parameters of a single sampling plan based on the criteria specified.

Stipulated Producer's Risk   Let's suppose the producer's risk α and its associated quality level p_1, which is the acceptable quality level (AQL), are specified. We desire single sampling plans that will accept lots of quality level p_1, $100(1-\alpha)\%$ of the time. Figure 10-9 shows the OC curves of sampling plans that meet this stipulated criteria. Note that several plans may satisfy this criteria. We want to find a sampling plan whose OC curve passes through the single point (AQL, $1-\alpha$). This criterion is not very restrictive; the OC curves of a variety of plans could pass through this point.

To find the appropriate sampling plan, first select an acceptance number c. As discussed previously, the Poisson distribution will be used to approximate the hypergeometric distribution when determining the lot acceptance probability. This is reasonable when the

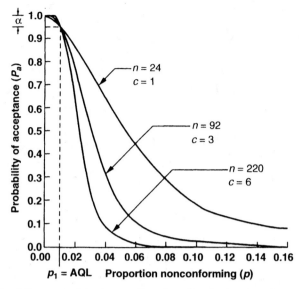

FIGURE 10-9 OC curves of single sampling plans for stipulated producer's risk and AQL.

sample size is a small fraction of the lot size and the lot proportion nonconforming p is small. The mean number of nonconforming items in the sample is given by $\lambda = np$. Hence, for a probability of lot acceptance P_a equal to $1-\alpha$ at $p = p_1$, the value of λ is found in Appendix A-2. Because $\lambda = np_1 = n$ (AQL), the sample size n is found by dividing the value of n(AQL) by AQL. Fractional computed values of the sample size are always rounded up to be conservative.

For a producer's risk α of 0.05, Table 10-5 lists the values of np_1 for an acceptance probability $P_a = 0.95$ and various values of c.

TABLE 10-5 **Values of np for a Producer's Risk of 0.05 and a Consumer's Risk of 0.10**

Acceptance Number, c	$P_a = 0.95, np_1$	$P_a = 0.10, np_2$	np_2/np_1
0	0.051	2.303	44.84
1	0.355	3.890	10.96
2	0.818	5.322	6.51
3	1.366	6.681	4.89
4	1.970	7.994	4.06
5	2.613	9.274	3.55
6	3.286	10.532	3.21
7	3.981	11.771	2.96
8	4.695	12.995	2.77
9	5.426	14.206	2.62
10	6.169	15.407	2.50
11	6.924	16.598	2.40
12	7.690	17.782	2.31
13	8.464	18.958	2.24
14	9.246	20.128	2.18
15	10.035	21.292	2.21

[a]*Source:* F.E. Grubbs, "On Designing Single Sampling Plans," *Annals of Mathematical Statistics*, XX: 256, 1949. Reprinted by permission of the Institute of Mathematical Statistics.

Example 10-7   Find a single sampling plan that satisfies a producer's risk of 5% for lots that are 1.5% nonconforming.

Solution   We are given $\alpha = 0.05$ and $p_1 = $ AQL $= 0.015$. If we choose an acceptance number $c = 1$, for which Table 10-5 gives $np_1 = 0.355$, the sample size is

$$n = \frac{0.355}{p_1} = \frac{0.355}{0.015} = 23.67 \simeq 24$$

If our acceptance number is 3, we have $np_1 = 1.366$, and the sample size is

$$n = \frac{1.366}{0.015} = 91.07 \simeq 92$$

If our acceptance number is 6, we have $np_1 = 3.286$, and the sample size is

$$n = \frac{3.286}{0.015} = 219.07 \simeq 220$$

Figure 10-9 shows the OC curves for these three sampling plans. Note that all three plans satisfy the producer's risk of 5% at the AQL value of 1.5%; However, they have varying degrees of protection against acceptance of poor quality lots, which would be of interest to the consumer. Of the three plans shown, $n = 220$, $c = 6$ provides the best protection to the consumer because it has the lowest probability of accepting poor quality lots. However, we must also consider the increased inspection costs associated with this plan, because the sample size for $c = 6$ is the largest of the three. (*Note*: We considered the values of c of 1, 3, and 6 for demonstration purposes. Other values of c could be selected as well.)

Stipulated Consumer's Risk   Let's suppose that the consumer's risk β and its associated quality level p_2, which is the limiting quality level (LQL), are given. We want to find sampling plans that will accept lots of quality level p_2, $100\beta\%$ of the time. Here again, a number of sampling plans will satisfy this criterion. Figure 10-10 shows the OC curves for three sampling plans that meet the criterion.

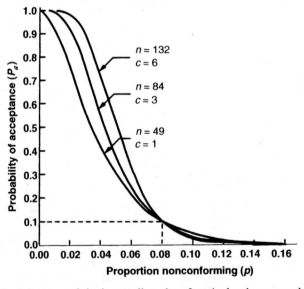

FIGURE 10-10   OC curves of single sampling plans for stipulated consumer's risk and LQL.

The procedure is similar to that used with producer's risk. A value of the acceptance number c is chosen. Based on the probability of acceptance of β, for lots of quality $p_2 = \text{LQL}$, the value of $\lambda = np_2$ is found in Appendix A-2. If the value of β is 0.10, we can use Table 10-5 to obtain the value of np_2. The sample size is calculated by dividing the value of np_2 by p_2.

Example 10-8   Find a single sampling plan that will satisfy a consumer's risk of 10% for lots that are 8% nonconforming.

Solution   We are given $\beta = 0.10$ and $p_2 = \text{LQL} = 0.08$. If we select an acceptance number of 1, Table 10-6 gives $np_2 = 3.890$. The sample size is

$$n = \frac{3.890}{p_2} = \frac{3.890}{0.08} = 48.62 \simeq 49$$

If the acceptance number is selected as 3, we have $np_2 = 6.681$, and the sample size is

$$n = \frac{6.681}{0.08} = 83.51 \simeq 84$$

If the acceptance number is 6, we have $np_2 = 10.532$, and the sample size is

$$n = \frac{10.532}{0.08} = 131.65 \simeq 132$$

Figure 10-10 shows the OC curves for these sampling plans. All three pass through the point (p_2, β), thus satisfying the consumer's stipulation. The degree of protection for extremely good batches, as far as the producer is concerned, is different. The plan $n = 132, c = 6$ will reject good batches (say, 1% nonconforming) the least frequently of the three plans. Of course, it has the largest sample size, which may cause the inspection cost to be high. Other values of the acceptance number could be selected as well.

Stipulated Producer and Consumer Risk   We desire sampling plans that satisfy a producer's risk α (given an associated quality level $p_1 = \text{AQL}$) and a consumer's risk β (given an associated quality level $p_2 = \text{LQL}$). Good lots, with quality level given by AQL, are to be rejected no more than $100\alpha\%$ of the time. Poor lots, with quality level specified by LQL, are to be accepted no more than $100\beta\%$ of the time.

Because the criteria are more stringent here than where only the producer's or the consumer's stipulation is satisfied, we may not have much flexibility in choosing the acceptance number and the associated sampling plans. It can be difficult to find a sampling plan that exactly satisfies both the producer's and consumer's stipulation.

Let's consider the plans shown in Figure 10-11. Two plans meet the producer's stipulation exactly and come close to meeting the consumer's stipulation. Two other plans meet the consumer's stipulation exactly and come close to meeting the producer's stipulation. Of these four plans, one must be selected based on additional criteria of concern to the user. It may be of interest, for example, to choose the plan with the smallest sample size to minimize inspection costs, or the one with the largest sample size to provide the most protection.

Alternatively, a preference to satisfy either the producer's or consumer's risk could be incorporated in the decision-making framework. That is, a user may desire the producer's stipulation to be satisfied exactly, with the consumer's stipulation satisfied as closely as possible, or vice versa. Such criteria will aid in selecting the appropriate sampling plan.

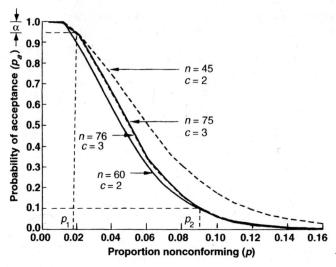

FIGURE 10-11   OC curves of sampling plans for stipulated producers's and consumer's risks.

Example 10-9   Find a single sampling plan that satisfies a producer's risk of 5% for lots that are 1.8% nonconforming, and a consumer's risk of 10% for lots that are 9% nonconforming.

Solution   We have $\alpha = 0.05$, $p_1 = AQL = 0.018$, $\beta = 0.10$, and $p_2 = LQL = 0.09$. First, we compute the ratio np_2/np_1, which is the ratio p_2/p_1 because n cancels out:

$$\frac{p_2}{p_1} = \frac{LQL}{AQL} = \frac{0.09}{0.018} = 5.00$$

For values of $\alpha = 0.05$ and $\beta = 0.10$, we use the last column in Table 10-5 to determine the possible acceptance numbers. We find that the ratio 5.00 falls between 6.51 and 4.89, corresponding to acceptance numbers of 2 and 3, respectively.

Two plans (one for $c = 2$ and one for $c = 3$) satisfy the producer's stipulation exactly: For $c = 2$, $np_1 = 0.818$, and the sample size is

$$n = \frac{np_1}{p_1} = \frac{0.818}{0.018} = 45.44 \simeq 45$$

For $c = 3$, $np_1 = 1.366$, and the sample size is

$$n = \frac{np_1}{p_1} = \frac{1.366}{0.018} = 75.88 \simeq 76$$

So, the plans $n = 45$, $c = 2$ and $n = 76$, $c = 3$ both satisfy the producer's stipulation exactly.

Next, we find that two plans ($c = 2$ and $c = 3$) satisfy the consumer's stipulation exactly: For $c = 2$, $np_2 = 5.322$, and the sample size is

$$n = \frac{np_2}{p_2} = \frac{5.322}{0.09} = 59.13 \simeq 60$$

For $c = 3$, $np_2 = 6.681$, and the sample size is

$$n = \frac{np_2}{p_2} = \frac{6.681}{0.09} = 74.23 \simeq 75$$

The plans $n = 60$, $c = 2$ and $n = 75$, $c = 3$ both satisfy the consumer's stipulation exactly. The four candidates are as follows:

Plan 1: $n = 45$, $c = 2$ Plan 3: $n = 60$, $c = 2$

Plan 2: $n = 76$, $c = 3$ Plan 4: $n = 75$, $c = 3$

Now let's see how close plans 1 and 2 (which satisfy the producer's stipulation) come to satisfying the consumer's stipulation. For a target value of the consumer's risk β of 0.10, we find the proportion nonconforming p_2 of batches that would be accepted $100\beta\%$ of the time. For $n = 45$ and $c = 2$ (plan 1), if $\beta = 0.10$, then $np_2 = 5.322$. Thus,

$$p_2 = \frac{np_2}{n} = \frac{5.322}{45} = 0.1183$$

For $n = 76$ and $c = 3$ (plan 2), if $\beta = 0.10$, then $np_2 = 6.681$. So

$$p_2 = \frac{np_2}{n} = \frac{6.681}{76} = 0.0879$$

Plan 1 accepts batches that are 11.83% nonconforming 10% of the time. On the other hand, plan 2 accepts batches that are only 8.79% nonconforming 10% of the time. Our goal is to find a plan that accepts batches that are 9% nonconforming 10% of the time.

Given that the target value of p_2 (the specified LQL) is 0.09, we find plan 2's value of $p_2 = 0.0879$ is closer to the target value than plan 1's value 0.1183. If our selection criteria calls for meeting the producer's stipulation exactly and closely meeting the consumer's stipulation, we would choose plan 2. Note that plan 2 is a little more stringent and therefore more conservative than our goal.

Now let's find a plan that satisfies the consumer's stipulation exactly and comes as close as possible to satisfying the producer's stipulation. For plans 3 and 4, we need to determine the proportion nonconforming p_1 of batches that would be accepted 95% of the time. This satisfies the producer's risk $\alpha = 0.05$.

For $n = 60$ and $c = 2$ (plan 3), if $\alpha = 0.05$, then $np_1 = 0.818$. So

$$p_1 = \frac{np_1}{n} = \frac{0.818}{60} = 0.0136$$

For $n = 75$ and $c = 3$ (plan 4), if $\alpha = 0.05$, then $np_1 = 1.366$. So

$$p_1 = \frac{np_1}{n} = \frac{1.366}{75} = 0.0182$$

Plan 3 rejects batches that are 1.36% nonconforming 5% of the time. On the other hand, plan 4 rejects batches that are 1.82% nonconforming 5% of the time. Since plan 4's value of $p_1 = 0.0182$ is closer to the target value of $p = 0.0180$, plan 4 is selected. Note that plan 4 is more stringent than our goal.

Another criterion we could use to select a sampling plan is to choose the one with the smallest sample size in order to minimize inspection costs. Of the four candidates plan 1 would be selected with $n = 45$, $c = 2$. This plan satisfies the producer's stipulation exactly.

Alternatively, we could select a plan with the largest sample size, which provides the most information. Here we would choose plan 2 with $n = 76$, $c = 3$. As discussed previously, this plan satisfies the producer's stipulation exactly and comes as close as possible to the consumer's stipulation.

Double Sampling Plans

The parameters of a double sampling plan, as explained previously, are as follows:

n_1:   size of the first sample
c_1:   acceptance number for the first sample
r_1:   rejection number for the first sample
n_2:   size of the second sample
c_2:   acceptance number for the second sample
r_2:   rejection number for the second sample

As noted earlier, for the same degree of protection (that is, the same probability of accepting lots of a given quality), double sampling plans may have a smaller average sample number (ASN) than corresponding single sampling plans. The size of the first sample, n_1, is always smaller than the sample size (n) of an equivalent single sampling plan. So, if a decision can usually be made on the first sample, the ASN will be lower for a double sampling plan. Or, if inspection can be curtailed during the second sample, the ASN will be reduced (most likely, less than that for a single sampling plan).

The OC Curve   The OC curve also measures the performance of double sampling plans. Calculations for double sampling plans are more involved than for single sampling plans.

Let P_{a1} denote the probability of accepting the lot on the first sample and P_{a2} the probability of lot acceptance on the second sample. The combined probability of acceptance is given by $P_a = P_{a1} + P_{a2}$. Let x_1 and x_2 denote the observed number of nonconforming items on the first and second samples, respectively. We have

$$
\begin{aligned}
P_{a1} &= P(x_1 \leq c_1) \\
P_{a2} &= P(x_1 = c_1 + 1)P(x_2 \leq c_2 - x_1) + P(x_1 = c_1 + 2)P(x_2 \leq c_2 - x_1) + \cdots \\
&\quad + P(x_1 = r_1 - 1)P(x_2 \leq c_2 - x_1) \\
P_a &= P_{a1} + P_{a2}
\end{aligned} \tag{10-11}
$$

In the following example we calculate the probability of lot acceptance for a given proportion nonconforming. The Poisson approximation to the hypergeometric distribution is used here. This procedure can be used to construct the OC curve.

Example 10-10   Let's consider a double sampling plan of lot size 3000 given by the following parameters: $n_1 = 40$, $c_1 = 1$, $r_1 = 5$, $n_2 = 80$, $c_2 = 5$, $r_2 = 6$. For a lot proportion nonconforming value of $p = 0.03$, find the probability of accepting such lots.

Solution   To find the probability of acceptance on the first sample (P_{a1}), we have $n_1 p = (40)(0.03) = 1.2$. The Poisson cumulative probability distribution tables in Appendix

A-2 give

$$P_{a1} = P(x_1 \le 1) = 0.663$$

We calculate P_{a2} as follows:

$$P_{a2} = P(x_1 = 2)P(x_2 \le 3) + P(x_1 = 3)P(x_2 \le 2) + P(x_1 = 4)P(x_2 \le 1)$$

Note that when calculating $P(x_2 \le 3)$ and other probabilities dealing with x_2, the value of $n_2 p = (80)(0.03) = 2.4$. So

$$P_{a2} = (0.216)(0.779) + (0.087)(0.570) + (.026)(0.308) = 0.2259$$

The combined probability of acceptance is

$$P_a = P_{a1} + P_{a2} = 0.663 + 0.2259 = 0.8889$$

Thus, about 66.3% of lots with a proportion nonconforming at 0.03 will be accepted on the first sample; for these lots a second sample will not be taken. For 22.59% of the lots, a second sample will be taken, and the lot will be accepted based on the evidence in both of the samples. Overall, 88.89% of the lots will be accepted by this sampling plan.

Repeating this computation for several values of the proportion nonconforming p allows us to construct the OC curve for the double sampling plan.

Figure 10-12 shows the OC curve for this double sampling plan. Both P_{a1}, the probability of acceptance on the first sample, and P_a, the probability of acceptance based on both samples, are shown. The difference between the ordinate values of the graphs for P_a and P_{a1} indicates the probability of lot acceptance on the second sample. For very small vames of p, the graphs of P_a and P_{a1} are close because a majority of the lots are accepted on the first sample. for very large values of p, the plots of P_a and P_{a1} are in close proximity because most

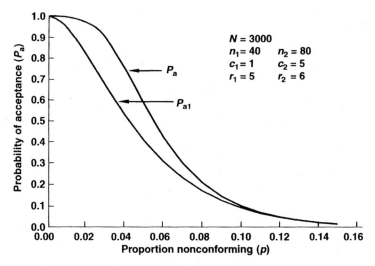

FIGURE 10-12 OC curve for the double sampling plan for Example 10-10.

of the lots are rejected on the first sample. In between these extremes, the increasing difference between the curves represents the increase in the probability of lot acceptance due to the second sample.

When judging the overall performance of the double sampling plan, the combined probability of acceptance P_a should be considered. The effectiveness of the sampling plan will be determined by its discriminatory power. Lots whose quality level is acceptable or better should be rejected infrequently, preferably with a probability no more than the producer's risk α. Lots of limiting quality level or poorer should be accepted infrequently, preferably with a probability no more than the consumer's risk β.

Average Sample Number and Average Total Inspection Curves   As noted earlier, the average sample number represents the average number of items inspected per lot, for a series of lots with a given proportion nonconforming p, that are needed to make a decision. For a double sampling plan without curtailed inspection, the average sample number was given in eq. (10-8).

A plot of ASN values for different values of p is known as an ASN curve (see Figure 10-7). One reason for selecting a double sampling plan over a single one is that the ASN for a double sampling plan is expected to be less. If the relevant range of the proportion nonconforming p is known, it is of interest if the ASN for a double sampling plan is less than that for an equivalent single sampling plan, at least for that range of p. Note that the ASN for a single sampling plan is equal to the sample size n.

The ASN curves for equivalent double and single sampling plans often plot as shown in Figure 10-13. Assuming that there is no curtailment of inspection for either plan, the single sampling plan will have a smaller ASN for values of the proportion nonconforming p between p_1 and p_2. Thus we need to know the average quality of the incoming batches. Such information will help us determine the usefulness of the double sampling plan. If the average proportion nonconforming is less than p_1 or greater than p_2, the double sampling plan is preferable.

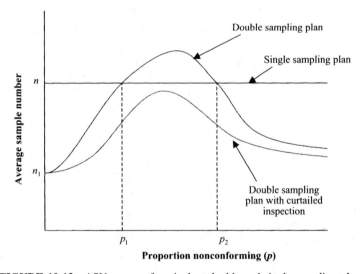

FIGURE 10-13   ASN curves of equivalent double and single sampling plans.

Curtailing inspection during the second sample of a double sampling plan reduces the average sample number. Thus, the ASN curve for a double sampling plan with curtailed inspection (shown in Figure 10-13) may be below that for a single sampling plan over the entire range of the proportion nonconforming values. In such a case, if reduction of inspection costs, were a criterion, the double sampling plan with curtailment would be preferable.

Another measure of performance, in conjunction with rectification inspection, is, the average total inspection, which is the average number of items inspected per lot for a series of lots with a given proportion nonconforming level p, assuming that rejected lots go through screening. For a double sampling plan, the ATI is given by eq. (10-7).

The significance and computation of the ATI curve for a single sampling plan were explained earlier. The ATI curve is a nondecreasing function of the proportion nonconforming p. When p is close to zero, ATI approaches n_1; for extremely large values of p, ATI approaches the lot size N. Given the quality level of the incoming product, minimizing the ATI reduces total costs of inspection and rectification.

Example 10-11 Find the average total inspection for lots with an incoming proportion nonconforming of 0.03 for a double sampling plan with lot size 3000 given by the following parameters: $n_1 = 40$, $c_1 = 1$, $r_1 = 5$, $n_2 = 80$, $c_2 = 5$, $r_2 = 6$.

Solution Usually we would first calculate P_{a1} and P_{a2}. However, we found these values in Example 10-10 for the same double sampling plan, so we will use those results:

$$P_{a1} = 0.663 \quad \text{and} \quad P_{a2} = 0.2259$$

Using these results, the average total inspection for incoming lots that are 3% non-conforming is

$$\text{ATI} = (40)(0.663) + (40 + 80)(0.2259) + (3000)(1 - 0.663 - 0.2259) = 386.93$$

So, the average total inspection per batch will be about 387 items if lots are 3% nonconforming. Recall that for a single batch, the total inspection will be either 40 or 120, respectively, if the lot is accepted on the first or second sample. If the lot is rejected on either sample, then the assumption of screening implies that all 3000 items in the lot will be inspected. An ATI curve for this sampling plan is shown in Figure 10-14.

Design of Double Sampling Plans One criterion for designing double sampling plans involves satisfying a specified level of the producer's risks α at an associated quality level $p_1 = \text{AQL}$ and meeting a consumer's risk β at a quality level $p_2 = \text{LQL}$. Thus, we want to find a plan where the OC curve passes through the two points (AQL, $1 - \alpha$) and (LQL, β).

Another sample procedure will help us find double sampling plans. Let's assume the sample sizes are either $n_2 = n_1$, or $n_2 = 2n_1$. We will use a pair of tables known as Grubbs' tables (Tables 10-6 and 10-7), named after their originator, Frank E. Grubbs, who proposed their use for constructing double sampling plans. Table 10-6 is used when $n_1 = n_2$, and Table 10-7 is used when $n_2 = 2n_1$. Both tables are based on a producer's risk α of 0.05 and a consumer's risk β of 0.10.

These tables are used to construct double sampling plans as follows. First, the ratio R of LQL to AQL (or p_2/p_1) is found. Using either Table 10-6 or 10-7, depending on the

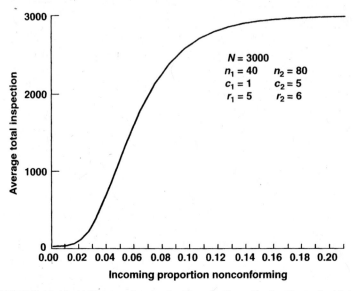

FIGURE 10-14   ATI curve for the double sampling plan for Example 10-11.

circumstances, the value closest to the calculated R-value is found. Next, the value of $n_1 p$ is read from the appropriate table. Dividing $n_1 p$ by p (which is either p_1 or p_2) yields the size of the first sample (n_1) The complete sampling plan, involving acceptance number for the two samples, is read from Tables 10-6 and 10-7.

TABLE 10-6   Grubbs Table I: Values for Constructing Double Sampling Plans Where $n_1 = n_2$ ($\alpha = 0.05$, $\beta = 0.10$)

Plan	$R = p_2/p_1$	Acceptance Numbers c_1	c_2	Approximate Values of $n_1 p$ $P_a = 0.95$	$P_a = 0.50$	$P_a = 0.10$	Approximate for ASN/ n_1 for $P_a = 0.95$
1	11.90	0	1	0.21	1.00	2.50	1.170
2	7.54	1	2	0.52	1.82	3.92	1.081
3	6.79	0	2	0.43	1.42	2.96	1.340
4	5.39	1	3	0.76	2.11	4.11	1.169
5	4.65	2	4	1.16	2.90	5.39	1.105
6	4.25	1	4	1.04	2.50	4.42	1.274
7	3.88	2	5	1.43	3.20	5.55	1.170
8	3.63	3	6	1.87	3.98	6.78	1.117
9	3.38	2	6	1.72	3.56	5.82	1.248
10	3.21	3	7	2.15	4.27	6.91	1.173
11	3.09	4	8	2.62	5.02	8.10	1.124
12	2.85	4	9	2.90	5.33	8.26	1.167
13	2.60	5	11	3.68	6.40	9.56	1.166
14	2.44	5	12	4.00	6.73	9.77	1.215
15	2.32	5	13	4.35	7.06	10.08	1.271
16	2.22	5	14	4.70	7.52	10.45	1.331
17	2.12	5	16	5.39	8.40	11.41	1.452

Source: Adapted from Chemical Corps Engineering Agency, *Manual 2: Master Sampling Plans for Single, Duplicate, Double, and Multiple Sampling,* Army Chemical Center, Edgewood Arsenal, MD, 1953.

TABLE 10-7   Grubbs' Table II: Values for Constructing Double Sampling Plans Where
$n_2 = 2n_1$ ($\alpha = 0.05$, $\beta = 0.10$)

Plan	$R = p_2/p_1$	Acceptance Numbers c_1	Acceptance Numbers c_2	Approximate Values of $n_1 p$ $P_a = 0.95$	Approximate Values of $n_1 p$ $P_a = 0.50$	Approximate Values of $n_1 p$ $P_a = 0.10$	Approximate for ASN/n_1 for $P_a = 0.95$
1	14.50	0	1	0.16	0.84	2.32	1.273
2	8.07	0	2	0.30	1.07	2.42	1.511
3	6.48	1	3	0.60	1.80	3.89	1.238
4	5.39	0	3	0.49	1.35	2.64	1.771
5	5.09	1	4	0.77	1.97	3.92	1.359
6	4.31	0	4	0.68	1.64	2.93	1.985
7	4.19	1	5	0.96	2.18	4.02	1.498
8	3.60	1	6	1.16	2.44	4.17	1.646
9	3.26	2	8	1.68	3.28	5.47	1.476
10	2.96	3	10	2.27	4.13	6.72	1.388
11	2.77	3	11	2.46	4.36	6.82	1.468
12	2.62	4	13	3.07	5.21	8.05	1.394
13	2.46	4	14	3.29	5.40	8.11	1.472
14	2.21	3	15	3.41	5.40	7.55	1.888
15	1.97	4	20	4.75	7.02	9.35	2.029
16	1.74	6	30	7.45	10.31	12.96	2.230

Source: Adapted from Chemical Corps Engineering Agency, *Manual 2: Master Sampling Plans for Single, Duplicate, Double, and Multiple Sampling,* Army Chemical Center, Edgewood Arsenal, MD, 1953.

Example 10-12   Find a double sampling plan for lot size 2500 where it is desired to reject batches that are 1.2% nonconforming no more than 5% of the time, while batches that are 7.5% nonconforming are accepted about 10% of the time. The sample sizes are equal, and the producer's stipulation must be satisfied exactly.

Solution   We have $N = 2500$, $\alpha = 0.05$, $p_1 = \text{AQL} = 0.012$, $\beta = 0.10$, $p_2 = \text{LQL} = 0.075$, and $n_1 = n_2$ Computing the ratio of p_2/p_1 have

$$R = \frac{p_2}{p_1} = \frac{\text{LQL}}{\text{AQL}} = \frac{0.075}{0.012} = 6.25$$

For $R = 6.25$, the closest value in Table 10-6 is $R = 6.79$ for plan 3, Its acceptance numbers are $c_1 = 0$ and $c_2 = 2$. If the producer's risk ($\alpha = 0.05$) is to be satisfied exactly, $P_a = 1 - 0.05 = 0.95$ for $p_1 = 0.012$. From Table 10-6 using $P_a = 0.95$, the value of $n_1 p$ is 0.43. The size of the first sample is thus

$$n_1 = \frac{0.43}{0.012} = 35.83 \simeq 36$$

Because the sample sizes are assumed to be equal, $n_2 = n_1 = 36$. So the double sampling plan is $n_1 = 36$, $c_1 = 0$, $n_2 = 36$, $c_2 = 2$. The values of r_1 and r_2 are both assumed to be 3.

The ASN corresponding to $P_a = 0.95$ can be found using Table 10-6. For plan 3, ASN/$n_1 = 1.340$. So ASN $= (1.340)(36) = 48.24$. For other values of p, the general formula for the ASN of a double sampling plan [eq. (10-8)] would have to be used.

Multiple Sampling Plans

A multiple sampling plan is an extension of a double sampling plan. The decision-making procedure is similar to the double sampling plan in that we can inspect two or more samples before we decide the fate of the lot.

For each sample, a sample size, acceptance number, and rejection number are specified. For example, the multiple sampling plan in the following table shows that a first sample of size 20 is chosen from the lot. If no nonconforming items are found in sample 1, the lot is accepted. If 2 or more nonconforming items are found, the lot is rejected. If 1 nonconforming item is found, a second sample of size 20 is chosen. If the combined number of non-conforming items is 1 or less in samples 1 and 2, the lot is accepted. If the combined number of nonconforming items is 3 or more, the lot is rejected. If the combined number of nonconforming items is 2, a third sample of size 20 is selected. This process continues until sample 4 of size 20 is chosen and the lot is either accepted or rejected at this final stage.

Sample	Cumulative Sample	Acceptance Number	Rejection Number
1	20	0	2
2	40	1	3
3	60	2	4
4	80	3	4

The average sample number is usually less for a multiple sampling plan than for an equivalent single or double sampling plan. If lots are exceptionally good or poor, a decision will usually be made at an earlier stage in the multiple sampling plan. Thus, we inspect fewer items, on average, to determine the disposition of the lot. The expression for the ASN for a multiple sampling was given previously by eq. (10-9).

For the example just described, the number of levels (k) is 4. The sample sizes are $n_1 = 20$, $n_2 = 20$, $n_3 = 20$, $n_4 = 20$. The probability of making a decision at the first level, P_1, is

$$P_1 = P(x_1 = 0) + P(x_1 \geq 2)$$

where x_1 represents the number of nonconforming items observed for the first sample. Similarly, the probability of making a decision on the second sample is

$$P_2 = P(x_1 = 1)P(x_2 = 0) + P(x_1 = 1)P(x_2 \geq 2)$$

where x_2 represents the number of nonconforming items found for the second sample.

Along the same lines, the probability of making a decision on the third sample is

$$P_3 = P(x_1 = 1)P(x_2 = 1)P(x_3 = 0) + P(x_1 = 1)P(x_2 = 1)P(x_3 \geq 2)$$

where x_3 represents the number of nonconforming units found for the third sample. Lastly, the probability of making a decision on the fourth sample is

$$P_4 = P(x_1 = 1)P(x_2 = 1)P(x_3 = 1)P(x_4 = 0) + P(x_1 = 1)P(x_2 = 1)P(x_3 = 1)P(x_4 \geq 1)$$

where x_4 represents the number of nonconforming items found for the fourth sample. Now that P_1, P_2, P_3, and P_4 can be evaluated, the ASN is calculated using eq. (10-9).

Because multiple sampling plans are more complex, their administrative costs are higher than equivalent single or double sampling plans. A single sampling plan is the easiest, and thus least costly, to administer.

Standard Sampling Plans

In the preceding sections we discussed several methods for determining sampling plans. Many organizations prefer to use existing plans, known as standardized sampling plans, rather than compute sampling plans of their own. They simply select a set of criteria and determine the standardized plans that best match this criteria. Although standardized plans use predefined criteria, companies can generally adjust their criteria to match the standardized plan. The advantage here is that plans can be selected with very little effort. Moreover, characteristics and performance measures of the plans are already calculated and tabulated.

There are two common lot-by-lot attribute sampling plans. The first is a plan developed by the American National Standards Institute, International Organization for Standardization and American Society for Quality. It is entitled (ANSI/ISO/ASQ Z1.4-2003) *Sampling Procedures and Tables for Inspection by Attributes.* This standard is the outgrowth of sampling plans developed during and after World War II, including Military Standard 105D (MIL-STD-105D). This standard emphasizes the system aspect of the sampling procedure using OC curves in the scheme framework. The tables and procedures of **MIL-STD-105E** are retained in the ANSI standard.

ANSI/ISO/ASQ Z1.4 is used as an acceptable quality level system. This means that the quality level of good lots should be rejected infrequently. If the process average proportion nonconforming is less than the AQL, the sampling plans in ANSI/ISO/ASQ Z1.4 are designed to accept the majority of the lots. However, protecting the consumer by not accepting poor lots (that is, the limiting quality level) was not a key criterion in ANSI/ISO/ASQ Z1.4 plans.

The Dodge-Romig system, developed by H. F. Dodge and H. G. Romig (1959), is designed to minimize the average total inspection while satisfying a consumer's risk β for batches with a given quality level specified by the limiting quality level. Other Dodge-Romig plans are designed to satisfy a given average outgoing quality limit while minimizing the average total inspection and are thus indexed by either the LQL or AOQL. The ANSI/ISO/ASQ Z1.4 plans are indexed by the AQL. The Dodge-Romig system is based on a rectifying sampling scheme where rejected lots go through 100% inspection, with nonconforming items replaced with acceptable ones.

A **sampling plan** determines the fate of a lot based on a certain sample size and acceptance criteria. For example, the previously described single sampling plans satisfy a producer's risk α at a certain quality level $p_1 = \text{AQL}$.

A **sampling scheme** is a set of sampling plans with rules provided for switching among them. A scheme is indexed by lot size and either AQL, LQL, or AOQL. A set of rules specifies the type of inspection to be used.

A **sampling system** is a collection of sampling schemes. It provides rules for the selection of an appropriate sampling plan. MIL-STD-105E and ANSI/ISO/ASQ Z1.4 are sampling systems.

Here we consider the Dodge-Romig plans. For the ANSI/ISO/ASQ Z1.4 plans, the reader may consult the appropriate references.

Dodge-Romig Plans   Dodge and Romig (1959) designed a set of plans based on achieving a certain overall level of quality for products sent to the consumer. Although ANSI/ISO/ASQ Z1.4 is a system based on AQL, it has little impact on the overall quality level because the sample sizes are quite small compared to the lot sizes and only the nonconforming items in the sample are detected. Dodge-Romig plans, however, are based on rectifying inspection.

They assume that lots rejected by the sampling plans go through 100% inspection and that nonconforming items are replaced by acceptable ones. This rectification process has an impact on the overall quality level of the product sent to the consumer.

Dodge-Romig plans can be single or double sampling plans. There are two sets of plans. One is based on satisfying a given limiting quality level (LQL) based on a consumer's risk β, the target value of which is 0.10. The other is based on meeting a certain value of the average outgoing quality limit. For both sets of plans, the objective is to minimize the average total inspection.

If the parameters of the plan under consideration (e.g., lot size and process average) are within certain ranges, the Dodge-Romig tables allow us to determine feasible plans very readily. Determining these plans from basic principles would take much more time. The trade-off is that the Dodge-Romig tables provide a sampling plan for a range of lot sizes and process averages. Thus, the plans listed may not be unique in minimizing ATI. Also, the plans minimize ATI only approximately because in listing a given plan, ranges are used for the lot size and process average.

Plans Based on LQL   These plans are used when protection is desired for the acceptance of individual lots of a certain quality level. The **Dodge-Romig LQL-based plans** accept lots with a quality level given by LQL 100β% of the time (a β of 0.10 was used to develop these plans). Plans exist for LQL values of 0.5, 1.0, 2.0, 3.0, 4.0, 5.0, 7.0, and 10.0% nonconforming. Both single and double sampling plans are listed. The choice between these two types is influenced by such factors as inspection cost per unit and administrative costs of operating the plan.

To use the Dodge-Romig tables, an estimate of the **process average** nonconforming $\bar{p}$ is necessary. Recent data from the process can be used to develop this estimate. If no data is available for the process, the largest value of the process average nonconforming found in the table can be used as a conservative estimate. As more information on the process becomes available, the value of the process average nonconforming rate should be updated; consequently, a new sampling plan may be found.

The Dodge-Romig table for a single sampling plan with an LQL of 5% is shown in Table 10-8. Note that the process average in Table 10-8 lists values to 2.5%. For values over 2.5% (which is half the LQL), sampling plans may not be preferable because 100% inspection becomes more economical. The table also provides a value of the AOQL (in percentage) for a given sampling plan.

Note from Table 10-8 that the plan is indexed by lot size. A range of lot sizes is provided, as well as a range of the process average nonconforming. As the lot size increases, the sample size also increases, but the relative sample size (as a function of lot size) decreases. Thus, total inspection costs become more economical for large lot sizes. Furthermore, as the process average increases, the average amount inspected also increases. Reducing the process average will help lower inspection costs.

Example 10-13   Find a Dodge-Romig plan when the lot size is 700, the LQL is 5%, and the process average is 1.30% nonconforming. A single sampling plan is desired.

Solution   Using Table 10-8 to index the lot size and process average, the single sampling plan is found to be $n = 130, c = 3$. For this sampling plan, the AOQL is 1.2%. This means that the worst average outgoing quality, regardless of incoming quality, will not exceed 1.2%. If the process average were not known, the maximum listed value of the process average would be used (in this case, the range 2.01 to 2.5%), and the plan would be $n = 200, c = 6$.

TABLE 10-8 Dodge-Romig Single Sampling Table for Limiting Quality Level of 5.0%ᵃ

Process Average

Lot Size	0–0.05%			0.06–0.50%			0.51–1.00%			1.01–1.50%			1.51–2.00%			2.01–2.50%		
	n	c	AOQL %	n	c	AOQL %	n	c	AOQL %	n	c	AOQL %	n	c	AOQL %	n	c	AOQL %
1–30	All	0	0	All	0	0	All	0	0	All	0	0	All	0	0	All	0	0
31–50	30	0	0.49	30	0	0.49	30	0	0.49	30	0	0.49	30	0	0.49	30	0	0.49
51–100	37	0	0.63	37	0	0.63	37	0	0.63	37	0	0.63	37	0	0.63	37	0	0.63
101–200	40	0	0.74	40	0	0.74	40	0	0.74	40	0	0.74	40	0	0.74	40	0	0.74
201–300	43	0	0.74	43	0	0.74	70	1	0.92	70	1	0.92	95	2	0.99	95	2	0.99
301–400	44	0	0.74	44	0	0.74	70	1	0.99	100	2	1.0	120	3	1.1	145	4	1.1
401–500	45	0	0.75	75	1	0.95	100	2	1.1	100	2	1.1	125	3	1.2	150	4	1.2
501–600	45	0	0.76	75	1	0.98	100	2	1.1	125	3	1.2	150	4	1.3	175	5	1.3
601–800	45	0	0.77	75	1	1.0	100	2	1.2	130	3	1.2	175	5	1.4	200	6	1.4
801–1,000	45	0	0.78	75	1	1.0	105	2	1.2	155	4	1.4	180	5	1.4	225	7	1.5
1,001–2,000	45	0	0.80	75	1	1.0	130	3	1.4	180	5	1.6	230	7	1.7	280	9	1.8
2,001–3,000	75	1	1.1	105	2	1.3	135	3	1.4	210	6	1.7	280	9	1.9	370	13	2.1
3,001–4,000	75	1	1.1	105	2	1.3	160	4	1.5	210	6	1.7	305	10	2.0	420	15	2.2
4,001–5,000	75	1	1.1	105	2	1.3	160	4	1.5	235	7	1.8	330	11	2.0	440	16	2.2
5,001–7,000	75	1	1.1	105	2	1.3	185	5	1.7	260	8	1.9	350	12	2.2	490	18	2.4
7,001–10,000	75	1	1.1	105	2	1.3	185	5	1.7	260	8	1.9	380	13	2.2	535	20	2.5
10,001–20,000	75	1	1.1	135	3	1.4	210	6	1.8	285	9	2.0	425	15	2.3	610	23	2.6
20,001–50,000	75	1	1.1	135	3	1.4	235	7	1.9	305	10	2.1	470	17	2.4	700	27	2.7
50,001–100,000	75	1	1.1	160	4	1.6	235	7	1.9	355	12	2.2	515	19	2.5	770	30	2.8

Source: H. F. Dodge and H. G. Romig, *Sampling Inspection Table: Single and Double Sampling.* 2nd ed. Copyright © 1959. Reprinted by permission of John Wiley & Sons, Inc.
ᵃn, size of sample; entry "All" indicates that each piece in a lot is to be inspected; c, acceptance number for sample; AOQL, average outgoing quality limit.

TABLE 10-9 **Dodge-Romig Single Sampling Table for AOQL of 3.0%**[a]

	Process Average																	
	0–0.06%			0.07–0.60%			0.61–1.20%			1.21–1.80%			1.81–2.40%			2.41–3.00%		
Lot Size	n	c	LQL %	n	c	LQL %	n	c	LQL %	n	c	LQL %	n	c	LQL %	n	c	LQL %
1–10	All	0	—	All	0	—	All	0	—	All	0	—	All	0	—	All	0	—
11–50	10	0	19.0	10	0	19.0	10	0	19.0	10	0	19.0	10	0	19.0	10	0	19.0
51–100	11	0	18.0	11	0	18.0	11	0	18.0	11	0	18.0	11	0	18.0	22	1	16.4
101–200	12	0	17.0	12	0	17.0	12	0	17.0	25	1	15.1	25	1	15.1	25	1	15.1
201–300	12	0	17.0	12	0	17.0	26	1	14.6	26	1	14.6	26	1	14.6	40	2	12.8
301–400	12	0	17.1	12	0	17.1	26	1	14.7	26	1	14.7	41	2	12.7	41	2	12.7
401–500	12	0	17.2	27	1	14.1	27	1	14.1	42	2	12.4	42	2	12.4	42	2	12.4
501–600	12	0	17.3	27	1	14.2	27	1	14.2	42	2	12.4	42	2	12.4	60	3	10.8
601–800	12	0	17.3	27	1	14.2	27	1	14.2	43	2	12.1	60	3	10.9	60	3	10.9
801–1,000	12	0	17.4	27	1	14.2	44	2	11.8	44	2	11.8	60	3	11.0	80	4	9.8
1,001–2,000	12	0	17.5	28	1	13.8	45	2	11.7	65	3	10.2	80	4	9.8	100	5	9.1
2,001–3,000	12	0	17.5	28	1	13.8	45	2	11.7	65	3	10.2	100	5	9.1	140	7	8.2
3,001–4,000	12	0	17.5	28	1	13.8	65	3	10.3	85	4	9.5	125	6	8.4	165	8	7.8
4,001–5,000	28	1	13.8	28	1	13.8	65	3	10.3	85	4	9.5	125	6	8.4	210	10	7.4
5,001–7,000	28	1	13.8	45	2	11.8	65	3	10.3	105	5	8.8	145	7	8.1	235	11	7.1
7,001–10,000	28	1	13.9	46	2	11.6	65	3	10.3	105	5	8.8	170	8	7.6	280	13	6.8
10,001–20,000	28	1	13.9	46	2	11.7	85	4	9.5	125	6	8.4	215	10	7.2	380	17	6.2
20,001–50,000	28	1	13.9	65	3	10.3	105	5	8.8	170	8	7.6	310	14	6.5	560	24	5.7
50,001–100,000	28	1	13.9	65	3	10.3	125	6	8.4	215	10	7.2	385	17	6.2	690	29	5.4

Source: H. F. Dodge and H. G. Romig. *Sampling Inspection Table:Single and Double Sampling.* 2nd ed. Copyright © 1959. Reprinted by permission of John Wiley & Sons. Inc.
[a]n, size of sample; entry "All" indicates that each piece in a lot is to be inspected; c, acceptance number for sample; LQL, limiting quality level corresponding to a consumer's risk (β) = 0.10.

Plans Based on AOQL When we need to provide a level of protection for the average quality level of a stream of batches, a plan based on the average outgoing quality limit is often appropriate. A specified value of AOQL is selected. The objective is to choose plans such that the worst average outgoing quality for a stream of lots, regardless of incoming quality, will not exceed this AOQL value. **Dodge-Romig AOQL-based plans** are designed to meet this criterion and also to minimize the average total inspection. The plans are tabulated for AOQL values of 0.10, 0.25, 0.50, 0.75, 1.00, 1.50, 2.00, 2.50, 3.00, 4.00, 5.00, 7.00, and 10.00%. Both single and double sampling plans are available for these AOQL values. As in the previous set of plans, the lot size and process average must be known in order to use the tables. The tables also provide the LQL values for a consumer's risk β of 0.10.

The Dodge-Romig table for a single sampling plan with an AOQL of 3.0% is shown in Table 10-9. Note that the process average is listed to a value of 3.0% (equal to the AOQL value of 3.0%). For process averages exceeding this value, 100% inspection becomes economical.

Example 10-14 Find a Dodge-Romig single sampling plan when the lot size is 1200, the average outgoing quality limit is 3%, and the process average is 1.4% nonconforming.

Solution From Table 10-9, indexing the lot size of 1200 and process average of 1.4% nonconforming, the single sampling plan is found to be $n = 65$, $c = 3$. For this plan, the LQL is 10.2%. This means that for individual lots with a nonconformance rate of 10.2%, the probability of accepting such lots would be 10%, the consumer's risk.

10-9 OTHER ATTRIBUTE SAMPLING PLANS

The acceptance sampling plans described in the preceding sections are, in a sense, general purpose plans. In this section we consider sampling plans that apply to special situations where the key criteria might be to reduce inspection time and effort, to simplify, or to provide better protection under special conditions.

Chain Sampling Plan

The chain sampling plan (ChSP-1) was proposed by Dodge (1955). It is used for tests that are costly or destructive. In these situations, the sample size must by necessity be small.

Sampling plans with small sample sizes usually have an acceptance number c that is zero. However, the OC curve for single sampling plans with $c = 0$ has an undesirable shape. The entire OC curve is convex; consequently, even for extremely good lots with a low proportion nonconforming, the probability of lot acceptance P_a decreases rapidly from 1.00. So, from a producer's point of view, there is a chance that good lots will be rejected more frequently than they should. Figure 10-15 shows the OC curve for single sampling plans with an acceptance number $c = 0$, as well as OC curves for plans with acceptance numbers of 1 and 2, which have high values of P_a for small values of p, a desirable feature.

The chain sampling plan achieves a desirable OC curve for $c = 0$ when the value of p is small. There are two parameters: First, the sample size n is the number of items to be selected at random from each lot. The second parameter i represents the number of preceding samples, the inspection results of which must be considered when deciding the fate of the current lot.

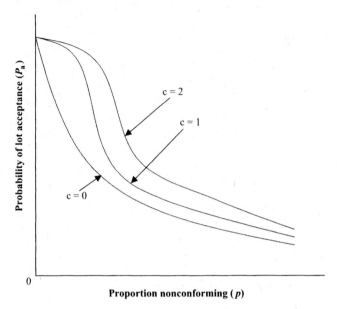

FIGURE 10-15   OC curves for $c = 0$, $c = 1$, and $c = 2$.

The plan works as follows. A random sample of n items is chosen from the lot. If no nonconforming items are found in the sample, the lot is accepted. If the number of nonconforming items is 2 or more, the lot is rejected. If the sample has 1 nonconforming item, the lot is accepted if the previous i samples each had 0 nonconforming items. The OC curve for this plan has a shape that is preferable to that for a single sampling plan with $c = 0$. Figure 10-16 shows the OC curves for chain sampling plans where the sample size is 10 and i

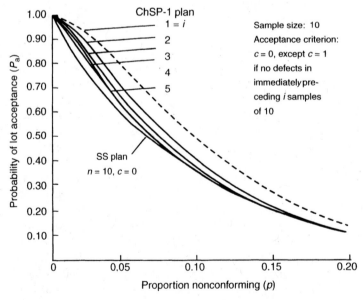

FIGURE 10-16   OC curves for ChSP-1 plans. [*Source*: H. F. Dodge (1955), "Chain Sampling Plans," *Industrial Quality Control*, 11(4): 10–13.]

varies from 1 to 5 and also the OC curve for a single sampling plan with $n = 10$ and $c = 0$. The chain sampling plan for $i = 1$ is shown as a dashed curve; it is not preferred because it deviates significantly from the OC curve for $n = 10$, $c = 0$ over the entire range of p.

Note that the OC curves for a chain sampling plan are influenced by the cumulative results of more than one sample. It is assumed that the quality level of the lot, as given by p, remains fixed over the accumulation period. In contrast, for ordinary lot-by-lot attribute inspection, decisions are based on information from each lot and are independent of the outcomes of previous lots.

The probability of lot acceptance P_a is calculated using the binomial distribution as an approximation to the hypergeometric distribution. It is given by

$$P_a = P(0, n) + P(1, n)[P(0, n)]^i \qquad (10\text{-}12)$$

where $P(0, n)$ and $P(1, n)$ represent the probabilities of obtaining 0 and 1 nonconforming items, respectively, from a sample of size n. The lot proportion nonconforming is assumed to be p.

Sequential Sampling Plan

A sequential sampling plan is similar to a multiple sampling plan in that the number of items required for sampling is influenced by the results of the sampling process itself. At each phase, based on the cumulative inspection results, a decision is made to either accept the lot, reject the lot, or continue sampling. Theoretically, the sampling process can continue indefinitely. However, if the number inspected is equal to approximately 3 times the number that would be inspected by an equivalent single sampling plan, a decision is made to terminate the plan and notify the vendor of the need to demonstrate an improved level of product before any further product can be accepted.

Sequential sampling is usually an item-by-item inspection process, even though it is possible to have groups of items at any given phase. Sequential sampling is used when it is desirable to arrive at a decision to either accept or reject the lot as soon as possible (i.e., when testing is expensive or destructive). The plan is based on the sequential probability ratio test developed by Wald (1947).

Figure 10-17 illustrates the concept behind item-by-item sequential sampling. There are two lines on the chart, an acceptance line and a rejection line. They determine the regions for acceptance and rejection and the continue sampling region. The equations of the acceptance and rejection lines are influenced by the producer's risk α and its associated quality level p_1, and by the consumer's risk β and its associated quality level p_2. The cumulative number of nonconforming items d is plotted versus the number of items inspected, n. If the plot of the cumulative number of nonconforming items touches the acceptance line or is below it, the lot is accepted. If it touches the rejection line or is above it, the lot is rejected. If the cumulative number of nonconforming items is within the region defined by the acceptance and rejection line, sampling is continued, and one more item is inspected.

The equation of the acceptance line (in slope–intercept form) is given by

$$d_a = -h_a + sn \qquad (10\text{-}13)$$

where $-h_a$ represents the intercept on the vertical axis and s represents the slope. The equation of the rejection line is given by

$$d_r = h_r + sn \qquad (10\text{-}14)$$

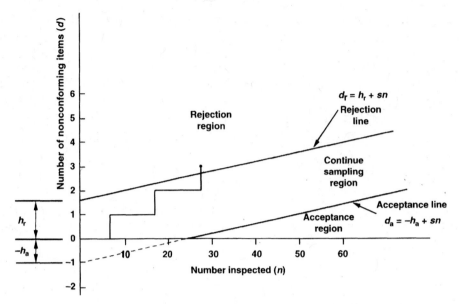

FIGURE 10-17   Item-by-item sequential sampling plan.

where h_r represents the intercept on the vertical axis and s represents the slope. These equations are based on satisfying both a given level of producer's risk at an associated quality level p_1 (the AQL) and a given level of consumer's risk at an associated quality level p_2 (the LQL). The expressions for the parameters of the acceptance line, h_a and s, and those for the rejection line, h_r and s, are as follows:

$$h_a = \frac{\ln[(1-\alpha)/\beta]}{k} \tag{10-15}$$

$$h_r = \frac{\ln[(1-\beta)/\alpha]}{k} \tag{10-16}$$

$$s = \frac{\ln[(1-p_1)/(1-p_2)]}{k} \tag{10-17}$$

where

$$k = \ln\left[\frac{p_2(1-p_1)}{p_1(1-p_2)}\right] \tag{10-18}$$

10-10 DEMING'S *kp* RULE

The objective of Deming's *kp* rule is to minimize the average total cost of inspection of incoming materials and final product for processes that are stable. It calls for 0% *or* 100% inspection. It can be shown that, if a process is stable, the distribution of the nonconforming items in a sample is independent of the distribution of the nonconforming items in the remainder of the lot. Items are submitted in lots, and a random sample is drawn from that lot to make a decision regarding the entire lot.

To use the *kp* rule, these assumptions are made. First, the inspection process is assumed to be completely reliable. This means that a nonconforming item, if inspected, will be labeled as such by the inspection process. Second, all items are inspected prior to moving forward to the next customer in the process. The next customer can be another department within the organization or a different company. The implication is that all nonconforming items will be detected prior to shipment to the succeeding customer. Third, the vendor provides the buyer with an extra supply of items in order to replace any nonconforming items that are found. The cost of providing these additional items is included in the vendor's charges and is considered as an overhead cost, which would thus be present regardless of the type of inspection plan used. Therefore, this cost is not included in the average cost function that is being minimized.

The following notation is necessary. Let

p: average proportion of nonconforming items in the lots

k_1:   cost of initial inspection of an item

k_2:   cost of repair or reassembly due to the usage of a nonconforming item

k: average cost to find a conforming item from the additional supply to replace a detected nonconforming item $= k_1/(1-p)$

x_i: 1 if item i is nonconforming; 0 otherwise

If item i is randomly selected from the lot, there is an initial inspection cost of k_1. If it is found to be nonconforming, the cost to replace it with a conforming one is k. Thus, the unit cost of initial inspection and replacement if nonconforming is given by

$$C_1 = \begin{cases} k_1 + kx_i & \text{if item } i \text{ is inspected} \\ 0 & \text{if item } i \text{ is not inspected} \end{cases} \qquad (10\text{-}19)$$

Similarly, the cost to repair and replace a nonconforming item is given by

$$C_2 = \begin{cases} (k_2 + k)x_i & \text{if item } i \text{ is not initially inspected} \\ 0 & \text{if item } i \text{ is initially inspected} \end{cases} \qquad (10\text{-}20)$$

Under the assumptions considered, C_1 and C_2 are mutually exclusive. If one is nonzero, the other is zero, and the total cost is given by $C = C_1 + C_2$. Thus, if item i is inspected, the total cost is $(k_1 + kx_i)$; if it is not inspected, the total cost is $(k_2 + k)x_i$. This concept can be extended to the average total cost per part for the entire lot. The parameter p represents the lot proportion nonconforming. So, the average total cost per part if items are inspected is $(k_1 + kp)$; if items are not inspected, it is $(k_2 + k)p$. The break-even value of p is obtained by equating the average total cost for inspection to that for no inspection. In other words,

$$k_1 + kp = (k_2 + k)p$$

or

$$p = \frac{k_1}{k_2} \qquad (10\text{-}21)$$

The *kp* rule is as follows:

1. If k_1/k_2 is greater than p, conduct no inspection. This situation occurs if the proportion nonconforming of incoming product is low, the unit cost of an incoming inspection is

high, and the cost of repairing a nonconforming item is low. Basically, this implies that there is very little risk associated with nonconforming items if the incoming quality is quite good.

2. If k_1/k_2 is less than p, conduct 100% inspection. This condition happens when the incoming proportion nonconforming is high, the unit cost of incoming inspection is low, and the cost of repairing a nonconforming item is high. In this situation, it is quite expensive for a nonconforming item to be allowed into production.

3. If k_1/k_2 equals p, then either no inspection or 100% inspection is conducted. Usually, if the estimated value of p is not very reliable, 100% inspection is conducted.

Note that conducting no inspection does not imply no information. Small samples should be drawn from lots, even when the rule suggests no inspection, in order to obtain information regarding the process. If 100% inspection is in operation, every effort must be made to upgrade the process quality level through process improvement procedures upon the initiative of management.

Example 10-15   In the manufacture of television sets, the cost to inspect a printed circuit board is $0.15. If a nonconforming circuit board is allowed in the assembly, the cost to subsequently disassemble it and replace the unit is $100. It is estimated that the proportion nonconforming of incoming circuit boards is 0.4%. What should the inspection policy be using Deming's *kp* rule?

Solution   We have $k_1 = 0.15$, $k_2 = 100$, and $p = 0.004$. The ratio $k_1/k_2 = 0.15/100 = 0.0015$. Since $k_1/k_2 < p$, the policy calls for 100% inspection of all incoming printed circuit boards.

Critique of the *kp* Rule

While Deming's *kp* rule provides an alternative to acceptance sampling, there are some practical limitations in its implementation (Vardeman 1986; Milligan 1991). The rule requires knowledge of the process average of nonconforming items, p. But, estimating p requires sampling. These samples may be large in size, depending on the desired accuracy. Thus, while we may wish to avoid acceptance sampling and move to an all-or-none scheme, we must sample to implement the all-or-none scheme.

Deming's *kp* rule assumes that the process average nonconformance rate, p, is stationary. This may not be true for processes over a long period of time. This means that samples will have to be taken periodically to estimate p and to modify the *kp* rule accordingly. Furthermore, even if the value of p is stationary, verification of this again requires sampling. Thus, a contradiction exists: Using Deming's *kp* rule, which advocates no inspection or 100% inspection, requires sampling.

Another drawback in implementing the *kp* rule is obtaining accurate estimates of k_1, unit inspection cost, and k_2, cost of repair or reassembly due to the usage of a nonconforming item. To estimate k_1, data must be collected on capital costs of equipment, the depreciation policy used, cost of obtaining capital, and the current value of the equipment. Additionally, operating costs that include inspection costs for labor, maintenance, utilities, rent, and other overhead items constitute a part of the unit cost k_1. Obtaining realistic estimates of k_2 is even more difficult because of both its tangible and intangible components. It is not easy to establish the cost of identifying nonconforming items later in the production process and the cost of their repair. And, there are the intangible costs: What is the cost associated with letting a nonconforming item slip by? This can involve warranty costs, product liability lawsuits, and costs of recall. How do we determine the cost of loss of goodwill due to a consumer experiencing a faulty product? These difficulties lead to crude estimates of k_2. Thus, the accuracy of the kp rule, which is derived from these estimates, becomes questionable.

The assumption of linearity in the cost structure is also subject to question. This applies to the unit cost of inspection and the cost to repair and replace a nonconforming item. Note that the level of the process where the nonconforming item is detected will influence the repair or replacement costs.

Observe that 100% inspection is not a viable policy when the inspection process involves destructive testing. Furthermore, there is no guarantee that 100% inspection will detect all nonconforming items because of such reasons as inspector fatigue and human error. While Deming's kp rule makes use of a simple decision framework, it has experienced limited usage because of the difficulties in obtaining accurate estimates of the model parameters.

10-11 SAMPLING PLANS FOR VARIABLES

A variable is a quality characteristic that is measured on a numerical scale, such as weight, pressure, temperature, viscosity, tensile strength, elasticity, resistance, and so on. The characteristic is often inherently a variable. Treating it as an attribute does not retain the precision of information offered by variables. If testing is destructive or costly, smaller sample sizes may be necessary. Acceptance sampling plans for variables are more suitable in these instances because they require smaller sample sizes than corresponding attribute sampling plans for the same degree of protection.

There are two basic types of variable sampling plans. The first deals with controlling a process parameter such as the mean or standard deviation. The desirable settings of the parameters are such that certain conditions regarding the mean lot quality and the corresponding probability of lot acceptance are satisfied. Plans can be designed for single or double specification limits and a process standard deviation that is known or unknown. Some plans in this category involve acceptance control charts, sequential sampling plans for variables, and hypothesis testing on process parameters (which is not treated in this chapter).

We examine acceptance control charts in the following sections. Sequential sampling plans for variables, which we do not discuss, are used when the quality characteristic is normally distributed and the process standard deviation is known. The procedure is similar to that for the sequential plans for attributes discussed earlier; in such plans the ordinate is the cumulative sum of the characteristic (ΣX), which is plotted as a function of the number of units inspected.

The second type of plan deals with controlling the proportion of the product that is nonconforming (i.e., does not satisfy specifications). Here as well, single or double specification limits and a process variability that is known or unknown can be used to determine the plans. In this chapter we focus on single sampling for both types of plans, although double or multiple sampling plans can also be designed.

Advantages and Disadvantages of Variable Plans

Some advantages of variable sampling plans over attribute plans are as follows:

1. For a comparable level of protection as specified by the producer's risk α, the acceptable quality level (AQL), the consumer's risk β, and the limiting quality level (LQL), sample sizes are smaller for a variable plan than for an attribute plan.

2. Variable sampling plans provide more information than attribute plans. Since a numerical value for the characteristic is specified, the extent to which the item is conforming or not is obtained. Attribute plans simply specify the item as conforming or not.

3. Variable sampling plans provide insight into the areas that deserve attention for quality improvement. The information from a variable plan may provide clues for remedial actions for improving process quality. Continual process improvement should be the goal of every organization; variable plans help achieve it.

Some disadvantages of variable sampling plans are as follows:

1. Each quality characteristic requires a separate sampling plan. Because the number of quality characteristics is usually large, this implies that several sampling plans must be monitored. With attribute sampling plans, several variables can be combined to form a single attribute plan.

2. The administrative and unit inspection costs are usually higher for variable plans than for attribute plans. The measuring instruments are more expensive because an exact measurement value is taken.

3. To make inferences from the variable sampling plans, we must know or estimate the distribution of the quality characteristic for the process under consideration.

10-12   VARIABLE SAMPLING PLANS FOR A PROCESS PARAMETER

Variable sampling plans for a process parameter are used when either the average quality of the product or process or the variability of the quality, is of concern. These plans can be used for items that are submitted in lots (e.g., in bags, boxes, drums, or bins). There are plans for situations in which the **process standard deviation** is known or unknown. The sampling plans in this section require that the distribution of the quality characteristic be normal. However, minor departures from normality may not significantly affect the tests.

Estimating Process Average: Single Specification Limit and Known Process Standard Deviation

Let's consider a single sampling plan to estimate the process average; a single specification limit is given, and the process standard deviation is known. The two parameters of the sampling plan are the sample size n and the **acceptance limit** $\bar{X}_a$. Let's suppose the lower specification limit for the density of a chemical is specified. A variable sampling plan may call for selecting a random sample of size n from the lot. The sample's average density is then found. If the sample average is less than the acceptance limit $\bar{X}_a$, the lot is rejected. Otherwise, the lot is accepted. Alternatively, if we specify an upper specification limit for the density, the lot is rejected if the sample average is greater than the corresponding acceptance limit. The process standard deviation σ is assumed to be known.

Now let's derive a variable sampling plan under the following conditions. We want to accept batches of good average quality, as denoted by $\bar{X}_1$, with a probability of $1 - \alpha$ and to accept batches of poor average quality, as denoted by $\bar{X}_2$, with a probability of β.

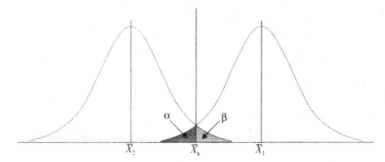

FIGURE 10-18   Relationship of $\bar{X}_1$ and $\bar{X}_2$ to $\bar{X}_a$ in the sampling distribution of $\bar{X}$.

Thus, we need to find a sampling plan whose OC curve passes through the points $(\bar{X}_1, 1 - \alpha)$ and $(\bar{X}_2, \beta)$. Our quality characteristic must also meet a minimum specification limit.

Figure 10-18 shows the relationship of $\bar{X}_1$ and $\bar{X}_2$ to $\bar{X}_a$ in the sampling distribution of the sample mean $\bar{X}$. Let Z_α denote the standard normal value corresponding to the tail area of α, and let Z_β denote the standard normal value corresponding to the tail area of β. We have

$$Z_\alpha = \frac{\bar{X}_a - \bar{X}_1}{\sigma/\sqrt{n}} \tag{10-22}$$

and

$$Z_\beta = \frac{\bar{X}_a - \bar{X}_2}{\sigma/\sqrt{n}} \tag{10-23}$$

Note that for the case shown in Figure 10-18, Z_α will be negative and Z_β will be positive. Equations (10-22) and (10-23) involve two unknowns, $\bar{X}_a$ and n. By solving them simultaneously, we get the sample size

$$n = \left[\frac{(Z_\beta - Z_\alpha)\sigma}{\bar{X}_1 - \bar{X}_2} \right]^2 \tag{10-24}$$

Using the form of n in eq. (10-24), eq. (10-23) gives the acceptance limit

$$\bar{X}_a = \frac{Z_\beta \bar{X}_1 - Z_\alpha \bar{X}_2}{Z_\beta - Z_\alpha} \tag{10-25}$$

Example 10-16   Ammonium nitrate is shipped in 500-kg bags; the lower specification for the concentration of nitrogen is 13%. The distribution of the concentration of nitrogen is known to be normal, with a standard deviation of 1.5%. Find a variable sampling plan that satisfies the following conditions:

1. Batches with a mean 2.5 standard deviations above the lower specification limit should be accepted with a probability of 0.95.

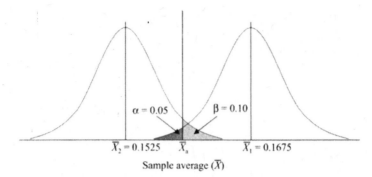

FIGURE 10-19   Process means $\bar{X}_1$ and $\bar{X}_2$ and their relationship to $\bar{X}_a$ for nitrogen concentration.

2. Batches with a mean 1.5 standard deviations above the lower specification limit should be accepted with a probability of 0.10.

Solution   First, we calculate the process average quality levels associated with the two conditions that are to be satisfied. Denoting the lower specification limit by LSL, we have

$$\bar{X}_1 = \text{LSL} + 2.5\sigma$$
$$= 0.13 + (2.5)(0.015) = 0.1675$$
$$\bar{X}_2 = \text{LSL} + 1.5\sigma$$
$$= 0.13 + (1.5)(0.015) = 0.1525$$

Thus, we wish to accept lots that have a mean quality level of 16.75% with a probability of 0.95. Additionally, we wish to accept lots that have a mean quality level of 15.25% with a probability of 0.10.

Figure 10-19 shows the relationship of these conditions with respect to the acceptance limit $\bar{X}_a$. We have $1 - \alpha = 0.95, \bar{X}_1 = 0.1675, \beta = 0.10, \bar{X}_2 = 0.1525$. From Appendix A-3, we have $Z_\alpha = Z_{0.05} = -1.645$ and $Z_\beta = Z_{0.10} = 1.282$. From eq. (10-24), the sample size is

$$n = \left\{ \frac{[1.282 - (-1.645)](0.015)}{0.1675 - 0.1525} \right\}^2$$
$$= 8.57 \simeq 9$$

From eq. (10-25), the acceptance limit is

$$\bar{X}_a = \frac{(1.282)(0.1675) - (-1.645)(0.1525)}{1.282 - (-1.645)} = 0.1591$$

The sampling plan works as follows. A random sample of size 9 is chosen from the lot, and the sample average concentration of nitrogen is found. If the sample average is less than 15.91%, the lot is rejected; otherwise, it is accepted.

Estimating Process Average: Double Specification Limits and Known Process Standard Deviation

In certain instances, the quality characteristic needs to be within a range of values, which necessitates the existence of double specification limits. For example, the diameter of an acceptable bearing should lie between upper and lower specification limits. Thus, the levels

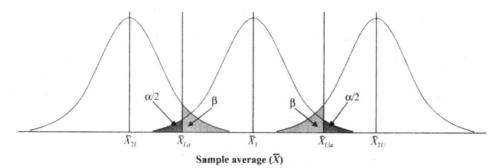

Sample average ($\bar{X}$)

FIGURE 10-20   Relationship between the various parameters for a plan using double specification limits.

of poor quality for the process average are given by two values, $\bar{X}_{2L}$ and $\bar{X}_{2U}$, and the associated consumer's risk is denoted by β. As in the previous section, the level of good quality of the process average, denoted by $\bar{X}_1$, is such that the producer's risk is given by α. Thus, the OC curve of the sampling plan should pass through the points $(\bar{X}_1, 1 - \alpha)$, $(\bar{X}_{2L}, \beta)$, and $(\bar{X}_{2U}, \beta)$. Accordingly, there will be two acceptance limits, $\bar{X}_{La}$ and $\bar{X}_{Ua}$, the lower and upper acceptance limits, respectively. The process standard deviation is assumed to be known and is represented by σ; the sample size is denoted by n.

Figure 10-20 shows the relationship between the various parameters of the sampling plan. It is assumed in this section that $\bar{X}_1$ is at the midpoint of $\bar{X}_{2L}$ and $\bar{X}_{2U}$. Assuming a normal distribution of the sample means, the following equations are obtained:

$$Z_{\alpha/2} = \frac{\bar{X}_{Ua} - \bar{X}_1}{\sigma/\sqrt{n}} \tag{10-26}$$

$$-Z_{\alpha/2} = \frac{\bar{X}_{La} - \bar{X}_1}{\sigma/\sqrt{n}} \tag{10-27}$$

$$Z_{\beta} = \frac{\bar{X}_{La} - \bar{X}_{2L}}{\sigma/\sqrt{n}} \tag{10-28}$$

$$-Z_{\beta} = \frac{\bar{X}_{Ua} - \bar{X}_{2U}}{\sigma/\sqrt{n}} \tag{10-29}$$

where $Z_{\alpha/2}$ and Z_{β} are the standard normal variate values when the right tail areas are $\alpha/2$ and β, respectively. Note that there are three unknowns: $\bar{X}_{La}$, $\bar{X}_{Ua}$, and n. One of the preceding equations can be eliminated as being redundant, and the unknown parameters of the sampling plan can be solved.

Example 10-17   The diameter of an axle must lie within a desirable upper and lower bound. Consequently, if the process average diameter is below 45 mm or above 47 mm, the desired probability of lot acceptance is 0.10. Let the producer's risk be 0.05 and the process standard deviation of the axle diameters be 0.6 mm. Find the variable acceptance sampling plan.

Solution   We have $\bar{X}_{2L} = 45$, $\bar{X}_{2U} = 47$, $\beta = 0.10$, $\alpha = 0.05$, and $\sigma = 0.6$. The value of $\bar{X}_1$ (midway between 45 and 47) is 46. Figure 10-21 shows the relationship between the various parameters, where $\bar{X}_{La}$ and $\bar{X}_{Ua}$ denote the lower and upper acceptance limits, respectively.

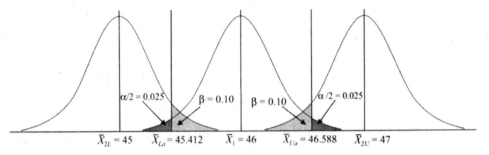

FIGURE 10-21   Relationship between various parameters for a sampling plan for axle diameter.

After finding the standard normal variates and using eqs. (10-26) to (10-29), we get

$$1.96 = \frac{\bar{X}_{Ua} - 46}{\sigma/\sqrt{n}} \tag{10-30}$$

$$-1.96 = \frac{\bar{X}_{La} - 46}{\sigma/\sqrt{n}} \tag{10-31}$$

$$1.282 = \frac{\bar{X}_{La} - 45}{\sigma/\sqrt{n}} \tag{10-32}$$

$$-1.282 = \frac{\bar{X}_{Ua} - 47}{\sigma/\sqrt{n}} \tag{10-33}$$

From eqs. (10-32) and (10-33) we get

$$\bar{X}_{La} + \bar{X}_{Ua} = 92$$

From eqs. (10-31) and (10-32), we have

$$\sqrt{n} = \frac{(1.282 + 1.96)\sigma}{46 - 45} = (1.282 + 1.96)(0.6) = 1.945$$

The sample size is computed as

$$n = (1.945)^2 = 3.784 \simeq 4$$

For this computed value of n, eq. (10-31) gives

$$\bar{X}_{La} = 46 - (1.96)\frac{0.6}{\sqrt{4}} = 45.412$$

Therefore, $\bar{X}_{Ua} = 46.588$.

The sampling plan operates as follows. A random sample of size 4 is chosen from the lot, and the average diameter of the axles is computed. If the sample average is less than 45.412 mm or greater than 46.588 mm, the lot is rejected. Otherwise, the lot is accepted.

Estimating Process Average: Single Specification Limit and Unknown Process Standard Deviation

In most cases, the process standard deviation of the quality characteristic is not known, and estimates of it are obtained from a sample. Assume that the distribution of the quality characteristic is normal. The sample variance s^2, for a sample of size n (as given previously) is

$$s^2 = \frac{\sum X_i^2 - (\sum X_i)^2/n}{n-1} \tag{10-34}$$

where X_i represents the value of the quality characteristic for the ith item in the sample. As before, suppose that the good quality level of the process is denoted by $\bar{X}_1$, such that the probability of accepting batches with this mean is $1 - \alpha$, where α denotes the producer's risk. For batches with a poor quality level, whose average is denoted by $\bar{X}_2$, the probability of acceptance is low and is given by β, the consumer's risk.

The sampling distribution of $(\bar{X} - \bar{X}_1)/(s/\sqrt{n})$ is approximately a t-distribution with $(n-1)$ degrees of freedom. The t-distribution is discussed in Chapter 4, and the t-tables are shown in Appendix A-4. The problem here is that because the sample size is not known, the t-statistic cannot be computed. One way to circumvent this is to estimate the process standard deviation, as denoted by $\hat{\sigma}$. After additional information is obtained, this estimate may be updated.

Neyman and Tobarska (1936) constructed OC curves for sampling plans with single specification limits based on the t-statistic for a producer's risk α of 0.05. Figure 10-22

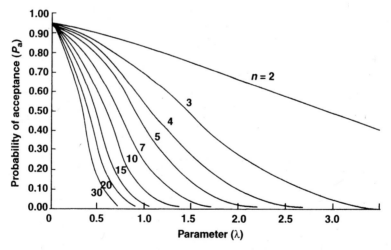

FIGURE 10-22 OC curves for sampling plans with a single specification limit when the process standard deviation is unknown for $\alpha = .05$. [*Source*: J. Neyman and B. Tobarska (1936), "Errors of the Second Kind in Testing Student's Hypothesis," *Journal of the American Statistical Association*, 31: 318–326. Reprinted with permission from the *Journal of the American Statistical Association*. Copyright © 1936 by the American Statistical Association. All rights reserved.]

shows these OC curves for different values of n. The abscissa represents the parameter λ, which is given by

$$\lambda = \frac{|\bar{X}_1 - \bar{X}_2|}{\hat{\sigma}} \tag{10-35}$$

If the probability of lot acceptance P_a is set equal to β, the sample size can be found from Figure 10-22.

Once the sample size is determined, the decision-making process is as follows. A random sample of size n is chosen, and the t-statistic is computed:

$$t = \frac{\bar{X} - \bar{X}_1}{s/\sqrt{n}} \tag{10-36}$$

where $\bar{X}$ and s represent the sample average and sample standard deviation, respectively. If a lower specification limit is given, the lot is rejected if the calculated value of $t < t_{\alpha,n-1}$, where $t_{\alpha,n-1}$ is the 100α percentile point of the t-distribution with $(n-1)$ degrees of freedom. Appendix A-4 is used to obtain the percentile points of the t-distribution. If an upper specification limit is given, the lot is rejected if $t > t_{1-\alpha,n-1}$, where $t_{1-\alpha,n-1}$ is the $100(1-\alpha)$ percentile point of the t-distribution within $(n-1)$ degrees of freedom.

Example 10-18   A soft-drink bottling company has a lower specification limit of 3.00 liters (L). Bottles with an average content of 3.08 L or more are accepted 95% of the time. Bottles with an average content of 2.97 L or less are to be accepted 10% of the time. The standard deviation of the bottle contents is unknown. However, management feels that a reasonable estimate is 0.2 L. Find a variable sampling plan.

Solution   We have $\bar{X}_1 = 3.08$, $\alpha = 0.05$, $\bar{X}_2 = 2.97$, $\beta = 0.10$, $\hat{\sigma} = 0.2$. The parameter λ given by eq. (10-35) is calculated as

$$\lambda = \frac{|3.08 - 2.97|}{0.2} = 0.55$$

From Figure 10-22, for $\alpha = 0.05$, $\lambda = 0.55$, and a probability of acceptance of 0.10, the sample size is approximately 30. From Appendix A-4, the t-statistic for a lower-tail area of $\alpha = 0.05$ and 29 degrees of freedom is -1.699.

The sampling plan operates as follows. A random sample of 30 bottles is chosen from the lot, and the sample average $\bar{X}$ and sample standard deviation s are computed. Given values of $\bar{X}$ and s, the t-statistic is computed as

$$t = \frac{\bar{X} - 3.08}{s/\sqrt{30}}$$

if $t < -1.699$, the lot is rejected. Otherwise, the lot is accepted.

10-13   VARIABLE SAMPLING PLANS FOR ESTIMATING THE LOT PROPORTION NONCONFORMING

Assume that the quality characteristic is normally distributed with a known standard deviation σ. A relationship can be established between the sample average $\bar{X}$, the process standard deviation σ, and the lot percent nonconforming. If a lower specification limit L is given, a standard normal deviate is found from

$$Z_L = \frac{\bar{X} - L}{\sigma} \qquad (10\text{-}37)$$

Alternatively, if the upper specification limit U is given, the standard normal deviate is calculated as

$$Z_U = \frac{U - \bar{X}}{\sigma} \qquad (10\text{-}38)$$

An estimate of the lot proportion nonconforming is the area under the portion of the normal curve that falls outside Z_L or Z_U. Note that as the sample mean moves farther away from the specification limit, the value of the corresponding deviate increases and the percent nonconforming decreases.

Two methods (Duncan 1986) are commonly used to estimate the proportion nonconforming of a lot: **Form 1** (the k-method) and **Form 2** (the M-method). With Form 1, the standard normal deviate, Z_L or Z_U, is compared to a critical value k. If Z_L or Z_U is greater than or equal to k, the lot is accepted; otherwise, the lot is rejected.

With Form 2, the lot percent nonconforming is estimated by first modifying the indices Z_L and Z_U to obtain the indices Q_L and Q_U which are unbiased and have minimum variance (Lieberman and Resnikoff 1955). If a lower specification limit L is given, Q_L is obtained as

$$Q_L = \frac{\bar{X} - L}{\sigma} \sqrt{\frac{n}{n-1}} = Z_L \sqrt{\frac{n}{n-1}} \qquad (10\text{-}39)$$

If an upper specification limit is given, Q_U is found as

$$Q_U = \frac{U - \bar{X}}{\sigma} \sqrt{\frac{n}{n-1}} = Z_U \sqrt{\frac{n}{n-1}} \qquad (10\text{-}40)$$

An estimate $\hat{p}$ of the lot percent nonconforming is obtained by finding the portion of the area under the standard normal distribution (Appendix A-3) that falls outside Q_L or Q_U. The **decision rule** is as follows. If $\hat{p}$ is greater than a **maximum allowable percent nonconforming** value M, the lot is rejected; otherwise, the lot is accepted.

Derivation of a Variable Sampling Plan with a Single Specification Limit and Known Process Standard Deviation

Let's suppose the following conditions must be satisfied. The probability of rejecting lots of a good quality level for which the proportion nonconforming is p_1 should be equal to α. The probability of accepting lots of a poor quality level for which the proportion nonconforming is p_2 should be equal to β. Therefore, the OC curve of the sampling plan must pass through the two points $(1 - \alpha, p_1)$ and (β, p_2). Let Z_α and Z_β denote the upper-tail percentile points of the standard normal distribution such that the tail areas are α and β, respectively. Suppose a lower specification limit is given.

Let μ_1 denote the process mean for which the proportion nonconforming is p_1, as shown in Figure 10-23a. Denoting the corresponding standard normal deviate as Z_1, we have

$$Z_1 = \frac{\mu_1 - L}{\sigma}$$

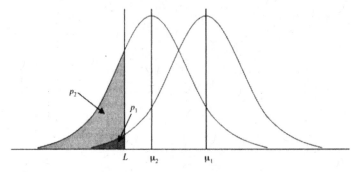

(a) Distribution of quality characteristic, X

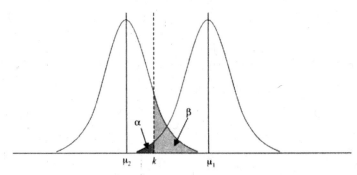

(b) Distribution of the sample average, $\bar{X}$

FIGURE 10-23   Distribution of X and $\bar{X}$.

where the tail area outside Z_1 is p_1. Similarly, we have

$$Z_2 = \frac{\mu_2 - L}{\sigma}$$

where the tail area outside Z_2 is p_2.

Note that the formulas for Z_1 and Z_2 differ from the usual definition of the standard normal deviate. Writing them in this format makes them positive; the values corresponding to the tail areas of the standard normal distribution that are outside the deviates Z_1 or Z_2 can be looked up. Given σ, we can determine the process mean locations μ_1 and μ_2.

Figure 10-23b shows the distribution of the sample averages; the relationship among μ_1, μ_2, α, and β is shown. Under Form 1, the lot is accepted if $(\bar{X} - L)/\sigma \geq k$, which is equivalent to the condition

$$\frac{\bar{X} - \mu}{\sigma/\sqrt{n}} \geq \left(k - \frac{\mu - L}{\sigma}\right)\sqrt{n}$$

Using the relation $Z_1 = (\mu_1 - L)/\sigma$ and the requirement of accepting lots of good quality (with process mean at μ_1) with a probability of $1 - \alpha$, we have

$$P\left[\frac{\bar{X} - \mu}{\sigma/\sqrt{n}} \geq (k - Z_1)\sqrt{n}\right] = 1 - \alpha$$

If $Z_{1-\alpha}$ is the standard normal deviate such that the probability of falling outside it is $1 - \alpha$, we have

$$(k - Z_1)\sqrt{n} = Z_{1-\alpha}$$

Note that $Z_{1-\alpha} = -Z_\alpha$, where Z_α represents the standard normal deviate such that the probability of falling outside it is α. For $\alpha < 0.50$, $Z_{1-\alpha}$ will be negative, and Z_α will be positive. Hence,

$$(k - Z_1)\sqrt{n} = -Z_\alpha \tag{10-41}$$

Using the relation $Z_2 = (\mu_2 - L)/\sigma$ and the requirement of accepting lots of poor quality (with process mean at μ_2) with a probability of β, we have

$$P\left(\frac{\bar{X} - \mu}{\sigma/\sqrt{n}} \geq \frac{k - Z_2}{\sqrt{n}}\right) = \beta$$

Denoting Z_β as the standard normal deviate such that the probability of falling outside it is β, we have

$$(k - Z_2)\sqrt{n} = Z_\beta \tag{10-42}$$

Solving eqs. (10-41) and (10-42) simultaneously, we get the sample size as

$$n = \left(\frac{Z_\alpha + Z_\beta}{Z_1 - Z_2}\right)^2 \tag{10-43}$$

The value of k can be found from either of the following equations:

$$k = Z_1 - \frac{Z_\alpha}{\sqrt{n}} \tag{10-44}$$

$$k = Z_2 + \frac{Z_\beta}{\sqrt{n}} \tag{10-45}$$

The use of eq. (10-44) or (10-45) is influenced by whether we desire to exactly achieve the value of α or β, respectively.

Using **Form 1**, the procedure for the variable sampling plan is as follows. A random sample of size n is selected. The sample mean $\bar{X}$ is found. Given a lower specification limit L, Z_L is found from eq. (10-37). The value of k is found from eq. (10-44) or (10-45), depending on whether we want to achieve the value of α or β. Sometimes, the value of k is chosen as the average of the values obtained using eqs. (10-44) and (10-45). If $Z_L \geq k$, the lot is accepted. Otherwise, the lot is rejected.

Using **Form 2**, if a lower specification limit L is given, a sample of size n is selected. The sample mean $\bar{X}$ is found. The value of Q_L is computed using eq. (10-39), and the percent of the distribution ($\hat{p}$) that exceeds this value is found from Appendix A-3. The maximum allowable percent nonconforming, M, is found from Appendix A-3 by determining the area that exceeds the standard normal value of $k\sqrt{n/(n-1)}$, where k is found from eq. (10-44) or (10-45), or the average of both of these values. If $\hat{p} \leq M$, the lot is accepted. Otherwise, the lot is rejected.

For Form 1, if an upper specification limit U is specified, a similar procedure is adopted. A random sample of size n is selected from a lot, where n is given by eq. (10-43). The sample average $\bar{X}$ is computed. Given the process standard deviation σ, Z_U is found from eq. (10-38). If $Z_U \geq k$, where k is found from eq. (10-44) or (10-45) or the average of these two values, the lot is accepted. Otherwise, the lot is rejected. The procedure for using Form 2 with an upper specification limit is similar to that for a lower specification limit.

Example 10-19   In the manufacture of heavy-duty utility bags for household use, the lower specification limit for the carrying load is 100 kg. We want to accept lots that are 2% nonconforming with a probability of 0.92. For lots that are 12% nonconforming, the desired probability of acceptance is 0.10. The standard deviation of the carrying load is known to be 8 kg. Find a variable sampling plan, and explain its operation.

Solution   We have $L = 100$, $p_1 = 0.02$, $\alpha = 0.08$, $p_2 = 0.12$, $\beta = 0.10$, and $\sigma = 8$. From the standard normal distribution tables in Appendix A-3, we get $Z_1 = 2.055$, $Z_2 = 1.175$, $Z_\alpha = 1.405$, and $Z_\beta = 1.282$. The sample size is

$$n = \left(\frac{1.405 + 1.282}{2.055 - 1.175} \right)^2 = 9.323 \simeq 10$$

If we want to satisfy the value of $\alpha = 0.08$ exactly, then k is

$$k = 2.055 - \frac{1.405}{\sqrt{10}} = 1.611$$

We select a sample of 10 bags and find the sample average of the carrying load to be 110 kg. We can use either Form 1 or Form 2.
Using Form 1, we calculate Z_L using eq. (10-37) as follows:

$$Z_L = \frac{110 - 100}{8} = 1.25$$

Since $Z_L < k$, the decision based on Form 1 is to reject the lot.
Using Form 2, we calculate Q_L using eq. (10-39) as follows:

$$Q_L = 1.25\sqrt{10/9} = 1.3176 \simeq 1.32$$

From Appendix A-3, the proportion that falls outside the standard normal value of 1.32 is 0.0934. This is the estimated proportion of nonconforming items ($\hat{p}$). The maximum allowable proportion nonconforming, M, is found by first calculating the standard normal value as

$$k\sqrt{n/(n-1)} = 1.611\sqrt{10/9} = 1.698 \simeq 1.70$$

From Appendix A-3, the area above this standard normal value of 1.70 is 0.0446, which is the maximum allowable proportion nonconforming M. Since $\hat{p} > M$, the decision is to reject the lot, which is consistent with our decision using Form 1.

Standardized Plans: ANSI/ISO/ASQ Z1.9 and MIL-STD-414

Military Standard 414 (MIL-STD-414) is a lot-by-lot acceptance sampling plan for variables. This standard was developed as an alternative to MIL-STD-105E, which uses attribute inspection. The civilian version of this plan, ANSI/ISO/ASQ Z1.9, was last updated in 2003.

ANSI/ISO/ASQ Z1.9 is an AQL-based plan in which it is assumed that the distribution of the quality characteristic is normal. Three types of inspection (normal, tightened, and reduced) are provided in the standard. Switching may take place between these types; the rules for switching are the same as those for ANSI/ISO/ASQ Z1.4. The recent quality history of the product determines the type of inspection.

Three general inspection levels, I, II, and III, and two special levels, S-3 and S-4 are used. The choice of general inspection level is determined by such factors as the unit inspection cost and the desired degree of discrimination between acceptable and unacceptable products. Inspection level I is used when less discrimination can be tolerated and the sample sizes are small. Level III is used when more discrimination is desired and the sample sizes are larger. Unless otherwise specified, level II is typically used.

The plans in ANSI/ISO/ASQ/Z1.9 can be used in conjunction with single or double specification limits. As noted earlier, two methods of decision making, namely Form 1 or Form 2, are used. For double specification limits, however, only Form 2 is used. Form 1 uses a critical value for the standardized distance, expressed in terms of the standard deviation, between the process mean and a given specification limit. If this standardized distance is greater than or equal to k, where k is a standard value tabulated in ANSI/ISO/ASQ Z1.9, the lot is accepted. Form 2, on the other hand, uses an estimate of the percentage of nonconfoming items that fall outside the specifications. If this estimate is less than or equal to M, where M is a standard value tabulated in the plan, the lot is accepted. For further details the reader should consult the appropriate references.

SUMMARY

In this chapter we have discussed a variety of acceptance sampling plans for attribute inspection. From the items selected for inspection, a count of the number of nonconformities or the number of nonconforming items, is made. This count, if less than or equal to a standard value (based on criteria that are to be satisfied by the sampling plan), determines acceptance of the lot. Alternatively, if the count exceeds the standard value of the acceptance number, the lot is rejected. This represents the simplest form of decision making associated with single sampling plans; this chapter has introduced double and multiple sampling plans, too. Much of the chapter is devoted to lot-by-lot attribute sampling plans, for which items are submitted for inspection in lots, or batches. Single and double sampling plan designs are based on such criteria as producer's risk and consumer's risk. In addition to lot-by-lot attribute inspection plans, some special plans such as chain sampling and sequential sampling have also been discussed. A discussion of Deming's kp rule for sampling, which provides a prescription for either 0% or 100% inspection, has been included.

This chapter has also introduced sampling plans for variables. Whereas attribute inspection is easier and usually less expensive, variable sampling plans yield more information. For the same degree of protection, the sample sizes for variable sampling plans are usually smaller than those for corresponding attribute sampling plans. Two main types of variable sampling plans have been presented. The first deals with controlling a process parameter. Plans for estimating the process average for single and double specification limits when the process standard deviation is known have been discussed. For cases in which the process standard deviation is unknown, plans that estimate the process average with a single specification limit are presented. The second type of variable sampling plan is designed to estimate the lot proportion nonconforming. Two methods of decision making, Forms 1 and 2, have been described. The case for which the process standard deviation is known and a single specification limit is given has been presented.

We have also presented Bayes' rule for updating estimates of probabilities, based on sample information. In this context, optimal decision-making concepts are discussed.

KEY TERMS

acceptability constant
acceptable quality level
acceptance limit
acceptance number
attribute sampling plans
average outgoing quality
average outgoing quality limit
average sample number
average total inspection
Bayes' rule
chain sampling plans
consumer's risk
decision rule
   Form 1
   Form 2
Deming's *kp* rule
double sampling plan
double specification limits
expected value of perfect
   information
inspection
   types of
    normal
    reduced
    tightened
limiting quality level
lot size
lot tolerance percent defective

maximum allowance percent
   nonconforming
multiple sampling plan
operating characteristic curve
   type A OC curve
   type B OC curve
process average
process standard deviation
producer's risk
rectifying inspection
rejectable quality level
sample size
sampling plans
   design
sampling scheme
sampling system
screening
sequential sampling plan
single sampling plan
single specification limit
standardized sampling plans
   ANIS/ISO/ASQ Standard Z1.4
   ANSI/ISO/ASQ Standard Z1.9
   Dodge–Romig plans
   MIL-STD-105E
   MIL-STD-414
variable sampling plan

EXERCISES

Discussion Questions

10-1 Discuss the advantages and disadvantages of sampling.

10-2 Distinguish between producer's risk and consumer's risk. In this context, explain the terms *acceptable quality level* and *limiting quality level*. Discuss instances for which one type of risk might be more important than the other.

10-3 What is the importance of the OC curve in the selection of sampling plans? Describe the impact of the sample size and the acceptance number on the OC curve. What is the disadvantage of having an acceptance number of zero?

10-4 Discuss the relative advantages and disadvantages of single, double, and multiple sampling plans.

10-5 Distinguish between average outgoing quality and acceptable quality level. Explain the meaning and importance of the average outgoing quality limit.

10-6   If you were interested in protection for acceptance of a single lot from a vendor with whom you do not expect to conduct much business, what criteria would you select, and why?

10-7   Explain the difference between average sample number and average total inspection. State any assumptions made.

10-8   If rectification inspection is used, discuss possible criteria to use in choosing sampling plans.

10-9   Discuss the context in which minimizing the average sample number would be a feasible criterion. Which type of sampling plan (single, double, or multiple) would be preferable, and what factors would influence your choice?

10-10   Compare and contrast chain sampling and sequential sampling plans. When are they used?

10-11   Discuss the assumptions made in Deming's kp rule. When would you use this rule?

10-12   What are the advantages and disadvantages of variable sampling plans over those for attributes?

10-13   What are the parameters of a variable sampling plan for which the process average quality is of interest? Explain the working procedure of such a plan when single and double specification limits are given.

10-14   Explain the difference between the decision-making procedure using Forms 1 and 2 for variable sampling plans that are designed to estimate the proportion of nonconforming items.

Problems

10-15   Consider a single sampling plan with a lot size of 1500, sample size of 150, and acceptance number of 3. Construct the OC curve. If the acceptable quality level is 0.05% nonconforming and the limiting quality level is 6% nonconforming, describe the protection offered by the plan at these quality levels.

10-16   Consider Exercise 10-15. Answer the same questions for the sampling plan $N = 1500$, $n = 200$, $c = 3$. Discuss the degree of protection of this plan compared to that in Exercise 10-15.

10-17   Suppose that desirable producer's risk is 3% and consumer's risk is 6%. Which of the plans described in Exercises 10-15 and 10-16 are preferable? Discuss your choice.

10-18   For the sampling plan $N = 1500$, $n = 150$, $c = 3$, construct the average outgoing quality curve. What is the AOQL? Interpret it.

10-19   Construct the ATI curve for the sampling plan $N = 1200$, $n = 50$, $c = 1$. Suppose that the process average nonconforming rate is 3%. Explain the value of ATI for that level of nonconformance.

10-20   For the double sampling plan $N = 2000$, $n_1 = 80$, $c_1 = 1$, $r_1 = 3$, $n_2 = 100$, $c_2 = 2$, $r_2 = 3$, construct and interpret the ASN curve. Suppose that process average nonconforming rate is 1.5%. Would you prefer the stated double sampling plan or a single sampling plan with $n = 100$, $c = 2$ in order to minimize ASN?

10-21   For the double sampling plan $N = 2200$, $n_1 = 60$, $c_1 = 1$, $r_1 = 5$, $n_2 = 120$, $c_2 = 4$, $r_2 = 5$, construct the ASN curve. Within what range of proportion nonconforming values would you prefer the stated double sampling plan over a single sampling plan with $n = 85$, $c = 2$ in order to minimize ASN?

10-22   A computer monitor manufacturer subcontracts its major parts to four vendors: A, B, C, and D. Past records show that vendors A and C provide 30% of the requirements each, vendor B provides 25%, and vendor D provides 15%. In a random sample of 10 parts, 4 were found to be nonconforming. It is known that vendors A and C operate at a level of 5% nonconformance rate, while vendors B and D are at 10% and 15%, respectively. What is the probability that the sample came from vendor A? From vendor B? From vendor C?

10-23   A manufacturer is considering replacement of an existing machine that performs an operation on a part. The variable costs are $0.38 per piece on the existing machine and $0.05 per piece on the new machine. The cost of the new machine is $40,000, while the existing machine can be scrapped at a value of $5400. It is known that the proportion nonconforming using the existing machine is 0.05. With the new machine, since the nonconformance rate will be influenced by the operator and the new machine, it cannot be specified exactly. Based on experience, the estimated probability distribution of the proportion nonconforming for the new machine is shown in Table 10-10. An order exists for 100,000 parts. Assume that the entire cost of the new machine can be assigned to this order. Note that costs may have to be calculated on the basis of good parts produced.

    (a) If a decision has to be made without sampling, what is that decision? What is the expected opportunity loss of the optimal decision?

    (b) Suppose that the seller of the new machine gives you the opportunity to sample the output from a new machine set up at another plant. You select a sample of 100 parts and find 4 nonconforming. What is your optimal decision?

TABLE 10-10

Proportion Nonconforming, p	Probability, $P(p)$
0.05	0.20
0.07	0.25
0.09	0.35
0.11	0.30

10-24   Determine the single sampling plans that will reject lots that are 1.3% nonconforming 8% of the time. Use acceptance numbers of 1, 3, and 5. From a consumer's point of view, which of these three plans would you choose?

10-25   Determine the single sampling plans that will accept lots that are 6% nonconforming 12% of the time. Use acceptance numbers of 1, 2, and 4. From a producer's point of view, which of these plans would you choose?

10-26   Determine single sampling plans that will accept lots that are 0.8% nonconforming with a probability of 0.96. Use acceptance numbers of 1, 3, and 4. If we desire batches that are 5% nonconforming to be accepted with a probability of no more than 0.04, which of the plans above would be preferable? Will the plan meet this criterion?

10-27 A sampling plan is desired to have a producer's risk of 0.05 at AQL $= 0.9\%$ and a consumer's risk of 0.10 at LQL $= 6.5\%$ nonconforming. Find the single sampling plan that meets the consumer's stipulation and comes as close as possible to meeting the producer's stipulation.

10-28 A sampling plan is desired to have a producer's risk of 0.05 at AQL $= 1.3\%$ nonconforming and a consumer's risk of 0.10 at LQL $= 7.1\%$ nonconforming. Find the single sampling plan that meets the producer's stipulation and comes as close as possible to meeting the consumer's stipulation.

10-29 A sampling plan is desired to have a producer's risk of 0.05 at AQL $= 2.0\%$ nonconforming and a consumer's risk of 0.10 at LQL $= 7\%$ nonconforming. Find the single sampling plan with the largest sample size. Find the single sampling plan with the smallest sample size.

10-30 Consider a double sampling plan given by the following parameters: $N = 1200$, $n_1 = 50$, $c_1 = 1$, $r_1 = 4$, $n_2 = 110$, $c_2 = 5$, $r_2 = 6$. Find the probability of accepting lots that are 4% nonconforming. What is the probability of rejecting a lot on the first sample?

10-31 Consider a double sampling plan given by the following parameters: $N = 2200$, $n_1 = 60$, $c_1 = 0$, $r_1 = 5$, $n_2 = 100$, $c_2 = 6$, $r_2 = 7$. Find the probability of accepting lots that are 3% nonconforming. What is the probability of accepting a lot on the first sample? What is the probability of making a decision on the first sample?

10-32 Refer to Exercise 10-30. What is the average sample number of incoming lots that are 2% nonconforming? What is the average total inspection for this quality level of 2% nonconforming?

10-33 A double sampling plan is desired that has a producer's risk of 0.05 at AQL $= 1.8\%$ nonconforming and a consumer's risk of 0.10 at LQL $= 8.5\%$ nonconforming. The lot size is 1500, and the sample sizes are assumed to be equal. Find the sampling plan if the producer's stipulation is to be satisfied exactly. Find the average sample number for lots that are 1.8% nonconforming.

10-34 It is desired to accept lots that are 9.5% nonconforming with a probability of 0.10 and to accept lots that are 2.3% nonconforming with a probability of 0.95. Find a double sampling plan for a lot size of 2000 if the second sample is to be twice as large as the first sample and the consumer's stipulation is to be satisfied exactly. Find the average sample number for lots that are 2.3% nonconforming.

10-35 Refer to Exercise 10-33. Find the double sampling plan if the second sample is to be twice as large as the first sample and the consumer's stipulation is to be satisfied exactly.

10-36 Refer to Exercise 10-33. Find the sampling plan if it is desired to accept batches that are 5% nonconforming with a probability of 0.5.

10-37 Find a Dodge–Romig single sampling plan if the lot size is 900, LQL $= 5\%$ nonconforming, and the process average is 0.8% nonconforming. What is the AOQL for the plan? Interpret it.

10-38 Find a Dodge–Romig single sampling plan if the lot size is 2200 and LQL $= 5.0\%$. Determine and interpret the AOQL for the plan.

10-39   Find a Dodge–Romig single sampling plan if the lot size is 600, the process average is 1.4% nonconforming, and AOQL = 3%. Determine and interpret the LQL for the plan.

10-40   A chain sampling plan is used with a sample size of 5 and a parameter i of 3. If lots have a proportion nonconforming of 0.06, find the probability of accepting such lots.

10-41   The equations for the acceptance and rejection lines for a sequential sampling plan are given as follows:

$$d_a = -0.95 + 0.06n$$
$$d_r = 1.22 + 0.06n$$

What is the first opportunity to reject? What is the first opportunity to accept?

10-42   A sequential sampling plan is to be used. It is desirable to have a producer's risk of 0.05 at AQL = 0.008 and a consumer's risk of 0.07 at LQL = 0.082. Determine the equations for the acceptance and rejection lines. What is the first opportunity to reject? What is the first opportunity to accept?

10-43   The initial inspection of transmission systems in automobiles is estimated to cost $0.50 per unit. If a nonconforming transmission is allowed in the assembly, the unit cost to eventually disassemble and replace it is $225. The estimated proportion nonconforming of the transmission systems is 0.003. Using Deming's kp rule, what inspection policy should be used?

10-44   In Exercise 10-43 if the initial inspection costs of the transmission systems is $1.00 per unit, what inspection policy should be followed using Deming's kp rule?

10-45   Refer to Exercise 10-43. Suppose that the monthly production is 3000 units. What is the average savings in total inspection costs per month when using the policy found from Deming's kp rule as opposed to no inspection?

10-46   Refer to Exercise 10-44. If the monthly production is 2000 units, what is the average savings in total inspection costs when using Deming's kp rule as opposed to 100% inspection?

10-47   In the construction industry, the initial inspection of tie beams is estimated to cost $0.20 per unit. If, however, a nonconforming beam is allowed for construction purposes, the unit cost of rectifying and replacing it is $50. What inspection policy should be followed, using Deming's kp rule, if the nonconformance rate of beams is 0.5%? What is the average savings in total inspection costs if 100% inspection is used as opposed to no inspection?

10-48   The upper specification limit for the resistance of coils is 30 Ω. The distribution of coil resistance is known to be normal with a standard deviation of 5 Ω. It is preferred to reject batches that have a mean of 2.3 standard deviations below the upper specification limit no more than 5% of the time. Also, batches that have a mean of 1 standard deviation below the upper specification should be accepted no more than 8% of the time. Find the parameters of a variable sampling plan and describe its operation.

10-49   The lower specification limit for the breaking strength of yarns is 25 g. The distribution of the breaking strength of yarns is normal with a variance of 6. It is desirable that lots with a mean such that 3% of the product is nonconforming be accepted 94% of the time. Lots with a mean such that 8% of the product is nonconforming should be accepted 7% of the time. Find the parameters of a variable sampling plan and describe its operation.

10-50   The tensile strength of an alloy has double specification limits. If the process average tensile strength is below $800\,kg/cm^2$ or above $1200\,kg/cm^2$, it is desired to accept such lots with a probability of 0.08. For lots with a process average of $1000\,kg/cm^2$, it is desired that the probability of acceptance be 0.96. The distribution of tensile strength is normal, with a standard deviation of $80\,kg/cm^2$. Find the variable sampling plan and describe its operation.

10-51   The length of connector pins has an upper specification limit of 45 mm and a lower specification limit of 40 mm. It is desirable that lots with a mean such that 8% of the product is nonconforming, either above the upper specification limit or below the lower specification limit, be accepted 6% of the time. We wish to accept lots with a mean of 42.5 mm with a probability of 0.94. The distribution of the length of the connector pin is normal, with a standard deviation of 0.8 mm. Find the parameters of a variable sampling plan and describe its operation.

10-52   The proportion of carbon monoxide in exhaust gases has an upper specification limit of 0.30. Emission control devices are being tested to meet such requirements. We wish that devices with an average carbon monoxide content of 0.15 or less be accepted 95% of the time. If the average carbon monoxide content is 0.34 or more, we wish the probability of acceptance to be 0.20. An estimate of the standard deviation of the proportion of carbon monoxide is 0.25. Find a variable sampling plan and explain its operation.

10-53   Refer to Exercise 10-52 regarding the proportion of carbon monoxide in exhaust gases, which has an upper specification limit of 0.30. If the average carbon monoxide content is 1 standard deviation below the upper specification limit, the devices should be rejected 5% of the time. If the average carbon monoxide content is 0.8 standard deviation above the upper specification limit, the probability of acceptance should be 0.20. The standard deviation of the proportion of carbon monoxide is unknown. Find a variable sampling plan and describe its operation.

10-54   Unleaded gasoline must meet certain federal standards. The octane number for a particular brand must be at least 89. The standard deviation of the octane number is estimated to be 4. It is preferred to accept shipments for which the average octane number is 94 about 95% of the time. Also, for those shipments that have an average octane number of 86, the probability of acceptance is desired to be 0.15. Find the parameters of a variable sampling plan.

10-55   A dairy has to control the amount of butterfat in its low-fat milk. The upper specification limit of the fat content is 4 g for 4-L containers. The standard deviation of the fat content for these containers is estimated to be 0.5 g. It is desired to accept 95% of the shipments when the proportion of containers that are nonconforming is 1%. Additionally, when 7% of the containers are nonconforming, the probability of acceptance is desired to be 8%. Find the parameters of a variable sampling plan and explain its operation, assuming that the first condition is to be satisfied exactly. A random sample of 13 containers yielded an average fat content of 3.05 g. What is the decision regarding the lot? Demonstrate the use of Forms 1 and 2.

10-56 The thickness of silicon wafers is an important characteristic in microelectronic circuits. The upper specification limit for the thickness is 0.015 mm. It is estimated that the standard deviation of the thickness of wafers is 0.0014 mm. We wish to accept lots that are 11% nonconforming with a probability of 5%. For lots that are 2% nonconforming, it is desired that the probability of rejection not exceed 4%. Find the parameters of a variable sampling plan, assuming the first of the two conditions is to be satisfied exactly. A random sample of 17 wafers yielded an average of 0.013 mm. Describe the decision regarding the lot based on Forms 1 and 2.

10-57 A cereal manufacturer who claims to meet certain mineral and vitamin requirements has a minimum specification of 25% for the iron content. The standard deviation of the iron content is estimated to be 3%. It is preferred to accept batches that are 1.5% nonconforming with a probability of 0.92. For batches that are 8% nonconforming, it is desired that the rejection probability be 0.88. Find the parameters of a variable sampling plan, assuming that the first condition is to be satisfied exactly. How would the sampling plan change if the second condition were to be satisfied exactly?

REFERENCES

ASQ (1993). ANSI/ASQ A3534-2. *Statistics—Vocabulary and Symbols—Statistical Quality Control*, Milwaukee, WI: American Society for Quality.

—(2003a). ANSI/ISO/ASQ Z1.4. *Sampling Procedures and Tables for Inspection by Attributes*, Milwaukee, WI: American Society for Quality.

—(2003b). ANSI/ISO/ASQ Z1.9. *Sampling Procedures and Tables for Inspection by Variables for Percent Nonconforming*, Milwaukee, WI: American Society for Quality.

Dodge, H.F. (1943). "A Sampling Inspection-Plan for Continuous Production," *Annals of Mathematical Statistics*, 14: 264–269.

—(1955). "Chain Sampling Inspection Plans," *Industrial Quality Control*, 11 (4): 10–13.

Dodge, H. F., and H. G. Romig (1959). *Sampling Inspection Tables: Single and Double Sampling*, 2nd ed. New York: Wiley.

Duncan, A.J. (1986). *Quality Control and Industrial Statistics,* 5th ed. Homewood, IL: Richard D. Irwin.

Grubbs, F. E. (1949). "On Designing Single Sampling Inspection Plans," *Annals of Mathematical Statistics*, XX: 256.

Lieberman, G.J., and G. J. Resnikoff (1955). "Sampling Plans for Inspection by Variables," *Journal of the American Statistical Association*, 50: 457–516.

Neyman, J., and B. Tobarska (1936). "Errors of the Second Kind in Testing Student's Hypothesis", *Journal of the American Statistical Association*, 31: 318–326.

Vardeman, S.B. (1986). "The Legitimate Role of Inspection in Modern SQC," *The American Statistician*, 40 (4): 325–328.

Wald, A. (1947). *Sequential Analysis*. New York: Wiley

PART V

PRODUCT AND PROCESS DESIGN

11

RELIABILITY

11-1 INTRODUCTION AND CHAPTER OBJECTIVES

In earlier chapters we dealt with quality assurance and control and explored the notion of quality during the manufacture of the product or the rendering of a service. Our discussion focused on a specific instant in time. In this chapter we examine the concept of reliability.

Reliability is a measure of the quality of the product over the long run. In here the concept of reliability is an extended time period over which the expected operation of the product is considered; that is, we expect the product will function according to certain expectations over a stipulated period of time.

To ensure customer satisfaction in the performance phase, we address measures to improve reliability in the design phase. The complex nature of products requires many components in their construction; thus, we need to be able to calculate system reliability. With the customer and warranty costs in mind, we must know the chances of successful operation of the product for at least a certain stipulated period of time. Such information helps the manufacturer to select the parameters of a warranty policy.

Our objective is to expose reliability calculations of systems, with a variety of components, in different configurations, such as series or parallel or combinations of both. We also introduce the concept of standby components and their impact on system reliability. Finally, our objective is to demonstrate use of reliability and life testing plans and develop parameter estimates through sampling plans.

11-2 RELIABILITY

Reliability is the probability of a product performing its intended function for a stated period of time under certain specified conditions. Four aspects of reliability are apparent from this

Fundamentals of Quality Control and Improvement, Third Edition, By Amitava Mitra
Copyright © 2008 John Wiley & Sons, Inc.

definition. First, reliability is a probability-related concept; the numerical value of this probability is between 0 and 1. Second, the functional performance of the product has to meet certain stipulations. Product design will usually ensure development of a product that meets or exceeds the stipulated requirements. For example, if the breaking strength of a nylon cord is expected to be 1000 kg, then in its operational phase/the cord must be able to bear weights of 1000 kg or more. Third, reliability implies successful operation over a certain period of time. Although no product is expected to last forever, the time requirement ensures satisfactory performance over at least a minimal stated period (say, two years). Fourth, operating or environmental conditions under which product use takes place are specified. Thus, if the nylon cord is specified to be used under dry conditions indoors, then satisfactory performance is expected for use under those conditions. In the context of these four aspects, the reliability of the nylon cord might be described as having a probability of successful performance of 0.92 in bearing loads of 1000 kg for two years under dry conditions.

11-3   LIFE-CYCLE CURVE AND PROBABILITY DISTRIBUTIONS IN MODELING RELIABILITY

Most products go through three distinct phases from product inception to wear-out. Figure 11-1 shows a typical **life-cycle curve** for which the failure rate λ is plotted as a function of time. This curve is often referred to as the **bathtub curve**; it consists of the debugging phase, the chance-failure phase, and the wear-out phase (Besterfield, 2003).

The **debugging phase**, also known as the **infant-mortality phase**, exhibits a drop in the failure rate as initial problems identified during prototype testing are ironed out. The **chance-failure phase**, between times t_1 and t_2, is then encountered; failures occur randomly and independently. This phase, in which the failure rate is constant, typically represents the useful life of the product. Following this is the **wear-out phase**, in which an increase in the failure rate is observed. Here, at the end of their useful life, parts age and wear out.

Probability Distributions to Model Failure Rate

Exponential Distribution   The life-cycle curve of Figure 11-1 shows the variation of the failure rate as a function of time. For the chance-failure phase, which represents the

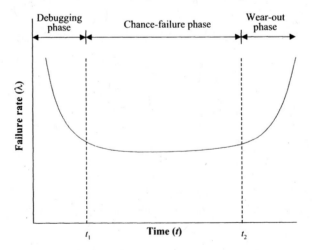

FIGURE 11-1   Typical life-cycle curve.

useful life of the product, the failure rate is constant. As a result, the **exponential distribution** can be used to describe the time to failure of the product for this phase. In Chapter 4 the exponential distribution was shown to have a probability density function given by

$$f(t) = \lambda e^{-\lambda t}, \quad t \geq 0 \qquad (11\text{-}1)$$

where λ denotes the failure rate. Figure 4-21 shows this density function.

The **mean time to failure** (MTTF) for the exponential distribution is given as

$$\text{MTTF} = \frac{1}{\lambda} \qquad (11\text{-}2)$$

Thus, if the failure rate is constant, the mean time to failure is the reciprocal of the failure rate. For repairable equipment, this is also equal to the **mean time between failures** (MTBF). There will be a difference between MTBF and MTTF only if there is a significant repair or replacement time upon failure of the product.

The reliability at time t, $R(t)$, is the probability of the product lasting up to at least time t. It is given by

$$R(t) = 1 - F(t) = 1 - \int_0^t e^{-\lambda t} dt = e^{-\lambda t} \qquad (11\text{-}3)$$

where $F(t)$ represents the cumulative distribution function at time t. Figure 11-2 shows the reliability function, $R(t)$, for the exponential failure distribution. At time 0, the reliability is 1, as it should be. Reliability decreases exponentially with time.

In general, the **failure-rate function** $r(t)$ is given by the ratio of the time-to-failure probability density function to the reliability function. We have

$$r(t) = \frac{f(t)}{R(t)} \qquad (11\text{-}4)$$

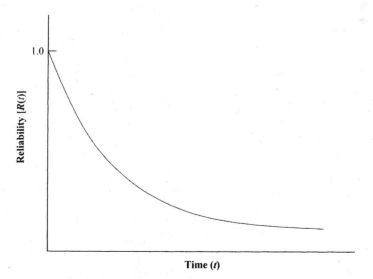

FIGURE 11-2   Reliability function for the exponential time-to-failure distribution.

For the exponential failure distribution

$$r(t) = \frac{\lambda e^{-\lambda t}}{e^{-\lambda t}} = \lambda$$

implying a **constant failure rate**, as mentioned earlier.

Example 11-1   An amplifier has an exponential time-to-failure distribution with a failure rate of 8% per 1000 hours. What is the reliability of the amplifier at 5000 hours? Find the mean time to failure.

Solution   The constant failure rate λ is obtained as

$$\lambda = 0.08/1000 \text{ hours} = 0.00008/\text{hour}$$

The reliability at 5000 hours is

$$R(t) = e^{-\lambda t} = e^{-(0.00008)(5000)} = e^{-0.4} = 0.6703$$

The mean time to failure is

$$\text{MTTF} = 1/\lambda = 1/0.00008 = 12,500 \text{ hours}$$

Example 11-2   What is the highest failure rate for a product if it is to have a probability of survival (i.e., successful operation) of 95% at 4000 hours? Assume that the time to failure follows an exponential distribution.

Solution   The reliability at 4000 hours is 0.95. if the constant failure rate is given by λ, we have

$$R(t) = e^{-\lambda t} \qquad or \quad 0.95 = e^{-\lambda(4000)}$$

This yields

$$\lambda = 0.0000128/\text{hour} = 12.8/10^6 \text{ hours}$$

Thus, the highest failure rate is $12.8/10^6$ hours for a reliability of 0.95 at 4000 hours.

Weibull Distribution   The **Weibull distribution** is used to model the time to failure of products that have a varying failure rate. It is therefore a candidate to model the debugging phase (failure rate decreases with time), or the wear-out phase (failure rate increases with time) (Henley and Kumamoto 1991). The Weibull distribution was introduced in Chapter 4. It is a three-parameter distribution whose density function is given in eq. (4-43) as

$$f(t) = \frac{\beta}{\alpha}\left(\frac{t-\gamma}{\alpha}\right)^{\beta-1} \exp\left[-\left(\frac{t-\gamma}{\alpha}\right)^{\beta}\right], \quad t \geq \gamma \tag{11-5}$$

The parameters are a location parameter, $\gamma(-\infty < \gamma < \infty)$, a scale parameter, $\alpha(\alpha > 0)$, and a shape parameter $\beta(\beta > 0)$. The probability density functions for $\gamma = 0$, $\alpha = 1$, and several values $\beta(\beta = 0.5, 1, 2, \text{ and } 4)$ are shown in Figure 4-22.

Appropriate parameters are used to model a wide variety of situations. If $\gamma = 0$ and $\beta = 1$, the Weibull distribution reduces to the exponential distribution. For reliability modeling, the location parameter will be zero. If $\alpha = 1$ and $\beta = 0.5$, for example, the failure rate decreases with time and can therefore be used to model components in the debugging phase. Similarly, if $\alpha = 1$ and $\beta = 3.5$, the failure rate increases with time and so can be used to model products in the wearout phase. In this case, the Weibull distribution approximates the normal distribution.

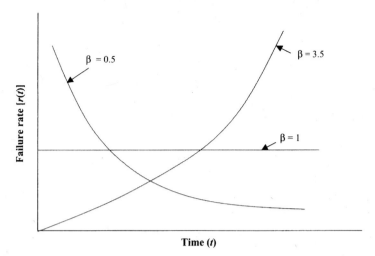

FIGURE 11-3   Failure-rate functions for the Weibull time-to-failure distribution for $\beta = 0.5, 1$, and 3.5.

The reliability function for the Weibull distribution is given by

$$R(t) = \exp\left[-\left(\frac{t}{\alpha}\right)^{\beta}\right] \tag{11-6}$$

The mean time to failure, as given by eq. (4-44), is

$$\text{MTTF} = \alpha\Gamma\left(\frac{1}{\beta} + 1\right) \tag{11-7}$$

The failure-rate function $r(t)$ for the Weibull time-to-failure probability distribution is

$$r(t) = \frac{f(t)}{R(t)} = \frac{\beta t^{\beta-1}}{\alpha^{\beta}} \tag{11-8}$$

Figure 11-3 shows the shape of the failure-rate function for the Weibull failure density function, for values of the parameter β of 0.5, 1, and 3.5.

Example 11-3   Capacitors in an electrical circuit have a time-to-failure distribution that can be modeled by the Weibull distribution with a scale parameter of 400 hours and a shape parameter of 0.2. What is the reliability of the capacitor after 600 hours of operation? Find the mean time to failure. Is the failure rate increasing or decreasing with time?

Solution   The parameters of the Weibull distribution are $\alpha = 400$ hours and $\beta = 0.2$. The location parameter γ is 0 for such reliability problems. The reliability after 600 hours of operation is given by

$$R(t) = \exp\left[-\left(\frac{t}{\alpha}\right)^{\beta}\right] = \exp\left[-\left(\frac{600}{400}\right)^{0.2}\right] = 0.3381$$

The mean time to failure is

$$\text{MTTF} = \alpha\Gamma\left(\frac{1}{\beta} + 1\right) = 400\Gamma\left(\frac{1}{0.2} + 1\right) = 48{,}000 \text{ hours}$$

The failure-rate function is

$$r(t) = \frac{0.2t^{0.2-1}}{(400)^{0.2}} = 0.0603t^{-0.8}$$

This function decreases with time. It would model components in the debugging phase.

Availability

The **availability** of a system at time t is the probability that the system will be up and running at time t. To improve availability, maintenance procedures are incorporated, which may include periodic or preventive maintenance or condition-based maintenance. An availability index is defined as

$$\text{availability} = \frac{\text{operating time}}{\text{operating time} + \text{downtime}}$$

Downtime may consist of active repair time, administrative time (processing of necessary paperwork), and logistic time (waiting time due to lack of parts). It is observed that maintainability is an important factor in influencing availability. Through design it is possible to increase the reliability and hence operational probability of a system. Further, downtime can be reduced through adequate maintenance plans. For a steady-state system, denoting the mean time to repair (MTTR) to include all the various components of downtime, we have (Blischke and Murthy 2000)

$$\text{availability} = \frac{\text{MTTF}}{\text{MTTF} + \text{MTTR}}$$

In the situation when the time-to-failure distribution is exponential (with a failure rate λ) and the time-to-repair distribution is also exponential (with a repair rate μ), the availability is given by $\mu/(\lambda + \mu)$.

11-4   SYSTEM RELIABILITY

Most products are made up of a number of components. The reliability of each component and the configuration of the system consisting of these components determines the **system reliability** (i.e., the reliability of the product). Although product design, manufacture, and maintenance influence reliability, improving reliability is largely the domain of design. One common approach for increasing the reliability of the system is through redundancy in design, which is usually achieved by placing components in parallel: As long as one component operates, the system operates. In this section we demonstrate how to compute system reliability for systems that have components in series, in parallel, or both.

Systems with Components in Series

Figure 11-4 shows a system with three **components** (A, B, and C) **in series**. For the system to operate, each component must operate. It is assumed that the components operate independent of each other (i.e., the failure of one component has no influence on the failure of any other component). In general, if there are n components in series, where the

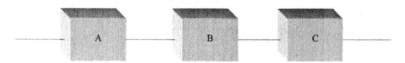

FIGURE 11-4   System with components in series.

reliability of the ith component is denoted by R_i, the system reliability is

$$R_s = R_1 \times R_2 \times \ldots \times R_n \tag{11-9}$$

The system reliability decreases as the number of components in series increases. Although overdesign in each component improves reliability, its impact would be offset by the number of components in series. Moreover, manufacturing capabilities and resource limitations restrict the maximum reliability of any given component. Product redesign that reduces the number of components in series is a viable alternative.

Example 11-4   A module of a satellite monitoring system has 500 components in series. The reliability of each component is 0.999. Find the reliability of the module. If the number of components in series is reduced to 200, what is the reliability of the module?

Solution   The system reliability for the module is

$$R_s = (0.999)^{500} = 0.606$$

Note that even with a high reliability of 0.999 for each component, the system reliability is only 60.6%. When the number of components in series is reduced to 200, the reliability of the module is

$$R_s = (0.999)^{200} = 0.819$$

Use of the Exponential Model   If the system components can be assumed to have a time to failure given by the exponential distribution and each component has a constant failure rate, we can compute the system reliability, failure rate, and mean time to failure. As noted earlier, when the components are in the chance-failure phase, the assumption of a constant failure rate should be justified.

Suppose that the system has n components in series, each with exponentially distributed time-to-failure with failure rates $\lambda_1, \lambda_2, \ldots, \lambda_n$. The system reliability is found as the product of the component reliabilities:

$$\begin{aligned} R_s &= e^{-\lambda_1 t} \times e^{-\lambda_2 t} \times \cdots \times e^{-\lambda_n t} \\ &= \exp\left[-\left(\sum_{i=1}^{n}\lambda_i\right)t\right] \end{aligned} \tag{11-10}$$

Equation (11-10) implies that the time to failure of the system is exponentially distributed with an equivalent failure rate of $\sum_{i=1}^{n}\lambda_i$. Thus, if each component that fails is replaced immediately by another that has the same failure rate, the mean time to failure for the system is given by

$$\text{MTTF} = \frac{1}{\sum_{i=1}^{n}\lambda_i} \tag{11-11}$$

When all components in series have an identical failure rate, say λ, the MTTF for the system [a special case of eq. (11-11)] is given by

$$MTTF = 1/(n\lambda)$$

Example 11-5   The automatic focus unit of a television camera has 10 components in series. Each component has an exponential time-to-failure distribution with a constant failure rate of 0.05 per 4000 hours. What is the reliability of each component after 2000 hours of operation? Find the reliability of the automatic focus unit for 2000 hours of operation. What is its mean time-to-failure?

Solution   The failure rate for each component is

$$\lambda = 0.05/4000 \text{ hours} = 12.5 \times 10^{-6}/\text{hour}$$

The reliability of each component after 2000 hours of operation is

$$R = \exp[-(12.5 \times 10^{-6})2000] = 0.975$$

The reliability of the automatic focus unit after 2000 hours of operation is

$$R_s = \exp[-(10 \times 12.5 \times 10^{-6})2000] = 0.779$$

The mean time to failure of the automatic focus unit is

$$MTTF = 1/(10 \times 12.5 \times 10^{-6}) = 8000 \text{ hours}$$

Example 11-6   Refer to Example 11-5 concerning the automatic focus unit of a television camera, which has 10 similar components in series. It is desired for the focus unit to have a reliability of 0.95 after 2000 hours of operation. What would be the mean time to failure of the individual components?

Solution   Let λ_s be the failure rate of the focus unit. Then, $\lambda_s = 10\lambda$, where λ represents the failure rate of each component. To achieve the reliability of 0.95 after 2000 hours, the value of λ_s is found from

$$0.95 = \exp[-\lambda_s(2000)] \qquad \text{or} \qquad \lambda_s = 0.0000256/\text{hour}$$

The failure rate for each component is

$$\lambda = \frac{\lambda_s}{10} = 0.00000256/\text{hour} = 2.56/10^6 \text{ hours}$$

The mean time to failure for each component would be

$$MTTF = \frac{1}{\lambda} = 390,625 \text{ hours}$$

Systems with Components in Parallel

System reliability can be improved by placing **components in parallel**. The components are redundant; the system operates as long as at least one of the components operates. The only time the system fails is when all the parallel components fail. Figure 11-5 demonstrates an example of a system with three components (A, B, and C) in parallel. All components are assumed to operate simultaneously. Examples of redundant components placed in parallel to improve the reliability of the system abound. For instance, the braking mechanism is a critical system in the automobile. Dual subsystems thus exist so that if one fails, the brakes still work.

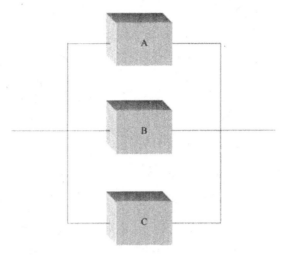

FIGURE 11-5   System with components in parallel.

Suppose that we have n components in parallel, with the reliability of the ith component denoted by R_i, $i = 1, 2, \ldots, n$. Assuming that the components operate randomly and independently of each other, the probability of failure of each component is given by $F_i = 1 - R_i$. Now, the system fails only if all the components fail. Thus, the probability of system failure is

$$
\begin{aligned}
F_s &= (1 - R_1)(1 - R_2) \cdots (1 - R_n) \\
&= \prod_{i=1}^{n} (1 - R_i)
\end{aligned}
$$

The reliability of the system is the complement of F_s and is given by

$$
\begin{aligned}
R_s &= 1 - F_s \\
&= 1 - \prod_{i=1}^{n} (1 - R_i)
\end{aligned}
\tag{11-12}
$$

Use of the Exponential Model   If the time to failure of each component can be modeled by the exponential distribution, each with a constant failure rate λ_i, $i = 1, \ldots, n$, the system reliability, assuming independence of component operation, is given by

$$
\begin{aligned}
R_s &= 1 - \prod_{i=1}^{n} (1 - R_i) \\
&= 1 - \prod_{i=1}^{n} (1 - e^{-\lambda_i t})
\end{aligned}
\tag{11-13}
$$

The time-to-failure distribution of the system is *not* exponentially distributed.
In the special case where all components have the same failure rate λ, the system reliability is given by

$$
R_s = 1 - (1 - e^{-\lambda t})^n
$$

Note that if the system reliability could have been expressed in the form $\exp(-\lambda_s t)$, only in that case could we have concluded that the time to failure of the system is exponentially distributed. A general definition of MTTF is given by

$$\text{MTTF} = \int_0^\infty R_s dt$$

$$= \int_0^\infty [1 - (1 - e^{-\lambda t})^n] dt$$

For this situation, after simplification, the mean time to failure for the system with n identical components in parallel, assuming that each failed component is immediately replaced by an identical component, is given by

$$\text{MTTF} = \frac{1}{\lambda}\left(1 + \frac{1}{2} + \frac{1}{3} + \cdots + \frac{1}{n}\right) \tag{11-14}$$

Example 11-7 Find the reliability of the system shown in Figure 11-5 with three components (A, B, and C) in parallel. The reliabilities of A, B, and C are 0.95, 0.92, and 0.90, respectively.

Solution The system reliability is

$$R_s = 1 - (1 - 0.95)(1 - 0.92)(1 - 0.90)$$
$$= 1 - 0.0004 = 0.9996$$

Note that the system reliability is much higher than that of the individual components. Designers can increase system reliability by placing more components in parallel, but the cost of the additional components necessitates a trade-off between the two objectives.

Example 11-8 For the system shown in Figure 11-5, determine the system reliability for 2000 hours of operation, and find the mean time to failure. Assume that all three components have an identical time-to-failure distribution that is exponential, with a constant failure rate of 0.0005/ hour. What is the mean time to failure of each component? If it is desired for the system to have a mean time to failure of 4000 hours, what should the mean time to failure be for each component?

Solution The failure rate of each component is $\lambda = 0.0005$/hour. For 2000 hours of operation, the system reliability is

$$R_s = 1 - \{1 - \exp[-(0.0005)(2000)]\}^3$$
$$= 1 - (0.63212)^3 = 0.7474$$

The mean time to failure for the system is

$$\text{MTTF} = \frac{1}{0.0005}\left(1 + \frac{1}{2} + \frac{1}{3}\right) = 3666.67 \text{ hours}$$

The mean time to failure for each component is

$$\text{MTTF} = \frac{1}{\lambda} = \frac{1}{0.0005} = 2000 \text{ hours}$$

By placing three identical components in parallel, the system MTTF has been increased by about 83.3%.

For a desired system MTTF of 4000 hours, we now calculate the required MTTF of the individual components. We have

$$4000 = \frac{1}{\lambda}\left(1 + \frac{1}{2} + \frac{1}{3}\right)$$

where λ is the failure rate for each component. Solving for λ, we get

$$\lambda = \frac{1.8333}{4000} = 0.00046/\text{hour}$$

Thus, the MTTF for each component would have to be

$$\text{MTTF} = \frac{1}{\lambda} = \frac{1}{0.00046} = 2173.91 \text{ hours}$$

Systems with Components in Series and in Parallel

Complex systems often consist of components that are both in series and in parallel. Reliability calculations are based on the concepts discussed previously, assuming that the components operate independently.

Example 11-9 Find the reliability of the eight-component system shown in Figure 11-6; some components are in series and some are in parallel. The reliabilities of the components are as follows: $R_{A_1} = 0.92$, $R_{A_2} = 0.90$, $R_{A_3} = 0.88$, $R_{A_4} = 0.96$, $R_{B_1} = 0.95$, $R_{B_2} = 0.90$, $R_{B_3} = 0.92$, and $R_{C_1} = 0.93$.

Solution We first find the reliabilities of each subsystem. For the subsystem with components A_1, A_2, A_3, and A_4, the reliability is

$$\begin{aligned} R_1 &= 1 - (1 - R_{A1}R_{A2})(1 - R_{A3}R_{A4}) \\ &= 1 - [1 - (0.92)(0.90)][1 - (0.88)(0.96)] = 0.9733 \end{aligned}$$

Similarly, the reliability of the subsystem with components B_1, B_2, and B_3 is

$$\begin{aligned} R_2 &= 1 - (1 - R_{B_1})(1 - R_{B_2})(1 - R_{B_3}) \\ &= 1 - (1 - 0.95)(1 - 0.90)(1 - 0.92) = 0.9996 \end{aligned}$$

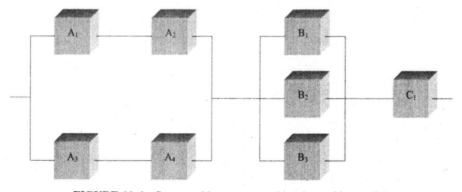

FIGURE 11-6 System with components in series and in parallel.

The system reliability is given by

$$R_s = R_1 \times R_2 \times R_{C_1}$$
$$= (0.9733)(0.9996)(0.93) = 0.9048$$

Use of the Exponential Model   If the time to failure for each component can be assumed to be exponentially distributed, the system reliability and mean time to failure can be calculated under certain conditions using procedures discussed previously.

Example 11-10   Find the system failure rate and the mean time to failure for the eight-component system shown in Figure 11-6. The failure rates (number of units per hour) for the components are as follows: $\lambda_{A_1} = 0.0006$, $\lambda_{A_2} = 0.0045$, $\lambda_{A_3} = 0.0035$, $\lambda_{A_4} = 0.0016$, $\lambda_{B_1} = 0.0060$, $\lambda_{B_2} = 0.0060$, $\lambda_{B_3} = 0.0060$, and $\lambda_{C_1} = 0.0050$.

Solution   First we compute failure rates for the subsystems. We have a failure rate of $(\lambda_{A_1} + \lambda_{A_2}) = 0.0051$ for the A_1/A_2 subsystem; for the A_3/A_4 subsystem, the failure rate is $(\lambda_{A_3} + \lambda_{A_4}) = 0.0051$. The mean time to failure for the $A_1/A_2/A_3/A_4$ subsystem is

$$\text{MTTF}_1 = \frac{1}{0.0051}\left(1 + \frac{1}{2}\right) = 294.118 \text{ hours}$$

The mean time to failure for the subsystem consisting of components B_1, B_2, and B_3 is

$$\text{MTTF}_2 = \frac{1}{0.0060}\left(1 + \frac{1}{2} + \frac{1}{3}\right) = 305.555 \text{ hours}$$

The system failure rate is

$$\lambda_s = \frac{1}{294.118} + \frac{1}{305.555} + 0.005 = 0.01167$$

The mean time to failure for the system is
$$\text{MTTF}_s = 85.69 \text{ hours}$$

Systems with Standby Components

In a standby configuration, one or more parallel components wait to take over operation upon failure of the currently operating component. Here, it is assumed that only one component in the parallel configuration is operating at any given time. Because of this, the system reliability is higher than for comparable systems with components in parallel. In parallel systems discussed previously, all components are assumed to operate simultaneously. Figure 11-7 shows a **standby system** with a basic component and two standby components in parallel. Typically, a failure-sensing mechanism triggers the operation of a standby when the component operating currently fails.

Use of the Exponential Model   If the time to failure of the components is assumed to be exponential with failure rate λ, the number of failures in a certain time t adheres to a Poisson distribution with parameter λt. Using the Poisson distribution introduced in Chapter 4, the probability of x failures in time t is given by

$$P(x) = \frac{e^{-\lambda t}(\lambda t)^x}{x!} \tag{11-15}$$

For a system that has a basic component in parallel with one standby component, the system will be operational at time t long as there is no more than one failure. Therefore, the system

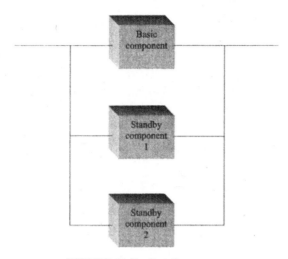

FIGURE 11-7   Standby system.

reliability would be

$$R = e^{-\lambda t} + e^{-\lambda t}(\lambda t)$$

For a standby system with a basic component and two standby components (shown in Figure 11-7), the system will be operational if the number of failures is less than or equal to 2. The system reliability is

$$R_s = e^{-\lambda t} + e^{-\lambda t}(\lambda t) + e^{-\lambda t}\frac{(\lambda t)^2}{2!}$$

In general, if there are n components on standby along with the basic component (for a total of $n+1$ components in the system), the system reliability is given by

$$R_s = e^{-\lambda t}\left[1 + \lambda t + \frac{(\lambda t)^2}{2!} + \frac{(\lambda t)^3}{3!} + \cdots + \frac{(\lambda t)^n}{n!}\right] \tag{11-16}$$

The mean time to failure for such a system is

$$\text{MTTF} = \frac{n+1}{\lambda} \tag{11-17}$$

Example 11-11   Find the reliability of the standby system shown in Figure 11-7, with one basic component and two standby components, each having an exponential time-to-failure distribution. The failure rate for each component is 0.004/hour and the period of operation is 300 hours. What is the mean time to failure?

   If the three components are in parallel (not in a standby mode), what is the reliability of the system? What is the mean time to failure in this situation?

Solution   The failure rate λ for each component is 0.004/hour. For 300 hours of operation, the system reliability is

$$\begin{aligned} R_s &= e^{-\lambda t}\left[1 + \lambda t + \frac{(\lambda t)^2}{2!}\right] \\ &= e^{-0.004(300)}\left\{1 + (0.004)(300) + \frac{[(0.004)(300)]^2}{2}\right\} \\ &= e^{-1.2}(1 + 1.2 + 0.72) = 0.879 \end{aligned}$$

The mean time to failure is

$$\text{MTTF} = \frac{n+1}{\lambda} = \frac{3}{0.004} = 750 \text{ hours}$$

If the system has all three components in parallel, the probability of failure of each component is

$$F_1 = 1 - e^{-\lambda t} = e^{-0.004(300)} = 0.6988$$

The system reliability is found as

$$R_s = 1 - (0.6988)^3 = 0.659$$

The mean time to failure for this parallel system is

$$\text{MTTF} = \frac{1}{\lambda}\left(1 + \frac{1}{2} + \frac{1}{3}\right)$$

$$= \frac{1}{0.004}\left(1 + \frac{1}{2} + \frac{1}{3}\right) = 458.333 \text{ hours}$$

Note that the system reliability and MTTF of the standby and parallel systems differ significantly.

11-5 OPERATING CHARACTERISTIC CURVES

We have discussed OC curves for acceptance sampling plans in previous chapters. In this section we discuss OC curves for life and reliability testing plans. A common life testing plan involves choosing a sample of items from the batch and observing their operation for a certain predetermined time. If the number of failures exceeds a stipulated acceptance number, the lot is rejected; if the number of failures is less than or equal to the acceptance number, the lot is accepted. Two options are possible. In the first option, an item that fails is replaced immediately by an identical item. In the second, failed items are not replaced.

In the calculations for OC curves, we assume that the time to failure of each item is exponentially distributed with a constant failure rate λ. The parameters of a life testing plan are the test time T, the sample size n, and the acceptance number c. The OC curve shows the probability of lot acceptance, P_a, as a function of the lot quality as indicated by the mean life (θ) or the mean time to failure of the item. Under the stated assumptions, the number of failures within a specified period adheres to the Poisson distribution. The Poisson distribution is used to calculate the probability of lot acceptance.

Example 11-12   The parameters of a life testing plan are as follows: time of test $T = 800$ hours, sample size $n = 12$, and acceptance number $c = 2$. Each item has an exponential time-to-failure distribution. When any item fails, it is immediately replaced by a similar item. Construct the OC curve for this plan.

Solution   The expected number of failures in the time period T is $nT\lambda$, where λ denotes the constant failure rate of the item. Denoting the mean life of the item by θ, we have $\lambda = 1/\theta$. From the Poisson distribution tables in Appendix A-2, we find the probability of lot

TABLE 11-1   Calculations for the OC Curve for the Life Testing Plan $n = 12$, $T = 800$, $c = 2$

Mean Life, θ	Failure Rate $\lambda = 1/\theta$	Expected Number of Failures, $nT\lambda$	Probability of Acceptance, P_a
1,000	0.001	9.6	0.0042
2,000	0.0005	4.8	0.1446
3,000	0.00033	3.2	0.3822
4,000	0.00025	2.4	0.5700
5,000	0.0002	1.92	0.6986
6,000	0.00017	1.6	0.7830
7,000	0.00014	1.37	0.8402
8,000	0.000125	1.2	0.8790
9,000	0.00011	1.07	0.9060
10,000	0.0001	0.96	0.9268
12,000	0.000083	0.8	0.9530
14,000	0.000071	0.69	0.9671
16,000	0.0000625	0.6	0.9770
20,000	0.00005	0.48	0.9872
30,000	0.000033	0.32	0.9952

acceptance for different values of the mean life θ. Suppose that the mean life of the item is 8000 hours, which leads to a failure rate λ of 1/8000. The expected number of failures is

$$nT\lambda = (12)(800)(1/8000) = 1.2$$

From Appendix A-2, for a mean number of failures of 1.2, the probability of 2 or fewer failures is, 0.879, which is the probability of lot acceptance. This represents one point on the OC curve. Table 11-1 presents the results of similar computations for values of the mean life θ between 1000 and 30,000 hours.

Figure 11-8 shows the OC curve for the life testing plan $n = 12$, $T = 800$, $c = 2$. Note that the OC curve shown in Figure 11-8 is valid for other life testing plans as long as the total number of item hours tested is 9600 and the acceptance number is 2. Note that the product of n and T represents the total item hours tested because failed items are being replaced. For instance, a life testing plan with parameters $n = 10$, $T = 960$, and $c = 2$ would have the same OC curve as one with parameters $n = 8$, $T = 1200$, and $c = 2$.

The notions of producer's risk α and consumer's risk β are also used in life testing plans. The risk of rejecting a good lot (products with a satisfactory mean life of θ_0) is the **producer's risk**. The risk of accepting a poor lot (products with an unsatisfactory mean life of θ_1) is the **consumer's risk.** These risks are illustrated in Figure 11-8. Suppose that items with a mean life of 20,000 hours are satisfactory ($\theta_0 = 20,000$); the associated producer's risk α is then $(1 - 0.9872) = 0.0128$, which is the probability of rejecting items with this mean life. Alternatively, suppose that items with a mean life θ_1 of 2000 hours are undesirable. The associated consumer's risk β would then be the probability of accepting such lots, which from Figure 11-8 is 0.1446.

An alternative variable for the horizontal axis of the OC curve could be θ/θ_0, the ratio of the actual mean life to the desired mean life θ_0 associated with good batches. For items with mean life θ_0, the probability of lot rejection is the producer's risk α. Thus, all OC curves would pass through the point given by $P_a = 1 - \alpha$ and $\theta/\theta_0 = 1$.

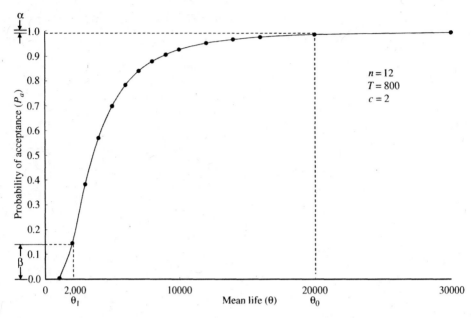

FIGURE 11-8 OC curve for the life testing plan in Example 11-12.

11-6 RELIABILITY AND LIFE TESTING PLANS

Plans for reliability and life testing are usually destructive in nature. They involve observing a sample of items until a certain number of failures occur, observing over a certain period of time to record the number of failures, or a combination of both. Such testing is usually done at the prototype stage, which can be expensive depending on the unit cost of the item. Although a longer accumulated test time is desirable for a precise estimation of product reliability or mean life, the cost associated with the testing plan is an important factor in its choice. Of course, testing is usually conducted under simulated conditions, but it should mimic the actual operating conditions as closely as possible. Several standardized plans have been established by the U.S. Department of Defense, such as **Handbook H-108,** and MIL-STD-690B. Plan H-108 will be introduced in this chapter.

Types of Tests

The three main types of tests are described here.

Failure-Terminated Test In **failure-terminated plans,** the tests are terminated when a preassigned number of failures occurs in the chosen sample. Lot acceptance is based on the accumulated test time of the items when the test is terminated. One acceptance criterion involves whether the estimated average life of the item exceeds a stipulated value. Let the sample size be denoted by n, the predetermined number of failures by r, and the stipulated mean life by C. From the test results, let's suppose that the accumulated test time of the items is $\hat{T}$, from which an estimate of the average life is

$$\hat{\theta} = \frac{\hat{T}}{r}$$

The lot is accepted if $\hat{\theta} \geq C$.

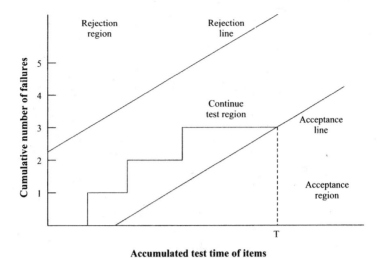

FIGURE 11-9   Sequential reliability test plan.

Time-Terminated Test   **A time-terminated test** is terminated when a preassigned time T is reached. Acceptance of the lot is based on the observed number of failures $\hat{r}$ during the test time. If the observed number of failures $\hat{r}$ exceeds a preassigned value r, the lot is rejected; otherwise, the lot is accepted.

Sequential Reliability Test   In sequential reliability testing, no prior decision is made as to the number of failures or the time to conduct the test. Instead, the accumulated results of the test are used to decide whether to accept the lot, reject the lot, or continue testing. Figure 11-9 shows a **sequential life testing plan.** The cumulative number of failures based on a chosen sample is plotted versus the accumulated test time of the items. Based on an acceptable mean life θ_0, an associated producer's risk α, a minimum mean life θ_1, and an associated consumer's risk β, equations for the acceptance line and the rejection lines are found. If the plot stays within the two lines, testing continues; if the plot falls in the acceptance region, the test is terminated and the lot is accepted. If the plot falls in the rejection region, the test is terminated and the lot is rejected.

In Figure 11-9, note that after three failures, when the accumulated test time reaches T, the test is terminated with an acceptance of the lot. The equations of the acceptance and rejection lines are similar to the sequential tests discussed for attribute sampling plans in Chapter 10. The main advantage of sequential test plans is that, for a similar degree of protection, the expected test time or the expected number of failures required to reach a decision is less than that for a time- or a failure-terminated plan.

For each of these tests, testing can take place **with** or **without replacement** of the failed items. In the situation for which the item is replaced, we assume that the replaced item has the same failure rate as the one it replaces. This assumption holds when the failure rate is constant, as in the chance-failure phase of the life of the product. Accordingly, the exponential distribution for the time to failure would be appropriate in this case.

Life Testing Plans Using the Exponential Distribution

The time to failure is assumed to have an exponential distribution with a constant failure rate. The mean life of the product is estimated when failure- or time-terminated tests are used. Both point estimates and confidence intervals for the mean life are obtained.

Failure-Terminated Test   Let the preassigned number of failures be denoted by r. Suppose that the failures occur at the following times, in rank order: $t_1 \leq t_2 \leq \ldots \leq t_r$. If the sample size is n, the accumulated life for the test items until the rth failure (T_r), assuming failed items are not replaced, is given by

$$T_r = \sum_{i=1}^{r} t_i + (n - r)t_r \tag{11-18}$$

If failed items are replaced with items having the same failure rate, we have

$$T_r = nt_r \tag{11-19}$$

An estimate of the mean life, in either case, is obtained from

$$\hat{\theta} = \frac{T_r}{r} \tag{11-20}$$

where the value of T_r is found using eq. (11-19) or (11-18), depending on whether the items are replaced or not, respectively.

Next, a confidence interval for the mean life θ is found. It is known that the statistic $2T_r / \theta$ has a chi-squared distribution with $2r$ degrees of freedom. Thus, a two-sided $100(1 - \alpha)\%$ confidence interval for the mean life is given by

$$\frac{2T_r}{\chi^2_{\alpha/2,2r}} < \theta < \frac{2T_r}{\chi^2_{1-\alpha/2,2r}} \tag{11-21}$$

where the chi-squared values with $2r$ degrees of freedom are found from Appendix A-5. This may be used for cases with and without replacement.

Example 11-13   Life testing is conducted for a sample of 15 transistors. The time to failure for each is exponentially distributed. The test is terminated after four failures, with no replacement of the failed items. The failure times (in hours) of the four transistors are 400, 480, 610, and 660. Estimate the mean life of the transistors and the failure rate. Find a 95% confidence interval for the mean life.

Solution   The accumulated life for the test items is

$$T_4 = (400 + 480 + 610 + 660) + (15 - 4)(660) = 9410 \text{ hours}$$

The estimated mean life (or mean time to failure) is $\hat{\theta} = 9410/4 = 2352.5$ hours. The estimated failure rate is

$$\lambda = \frac{1}{\hat{\theta}} = \frac{1}{2352.5} = 0.000425/\text{hour}$$

A 95% confidence interval for the mean life is

$$\frac{2T_r}{\chi^2_{0.025,8}} < \theta < \frac{2T_r}{\chi^2_{0.975,8}}, \quad \frac{(2)(9410)}{17.53} < \theta < \frac{(2)(9410)}{2.18}$$

$$1073.588 < \theta < 8633.027$$

Example 11-14   Refer to Example 11-13 concerning the life testing of a sample of 15 transistors. The test is terminated after four failures. Assume that each item that fails is replaced by an identical item. The failure times of the four transistors are 420, 490, 550, and 580 hours. Estimate the mean life of the transistors and the failure rate. Find a 95% confidence interval for the mean life.

Solution   The accumulated life on the test items is

$$T_4 = (15)(580) = 8700 \text{ hours}$$

The estimated mean life is found to be $\hat{\theta} = 8700/4 = 2175$ hours. The estimated failure rate is

$$\hat{\lambda} = \frac{1}{\hat{\theta}} = \frac{1}{2175} = 0.00046/\text{hour}$$

A 95% confidence interval for the mean life is

$$\frac{(2)(8700)}{17.53} < \theta < \frac{(2)(8700)}{2.18}, \quad \text{or} \quad 992.584 < \theta < 7981.651$$

Time-Terminated Test   Let the preassigned time to terminate the test be denoted by T for a sample of size n. Let t_i denote the time of failure of the ith item. In this case, the observed number of failures is a random variable. If failed items are not replaced, the accumulated life for the test items is given by

$$T_x = \sum_{i=1}^{x} t_i + (n - x)T \tag{11-22}$$

where x represents the observed number of failures.

When failed items are replaced with similar items, the accumulated life of the test items is

$$T_x = nT \tag{11-23}$$

An estimate of the mean life may be obtained from

$$\hat{\theta} = \frac{T_x}{x}$$

For both situations, an approximate $100(1 - \alpha)$% confidence interval for the mean life is given by

$$\frac{2T_x}{\chi^2_{\alpha/2, 2(x+1)}} < \theta < \frac{2T_x}{\chi^2_{1-\alpha/2, 2(x+1)}} \tag{11-24}$$

where the chi-squared values with $2(x + 1)$ degrees of freedom are found in Appendix A-5.

Example 11-15   A sample of 12 electronic components is tested for 1000 hours with no replacement of failed components. The time to failure is exponentially distributed. Three components failed within the prescribed test time, the failure times being 650, 680, and 720 hours. Estimate the mean time to failure and the failure rate. Find a 90% confidence interval for the mean time to failure.

Solution   The accumulated life for the test items is

$$T_3 = (650 + 680 + 720) + (12 - 3)(1000) = 11{,}050 \text{ hours}$$

An estimate of the mean time to failure is $\hat{\theta} = 11{,}050/3 = 3683.333$ hours.

An estimate of the failure rate is

$$\lambda = \frac{1}{\hat{\theta}} = \frac{1}{3683.333} = 0.00027/\text{hour}$$

A 90% confidence interval for the mean time to failure is given by

$$\frac{(2)(11,050)}{\chi^2_{0.05,8}} < \theta < \frac{(2)(11,050)}{\chi^2_{0.95,8}}, \quad \frac{2(11,050)}{15.51} < \theta < \frac{2(11,050)}{2.73}$$

$$1424.887 < \theta < 8095.238$$

Standard Life Testing Plans Using Handbook H-108

The Quality Control and Reliability Handbook H-108 (U.S. Department of Defense, 1960) was developed by the Bureau of Naval Weapons, U.S. Department of the Navy. Life testing plans in the handbook are based on a time-to-failure distribution that is exponential, and all three types of plans (failure-terminated, time-terminated, and sequential life testing) are included. For each plan, provision is made for situations with and without replacement of failed units. We demonstrate some selected applications of failure- and time-terminated tests here. Tables from the handbook used to determine the life testing plans are reproduced here as necessary.

Failure-Terminated Plans   A sample of n items is selected from the lot and tested until the occurrence of the rth failure. If the estimated mean life $\hat{\theta}$ is greater than or equal to a criterion value C given by the plan, the lot is accepted. The producer's risk α is the probability of rejecting a lot with a satisfactory mean life θ_0.

Example 11-16   A life testing plan is to be terminated upon occurrence of the sixth failure. The plan should accept a lot having an acceptable mean life of 1200 hours with a probability 0.95. Twenty items are placed on test. The six failures occur at the following times (in hours): 480, 530, 560, 600, 640, and 670. Failed items are not replaced. Using *Handbook H-108*, determine whether the lot should be accepted.

Solution   The following parameters are given: $r = 6$, $n = 20$, $\theta_0 = 1200$, $\alpha = 0.05$. An estimate of the mean life is

$$\hat{\theta} = \frac{480 + 530 + 560 + 600 + 640 + 670 + (20 - 6)(670)}{6}$$

$$= 2143.333 \text{ hours}$$

Table 11-2 shows the values of C/θ_0 for a given producer's risk α and the predetermined number of failures r given in *Handbook H-108*. For $r = 6$ and $\alpha = 0.05$, the code is B–6, and C/θ_0 is 0.436. The acceptability criterion C is

$$C = \theta_0 \ (C/\theta_0) = (1200)(0.436) = 523.2$$

The estimated mean life $\hat{\theta}$ of 2143.33 hours exceeds the criterion value C. Therefore, the lot should be accepted.

Time-Terminated Plans   For time-terminated plans using *Handbook H-108*, the preassigned test time is denoted by T. An acceptable mean lot life is denoted by θ_0,

TABLE 11-2 Master Table for Life Tests Terminated upon Occurrence of Preassigned Number of Failures

Rejection Number, r	Producer's Risk (α)					
	0.01		0.05		0.10	
	Code	C/θ_0	Code	C/θ_0	Code	C/θ_0
1	A–1	0.010	B–1	0.052	C–1	0.106
2	A–2	0.074	B–2	0.178	C–2	0.266
3	A–3	0.145	B–3	0.272	C–3	0.367
4	A–4	0.206	B–4	0.342	C–4	0.436
5	A–5	0.256	B–5	0.394	C–5	0.487
6	A–6	0.298	B–6	0.436	C–6	0.525
7	A–7	0.333	B–7	0.469	C–7	0.556
8	A–8	0.363	B–8	0.498	C–8	0.582
9	A–9	0.390	B–9	0.522	C–9	0.604
10	A–10	0.413	B–10	0.543	C–10	0.622
15	A–11	0.498	B–11	0.616	C–11	0.687
20	A–12	0.554	B–12	0.663	C–12	0.726
25	A–13	0.594	B–13	0.695	C–13	0.754
30	A–14	0.625	B–14	0.720	C–14	0.774
40	A–15	0.669	B–15	0.755	C–15	0.803
50	A–16	0.701	B–16	0.779	C–16	0.824
75	A–17	0.751	B–17	0.818	C–17	0.855
100	A–18	0.782	B–18	0.841	C–18	0.874

where the probability of rejection is α, the producer's risk, and a minimum mean life is denoted by θ_1, where the probability of acceptance is β, the consumer's risk. The rejection criterion number r is obtained from the tables. If the observed number of failures within the preassigned test time is greater than or equal to r, the lot is rejected; otherwise, the lot is accepted. The following examples demonstrate the use of tables from *Handbook H-108*. The first example shows the determination of a plan given the producer's and consumer's risks, their associated mean lives, and the sample size.

Example 11-17   Find a time-terminated life testing plan (using *Handbook H-108*) that rejects lots with a mean life of 1250 hours with a probability of 0.05 and accepts lots with a mean life of 400 hours with a probability of 0.10. Items that fail during the test are not replaced.

Solution   We have $\theta_0 = 1250$, $\alpha = 0.05$, $\theta_1 = 400$, $\beta = 0.10$. Hence,

$$\frac{\theta_1}{\theta_0} = \frac{400}{1250} = 0.32$$

Table 11-3 shows the life test sampling plan code designation for various values of α, β, and θ_1/θ_0 given in *Handbook H-108*. Using Table 11-3, we select the table value of θ_1/θ_0 that matches the calculated value. If an exact match is not found, the next larger value of θ_1/θ_0 in the table is used. For this example, the code is B–8.

For each code letter, the *Handbook H-108* lists a table for determining the rejection number r and the value of T/θ_0. Table 11-4 shows the value of T/θ_0 for code letter B when

TABLE 11-3 Life Testing Plan Code Designation

$\alpha = 0.01$ $\beta = 0.01$		$\alpha = 0.05$ $\beta = 0.10$		$\alpha = 0.10$ $\beta = 0.10$		$\alpha = 0.25$ $\beta = 0.10$		$\alpha = 0.50$ $\beta = 0.10$	
Code	θ_1/θ_0	Code	θ_1/θ_0	Code	θ_1/θ_0	Code	θ_1/θ_0	Code	θ_1/θ_0
A–1	0.004	B–1	0.022	C–1	0.046	D–1	0.125	E–1	0.301
A–2	0.038	B–2	0.091	C–2	0.137	D–2	0.247	E–2	0.432
A–3	0.082	B–3	0.154	C–3	0.207	D–3	0.325	E–3	0.502
A–4	0.123	B–4	0.205	C–4	0.261	D–4	0.379	E–4	0.550
A–5	0.160	B–5	0.246	C–5	0.304	D–5	0.421	E–5	0.584
A–6	0.193	B–6	0.282	C–6	0.340	D–6	0.455	E–6	0.611
A–7	0.221	B–7	0.312	C–7	0.370	D–7	0.483	E–7	0.633
A–8	0.247	B–8	0.338	C–8	0.396	D–8	0.506	E–8	0.652
A–9	0.270	B–9	0.361	C–9	0.418	D–9	0.526	E–9	0.667
A–10	0.291	B–10	0.382	C–10	0.438	D–10	0.544	E–10	0.681
A–11	0.371	B–11	0.459	C–11	0.512	D–11	0.608	E–11	0.729
A–12	0.428	B–12	0.512	C–12	0.561	D–12	0.650	E–12	0.759
A–13	0.470	B–13	0.550	C–13	0.597	D–13	0.680	E–13	0.781
A–14	0.504	B–14	0.581	C–14	0.624	D–14	0.703	E–14	0.798
A–15	0.554	B–15	0.625	C–15	0.666	D–15	0.737	E–15	0.821
A–16	0.591	B–16	0.658	C–16	0.695	D–16	0.761	E–16	0.838
A–17	0.653	B–17	0.711	C–17	0.743	D–17	0.800	E–17	0.865
A–18	0.692	B–18	0.745	C–18	0.774	D–18	0.824	E–18	0.882

TABLE 11-4 Values of T/θ_0 for $\alpha = 0.05$: Time Terminated, Testing Without Replacement, Code Letter B

Code	r	Sample Size									
		$2r$	$3r$	$4r$	$5r$	$6r$	$7r$	$8r$	$9r$	$10r$	$20r$
B–1	1	0.026	0.017	0.013	0.010	0.009	0.007	0.006	0.006	0.005	0.003
B–2	2	0.104	0.065	0.048	0.038	0.031	0.026	0.023	0.020	0.018	0.009
B–3	3	0.168	0.103	0.075	0.058	0.048	0.041	0.036	0.031	0.028	0.014
B–4	4	0.217	0.132	0.095	0.074	0.061	0.052	0.045	0.040	0.036	0.017
B–5	5	0.254	0.153	0.110	0.086	0.071	0.060	0.052	0.046	0.041	0.020
B–6	6	0.284	0.170	0.122	0.095	0.078	0.066	0.057	0.051	0.045	0.022
B–7	7	0.309	0.185	0.132	0.103	0.084	0.072	0.062	0.055	0.049	0.024
B–8	8	0.330	0.197	0.141	0.110	0.090	0.076	0.066	0.058	0.052	0.025
B–9	9	0.348	0.207	0.148	0.115	0.094	0.080	0.069	0.061	0.055	0.027
B–10	10	0.363	0.216	0.154	0.120	0.098	0.083	0.072	0.064	0.057	0.028
B–11	15	0.417	0.246	0.175	0.136	0.112	0.094	0.082	0.072	0.065	0.032
B–12	20	0.451	0.266	0.189	0.147	0.120	0.102	0.088	0.078	0.070	0.034
B–13	25	0.475	0.280	0.199	0.154	0.126	0.107	0.093	0.082	0.073	0.036
B–14	30	0.493	0.290	0.206	0.160	0.131	0.111	0.096	0.085	0.076	0.037
B–15	40	0.519	0.305	0.216	0.168	0.137	0.116	0.101	0.089	0.079	0.039
B–16	50	0.536	0.315	0.223	0.173	0.142	0.120	0.104	0.092	0.082	0.040
B–17	75	0.564	0.331	0.235	0.182	0.149	0.126	0.109	0.096	0.086	0.042
B–18	100	0.581	0.340	0.242	0.187	0.153	0.130	0.112	0.099	0.089	0.043

TABLE 11-5   Values of T/θ_0 for $\alpha = 0.05$: Time Terminated, Testing with Replacement, Code Letter B

Code	r	Sample Size									
		$2r$	$3r$	$4r$	$5r$	$6r$	$7r$	$8r$	$9r$	$10r$	$20r$
B–1	1	0.026	0.017	0.013	0.010	0.009	0.007	0.006	0.006	0.005	0.003
B–2	2	0.089	0.059	0.044	0.036	0.030	0.025	0.022	0.020	0.018	0.009
B–3	3	0.136	0.091	0.068	0.055	0.045	0.039	0.034	0.030	0.027	0.014
B–4	4	0.171	0.114	0.085	0.068	0.057	0.049	0.043	0.038	0.034	0.017
B–5	5	0.197	0.131	0.099	0.079	0.066	0.056	0.049	0.044	0.039	0.020
B–6	6	0.218	0.145	0.109	0.087	0.073	0.062	0.054	0.048	0.044	0.022
B–7	7	0.235	0.156	0.117	0.094	0.078	0.067	0.059	0.052	0.047	0.023
B–8	8	0.249	0.166	0.124	0.100	0.083	0.071	0.062	0.055	0.050	0.025
B–9	9	0.261	0.174	0.130	0.104	0.087	0.075	0.065	0.058	0.052	0.026
B–10	10	0.271	0.181	0.136	0.109	0.090	0.078	0.068	0.060	0.054	0.027
B–11	15	0.308	0.205	0.154	0.123	0.103	0.088	0.077	0.068	0.062	0.031
B–12	20	0.331	0.221	0.166	0.133	0.110	0.095	0.083	0.074	0.066	0.033
B–13	25	0.348	0.232	0.174	0.139	0.116	0.099	0.087	0.077	0.070	0.035
B–14	30	0.360	0.240	0.180	0.144	0.120	0.103	0.090	0.080	0.072	0.036
B–15	40	0.377	0.252	0.189	0.151	0.126	0.108	0.094	0.084	0.075	0.038
B–16	50	0.390	0.260	0.195	0.156	0.130	0.111	0.097	0.087	0.078	0.039
B–17	75	0.409	0.273	0.204	0.164	0.136	0.117	0.102	0.091	0.082	0.041
B–18	100	0.421	0.280	0.210	0.168	0.140	0.120	0.105	0.093	0.084	0.042

testing is conducted without replacement of the failed items; a similar table for testing with replacement of failed items is shown in Table 11-5.

Using Table 11-4 and a code of B–8, the rejection number is 8. Note that the sample size is indicated as multiples of the rejection number. Increasing the sample size reduces the average time needed to reach a decision at the expense of added costs because more items must be tested.

Suppose we use a sample size that is a multiple, $4r$, of the rejection number. From Table 11-4, the value of T/θ_0 is 0.141. Thus,

$$T = (0.141)(1250) = 176.25 \text{ hours}$$

The life testing plan can now be specified. A random sample of 32 items is selected from the lot and tested simultaneously. If the eighth failure occurs before the test termination time of 176.25 hours, the lot is rejected; otherwise, the lot is accepted.

The next example illustrates the determination of a plan given the producer's risk, the associated mean life, the rejection number, and the sample size.

Example 11-18   Find a time-terminated life testing plan that rejects lots with a mean life of 1800 hours with a probability of 0.05. The rejection number is 5, and the sample size is 30. Items that fail during the test are replaced. Determine the plan using the *Handbook H-108*.

Solution   We are given the following parameters: $\theta_0 = 1800$, $\alpha = 0.05$, $r = 5$, and $n = 30$. For $\alpha = 0.05$ and life testing plans with replacement, we use Table 11-5 and a rejection number r of 5 and a sample size of $6r$ to get $T/\theta_0 = 0.066$. The test termination time is

$$T = (0.066)(1800) = 118.8 \text{ hours}$$

TABLE 11-6 Life Tests for Predetermined Time Based on α, β, θ_1/θ_0, and T/θ_0 (Without Replacement)

θ_1/θ_0	$\alpha=0.01$ r	T/θ_0 $\frac{1/3}{n}$	$\frac{1/5}{n}$	$\frac{1/10}{n}$	$\frac{1/20}{n}$	$\alpha=0.05$ r	T/θ_0 $\frac{1/3}{n}$	$\frac{1/5}{n}$	$\frac{1/10}{n}$	$\frac{1/20}{n}$	$\alpha=0.10$ r	T/θ_0 $\frac{1/3}{n}$	$\frac{1/5}{n}$	$\frac{1/10}{n}$	$\frac{1/20}{n}$
			$\beta=0.01$					$\beta=0.01$					$\beta=0.01$		
2/3	136	403	622	1179	2275	95	289	447	843	1639	77	238	369	699	1358
1/2	46	119	182	340	657	33	90	138	258	499	26	73	112	210	407
1/3	19	41	61	113	216	13	30	45	83	160	11	27	40	75	145
1/5	9	15	22	39	74	7	13	20	36	69	5	10	14	26	51
1/10	5	6	9	15	28	4	6	9	15	29	3	5	7	12	23
			$\beta=0.05$					$\beta=0.05$					$\beta=0.05$		
2/3	101	291	448	842	1632	67	198	305	575	1116	52	156	242	456	886
1/2	35	87	132	245	472	23	59	90	168	326	18	48	73	137	265
1/3	15	30	45	82	157	10	21	32	59	113	8	18	27	50	97
1/5	8	13	18	33	62	5	8	12	22	41	4	7	10	19	36
1/10	4	4	6	10	18	3	4	5	9	17	2	2	3	6	11
			$\beta=0.10$					$\beta=0.10$					$\beta=0.10$		
2/3	83	234	359	675	1307	55	159	245	462	895	41	121	186	351	681
1/2	30	72	109	202	390	19	47	72	134	258	15	39	59	110	213
1/3	13	25	37	67	128	8	16	24	43	83	6	12	18	34	66
1/5	7	11	15	26	50	4	6	9	15	29	3	5	7	12	23
1/10	4	4	6	10	18	3	4	5	9	17	2	2	3	6	11

TABLE 11-7 Life Tests for Predetermined Time Based on α, β, θ_1/θ_0, and T/θ_0 (with Replacement)

θ_1/θ_0	$\alpha=0.01$ r	T/θ_0 1/3 n	T/θ_0 1/5 n	T/θ_0 1/10 n	T/θ_0 1/20 n	$\alpha=0.05$ r	T/θ_0 1/3 n	T/θ_0 1/5 n	T/θ_0 1/10 n	T/θ_0 1/20 n	$\alpha=0.10$ r	T/θ_0 1/3 n	T/θ_0 1/5 n	T/θ_0 1/10 n	T/θ_0 1/20 n
		$\alpha=0.01$		$\beta=0.01$			$\alpha=0.05$		$\beta=0.01$			$\alpha=0.10$		$\beta=0.01$	
2/3	136	331	551	1103	2207	95	238	397	795	1591	77	197	329	659	1319
1/2	46	95	158	317	634	33	72	120	241	483	26	59	98	197	394
1/3	19	31	31	103	206	13	23	38	76	153	11	21	35	70	140
1/5	9	10	10	35	70	7	9	16	32	65	5	7	12	24	48
1/10	5	4	6	12	25	4	4	6	13	27	3	3	5	11	22
		$\alpha=0.01$		$\beta=0.05$			$\alpha=0.05$		$\beta=0.05$			$\alpha=0.10$		$\beta=0.05$	
2/3	101	237	395	790	1581	67	162	270	541	1082	52	128	214	429	859
1/2	35	68	113	227	454	23	47	78	157	314	18	38	64	128	256
1/3	15	22	37	74	149	10	16	27	54	108	8	13	23	46	93
1/5	8	8	14	29	58	5	6	10	19	39	4	5	8	17	34
1/10	4	3	4	8	16	3	3	4	8	16	2	2	3	5	10
		$\alpha=0.01$		$\beta=0.10$			$\alpha=0.05$		$\beta=0.10$			$\alpha=0.10$		$\beta=0.10$	
2/3	83	189	316	632	1265	55	130	216	433	876	41	99	165	330	660
1/2	30	56	93	187	364	19	37	62	124	248	15	30	51	102	205
1/3	13	18	30	60	121	8	11	19	39	79	6	9	15	31	63
1/5	7	7	11	23	46	4	4	7	13	27	3	4	6	11	22
1/10	2	4	4	8	16	3	3	4	8	16	2	2	2	5	10

Source: Handbook H-108, pp. 2.53–2.54.

The plan is as follows. A random sample of 30 items is selected from the lot and testing is conducted for 118.8 hours. If the fifth failure occurs before the termination time, the lot is rejected; if the fifth failure has not occurred within 118.8 hours, the lot is accepted. Items that fail are replaced.

In the following example we determine a time-terminated plan given the producer's and consumer's risks, their associated mean lives, and the test time.

Example 11-19   Find a time-terminated life testing plan that accepts a lot with a mean life of 5000 hours with a probability of 0.90 but rejects a lot with a mean life of 1000 hours with a probability of 0.95. The test should be terminated by 500 hours. Items that fail are not replaced. Determine the plan assuming the time to failure is exponentially distributed.

Solution   We are given the following parameters: $\theta_0 = 5000$, $\alpha = 0.01$, $\theta_1 = 1000$, $\beta = 0.05$, $T = 500$. The following ratios are calculated:

$$\frac{\theta_1}{\theta_0} = \frac{1000}{5000} = \frac{1}{5}; \quad \frac{T}{\theta_0} = \frac{500}{5000} = \frac{1}{10}$$

Table 11-6 lists the values of the rejection number r and the sample size n, knowing α, β, θ_1/θ_0, and T/θ_0 for time-terminated plans, without replacement of failed units. Table 11-7 lists values of the rejection number r and the sample size n, given α, β, θ_1/θ_0, and T/θ_0, for time-terminated plans with replacement of failed items.

We use Table 11-6. For $\alpha = 0.10$, $\beta = 0.05$, $\theta_1/\theta_0 = 1/5$, and $T/\theta_0 = 1/10$, we have $r = 4$ and $n = 19$. Therefore, the life testing plan is to select a sample of 19 items from the lot. If the fourth failure occurs before the termination time of 500 hours, the lot is rejected; otherwise, the lot is accepted.

SUMMARY

This chapter introduced the concept of reliability and methods for its computation. Reliability implies the successful operation of a product over a certain period of time under stipulated environmental conditions. Reliability is needed in the performance phase of quality, but plans for its achievement are actually addressed in the design phase. Procedures for improving the reliability of a system by means of adding components in parallel (redundant systems) or through standby systems have been discussed. The concept of availability is also presented. The life-cycle curve of a product was studied to identify the period of useful life of the product. Probability distributions that model different phases of the life-cycle curve have been explored. Some tests for reliability and life testing have been covered in this chapter. The exponential distribution for time to failure is used as a basis for these tests. Failure-terminated, time-terminated, and sequential tests for life testing have been introduced. Standardized tests developed by the government have been treated in some detail.

KEY TERMS

accumulated life

availability

bathtub curve

chance-failure phase

chi-squared random variable

consumer's risk

debugging phase

failure-rate curve

failure-rate function	operating characteristic curve
constant failure rate	probability distributions
infant-mortality phase	exponential distribution
life-cycle curve	Weibull distribution
life testing plans	Poisson process
failure-terminated	producer's risk
Handbook H-108	reliability
sequential	system reliability
time-terminated test	systems
with replacement	complex
without replacement	components in parallel
mean time between failures	components in series
mean time to failure	standby
mean time to repair	wearout phase

EXERCISES

Discussion Questions

11-1 Define reliability. Explain its role in quality control and improvement.

11-2 Describe the life cycle of a product. What probability distributions would you use to model each phase?

11-3 Explain procedures that might improve the reliability of a system. How would you increase the availability of a system? Distinguish between a system with components in parallel and another with standby components.

11-4 Distinguish between failure-, time-terminated, and sequential tests for reliability and life testing.

Problems

11-5 A transistor has an exponential time-to-failure distribution with a constant failure rate of 0.00006/hour. Find the reliability of the transistor after 4000 hours of operation. What is the mean time to failure? If the repair rate is 0.004/hour, find the availability.

11-6 An electronic component in a video recorder has an exponential time-to-failure distribution. What is the minimum mean time to failure of the component if it is to have a probability of 0.92 of successful operation after 6000 hours of operation?

11-7 An optical sensor has a Weibull time-to-failure distribution with a scale parameter of 300 hours and a shape parameter of 0.5. What is the reliability of the sensor after 500 hours of operation? Find the mean time to failure.

11-8 A remote control unit has 40 components in series. The reliability of each component is 0.9994. What is reliability of the remote control unit? If a redesign has 25 components in series, what is the reliability of the unit?

11-9 Refer to Exercise 11-8 concerning the redesigned remote control unit with 25 components in series. If it is desired that the remote unit has a reliability of 0.996 for 3000 hours of operation, what should the failure rate be for each component? What should the mean time to failure be for each component? Assume that the time to failure for each component is exponentially distributed.

11-10   Four components A, B, C, and D are placed in parallel to make a subassembly in a circuit board. The reliabilities of A, B, C, and D are 0.93, 0.88, 0.95, and 0.92, respectively. Find the reliability of the subassembly.

11-11   Refer to Exercise 11-10. Each component has a time to failure that is exponentially distributed, with a mean time to failure of 3000 hours. Find the reliability of the subassembly for 2500 hours of operation. What is the mean time to failure of the subassembly? If it is desired that the subassembly has a mean time to failure of 6600 hours, what would have to be the mean time to failure of the components?

11-12   Consider the seven-component system shown in Figure 11-10. The reliabilities of the components are as follows: $R_A = 0.96$, $R_B = 0.92$, $R_C = 0.94$, $R_D = 0.89$, $R_E = 0.95$, $R_F = 0.88$, $R_G = 0.90$. Find the reliability of the system. If you had a choice of improving system reliability by modifying any two components, how would you proceed?

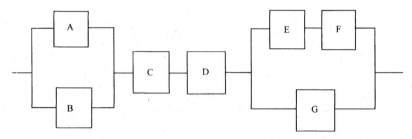

FIGURE 11-10   System with seven components.

11-13   Consider the seven-component system shown in Figure 11-10. Assume that the time to failure for each component has an exponential distribution. The failure rates are as follows: $\lambda_A = 0.0005$/hour, $\lambda_B = 0.0005$/hour, $\lambda_C = 0.0003$/h, $\lambda_D = 0.0008$/hour, $\lambda_E = 0.0004$/hour, $\lambda_F = 0.006$/hour, and $\lambda_G = 0.0064$/hour. Find the reliability of the system after 1000 hours. What is the mean time to failure of the system?

11-14   A standby system has a basic unit with four standby components. The time to failure of each component has an exponential distribution with a failure rate of 0.008/hour. For a 400 hour operation period, find the reliability of the standby system. What is its mean time to failure? Suppose that all five components are operating simultaneously in parallel. What would the system reliability be in that situation? What would the mean time to failure be?

11-15   Refer to Exercise 11-13 and the system shown in Figure 11-10. Suppose that component B is a standby component. Find the reliability of the system after 1000 hours. What is the mean time to failure?

11-16   Construct the OC curve for the life testing plan $n = 6$, $T = 900$ hours, $c = 3$. For a producer's risk of 0.05, what is the associated quality of batches as indicated by their mean life? For a consumer's risk of 0.10, what is the associated quality level of batches as indicated by their mean life?

11-17   A sample of 20 diodes is chosen for life testing. The time to failure of the diodes is exponentially distributed. The test is terminated after six failures, with no replacement of the failed items. The failure times (in hours), of the six diodes are 530, 590,

670, 700, 720, and 780. Estimate the mean time to failure of the diodes as well as the failure rate. Find a 95% confidence interval for the mean life.

11-18   Refer to Exercise 11-17. Assume that each failed item is replaced with an identical unit. Estimate the mean time to failure and the failure rate. Find a 90% confidence interval for the mean time to failure.

11-19   A sample of 25 relays is chosen for life testing. The time to failure of a relay is exponentially distributed. The test is terminated after 800 hours, with 5 failures being observed at times 610, 630, 680, 700, and 720 hours. Failed items are not replaced. Estimate the mean life and the failure rate of the relays. Find a 95% confidence interval for the mean life.

11-20   A life testing plan is to be terminated after the eighth failure. It should reject a lot that has an acceptable mean life of 900 hours with a probability of 0.10. Items that fail during the test are not replaced. A sample of 15 items is placed on test with the eight failures occurring at the following times (in hours): 400, 430, 435, 460, 460, 490, 520, 530. Based on *Handbook H-108*, should the lot be accepted?

11-21   Refer to Exercise 11-20. Assume that failed items are immediately replaced during the test. Using *Handbook H-108*, what is your recommendation on the lot?

11-22   A life testing plan is to be terminated after the third failure. It should accept a lot that has an acceptable mean life of 600 hours with a probability of 0.99. Failed items are not replaced during the test. A sample of eight items is chosen, and three failures are observed at the following times (in hours): 200, 240, 250. Using *Handbook H-108*, should the lot be accepted?

11-23   A time-terminated life testing plan is to be found that will reject lots with a mean life of 1500 hours with a probability 0.05 and accept lots with a mean life of 600 hours with a probability of 0.10. Items that fail during the test are not replaced. Determine the plan, using *Handbook H-108*, if the sample size is to be five times the rejection number.

11-24   Refer to Exercise 11-23. Suppose that items that fail during the test are immediately replaced with similar items. Determine the life testing plan using *Handbook H-108.*

11-25   A time-terminated life testing plan is to be found that will reject lots that have a mean life of 1400 hours with a probability of 0.05. The rejection number is 7, with a sample size of 35. Determine the plan using *Handbook H-108*, if the time to failure is exponentially distributed. Items that fail during the test are not replaced.

11-26   Refer to Exercise 11-25. If failed items are replaced with similar items, determine a life testing plan using *Handbook H-108*.

11-27   A time-terminated life testing plan is desired that will accept lots with a mean life of 6000 hours with a probability of 0.99, and will accept lots with a mean life of 2000 hours with a probability of 0.10. The test should be terminated by 1200 hours. Items that fail will be replaced immediately. Determine the plan using *Handbook H-108,* assuming that the time to failure is exponential.

11-28   Refer to Exercise 11-27. Find the appropriate plan using *Handbook H-108*, if failed items are not replaced during the test.

11-29   In a time-terminated life testing plan, it is desired to reject lots with a mean life of 1500 hours with a probability of 0.95 and also to reject lots with a mean life of 7500 hours with a probability of 0.10. The test is to be terminated by 2500 hours. Items that fail during the test will be replaced. Assuming the time-to-failure distribution to be exponential, determine the plan using *Handbook H-108*.

11-30   Refer to Exercise 11-29. If failed items are not replaced during the test, determine the plan using *Handbook H-108*.

REFERENCES

Besterfield D. H. (2003). *Quality Control*, 7th ed. Upper Saddle River, NJ: Prentice Hall.

Blischke, W. R., and D. N. P. Murthy (2000). *Reliability Modeling, Prediction and Optimization*, New York: Wiley.

Henley, E. J., and H. Kumamoto (1991). *Probabilistic Risk Assessment: Reliability Engineering, Design, and Analysis*, reprint ed., Piscataway, NJ: Institute of Electrical and Electronic Engineers.

U.S. Department of Defense, Office of the Assistant Secretary of Defense (1960). *Sampling Procedures and Tables for Life and Reliability Testing Based on Exponential Distribution*, Quality Control and Reliability Handbook H108, Washington, DC: Government Printing Office.

U.S. Department of Defense (1968). *Failure Rate Sampling Plans and Procedures*, MIL-STD-690B, Washington, DC: Government Printing Office.

12

EXPERIMENTAL DESIGN AND THE TAGUCHI METHOD

12-1 INTRODUCTION AND CHAPTER OBJECTIVES

Quality should be designed into a product. Principles of experimental design, which aid in selecting optimal product and process parameters, facilitate this objective. We have emphasized the neverending cycle of quality improvement throughout this text. The quality control phase is followed by quality improvement, which leads to the creation of a better product or service.

We have examined the principles of quality control in the chapters dealing with control charts. Control chart techniques deal with on-line control methods, in which it is necessary to determine whether the process is in a state of statistical control. Adjustments are made on the process, preferably in real time, as data from the process are collected and analyzed to determine its state. Data collected from the process are used to make changes in the process. Although these methods, sometimes collectively grouped under the heading of statistical process control (SPC), are no doubt useful, they nevertheless provide an action-taking framework in the product-manufacturing phase or in the operational phase of a service that is somewhat downstream in the cycle.

In this chapter we discuss a class of procedures known as off-line control methods, which are used in the design phase of the product/process or service, and which are quite beneficial. The Taguchi method of experimental design is introduced in this context. These procedures are used during initial design and prototype testing to determine optimal settings for the product and process parameters. Analysis is thus conducted earlier in the cycle, which allows us to address problems that might appear in the production phase. The product or process is then designed to minimize any detrimental effects. The proper choice in the design phase of parameter settings for both the product and process can reduce costs and improve quality.

We introduce several common experimental designs such as the completely randomized design, the randomized block design, and the Latin square design in this chapter. The details of the analysis associated with these designs, such as the analysis of variance, are not extensively discussed. For a thorough treatment of the relevant analytical procedures, consult the references provided (Box et al., 1978; Montgomery 2004b; Peterson 1985; Wadsworth et al. 2001.). Our objective will be on the fundamental principles of experimental design.

There are several advantages to experimental design. First, it can be useful in identifying the key decision variables that may not only keep a process in control but improve it. Second, in the development of new processes for which historical data are not available, experimental design used in the developmental phase can identify important factors and their associated levels that will maximize yield and reduce overall costs. This approach cuts down on the lead time between design and manufacturing and produces a design that is robust in its insensitivity to uncontrollable factors. The principles of factorial experiments are also presented (Box et al., 2005; Hunter 1985; Raghavarao 1971; Ryan 2000.). These help us determine the more critical factors and the levels at which the chosen factors should be maintained (Box and Wilson 1951; Box and Draper 1986). Such experiments are usually conducted under controlled conditions (in a laboratory, for example). They can provide crucial information for choosing the appropriate parameters (Gunter 1990d) and levels for the manufacturing phase or for the delivery of services.

12-2 EXPERIMENTAL DESIGN FUNDAMENTALS

An **experiment** is a planned method of inquiry conducted to support or refute a hypothesized belief or to discover new information about a product, process, or service. It is an active method of obtaining information, as opposed to passive observation. It involves inducing a response to determine the effects of certain controlled parameters. Examples of controllable **parameters**, often called **factors**, are the cutting tool, the temperature setting of an oven, the selected vendor, the amount of a certain additive, or the type of decor in a restaurant. The factors may be **quantitative** or **qualitative** (discrete). For quantitative factors, we must decide on the range of values, how they are to be measured, and the levels at which they are to be controlled during experimentation. For example, we might be interested in the effect of temperature on the output of a chemical compound. Three selected levels of the factor (temperature in this case) could be 100, 200, and 300°C. The **response variable** in this case is the output of the chemical compound as measured in, say, kilograms. Qualitative factors, such as the vendor of an input raw material, are arbitrarily assigned codes representing discrete levels.

In the terminology of experimental design, a **treatment** is a certain combination of **factor levels** whose effect on the response variable is of interest. For example, suppose we have two factors—oven temperature and the selected vendor of a certain raw material—that are of interest in determining the effect on the output of the response variable, which is the yield of a chemical compound. The temperature can be set at 100,200, or 300°C, and the raw material can be purchased from vendor A or B. The selected vendor is then a qualitative factor. The treatments for this experiment are the various combinations of these two factor levels. The case where vendor A is chosen and the oven temperature is set at 100°C is one treatment. A total of six treatments could possibly be considered in this situation.

An **experimental unit** is the quantity of material (in manufacturing) or the number served (in a service system) to which one trial of a single treatment is applied. For the chemical compound, if production is in batches of 2000 kg and each treatment is applied to a batch, the experimental unit is the batch. A **sampling unit**, on the other hand, is a fraction of the experimental unit on which the treatment effect is measured. If 100 g of the mixture is chosen for analysis, then 100 g is the sampling unit.

The variation in experimental units that have been exposed to the same treatment is attributed to **experimental error**. This variability is due to uncontrollable factors, or **noise factors**.

Experimental design is used to determine the impact of factors on the response variable. With quantitative factors, which vary on a **continuous** scale, we can obtain information about the variable's behavior even for factor levels that have not been experimentally determined. This concept is illustrated in Figure 12-1, where the quantitative factor is the amount of catalyst used in a process and the response variable is the reaction time. Six experiments are conducted for catalyst levels of 10, 20, 30, 30, 40, and 40 g. Note that as the amount of catalyst increases, the reaction time increases at a decreasing rate. Assuming that the behavior of the response variable is representative of the observed values, a smooth curve is drawn to depict its relationship with the factor of interest. This relationship is then used to interpolate the performance of the response variable for values of the amount of catalyst (say, 15 or 25 g) for which experiments have not been performed.

Such interpolation may not be feasible for qualitative factors with **discrete** levels. For example, let's say our response variable is the elasticity of a rubber compound and the factor is the vendor of raw material. We have four vendors: A, B, C, and D. Observations of the

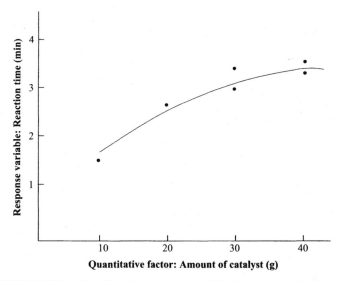

FIGURE 12-1 Behavior of a response variable for a quantitative factor.

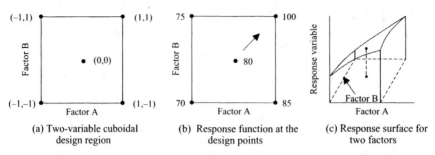

(a) Two-variable cuboidal design region

(b) Response function at the design points

(c) Response surface for two factors

FIGURE 12-2 Experimental design for more than one variable.

response variable are available, say, for vendors A, B, and D. From this information, we cannot interpolate the elasticity of the compound for raw material provided by vendor C.

Frequently, more than one factor is considered in experimental analysis. If more than one factor is under consideration, experiments do not usually involve changing one variable at a time. Some disadvantages of changing one variable at a time are discussed later. With two variables, we get a design known as the cuboidal design, shown in Figure 12-2a. Each factor can be set at two levels, denoted by −1 and 1. The point (0, 0) represents the current setting of the factor levels. In this notation, the first coordinate represents the level of factor A, and the second coordinate represents the level of factor B. A level of −1 represents a value below the current setting of 0, and a level of 1 represents a value above the current setting. Thus, four new design points, (−1, −1), (1, −1), (−1,1), and (1,1), can be investigated. The value of the response variable at these design points determines the next series of experiments to be performed.

Let's suppose that we have obtained the response function value at the experimental design points shown in Figure 12-2b and the response surface shown in Figure 12-2c. If our objective is to maximize the response function, we will consider the direction in which the gradient is steepest, as indicated by the arrow. Our next experiment could be conducted in a region with the center at (2,2) with respect to the current origin. These same principles apply to more than two variables as well (Gunter 1989,1990a,c).

Experiments with two or more variables usually avoid changing only one variable at a time—that is, keeping the other variables constant. There are reasons for this. Figure 12-3a shows the design points for such an approach, with (0, 0) representing the current state. If the value of factor A is fixed, the two design points are (0, −1) and (0, 1). Likewise, if factor B is fixed, the two additional design points are (−1, 0) and (1, 0). Keep in mind the response function shown in Figure 12-2b for the same factors. Figure 12-3b shows the response function value at the design points. Note that to maximize the response function in this experiment, we would move in the direction of the arrow shown along the major axis of factor A. The next experiment would be conducted in a region with the center at (2, 0) with respect to the current origin. This, however, does not lead to the proper point for maximizing the response function indicated in Figure 12-2c. These concepts also apply to the case with more than two variables. Experiments for one variable at a time are not as cost-effective.

Another drawback to the one-variable-at-a-time approach is that it cannot detect **interactions** between two or more factors (Gunter 1990b). Interactions exist when the nature of the relationship between the response variable and a certain factor is influenced by the level of some other factor. This concept is illustrated graphically in Figure 12-4.

Let's suppose that factor A represents the raw material vendor and factor B represents the amount of catalyst used in a chemical process. The response variable is the elasticity of the output product, the values of which are shown beside the plotted points. In Figure 12-4a,

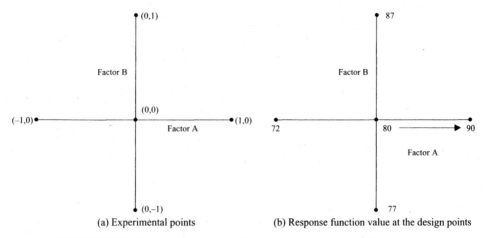

(a) Experimental points   (b) Response function value at the design points

FIGURE 12-3   Design approach involving changes in one variable at a time.

the functional relationship between the response and factor A stays the same regardless of the level of factor B. This depicts the case for which there is no interaction between factors A and B. The factors are said to have an additive effect on the response variable. The rate of change of the response as the level of factor A changes from -1 to 1 is the same for each level of B.

Now observe Figure 12-4b, where interaction does exist between factors A and B. The rate of change of the response as a function of factor A's level changes as B's level changes from -1 to 1 The response has a steeper slope when B is 1. Hence, B's level does influence the relationship between the response function and factor A. In such a situation, interaction is said to exist between factors A and B.

Interactions depict the joint relationship of factors on the response function. Such effects should be accounted for in multifactor designs because they better approximate real-world events. For example, a new motivational training program for employees (factor B) might impact productivity (response variable) differently for different departments (factor A) within the organization. Another way to represent interaction is through **contour plots** of the

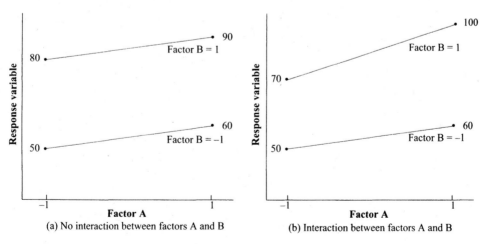

(a) No interaction between factors A and B   (b) Interaction between factors A and B

FIGURE 12-4   Interaction between factors.

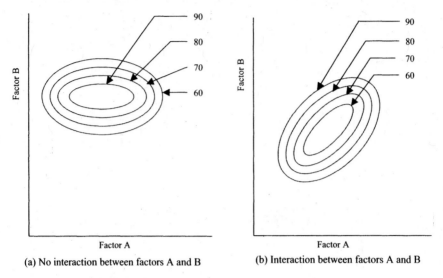

(a) No interaction between factors A and B

(b) Interaction between factors A and B

FIGURE 12-5 Contour plots of the response function.

response function. Figure 12-5a shows contour plots of the response function with factors A and B when there is no interaction between the factors. Here, the axes of the plotted contours are parallel to the axes of the factors. In Figure 12-5b, the response surface has a form of curvature. The natural axes of the contours are rotated with respect to the factor coordinate axes. Interaction between factors A and B exists in these circumstances.

Features of Experimentation

Experimental design incorporates features of replication, randomization, and blocking. **Replication** involves a repetition of the experiment under similar conditions (that is, similar treatments). It allows us to obtain an estimate of the experimental error, the variation in the experimental units under identically controlled conditions. The experimental error forms a basis for determining whether differences in the statistics found from the observations are significant. It serves as a benchmark for determining what factors or interactions are significant in influencing the response variable.

Replication serves another important function. If the sample mean is used to estimate the effect of a factor, replication increases the precision by reducing the standard deviation of the mean. In Chapter 4 we found the following from the **central limit theorem**. The variance of the sample mean $\bar{y}$ is given by

$$\sigma_{\bar{y}}^2 = \frac{\sigma^2}{n} \tag{12-1}$$

where σ^2 represents the **variance** of the observations and n is the number of replications used to determine the sample mean. As the number of replications increases, the standard deviation of the mean ($\sigma_{\bar{y}}$) decreases.

Replication can also provide a broader base for decision making. With an increase in the number of replications, a variety of experimental units can be used, permitting inferences to be drawn over a broader range of conditions.

A second feature of experimentation deals with the concept of randomization. **Randomization** means that the treatments should be assigned to the experimental units in such a way that every unit has an equal chance of being assigned to any treatment. Such a process eliminates **bias** and ensures that no particular treatment is favored. Random assignments of the treatments and the order in which the experimental trials are run ensures that the observations, and therefore the experimental errors, will be independent of each other. Randomization "averages out" the effect of uncontrolled factors.

Let's consider a situation in which customer reaction and satisfaction are to be tested for three types of loan application in a financial institution. The factor is the type of loan application, with three levels. Customers who are given these applications should be randomized so that the effects of all other factors are averaged out. Any differences in the observed customer satisfaction levels will thus be attributed to differences in the loan applications. Hence, care should be taken to ensure that the different loan applications are given to customers of various financial status or educational backgrounds, not merely to customers with similar backgrounds.

The third feature of experimentation is the principle of **blocking**, or local control. The idea is to arrange similar experimental units into groups, or blocks. Variability of the response function within a block can then be attributed to the differences in the treatments because the impact of other extraneous variables has been minimized. Treatments are assigned at random to the units within each block. Variability between blocks is thus eliminated from the experimental error, which leads to an increase in the precision of the experiment.

Let's consider a situation in which the effect of four different machine tools on the surface finish of a machined component is to be evaluated. We could create blocks consisting of different feed rates associated with the machining operation. The effect on surface finish of the four machine tools could then be compared within blocks. Assessing differences between blocks due to the variabilities in the feed rates would be distinct from determining the effect of the treatment—in this case, the type of machine tool. Blocking allows treatments to be compared under nearly equal conditions because the blocks consist of uniform units.

12-3 SOME EXPERIMENTAL DESIGNS

The design of an experiment accounts for the following:

- The set of treatments to be considered
- The set of experimental units to be included
- The procedures through which the treatments are to be assigned to the experimental units (or vice versa)
- The response variable(s) to be observed
- The measurement process to be used

A manageable number of treatments is chosen from a larger set based on their relative degree of impact on the response variable. Economic considerations (such as the cost associated with the application of a particular treatment) are also a factor in selecting treatments. Ideally, the goal is to optimize the response variable in the most cost-effective manner possible (Box and Draper 1986).

The distinguishing feature of the experimental designs considered in this section is the manner in which treatments are assigned. The selection of an appropriate measurement process is critical to all experimental designs. **Measurement bias**, which occurs when differences in the evaluation and observation processes go unrecognized, can cause serious problems in the analysis. Suppose we want to determine the effectiveness of a new drug on lowering blood cholesterol levels. The experimental subjects are randomly selected patients while the evaluator is a doctor or medical technician who monitors cholesterol levels. Patients, in this experiment, are randomly assigned either the "old" or the "new" drug, which are the treatments. Measurement bias can be eliminated by not divulging the assignment of treatment to the experimental subject and the evaluator. Any personal bias that the evaluator might have toward the new drug is eliminated since the evaluator does not know the type of drug received by each patient. Such an experiment, where the experimental subjects and the evaluator are not cognizant of the treatment assigned to each subject, is known as a **double-blind study**. When the knowledge of the treatment assignment is not disclosed to the experimental subject or the evaluator, the study is said to be **single-blind**.

For the experimental designs in this chapter, we consider situations that use **fixed effects models**, where the treatment effects are considered to be fixed. This means that the inferences drawn from the analysis pertain only to the treatments selected in the study. Alternatively, if the selected treatments are randomly chosen from a population of treatments, a **random effects model** is used to make inferences about the whole population of treatments.

To illustrate these models, let's consider an experiment in which the treatments are the amount of a certain compound used in a chemical process: 100, 200, and 300 g of the compound. The response variable is the reaction time. A fixed effects model considers the effect of *only* these three treatments on the response. That is, we are only interested in significant differences between these three treatments.

Now suppose the treatment (the amount of a certain compound) is assumed to be a normal random variable with a mean of 200 g and a standard deviation of 100 g. We could conceivably have an infinite number of treatments. From this population of available treatments, we randomly select three: 100, 200, and 300 g of the compound. These are the same amounts that we used in the fixed effects model; however, here we seek to make inferences on a population of treatments (the amount of chemical compound used) that is normally distributed. Specifically, the null hypothesis is that the variance of the treatments equals zero. This is a random effects model.

Completely Randomized Design

A completely randomized design is the simplest and least restrictive design. Treatments are randomly assigned to the experimental units, and each unit has an equal chance of being assigned to any treatment. One way to accomplish this is through the use of uniform random number tables.

Suppose that we have four treatments, A, B, C, and D, that we would like to assign to three experimental units, making a total of 12 experimental units. In the terminology of experimental design, each treatment is replicated three times; thus, we need 12 random numbers. Starting at any point in a random number table and going across a row or down a column, we obtain the following 12 three-digit random numbers sequentially: 682, 540, 102, 736, 089, 457, 198, 511, 294, 821, 304, and 658. Treatment A is assigned to the first three numbers, treatment B to the second three numbers, and so forth. These numbers are then ranked from smallest to largest, with the ranks corresponding to the experimental unit numbers. The scheme of assignment is depicted in Table 12-1. From Table 12-1 we find that experimental unit 1 is assigned treatment B, experimental unit 2 is assigned treatment A, and so on. A table of uniform random numbers is given in Appendix A-8. The rank order of the random number should be the time order of the data collection and experimentation.

TABLE 12-1   Assignment of Treatments to Experimental Units in a Completely Randomized Design

Random Number	Rank of Random Number (Experimental Unit)	Treatment
682	10	A
540	8	A
102	2	A
736	11	B
089	1	B
457	6	B
198	3	C
511	7	C
294	4	C
821	12	D
304	5	D
658	9	D

There are several advantages to a completely randomized design. First, any number of treatments or replications can be used. Different treatments need not be replicated the same number of times, making the design flexible. Sometimes it is not feasible to replicate treatments an equal number of times. In a situation where the treatments represent different vendors, vendor A, because of capacity and resource limitations, may not be able to supply as much raw material as vendors B and C. When all treatments are replicated an equal number of times, the experiment is said to be **balanced**; otherwise, the experiment is **unbalanced**. The statistical analysis is simple even for the unbalanced case, although comparisons between the treatment effects are more precise in the balanced case.

Another advantage is that the completely randomized design provides the most **degrees of freedom** for the experimental error. This ensures a more precise estimation of the experimental error. Recall that the experimental error provides a basis for determining which **treatment effects** are significantly different.

One disadvantage of a completely randomized design is that its precision may be low if the experimental units are not uniform. This problem is overcome by means of blocking, or grouping, similar homogeneous units. Designs utilizing the principle of blocking will be discussed later.

Analysis of Data   The statistical procedure used most often to analyze data is known as the **analysis of variance** (ANOVA). This technique determines the effects of the treatments, as reflected by their means, through an analysis of their variability. Details of this procedure are found in the listed references (Box et al., 2005; Peterson 1985). The total variability in the observations is partitioned into two components: the variation among the treatment means (also known as the **treatment sum of squares**) and the variation among the experimental units within treatments (also known as the **error sum of squares**). We have

$$\text{total sum of squares (SST)} = \text{treatment sum of squares (SSTR)} \qquad (12\text{-}2)$$
$$+ \text{error sum of squares (SSE)}$$

The **mean squares for treatments** and for error are obtained by dividing the corresponding sum of squares by the appropriate number of degrees of freedom. This number is 1 less than the number of observations in each source of variation. For a balanced design with p treatments, each with r replications, the total number of observations is rp. The total variability, therefore, has $rp - 1$ degrees of freedom. The number of degrees of freedom for the treatments is $p - 1$. For each treatment, there are r observations, so $r - 1$ degrees of freedom apply toward the experimental error. The total number of degrees of freedom for the experimental error is, therefore, $p(r - 1)$. We have the following notation:

p: number of treatments

r: number of replications for each treatment

y_{ij}:   response variable value of the jth experimental unit that is assigned treatment i, $i = 1, 2, \ldots, p; j = 1, 2, \ldots, r$

$y_{i\cdot}$:   sum of the responses for the ith treatment; that is, $\sum_{j=1}^{r} y_{ij}$

$\bar{y}_{i\cdot}$:   mean response of the ith treatment; that is, $y_{i\cdot}/r$

$y_{\cdot\cdot}$:   grand total of all observations; that is, $\sum_{i=1}^{p} \sum_{j=1}^{r} y_{ij}$

$\bar{\bar{y}}$:   grand mean of all observations; that is, $y_{\cdot\cdot}/rp$

The notation, consisting of r observations for each of the p treatments, is shown in Table 12-2. The computations of the sum of squares are as follows. A correction factor C is first computed as

$$C = \frac{y_{\cdot\cdot}^2}{rp} \tag{12-3}$$

The total sum of squares is

$$SST = \sum_{i=1}^{P} \sum_{j=1}^{r} y_{ij}^2 - C \tag{12-4}$$

The treatment sum of squares is determined from

$$SSTR = \frac{\sum_{i=1}^{P} y_{i\cdot}^2}{r} - C \tag{12-5}$$

TABLE 12-2   Notation for the Comletely Randomized Design in the Balanced Case

Treatment	Replication				Sum	Mean
	1	2	$\ldots$	r		
1	y_{11}	y_{12}	$\ldots$	y_{1r}	$y_{1\cdot}$	$\bar{y}_{1\cdot}$
2	y_{21}	y_{22}	$\ldots$	y_{2r}	$y_{2\cdot}$	$\bar{y}_{2\cdot}$
$\vdots$	$\vdots$	$\vdots$	$\ddots$	$\vdots$	$\vdots$	$\vdots$
p	y_{p1}	y_{p2}	$\ldots$	y_{pr}	$y_{p\cdot}$	$\bar{y}_{p\cdot}$
					$y_{\cdot\cdot}$	$\bar{\bar{y}}$

Finally, the error sum of squares is

$$SSE = SST - SSTR \qquad (12\text{-}6)$$

Next, the **mean squares** are found by dividing the sum of squares by the corresponding number of degrees of freedom. So, the mean squares for treatment are

$$MSTR = \frac{SSTR}{p-1} \qquad (12\text{-}7)$$

The mean square error is given by

$$MSE = \frac{SSE}{p(r-1)} \qquad (12\text{-}8)$$

Test for Differences Among Treatment Means   It is desirable to test the null hypothesis that the treatment means are equal against the alternative hypothesis that at least one treatment mean is different from the others. Denoting the treatment means, $\mu_1, \mu_2, \ldots, \mu_p$ we have the hypotheses

H_0 : $\mu_1 = \mu_2 = \cdots = \mu_p$
H_a :   At least one μ_i is different from the others

The test procedure involves the **F-statistic**, which is the ratio of the mean squares for treatment to the mean squares for error. The **mean square error** (MSE) is an unbiased estimate of σ^2, the variance of the experimental error. The test statistic is

$$F = \frac{MSTR}{MSE} \qquad (12\text{-}9)$$

with $(p - 1)$ degrees of freedom in the numerator and $p(r - 1)$ degrees of freedom in the denominator. For a chosen level of significance α, the critical value of F, which is found from the tables in Appendix A-6, is denoted by $F_{\alpha, p-1, p(r-1)}$. If the computed test statistic $F > F_{\alpha, p-1, p(r-1)}$, the null hypothesis is rejected, and we conclude that the treatment means are not all equal at the chosen level of significance. This computational procedure is known as analysis of variance; it is shown in tabular format in Table 12-3.

If the null hypothesis is not rejected, there is no significant difference among the treatment means. The practical implication is that as far as the response variable is concerned, it does not matter which treatment is used. Whichever treatment is the most cost-effective would be selected.

TABLE 12-3   ANOVA Table for the Completely Randomized Design for a Balanced Experiment

Source of Variation	Degrees of Freedom	Sum of Squares	Mean Square	F-Statistic
Treatments	$p - 1$	SSTR	$MSTR = \dfrac{SSTR}{p-1}$	$F = \dfrac{MSTR}{MSE}$
Error	$p(r - 1)$	SSE	$MSE = \dfrac{SSE}{p(r-1)}$	
Total	$rp - 1)$	SST		

If the null hypothesis is rejected, management might be interested in knowing which treatments are preferable. The sample treatment means would be ranked; depending on whether a large or small value of the response variable is preferred, the desirable treatments would then be identified. The sample treatment means serve as point estimates of the population treatment means. If interval estimates are desired, the following relation can be used to find a $100(1 - \alpha)\%$ confidence interval for the treatment mean μ_i:

$$L(\mu_i) = \bar{y}_i \pm t_{\alpha/2, p(r-1)} \sqrt{\frac{MSE}{r}} \tag{12-10}$$

where $t_{\alpha/2, p(r-1)}$ is the t-value for a right-tail area of $\alpha/2$ and $p(r - 1)$ degrees of freedom; it is found from Appendix A-4. The quantity $\sqrt{MSE/r}$ represents the standard deviation of the sample treatment mean $\bar{y}_i$.

Perhaps management wants to estimate the difference between two treatment means. Such an inference would indicate whether one treatment is preferable to the other. A $100(1 - \alpha)\%$ confidence interval for the difference of two treatment means, $\mu_{i1} - \mu_{i2}$, is given by

$$L(\mu_{i1} - \mu_{i2}) = (\bar{y}_{i1} - \bar{y}_{i2}) \pm t_{\alpha/2, p(r-1)} \sqrt{\frac{2MSE}{r}} \tag{12-11}$$

where $t_{\alpha/2, p(r-1)}$ is the t-value for a right-tail area of $\alpha/2$ and $p(r - 1)$ degrees of freedom. The quantity $\sqrt{2MSE/r}$ represents the standard deviation of the difference between the sample treatment means, $\bar{y}_{i1} - \bar{y}_{i2}$.

Example 12-1   Three adhesives are being analyzed for their impact on the bonding strength of paper in a pulp and paper mill. The adhesives are each randomly applied to four batches. The data are shown in Table 12-4. Here, the three treatments are the adhesives ($p = 3$), and the number of replications for each treatment is 4 ($r = 4$). The completely randomized design is a balanced one.

   (a) Is there a difference among the adhesives in terms of the mean bonding strength? Test at the 5% level of significance.

Solution   To test for differences among the adhesive mean bonding strengths, we perform analysis of variance using Minitab. We select **Stat** > **ANOVA** > **Oneway**. We first enter the data by listing the value of the response variable (bonding strength in this case) in one column and the corresponding treatment or factor (adhesive type) in a second column.

TABLE 12-4   Bonding Strength Data

Adhesive Type	Bonding Strength (kg)				Sum	Mean
1	10.2	11.8	9.6	12.4	44.0	11.000
2	12.8	14.7	13.3	15.4	56.2	14.050
3	7.2	9.8	8.7	9.2	34.9	8.725
					135.1	11.258

One-way ANOVA: Bonding Strength versus Adhesive

```
Source    DF    SS    MS     F     P
Adesive    2   57.11 28.56 19.36 0.001
Error      9   13.28  1.48
Total     11   70.39
s = 1.215   R-Sq = 81.14%   R-Sq(adj) = 76.95
                    Individual 95% CIs For Mean Based on
                    Pooled StDev
Level     N  Mean   StDev -+---------+---------+---------+-----
1         4 11.000 1.317              (----*----)
2         4 14.050 1.207                         (----*-----)
3         4  8.725 1.112 (-----*----)
                         -+---------+---------+---------+-----
                         7.5      10.0      12.5      15.0
Pooled StDev = 1.215
```

FIGURE 12-6   Minitab's ANOVA table for bonding strength data.

The column number of the response variable is input in *Response* and that of the factor in *Factor* and we then click **OK**. The ANOVA table is shown in Figure 12-6.

To test the hypothesis

$H_0:$ $\mu_1 = \mu_2 = \mu_3$

$H_a:$ At least one μ_i is different from the others

the F-statistic (shown in Figure 12-6) is computed as

$$F = \frac{\text{MSTR}}{\text{MSE}} = \frac{28.56}{1.48} = 19.36$$

From Appendix A-6, for a chosen significance level α of 0.05, $F_{0.05,2,9} = 4.26$. Since the calculated value of F exceeds 4.26, we reject the null hypothesis and conclude that there is a difference in the mean bonding strengths of the adhesives. Note also from Figure 12-6 that the p-value associated with the calculated F-value is $0.001 < 0.05$. So we reject the null hypothesis.

(b) Find a 90% confidence interval for the mean bonding strength of adhesive 1.

Solution   A 90% confidence interval for the mean bonding strength of adhesive 1 is given by

$$L(\mu_1) = \bar{y}_1 \pm t_{0.05,9} \sqrt{\frac{\text{MSE}}{4}}$$

$$= 11.000 \pm 1.833 \sqrt{\frac{1.48}{4}}$$

$$= 11.000 \pm 1.115 = (9.885, 12.115)$$

Note that Figure 12-6 also shows 95% confidence intervals for the treatment means.

(c) Find a 95% confidence interval for the difference in the mean bonding strengths of adhesives 1 and 3. Is there a difference in the means of these two adhesives?

Solution A 95% confidence interval for the difference in the mean bonding strengths of adhesives 1 and 3 is given by

$$L(\mu_1 - \mu_3) = (\bar{y}_1 - \bar{y}_3) \pm t_{0.025,9} \sqrt{\frac{2\text{MSE}}{4}}$$

$$= (11.00 - 8.725) \pm 2.262 \sqrt{\frac{(2)(1.48)}{4}}$$

$$= 2.275 \pm 1.945 = (0.33, \ 4.22)$$

Since the confidence interval does not include the value of zero, we conclude that there is a difference in the means of adhesives 1 and 3 at the 5% level of significance.

Randomized Block Design

Blocking involves grouping experimental units that are exposed to similar sources of variation. The idea is to eliminate the effects of extraneous factors within a block such that the impact of the individual treatments, which is our main concern, can be identified. Experimental units within a block should therefore be as homogeneous as possible.

To illustrate, let's consider a situation in which we want to investigate the differences between the raw materials from vendors A, B, and C. Processing will take place in either of two machines, M1 or M2. The treatments of interest are the vendors. Extraneous factors are those due to differences caused by M1 and M2. If we randomly assign raw materials from the vendors to the machines, we will not be able to claim that any differences in the output are due to the vendors, because the variability of the machines will not have been isolated.

Now let's use the concept of blocking in the experimental design. The blocking variable is the type of machine (M1 or M2). For a given machine, say M1 (which represents a block), the raw materials from vendors A, B, and C are randomly used. Differences in the output from this block now represent the impact of the treatments, not the variability of the machines. This scheme would be repeated for the remaining block, machine M2. Differences in the outputs between the two blocks will measure the impact of the blocking variable (that is, whether blocking is effective).

The criterion on which we base blocking is identified before we conduct the experiment. Successful blocking minimizes the variation among the experimental units *within* a block while maximizing variation *between* blocks. An advantage of blocking is that differences between blocks can be eliminated from the experimental error, thereby increasing the precision of the experiment. However, blocking is effective only if the variability within blocks is smaller than the variability over the entire set. Precision usually decreases as the block size (the number of experimental units per block) increases.

In a randomized block design, the experimental units are first divided into blocks. Within a block, the treatments are randomly assigned to the experimental units as in a completely randomized design. If we have p treatments and r blocks in a completely randomized block design, each of the p treatments is applied once in each block, resulting in a total of rp experimental units. Thus, r blocks, each consisting of p units, are chosen such that the units within blocks are homogeneous. Next, the p treatments are randomly assigned to the units within the blocks, with each treatment occurring only once within each block. The blocks serve as replications for the treatments. We assume that there is no interaction between the treatments and blocks.

Table 12-5 shows the layout for an example in which there are four treatments, A, B, C, and D, and three blocks, I, II, and III (hence, $p = 4$ and $r = 3$). Within each block, the four

TABLE 12-5   Assignment of Treatments to Units Within Blocks in a Randomized Block Design

Block I	Block II	Block III
B	C	C
A	A	B
C	D	D
D	B	A

treatments are randomly applied to the experimental units. Each treatment appears only once in each block, The treatments can also be assigned as for the completely randomized design. A random number table would then be used, with the scheme depicted in Table 12-1 adopted.

There are several advantages to using the randomized block design. First and foremost is that it increases the precision by removing a source of variation (variability between blocks) from the experimental error. The design can accommodate any number of treatments and blocks. The statistical analysis is relatively simple and can easily accommodate the dropping of an entire treatment or block, due to reasons such as faulty observations. One disadvantage is that missing data can require complex calculations. The number of degrees of freedom for the experimental error is not as large as that for the completely randomized design. In fact, if there are r blocks, then there are $r-1$ fewer degrees of freedom in the experimental error. Furthermore, if the experimental units are homogeneous, the completely randomized design is more efficient.

Analysis of Data   Let's define the following notation:

- p : number of treatments
- r : number of blocks
- y_{ij} : response variable value of the ith treatment applied to the jth block, $i = 1, 2, \ldots p$; $j = 1, 2, \ldots, r$
- $y_{i\cdot}$: $\sum_{j=1}^{r} y_{ij} = $ sum of responses for the ith treatment
- $\bar{y}_{i\cdot}$: $y_{i\cdot}/r = $ mean response of the ith treatment
- $y_{\cdot j}$: $\sum_{i=1}^{p} y_{ij} = $ sum of the responses for the jth block
- $\bar{y}_{\cdot j}$: $y_{\cdot j}/p = $ mean response of the jth block
- $y_{\cdot\cdot}$: $\sum_{i=1}^{p} \sum_{j=1}^{r} y_{ij} = $ grand total of all observations
- $\bar{\bar{y}}$: $y_{\cdot\cdot}/rp = $ grand mean of all observations

The notation, consisting of the p treatments assigned to each of the r blocks, is shown in Table 12-6.

Computations for the sum of squares are as follows. The correction factor C, is given by eq. (12-3), the total sum of squares is given by eq. (12-4), and the treatment sum of squares is given by eq. (12-5).

For a randomized block design, the **block sum of squares** is

$$\text{SSB} = \frac{\sum_{j=1}^{r} y_{\cdot j}^{2}}{p} - C \tag{12-12}$$

TABLE 12-6   Notation for the Randomized Block Design

Treatment	Block 1	2	...	r	Sum	Mean
1	y_{11}	y_{12}	...	y_{1r}	$y_1.$	$\bar{y}_1.$
2	y_{21}	y_{22}	...	y_{2r}	$y_2.$	$\bar{y}_2.$
$\vdots$	$\vdots$	$\vdots$	$\ddots$	$\vdots$	$\vdots$	$\vdots$
p	y_{p1}	y_{p2}	...	y_{pr}	$y_p.$	$\bar{y}_p.$
Sum	$y_{.1}$	$y_{.2}$	...	$y_{.r}$	$y_{..}$	
Mean	$\bar{y}_{.1}$	$\bar{y}_{.2}$	...	$\bar{y}_{.r}$		$\bar{\bar{y}}$

The error sum of squares is

$$SSE = SST - SSTR - SSB$$

The number of degrees of freedom for the treatments is $p - 1$ and for the blocks is $r - 1$, 1 less than the number of blocks. The number of degrees of freedom for the error is $(p - 1)(r - 1)$, and the total number is $rp - 1$. The mean squares for the treatments, blocks, and experimental error are found on division of the respective sum of squares by the corresponding numbers of degrees of freedom.

Test for Difference Among Treatment Means   Denoting the treatment means by $\mu_1, \mu_2, \ldots,$ μ_p, one set of hypotheses to test would be

H_0 : $\mu_1 = \mu_2 = \cdots \mu_p$
H_a :   At least one μ_i is different from the others

This test is conducted by finding the F-statistic (the ratio of the mean squares for treatment to the mean square error), that is,

$$F = \frac{MSTR}{MSE}$$

The calculated F-statistic is compared to the critical value of F from Appendix A-6 for a chosen significance level α, with $p - 1$ degrees of freedom in the numerator and $(p - 1)(r - 1)$ in the denominator. If the computed test statistic $F > F_{\alpha, p-1, (p-1)(r-1)}$, the null hypothesis is rejected, and we conclude that the treatment means are not all equal at the chosen level of significance. The ANOVA table is shown in Table 12-7.

Note that blocking is typically used as a means of error control. The variability between blocks is removed from the experimental error. Whereas the treatments are replicated (each treatment being used once in each block), the blocking classifications are not. Therefore, we do not have a true measure of the experimental error for testing the difference between block means.

Sometimes, an F-statistic given by the ratio of the mean squares for blocks to the mean squares for error; that is,

$$F_B = \frac{MSB}{MSE} \tag{12-13}$$

TABLE 12-7   ANOVA Table for the Randomized Block Design

Source of Variation	Degrees of Freedom	Sum of Squares	Mean Square	F-Statistic
Treatments	$p-1$	SSTR	$MSTR = \dfrac{SSTR}{p-1}$	$F = \dfrac{MSTR}{MSE}$
Blocks	$r-1$	SSB	$MSB = \dfrac{SSB}{r-1}$	
Error	$(p-1)(r-1)$	SSE	$MSE = \dfrac{SSE}{(p-1)(r-1)}$	
Total	$rp-1$	SST		

is used to determine whether blocking has been effective in reducing the experimental error. The computed value of F_B from eq. (12-13) is compared to $F_{\alpha,\,r-1,\,(p-1)(r-1)}$ for a level of significance α. If $F_B > F_{\alpha,\,r-1,\,(p-1)(r-1)}$, we conclude that blocking has been effective; technically, though, a true significance level cannot be assigned to the test because we intentionally try to select the blocks to maximize the variability between them.

If $F_B < F_{\alpha,\,r-1,\,(p-1)(r-1)}$, we infer that blocking has not been effective. In such a case, the efficiency of the experiment has decreased relative to the completely randomized design, because we lost $r-1$ degrees of freedom from the error without a sufficiently compensating reduction in the mean square error. This means that even though the sum of squares for error (SSE) decreased due to blocking, the MSE did not decrease because the reduction in the number of degrees of freedom was not adequately compensated for. Note that, since we strive to maximize the variability between blocks, it is not necessary to test the null hypothesis that the block means are equal.

Interval estimates for a single treatment mean μ_i are obtained as before. A $100(1-\alpha)\%$ confidence interval for μ_i is given by

$$L(\mu_i) = \bar{y}_{i\cdot} \pm t_{\alpha/2,(p-1)(r-1)} \sqrt{\frac{MSE}{r}} \tag{12-14}$$

where $t_{\alpha/2,(p-1)(r-1)}$ is the t-value for a right-tail area of $\alpha/2$ and $(p-1)(r-1)$ degrees of freedom and is found from Appendix A-4.

Usually, it would be of interest to estimate the difference between two treatment means. A $100(1-\alpha)\%$ confidence interval for the difference between two treatment means, $\mu_{i1} - \mu_{i2}$, is given by

$$L(\mu_{i1}-\mu_{i2}) = (\bar{y}_{i1} - \bar{y}_{i2}) \pm t_{\alpha/2,(p-1)(r-1)} \sqrt{\frac{2MSE}{r}} \tag{12-15}$$

where $t_{\alpha/2,(p-1)(r-1)}$ is the t-value for a right-tail area of $\alpha/2$ and $(p-1)(r-1)$ degrees of freedom.

Example 12-2   A construction company intends to test the efficiency of three different insulators. Since the area where the company builds has varying temperature differentials, the following experimental procedure has been planned. The company has divided the area into four geographical regions, based on climatic differences. Within each geographical region, it randomly uses each of the three insulators and records the energy loss as an index. Smaller values of the index correspond to lower losses. Table 12-8 shows the energy loss data.

TABLE 12-8   Energy Loss Data for Example 12-2

| Insulator | Geographical Region | | | | Sum | Mean |
	I	II	III	IV		
1	19.2	12.8	16.3	12.5	60.8	15.2
2	11.7	6.4	7.3	6.2	31.6	7.9
3	6.7	2.9	4.1	2.8	16.5	4.125
Sum	37.6	22.1	27.7	21.5	108.9	
Mean	12.533	7.367	9.233	7.167		9.075

(a) Is there a difference in the mean energy loss for the three insulators? Test at a significance level of 10%.

Solution   Here, the treatments are the insulators, and the blocks are the geographical regions. Using the raw data in Table 12-8, we calculate the sample treatment sums and means and the block sums and means. The number of treatments p is 3, and the number of blocks r is 4. These are also shown in Table 12-8.

We use Minitab to obtain the analysis of variance by selecting **Stat > ANOVA > Balanced ANOVA**. Data is first entered by listing the value of the response variable (energy loss index) in one column, the corresponding treatment (type of insulator) in a second column, and the blocking variable (geographical region) in a third column. The model used is to predict the response variable using the main effects of the treatments and blocks. The resulting ANOVA table is shown in Figure 12-7.

To test whether there is a difference in the mean energy loss index between the three insulators, the F-statistic is calculated as the ratio of the mean squares for insulators to the mean squares for error.

From Figure 12-7, note that the value of the F-statistic is 170.90, and the p-value [which is $P(F > 170.90)$] is 0.000. Since this p-value < 0.10, the chosen level of significance, we reject the null hypothesis and conclude that at least one of the mean energy loss indices, using the three insulators, is different from the others.

(b) Find a 99% confidence interval for the mean energy loss index for insulator 3.

Solution   A 99% confidence interval for the mean energy loss index for insulator 3 is

$$L(\mu_3) = \bar{y}_3. \pm t_{0.005,6}\sqrt{\frac{MSE}{4}}$$

$$= 4.125 \pm (3.707)\sqrt{\frac{0.742}{4}}$$

$$= 4.125 \pm 1.597 = (2.528, 5.722)$$

Analysis of Variance for Energy Loss

Source	DF	SS	MS	F	P
Insulator	2	253.595	126.797	170.90	0.000
Region	3	55.636	18.545	25.00	0.001
Error	6	4.452	0.742		
Total	11	313.683			

FIGURE 12-7   Minitab's ANOVA Table for Energy Loss Data.

(c) Find a 90% confidence interval for the difference in the mean energy loss index of insulators 2 and 3. Is there a difference in these two means?

Solution   A 90% confidence interval for the difference in the mean energy loss index of insulators 2 and 3 is

$$L(\mu_2 - \mu_3) = (\bar{y}_2. - \bar{y}_3.) \pm t_{0.05,6} \sqrt{\frac{2MSE}{4}}$$

$$= (7.9 - 4.125) \pm 1.943 \sqrt{\frac{(2)(0.742)}{4}}$$

$$= 3.775 \pm 1.183 = (2.592, 4.958)$$

Since the confidence interval does not include the value of zero, we conclude that there is a difference in the mean energy loss index of insulators 2 and 3 at the 10% level of significance.

(d) Which insulator should the company choose?

Solution   Since smaller values of the energy loss index correspond to low losses, insulator 3 would be chosen, as it has the lowest mean (4.125). Part (c) showed that the mean energy loss of insulator 3 is different from that of insulator 2. Since the mean energy loss of insulator 1 is even higher than that of insulator 2, the mean energy loss of insulator 3 will be significantly different from that of insulator 1.

Latin Square Design

In the randomized block design, we control one source of variation. Experimental units are grouped into blocks consisting of homogeneous units based on a criterion that reduces the variability of the experimental error. Variability between blocks is thus separated from the experimental error, and each treatment appears an equal number of times in each block. The Latin square design controls two sources of variation; in effect, it blocks in two variables. The number of groups, or blocks, for each of the two variables is equal, which in turn is equal to the number of treatments. Suppose we denote one of the blocking variables as the row variable and the other as the column variable. The treatments are assigned so that each treatment appears once and only once in each row and column. The number of experimental units required is the square of the number of treatments.

Let's consider an example in which blocking in two variables is appropriate. We will assume that the productivity of bank tellers varies from one person to another and that productivity changes as a function of the day of the week. Bank management is interested in determining whether the type of break, as characterized by the frequency and duration, affects productivity. They decide to test five types of breaks (A, B, C, D, and E). Type A represents short breaks (say, 5 min) that occur once each hour. Type B represents longer breaks (10 min) that occur once every two hours, and so forth. In the Latin square design, five tellers (I, II, III, IV, and V) are randomly selected and comprise the first blocking variable (the row classification). The five days of the week (1, 2, 3, 4, and 5) comprise the column classification. The treatments are the types of break (A, B, C, D, and E). Treatments are always denoted by Latin letters, hence the name **Latin square design**. The design is square because the number of row and column classifications is equal (in this case, 5), which is also

TABLE 12-9   Latin Square Design for Five Treatments

Teller	Day of Week				
	1	2	3	4	5
I	A	B	C	D	E
II	B	C	D	E	A
III	C	D	E	A	B
IV	D	E	A	B	C
V	E	A	B	C	D

equal to the number of treatments. Each break type is assigned to each teller and to each day of the week only once.

One particular assignment of break type (treatments) is shown in Table 12-9. For instance, type A is assigned to teller I for the first day of the week (day 1). Similarly, type B is assigned to teller I for the second day of the week (day 2), and so on. Note that each treatment occurs only once in each row and in each column. For five treatments, 25 experimental units are needed for the Latin square design. The assignment strategy shown in Table 12-9 is one of several that could satisfy the preceding criteria.

The Latin square design is an example of an **incomplete block design**, in which not all treatments are necessarily assigned to each block. In the example just considered, we actually have 25 blocks, which represent the different combinations of tellers and day of the week (that is, block 1: teller I, day 1; block 2: teller I, day 2; and so on). For a randomized block design, in which each treatment appears once in each block, we would need 125 experimental units. In the Latin square design we can get by with only 25 experimental units. In general, if we have p treatments, we need p^2 experimental units; each of the two blocking variables should have p classes.

The main advantage of the Latin square design is that it allows the use of two blocking variables. This usually results in greater reductions in the variability of experimental error than can be obtained by using either of the two blocking variables individually.

There are some disadvantages to the Latin square design, however. First, since the number of classes of each blocking variable must equal the number of treatments, very few degrees of freedom are left for the experimental error if the number of treatments is small. When the number of treatments is large, the number of degrees of freedom for the experimental error may be larger than necessary. Since the number of experimental units is equal to the square of the number of treatments, the number of treatments has to be limited (usually to 10 or less) for practical studies. Also, the model is restrictive because it assumes that there is no interaction between either the blocking variable and the treatment or between the blocking variables themselves. Another restriction is that the number of classes for the two blocking variables must be the same.

Randomization of the Latin Square Design   For a given number of treatments, various Latin squares are possible. For example, for three treatments denoted by A, B, and C ($p = 3$), four possible Latin squares are shown in Table 12-10. As p increases, the number of possible arrangements increases dramatically. The idea behind randomization is to select one out of all possible Latin squares for a given number of treatments in such a way that each square has an equal chance of being selected.

TABLE 12-10   Four Possible Latin Square Designs for Three Treatments

	Blocking variable I				Blocking variable I		
	A	B	C		A	C	B
Blocking variable II	B	C	A	Blocking variable II	B	A	C
	C	A	B		C	B	A
	Blocking variable I				Blocking variable I		
	B	A	C		C	B	A
Blocking variable II	C	B	A	Blocking variable II	A	C	B
	A	C	B		B	A	C

Analysis of Data   The analysis of variance is similar to the previously described procedures. The following notation is used:

p :   number of treatments

$y_{ij(k)}$:   response variable value of the experimental unit in the ith row and jth column subjected to the kth treatment

$y_{i\cdot}$: $\sum_{j=1}^{p} y_{ij} =$ sum of the responses in the ith row

$y_{\cdot j}$: $\sum_{i=1}^{p} y_{ij} =$ sum of the responses in the jth column

$y_{\cdot\cdot}$: $\sum_{i=1}^{p} \sum_{j=1}^{p} y_{ij} =$ grand total of all observations

T_k :   sum of the responses for the kth treatment

$\bar{y}_k$: T_k/p sample mean response for the kth treatment

The notation for the Latin square design is shown in Table 12-11. The sum of squares is calculated as follows.

A correction factor C is computed as

$$C = \frac{y_{\cdot\cdot}^2}{p^2} \qquad (12\text{-}16)$$

The total sum of squares is

$$\text{SST} = \sum_{i=1}^{p} \sum_{j=1}^{p} y_{ij}^2 - C \qquad (12\text{-}17)$$

TABLE 12-11   Notation for the Latin Square Design

	Column				
Row	1	2	$\cdots$	p	Sum
1	y_{11}	y_{11}	$\cdots$	y_{1p}	$y_{1\cdot}$
2	y_{21}	y_{22}	$\cdots$	y_{2p}	$y_{2\cdot}$
$\vdots$	$\vdots$	$\vdots$	$\vdots$	$\vdots$	$\vdots$
p	y_{p1}	y_{p2}	$\cdots$	y_{pp}	$y_{p\cdot}$
Sum	$y_{\cdot 1}$	$y_{\cdot 2}$	$\cdots$	$y_{\cdot p}$	$y_{\cdot\cdot}$
	Treatment				
	1	2	$\cdots$	p	
Sum	T_1	T_2	$\cdots$	T_p	
Mean	$\bar{y}_1$	$\bar{y}_2$	$\cdots$	$\bar{y}_p$	

The sum of squares for the row blocking variable is

$$SSR = \frac{\sum_{i=1}^{p} y_{i.}^2}{p} - C \tag{12-18}$$

The sum of squares for the column blocking variable is

$$SSC = \frac{\sum_{j=1}^{p} y_{.j}^2}{p} - C \tag{12-19}$$

The sum of squares for the treatments is

$$SSTR = \frac{\sum_{k=1}^{p} T_k^2}{p} - C \tag{12-20}$$

Finally, the error sum of squares is

$$SSE = SST - SSR - SSC - SSTR \tag{12-21}$$

The number of degrees of freedom for the row blocking variable as well as for the column blocking variable is $p - 1$. The number of degrees of freedom for the treatments is $(p - 1)$. The total number of degrees of freedom is $p^2 - 1$. The number of degrees of freedom for the experimental error is found as follows:

$$\text{error degrees of freedom} = (p^2 - 1) - (p - 1) - (p - 1) - (p - 1)$$
$$= (p - 1)(p - 2)$$

Next, the mean squares are found by dividing the sum of squares by the appropriate number of degrees of freedom. A tabular format of the ANOVA calculations is shown in Table 12-12.

TABLE 12-12   ANOVA Table for the Latin Square Design

Source of Variation	Degrees of Freedom	Sum of Squares	Mean Square	F-Statistic
Rows (blocking variable I)	$p - 1$	SSR	$MSR = \dfrac{SSR}{p-1}$	
Columns (blocking variable II)	$p - 1$	SSC	$MSC = \dfrac{SSC}{p-1}$	
Treatments	$p - 1$	SSTR	$MSTR = \dfrac{SSTR}{p-1}$	$F = \dfrac{MSTR}{MSE}$
Error	$(p - 1)(p - 2)$	SSE	$MSE = \dfrac{SSE}{(p-1)(p-2)}$	
Total	$p^2 - 1$	SST		

Test for Difference Among Treatment Means   If the treatment means are denoted by μ_1, $\mu_2, \ldots, \mu_p$, one set of hypotheses to test is

H_0: $\mu_1 = \mu_2 = \ldots, = \mu_p$

H_a: At least one μ_i is different from the others

The test is conducted by calculating the *F*-statistic as the ratio of the mean squares for treatments to the mean square for error, that is,

$$F = \frac{\text{MSTR}}{\text{MSE}}$$

If the calculated *F*-statistic $> F_{\alpha,(p-1),(p-1)(p-2)}$, which is found in Appendix A-6, for a right-tail area of α with $p-1$ degrees of freedom in the numerator and $(p-1)(p-2)$ in the denominator, the null hypothesis is rejected. In that case, we conclude that at least one of the treatment means is different from the others at a chosen level of significance α.

Note that treatments are replicated in the Latin square (as in the randomized block design), but the row and column classifications of the two blocking variables are not. Thus, an *F*-statistic for determining the effectiveness of the row blocking variable calculated as

$$F_R = \frac{\text{MSR}}{\text{MSE}}$$

is used in an approximate significance test to determine whether the row classification is effective in reducing the experimental error. Similarly, for testing the effectiveness of the column blocking variable, an *F*-statistic calculated as

$$F_C = \frac{\text{MSC}}{\text{MSE}}$$

is used in an approximate significance test for the column blocking variable. Each of these two statistics can be compared with $F_{\alpha,p-1,(p-1)(p-2)}$ to make such inferences.

A $100(1 - \alpha)\%$ confidence interval for a single treatment mean μ_k is given by

$$L(\mu_k) = \bar{y}_k \pm t_{\alpha/2,(p-1)(p-2)} \sqrt{\frac{\text{MSE}}{p}} \qquad (12\text{-}22)$$

where $t_{\alpha/2,(p-1)(p-2)}$ is the *t*-value for a right-tail area of $\alpha/2$ and $(p-1)(p-2)$ degrees of freedom and is found from Appendix A-4.

As with the previous designs, it is usually of interest to estimate the difference between two treatment means. A $100(1 - \alpha)\%$ confidence interval for the difference between two treatment means, $\mu_{k1} - \mu_{k2}$, is given by

$$L(\mu_{k1} - \mu_{k2}) = (\bar{y}_{k1} - \bar{y}_{k2}) \pm t_{\alpha/2,(p-1)(p-2)} \sqrt{\frac{2\text{MSE}}{p}} \qquad (12\text{-}23)$$

where $t_{\alpha/2,(p-1)(p-2)}$ is the *t*-value for a right-tail area of $\alpha/2$ and $(p-1)(p-2)$ degrees of freedom.

TABLE 12-13   Pricing Policy Data

Sales Volume Class	Geographical Location								Sum
	Northeast		East		Midwest		Southeast		
1	A	19.2	B	15.4	C	6.6	D	10.5	51.7
2	B	13.2	C	5.3	D	8.2	A	16.8	43.5
3	C	4.2	D	9.4	A	14.6	B	8.5	36.7
4	D	8.4	A	13.3	B	7.6	C	6.2	35.5
		45.0		43.4		37.0		42.0	167.4

From a practical point of view, the number of treatments in a randomized Latin square design that can be analyzed is somewhat restricted. Note that the number of error degrees of freedom is given by $(p - 1)(p - 2)$, where p is the number of treatments. If p is very small, the number of error degrees of freedom is small, which will cause the mean square error to be large. The number of treatments cannot be less than 3, because if $p = 2$, the number of error degrees of freedom is zero. As p becomes large, the number of experimental units required (p^2) becomes too large and may not be feasible from a cost and operational standpoint.

Example 12-3   A retail company is interested in testing the impact of four different pricing policies (A, B, C, and D) on sales. The company suspects that variation in sales could be affected by factors other than the pricing policy, such as store location and sales volume. The company has four location classifications: Northeast, East, Midwest, and Southeast. It has four classes of sales volume: 1, 2, 3, and 4, with class 1 representing the largest volume, and the others representing decreasing volume in succession. Each pricing policy is applied in each geographic region and each sales volume class exactly once. Table 12-13 shows the sales (in thousands of dollars) for a three-month period, with the pricing policy shown appropriately.

(a) Is there a difference in the pricing policies in terms of mean sales? Test at a level of significance of 5%.

Solution   Table 12-13 also shows the row and column sums for the blocking variables sales volume class and geographical location. The sum of squares for each of the two blocking variables as well as for the treatments (the pricing policies) are found as follows. The correction factor is

$$C = \frac{y_{..}^2}{p^2} = \frac{(167.40)^2}{4^2} = 1751.4225$$

From eq. (12-17), the total sum of squares is

$$SST = \sum_{i=1}^{p}\sum_{j=1}^{p} y_{ij}^2 - C$$

$$= 2046.48 - 1751.4225 = 295.0575$$

TABLE 12-14   Sums and Means of Sales for Pricing Policy Data

	Pricing Policy			
	A	B	C	D
Sum	63.9	44.7	22.3	36.5
Mean	15.975	11.175	5.575	9.125

The sum of squares of the row blocking variable (sales volume class) is found from eq. (12-18) as

$$SSR = \frac{\sum_{i=1}^{p} y_{i\cdot}^2}{p} - C$$

$$= \frac{(51.7)^2 + (43.5)^2 + (36.7)^2 + (35.5)^2}{4} - 1751.4225 = 41.6475$$

The sum of squares for the column blocking variable (geographical location) is found from eq. (12-19) as

$$SSC = \frac{\sum_{j=1}^{p} y_{\cdot j}^2}{p} - C$$

$$= \frac{(45.0)^2 + (43.4)^2 + (37.0)^2 + (42.0)^2}{4} - 1751.4225 = 8.9675$$

Table 12-14 shows the sums of sales and their means for the pricing policies. The sum of squares due to the pricing policies is found from eq. (12-20) as

$$SSTR = \frac{\sum_{k=1}^{p} T_k^2}{p} - C$$

$$= \frac{(63.9)^2 + (44.7)^2 + (22.3)^2 + (36.5)^2}{4} - 1751.4225 = 226.2875$$

The error sum of squares is

$$SSE = SST - SSR - SSC - SSTR$$
$$= 295.0575 - 41.6475 - 8.9675 - 226.2875 = 18.155$$

The ANOVA table is shown in Table 12-15. To test the null hypothesis that the mean sales for the four pricing policies (A, B, C, and D) are equal against the alternative hypothesis that at least one mean is different from the others, the F-statistic is

$$F = \frac{MSTR}{MSE} = \frac{75.4292}{3.0258} = 24.9287$$

For a level of significance α of 5%, the value of F from Appendix A-6 is $F_{0.05,3,6} = 4.76$. Since the calculated value of $F = 24.9287 > 4.76$, we reject the null hypothesis and conclude that at least one mean sales value differs from the others at this level of significance.

TABLE 12-15 ANOVA Table for Pricing Policy Data

Source of Variation	Degrees of Freedom	Sum of Squares	Mean Square	F-Statistic
Sales volume class (rows)	3	41.6475	13.8825	
Geographic location (columns)	3	8.9675	2.9892	
Pricing policy (treatments)	3	226.2875	75.4292	24.9287
Error	6	18.1550	3.0258	
	15	295.0575		

(b) Find a 90% confidence interval for the mean sales using pricing policy A.

Solution   A 90% confidence interval for policy A's mean sales is

$$L(\mu_A) = \bar{y}_A \pm t_{0.05,6} \sqrt{\frac{MSE}{4}}$$

$$= 15.975 \pm 1.943 \sqrt{\frac{3.0258}{4}} = (14.285, 17.665)$$

(c) Find a 95% confidence interval for the difference in mean sales for pricing policies A and C. Is there a difference between these two policies?

Solution   A 95% confidence interval for the difference in policies A and C's mean sales is

$$L(\mu_A - \mu_C) = (\bar{y}_A - \bar{y}_C) \pm t_{0.025,6} \sqrt{\frac{2MSE}{4}}$$

$$= (15.975 - 5.575) \pm (2.447) \sqrt{\frac{(2)(3.0258)}{4}} = (7.390, 13.410)$$

Since the confidence interval does not include zero, at the 5% level of significance we conclude that the mean sales of pricing policies A and C differ.

(d) Which pricing policy should be used?

Solution   We have concluded that there is a difference between the mean sales of the four pricing policies. From Table 12-14 we find that pricing policy A yields the highest mean sales of 15.975. When we check if mean sales of policies A and B (the two highest values) differ at the 5% level of significance, we find

$$\bar{y}_A - \bar{y}_B = 15.975 - 11.175 = 4.8$$

Using the calculations from part c, a 95% confidence interval for $\mu_A - \mu_B$ is

$$4.8 \pm 3.010 = (1.790, 7.810)$$

Since the confidence interval does not include zero, we conclude that the mean sales for policy A differs significantly from policy B at the 5% level of significance. Hence, if our objective is to maximize sales, pricing policy A is preferable. If there were no significant differences between policies A and B, we would then look at the costs of implementing them.

Let's investigate the effectiveness of blocking in two variables; sales volume and geographic location. For the row classification of sales volume, the calculated F-statistic (from Table 12-15) is

$$F_R = \frac{\text{MSR}}{\text{MSE}} = \frac{13.8825}{3.0258} = 4.588$$

Since the table value of F at the 5% level of significance is $F_{0.05,3,6} = 4.76$, which exceeds the calculated value of F_R, we cannot conclude that blocking through the classification of sales volume has been effective at this significance level. Similarly, to test the effectiveness of the column blocking variable, geographic location, we calculate the F-statistic as follows:

$$F_C = \frac{\text{MSC}}{\text{MSE}} = \frac{2.9892}{3.0258} = 0.988$$

The value of $F_C = 0.988 < F_{0.05,3,6} = 4.76$. So, we cannot conclude that blocking by means of geographic location has been effective.

For future experiments, either the completely randomized design or the randomized block design with sales volume class as the blocking variable could be investigated. If the Latin square design were to be run, we should select some alternative blocking variables, the inclusion of which would effectively reduce the error mean squares.

12-4 FACTORIAL EXPERIMENTS

When there are two or more factors, each at two or more levels, a treatment is defined as a combination of the levels of each factor. In a factorial experiment, all possible combinations of the factors (i.e., all treatments) are represented for each complete replication of the experiment. The number of treatments is equal to the product of the number of factor levels and can therefore become large when either the factors or the levels are numerous.

Let's consider the assembly of television sets. Our response variable is the assembly time and the factors are the experience of operators (six levels), the training program used (four levels), and the type of lighting used in the factory (three levels). A factorial experiment with these three factors would require 72 ($6 \times 4 \times 3$) treatments. Thus, a single replication of the treatments, with each treatment used only once, would require 72 trials, which could be cumbersome. Moreover, treatments usually need to be replicated because the purpose of factorial experiments is to ensure that any possible interaction effects between factors can be estimated. This further increases the total number of required trials. Note that the factorial set of treatments can be used with any of the previously discussed design types (the completely randomized design, randomized block design, or Latin square design).

Factorial experiments offer the ability to estimate the interaction effects between factors, which is not possible with the one-variable-at-a-time approach. In the presence of two or more factors, knowledge of **interaction effects** is of prime importance to the experimenter. When interaction between two factors is significant, the difference in the response variable to changes in the levels of one factor is not the same at all levels of the other factor. If the effect of both factors on the response variable is to be determined, interaction effects cannot be ignored. Interaction effects, if present, mask the **main effects** of the factors, so testing for the main effects if the interaction effects are significant is inappropriate. Only when the factors are independent and interaction is not present are the main effects of the factors of interest. In such cases, fewer trials are required to estimate the main effects to the same degree of precision. Usually, single-factor experiments require more trials to achieve the same degree of precision as factorial experiments.

Factorial experiments play an important role in exploratory studies to determine which factors are important. Additionally, by conducting experiments for a factor at several levels of other factors, the inferences from a factorial experiment are valid over a wide range of conditions. As noted earlier, a disadvantage of factorial experiments is the exponential increase in experiment size as the number of factors and/or their number of levels increase. The cost to conduct the experiments can become prohibitive. Another drawback is the feasibility of finding blocks large enough to complete a full replication of treatments, assuming that each treatment is used once in each block. Also, higher-order interactions need interpreting, and this can be complex.

We discuss the simplest type of factorial experiment, involving two factors, with a certain number of levels, for completely randomized and randomized block designs.

Two-Factor Factorial Experiment Using a Completely Randomized Design

Let the two factors be denoted by A and B, with factor A having a levels and B having b levels. The ab treatments are randomly assigned to the experimental units. Suppose the experiment is replicated n times, yielding a total of abn observations. This model is described by

$$y_{ijk} = \mu + \alpha_i + \beta_j + (\alpha\beta)_{ij} + \varepsilon_{ijk}$$
$$i = 1, 2, \ldots, a \quad j = 1, 2, \ldots, b \quad k = 1, 2, \ldots, n \qquad (12\text{-}24)$$

where y_{ijk} represents the response for the ith level of factor A, the jth level of factor B, and the kth replication; μ is the overall mean effect; α_i is the effect of the ith level of factor A; β_j is the effect of the jth level of factor B; $(\alpha\beta)_{ij}$ is the effect of the interaction between factors A and B; and ε_{ijk} is the random error component (assumed to be normally distributed with mean zero and constant variance σ^2). The following notation is used:

$y_{i..}$: $\sum_{j=1}^{b} \sum_{k=1}^{n} y_{ijk} = $ sum of the responses for the ith level of factor A

$\bar{y}_{i..}$: $y_{i..}/(bn) = $ average response for the ith level of factor A

$y_{.j.}$: $\sum_{i=1}^{a} \sum_{k=1}^{n} y_{ijk} = $ sum of the responses for the jth level of factor B

$\bar{y}_{.j.}$: $y_{.j.}/an = $ average response for the jth level of factor B

$y_{ij.}$: $\sum_{k=1}^{n} y_{ijk} = $ sum of the responses for the ith level of factor A and jth level of factor B

$\bar{y}_{ij.}$: $y_{ij.}/n = $ average response for the ith level of factor A and jth level of factor B

$y_{...}$: $\sum_{i=1}^{a} \sum_{j=1}^{b} \sum_{k=1}^{n} y_{ijk} = $ grand total of all observations

$\bar{y}_{...}$: $y_{...}/abn = $ grand mean of all observations

The notation is shown in Table 12-16.

Computations of the sum of squares needed for the analysis of variance are as follows. The correction factor is calculated as

$$C = \frac{y_{...}^2}{abn} \qquad (12\text{-}25)$$

TABLE 12-16 Notation for Factorial Experiments Using a Completely Randomized Design

Factor A	Factor B 1	2	...	b	Sum	Average
1	$y_{111}, y_{112}, \ldots, y_{11n}$ Sum $= y_{11\cdot}$	$y_{121}, y_{122}, \ldots, y_{12n}$ Sum $= y_{12\cdot}$	$\cdots$	$y_{1b1}, y_{1b2}, \ldots, y_{1bn}$ Sum $= y_{1b\cdot}$	$y_{1\cdot\cdot}$	$\bar{y}_{1\cdot\cdot}$
2	$y_{211}, y_{212}, \ldots, y_{21n}$ Sum $= y_{21\cdot}$	$y_{221}, y_{222}, \ldots, y_{22n}$ Sum $= y_{22\cdot}$	$\cdots$	$y_{2b1}, y_{2b2}, \ldots, y_{2bn}$ Sum $= y_{2b\cdot}$	$y_{2\cdot\cdot}$	$\bar{y}_{2\cdot\cdot}$
$\vdots$	$\vdots$	$\vdots$	$\vdots$	$\vdots$	$\vdots$	$\vdots$
a	$y_{a11}, y_{a12}, \ldots, y_{a1n}$ Sum $= y_{a1\cdot}$	$y_{a21}, y_{a22}, \ldots, y_{a2n}$ Sum $= y_{a2\cdot}$	$\cdots$	$y_{ab1}, y_{ab2}, \ldots, y_{abn}$ Sum $= y_{ab\cdot}$	$y_{a\cdot\cdot}$	$\bar{y}_{a\cdot\cdot}$
Sum	$y_{\cdot1\cdot}$	$y_{\cdot2\cdot}$	$\cdots$	$y_{\cdot b\cdot}$	$y_{\cdots}$	
Average	$\bar{y}_{\cdot1\cdot}$	$\bar{y}_{\cdot2\cdot}$	$\cdots$	$\bar{y}_{\cdot b\cdot}$		$\bar{y}_{\cdots}$

The total sum of squares is given by

$$\text{SST} = \sum_{i=1}^{a} \sum_{j=1}^{b} \sum_{k=1}^{n} y_{ijk}^2 - C \tag{12-26}$$

The sum of squares for the main effects of factor A is found from

$$\text{SSA} = \frac{\sum_{i=1}^{a} y_{i\cdot\cdot}^2}{bn} - C \tag{12-27}$$

The sum of squares for the main effects of factor B is found from

$$\text{SSB} = \frac{\sum_{j=1}^{b} y_{\cdot j\cdot}^2}{an} - C \tag{12-28}$$

The sum of squares for the subtotals between the cell totals is given by

$$\text{SS}_{\text{subtotal}} = \frac{\sum_{i=1}^{a} \sum_{j=1}^{b} y_{ij\cdot}^2}{n} - C \tag{12-29}$$

The sum of squares due to the interaction between factors A and B is given by

$$\text{SSAB} = \text{SS}_{\text{subtotal}} - \text{SSA} - \text{SSB} \tag{12-30}$$

Finally, the error sum of squares is

$$\text{SSE} = \text{SST} - \text{SS}_{\text{subtotal}} \tag{12-31}$$

There are $a - 1$ degrees of freedom for factor A, 1 less than number of levels of A. Similarly, there are $b - 1$ degrees of freedom for factor B, 1 less the number of levels of B. The number of degrees of freedom for the interaction between factors A and B is the product of the number of degrees of freedom for each—that is, $(a - 1)(b - 1)$. Since the total number of degrees of freedom *is* $abn - 1$, by subtraction, the number of degrees of freedom for the experimental error is $ab(n - 1)$. The mean squares for each category are found by dividing the sum of squares by the corresponding number of degrees of freedom. To test for the significance of the factors and the interaction, the F-statistic is calculated. Table 12-17 shows the ANOVA table.

TABLE 12-17   ANOVA Table for the Two-Factor Factorial Experiment Using a Completely Randomized Design

Source of Variation	Degrees of Freedom	Sum of Squares	Mean Square	F-statisic
Factor A	$a - 1$	SSA	$MSA = \frac{SSA}{a-1}$	$F_A = \frac{MSA}{MSE}$
Factor B	$b - 1$	SSB	$MSB = \frac{SSB}{b-1}$	$F_B = \frac{MSB}{MSE}$
Interaction between A and B	$(a - 1)(b - 1)$	SSAB	$MSAB = \frac{SSAB}{(a-1)(b-1)}$	$F_{AB} = \frac{MSAB}{MSE}$
Error	$ab(n - 1)$	SSE	$MSE = \frac{SSE}{ab(n-1)}$	
Total	$abn - 1$	SST		

Tests for Significance of Factors   In a factorial experiment, the first set of hypotheses to be tested usually deals with the interaction between factors. The hypothesis of interest is thus

H_0 : $(\alpha\beta)_{ij} = 0$ for all i, j; $i = 1, 2, \ldots, a$; $j = 1, 2, \ldots, b$
H_a :   At least one $(\alpha\beta)_{ij}$ is different from zero

The test statistic is the F-statistic, which is shown in Table 12-17 and is calculated as

$$F_{AB} = \frac{MSAB}{MSE} \tag{12-32}$$

which is the ratio of the mean square due to interaction to the mean square error. If the calculated value of $F_{AB} > F_{\alpha,(a-1)(b-1),ab(n-1)}$, the null hypothesis is rejected, and we conclude that the interactions between factors are significant at the level α.

The presence of significant interaction effects masks the main effects of the factors, in which case the main effects are usually not tested. When interaction exists, the value of the response variable due to changes in factor A depends on the level of factor B. In such a situation, the mean for a level of factor A, averaged over all levels of factor B, does not have any practical significance. What is of interest under these circumstances is the treatment means along with their standard deviations. A plot of the treatment means may identify the preferable levels of the factors.

The standard deviation of the treatment mean, which is the mean of any combination of factors A and B, is given by

$$s_{\bar{y}(AB)} = \sqrt{\frac{MSE}{n}} \tag{12-33}$$

Thus, a $100(1 - \alpha)\%$ confidence interval for a treatment mean when factor A is at level i and factor B is at level j is given by

$$\bar{y}_{ij.} \pm t_{\alpha/2,ab(n-1)} \sqrt{\frac{MSE}{n}} \tag{12-34}$$

where $t_{\alpha/2,ab(n-1)}$ is the t-value found from Appendix A-4 for a right-tail area of $\alpha/2$ and $ab(n - 1)$ degrees of freedom. For inferences on the difference d of two treatment means, the standard deviation is

$$s_{d(AB)} = \sqrt{\frac{2MSE}{n}} \tag{12-35}$$

Let's consider the treatments for which factor A is at level i and factor B is at level j, which yields a sample average of $\bar{y}_{ij.}$. The other treatment has factor A at level i' and factor B at level j', the sample average being $\bar{y}_{i'j'.}$. A $100(1-\alpha)\%$ confidence interval for the difference between the two treatment means is given by

$$(\bar{y}_{ij.} - \bar{y}_{i'j'.}) \pm t_{\alpha/2,ab(n-1)} \sqrt{\frac{2MSE}{n}} \tag{12-36}$$

When the interaction effects are not significant, we can test for the effects of the two factors. To test whether factor A is significant, the F-statistic, as shown in Table 12-17 is calculated as the ratio of the mean squares due to factor A to the mean squares for error, that is

$$F_A = \frac{MSA}{MSE} \tag{12-37}$$

This calculated value of F_A is then compared to $F_{\alpha,a-1,ab(n-1)}$, the critical value of F found from Appendix A-6 for a right-tail area of α, $(a-1)$ numerator degrees of freedom and $ab(n-1)$ denominator degrees of freedom. If $F_A > F_{\alpha,a-1,ab(n-1)}$, we conclude that the factor means at the different levels of A are not all equal.

A $100(1-\alpha)\%$ confidence interval for the mean of factor A when at level i is given by

$$\bar{y}_{i..} \pm t_{\alpha/2,ab(n-1)} \sqrt{\frac{MSE}{bn}} \tag{12-38}$$

Also, a $100(1-\alpha)\%$ confidence interval for the difference between the factor A means that are at levels i and i' is found from

$$(\bar{y}_{i..} - \bar{y}_{i'..}) \pm t_{\alpha/2,ab(n-1)} \sqrt{\frac{2MSE}{bn}} \tag{12-39}$$

The inferences described for factor A can be applied to factor B as well. To determine the significance of factor B—as to whether the means at all levels are equal—the test statistic is

$$F_B = \frac{MSB}{MSE} \tag{12-40}$$

If $F_B > F_{\alpha,b-1,ab(n-1)}$, the means of factor B are not all equal for the different levels. A $100(1-\alpha)\%$ confidence interval for the mean of factor B when at level j is given by

$$\bar{y}_{.j.} \pm t_{\alpha/2,ab(n-1)} \sqrt{\frac{MSE}{an}} \tag{12-41}$$

Similarly, a $100(1-\alpha)\%$ confidence interval for the difference between two factor B means that are at levels j and j' is determined from

$$(\bar{y}_{.j.} - \bar{y}_{.j'.}) \pm t_{\alpha/2,ab(n-1)} \sqrt{\frac{2MSE}{an}} \tag{12-42}$$

Two-Factor Factorial Experiment Using a Randomized Block Design

The randomized block design has been illustrated for the one-factor case. The same principles apply to the two-factor case. A unique feature, however, is the potential interaction between the two factors.

Suppose that we have two factors, A and B; factor A has a levels, and factor B has b levels. Let the number of blocks be r, with each block containing ab units. The ab treatments are randomly assigned to the ab units within each block. The units within each block are chosen to be homogeneous with respect to the blocking variable. We want to determine the impact of two factors, temperature and pressure, on the viscosity of a compound. The possible levels of temperature are $75°$, $150°$, $200°$ and $250°$C, and the levels of pressure are 50, 75, 100, 125, and $150 \, kg/cm^2$. Because the batches of incoming material vary, the batch is selected as the blocking variable, and we select 10 batches. Within each batch, we randomly apply the 20 treatments to selected units. Factor A denotes temperature and factor B denotes pressure, so we have $a = 4$, $b = 5$, $r = 10$.

The model for the two-factor factorial experiment using a randomized block design is

$$y_{ijk} = \mu + \alpha_i + \beta_j + \rho_k + (\alpha\beta)_{ij} + \varepsilon_{ijk}$$
$$i = 1, 2, \ldots, a \quad j = 1, 2, \ldots, b \quad k = 1, 2, \ldots, r \tag{12-43}$$

where y_{ijk} represents the response for the ith level of factor A, the jth level of factor B, and the kth block; μ is the overall mean effect; α_i is the effect of the ith level of factor A; β_j is the effect of the jth level of factor B; and $(\alpha\beta)_{ij}$ is the interaction effect between factors A and B. The quantity ρ_k represents the effect of block k, and ε_{ijk} represents the random error component, assumed to be distributed normally with mean zero and constant variance σ^2. We use the following notation:

$y_{i..}$: $\sum_{j=1}^{b} \sum_{k=1}^{r} y_{ijk}$ = sum of the responses for the ith level of factor A

$\bar{y}_{i..}$: $y_{i..}/(br)$ = average response for the ith level of factor A

$y_{.j.}$: $\sum_{i=1}^{a} \sum_{k=1}^{r} y_{ijk}$ = sum of the responses for the jth level of factor B

$\bar{y}_{.j.}$: $y_{.j.}/(ar)$ = average response of the jth level of factor B

$y_{ij.}$: $\sum_{k=1}^{r} y_{ijk}$ = sum of the responses for the treatment if factor A is at the ith level and factor B is at the jth level over the blocks

$\bar{y}_{ij.}$: $y_{ij.}/r$ = average response for the treatment consisting of the ith level of factor A and jth level of factor B over the blocks

$y_{..k}$: $\sum_{i=1}^{a} \sum_{j=1}^{b} y_{ijk}$ = sum of the responses for the kth block

$y_{...}$: $\sum_{i=1}^{a} \sum_{j=1}^{b} \sum_{k=1}^{r} y_{ijk}$ = grand total of all observations

The computations for the sum of squares are similar to those for the one-factor case, except that we have an additional factor and that factor interactions need to be considered. The correction factor is

$$C = \frac{y_{...}^2}{abr} \tag{12-44}$$

The total sum of squares is

$$SST = \sum_{i=1}^{a} \sum_{j=1}^{b} \sum_{k=1}^{r} y_{ijk}^2 - C \tag{12-45}$$

The sum of squares due to the blocks is

$$\text{SSBL} = \frac{\sum\limits_{k=1}^{r} y_{\cdot\cdot k}^2}{ab} - C \tag{12-46}$$

The sum of squares for the main effects of factor A is

$$\text{SSA} = \frac{\sum\limits_{i=1}^{a} y_{i\cdot\cdot}^2}{br} - C \tag{12-47}$$

The sum of squares for the main effects of factor B is

$$\text{SSB} = \frac{\sum\limits_{j=1}^{b} y_{\cdot j\cdot}^2}{ar} - C \tag{12-48}$$

The sum of squares for the interaction between factors A and B is given by

$$\text{SSAB} = \frac{\sum\limits_{i=1}^{a}\sum\limits_{j=1}^{b} y_{ij\cdot}^2}{r} - C - \text{SSA} - \text{SSB} \tag{12-49}$$

Finally, the error sum of squares is

$$\text{SSE} = \text{SST} - \text{SSBL} - \text{SSA} - \text{SSB} - \text{SSAB} \tag{12-50}$$

The number of degrees of freedom is $a - 1$ for factor A, $b - 1$ for factor B, $r - 1$ for the blocks, and $(a - 1)(b - 1)$ for the interaction between factors A and B. Since the total number of degrees of freedom is $abr - 1$, 1 less than the total number of observations, the number of error degrees of freedom upon subtraction is $(ab - 1)(r - 1)$. The complete ANOVA table is shown in Table 12-18.

Tests for Significance of Factors   As previously mentioned for factorial experiments, the significance of interaction effects between the factors is usually tested first. The test statistic, as shown in Table 12-18, is the F-statistic calculated as the ratio of the mean square due to

TABLE 12-18   ANOVA Table for the Two-Factor Factorial Experiment Using a Randomized Block Design

Source of Variation	Degrees of Freedom	Sum of Squares	Mean Sqaure	F-Statistic
FactorA	$a - 1$	SSA	$\text{MSA} = \dfrac{\text{SSA}}{a-1}$	$F_A = \dfrac{\text{MSA}}{\text{MSE}}$
Factor B	$b - 1$	SSB	$\text{MSB} = \dfrac{\text{SSB}}{b-1}$	$F_B = \dfrac{\text{MSB}}{\text{MSE}}$
Interaction between A and B	$(a - 1)(b - 1)$	SSAB	$\text{MSAB} = \dfrac{\text{SSAB}}{(a-1)(b-1)}$	$F_{AB} = \dfrac{\text{MSAB}}{\text{MSE}}$
Blocks	$r - 1$	SSBL	$\text{MSBL} = \dfrac{\text{SSBL}}{r-1}$	$F_{BL} = \dfrac{\text{MSBL}}{\text{MSE}}$
Error	$(ab - 1)(r - 1)$	SSE	$\text{MSE} = \dfrac{\text{SSE}}{(ab-1)(r-1)}$	
Total	$abr - 1$	SST		

interaction to the mean square error, that is,

$$F_{AB} = \frac{MSAB}{MSE}$$

If $F_{AB} > F_{\alpha, (a-1)(b-1), (ab-1)(r-1)}$, which is the table value found from Appendix A-6 for a level of significance α and $(a-1)(b-1)$ degrees of freedom in the numerator and $(ab-1)(r-1)$ in the denominator, we conclude that the interaction effects between factors A and B are significant. Since the individual factor mean effects are not important to test in the presence of interaction, a $100(1-\alpha)\%$ confidence interval for the treatment mean is found from

$$\bar{y}_{ij\cdot} \pm t_{\alpha/2, (ab-1)(r-1)} \sqrt{\frac{MSE}{r}} \tag{12-51}$$

where $t_{\alpha/2, (ab-1)(r-1)}$ is the t-value found from Appendix A-4 for a right-tail area of $\alpha/2$ and $(ab-1)(r-1)$ degrees of freedom.

Inferences about the difference between two treatment means through confidence intervals with a level of confidence of $(1-\alpha)$ may be found as follows:

$$(\bar{y}_{ij\cdot} - \bar{y}_{i'j'\cdot}) \pm t_{\alpha/2, (ab-1)(r-1)} \sqrt{\frac{2MSE}{r}} \tag{12-52}$$

When the interaction effects are not significant, tests for the individual factor means can be conducted. The significance of the means of factor A is tested by calculating the F-statistic, shown in Table 12-18

$$F_A = \frac{MSA}{MSE}$$

At a level of significance α, if $F_A > F_{\alpha,(a-1),(ab-1)(r-1)}$, we conclude that the means of factor A are not all equal at the different levels.

In making an inference about an individual mean for a certain level of factor A, a $100(1-\alpha)\%$ confidence interval is given by

$$\bar{y}_{i\cdot\cdot} \pm t_{\alpha/2, (ab-1)(r-1)} \sqrt{\frac{MSE}{br}} \tag{12-53}$$

Similarly, a $100(1-\alpha)\%$ confidence interval for the difference between the factor A means, at levels i and i', is found from

$$(\bar{y}_{i\cdot\cdot} - \bar{y}_{i'\cdot\cdot}) \pm t_{\alpha/2, (ab-1)(r-1)} \sqrt{\frac{2MSE}{br}} \tag{12-54}$$

The inference-making procedure described for factor A can be applied in a similar manner to factor B. First, to determine the significance of factor B concerning whether the means at all levels are equal, the test statistic (as shown in Table 12-18) is

$$F_B = \frac{MSB}{MSE}$$

If $F_B > F_{\alpha,b-1,(ab-1)(r-1)}$, we conclude that the factor B means are not all equal at a level of significance α. As with factor A, a $100(1-\alpha)\%$ confidence interval for the mean of factor B at level j is given by

$$\bar{y}_{\cdot j \cdot} \pm t_{\alpha/2, (ab-1)(r-1)} \sqrt{\frac{MSE}{ar}} \tag{12-55}$$

A $100(1-\alpha)\%$ confidence interval for the difference between two factor B means that are at levels j and j' is found from

$$(\bar{y}_{\cdot j \cdot} - \bar{y}_{\cdot j' \cdot}) \pm t_{\alpha/2, (ab-1)(r-1)} \sqrt{\frac{2MSE}{ar}} \tag{12-56}$$

Example 12-4 In a study on the effectiveness of synthetic automobile fuels, two factors are of importance. Factor A is an additive that is to be tested at three levels of use in successively increasing amounts. Factor B is a catalyst, for which three levels of use (1, 2, and 3, representing successively increasing amounts) are also to be tested. Forty-five automobiles are randomly selected for the study, and each of the nine treatments is randomly used in five different automobiles. Table 12-19 shows the efficiency ratings, in percentage, of the treatments.

(a) At a level of significance of 5%, what can we conclude about the significance of the factors

Solution The problem involves two factors using a completely randomized design. Using the raw data a given in Table 12-19, the summary information shown in Table 12-20 is computed.

We conduct the analysis of variance using Minitab by selecting **Stat > ANOVA > Balanced ANOVA**. Under **Model**, input Additive Catalyst Additive * Catalyst, indicating that the main effects and interaction effects should be considered. Data are entered by listing

TABLE 12-19   Efficiency Ratings, in Percentage, of a Synthetic Fuel Involving an Additive and a Catalyst

Additive Level	Catalyst Level		
	1	2	3
I	75	64	43
	72	62	48
	66	58	42
	74	56	46
	65	60	46
II	50	73	58
	58	70	62
	46	67	54
	53	68	60
	55	70	62
III	67	82	44
	60	76	43
	60	80	38
	62	84	42
	63	75	44

TABLE 12-20 Summary Information for Synthetic Fuel Data

Additive Level	Catalyst Level			Sum	Average
	1	2	3		
I	Sum $(y_{11.}) = 352$ Average $(\bar{y}_{11.}) = 70.4$	Sum $(y_{12.}) = 300$ Average $(\bar{y}_{12.}) = 60$	Sum $(y_{13.}) = 225$ Average $(\bar{y}_{13.}) = 45$	$y_{1..} = 877$	$\bar{y}_{1..} = 58.467$
II	Sum $(y_{21.}) = 262$ Average $(\bar{y}_{21.}) = 52.4$	Sum $(y_{22.}) = 348$ Average $(\bar{y}_{22.}) = 69.6$	Sum $(y_{23.}) = 296$ Average $(\bar{y}_{23.}) = 59.2$	$y_{2..} = 906$	$\bar{y}_{2..} = 60.400$
III	Sum $(y_{31.}) = 312$ Average $(\bar{y}_{31.}) = 62.4$	Sum $(y_{32.}) = 397$ Average $(\bar{y}_{32.}) = 79.4$	Sum $(y_{33.}) = 211$ Average $(\bar{y}_{33.}) = 42.2$	$y_{3..} = 920$	$\bar{y}_{3..} = 61.333$
Sum	$y_{.1.} = 926$	$y_{.2.} = 1045$	$y_{.3.} = 732$	$y_{...} = 2703$	
Average	$\bar{y}_{.1.} = 61.733$	$\bar{y}_{.2.} = 69.667$	$\bar{y}_{.3.} = 48.800$		$\bar{y}_{...} = 60.067$

the value of the response variable (efficiency rating) in one column, the corresponding first factor (additive level) in a second column, and the second factor (catalyst level) in a third column. Note that there are five replications at each factor combination. The Minitab ANOVA table is shown in Figure 12-8.

The value of F for testing the significance of the interaction effects between additive and catalyst is shown in Figure 12-8 to be 54.33, with the p-value as .000. Since the p-value is less than .05, the chosen level of significance, we reject the null hypothesis and conclude that the interaction effects between additive and catalyst are significant. Although Figure 12-8 shows the values of F for testing the significance of the additive (factor A) means and the catalyst (factor B) means, they are not necessary because the interaction effect is significant.

(b) What level of additive and the level of catalyst should be used?

Solution Since we found the interaction effects to be significant in part a, we need to take into account the joint effect of both factors on the efficiency rating. Let's consider the mean efficiency for different levels of the catalyst at each level of the additive. We use the summary information given in Table 12-20 to construct the plot shown in Figure 12-9 where we find that the mean efficiency does not change uniformly with catalyst levels for each additive level. For additive levels 2 and 3, as the catalyst level increases from 1 to 2, the mean

---

Analysis of Variance (Balanced Designs)

Analysis of Variance for Efficiency

Source	DF	SS	MS	F	P
Additive	2	64.13	32.07	2.76	0.076
Catalyst	2	3328.13	1664.07	143.45	0.000
Additive*Catalyst	4	2520.93	630.23	54.33	0.000
Error	36	417.60	11.60		
Total	44	6330.80			

FIGURE 12-8 Minitab's ANOVA table for the synthetic fuel data using a completely randomized design.

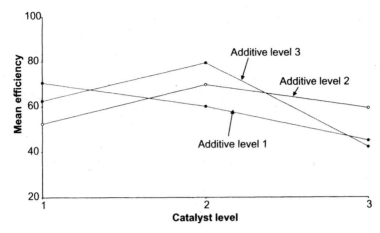

FIGURE 12-9   Plot of treatment means for the synthetic fuel data.

efficiency increases, but as the catalyst level increases from 2 to 3, the mean efficiency decreases, and the decline is more drastic for additive level 3. For additive level 1, the mean efficiency decreases as the catalyst level increases from 1 to 2 and then to 3. If maximizing efficiency is our objective, we will choose additive level 3 and catalyst level 2.

(c) Is there a difference between the means of catalyst levels 1 and 3 at a 5% level of significance?

Solution   We find a 95% confidence interval for the difference between the mean efficiencies of catalyst levels 1 and 3. This interval is given by

$$(\bar{y}_{.1.} - \bar{y}_{.3.}) \pm t_{0.025,36} \sqrt{\frac{2\text{MSE}}{(3)(5)}}$$

$$= (61.733 - 48.80) \pm (2.028) \sqrt{\frac{(2)(11.6)}{15}}$$

$$= 12.933 \pm 2.522$$

Observe that the confidence interval does not include zero, indicating a difference between the means at this significance level. Note, however, as discussed in parts a and b, that because the interaction effect is significant, we cannot consider one factor effect separate from the other.

(d) Find a 95% confidence interval for the difference between the treatment means for additive level 1, catalyst level 1 and for additive level 2, catalyst level 3.

Solution   We have

$$(\bar{y}_{11.} - \bar{y}_{23.}) \pm t_{0.025,36} \sqrt{\frac{2\text{MSE}}{5}}$$

$$= (70.4 - 49.2) \pm (2.028) \sqrt{\frac{(2)(11.6)}{5}} = 11.2 \pm 4.368$$

Since this confidence interval does not include zero, we conclude that there is a difference between these two treatment means at the 5% level of significance.

Role of Contrasts

In factorial experiments involving multiple treatments, we often want to test hypotheses on a linear combination of the treatment means or to partition the total sum of squares of all treatments into component sums of squares of certain desirable treatments. A contrast can be helpful in such circumstances. A **contrast** of treatment means is a linear combination of the means such that the sum of the coefficients of the linear combination equals zero. Suppose that we have p treatments with means $\mu_1, \mu_2, \ldots, \mu_p$. Then, a contrast of means is defined as

$$L = k_1\mu_1 + k_2\mu_2 + \ldots + k_p\mu_p$$
$$= \sum_{i=1}^{p} k_i\mu_i \tag{12-57}$$

where k_i are constants such that $\sum_{i=1}^{p} k_i = 0$

Let's consider an example in which four selling methods for computer salespeople are of interest. Let the mean sales for each method be denoted by μ_1, μ_2, μ_3 and μ_4. We desire to test the null hypothesis that the average sales of people using the first two methods is the same as the average sales of people using the last two methods. The null hypothesis is

$$H_o: \quad L = 0$$

where

$$L = \frac{\mu_1 + \mu_2}{2} - \frac{\mu_3 + \mu_4}{2}$$

The quantity L is a contrast, its coefficients are given by $k_1 = \frac{1}{2}, k_2 = \frac{1}{2}, k_3 = -\frac{1}{2}$, and $k_4 = -\frac{1}{2}$. Note that the sum of the coefficients equals zero. To test the null hypothesis, we need a sample estimate of L and the sample variance of the estimate of L. The sample estimate of L is given by

$$\hat{L} = k_1\bar{y}_1 + k_2\bar{y}_2 + \cdots + k_p\bar{y}_p = \sum_{i=1}^{p} k_i\bar{y}_i \tag{12-58}$$

where $\bar{y}_i$ is the sample mean of treatment i. The variance of $\hat{L}$ is found from

$$\text{Var}(\hat{L}) = \left(\frac{k_1^2}{r_1} + \frac{k_2^2}{r_2} + \cdots + \frac{k_p^2}{r_p}\right)\text{MSE} = \left(\sum_{i=1}^{p} \frac{k_i^2}{r_i}\right)\text{MSE} \tag{12-59}$$

where r_i represents the number of observations from which the ith sample mean is computed and MSE is the mean square error for the design used in the experiment. The test statistic, then, is the t-statistic given by

$$t = \frac{\hat{L}}{\sqrt{\text{Var}(\hat{L})}} \tag{12-60}$$

At a level of significance α, if $|t| > t_{\alpha/2,\nu}$ where ν is the number of degrees of freedom of the experimental error and $\alpha/2$ is the right-tail area of the t-distribution, the null hypothesis is rejected. If the null hypothesis is rejected, we may need to find an interval estimate for the contrast L. A 100 $(1 - \alpha)$% confidence interval for L is given by

$$\hat{L} \pm t_{\alpha/2, \nu} \sqrt{\mathrm{Var}(\hat{L})} \qquad (12\text{-}61)$$

where $t_{\alpha/2, \nu}$ represents the t-value, found from Appendix A-4, for a right-tail area of $\alpha/2$ and ν degrees of freedom.

Example 12-5   Table 12-21 shows the total computer sales, in thousands of dollars, of salespeople using five different methods. The number of people chosen from each method is also indicated. A completely randomized design is used. The mean square error is 30.71.

(a) Is the mean sales using method 2 different from the average of the mean sales of methods 1 and 3? Test at a significance level of 5%.

Solution   The total number of experimental units is 21, yielding a total of 20 degrees of freedom. Since there are five selling methods (treatments), there are 4 degrees of freedom for the treatments. Upon subtraction, the number of degrees of freedom for the error is 16.

If μ_i represents the mean sales using method i, the null hypothesis to be tested is

$$H_o : \mu_2 - \frac{\mu_1 + \mu_3}{2} = 0$$

The corresponding contrast L can be written as $L = 2\mu_2 - \mu_1 - \mu_3$. Note that L is a contrast, with its estimate $\hat{L}$ obtained as $\hat{L} = 2\bar{y}_2 - \bar{y}_1 - \bar{y}_3$, where $\bar{y}_i$ represents the sample mean for method i. We have

$$\hat{L} = 2\left(\frac{239}{4}\right) - \frac{80}{3} - \frac{255}{5} = 119.5 - 26.667 - 51.0 = 41.833$$

The variance of $\hat{L}$ is found from eq. (12-59) as

$$\mathrm{Var}(\hat{L}) = \left[\frac{(2)^2}{4} + \frac{(-1)^2}{3} + \frac{(-1)^2}{5}\right](30.71) = 47.078$$

For testing the specified null hypothesis, the test statistic given by eq. (12-60) is

$$t = \frac{\hat{L}}{\sqrt{\mathrm{Var}(\hat{L})}} = \frac{41.833}{\sqrt{47.078}} = 6.097$$

TABLE 12-21   Computer Sales for Five Different Methods (thousands)

	Selling Method				
	1	2	3	4	5
Number of salespeople	3	4	5	4	5
Total sales	80	239	255	249	367

At the 5% level of significance, with 16 error degrees of freedom, the t tables in Appendix A-4 give $t_{.025,16} = 2.120$. The test statistic of 6.097 is greater than 2.120, so we reject the null hypothesis.

Thus, the mean sales using method 2 is different from the average of the mean sales using methods 1 and 3.

(b) Find a 95% confidence interval for the contrast in part (a).

Solution   A 95% confidence interval for $L = 2\mu_2 - \mu_1 - \mu_3$ is given by

$$\hat{L} \pm t_{0.025,16} \sqrt{\text{Var}(\hat{L})} = 41.833 \pm (2.120)\sqrt{47.078}$$
$$= 41.833 \pm 14.546 = (27.287, 56.379)$$

(c) Can we conclude that the average of the mean sales using methods 4 and 5 is different from that using methods 2 and 3? Test at the 1% level of significance.

Solution   The null hypothesis to be tested is

$$H_0: \quad \frac{\mu_4 + \mu_5}{2} - \frac{\mu_2 + \mu_3}{2} = 0$$

The corresponding contrast L is given by $L = \mu_4 + \mu_5 - \mu_2 - \mu_3$.

The estimate of L is

$$\hat{L} = \bar{y}_4 + \bar{y}_5 - \bar{y}_2 - \bar{y}_3$$

$$= \frac{249}{4} + \frac{367}{5} - \frac{239}{4} - \frac{255}{5} = 62.25 + 73.4 - 59.75 - 51.0 = 24.9$$

The variance of $\hat{L}$ is given by

$$\text{Var}(\hat{L}) = \left[\frac{(1)^2}{4} + \frac{(1)^2}{5} + \frac{(-1)^2}{4} + \frac{(-1)^2}{5} \right](30.71) = 27.639$$

The test statistic is $t = \dfrac{\hat{L}}{\sqrt{\text{Var}(\hat{L})}} = \dfrac{24.9}{\sqrt{27.639}} = 4.736$.

For a two-tailed test at a 1% level of significance, the t-value for 16 degrees of freedom, from Appendix A-4, is $t_{0.005,16} = 2.921$. Since the test statistic of 4.736 is greater than 2.921, we reject the null hypothesis and conclude that the average of the mean sales using methods 4 and 5 is not equal to that using methods 2 and 3.

Orthogonal Contrasts   For a given set of treatments, it is possible to form more than one meaningful contrast. Given p treatments, we can construct $p - 1$ contrasts that are statistically independent of each other. Two contrasts L_1 and L_2 are independent if they are orthogonal to each other. Suppose, we have the two following contrasts:

$$L_1 = k_{11}\mu_1 + k_{12}\mu_2 + \cdots + k_{1p}\mu_p$$
$$L_2 = k_{21}\mu_1 + k_{22}\mu_2 + \cdots + k_{2p}\mu_p$$

The contrasts L_1 and L_2 are orthogonal if and only if

$$\sum_{j=1}^{p} k_{1j}k_{2j} = 0$$

that is, if the sum of the products of the corresponding coefficients associated with the treatment means is zero.

To illustrate, let's consider the following two contrasts involving three treatment means:

$$L_1 = \mu_1 - \mu_3$$
$$L_2 = -\mu_1 + 2\mu_2 - \mu_3$$

Thus, L_1 is used to test the null hypothesis of the equality of treatment means 1 and 3, and L_2 is used to test the null hypothesis that the mean of treatments 1 and 3 equals the mean of treatment 2. Note that the contrasts L_1 and L_2 are orthogonal, because the sum of the products of the corresponding coefficients of the means is $1(-1) + 0(2) + (-1)(-1) = 0$.

Contrasts of Totals Similar to contrasts of means, contrasts of totals are also useful in analysis of variance. Such contrasts aid in partitioning the treatment sum of squares into components such that each component is associated with a contrast of interest. Let r_j represent the number of replications for the jth treatment (out of p treatments). Let

$$T_i = y_{i \cdot} = \sum_{j=1}^{r_i} y_{ij}$$

be the sum of the responses of the ith treatment. Then $\hat{L}$ is a contrast of totals given by

$$\hat{L} = k_1 T_1 + k_2 T_2 + \cdots + k_p T_p \tag{12-62}$$

if and only if $r_1 k_1 + r_2 k_2 + \cdots + r_p k_p = 0$. If the number of replications is given by r for each of the treatments, $\hat{L}$ is a contrast if

$$\sum_{i=1}^{p} k_i = 0$$

The variance of the contrast of totals is obtained from

$$\mathrm{Var}(\hat{L}) = (r_1 k_1^2 + r_2 k_2^2 + \cdots + r_p k_p^2)\mathrm{MSE} \tag{12-63}$$

where MSE is the mean square error for the experimental design considered. Two contrasts of totals given by

$$\hat{L}_1 = k_{11}T_1 + k_{12}T_2 + \cdots + k_{1p}T_p$$
$$\hat{L}_2 = k_{21}T_1 + k_{22}T_2 + \cdots + k_{2p}T_p$$

are orthogonal if and only if $\sum_{j=1}^{p} k_{1j}k_{2j} = 0$.

This condition is similar to the condition for the contrasts for means.

Orthogonal contrasts for totals are important because they can be used to decompose the treatment sum of squares in analysis of variance and thereby test hypotheses on the effectiveness of selected treatments. If L_1 is a contrast of totals, then a single degree-of-freedom component for that contrast of the treatment sum of squares (SSTR) is

$$S_1^2 = \frac{\hat{L}_1^2}{D_1} \tag{12-64}$$

where $D_1 = r_1 k_{11}^2 + r_2 k_{12}^2 + \cdots r_p k_{1p}^2 = \sum_{j=1}^{p} r_j k_{1j}^2$.

Similarly if $\hat{L}_1$ and $\hat{L}_2$ are orthogonal, the sum of squares for the contrast $\hat{L}_2$ is given by $S_2^2 = \hat{L}_2^2/D_2$, where $D_2 = r_1 k_{21}^2 + r_2 k_{22}^2 + \cdots + r_p k_{2p}^2$. In fact, S_2^2 is a component of $SSTR - S_1^2$.

Similarly, if $L_1, L_2, \ldots, L_{p-1}$ are mutually orthogonal contrasts of totals, then

$$S_1^2 + S_2^2 + \cdots + S_{p-1}^2 = SSTR$$

Example 12-6   A regional bank is considering various policies to stimulate the local economy and promote home improvement loans. One factor is the interest on the mortgage loan, which the bank has chosen to be of two types: fixed rate or variable rate. Another factor is the payback period, which the bank has selected to be 10 or 20 years. For each of the four treatments, the bank has randomly selected four loan amounts.

The four treatments are represented by the notation shown in Table 12-22. In factorial experiments in which each factor has two levels–low and high–a factor at the high level is denoted by a lowercase letter. A factor at the low level is indicated by the absence of this letter. Additionally, the notation (1) denotes the treatment where all factors are at the low level. Thus, Table 12-23 shows treatment i that indicates that the interest type is at the high level (variable rate) and the payback period is at the low level (10 years). Table 12-23 shows the loan amounts (in thousands of dollars).

We will use a set of three orthogonal contrasts of totals, the coefficients of which are shown in Table 12-24 . Note that capital letters denote contrasts, whereas lowercase letters denote treatments. Contrast I compares the totals for the treatments when the interest rate is variable

TABLE 12-22   Notation for Treatments Comprising the Interest Rate and Payback Period

Treatment	Type of Interest Rate and Payback Period
(1)	Fixed-rate interest and 10-year payback period
i	Variable-rate interest and 10-year payback period
p	Fixed-rate interest and 20-year payback period
ip	Variable-rate interest and 20-year payback period

TABLE 12-23   Loan Amounts for Types of Interest Rate and Payback Period (hundreds)

	Treatment		
(1)	i	p	ip
16	22	38	56
12	18	45	58
14	25	40	54
12	20	42	59
54	85	165	227

TABLE 12-24   Coefficients of a Set of Three Orthogonal Contrasts of Totals for the Bank Loan Amounts

	Treatment			
Contrast	(1)	i	p	ip
I	−1	+1	−1	+1
P	−1	−1	+1	+1
IP	+1	−1	−1	+1

versus when it is fixed. Using Table 12-24 and eq. (12-62), contrast I is given by

$$\text{constrast } I = -T_1 + T_2 - T_3 + T_4$$

where $T_1 = $ Sum for treatment (1)

   $T_2 = $ Sum for treatment i

   $T_3 = $ Sum for treatment p

   $T_4 = $ Sum for treatment ip

In experimental design notation, this same relation is expressed as

$$I = (-1) + i - p + ip$$

where the quantities on the right-hand side of the equation represent the sum of the response variables for the corresponding treatments. The other two contrasts can be similarly expressed using Table (12-24). Thus, we have

$$\text{contrast } P = (-1) - i + p + ip$$

which compares the totals for treatments when the payment period is 20 years versus when it is 10 years. We also have

$$\text{contrast } IP = (1) - i - p + ip$$

which compares two sets of totals. The first set includes treatments for fixed-rate interest and a 10-year payback period and for variable-rate interest and a 20-year payback period. The second set includes treatments for variable-rate interest and a 10-year payback period and for fixed-rate interest and a 20-year payback period. Based on the previously discussed concepts, contrast I represents the main effect of the interest rate factor; contrast P represents the main effect of the payback period factor; the contrast IP represents the interaction effects between the type of interest and the payback period. Now let's compute the sum of squares for these orthogonal contrasts of totals. For contrast I, an estimate is

$$\hat{I} = (-1)(54) + (1)(85) + (-1)(165) + (1)(227) = 93$$

This estimate is found by using the sum of the treatments from Table 12-23, the contrast coefficients from Table 12-24 and eq. (12-62). The sum of squares due to contrast I is obtained using eq. (12-64) as follows:

$$S_1^2 = \frac{(93)^2}{(4)(4)} = 540.5625$$

Note that the number of replications for each treatment is 4 ($r_1 = r_2 = r_3 = r_4 = 4$). Similarly, the estimate of contrast P is

$$\hat{P} = (-1)(54) + (-1)(85) + (1)(165) + (1)(227) = 253$$

The sum of squares due to contrast P is

$$S_2^2 = \frac{(253)^2}{(4)(4)} = 4000.2625$$

The estimate of contrast IP is

$$\hat{IP} = (1)54 + (-1)(85) + (-1)(165) + (1)(227) = 31$$

The sum of squares due to contrast IP is

$$S_3^2 = \frac{(31)^2}{(4)(4)} = 60.0625$$

Since I, P, and IP are *three mutually orthogonal contrasts*, of the four treatments, the total treatment sum of squares is therefore the sum of these three component sums of squares. Hence, the treatment sum of squares is

$$\text{SSTR} = S_1^2 + S_2^2 + S_3^2 = 540.5625 + 4000.5625 + 60.0625 = 4601.1875$$

Since the given design is a completely randomized one, the same results can also be obtained using the formulas in Section 12-3. The use of orthogonal contrasts helps us test the significance of specific factor configurations.

The 2^k Factorial Experiment

Experimentation involves first determining the more important factors (in terms of impact on the response variable) while taking into account the number of factors that can be dealt with feasibly. Next, the desirable levels of the selected factors are identified. Finally, it may be of interest to find the relationship between the factor levels, the corresponding responses, and the physical and economic constraints that are imposed. Although different types of designs can be used at each of these stages, multifactor experiments are usually employed. One such multifactor experiment used in the exploratory stage is the 2^k factorial experiment, which involves k factors, each at two levels. The total number of treatments is 2^k. Thus, if we have two factors, each at two levels, there are 2^2 or 4 treatments.

Usually, the level of each factor is thought of as either low or high. A treatment is represented by a series of lowercase letters; the presence of a particular letter indicates that the corresponding factor is at the high level. The absence of this letter indicates that the factor is at the low level. The notation (1) is used to indicate the treatment for which all factors are at the low level. Suppose that we have three factors, A, B, and C, each at two levels. The full factorial set of eight treatments is designated in the following order:

1. (1)   5. c
2. a   6. ac
3. b   7. bc
4. ab   8. abc

Figure 12-10 shows the layout of a 2^3 experiment with three factors, A, B, and C. The high levels are denoted by 1 and the low levels by -1 for each factor.

To estimate the main effects and interactions, the concept of contrasts, introduced earlier, is used. For example, to determine the main effect of factor A, we average the observations

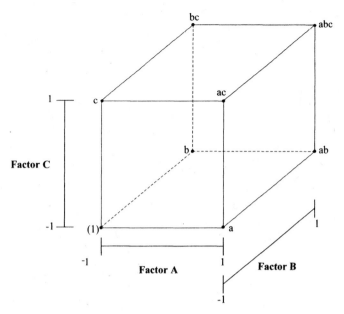

FIGURE 12-10 Layout of a 2^3 experiment with three factors.

for which A is at the high level and subtract from this value the average of the observations for which A is at the low level. From Figure 12-10, if there are r replications for each treatment, the main effect of factor A is

$$A = \frac{a + ab + ac + abc}{4r} - \frac{(1) + b + c + bc}{4r}$$

$$= \frac{1}{4r}[a + ab + ac + abc - (1) - b - c - bc] \qquad (12\text{-}65)$$

where the lowercase letters denote the sum of the responses for the corresponding treatments. The quantity inside the square brackets in eq. (12-65) is a valid contrast of totals.

Similarly, the main effect of factor B is

$$B = \frac{1}{4r}[b + ab + bc + abc - (1) - a - c - ac] \qquad (12\text{-}66)$$

The quantity inside the square brackets of eq. (12-66) is also a valid contrast. Note that the contrasts for factors A and B obtained from eqs. (12-65) and (12-66) are also orthogonal contrasts of totals. If we can come up with a set of orthogonal contrasts to estimate the main effects and interactions, then the methodology described earlier can be used to determine the corresponding sum of squares for the treatment components. Table 12-25 is such a table of coefficients for orthogonal contrasts in a 2^3 factorial experiment. Since the magnitude of each coefficient is 1, it has been omitted.

Note that the contrasts for factors A and B match those obtained from eqs. (12-65) and (12-66). We constructed Table 12-25 as follows. For each main effect contrast, we use a + for each treatment combination in which the letter for the effect occurs. All other combinations receive a −. The signs for the interaction contrasts are the product of the signs of the main

TABLE 12-25   Coefficients for Orthogonal Contrasts in a 2^3 Factorial Experiment

Contrast	(1)	a	b	ab	c	ac	bc	abc
				Treatment				
A	−	+	−	+	−	+	−	+
B	−	−	+	+	−	−	+	+
AB	+	−	−	+	+	−	−	+
C	−	−	−	−	+	+	+	+
AC	+	−	+	−	−	+	−	+
BC	+	+	−	−	−	−	+	+
ABC	−	+	+	−	+	−	−	+

effects whose letters appear in the interaction. For example, consider the contrast AB. Its sign under treatment column a is −, which is the product of the sign of A in the treatment column a (+) and the sign of B in the treatment column a (−). The contrast for any of the main effects or interactions can thus be obtained from Table 12-25 For instance, the interaction AC is estimated as

$$AC = \frac{1}{4r}[(1) + b + ac + abc - a - ab - c - bc]$$

The sum of squares for the set of orthogonal contrasts for the totals is found from eq. (12-64). If the number of replications for each treatment is r, then (because the contrast coefficients are unity in magnitude), we use eq. (12-64) for a 2^k factorial experiment to get the sum of squares for an effect:

$$SS = \frac{(\text{contrast})^2}{r2^k} \tag{12-67}$$

Also, the following relation is used to estimate the effect:

$$\text{effect} = \frac{\text{contrast}}{r2^{k-1}} \tag{12-68}$$

Note that eq. (12-65), derived to estimate the main effect of A, matches the general expression given by eq. (12-68).

Example 12-7   In the furniture industry, the quality of surface finish of graded lumber is an important characteristic. Three factors are to be tested, each at two levels, for their impact on the surface finish. Factor A is the type of wood: oak (level −1) or pine (level 1). Factor B is the rate of feed: 2 m/min (level −1) or 4 m/min (level 1). Factor C is the depth of cut: 1 mm (level −1) or 3 mm (level 1). For each treatment combination, three replications are carried out using a completely randomized design. Table 12-26 shows the surface finish for the eight treatments. The larger the value, the rougher the surface finish.

(a) Find the main effects and the interaction effects.

Solution   Since we have a 2^3 factorial experiment, the orthogonal contrasts and their coefficients corresponding to each treatment combination are as shown in Table 12-25. To find the effect of each contrast, we use eq. (12-68) and the coefficients from Table 12-25.

TABLE 12-26 Surface Finish of Wood Based on Type of Wood, Rate of Feed, and Depth of Cut

Treatment	Surface Finish			Sum
(1)	6	8	9	23
a	10	16	15	41
b	18	12	15	45
ab	12	9	10	31
c	20	26	29	75
ac	34	28	32	94
bc	36	44	46	126
abc	25	22	24	71

For example, the main effect of factor A, using the data from Table 12-26, is

$$A = \frac{1}{3(4)}[a + ab + ac + abc - (1) - b - c - bc]$$

$$= \frac{1}{12}[41 + 31 + 94 + 71 - 23 - 45 - 75 - 126]$$

$$= -2.667$$

Similarly, the other main effects and interaction effects are found as shown in Table 12-27

(b) Find the sum of squares for each of the main effects and interaction effects.

Solution The sum of squares for the factorial effects are found from eq. (12-67). For example, the sum of squares for A is

$$SS_A = \frac{(-32)^2}{3(8)} = 42.667$$

Table 12-27 shows the sum of squares for the factorial effects. The total sum of squares is found from eq. (12-4) as SST = 3025.833.

The error sum of squares, found by subtracting the sum of squares of the factorial effects from the total sum of squares, is 169.333.

(c) Which effects are significant? Test at the 5% level of significance.

TABLE 12-27 ANOVA Table for the Contrasts in the Surface Finish of Wood

Source of Variation	Effect	Degrees of Freedom	Sum of Squares	Mean Square	F-Statistic
A	-2.667	1	42.666	42.667	4.032
B	3.333	1	66.666	66.667	6.299
C	18.833	1	2128.167	2128.167	201.093
AB	-8.833	1	468.167	468.167	44.238
AC	-3.333	1	66.667	66.667	6.299
BC	1.333	1	10.667	10.667	1.008
ABC	-3.500	1	73.500	73.500	6.945
Error		16	169.333	10.583	
		23	3025.833		

Solution From Table 12-27 which shows the sum of squares and the degrees of freedom of the effects, the mean square for each effect is found. The F-statistic for testing the significance of the factorial effects is given by the ratio of the mean square for that effect to the mean square error. At the 5% level of significance, the critical value of F from Appendix A-6 is found after interpolation to be $F_{0.05,1,16} = 4.497$. Comparing the calculated value of the F-statistic with this critical value, we find that the interactions effects AB, AC, and ABC are significant. The F-statistics for the main effects B and C also exceed the critical value; however, because the interaction effects AB, AC, and ABC are significant, we cannot make any definite inferences about the main effects.

Confounding in 2^k Factorial Experiments

A basic tenet of experimental design is to group the experimental units into blocks consisting of homogeneous units. Treatments are then randomly assigned to experimental units within blocks. The idea behind blocking is to minimize the variation within blocks and thereby compare the treatments under conditions that are as uniform as possible.

The experimental error variance usually increases as the block size increases. In multifactor factorial experiments, the number of treatments increases dramatically with the number of factors. For instance, for 6 factors with 2 levels each, the number of treatments is $2^6 = 64$. Thus, for a complete replication of the treatments in one block, 64 experimental units are needed. It is often difficult to obtain a large number of homogeneous units per block for complete replication of all treatments.

To increase the precision of the experiment by blocking, the block size must be kept as small as possible. One way to achieve this is through **confounding,** where the estimation of a certain treatment effect is "confounded" (made not distinguishable) with the blocks. In confounding, not all treatments are replicated in each block. The block contrast is then equivalent to the treatment contrast that is confounded. Thus, it is not possible to estimate the effects between blocks because estimating one treatment effect essentially matches it. Thus, we are not able to conclude whether any observed differences in the response variable between blocks is solely due to differences in the blocks. Such an effect cannot be isolated from that of one of the treatments, implying that differences between blocks could be due to this particular treatment (that is confounded with the blocks), or the blocks themselves, or both. Although the treatment that is confounded with the blocks cannot be estimated, this sacrifice is made with the hope of improving the precision of the experiment through reduced block sizes. We confound the 2^k experiment into blocks of size 2^p, where $p < k$.

Let's consider a 2^3 factorial experiment. For a complete replication, 8 experimental units are needed, which then influences the block size. Suppose we want to base blocking on the raw material vendor. Because of the limited supply of raw material from each vendor, we can only obtain 4 experimental units from each vendor rather than the 8 that we need for a complete replication. The question is, How are the eight treatments to be divided into two blocks of size 4, since only 4 experimental units can be obtained from a vendor?

For the answer to this, let's refer to the coefficients for orthogonal contrasts in a 2^3 factorial experiment shown in Table 12-25. The confounding contrast is typically selected to be a high-order interaction. Look at ABC, the three-way interaction between factors A, B, and C, whose coefficients are given in Table 12-25. The contrast for ABC is given by

$$\text{contrast}(ABC) = a + b + c + abc - (1) - ab - ac - bc$$

Block 1	Block 2
a	(1)
b	ab
c	ac
abc	bc

FIGURE 12-11   Confounding in a 2^3 experiment with two blocks.

We group the plus-sign terms in one group *(a, b, c, abc)* and the minus-sign terms in another group [(1), *ab, ac, bc*]. Next, we construct two blocks of size 4 each. In block 1, the applied treatments are *a, b, c,* and *abc*. In block 2, the applied treatments are (1), *ab, ac,* and *bc*. This is demonstrated in Figure 12-11.

An estimate of the block effect is obtained by determining the difference in the average responses obtained from the two blocks. Note that this is identical to estimating the effect of the treatment *ABC*. Since the four positive coefficients in the contrast for *ABC* are assigned to block 1 and the four negative coefficients are assigned to block 2, the block effect and the *ABC* interaction effects are identical; that is, *ABC* is confounded with blocks. In other words, the effect of *ABC* cannot be distinguished from the block effects. Thus, the contrast *ABC* is the confounding contrast.

Now let's estimate the main effect of factor A as given by the contrast (from Table 12-25):

$$\text{contrast}(A) = a + ab + ac + abc - (1) - b - c - bc$$

Note that there are four runs from each block. Also, from each block there is one treatment whose effect is added and another whose effect is subtracted. For example, from block 1, the response variable is added for treatments *a* and *abc*, and it is subtracted for treatments *b* and *c*. So, any differences between the two blocks cancels out, and estimating the main effect of factor A is not affected by blocking. The same is true for the remaining main effects and two-way interactions. The scheme for blocking previously demonstrated can be used for any chosen contrast. Usually, the factors chosen to confound with blocks are the higher-order interactions, which are normally not as important as the main effects and the low-order interactions.

Fractional Replication in 2^k Experiments

The number of runs in a factorial experiment increases exponentially with the number of factors. Cost concerns often require using fewer experiments than called for by the full factorial. Fractional replication of factorial experiments achieves this purpose. Let's consider a 2^6 factorial experiment, which requires 64 experimental runs for a full replication. There are 63 degrees of freedom for estimating the treatment effects. There are 6 main effects, 15 two-way interactions (C_2^6), 20 three-way interactions, 15 four-way interactions, 6 five-way interactions, and 1 six-way interaction. If only the main effects and two-way interactions (which require 21 degrees of freedom for estimation) are considered important, we might be able to get by with fewer than 64 experimental runs.

Half-Fraction of the 2^k Experimental   A half-fraction of the 2^k experiment consists of 2^{k-1} experimental runs. To determine the treatment combinations that will be represented in the experiment, we make use of a **defining contrast**, or a **generator**. This usually represents a high-order interaction, which may be unimportant or negligible and whose estimation we might not be interested in. For example, let's consider the 2^3 factorial design with factors A, B, and C and the table of coefficients of orthogonal contrasts shown in Table 12-25. We select the defining contrast I to be associated with the three-way interaction ABC; that is, $I = ABC$. For a 2^{3-1} fractional experiment, we could select only those treatments associated with plus signs in the defining contrast. From Table 12-25, these would be a, b, c, and abc, which involve only 4 runs of the experiment. Alternatively, we could have selected the four treatments with minus signs in the defining contrast. These would have been the treatments (1), ab, ac, and bc, and the defining contrast would have been $I = -ABC$. The fraction with the plus sign in the defining contrast is called the **principal fraction**, and the other is called the **alternate fraction**.

The disadvantage of a fractional replication of a full factorial experiment is that certain effects cannot be separately estimated. Let's demonstrate this with an example. Suppose that we are dealing with a 2^{3-1} experiment with generator $I = ABC$. With four runs (a, b, c, and abc), there will be 3 degrees of freedom to estimate the main effects. Using Table 12-25 and noting that only four of the eight possible treatments are run, it would seem that the estimates of the main effects are given by

$$A = \frac{1}{2}(a + abc - b - c), \quad B = \frac{1}{2}(b + abc - a - c), \quad C = \frac{1}{2}(c + abc - a - b)$$

where the lowercase letters represent the sum of the responses in the corresponding treatments. However, we will see that in a fractional factorial experiment, these expressions, strictly speaking, represent the combined effect of the factor shown and its alias. Also using Table 12-25, and noting that the treatments run are (a, b, c, and abc), it would seem that the estimates of the two-factor interactions are given by

$$AB = \frac{1}{2}(c + abc - a - b), \quad AC = \frac{1}{2}(b + abc - a - c), \quad BC = \frac{1}{2}(a + abc - b - c)$$

Again, we will see that strictly speaking, these expressions also represent the combined effect of the factor shown and its alias. The linear combination of the sum of the treatment responses that estimates the main effect A also estimates the interaction BC. In this case, A and BC are said to be **aliases**. If the effect of the contrast is significant, we cannot conclude whether it is due to the main effect of A, the interaction BC, or a mixture of both. Similarly, B and AC are aliases, and so are C and AB. To find the aliases of a given contrast, we find its interaction with the defining contrast as follows. The letters appearing in both contrasts are combined, and any letter that appears twice is deleted. For example, the alias of A, using the defining contrast as ABC, is found as follows:

$$\text{alias}(A) = A * ABC = BC$$

Similarly, the aliases for B and C are alias $(B) = B * ABC = AC$, alias $(C) = C * ABC = AB$.

Since the contrasts *A* and *BC* are aliases, if only four treatments *a, b, c,* and *abc* are run, the *combined effect* of both contrasts is obtained from $\frac{1}{2}(a + abc - b - c)$. Therefore, using the principal fraction, an estimate of $(A + BC)$ is obtained from $\frac{1}{2}(a + abc - b - c)$; an estimate of *(B + AC)* is obtained from $\frac{1}{2}(b + abc - a - c)$; and that of *(C + AB)* is obtained from $\frac{1}{2}(c + abc - a - b)$. If the two-factor interactions can be assumed to be negligible, the main effects *A, B,* and *C* can be estimated.

Alternatively, if we are not sure whether the two-factor interactions are significant, then, by running the alternate fraction [that is, the treatments (1), *ab, ac,* and *bc*], we can estimate the main effects and the interactions by combining the results of the principal fraction and the alternate function. If we had selected the defining contrast $I = -ABC$, the runs would have consisted of the alternate fraction of treatments (1), *ab, ac,* and *bc*. The alias of *A* would be $A* -ABC = -BC$, the alias of *B* would be $-AC$, and the alias of *C* would be $-AB$. Thus, the estimate of *(A − BC)* would be obtained from $\frac{1}{2}[ab + ac - bc - (1)]$, the estimate of *(B − AC)* would be obtained from $\frac{1}{2}[ab + bc - ac - (1)]$, and that of *(C − AB)* would be obtained from $\frac{1}{2}[ac + bc - ab - (1)]$. Upon combining the results of the principal and alternate fractions, the estimate of the main effect of factor A is

$$A = \frac{1}{2}(A + BC) + \frac{1}{2}(A - BC)$$

Similarly, the other two main effects are given by

$$B = \frac{1}{2}(B + AC) + \frac{1}{2}(B - AC), \quad C = \frac{1}{2}(C + AB) + \frac{1}{2}(C - AB)$$

The interaction effects can also be obtained by combining the results of the two fractions as follows:

$$AB = \frac{1}{2}(C + AB) - \frac{1}{2}(C - AB), AC = \frac{1}{2}(B + AC) - \frac{1}{2}(B - AC), BC = \frac{1}{2}(A + BC) - \frac{1}{2}(A - BC)$$

This process—combining the results of two fractional factorial experiments to isolate and estimate main effects and interactions that ordinarily cannot be estimated from either fraction separately—creates a very desirable situation. If we cannot run a factorial experiment in a full replication, we can run sequences of small efficient experiments and then combine the information as we continue with the process of experimentation. The knowledge that we accumulate from the ongoing sequence of experiments can be used to update or modify the experiments that follow. For instance, if a fractional replication suggests the nonsignificance of certain factors, we can drop those factors in future experiments, which may lead to a full factorial experiment in the remaining factors.

Design Resolution   One method of categorizing fractional factorial experiments is based on the alias patterns produced. Obviously, we prefer that the main effects not be aliased with each other so that we can estimate them independently. The degree to which the main effects are aliased with the interaction terms (two-factor or higher-order) is represented by the resolution of the corresponding design. In particular, designs of resolutions III, IV, and V are important.

Resolution III Designs   Here, main effects are not aliased with each other, but they are aliased with two-factor interactions. Also, two-factor interactions may be aliased with each other. The 2^{3-1} design with the defining contrast given by $I = ABC$ is of resolution III. In such a design, the alias structure is given by $A = BC, B = AC,$ and $C = AB$. The main effects can be estimated only if the two-factor and higher interactions are not present.

Resolution IV Designs   Here, main effects are not aliased with each other or with any two-factor interactions. Two-factor interactions are aliased with each other. Thus, the main effects can be estimated regardless of the significance of the two-factor interactions as long as three-factor and higher order interactions are not present. An example of a resolution IV design is a 2^{4-1} design with the defining contrast given by $I = ABCD$. It can be shown that the alias structure is $A = BCD$, $B = ACD$, $C = ABD$, $D = ABC$, $AB = CD$, $AC = BD$, and $AD = BC$.

Resolution V Designs   Here, no main effect or two-factor interaction is aliased with any other main effect or two-factor interaction. Two-factor interactions, however, are aliased with three-factor interactions. An example of such a design is the 2^{5-1} design with a defining contrast given by $I = ABCDE$.

The 2^{k-p} Fractional Factorial Experiment   The 2^{k-1} fractional factorial experiment

is a half-replicate of the 2^k experiment, but it is possible to find experiments of smaller size that provide almost as much information at a greatly reduced cost. The 2^{k-p} experiment is a $1/2^p$ fraction of the 2^k factorial experiment. Therefore, a 2^{k-2} experiment employs 1/4 of the total number of treatments.

Let's consider a factorial experiment with 6 factors, each at 2 levels. A 2^{6-1} fractional factorial experiment requires 32 experimental runs, which gives us 31 degrees of freedom to estimate effects. Here, we have 6 main effects and 15 two-factor interactions. Thus, 22 runs will suffice if only main effects and two-factor interactions are deemed important, making the 2^{6-1} experiment inefficient because it requires 32 runs. However, a 2^{6-2} fractional factorial experiment needs 16 runs and has 15 degrees of freedom to estimate effects. We could estimate all 6 of the main effects and some (not more than 9) of the two-factor interactions. If this meets our needs, the 2^{6-2} experiment will be more cost-efficient than the 2^{6-1} experiment.

To generate the treatment combinations for a 2^{k-p} fractional factorial experiment, we use p defining contrasts, or generators. Using the first contrast, we create a 2^{k-1} fractional experiment as demonstrated previously. Starting with the treatments selected in the 2^{k-1} experiment and using the second contrast, we create a half-replicate of the 2^{k-1} experiment yielding a 2^{k-2} fractional experiment. This procedure continues until the *p*th contrast is used on the treatment combinations in the 2^{k-p+1} experiment to yield the 2^{k-p} fractional experiment. The aliases for the contrasts can be found using the principle explained previously.

To illustrate, let's generate a 2^{5-2} fractional factorial. The coefficients for orthogonal contrasts in a 2^5 factorial experiment with factors A, B, C, D, and E are obtained in a manner similar to that used for the coefficients shown in Table 12-25. Table 12-28 shows a portion of this table for selected contrasts (*A, B, C, D, E, AB,* and *AC*). The full factorial experiment has 32 treatment combinations. We choose $I = AB$ as our first defining contrast. From the 32 treatments, we identify from Table 12-28 those with a plus sign in the defining contrast AB. So, a half-replicate of the 2^5 experiment is given by the 2^{5-1} fractional factorial, which consists of the following treatments: (1), *ab, c, abc, d, abd, cd, abcd, e, abe, ce, abce, de, abde, cde,* and *abcde.* If we choose AC of the contrasts in the 2^{5-1} experiment as our second defining contrast, we select those (from Table 12-28 that also have a plus sign in the contrast AC. The following eight treatments are obtained for the 2^{5-2} fractional factorial experiment using the defining contrasts AB and AC: (1), *abc, d, abcd, e, abce, de,* and *abcde.*

TABLE 12-28 Coefficients for Some Orthogonal Contrasts in a 2^5 Factorial Experiment

Contrast	Treatment										
	(1)	a	b	ab	c	ac	bc	abc	d	ad	bd
A	−	+	−	+	−	+	−	+	−	+	−
B	−	−	+	+	−	−	+	+	−	−	+
C	−	−	−	−	+	+	+	+	−	−	−
D	−	−	−	−	−	−	−	−	+	+	+
E	−	−	−	−	−	−	−	−	−	−	−
AB	+	−	−	+	+	−	−	+	+	−	−
AC	+	−	+	−	−	+	−	+	+	−	+
	abd	cd	acd	bcd	abcd	e	ae	be	abe	ce	ace
A	+	−	+	−	+	−	+	−	+	−	+
B	+	−	−	+	+	−	−	+	+	−	−
C	−	+	+	+	+	−	−	−	−	+	+
D	+	+	+	+	+	−	−	−	−	−	−
E	−	−	−	−	−	+	+	+	+	+	+
AB	+	+	−	−	+	+	−	−	+	+	−
AC	−	−	+	−	+	+	−	+	−	−	+
	bce	abce	de	ade	bde	abde	cde	acde	bcde	abcde	
A	−	+	−	+	−	+	−	+	−	+	
B	+	+	−	−	+	+	−	−	+	+	
C	+	+	−	−	−	−	+	+	+	+	
D	−	−	+	+	+	+	+	+	+	+	
E	+	+	+	+	+	+	+	+	+	+	
AB	−	+	+	−	−	+	+	−	−	+	
AC	−	+	+	−	+	−	−	+	−	+	

Now let's find the aliases of the contrasts for this 2^{5-2} fractional factorial experiment using the defining contrasts AB and AC. The generalized interaction of the two also acts as a defining contrast (BC). The following alias structure is obtained:

$$I = AB = AC = BC$$
$$A = B = C = ABC$$
$$D = ABD = ACD = BCD$$
$$E = ABE = ACE = BCE$$
$$AD = BD = CD = ABCD$$
$$AE = BE = CE = ABCE$$
$$DE = ABDE = ACDE = BCDE$$
$$ADE = BDE = CDE = ABCDE$$

Note that three of the five main effects (A, B, and C) are aliases of each other. The main effects D and E have three-factor interactions as aliases. Of the two-factor interactions, AB, AC, and BC are aliases; AD, BD, and CD are aliases; and so are AE, BE, and CE. The two-factor interaction DE is an alias of the four-factor interactions $ABDE$, $ACDE$, and $BCDE$. Thus, if estimating each main effect is of interest, selecting AB and AC as the defining contrasts may not be appropriate. We would not be able to isolate the effects of A, B, and C using the generated 2^{5-2} experiment. Furthermore, if estimating two-factor interactions is desired, we will encounter difficulty because several two-factor interactions are aliases of each other.

The choice of defining contrasts, therefore, is influential in determining the **alias structure**. The generating contrasts should be selected based on the factors that are of interest in estimation. Suggestions for choosing defining contrasts for fractional factorial experiments are found in books on experimental design listed in the references (Box et al., 1978; Peterson 1985; Raghavarao 1971).

Note that principles of confounding and fractional replications differ. In confounding, the aim is to reduce the block size in order to increase the precision of the experiment without necessarily reducing the treatment combinations in the experiment. It is possible to have all the treatment combinations that occur in a full factorial experiment and still keep the block size small. The only sacrifice made is that a factorial contrast is confounded with the block contrast and so cannot be estimated—the remaining factorial contrasts can be estimated. But, if the confounding contrast is chosen so that it is one in which we have no interest, then we do not lose anything. On the other hand, in a fractional factorial experiment, the objective is to reduce the size of the experiment—that is, the number of treatments. Depending on the degree of fractionalization, contrasts will have aliases. Thus, it will not be possible to estimate all of the factorial effects separately.

Example 12-8   Refer to Example 12-7 concerning the quality of surface finish of graded lumber. The three factors, each at two levels, are the type of wood (factor A), the rate of feed (factor B), and the depth of cut (factor C). The raw data, consisting of three replications for each treatment combination, are shown in Table 12-26. We decide to use a 2^{3-1} fractional factorial experiment, with the defining contrast being $I = ABC$.

From Table 12-25, for contrast ABC, the following treatments have a plus sign: a, b, c, and abc. These four treatments are included in the 2^{3-1} fractional factorial experiment.

Now we determine the alias structure. The main effects have the following aliases:

$$A = A * ABC = BC$$
$$B = B * ABC = AC$$
$$C = C * ABC = AB$$

Thus, each main effect is an alias of a two-factor interaction. This design is therefore of resolution III. To estimate the effects of their aliases, we use Tables 12-25 and 12-26:

$$A + BC = \frac{1}{3(2)}(a + abc - b - c)$$
$$= \frac{1}{6}(41 + 71 - 45 - 75) = -1.333$$

$$B + AC = \frac{1}{3(2)}(b + abc - a - c)$$
$$= \frac{1}{6}(45 + 71 - 41 - 75) = 0$$

$$C + AB = \frac{1}{3(2)}(c + abc - a - b)$$
$$= \frac{1}{6}(75 + 71 - 41 - 45) = 10.000$$

We find that the effect of $C + AB$ exceeds that of the others.

TABLE 12-29   ANOVA Table for the 2^{3-1} Fractional Factorial Experiment for the Surface Finish of Wood Example

Source of Variation	Effect	Degrees of Freedom	Sum of Squares	Mean Square	F-Statistic
A + BC	−1.333	1	5.333	5.333	0.500
B + AC	0.	1	0.	0.	0.
C + AB	10.000	1	300.000	300.000	28.124
Error		8	85.334	10.667	
		11	390.667		

To determine the sum of squares for the effects, we proceed in the same manner as we did with factorial experiments. Thus, the sum of squares for $A + BC$ is given by

$$\mathrm{SS}_{A+BC} = \frac{(-8)^2}{3(4)} = 5.333$$

The sum of squares for the remaining effects are shown in Table 12-29. The total sum of squares is found from eq. (12-4); the sum of squares for error is found by subtraction of the sum of squares of the treatments from the total sum of squares.

We find from Table 12-29 that the effect of $C + AB$ is highly significant. The calculated F-statistic value of 28.124 far exceeds the critical F value of 5.32 from Appendix A-6 for a 5% level of significance with 1 numerator and 8 denominator degrees of freedom. To isolate the effect of C, we would need to perform more experiments using the defining contrast $I = -ABC$, which would yield the alternate fraction. Alternatively, if the complete 2^3 experiment is run using all 8 treatment combinations, the effect of C would be estimated separately from that of AB. Example 12-7 deals with this case.

12-5   THE TAGUCHI METHOD

In the following sections we deal with the philosophy and experimental design principles devised by Genichi Taguchi, a Japanese engineer whose ideas in quality engineering have been used for many years in Japan. The underlying theme of the neverending cycle of quality improvement is supported by Taguchi's work (Taguchi and Wu 1980; Taguchi 1986, 1987). **Quality engineering** has the objective of designing quality into every product and corresponding process. It directs quality improvement efforts upstream from the manufacturing process to the design phase and is therefore referred to as an **off-line quality control** method. The techniques of statistical process control discussed in preceding chapters are known as on-line quality control methods. Taguchi's off-line methods are effective in improving quality and cutting down costs at the same time (Taguchi et al. 1989). Off-line methods improve product manufacturability and reduce product development and lifetime costs.

In Taguchi's method, quality is measured by the deviation of a characteristic from its target value. A loss function is developed for this deviation. Uncontrollable factors, known as noise, cause such deviation and thereby lead to loss. Since the elimination of noise factors is impractical and often impossible, the **Taguchi method** seeks to minimize the effects of noise and to determine the optimal level of the important controllable factors based on the concept of **robustness** (Dehnad 1989; Nair 1992). The objective is to create a product/process design that is insensitive to all possible combinations of the uncontrollable noise factors and is at the same time effective and cost-efficient as a result of setting the key controllable factors at certain levels. Whereas the foundations of Taguchi's quality engineering seem to be well accepted, there are criticisms, which we will discuss later, of the statistical aspects of his experimental designs.

12-6 THE TAGUCHI PHILOSOPHY

According to Taguchi, "Quality is the loss imparted to society from the time a product is shipped." Taguchi attaches a monetary value to quality because he feels that this will make quality improvement understood by all—technical personnel as well as management. Typical examples of loss to society include failure to meet the customer's requirements and unsatisfactory performance that leads to loss of goodwill and reduced market share. The Taguchi concept of quality loss has been extended by Kackar (1986) to include the loss to society while the product is being manufactured. For instance, the raw materials and labor depleted when making the product or rendering the service may also be considered societal losses. Similarly, the cost of environmental pollution falls into this category.

The purpose of quality improvement, then, is to discover innovative ways of designing products and processes that will save society more than they cost in the long run. All products ultimately cause loss because they break and then need to be repaired or replaced or because they wear out after an adequate performance in their functional life. The degree to which the product or service meets consumers' expectations affects the magnitude of loss. Taguchi contends that the loss due to a product's variation in performance is proportional to the square of the performance characteristic's deviation from its target value.

Figure 12-12 shows an example of the quadratic loss function. This function is an example of the target-is-best condition. If the quality characteristic y is at the target value m, then the loss $L(y)$ is minimized. As the characteristic deviates from the **target** (or **nominal**) **value** in

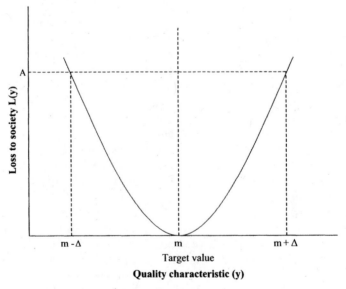

FIGURE 12-12 Loss function for a situation in which the target value is best.

either direction, the loss increases in a quadratic manner. Examples include the dimensions of a part (length, thickness, diameter, etc.) and the operating pressure in a press.

The quality and cost of a manufactured product is influenced by the engineering design of the product as well as the process. By concentrating on product and process designs that are robust i.e., less sensitive to uncontrollable factors such as temperature, humidity, and manufacturing variation (the noise factors), the Taguchi concept attempts to reduce the impact of noise rather than eliminate it. Frequently, the elimination of noise factors is neither practical (because it is very costly) nor feasible. Manufacturing variability cannot be totally eliminated; incoming components and raw material inherently exhibit variation. By dampening the impact of the noise factors and by selecting the controllable factor levels that force the desirable quality characteristics to stay close to target values, a robust design of the product, and thereby of the process, is achieved.

Taguchi advocates a three-stage design operation to determine the target values and tolerances for relevant parameters in the product and the process: system design, parameter design, and tolerance design. In **system design**, scientific and engineering principles and experience are used to create a prototype of the product that will meet the functional requirements and also to create the process that will build it.

Parameter design involves finding the optimal settings of the product and process parameters in order to minimize performance variability. Taguchi defines a performance measure known as the **signal-to-noise (S/N) ratio** and tries to select the parameter levels that maximize this ratio. The term **signal** represents the square of the mean value of the quality characteristic, whereas noise is a measure of the variability (as measured by the variance) of the characteristic.

As mentioned previously, the uncontrollable factors are known as noise factors. They cause a quality characteristic to deviate from the target value and hence create a loss in quality. In general, noise arises from two sources: **internal** and **external noise**. Internal sources of noise are a result of settings of the product and process parameters. Examples include manufacturing variations and product deterioration over time due to wear and tear. Manufacturing imperfections occur because of the inevitable uncertainties in a manufacturing process. They create product-to-product noise, or variability. External sources of noise are those variables that are external to the product and that affect its performance. Examples include changes in such environmental conditions as humidity, temperature, and dust.

In the parameter design phase, not all sources of noise can be accounted for because of the lack of knowledge concerning the various factors that affect the product's performance. Moreover, physical restrictions—based on the number of experiments to be performed and the number of factors that may be included in the analysis—also limit the choice of factors. In the parameter design phase, only those sources of noise that are included in the study are termed noise factors. Thus, noise factors should be judiciously chosen to represent those that most impact the product's performance.

Suppose we are producing collapsible tubes for which the important quality characteristic Y is the diameter. One of the controllable factors X found to have an effect on the diameter is the mold pressure. Figure 12-13 shows the relationship between the diameter and the mold pressure. Our target value of the diameter is y_0, which can be achieved by setting the mold pressure at x_0. Any deviation of the mold pressure from x_0 causes a large variability in the diameter. However, if the mold pressure is set at x_1, the variability in the diameter is much less even if noise factors cause the mold pressure to deviate around x_1. One way to achieve the small variability in the diameter is to use an expensive press that has little variability around

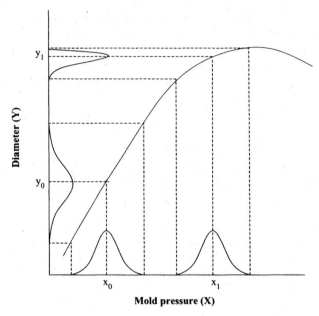

FIGURE 12-13   Relationship between diameter and mold pressure.

the set value x_0. Alternatively, a more economical solution that achieves the same level of quality involves the following strategy. We would use a less expensive press and keep the pressure setting at x_1, which yields a small variability in the output diameter in the presence of uncontrollable factors. Since the output diameter y_1 is far removed from the target value y_0 in this situation, we would look for other parameters that have a linear effect on the diameter and whose setting can be adjusted in a cost-efficient manner.

One such parameter may be the pellet size S, whose effect is shown in Figure 12-14. Since tubes are extruded from pellets, the pellet size has a linear relationship with tube diameter. We can control the pellet size to adjust the mean value from y_1 to y_0. Such a parameter is sometimes called an **adjustment factor**. By choosing a pellet size of s_0, we can achieve the desired target of y_0. Of course, we are assuming that because the relationship between pellet

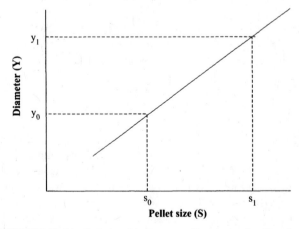

FIGURE 12-14   Relationship between diameter and pellet size.

size and diameter is linear, no increase in the variability of the diameter will result when the pellet size is set at s_0.

In general, it is much easier to adjust the mean value of a performance characteristic to its target value than to reduce the performance variability. The preceding example demonstrates how we can use the nonlinear effects of product and process parameters to reduce the variability of the output and thereby minimize the impact of the sources of variation. This is the key to achieving robust designs in the Taguchi method. This method is cost-efficient in that it reduces the performance variability by minimizing the effects of factors we cannot control instead of trying to control or eliminate them.

Our example also illustrates the fundamental idea of the *two-step procedure* used in the parameter design phase. In the first step, we choose parameter levels that minimize the variability of the output quality characteristic. This step makes use of the nonlinear effects of the parameters on the response variable and creates a design that is robust with respect to the uncontrollable sources of variation. In the second step, we identify parameters that have a linear effect on the average value of the response but do not impact variability. These parameters, known as adjustment factors, are used to drive the quality characteristic to the target value without increasing the variability.

After the system design and parameter design stages comes the third stage, **tolerance design**. In this step we set tolerances (i.e., a range of admissible values) around the target values of the control parameters identified in the parameter design phase. We do this only if the performance variation achieved by the settings identified in the parameter design stage is not acceptable. For instance, in the example concerning the diameter of collapsible tubes, tolerances for the mold pressure could be $x_1 \pm \Delta x_1$, or those for the pellet size could be $s_0 \pm \Delta s_0$, where Δx_1 and Δs_0 represent the permissible variabilities for mold pressure and pellet size, respectively.

Tolerances that are too rigid or tight will increase manufacturing costs, and tolerances that are too wide will increase performance variation, which in turn will increase the customer's loss. Our objective is to find the optimal trade-off between these two costs. Usually, after the parameter design stage is completed and the parameter control settings are adjusted, **confirmation** (or **verification**) **experiments** are conducted. These experiments reaffirm the degree of improvement realized when the chosen parameter settings are used. It is possible that the observed degree of variability of the performance characteristic is more than what we have in mind. If this is the case, tolerance design may be needed to reduce the variability of the input parameters, which will thereby reduce output variation.

Typical questions at this stage concern which parameters are to be tightly controlled. Cost is always a factor here. In most cases, the costs of tightening the variability of control factors differ. Here again, a trade-off will be made on the effect of a parameter on the output characteristic and the cost of tightening it to desirable bounds.

12-7 LOSS FUNCTIONS

As noted earlier, in the Taguchi philosophy, quality is the loss imparted to society from the time a product is shipped. In the preceding section we extended this definition to include the loss incurred while the product is being manufactured. The components of loss include the expense, waste, and lost opportunity that result when a product fails to meet the target value exactly. During production, costs such as inspection, scrap, and rework contribute to loss. Although those costs are easy to account for, there are others that are much more difficult to measure, such as the loss associated with customer dissatisfaction that arises from variation in products and services.

Profit-increasing measures such as productivity improvement and waste reduction are important, but they are bounded by such factors as labor and material costs and technology. Real growth in market share, however, is very much influenced by society—that is, by customer satisfaction. Price and quality are critical factors in this context. Any cost incurred by the customer or any loss resulting from poor quality will have a significant negative impact. On the other hand, a satisfied customer who realizes savings from using the product or service will turn it over many times to greatly improve market share. These ideas lead us to conclude that variability is the key concept that relates product quality to dollars. We can express this variability through a loss function.

The quality loss function is stated in financial terms, which provides a common language for various entities within an organization, such as management, engineering, and production. It can also be related to performance measures such as the signal-to-noise ratio, which is used in the parameter design phase. The loss function is assumed to be proportional to the square of the deviation of the quality characteristic from the target value. The traditional notion of the loss function is that as long as the product's quality characteristic is within certain specification limits or tolerances, no loss is incurred. Outside these specifications, the loss takes the form of a step function with a constant value. Figure 2-8 shows this traditional loss function indicated by L_0.

There are drawbacks inherent to this notion of loss function. Obviously, the function makes no distinction between a product whose characteristic is exactly on target (at m) versus one whose characteristic is just below the upper specification limit (USL) or just above the lower specification limit (LSL). There will be performance differences in these products, but the traditional loss function does not reflect these differences. Furthermore, it is difficult to justify the sudden step increase in loss as the quality characteristic just exceeds the specification limits. Is there a great functional difference between a product with a quality characteristic value of $USL - \delta$ and one with a value of $USL + \delta$, where δ is very small and approaches zero? Most likely, the answer is no. Additionally, it is unreasonable to say that the loss remains constant, say at a value L_0, for all values of the characteristic beyond the specification limits. There will obviously be a significant functional difference between products that are barely above the USL and those that are far above the USL. These differences would cause a performance variation, so the loss to society would be different.

The Taguchi loss function overcomes these deficiencies, because the loss increases quadratically with increasing deviation from the target value. In the following subsections we discuss expressions for the loss functions based on three situations: target is best, smaller is better, and larger is better.

Target Is Best

Let's consider characteristics for which a target (or nominal) value is appropriate. As the quality characteristic deviates from the target value, the loss increases in a quadratic manner, as shown in Figure 12-12. Examples of quality characteristics in this category include product dimensions such as length, thickness, and diameter; a product characteristic such as the viscosity of an oil; and a service characteristic such as the degree of management involvement. Taguchi's loss function is given by

$$L(y) = k(y - m)^2 \tag{12-69}$$

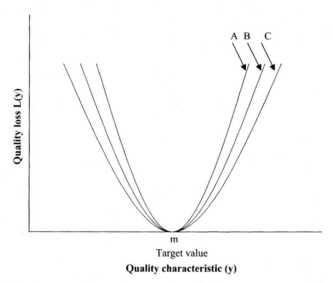

FIGURE 12-15   Taguchi loss function indicating the degree of financial importance.

where k is a proportionality constant, m is the target value, and y is the value of the quality characteristic. Note that when $y = m$ (i.e., when the value of the quality characteristic is on target), the loss is zero.

The constant k can be evaluated if the loss $L(y)$ is known for any particular value of the quality characteristic; it is influenced by the financial importance of the quality characteristic. For instance, if a critical dimension of the braking mechanism in an automobile deviates from a target value, the loss may increase dramatically. Figure 12-15 shows three quality loss functions with different degrees of financial importance. The steeper the slope, the more important the loss function. In Figure 12-15, loss function A is more important than loss functions B and C.

To determine the value of the constant k, suppose the functional tolerance range of the quality characteristic is $(m - \Delta, m + \Delta)$, as shown in Figure 12-12. This represents the maximum permissible variation, beyond which the average product does not function satisfactorily. An average customer viewpoint is represented. Suppose the consumer's average loss is A when the quality characteristic is at the limit of the functional tolerance. This loss represents costs to the consumer for repair or replacement of the product, with the associated dissatisfaction. Using eq. 12-69, we find the proportionality constant k as

$$A = k\Delta^2 \quad \text{or} \quad k = \frac{A}{\Delta^2}$$

The loss function given by eq. 12-69 can be rewritten as

$$L(y) = \frac{A}{\Delta^2}(y - m)^2 \tag{12-70}$$

The **expected loss** is the mean loss over many instances of the product. The expectation is taken with respect to the distribution of the quality characteristic Y. We have

$$\begin{aligned}
E[L(y)] &= E[k(y - m)^2] \\
&= k(\text{variance of } y + \text{squared bias of } y) \\
&= k[\text{Var}(y) + (\mu - m)^2] = k(\text{MSD})
\end{aligned} \tag{12-71}$$

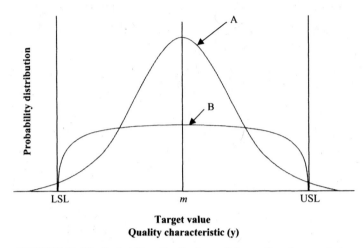

FIGURE 12-16   Probability distributions of the quality characteristic.

Here, MSD represents the **mean square deviation** of y and is estimated as

$$\text{MSD} = \frac{\sum\limits_{i=1}^{n}(y_i - m)^2}{n}$$

over a sample of n items. In this expression, μ and $\text{Var}(y)$ represent the mean and variance of y, respectively. Note that $\text{MSD}(y) = \text{Var}(y) + (\mu - m)^2$; that is, the mean square deviation of a quality characteristic is the sum of the variance of the characteristic and the squared bias.

   With the objective of minimizing the expected loss, note that if the mean of the characteristic is at the target value m, the mean square deviation is just the variance of the characteristic. As with parameter design, the variance of the response variable can be minimized by selecting appropriate levels of the controllable factors to minimize the performance variation. Additionally, the levels of the control factors that have a linear relationship with the response variable can be changed to move the average of the response variable to the target value.

   The form of the distribution of the quality characteristic also influences the expected loss. Figure 12-16 shows two probability distributions of a quality characteristic. Distribution A is a normal distribution with most of the product tightly clustered around the target value. A small fraction of the product lies outside the specification limits. Distribution B, on the other hand, is close to a uniform distribution. The product is uniformly distributed over the range of the specification limits, with none of the product lying outside. According to the traditional view, all of the product related to distribution B will be considered acceptable. However, if we consider the more appropriate quadratic loss function, the expected loss for the product from distribution A will be less than that for distribution B. This is due to the fact that much more of the product from distribution A is close to the target value, resulting in a lower expected loss.

Determination of Manufacturing Tolerances   The importance of realistic tolerances based on customer satisfaction and thereby loss has been stressed previously. In the preceding subsection we computed the proportionality constant k of the loss function in eq. (12-70). Now suppose the question is, how much should a manufacturer spend to repair a product

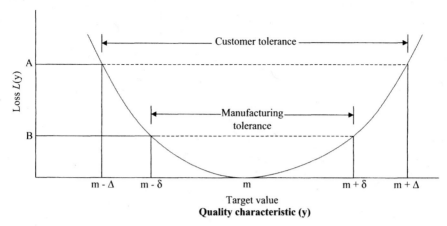

FIGURE 12-17   Manufacturing tolerance for a situation in which target is best.

before shipping it to the customer in order to avoid losses due to the product not meeting customer tolerances? This amount represents the break-even point, in terms of loss, between the manufacturer and the customer. Figure 12-17 shows the customer tolerance $m \pm \Delta$, the associated loss A, and the manufacturing tolerance $m \pm \delta$ associated with a cost B to the manufacturer. This cost, B, is the cost of repairing an item, prior to shipping such that the item will exceed the customer's tolerance limits. The idea is that fixing a potential problem prior to shipment reduces loss in the long run due to wear and tear of the product in its usage. We would thus like to determine δ.

Based on the fact that the loss is A for customer tolerances of $m \pm \Delta$, the value of the proportionality constant k of the loss function is found to be $k = A/\Delta^2$. The loss function is given by eq. (12-70). Now what happens to the manufacturing tolerance of $m \pm \delta$ and the associated cost B? Using the loss function from eq. (12-70), we get

$$ B = \frac{A}{\Delta^2} \delta^2 \quad \text{or} \quad \delta = \left(\frac{B}{A} \right)^{1/2} \Delta \qquad (12\text{-}72) $$

The manufacturing tolerances are then found as $m \pm \delta$. As long as the quality characteristic is within δ units of the target value m, the manufacturer will not spend any extra money.

When the characteristic is at δ units from m, the manufacturer spends an amount equal to B, on average, to repair the product. As the quality characteristic deviates farther from the target value, the customer's loss increases beyond acceptable limits. In the long run, the manufacturer would be wise to spend an amount B in this situation, since the customer's loss would otherwise be much more, thereby increasing the societal loss. The manufacturing tolerances are the limits for shipping the product.

Smaller Is Better

Now let's consider quality characteristics that are nonnegative and for which the ideal value is zero (i.e., smaller values are better). Characteristics such as the fuel consumption of an aircraft or automobile, the wear of a bearing, an impurity in water, the shrinkage of a gasket, and customer waiting time in a bank are examples of smaller is better.

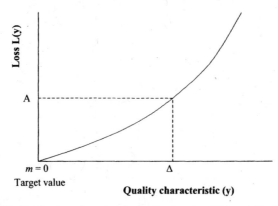

FIGURE 12-18   Loss function for a situation in which smaller is better.

Figure 12-18 shows the loss function for this case, which is given by

$$L(y) = ky^2 \qquad (12\text{-}73)$$

If the loss caused by exceeding the customer's tolerance level Δ is A, the value of the proportionality constant k in eq. (12-73) is $k = A/\Delta^2$.

The expected, or average, loss can be found, as before, as the expectation with respect to the probability distribution of the quality characteristic. Thus, the expected loss over many produced items is given by

$$E[L(y)] = kE(y^2)$$
$$= \frac{A}{\Delta^2}[\text{Var}(y) + \mu^2] = \frac{A}{\Delta^2}\text{MSD} \qquad (12\text{-}74)$$

where MSD represents the mean square deviation and is estimated as $(\sum_{i=1}^{n} y_i^2)/n$ for a sample of n items; Var(y) represents the variance of y, and μ is the mean value of y.

Larger Is Better

Finally, we have quality characteristics that are nonnegative and have an ideal target value that is infinite (i.e., larger values are better). Examples include the weld strength in bridge beams, the amount of employee involvement in quality improvement, and the customer acceptance rate of a product. Here, no target value is predetermined.

Figure 12-19 shows the loss function for this case, which is given by

$$L(y) = k\frac{1}{y^2} \qquad (12\text{-}75)$$

If the loss is A when the quality characteristic falls below the customer's tolerance level Δ, the value of the proportionality constant k is given by $k = A\Delta^2$. The expected, or average, loss is given by the expectation with respect to the probability distribution of the quality characteristic. The expected loss is

$$E[L(y)] = kE\left(\frac{1}{y^2}\right) = A\Delta^2 E\left(\frac{1}{y^2}\right) \qquad (12\text{-}76)$$

The quantity $E(1/y^2)$ can be estimated from a sample of n items as $\left(\sum_{i=1}^{n} 1/y_i^2\right)/n$.

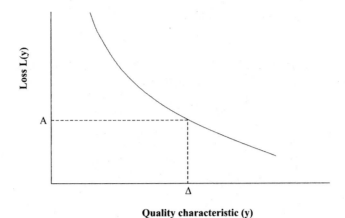

FIGURE 12-19 Loss function for a situation in which larger is better.

Example 12-9 Customer tolerances for the height of a steering mechanism are 1.5 ± 0.020 m. For a product that just exceeds these limits, the cost to the customer for getting it fixed is $50. Ten products are randomly selected and yield the following heights (in meters): 1.53, 1.49, 1.50, 1.49, 1.48, 1.52, 1.54, 1.53, 1.51, and 1.52. Find the average loss per product item.

Solution   Note that this is a situation in which target is best. The target value m is 1.5 m. The loss function is given by $L(y) = k(y-m)^2$, where y represents the height of the steering mechanism and k is a proportionality constant. Given the information

$$k = \frac{A}{\Delta^2} = \frac{50}{(0.02)^2} = 125,000$$

The expected loss per item is given by $E[L(y)] = 125{,}000E(y-1.5)^2$, where $E(y-1.5)^2$ is estimated as follows:

$$\frac{\sum_{i=1}^{10}(y_i - 1.5)^2}{10} = \frac{[(1.53-1.5)^2 + (1.49-1.5)^2 + (1.50-1.5)^2 + \ldots + (1.52-1.5)^2]}{10}$$
$$= 0.0049/10 = 0.00049$$

Hence, the expected loss per item is $125{,}000(0.00049) = \$61.25$.

Example 12-10   Refer to Example 12-9 concerning the height of a steering mechanism. The manufacturer is considering changing the production process to reduce the variability in the output. The additional cost for the new process is estimated to be $5.50/item. The annual production is 20,000 items. Eight items are randomly selected from the new process, yielding the following heights: 1.51, 1.50, 1.49, 1.52, 1.52, 1.50, 1.48, 1.51. Is the new process cost-efficient? If so, what is the annual savings?

Solution   The loss function from Example 12-9 is $L(y) = 125{,}000(y-1.5)^2$. For the new process, we estimate the mean square deviation around the target value of 1.5:

$$\frac{\sum_{i=1}^{8}(y_i-1.5)^2}{8} = \frac{(1.51-1.5)^2 + (1.50-1.5)^2 + \ldots + (1.51-1.5)^2}{8}$$
$$= 0.0015/8 = 0.0001875$$

The expected loss per item for the new process is

$$(125,000)(0.0001875) = \$23.44$$

The expected loss per item for the former process is \$61.25, which represents a savings of \$61.25 − \$23.44 = \$37.81/item, and the added cost is only \$5.50/item. Thus, the net savings per item by using the new process is \$37.81 − \$5.50 = \$32.31, making it cost-efficient.

The net annual savings compared to the former process is $(20,000)(32.31) = \$646,200$.

Example 12-11   Refer to Example 12-9 concerning the height of a steering mechanism. The manufacturer decides to rework the height, prior to shipping the product, at a cost of \$3.00 per item. What should the manufacturer's tolerance be?

Solution   The loss function from Example 12-9 is $L(y) = 125,000(y-1.5)^2$. Let the manufacturer's tolerance be given by $1.5 \pm \delta$. We have

$$3 = 125,000\delta^2 \quad \text{or} \quad \delta = \left(\frac{3}{125,000}\right)^{1/2} = 0.0049$$

Thus, the manufacturer's tolerance is 1.5 ± 0.0049 m. For products with heights that equal or exceed these limits, the manufacturer should rework the items at the added cost of \$3.00 per item to provide cost savings in the long run. Otherwise, these items will incur a loss to the customer at a rate of \$50 per item.

12-8 SIGNAL-TO-NOISE RATIO AND PERFORMANCE MEASURES

To determine the effectiveness of a design, we must develop a measure that can evaluate the impact of the design parameters on the output quality characteristic. An acceptable performance measure of the output characteristic should incorporate both the desirable and undesirable aspects of performance. As noted earlier, the term *signal*, or average value of the characteristic, represents the desirable component, which will preferably be close to a specified target value. The term *noise* represents the undesirable component and is a measure of the variability of the output characteristic, which will preferably be as small as possible. Taguchi has combined these two components into one measure known as the signal-to-noise (S/N) ratio. Mathematical expressions for the S/N ratio are dependent on the three situations (target is best, smaller is better, and larger is better). A performance measure should have the property that when it is maximized, the expected loss will be minimized.

Target Is Best

Equation (12-71) gives the expected loss associated with the output characteristic. It is preferable for the output characteristic to be on target and at the same time for the variance to be as small as possible.

Variance not Related to the Mean   Note that the response variable *Y* is influenced by the settings of the product and process parameters $(x_1, x_2 \ldots, x_p)$. In the Taguchi method, we determine the parameter settings that optimize the performance measure associated with *Y*. In certain situations, the mean and the variance are functionally independent of each other, This means that there are certain parameters known as adjustment parameters (noted earlier), whose levels can be adjusted to change the average output characteristic level without influencing the variability in performance. The adjustment parameter can therefore be set to a value such that the mean output is on target, making the bias equal to zero, without affecting the variability.

In this situation, our main concern is to reduce the variance of the output characteristic. A possible performance measure is given by

$$\xi = -\log(\text{variance of } y)$$

an estimate of which is the **performance statistic** given by

$$Z = -\log(s^2) \tag{12-77}$$

where s^2 is the sample variance of y.

Note that the larger the performance statistic Z, the smaller the variability. The performance statistic given by eq. (12-77) is not a Taguchi performance statistic. One advantage of this performance statistic is that the logarithmic transformation on the data often reduces the dependence of the sample variance s^2 on the sample mean $\bar{y}$, thus satisfying the assumption that the variance is not related to the mean.

Variance Related to the Mean   Here we assume that the standard deviation of the output characteristic is linearly proportional to the mean and that the bias can be reduced independently of the coefficient of variation, which is the ratio of the standard deviation to the mean. Thus, knowledge of the standard deviation by itself may not provide a measure of goodness of performance. For instance, suppose that the standard deviation of a part length is 1 mm. Is this a superior process? We cannot tell. However, if we know the mean length, we are in a position to make a judgment. A mean length of 15 mm might not reflect a good process, whereas a mean length of 100 mm might. The ratio of the mean to the standard deviation could be a good performance measure. Actually, the signal-to-noise measure proposed by Taguchi is proportional to the square of the ratio of the mean to the standard deviation. An objective would be to maximize this ratio.

Let's consider an example in which one of three designs must be selected. Figure 12-20 shows the levels of the mean and standard deviation of the output quality characteristic for

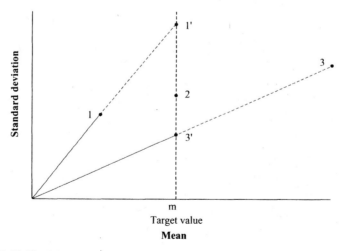

FIGURE 12-20   Mean and standard deviation of output characteristic for three designs.

the three designs (indicated by points 1, 2, and 3), as well as the target value m. Considering only the standard deviation, we might be tempted to choose design 1, which has the smallest standard deviation. However, it is offset from the target value of m. So, if an adjustment parameter is identified, its setting can be changed to alter the mean of the output to the target value. In the process of changing the mean to the target value, the standard deviation of the output variable will increase linearly. The point indicated by 1′ in Figure 12-20 represents the adjusted mean and standard deviation.

Similarly, for design 3, a downward adjustment in the mean value, point 3′, could be made, corresponding to a reduction in the standard deviation. For design point 2, whose mean is at the target value, no adjustment is necessary. Comparing the three design points whose means are at the target value, point 3′ has the smallest standard deviation. Therefore, design point 3 should be selected because it maximizes the performance measure, which is the ratio of the mean to the standard deviation. Even after the mean has been adjusted to the target value, design point 3′ will still have the maximum value for the performance measure.

The Taguchi method of parameter design involves trying to identify adjustment parameters and their ranges for which the S/N ratio (which is proportional to the ratio of the mean to the standard deviation) is constant. Figure 12-21 shows one such parameter where the S/N ratio is nearly constant within the range of (a, b). The effect of a change in the mean is linear as the level of the parameter is varied. Thus, to adjust the mean to the target value m, we change the parameter setting to level c, and we expect that the new standard deviation will remain at a minimum, if, in fact, it were minimum before adjustment of the mean, since the S/N ratio is assumed to be constant.

Taguchi's S/N ratio for this case is given by $\xi = \mu^2 / \sigma^2$ where μ and σ represent the mean and standard deviation of the output characteristic y, respectively. A related performance statistic is

$$z = 10 \log \frac{\bar{y}^2}{s^2} \tag{12-78}$$

where $\bar{y}$ and s denote the sample mean and sample standard deviation of the output characteristic y, respectively. This statistic is maximized in the parameter design stage, and the associated parameter levels are determined.

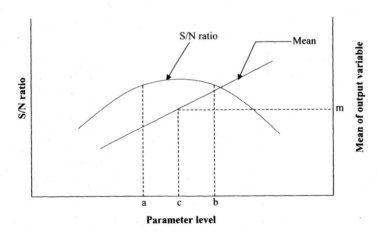

FIGURE 12-21   Effect of a parameter setting on the S/N ratio and the mean of the output characteristic.

Smaller Is Better

The loss function in the case for which smaller values are better is given by eq. (12-73), and the expected loss is represented by eq. (12-74). Since these characteristics have an ideal target value of zero, they can be considered as noise themselves. The performance measure is the mean square deviation from zero; and the performance statistic that should be maximized in the parameter design stage is given by

$$Z = -10 \log \frac{\sum_{i=1}^{n} y_i^2}{n} \qquad (12\text{-}79)$$

In some instances, a characteristic for which a smaller value is better is incorporated into another characteristic in the experimental design and analysis phase. For instance, the surface roughness on a bearing is a smaller-is-better characteristic by itself, but it could be treated as a noise variable in a design where the output characteristic is the diameter of the bearing, which is a target-is-best characteristic.

Larger Is Better

The larger-is-better case is merely the inverse of the smaller-is-better case. The loss function is given by eq. (12-75), and the expected loss is represented by eq. (12-76). The performance statistic is given by

$$Z = -10 \log \frac{\sum_{i=1}^{n} 1/y_i^2}{n} \qquad (12\text{-}80)$$

The goal is to determine the parameter settings that will maximize this performance statistic.

12-9   CRITIQUE OF S/N RATIOS

To minimize expected loss, a good performance measure incorporates engineering and statistical knowledge about the product and the process. Taguchi has proposed more than 60 signal-to-noise ratios for particular applications. Three such ratios are given by eqs.(12-78), (12-79) , and (12-80); Taguchi advocates maximizing these ratios in experimental analysis, as a means of minimizing variability. If variability is minimized, the design will be robust, and one of the major goals of Taguchi's method will be met.

Taguchi's S/N ratios have met with criticism, however. Let's consider the ratio for the target-is-best case given in eq. (12-78). This can be rewritten as

$$Z = 10 \log(\bar{y}^2) - 10 \log(s^2)$$

Thus, maximizing the performance statistic Z can be achieved through either a large positive value or a large negative value of the mean $\bar{y}$, by a small value of the sample variance s^2, or by both. Taguchi's S/N ratio measures the simultaneous effect of factors on the mean response and the variability. If our objective is to reduce variability (as in the first step of Taguchi's two-step parameter design procedure), there is no guarantee that maximizing Taguchi's S/N ratio will achieve it. It is not clear that the transformation $10 \log(\bar{y}^2/s^2)$ separates the dispersion and location effects; hence, Taguchi's two-step parameter design approach is not fully justifiable on theoretical grounds.

Additionally, if Taguchi's S/N ratio is maximized, it is assumed that having a unit increase in $\log(\bar{y}^2)$ and having a unit decrease in $\log(s^2)$ are equally desirable. This assumption may not be justified in all circumstances. In certain situations, a reduction of variability could overshadow the closeness to the target value.

One recommendation is to minimize $\log(s^2)$ or to maximize $-\log(s^2)$ in the first step of the parameter design procedure. This performance measure is given in eq. (12-77). If the mean and variability can be isolated, then the performance measure $\log(s^2)$ can be minimized in the first step by choosing appropriate parameter settings. In the next step, the performance measure is the sample mean $\bar{y}$, which is used to determine the settings of the adjustment factors such that the mean response is close to the target value. This approach provides us with a better understanding of the process in terms of factors that influence process variability.

Hunter (1985) points out that one way to maximize the S/N ratio $10\log(\bar{y}^2/s^2)$ is to take the logarithm of the y-values and then identify the factor-level settings that minimize the value of s^2. Box (1986) argues that this S/N ratio can be justified if a logarithmic transformation is needed to make the average and variance independent as well as to satisfy other assumptions such as normality of the error distribution and constancy of the error variance. If s^2 [or a function of it, such as $\log(s^2)$] is to be used as a performance statistic in the first step, it is preferable for the variance to be unrelated to the mean, because the mean is adjusted in the second step.

The S/N ratios for the smaller-is-better and larger-is-better cases have similar disadvantages. It has been demonstrated in several simulation studies (Schmidt and Boudot 1989), that the S/N ratios given by eqs. (12-79) and (12-80) may not be effective in identifying dispersion effects; however, they do reveal location effects that are due to the mean. The S/N ratio given by eq. (12-79) can be shown to be a function of the mean and the variance because

$$\frac{\sum_{i=1}^{n} y_i^2}{n} = \bar{y}^2 + \frac{n-1}{n}s^2$$

Thus, the location and dispersion effects are confounded in the S/N ratio.

An alternative measure designed to overcome some of the drawbacks of Taguchi's S/N ratio has been proposed by Leon et al. (1987). The study recommends a performance measure that is independent of adjustment known as PERMIA. It involves a two-step procedure as suggested by Taguchi. In the first step, PERMIA is used instead of the S/N ratio to determine the parameter settings of the nonadjustment factors. The selection of the appropriate response variable at this stage is important because this response variable should not depend on the values of the adjustment factors. The second step is similar to Taguchi's, in that levels of the adjustment factors are determined in order to drive the mean response toward the target value.

12-10   EXPERIMENTAL DESIGN IN THE TAGUCHI METHOD

At the core of product and process design is the concept of experimental design. How we design our experiments guides us in selecting combinations of the various factor levels that enable us to determine the output characteristic and thereby calculate the performance statistic. The matrix that designates the settings of the controllable factors (design parameters) for each run, or experiment, is called an **inner array** by Taguchi; the matrix that designates the setting of the uncontrollable or noise factors is called an **outer array**. Each run consists of a setting of the design parameters and an associated setting of the noise factors. The inner and outer arrays are designated as the **design** and **noise matrices** respectively.

For design factors that are quantitative, three levels are necessary to estimate the quadratic (or nonlinear) effect, if any. If only two levels of a factor are tested, then only its linear effects on the response variable can be estimated. One disadvantage of selecting three levels for each design factor is that the number of experiments to be performed increases, which in turn increases the cost of design.

Orthogonal Arrays and Linear Graphs

Once the design and noise factors and the number of settings have been selected, a series of experiments is run to determine the optimal settings of the design parameters. Taguchi's two-step procedure for the parameter design phase is used here as well. Rather than run experiments at all possible combinations of design and noise factor levels, Taguchi relies on running only a portion of the total number of possible experiments, using the concept of an orthogonal array. The intent is to concentrate on the vital few rather than the trivial many within the confines of engineering and cost constraints.

An orthogonal array represents a matrix of numbers. Each row represents the levels, or states, of the chosen factors, and each column represents a specific factor whose effects on the response variable are of interest. Orthogonal arrays have the property that every factor setting occurs the same number of times for every test setting of all other factors. This allows us to make a balanced comparison among factor levels under a variety of conditions. In addition, any two columns of an orthogonal array form a two-factor complete factorial design. Using an orthogonal array minimizes the number of runs while retaining the pairwise balancing property.

Orthogonal arrays are generalized versions of Latin square designs. We have seen that, in a Latin square design, a blocking of two factors is accomplished. A **Graeco-Latin square design** blocks in three factors. The concept of blocking in more than three factors may be conducted through a generalized Graeco-Latin square design, of which orthogonal arrays are a subset. In fact, all common fractional factorial designs are orthogonal arrays (see Table 12-28). Several methods for constructing orthogonal arrays are provided by Raghavarao (1971). Taguchi (1986, 1987) recommends the use of orthogonal arrays for the inner array (design matrix) and the outer array (noise matrix).

Taguchi uses linear graphs with the orthogonal arrays. These graphs show the assignment of factors to the columns of the orthogonal array and help us visualize interactions between factors. Table 12-30 shows several orthogonal arrays and their associated linear graphs.

The notation used in Table 12-30 is as follows. The orthogonal array represented by the L_4 design is a half-replicate of a 2^3 full factorial experiment, in which three factors, each at two levels, are considered. In the L_4 design, four experiments are conducted (one-half of 2^3, the full factorial experiment). The two levels of each factor are denoted by 1 and 2. Thus, in the first experiment, each of the three factors is at level 1. In the fourth experiment, factors 1 and 2 are at level 2 and factor 3 is at level 1. The linear graph for the L_4 table has two nodes, labeled 1 and 2, and an arc joining them labeled as 3. These numbers correspond to the columns of the orthogonal array.

If the main effects of two of the factors are to be estimated, they should be placed in columns 1 and 2 of the orthogonal array. Column 3 represents the interaction effect of the factors assigned to columns 1 and 2. The implication is that such a design will not be able to estimate the main effect of a third separate factor independent of the interaction effects

TABLE 12-30A   Orthogonal Arrays and Associated Linear Graphs: Legend for Linear Graphs

Symbol	Group	Degree of Difficulty
○	1	Easy
◎	2	Moderately easy
◉	3	Fairly difficult
●	4	Difficult

TABLE 12-30C   Orthogonal Arrays and Associated Linear Graphs: L_8 (2^7) Series

L_8 (2^7) Series

	Factor						
Experiment	1	2	3	4	5	6	7
1	1	1	1	1	1	1	1
2	1	1	1	2	2	2	2
3	1	2	2	1	1	2	2
4	1	2	2	2	2	1	1
5	2	1	2	1	2	1	2
6	2	1	2	2	1	2	1
7	2	2	1	1	2	2	1
8	2	2	1	2	1	1	2
	Group 1	Group 2		Group 3			

Linear graphs for L_8 table.

TABLE 12-30B   Orthogonal Arrays and Associated Linear Graphs: L_4 (2^3) Series

L_4 (2^3) Series

	Factor		
Experiment	1	2	3
1	1	1	1
2	1	2	2
3	2	1	2
4	2	2	1
	Group 1	Group 2	

Linear graph for L_4 table.

TABLE 12-30D   Orthogonal Arrays and Associated Linear Graphs: L_9 (3^4) Series

L_9 (3^4) Series

	Factor			
Experiment	1	2	3	4
1	1	1	1	1
2	1	2	2	2
3	1	3	3	3
4	2	1	2	3
5	2	2	3	1
6	2	3	1	2
7	3	1	3	2
8	3	2	1	3
9	3	3	2	1
	Group 1	Group 2		

Linear graph for L_9 table:

TABLE 12-30E   Orthogonal Arrays and Associated Linear Graphs: L_{16} (2^{15}) Series

	L_{16} (2^{15}) Series														
	Factor														
Experiment	1	2	3	4	5	6	7	8	9	10	11	12	13	14	15
1	1	1	1	1	1	1	1	1	1	1	1	1	1	1	1
2	1	1	1	1	1	1	1	2	2	2	2	2	2	2	2
3	1	1	1	2	2	2	2	1	1	1	1	2	2	2	2
4	1	1	1	2	2	2	2	2	2	2	2	1	1	1	1
5	1	2	2	1	1	2	2	1	1	2	2	1	1	2	2
6	1	2	2	1	1	2	2	2	2	1	1	2	2	1	1
7	1	2	2	2	2	1	1	1	1	2	2	2	2	1	1
8	1	2	2	2	2	1	1	2	2	1	1	1	1	2	2
9	2	1	2	1	2	1	2	1	2	1	2	1	2	1	2
10	2	1	2	1	2	1	2	2	1	2	1	2	1	2	1
11	2	1	2	2	1	2	1	1	2	1	2	2	1	2	1
12	2	1	2	2	1	2	1	2	1	2	1	1	2	1	2
13	2	2	1	1	2	2	1	1	2	2	1	1	2	2	1
14	2	2	1	1	2	2	1	2	1	1	2	2	1	1	2
15	2	2	1	2	1	1	2	1	2	2	1	2	1	1	2
16	2	2	1	2	1	1	2	2	1	1	2	1	2	2	1
	1	2			3						4				

Groups

Linear graphs for L_{16} table:

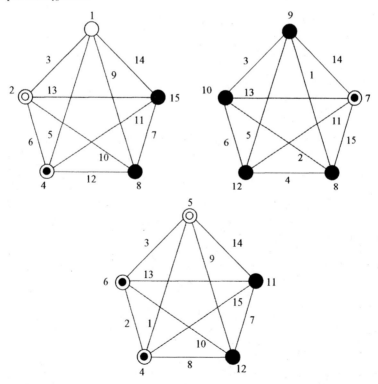

TABLE 12-30F Orthogonal Arrays and Associated Linear Graphs: L_{27} (3^{13}) Series

L_{27} (3^{13}) Series

Experiment	Factor												
	1	2	3	4	5	6	7	8	9	10	11	12	13
1	1	1	1	1	1	1	1	1	1	1	1	1	1
2	1	1	1	1	2	2	2	2	2	2	2	2	2
3	1	1	1	1	3	3	3	3	3	3	3	3	3
4	1	2	2	2	1	1	1	2	2	2	3	3	3
5	1	2	2	2	2	2	2	3	3	3	1	1	1
6	1	2	2	2	3	3	3	1	1	1	2	2	2
7	1	3	3	3	1	1	1	3	3	3	2	2	2
8	1	3	3	3	2	2	2	1	1	1	3	3	3
9	1	3	3	3	3	3	3	2	2	2	1	1	1
10	2	1	2	3	1	2	3	1	2	3	1	2	3
11	2	1	2	3	2	3	1	2	3	1	2	3	1
12	2	1	2	3	3	1	2	3	1	2	3	1	2
13	2	2	3	1	1	2	3	2	3	1	3	1	2
14	2	2	3	1	2	3	1	3	1	2	1	2	3
15	2	2	3	1	3	1	2	1	2	3	2	3	1
16	2	3	1	2	1	2	3	3	1	2	2	3	1
17	2	3	1	2	2	3	1	1	2	3	3	1	2
18	2	3	1	2	3	1	2	2	3	1	1	2	3
19	3	1	3	2	1	3	2	1	3	2	1	3	2
20	3	1	3	2	2	1	3	2	1	3	2	1	3
21	3	1	3	2	3	2	1	3	2	1	3	2	1
22	3	2	1	3	1	3	2	2	1	3	3	2	1
23	3	2	1	3	2	1	3	3	2	1	1	3	2
24	3	2	1	3	3	2	1	1	3	2	2	1	3
25	3	3	2	1	1	3	2	3	2	1	2	1	3
26	3	3	2	1	2	1	3	1	3	2	3	2	1
27	3	3	2	1	3	2	1	2	1	3	1	3	2

	1		2						3				

Groups

Linear graphs for L_{27} table:

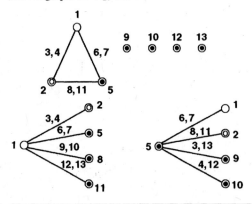

TABLE 12-30G   Orthogonal Arrays and Associated Linear Graphs: L_{32} ($2^1 \times 4^9$) Series

L_{32} ($2^1 \times 4^9$) Series

Experiment	Factor									
	1	2	3	4	5	6	7	8	9	10
1	1	1	1	1	1	1	1	1	1	1
2	1	1	2	2	2	2	2	2	2	2
3	1	1	3	3	3	3	3	3	3	3
4	1	1	4	4	4	4	4	4	4	4
5	1	2	1	1	2	2	3	3	4	4
6	1	2	2	2	1	1	4	4	3	3
7	1	2	3	3	4	4	1	1	2	2
8	1	2	4	4	3	3	2	2	1	1
9	1	3	1	2	3	4	1	2	3	4
10	1	3	2	1	4	3	2	1	4	3
11	1	3	3	4	1	2	3	4	1	2
12	1	3	4	3	2	1	4	3	2	1
13	1	4	1	2	4	3	3	4	2	1
14	1	4	2	1	3	4	4	3	1	2
15	1	4	3	4	2	1	1	2	4	3
16	1	4	4	3	1	2	2	1	3	4
17	2	1	1	4	1	4	2	3	2	3
18	2	1	2	3	2	3	1	4	1	4
19	2	1	3	2	3	2	4	1	4	1
20	2	1	4	1	4	1	3	2	3	2
21	2	2	1	4	2	3	4	1	3	2
22	2	2	2	3	1	4	3	2	4	1
23	2	2	3	2	4	1	2	3	1	4
24	2	2	4	1	3	2	1	4	2	3
25	2	3	1	3	3	1	2	4	4	2
26	2	3	2	4	4	2	1	3	3	1
27	2	3	3	1	1	3	4	2	2	4
28	2	3	4	2	2	4	3	1	1	3
29	2	4	1	3	4	2	4	2	1	3
30	2	4	2	4	3	1	3	1	2	4
31	2	4	3	1	2	4	2	4	3	1
32	2	4	4	2	1	3	1	3	4	2
Groups	1	2	3							

Linear graph for L_{32} table:

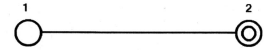

1   2

TABLE 12-30H Orthogonal Arrays and Associated Linear Graphs: L_{18} ($2^1 \times 3^7$) Series

L_{18} ($2^1 \times 3^7$) Series

Experiment	Factor							
	1	2	3	4	5	6	7	8
1	1	1	1	1	1	1	1	1
2	1	1	2	2	2	2	2	2
3	1	1	3	3	3	3	3	3
4	1	2	1	1	2	2	3	3
5	1	2	2	2	3	3	1	1
6	1	2	3	3	1	1	2	2
7	1	3	1	2	1	3	2	3
8	1	3	2	3	2	1	3	1
9	1	3	3	1	3	2	1	2
10	2	1	1	3	3	2	2	1
11	2	1	2	1	1	3	3	2
12	2	1	3	2	2	1	1	3
13	2	2	1	2	3	1	3	2
14	2	2	2	3	1	2	1	3
15	2	2	3	1	2	3	2	1
16	2	3	1	3	2	3	1	2
17	2	3	2	1	3	1	2	3
18	2	3	3	2	1	2	3	1
Groups	1	2	3					

Linear graph for L_{18} table:

TABLE 12-30I Orthogonal Arrays and Associated Linear Graphs: L_{16} (4^5) Series

L_{16} (4^5) Series

Experiment	Factor				
	1	2	3	4	5
1	1	1	1	1	1
2	1	2	2	2	2
3	1	3	3	3	3
4	1	4	4	4	4
5	2	1	2	3	4
6	2	2	1	4	3
7	2	3	4	1	2
8	2	4	3	2	1
9	3	1	3	4	2
10	3	2	4	3	1
11	3	3	1	2	4
12	3	4	2	1	3
13	4	1	4	2	3
14	4	2	3	1	4
15	4	3	2	4	1
16	4	4	1	3	2
Groups	1	2			

Linear graph for L_{16} table:

*Source: G. Taguchi. *System of Experimental Design-Engineering Methods to Optimize Quality and Minimize Costs*, Vol. 2, 1987. Reprinted by permission of American Supplier Institute, Inc. Bingham Farms, Michigan (U.S.A.).

between the first two factors. Alternatively, if we have only two factors and we would like to estimate their main effects and their interaction, then the two factors should be assigned to columns 1 and 2, and their interaction should be assigned to column 3.

The linear graphs shown in Table 12-30 have different types of points, such as circles, double circles, dots with circles, and completely shaded circles. These represent the degree of difficulty associated with the number of changes that must be made in the level of a factor assigned to the corresponding column. Table 12-30 shows such groups (groups 1 through 4) and the associated difficulty for adjusting the factor levels, ranging from easy to difficult.

Let's consider a situation where it is either very difficult or expensive to adjust the factor level, in which case we prefer that the factor level not be changed any more than is necessary. An example might be the octane level of a gasoline; it is not feasible to change octane levels frequently because a certain minimum quantity must be produced. Another example is the setting of a pneumatic press, where the setup time is high. Frequent adjustments of settings requires an unusual amount of downtime, making it cost-ineffective.

In the first column of the L_4 design, if the most difficult parameter is placed in column 1 and the experiments are conducted in the order specified in the orthogonal array, then this factor is changed only once. The first two experiments are conducted at the first level of the parameter, and the last two experiments are conducted at the second level of the parameter. On the other hand, the factor assigned to column 2 requires three changes. Column 1 has been assigned the "easy" degree of difficulty (group 1) because it corresponds to the fewest number of factor level changes. Columns 2 and 3 are placed in group 2, where the number of required changes in levels is not as easy as that for column 1.

Similar interpretations hold for the other orthogonal arrays shown in Table 12-30. The factor assigned to column 1 changes levels only once for each level at which it is tested. Factors that are assigned to columns of increasing index need more setting changes to satisfy the required levels of the orthogonal array. The linear graphs thus provide some guidance to performing experiments in a cost-effective manner. Factors whose levels are difficult or costly to change should be placed in columns with low indices. Linear graphs also assist in estimating interaction effects between factors. Desirable interactions between certain factors should be assigned to appropriate columns in the orthogonal array, as found from the associated linear graph.

The L_8 (2^7) orthogonal array indicates the factor assignment levels for seven factors, each at two levels. This is a fractional replicate of the full factorial experiment that requires a total of 2^7, or 128, experiments. The L_8 orthogonal array needs only 8 experiments, making it cost-effective relative to the full factorial experiment. Note that there are two linear graphs for the L_8 orthogonal array. From either of these graphs, we find that four main effects can be estimated independently of the others if assigned to columns 1, 2, 4, and 7. If the interaction effect between factors 1 and 2 is desired, it should be assigned to column 3.

The L_9 (3^4) orthogonal array represents the design for which four factors are varied at three levels each. It is a fractional replicate of the full factorial experiment that needs 3^4, or 81, experiments. The linear graph for the L_9 orthogonal array shows that the main effects of two factors can be estimated independently of the others if assigned to columns 1 and 2. The main effects of the two other factors cannot be estimated independently of the effects of the remaining factors. They will be confounded by the interaction effects between the factors assigned to the first two columns.

Table 12-30 also shows the orthogonal arrays and the associated linear graphs for several other experiments. The array for the L_{16} (2^{15}) series shows the experimental layout for estimating 15 factors, each at two levels. It requires 16 experiments. Three linear graphs for the L_{16} array are also shown. The L_{18} ($2^1 \times 3^7$) array shows the design for one factor at two levels and 7 factors at three levels each. A total of 8 factors are used, requiring 18 experiments. The main effects of two factors can be estimated independently of the others if assigned to columns 1 and 2. Other designs shown are the L_{27} (3^{13}) array for estimating 13 factors each at three levels, requiring 27 experiments; the L_{32} ($2^1 \times 4^9$) array for estimating 10 factors, 1 factor at two levels and 9 others at four levels each, requiring a total of 32 experiments; and the L_{16} (4^5) array for estimating 5 factors at four levels each, requiring 16 experiments.

Note that all possible interactions between factors cannot be estimated from the orthogonal array; a full factorial experiment would be needed. Taguchi suggests that interactions be ignored during the initial experiments, his rationale being that it is difficult to identify significant interactions without some prior knowledge or experience. Unassigned columns in the orthogonal array can be assigned to the possible interactions after assigning the factors whose main effects are desirable.

Example 12-12   A consumer magazine subscription service has four factors, A, B, C, and D, each to be analyzed at two levels. Also of interest are the interactions of B × C, B × D, and C × D. Show the experimental design for this case.

Solution   Since we have a total of seven factors (including the interactions, which are treated like factors), each of two levels, we check the L_8 (2^7) orthogonal array to see whether it can be used. From the first linear graph for L_8 shown in Table 12-30, we assign factors to the corresponding columns as shown in the linear graph in Table 12-31. Thus, the L_8 array is the appropriate choice.

TABLE 12-31   Allocation of Factors to Columns in an L_8 Orthogonal Array

Experiment	Factor						
	B	C	B × C	D	B × D	C × D	A
1	1	1	1	1	1	1	1
2	1	1	1	2	2	2	2
3	1	2	2	1	1	2	2
4	1	2	2	2	2	1	1
5	2	1	2	1	2	1	2
6	2	1	2	2	1	2	1
7	2	2	1	1	2	2	1
8	2	2	1	2	1	1	2

Linear graph:

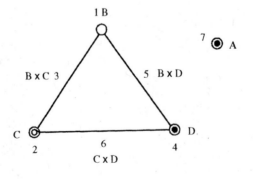

The linear graphs can also be modified to estimate certain interaction effects. In the L_8 orthogonal array, the estimation of seven factor effects is possible. The following example shows an experimental design in which five main effects and two desirable interactions can be estimated.

Example 12-13   The rapid transit authority in a large metropolitan area has identified five factors, A, B, C, D, and E, each to be investigated at two levels. Interactions A × C and A × D will also be estimated. Determine an appropriate experimental design for estimating the main effects and the two interactions.

Solution   With a total of seven factors to estimate, each at two levels, we again consider the $L_8 (2^7)$ orthogonal array and its second linear graph, which is shown in Table 12-30, The linear graph shows the assignment of three interaction terms. However, we need to estimate only two interactions. For the interaction that is not used, we remove the associated two column numbers and treat them as individual points to be assigned to the main effects. The linear graph shown in Table 12-32 shows our approach.

TABLE 12-32   Modifying L_8's Linear Graph to Estimate Interaction Effects

Experiment	Factor						
	A	C	A × C	D	A × D	B	E
1	1	1	1	1	1	1	1
2	1	1	1	2	2	2	2
3	1	2	2	1	1	2	2
4	1	2	2	2	2	1	1
5	2	1	2	1	2	1	2
6	2	1	2	2	1	2	1
7	2	2	1	1	2	2	1
8	2	2	1	2	1	1	2

Modification of linear graph:

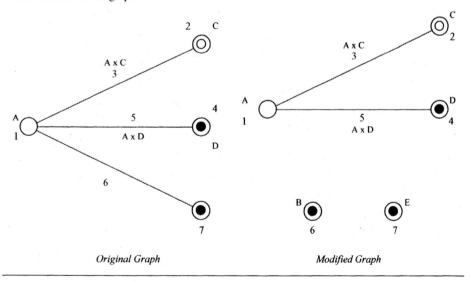

Original Graph   *Modified Graph*

We initially assign factor A to column 1, factor C to column 2, and factor D to column 4. Then, the interaction $A \times C$ is assigned to column 3 and the interaction $A \times D$ to column 5. Since we don't need to estimate a third interaction, we remove column numbers 6 and 7 from the original graph and treat them as separate points. We assign the main effects of B and E to columns 6 and 7, respectively, as shown in the modified graph. Table 12-32 also shows the final assignment and the experimental design for estimating the required effects.

We must be careful, however, when we assign an interaction column to an independent main effect. The interaction effect that would have occurred in that column will still be present and will be confounded with the main effect. For this example, column 6, which originally corresponded to the interaction between the factors assigned to columns 1 and 7, (i.e., $A \times E$) will be confounded with B, which has been assigned to that column. Only under the assumption that the interaction between $A \times E$ is insignificant can we estimate the effect of B.

As mentioned previously, Taguchi recommends verification experiments after the optimal parameter levels have been identified. If the results associated with a main-effect factor assigned to an interaction column cannot be repeated, it might be a clue to analyze interaction terms. The whole experiment may have to be reconducted, taking into account the interaction term in the design.

Another feature of linear graphs is their applicability to the construction of **hybrid orthogonal arrays**. The number of levels of each factor we want to test may not always be the same. For instance, an automobile insurance company may be interested in the impact of several factors — the age of the applicant, the type of work performed, and the number of miles commuted to work on a weekly basis — to determine the applicant's probability of being involved in an accident. Certainly, each factor need not be considered at the same number of levels. For instance, the company might select five age classification groups, three work classification groups, and four mileage classification groups. We would thus need a design that incorporates a different number of factor levels. The following example demonstrates a procedure that allows us to do this.

Example 12-14   A commercial bank has identified five factors (A, B, C, D, and E) that have an impact on its volume of loans. There are four levels for factor A, and each of the other factors is to be tested at two levels. Determine an appropriate experimental design.

Solution   We first calculate the number of degrees of freedom associated with the factors. The number of degrees of freedom for each factor is 1 less than the number of levels. The total number of degrees of freedom associated with an orthogonal array is 1 less than the number of experiments. The number required for this study is 7 (3 for A and 1 each for B, C, D, and E). The number of degrees of freedom associated with the L_8 orthogonal array is also 7, 1 less than the total number of experiments. Thus, we can use the L_8 array to select a suitable hybrid design.

Each column in the L_8 array can handle a two-level factor, which means that it has 1 degree of freedom. In our study, factor A is to be studied at four levels, needing 3 degrees of freedom. We thus allocate three columns to factor A, after they have been combined appropriately. Factors, B, C, D, and E, each with 1 degree of freedom, are assigned to the remaining four columns. Which three columns should we allocate to A, and how do we combine them? We use the linear graph to determine this. We must identify a line in the graph that can be removed without affecting the rest of the design (to the extent possible). Let's consider the line segment joining nodes 1 and 2 in the first linear graph for L_8 shown in Table 12-30 (also shown in Table 12-33. Removing the line segment and nodes 1 and 2 also removes node 3.

TABLE 12-33   Creation of a Hybrid Orthogonal Array

Experiment	Factor					Columns 1,2,3			New Assignment (Factor A)
	A	B	C	D	E				
1	1	1	1	1	1	1	1	1	1
2	1	2	2	2	2	1	1	1	1
3	2	1	1	2	2	1	2	2	2
4	2	2	2	1	1	1	2	2	2
5	3	1	2	1	2	2	1	2	3
6	3	2	1	2	1	2	1	2	3
7	4	1	2	2	1	2	2	1	4
8	4	2	1	1	2	2	2	1	4

Original linear graph: Modified graph:

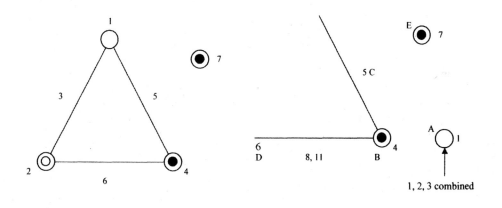

We then sequentially assign values to each unique combination generated by the removed columns, 1, 2, and 3. A new column is created that combines the effects of these three columns. For example, the combination (1,1,1) is assigned a level of 1, (1,2,2) is assigned a level of 2, and so on. Table 12-33 shows the assignment of the levels of this new column, which is now assigned to factor A. Note that it has four levels. The remaining factors B, C, D, and E are assigned to columns 4, 5, 6, and 7 of the original L_8 orthogonal array. In the hybrid orthogonal design in Table 12-33, note that factor A has four levels, whereas B, C, D, and E have two levels each.

Remember that interaction effects are assumed to be insignificant in the construction of this design. Thus, since we have assigned factor C to column 5 of the original L_8 array, the interaction between the factors in columns 1 and 4 (in this case, A × B) is confounded with the main effect of C. The main effect of factor C can therefore be estimated only under the assumption that the interaction effect A × B is insignificant. Similar conclusions may be drawn for the remaining factor assignments.

Estimation of Effects

Once the experimental design is selected, the factors are set at the levels indicated by the chosen design, and experiments are conducted to observe the value of the response variable Y.

TABLE 12-34   Experimental Layout Using an L_9 Inner Array and an L_4 Outer Array

				E	1	1	2	2		
	Noise Factors			F	1	2	1	2		
	(L_4 Outer Array)			G	1	2	1	1		
	Design Factors (L_9 Inner Array)									
Experiment	A	B	C	D					Mean Response, $\bar{y}$	S/N Ratio, Z
1	1	1	1	1	y_{11}	y_{12}	y_{13}	y_{14}	$\bar{y}_1$	Z_1
2	1	2	2	2	y_{21}	y_{22}	y_{23}	y_{24}	$\bar{y}_2$	Z_2
3	1	3	3	3	.	.	.	.	.	.
4	2	1	2	3	.	.	.	.	.	.
5	2	2	3	1	.	.	.	.	.	.
6	2	3	1	2	.	.	.	.	.	.
7	3	1	3	2	.	.	.	.	.	.
8	3	2	1	3	.	.	.	.	.	.
9	3	3	2	1	y_{91}	y_{92}	y_{93}	y_{94}	$\bar{y}_9$	Z_9

The control factors are placed in an inner array (a selected orthogonal array) and the noise factors in an outer array. The layout of such a design is shown in Table 12-34 in which four control factors, A, B, C, and D, each at three levels, are placed in an L_9 inner array, and three noise factors, E, F, and G, each at two levels, are placed in an L_4 outer array. The purpose of including noise factors in the experiment is to determine the levels of the design factors that are least sensitive to noise. A robust design, which is one of the principal objectives of the Taguchi method, would thus be obtained.

The different combination levels of the noise factors are essentially treated as replications for a given setting of the design factors. In Table 12-34 there are nine settings of the design factors, based on the L_9 orthogonal array. For each of these settings, there are four replications corresponding to the setting of the three noise factors, based on an L_4 orthogonal array. For the example in Table 12-34 a total of 36 experiments are conducted. For each setting of the design factors, a measure of the mean and variability of the output characteristic can be calculated. For instance, for experiment 1, where the design factors A, B, C, and D are each at level 1, let the observed value of the response variable at the four noise factor settings be denoted by y_{11}, y_{12}, y_{13}, and y_{14}. The mean of these four values is denoted by $\bar{y}_1$, and the value of the associated S/N ratio is denoted by Z_1. This computation is then repeated for the remaining runs. The summary values of the mean response and the S/N ratio are then used in the analysis that occurs in the parameter design phase.

Taguchi recommends analyzing the means and S/N ratios through simple plots and summary measures to keep the analysis simple. The ANOVA procedures described previously can also be used to determine significant factor effects (Minitab 2007; Peterson 1985). However, we will only demonstrate a simple analysis using averages and plots.

First, we calculate the **main effect** of a factor by determining the average of the response variable over all replications for a given level of the factor. For instance, the mean response when factor A is at level 1 can be found for the example in Table 12-34 as follows:

$$\bar{A}_1 = \frac{\bar{y}_1 + \bar{y}_2 + \bar{y}_3}{3}$$

Similarly, the mean responses when factor A is at levels 2 and 3, respectively, are

$$\bar{A}_2 = \frac{\bar{y}_4 + \bar{y}_5 + \bar{y}_6}{3}, \quad \bar{A}_3 = \frac{\bar{y}_7 + \bar{y}_8 + \bar{y}_9}{3}$$

In the same manner, the main effects of the remaining factors at each level are found. For example, the mean response when factor B is at level 3 is

$$\bar{B}_3 = \frac{\bar{y}_3 + \bar{y}_6 + \bar{y}_9}{3}$$

The main effects are then plotted to determine the significance of the factor. In this case, the values $\bar{A}_1$, $\bar{A}_2$, and $\bar{A}_3$ are plotted versus the three levels of factor A. If the plot is close to a horizontal line, the implication is that factor A does not have a significant impact. Figure 12-22 shows various possible effects of factor A on the average response. Figure 12-22a depicts a situation where factor A's main effect is not significant. Any level of the factor produced about the same average response. To aid in selecting the level of factor A here, additional criteria such as cost could be used.

On the other hand, if the plot resembles a nonlinear function, the region where the curve is flat is used to select the level of A that produces minimum variability in the response variable. Figure 12-22b demonstrates this situation. To create a design that is robust to the noise factors, we would choose level 2 of factor A, as it will lead to the smallest variability in the average response. If factor A has a linear relationship with the average response, as shown in Figure 12-22c, it could be used as an adjustment factor in the second step of the parameter

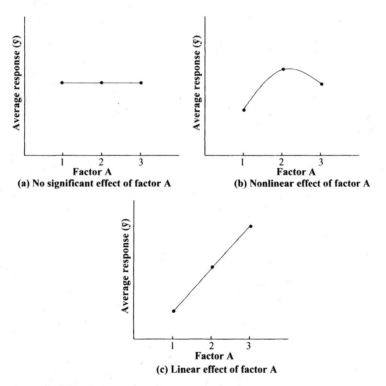

FIGURE 12-22   Effect of factor on response variable.

design procedure. An adjustment factor is used to shift the average response toward a target value without influencing its variability.

The preceding conclusions relating to the main effect of a factor are based on the assumption that interaction effects between factors are insignificant. If interaction effects are significant, it may not make sense to determine the significance of the main effects independently of each other.

The **interaction effects** between factors may also be estimated using a procedure similar to that for the main effects of individual factors. The assumption is that the interaction effect has been assigned an appropriate column using the linear graph and the orthogonal array. Let's consider the L_9 orthogonal array and its associated linear graph. We assume that factor C in Table 12-34 represents the interaction between factors A and B. Note that C, which is A $\times$ B, is assigned to column 3 of the L_9 array, which is in agreement with L_9's linear graph. We can estimate the interaction effect between A and B by treating C like any other factor. Thus, using Table 12-34 the average response when A $\times$ B is at level 1 is

$$(\overline{A \times B})_1 = \frac{\bar{y}_1 + \bar{y}_6 + \bar{y}_8}{3}$$

The average responses at the other levels of A $\times$ B are

$$(\overline{A \times B})_2 = \frac{\bar{y}_2 + \bar{y}_4 + \bar{y}_9}{3}, \quad (\overline{A \times B})_3 = \frac{\bar{y}_3 + \bar{y}_5 + \bar{y}_7}{3}$$

These three averages are plotted against the levels of A $\times$ B. If the plot is a horizontal line, it implies that the interaction effect between A and B is not significant. In that case, we would investigate the main effects of A and B to determine the optimal settings of the factor levels. For significant interaction effects, the optimal level of each factor would be determined based on the joint effect of both.

Example 12-15   Various components of a drug for lung cancer have positive and negative effects depending on the amount used. Scientists have identified four independent factors that seem to affect the performance of the drug. Each factor can be tested at three levels. Testing is expensive. Determine an experimental design that will test the impact of these factors in a cost-effective manner.

Solution   We have four factors, A, B, C, and D, each at three levels, so the $L_9(3^4)$ orthogonal array is appropriate. This design can test the effects of four factors, each at three levels, in only nine experiments. However, because we have four independent factors whose effects are to be estimated, it will be necessary to assume that the interaction effects between factors are insignificant. For instance, if we consider the linear graph for the L_9 array, the interaction effect between the factors assigned to columns 1 and 2 will be confounded with the main effects of those factors assigned to columns 3 and 4. Our experimental design is to assign the following: factor A to column 1, B to column 2, C to column 3, and D to column 4. The interaction effect A $\times$ B will then be confounded with the main effect of C, which is assigned to column 3. Also, interaction effect A $\times$ B will be confounded with the main effect of D, which is assigned to column 4.

We select the L_9 orthogonal array and assign the four factors A, B, C, and D to columns 1, 2, 3, and 4, respectively. The response variable, measuring the impact of the drug, is recorded on a coded scale for the nine experiments. The target value is zero, with the observed coded

TABLE 12-35   Coded Response for Effect of Drug Due to Four Factors

Experiment	Factor				Coded Response
	A	B	C	D	
1	1	1	1	1	−3.5
2	1	2	2	2	7.3
3	1	3	3	3	1.8
4	2	1	2	3	−4.4
5	2	2	3	1	9.5
6	2	3	1	2	−6.2
7	3	1	3	2	−4.0
8	3	2	1	3	2.4
9	3	3	2	1	−2.5

responses being positive and negative. The experimental design and the values of the coded response variable are shown in Table 12-35. Only one replication is conducted for each setting of the design factors, so the outer array of the noise factors is not shown. Determine the main effects, and plot the average response curves. What are the optimal settings of the design parameters?

We calculate the main effects of each factor at the three levels, beginning with factor A, where the average responses at levels 1, 2, and 3 are

$$\bar{A}_1 = (-3.5 + 7.3 + 1.8)/3 = 1.867$$
$$\bar{A}_2 = (-4.4 + 9.5 - 6.2)/3 = -0.367$$
$$\bar{A}_3 = (-4.0 + 2.4 - 2.5)/3 = -1.367$$

For factors B, C, and D, the average responses are

$$\bar{B}_1 = (-3.5 - 4.4 - 4.0)/3 = -3.967$$
$$\bar{B}_2 = (7.3 + 9.5 + 2.4)/3 = 6.400$$
$$\bar{B}_3 = (1.8 - 6.2 - 2.5)/3 = -2.300$$
$$\bar{C}_1 = (-3.5 - 6.2 + 2.4)/3 = -2.433$$
$$\bar{C}_2 = (7.3 - 4.4 - 2.5)/3 = 0.133$$
$$\bar{C}_3 = (1.8 + 9.5 - 4.0)/3 = 2.433$$
$$\bar{D}_1 = (-3.5 + 9.5 - 2.5)/3 = 1.167$$
$$\bar{D}_2 = (7.3 - 6.2 - 4.0)/3 = -0.967$$
$$\bar{D}_3 = (1.8 - 4.4 + 2.4)/3 = -0.067$$

The average response for each factor is then plotted against the factor levels, and the corresponding graphs are shown in Figure 12-23. First, factors B and D clearly have a nonlinear impact on the average response. To minimize the variability in the average response and thus create a robust design, factors B and D are set at level 2. At these levels, the response curve is approximately flat, making it less sensitive to variations resulting from noise factors.

In the second step, to move the average response close to the target value of zero, we manipulate the levels of factors that are linearly related to the average response, thereby not affecting the variability that we attained in the first step. We find that factors A and C each has a nearly linear relationship with the average response. To move the average response close to the target value of zero, adjustment factors A and C are each set at level 2. Hence, the optimal

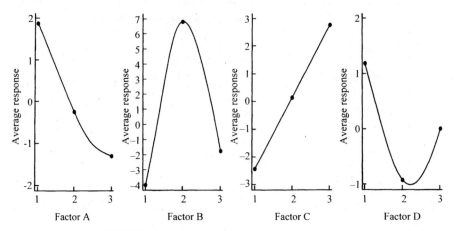

FIGURE 12-23   Average response curve for each factor.

settings are factor A at level 2, factor B at level 2, factor C at level 2, and factor D at level 2. In practice, confirmation experiments would be run at these levels to validate the results and/or to come up with more refined settings that would reduce variability further.

12-11 PARAMETER DESIGN IN THE TAGUCHI METHOD

The focal point of the Taguchi method is an economical quality design. Figure 12-24 shows the three phases of the Taguchi method discussed previously. In the system design phase, the design engineer uses practical experience coupled with scientific and engineering principles to create a functional design. Raw materials and components are identified, the sequential steps in the process through which the product is to be manufactured are proposed and analyzed, tooling requirements are studied, production constraints related to capacity are investigated, and all other issues related to the creation and production of a feasible design are dealt with.

The second phase is parameter design, which involves determining influential parameters and their settings. Usually, a subset of all possible parameters is selected and analyzed in an experimental framework. The experiments may be conducted physically or through computer simulation. The latter is more cost-effective. However, in order to use computer simulation, we must first establish mathematical models that realistically describe the relationships between the parameters and the output characteristic. In step 1 of this phase, the levels of the selected design parameters that maximize a performance statistic such as the signal-to-noise ratio are determined. The nonlinearity in the relationship between the S/N ratio and the levels of the design parameters is used to determine the optimal settings. The idea is to reduce the performance variability and create a design that is robust to noise factors (such as those due to variation in incoming raw materials and components, manufacturing, and product use). For example, the settings of design parameters (such as depth of cut, rate of feed of tool, amount of catalyst, and number of servers in a fast-food restaurant) that maximize the performance statistic could be determined. When a nonlinear relation exists between the parameter level and the performance statistic, this approach will minimize the sensitivity of the performance statistic to input variations.

The second step is to identify parameters that have a linear relationship with the mean response. Settings for these parameters are chosen to adjust the average response to the

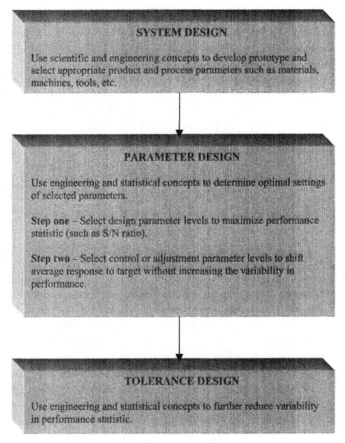

FIGURE 12-24   The three phases of the Taguchi method.

target value. Such parameters are also known as control, or adjustment, factors. By exploiting the linearity of the relationship, it is hoped that adjustment of the factor levels will not affect the variability. Examples include the specific gravity of a liquid, the time that a hot specimen is immersed in a bath, the concentration of a solution, and so on. The factors will depend on the product and the process. Following selection of the optimal parameter and adjustment factor levels, Taguchi recommends running confirmation experiments to verify the chosen settings. The results can also be used to identify factors that are not very sensitive to the performance statistic. The settings of these factors can also be chosen to minimize cost.

The purpose of the third phase, tolerance design, is to determine tolerances, or permissible ranges, for the parameter and factor settings identified in the parameter design phase. The phase is utilized only if the variability of the performance statistic exceeds our desirable levels. Setting such tolerances around the target values helps stabilize the performance measure. The objective of this phase is to further reduce the variability of the performance measure from what was achieved in the parameter design phase. Along the same lines, the tolerances of parameters that have no significant influence on the performance statistic can be relaxed in order to reduce costs.

The parameter design phase is emphasized in the Taguchi approach. Although many companies spend more than 60% of their design activity on system design, they ignore parameter design. Reducing cost and achieving quality at the same time is more likely through careful parameter design. System design may come up with a multitude of designs, but it does not test for the sensitivity of the desired output to the input factors in order to come up with a cost-effective design.

Let's consider, for example, a situation in which a transportation company is expected to meet growing demand and ship goods with as small a deviation from the target date as possible. An important factor is the type of truck, which has three possible levels. Using a new-model truck, which costs more than existing ones but is dependable and efficient, becomes level 1. Downtime of the new trucks is small, and fuel efficiency is high. The existing-model trucks, level 2, cost less than the new trucks but break down more frequently and get poorer gas mileage. Subcontracting the major routes, where the level of profit margin is low, is assigned to level 3. In the parameter design phase, we might find that level 1 maximizes a chosen performance characteristic (say, total profit), minimizes its variability, and is also cost-effective.

Application to Attribute Data

Our discussion of design and analysis using the Taguchi method has dealt with variable data. Variable data have obvious advantages over attribute data in that they provide a quantitative measure of the degree to which the response variable comes close to a target value. For instance, if we are measuring the carbon monoxide concentration in car exhaust, variable data will show the precise concentration as a numerical value, say 0.25. On the other hand, attribute data may indicate levels of carbon monoxide concentration grouped into such categories as poor, acceptable, and highly desirable. The subjective judgment of the experimenter often influences the category to which the response variable is assigned. For example, consider a food taster in a restaurant, who has to classify a menu item as undesirable, fair, good, or excellent. Based on the qualitative opinion of the taster, which could be influenced by personal bias and past experiences, a rating would be assigned. The same food item could get different ratings from different tasters.

The design and experimental layout when the response variable is an attribute is similar to when data are variable. The difference is in the observation of the output variable, which is now classified as an attribute, and its analysis. To keep the analysis simple, the *accumulation type method* to analyze the results is discussed. Using summary statistics from this accumulation procedure, the optimal level of the parameters may be selected. The following example illustrates the procedure.

Example 12-16   A financial analyst wishes to investigate the effect of four factors on a customer's ability to repay loans. The annual income level (factor A) is represented by three levels: less than $40,000 (level 1), between $40,000 and $60,000 (level 2), and over $60,000 (level 3). The number of years of job experience (factor B) is represented by three levels: less than one year (level 1), between one and four years (level 2), and over four years (level 3). The amount of outstanding debt (factor C) is represented by three levels: less than $20,000 (level 1), between $20,000 and $40,000 (level 2), and over $40,000 (level 3). The number of dependents (factor D) is represented by three levels: less than three (level 1), between three and five (level 2), and more than five (level 3). The ability to repay loans is measured by outcomes classified into three categories: loan repaid (category R), loan payment disrupted but eventually repaid (category E), and loan defaulted (category F).

TABLE 12-36   Experimental Layout and Outcomes Classified as Attributes

| | Factor | | | | Outcome | | |
Experiment	A	B	C	D	Loan Repaid R	Loan Payment Disrupted but Eventually Paid (E)	Loan Defaulted (F)
1	1	1	1	1	0	1	0
2	1	2	2	2	1	0	0
3	1	3	3	3	0	0	1
4	2	1	2	3	0	1	0
5	2	2	3	1	0	0	1
6	2	3	1	2	1	0	0
7	3	1	3	2	0	1	0
8	3	2	1	3	1	0	0
9	3	3	2	1	1	0	0

Solution   Given the four factors at three levels each, an L_9 (3^4) orthogonal array is selected for the parameter design of the controllable factors. Nine experiments are conducted, and the results of each experiment are shown in Table 12-36. The outcome for each experiment is denoted by a 1 or 0 under the categories R, E, and F.

An accumulation analysis is conducted as follows. For each factor at each level, we determine the total number of times the outcomes are in one of the three categories (R, E, and F) and calculate the percentage of time each of them occurs. Table 12-37 shows the summarized accumulation results. For example, when A is at level 1, denoted by A_1, one loan was repaid (R), one had payment disrupted but eventually repaid (E), and one had defaulted (F). For a total of three loans studied at this level, 33.333% occurred in each of the outcome categories R, E, and F. Similar analyses are conducted for all of the factors at each of their levels.

The choice of factor levels would be influenced by the degree to which we tolerate the defaulting loan payments. For example, suppose that borrowers who default are not acceptable at all, whereas the financial analyst is somewhat tolerant of borrowers who disrupt

TABLE 12-37   Accumulation Analysis of Observed Attribute Outcomes

| | Accumulated Outcomes | | | Percentage | | |
Factor Level	R	E	F	R	E	F
A_1	1	1	1	33.333	33.333	33.333
A_2	1	1	1	33.333	33.333	33.333
A_3	2	1	0	66.667	33.333	0.0
B_1	0	3	0	0.0	100.0	0.0
B_2	2	0	1	66.667	0.0	33.333
B_3	2	0	1	66.667	0.0	33.333
C_1	2	1	0	66.667	33.333	0.0
C_2	2	1	0	66.667	33.333	0.0
C_3	0	1	2	0.0	33.333	66.667
D_1	1	1	1	33.333	33.333	33.333
D_2	2	1	0	66.667	33.333	0.0
D_3	1	1	1	33.333	33.333	33.333

payment but eventually repay their loans. We find that the chosen level of factor A is A_3, which yields the highest percentage of loans that are repaid, and no loan defaults, meeting the specified requirements. The financial manager would therefore select candidates whose income level exceeds \$60,000 (level A_3) for consideration. For factor B we find that 66.667% of loans are repaid for levels B_2 and B_3, the highest in that category. However, each of these two levels also results in 33.333% defaulted loans. If the analyst is intolerant of borrowers who default, the chosen level of factor B may be B_1. At this level of B_1, even though none of the loans are repaid without interruption, none of the loans are defaulted. Based on similar reasoning, for factor C, either of levels C_1 or C_2 is preferable; the final choice would be based on other considerations. Finally, for factor D, level D_2 would satisfy the desired requirements. Thus, the factor settings for each of the four factors are determined.

12-12 CRITIQUE OF EXPERIMENTAL DESIGN AND THE TAGUCHI METHOD

Several criticisms of the Taguchi method (Montgomery 2004a; Nair 1992; Tsui 1992) are of interest. As discussed earlier, Taguchi advocates the use of orthogonal arrays, several of which are shown in Table 12-30 in his experimental design. Examples include two-, three-, and four-level fractional factorial experiments. For example, the L_8 array is really a 2^{7-4} fractional factorial experiment. The disadvantage of an orthogonal array is that the **alias structure** is not readily apparent from the design. It can be shown that the alias structure for this L_8 design is such that main effects are confounded with two-way interactions. This could lead to inaccurate conclusions because an inference on the significance of a main effect depends on the interaction effect. Only if the interaction effect is insignificant will our conclusions drawn on the main effects hold.

Additionally, the alias structure for orthogonal arrays with three or more levels is messy. In fact, it is possible to find fractional factorial experiments with more desirable alias structures. For example, we can find a 2^{4-1} fractional factorial experiment that will have a total of eight experiments (as in the L_8 design) such that the main effects will be confounded with three-factor interactions. This is preferable to the L_8 design where main effects are confounded with two-factor interactions.

Three types of interactions are possible in parameter design experiments: among design parameters, between design parameters and **noise parameters**, and among noise parameters. Taguchi, even though he recognizes interactions between design parameters, downplays their importance relative to the main effects. He proposes the inclusion of as many design parameters as possible, based on the number of experiments chosen. For example, using an L_9 orthogonal array, for which eight factors can be estimated, Taguchi advocates testing eight design parameters rather than a few important design parameters and possible two-way interactions between some of them. This is a potential drawback. Taguchi prefers using three or more levels of the factors to estimate curvature (nonlinear effects) rather than investigating potential interactions.

An alternative procedure could be to identify possible main effects and important interactions and then to consider curvature in the important variables. This may lead to a simpler interpretation of the data and better understanding of the process.

Perhaps a more important concern is the inclusion of the correct variables in the experiment. Out of all possible parameters, which are the key ones? Have they all been included in the design? In response to these issues, Taguchi's parameter design procedure is a form of **screening design**. Critiques suggest that other forms of design such as **response surface designs** are equally effective (Box and Draper 1986). Response surface methodology, originally developed by Box and Wilson (1951), attempts to determine the shape of the response function and its sensitivity to the parameter design factors. It determines points of maxima and minima and the region in the design parameter space where these occur. The advantage of these response surface methods is that they help us understand the nature of the response function and thereby the process. Improved product and process designs are thus facilitated. Knowledge of the model, relating the response variable or performance statistic to the design factors and/or their function, is important.

The Taguchi method attempts to identify the factors that most influence variability of the performance measure. It does not focus on the reason why this happens. In Taguchi's method, the focus is on whether the variability is most influenced by the main effects, by interactions, or by curvature. On the other hand, alternative designs using fractional factorial experiments attempt to identify which components cause the variability to happen and the nature in which they contribute to the variability. The result is that we understand the underlying causes rather than merely identify the significant factors.

Another criticism of the Taguchi method of experimental design is that the adoption of the inner- and outer-array structure leads to a larger number of experiments. Furthermore, sometimes we cannot even estimate the two-factor interactions between the design parameters. The L_9 array allows us to estimate only four factor effects. Alternative experimental designs using fractional factorials may be found that are superior to the one proposed by Taguchi. One example is the 2^{7-2} fractional factorial experiment, which accommodates seven factors, A through G. If the defining contrasts are $I = ABCDF = ABDEG$, none of the main effects are aliased with two-factor interactions. Moreover, only 32 runs are required.

Some researchers maintain that, in general, the inner- and outer-array approach is unnecessary. They prefer a strategy that uses a single array that accommodates both design and noise factors. If defining contrasts are properly chosen, not only the main effects but also desirable interactions can be estimated, and at a lower cost because of fewer experiments that are required.

In the context of experimental design, the linear graphs provided by Taguchi seem to have been developed heuristically. Their use may lead to inefficient designs, as we have seen in the case where main effects are confounded with two-way interactions. A better approach could be to use a design that specifies the complete alias structure.

The Taguchi method uses marginal averages to determine optimal parameter levels. However, Taguchi's procedure for computing marginal averages may not, in general, identify the optimal settings because it ignores the interaction effects between the parameters. However, when interaction effects are not pronounced, Taguchi's approach allows the optimal setting to be approximated. The use of marginal averages is analogous to the one-factor-at-a-time approach, the disadvantages of which we have previously discussed. Furthermore, marginal averages implicitly assign values for factor-level combinations that are not included in the orthogonal array. The method of assignment forces the interactions to be approximately zero, which may not be the case in practice. Obviously, it is difficult to estimate the value of the response variable for missing factor combinations without knowing the response surface.

In Taguchi's parameter design procedure, it is implicitly assumed in using the S/N ratio as a performance measure that the standard deviation of the response is linearly proportional to its mean. This may not be valid in all situations. Furthermore, the knowledge gained through sequential experimentation is not used. In maximizing the S/N ratio, we need to be aware that the maximization is occurring only over the design points considered in the design and not over the entire feasible region defined by the design parameters. In practice, optimizing a nonlinear function, such as the S/N ratio, subject to constraints on the bounds of the factor levels might be more appropriate, in which case, nonlinear programming techniques would be used.

Although these views represent the major criticisms of the Taguchi method, there is still an important place in the field of quality control and improvement for this technique. Taguchi's ideas concerning the loss function have changed the way we think in terms of the definition of quality. The challenge of meeting zero defects, whereby all products are within certain stated specifications, is not enough. Being close to the target value with minimum variability is the real goal. Quality engineering principles that are based on developing product and process designs that are insensitive to environmental factors, are on target with minimum variability, and have the lowest possible cost are enduring ideas.

Taguchi's ideas have motivated many practitioners to focus on achieving quality through design by conducting and analyzing experimental results. Creating a loss function in financial terms that is easily understood by operating personnel and management is a major impact of his philosophy and design procedure. Everyone relates to a bottom line, for example expected loss, when stated in dollars. Taguchi believes in a simplistic and economical design of experiments where the goal is to identify a combination of design parameters that achieves a better level of the performance criteria, which may not necessarily be optimal. However, the added cost of obtaining an optimal design may be prohibitive relative to the gain in the performance criteria. Thus, what Taguchi offers is a good and close-to-optimal solution at a reduced cost. Manufacturing and service industries can always view his method in this light and use it as a stepping stone for subsequent improvement through refined analyses.

SUMMARY

This chapter has examined some fundamental concepts of experimental design. With the objective of achieving quality at a reduced cost, this chapter has discussed procedures through which the impact of design factors on the performance of the product or process can be analyzed. Traditional experimental designs such as the completely randomized design, randomized block design, and Latin square design have been presented. In an effort to achieve a desired combination of parameter levels, the purpose of assigning a certain level to each parameter is to ensure that the process of estimating the effects of these parameters is not biased. The notion of determining the significance of a factor in terms of the mean square of the error component is a central concept in the analysis of variance. The basis for the evaluation process using ANOVA procedures has been illustrated.

Factorial and fractional factorial experiments have been discussed. A full factorial experiment represents all possible combinations of the various factors. A fractional factorial experiment represents a subset of the full factorial experiment, with the experiments selected in such a manner that desirable effects can be estimated. For factorial experiments, the role of contrasts in estimating the treatment effects has been discussed. The procedure of blocking to increase the precision of the experiment, as achieved through the principle of confounding, has been demonstrated. Procedures for selecting fractional replicates of the full design, in which it is desirable to reduce the size of the experiment, have also been presented. This chapter has provided a foundation for product and process design by introducing some common designs and methods for analyzing data using those designs. Inferences from such analysis help us determine treatments that can favorably affect the response.

This chapter has also presented concepts and procedures associated with the Taguchi method of design and analysis. Taguchi's philosophy has been described, and his contributions to the area of quality improvement through design have been discussed. The Taguchi loss function, which is a measure of the deviation of an output characteristic from a target value, has been explained; the three phases of the Taguchi method—system design, parameter design, and tolerance designs—have been presented.

Since parameter design is the important phase in this sequence, it has been discussed in more detail. The objective is to create a design that is robust to the uncontrollable factors and will also achieve a desired target value with the least variability. Performance measures such as the signal-to-noise ratio have been introduced. Cost is a key consideration in the Taguchi approach, with the idea of obtaining the best possible design at the lowest possible cost.

KEY TERMS

accumulation analysis
adjustment factor
alias
alias structure
alternate fraction
analysis of variance
array
   inner
   outer
balanced experiment
bias
blocking
central limit theorem
completely randomized design
confirmation experiments
confounding
contour plot
contrast
   orthogonal
defining contrast
degrees of freedom
design matrix
design resolution
double-blind study
effects
   interaction
   main
expected loss
experiment

experimental design
experimental error
experimental unit
F-statistic
factor
   qualitative
   quantitative
factor levels
   continuous
   discrete
factorial experiment
fixed effects model
fractional factorial experiment
generator
hybrid orthogonal arrays
incomplete block design
interaction
Latin square design
linear graph
loss function
   target is best
   smaller is better
   larger is better
manufacturing tolerance
mean square
   error
   treatment
mean square deviation
measurement bias

noise

  matrix

noise factors

  external

  internal

nominal value

off-line quality control

orthogonal array

parameter

  design

  noise

performance

  measure

  statistic

principal fraction

quality engineering

random effects model

randomization

randomized block design

replication

response surface designs

response variable

robustness

sampling unit

screening design

signal

signal-to-noise ratio

single-blind study

sum of squares

  block

  error

  total

  treatment

system design

Taguchi

  philosophy

  method

target value

tolerance design

treatment

treatment effect

  fixed

unbalanced experiment

variance

verification experiments

EXERCISES

Discussion Questions

12-1 Distinguish between factor, treatment, and treatment levels in the context of a health care facility.

12-2 Explain the importance of experimental design in quality control and improvement for a financial institution.

12-3 Explain the principles of replication, randomization, and blocking, and discuss their roles in experimental design in a semiconductor manufacturing company.

12-4 Explain the concept of interaction between factors, and give some examples in the entertainment industry.

12-5 What is the difference between qualitative and quantitative variables? Give examples of each in the transportation industry. Which of these two classes permit interpolation of the response variable?

12-6 What is the difference between a fixed effects model and a random effects model? Give some examples in the logistics area.

12-7 Explain the difference between the completely randomized design and the randomized block design. Discuss in the context of a gasoline refining process. Under what conditions would you prefer to use each design?

12-8 Distinguish between a randomized block design and a Latin square design. What are the advantages and disadvantages of a Latin square design?

12-9   Explain why it does not make sense to test for the main effects in a factorial experiment if the interaction effects are significant.

12-10   What is the utility of contrasts in experimental design? What are orthogonal contrasts, and how are they helpful?

12-11   What are the features of a 2^k factorial experiment? What are the features of a 2^{k-2} fractional factorial experiment, and how is it constructed?

12-12   Clearly distinguish between the principles of confounding and fractionalization.

12-13   What is the role of a defining contrast, or generator, in fractional factorial experiments? Distinguish between a principal fraction and an alternate fraction.

12-14   Discuss Taguchi's philosophy for quality improvement. Discuss his loss function and its contributions.

12-15   Compare and contrast Taguchi's loss functions for the situations target is best, smaller is better, and larger is better. Give examples in the hospitality industry.

12-16   Discuss the signal-to-noise ratio. How is it used in the Taguchi method? What is an adjustment parameter, and how is it used?

Problems

12-17   A large retail company has to deliver its goods to distributors throughout the country. It has been offered trial-run services by three transportation companies. To test the efficiency of these three companies, it randomly assigns its outgoing product shipments to these transporters and determines the degree of lateness as a proportion of the time allocated for delivery. Table 12-38 shows the degree of lateness of the three companies for five shipments,
(a) Is there a difference in the mean degree of lateness of the three transportation companies? Test at the 10% level of significance.
(b) Find a 95% confidence interval for the mean degree of lateness of company 1.
(c) Find a 90% confidence interval for the difference in the mean degree of lateness of companies 2 and 3. Can we conclude that there is a difference in their means at the 10% level of significance?
(d) Which company would you choose? Why?

TABLE 12-38

Transportation Company	Degree of Lateness Values				
1	0.04	0.00	0.02	0.03	0.02
2	0.15	0.11	0.07	0.09	0.12
3	0.06	0.03	0.04	0.04	0.05

12-18   An airline is interested in selecting a software package for its reservation system. Even though the company would like to maximize use of its available seats, it prefers to bump as few passengers as possible. It has four software packages to choose from. The airline randomly chooses a package and uses it for a month. Over the span of a year, the company uses each package three times. The number of passengers bumped for each month for a given software package is shown in Table 12-39.

(a) Is there a difference in the software packages as measured by the mean number of passengers bumped? Test at the 5% level of significance.

(b) Find a 90% confidence interval for the mean number of passengers bumped using software package 2.

(c) Find a 95% confidence interval for the difference in the mean number of passengers bumped using software packages 1 and 2. Is there a difference in their means at the 5% level of significance?

(d) Which software package would you choose?

TABLE 12-39

Software Package	Passengers Bumped		
1	12	14	9
2	2	4	3
3	10	9	6
4	7	6	7

12-19 Three training programs are being considered for auditors in an accounting firm. The success of the training programs is measured on a rating scale from 0 to 100, with higher values indicating a desirable program. The company has categorized its auditors into three groups, depending on their number of years of experience. Three auditors are selected from each group and are randomly assigned to the training programs. Table 12-40 shows the ratings assigned to the auditors after they completed the training program.

(a) Is there a difference in the training programs as indicated by the mean evaluation scores? Test at the 10% level of significance.

(b) Find a 90% confidence interval for the difference in the mean evaluation scores of training programs 1 and 3. Is there a difference between these two programs?

TABLE 12-40

Years of Experience	Training Programs		
	1	2	3
0–5	70	65	81
5–10	75	80	87
10–15	87	82	94

12-20 A doctor is contemplating four types of diet to reduce the blood sugar levels of patients. Because of differences in the metabolism of patients, the doctor categorizes the patients into five age groups. From each age group, four patients are selected and randomly assigned to the diets. After two months on the diet, the reduction in patients' blood sugar levels is found. The results are shown in Table 12-41.

(a) Is there a difference in the diet types as evidenced by the mean reduction in blood sugar level? Test at the 5% level of significance.

TABLE 12-41

Patient Age Group	Diet Type			
	1	2	3	4
20–30	40	70	35	40
30–40	60	80	65	60
40–50	65	60	60	50
50–60	30	50	40	20
60–70	45	55	50	30

(b) Find a 90% confidence interval for the mean reduction in blood sugar level for diet type 1.

(c) Find a 95% confidence interval for the difference in the mean reduction in blood sugar levels between diet types 2 and 4. Is there a difference between the two diet types?

(d) Which diet type would you prefer?

12-21　A consulting firm wishes to evaluate the performance of four software packages (A, B, C, and D) as measured by the computational time. The experimenter wishes to control for two variables: the problem type and the operating system configuration used. Four classes of each of these two variables are identified. Each software package is used only once on each problem type and only once on each operating system configuration. Table 12-42 shows the computational times (ms) for the four software packages under these controlled conditions.

(a) Test at the 5% level of significance for the equality of the mean computational times of the four software packages.

(b) Find a 95% confidence interval for the mean computational time using software package C.

(c) Find a 95% confidence interval for the difference in the mean computational times of software packages A and C. Is there a difference between the two?

(d) Which software package would you choose, and why?

(e) Was blocking of the two variables, problem type and operating system configuration, effective?

(f) If you had to rerun the experiment, what type of design would you use?

TABLE 12-42

Problem Type	Operating System Configuration							
	I		II		III		IV	
1	B	14.2	A	37.3	D	24.3	C	22.6
2	D	18.5	C	18.1	B	27.4	A	39.6
3	A	36.1	B	16.8	C	13.2	D	20.3
4	C	17.5	D	24.6	A	36.7	B	19.5

12-22　Two controllable factors, temperature and pressure, are each kept at three levels to determine their impact on the ductility of an alloy being produced. The temperature levels are 150, 250, and 300 °C, respectively. Pressure is controlled at 50,100, and

150 kg/cm^2. Each of the nine treatments is replicated five times. The ductility of the produced alloy is shown in Table 12-43. The higher the value, the more ductile the alloy.

(a) Test whether the interaction effects between temperature and pressure are significant at the 5% level of significance.

(b) What are the desirable settings of the process temperature and pressure if a ductile alloy is preferred?

(c) Find a 90% confidence interval for the mean ductility when the process temperature is set at 150 °C and the pressure at 150 kg/cm^2.

(d) Find a 95% confidence interval for the difference in mean ductility for a process temperature of 250 °C and pressure of 150 kg/cm^2 versus a process temperature of 300 °C and pressure of 100 kg/cm^2.

TABLE 12-43

Temperature (C°)	Pressure (kg/cm^2)		
	50	100	150
150	50	44	75
	65	49	80
	70	52	81
	73	40	76
	68	45	82
250	40	25	86
	45	30	88
	56	20	82
	50	33	81
	52	34	76
300	72	62	60
	60	71	62
	75	65	55
	73	68	52
	70	72	50

12-23   Consider Example 12-4 concerning the efficiency of synthetic fuel for which the factors are an additive and a catalyst. The original data are given in Table 12-19. The experiment is conducted using only five randomly chosen automobiles (A, B, C, D, and E). Each treatment is used on each automobile. The reason for choosing only five automobiles and replicating the experiment is to eliminate any variation in the fuel efficiencies that might arise due to differences in the automobiles. The data in Table 12-19 is interpreted as follows. For additive level I and catalyst level 1, the response value is 75 for automobile A, 72 for automobile B, and so forth. Similar interpretations hold for the other treatments.

(a) What is your conclusion regarding the significance of the factors or their interactions? Test at the 5% level of significance.

(b) Find a 95% confidence interval for the mean efficiency when the additive is at level I and the catalyst is at level 2.

(c) Find a 95% confidence interval for the difference in the treatment means when the additive is at level I and the catalyst is at level 3 versus when the additive is at level II and the catalyst is at level 3. What is your conclusion?

(d) What levels of additive and catalyst would you prefer to choose?

(e) Was it beneficial to select only five automobiles and replicate the treatments on only those automobiles? Test at the 5% level of significance.

12-24   Refer to Exercise 12-17.

(a) Is there a difference between the mean degree of lateness of company 3 and that of the averages of companies 1 and 2? Test at the 5% level of significance.

(b) Find a 90% confidence interval for the contrast defined in part (a).

12-25   Refer to Exercise 12-18.

(a) Is there a difference in the mean number of passengers bumped using software packages 1 and 2 from that using software packages 3 and 4? Test at the 10% level of significance.

(b) Find a 95% confidence interval for the contrast that tests for the difference in the mean number of passengers bumped using software package 3 and the average of that using software packages 1 and 2.

12-26   Refer to Exercise 12-19. Find a 90% confidence interval for the difference in the mean effectiveness of program 1 and the average of that using programs 2 and 3.

12-27   Refer to Exercise 12-19. Consider the following two contrasts of totals: (1) difference between the totals for training programs 1 and 3; (2) difference between the totals for the sum of training programs 1 and 3 and twice that of training program 2.

(a) Find the sum of squares due to each of these two contrasts.

(b) What null hypothesis is indicated by contrast (2)? Test the hypothesis at a level of significance of 0.05, and explain your conclusions.

12-28   Refer to Exercise 12-20. Consider the following three contrasts: (1) difference between the sum of the reduction in blood sugar levels using diet types 1 and 3 from that using diet types 2 and 4; (2) difference between the sum of the reduction in blood sugar level using diet types 1 and 2 from that using diet types 3 and 4; (3) difference in the sum of the reduction in blood sugar level using diet types 2 and 3 from that using diet types 1 and 4.

(a) Find the sum of squares due to each of the contrasts, and discuss their contribution to the treatment sum of squares.

(b) What null hypothesis is indicated by contrast (1)? Test the hypothesis at a 5% level of significance.

(c) What type of hypothesis is indicated by contrast (2)? Test the hypothesis at the 10% level of significance.

12-29   Write out the treatment combinations for a 2^4 factorial experiment.

12-30   In the search for a lower-pollution synthetic fuel, researchers are experimenting with three different factors, each controlled at two levels, for the processing of such a fuel. Factor A is the concentration of corn extract at 5% and 10%, factor B is the concentration of an ethylene-based compound at 15% and 25%, and factor C is the distillation temperature at 120 °C and 150 °C. The levels of undesirable emission of the fuel are shown in Table 12-44 for three replications of each treatment; each level

TABLE 12-44

Treatment	Degree of Undesirable Emission Level (ppm)		
(1)	30	24	26
a	18	22	24
b	30	32	25
ab	43	47	41
c	28	24	22
ac	54	49	46
bc	58	48	50
abc	24	20	22

is randomly assigned to a treatment. The larger the level of emission, the worse the impact on the environment.

(a) Find all of the main effects and the interaction effects.

(b) Find the sum of squares for each of the effects and the interaction effects.

(c) At the 5% level of significance, which effects are significant? Interpret your inferences.

12-31   Consider a 2^4 factorial experiment. Set up a table of the coefficients for orthogonal contrasts similar to Table 12-25. Write down the contrasts for estimating the main effects and the two-factor interactions. If the four-way interaction effect $ABCD$ is not significant, use that as a basis to confound the design into two blocks.

12-32   In Exercise 12-31, use AB as the confounding factor to divide the experiment into two blocks. How would you estimate the effect of factor A?

12-33   Consider a 2^4 factorial experiment. Using BC as the defining contrast, find the treatment combinations in a 2^{4-1} fractional factorial experiment. Find the aliases of the contrasts. How would it be possible to estimate the effect of the contrast BC? If AD is used as a second defining contrast, determine the treatment combinations in a 2^{4-2} fractional factorial experiment. What is the alias structure now?

12-34   In a 2^{5-2} fractional factorial experiment, using CDE and AB as the generators, find the treatment combinations. Find the aliases of the contrasts.

12-35   Refer to Exercise 12-30. Factor A is the concentration of corn extract, factor B is the concentration of an ethylene based compound, and factor C is the distillation temperature. Each factor will be controlled at two levels. Suppose the experimenter runs a 2^{3-1} fractional factorial experiment, with the defining contrast being $I = ABC$. Using the data in Table 12-44, perform an analysis to determine the significance of effects. Which effects cannot be estimated? What is the alias structure? Comment on your inferences if the level of significance is 5%.

12-36   A manufacturer of magnetic tapes is interested in reducing the variability of the thickness of the coating on the tape. It is estimated that the loss to the consumer is $10 per reel if the thickness exceeds 0.005 ± 0.0004 mm. Each reel has 200 m of tape. A random sample of 10 yields the following thickness (in millimeters): 0.0048, 0.0053, 0.0051, 0.0051, 0.0052, 0.0049, 0.0051, 0.0047, 0.0054, 0.0052. Find the average loss per reel.

12-37   Refer to Exercise 12-36. The manufacturer is considering adopting a new process to reduce the variability in the thickness of coating. It is estimated that the additional cost for this improvement is $0.03 per linear meter. The annual production is 10,000 reels. Each reel has 200 m of tape. A random sample of size 8 from the new process yielded the following thickness (in millimeters): 0.0051, 0.0048, 0.0049, 0.0052, 0.0052, 0.0051, 0.0050, 0.0049. Is it cost-effective to use the new process? What is the annual savings or loss?

12-38   Refer to Exercise 12-36. Suppose that the manufacturer can rework the thickness prior to shipping the product at a cost of $2.00 per reel. What should the manufacturer's tolerance be?

12-39   Refer to Exercise 12-36. Suppose the manufacturer has the ability to center the process such that the average thickness of the coating is at 0.005 mm, which is the target value. In doing so, the manufacturer estimates that the standard deviation of the process will be 0.018 mm. The cost of making this change in the process is estimated to be $1.50 per reel. Would it be cost-effective to make this change, compared to the original process? What would the annual savings or loss be if the annual production is 10,000 reels?

12-40   A restaurant believes that two of the most important factors that help it attract and retain customers are the price of the item and the time taken to serve the customer. Based on the price for similar items in other neighboring restaurants, it is estimated that the customer tolerance limit for price is $8, and the associated customer loss is estimated to be $50. Similarly, the customer tolerance limit for the service time is 10 minutes for which the associated customer loss is $40. A random sample of size 10 yields the following values of price: 6.50, 8.20, 7.00, 8.50, 5.50, 7.20, 6.40, 5.80, 7.40, 8.30. The sample service times (in minutes) are 5.2, 7.5, 4.8, 11.4, 9.8, 10.5, 8.2, 11.0, 12.0, 8.5. Find the total expected loss per customer. If the restaurant expects 2000 customers monthly, what is the expected monthly loss?

12-41   Refer to Exercise 12-40. The restaurant is thinking of hiring more personnel to cut down the service time. However, the additional cost of increasing personnel is estimated to be $0.50 per customer. The results of sampling with the added personnel yields the following waiting times (in minutes): 8.4, 5.6, 7.8, 6.8, 8.5, 6.2, 6.5, 5.9, 6.4, 7.5. Is it cost-effective to add personnel? What is the total expected monthly loss?

12-42   The Environmental Protection Agency has identified four factors (A, B, C, and D), each at two levels, that are significant in their effect on the air pollution level at a photographic film production facility. The agency also feels that the interaction effects $A \times C$, $A \times B$, and $B \times C$ are important. Show an experimental design that can estimate these effects using a minimal number of experiments.

12-43   A baseball team manager believes that five factors (A, B, C, D, and E), each at two levels, are significant in affecting runs batted in. The manager believes that the interactions $B \times C$ and $B \times E$ are important. Show an experimental design using an orthogonal array that can estimate these effects.

12-44   The tourism board of a large metropolitan area is seeking ways to promote tourism. They have identified five factors (A, B, C, D, and E) that they feel have an impact on tourist satisfaction. Factor C has four levels, and each of the other four factors has two levels. Show an experimental design, using an orthogonal array, that could estimate the factor effects.

12-45   A city library has established three factors (A, B, and C), each at three levels, that influence the satisfaction of their patrons. The library governance committee also believes that the interaction $B \times C$ is important. Using an orthogonal array, set up an appropriate experimental design.

12-46   In a drilling operation, four factors (A, B, C, and D), each at three levels, are thought to be of importance in influencing the volume of crude oil pumped. Using an L_9 orthogonal array, the factors A, B, C, and D are assigned to columns 1, 2, 3, and 4, respectively. The response variable showing the number of barrels (in thousands) pumped per day for each of the nine experiments is as follows:

Experiment	1	2	3	4	5	6	7	8	9
Barrels per Day (thousands)	6.8	15.8	10.5	5.2	17.1	3.4	5.9	12.2	8.5

Show the experimental design and the response variable for the corresponding experiments. Determine the main effects. Plot the average response curves. What are the optimal settings of the design parameters?

12-47   Consider Exercise 12-46. With the assignment of factors A, B, C, and D to columns 1, 2, 3, and 4, respectively, of an L_9 orthogonal array, the output is as follows for another replication of the nine experiments:

Experiment	1	2	3	4	5	6	7	8	9
Barrels per Day (thousands)	12.2	18.3	13.5	8.3	17.2	7.5	7.9	15.7	14.8

Determine the main effects. Plot the average response curves. What are the optimal settings of the design parameters?

12-48   In a food processing plant, four design parameters, A, B, C, and D, each at three levels, have been identified as having an effect on the moisture content in packaged meat. Three noise factors, E, F, and G, each at two levels, are also to be investigated in the experiment. In the inner array, an L_9 orthogonal array is used with the factors A, B, C, and D assigned to columns 1, 2, 3, and 4, respectively. For the outer array, an L_4 orthogonal array is used with the noise factors E, F, and G assigned to columns 1, 2, and 3, respectively. Table 12-45 shows the moisture content in a coded scale for four replications of each of the nine experiments in the inner array. It is preferable that the moisture content be around a coded value of 20. Show the complete parameter design layout using the Taguchi method. Calculate the mean response and the appropriate signal-to-noise ratio. Plot the average S/N ratios and the average responses, and discuss how the design factor levels are to be selected. Use the average response plots to determine the existence of possible interactions $B \times E$ and $C \times F$. What are the optimal settings of the design parameters using Taguchi's parameter design approach?

12-49   Consider Exercise 12-48. The design factors are A, B, C, and D, each at three levels. These are assigned to an orthogonal array (inner array) with the factors A, B, C, and D assigned to columns 1, 2, 3, and 4, respectively. Suppose that in addition to the

TABLE 12-45

Inner-Array	Outer-Array Experiment			
Experiment	1	2	3	4
1	18.5	21.2	20.5	19.3
2	16.8	17.3	20.9	18.5
3	21.1	21,8	20.6	19.4
4	20.2	17.7	19.8	20.8
5	16.2	21.5	21.2	21.4
6	18.3	18.5	17.8	17.2
7	20.6	21.4	16.8	19.5
8	17.5	20.0	21.0	20.4
9	20.4	18.8	19.6	18.3

three main effects of the noise factors E, F, and G, it is felt that the interaction effects E × F and E × G should be investigated. What type of design would you use for the outer array?

Suppose the outer array selected is L_8, with the assignments as follows: E to column 1, F to column 2, E × F to column 3, G to column 4, and E × G to column 5. The remaining two columns are assigned to the experimental error. Table 12-46 shows the moisture content of the eight replications for each of the nine experiments in the inner array. The coded target value is 20.

Show the complete parameter design layout using the Taguchi method. Calculate the mean response and the appropriate signal-to-noise ratios. Plot the average S/N ratios and the average responses, and discuss how the design factor levels are to be selected. Use the average response plots to determine the existence of the interaction effects A × E, B × F, E × F, and E × G. What are the optimal settings of the design parameters using Taguchi's parameter design approach?

TABLE 12-46

Inner-Array	Outer-Array Experiment							
Experiment	1	2	3	4	5	6	7	8
1	19.3	20.2	19.1	18.4	21.1	20.6	19.5	18.7
2	20.6	18.5	20.2	19.4	20.1	16.3	17.2	19.4
3	18.3	20.7	19.4	17.6	20.4	17.3	18.2	19.2
4	20.8	21.2	20.2	19.9	21.7	22.2	20.4	20.6
5	18.7	19.8	19.4	17.2	18.5	19.7	18.8	18.4
6	21.1	20.2	22.4	20.5	18.7	21.4	21.8	20.6
7	17.5	18.3	20.0	18.8	20.2	17.7	17.9	18.2
8	20.4	21.2	22.4	21.9	21.5	20.8	22.5	21.7
9	18.0	20.2	17.6	22.4	17.2	21.6	18.5	19.2

12-50 In a textile processing plant the quality of the output fabric is believed to be influenced by four factors (A, B, C, and D), each of which can be controlled at three levels. The fabric is classified into three categories: acceptable, second-class, or reject. An L_9 orthogonal array is selected for the design factors, with the factors A,

B, C, and D assigned to columns 1, 2, 3, and 4, respectively. The observations are shown in Table 12-47. Management wants to eliminate reject product altogether. Conduct an accumulation analysis, and determine the optimal settings of the design factors.

TABLE 12-47

	Factor				Quality of Fabric		
Experiment	A	B	C	D	Acceptable	Second Class	Reject
1	1	1	1	1	0	1	0
2	1	2	2	2	1	0	0
3	1	3	3	3	0	1	0
4	2	1	2	3	0	1	0
5	2	2	3	1	0	0	1
6	2	3	1	2	1	0	0
7	3	1	3	2	0	0	1
8	3	2	1	3	0	1	0
9	3	3	2	1	1	0	0

12-51   Consider Exercise 12-50. Four factors (A, B, C, and D), each at three levels, are controlled in an experiment using an L_9 orthogonal array. The output quality is classified as acceptable or unacceptable; unacceptable includes both the second-class and reject classes. Consider the data shown in Table 12-47. Combine the second-class and reject categories into one class and label it as unacceptable, and accumulate the data accordingly. From the accumulation analysis, determine the optimal settings of the design factors.

REFERENCES

Box, G.E.P. (1986). *Studies in Quality Improvement: Signal to Noise Ratios, Performance Criteria and Statistical Analysis: Part I*, Report 11. Center for Quality and Productivity Improvement. Madison, WI: University of Wisconsin.

Box, G.E.P., and N.R. Draper (1986). *Empirical Model Building and Response Surfaces*. New York: Wiley.

Box, G.E.P., and K.B. Wilson (1951). "On the Experimental Attainment of Optimum Conditions," *Journal of the Royal Statistical Society, Series B*, 13 (1): 1–45 (with discussion).

Box, G.E.P., W.G. Hunter, and J.S. Hunter (2005). *Statistics for Experimenters,* 2nd ed. New York: Wiley.

Byrne, D.M., and S. Taguchi (1987). "The Taguchi Approach to Parameter Design," *Quality Progress*, 20 (12): 19–26.

Dehnad, K. (1989). *Quality Control, Robust Design, and the Taguchi Method*. Pacific Grove, CA: Wadsworth & Brooks/Cole.

Gunter, B. (1989). "Statistically Designed Experiments: 1. Quality Improvement, the Strategy of Experimentation, and the Road to Hell," *Quality Progress*, 22: 63–64.

— (1990a). "Statistically Designed Experiments: 2. The Universal Structure Underlying Experimentation," *Quality Progress*, 23: 87–89.

— (1990b). "Statistically Designed Experiments: 3. Interaction," *Quality Progress*, 23: 74–75.

— (1990c). "Statistically Designed Experiments: 4. Multivariate Optimization," *Quality Progress*, 23: 68–70.

— (1990d). "Statistically Designed Experiments: 5. Robust Process and Product Design and Related Matters," *Quality Progress*, 23: 107–108.

Hunter, J.S. (1985). "Statistical Design Applied to Product Design," *Journal of Quality Technology*, 17 (4): 210–221.

Kackar, R.N. (1986). "Taguchi's Quality Philosophy: Analysis and Commentary," *Quality Progress*, 19 (12): 21–29.

Leon, R.V., A.C. Shoemaker, and R.N. Kackar (1987). "Performance Measures Independent of Adjustments," *Technometrics*, 29 (3): 253–265; (discussion) 266–285.

Minitab, Inc.(2007). *Release 15*. State College, PA: Minitab.

Montgomery D.C., (2004a). *Design and Analysis of Experiments*. 6th ed., Hoboken, NJ: Wiley.

— (2004b). *Introduction to Statistical Quality Control*. 5th ed., Hoboken, NJ: Wiley.

Nair, V.N. (1992). "Taguchi's Parameter Design: A Panel Discussion," *Technometrics*, 34 (2): 127–161.

Peterson, R.G. (1985). *Design and Analysis of Experiments*. New York: Marcel Dekker.

Raghavarao, D. (1971). *Constructions and Combinatorial Problems in Design of Experiments*. New York: Wiley.

Ryan, T.P. (2000). *Statistical Methods for Quality Improvement*. 2nd ed., Hoboken, NJ: Wiley.

Schmidt, S.R., and J.R. Boudot (1989). "A Monte Carlo Simulation Study Comparing Effectiveness of Signal- to-Noise Ratios and Other Methods for Identifying Dispersion Effects," presented at the 1989 Rocky Mountain Quality Conference.

Taguchi, G. (1986). *Introduction to Quality Engineering: Designing Quality into Products and Processes*. Tokyo, Japan: Asian Productivity Organization.

— (1987). *System of Experimental Designs: Engineering Methods to Optimize Quality and Minimize Costs*, Vol. 1 and 2. Dearborn, MI: American Supplier Institute.

Taguchi, G., and Y. Wu (1980). *Introduction to Off-Line Quality Control System*. Central Japan Quality Control Association. Nagoya, Japan: Available from American Supplier Institute, Bingham Farms, MI.

Taguchi, G., A. Elsayed, and T. Hsiang (1989). *Quality Engineering in Production Systems*. New York: McGraw-Hill.

Tsui, K.L. (1992). "An Overview of Taguchi Method and Newly Developed Statistical Methods for Robust Design," *IIE Transactions*, 24 (5): 44–57.

Wadsworth, H.M., K.S. Stephens, and A.B. Godfrey (2001). *Modern Methods for Quality Control and Improvement*. 2nd ed., Hoboken, NJ: Wiley.

APPENDIXES

APPENDIX A-1   Cumulative Binomial Distribution

		\multicolumn{10}{c}{$p =$ Probability of Occurrence}
n	X	0.05	0.10	0.15	0.20	0.25	0.30	0.35	0.40	0.45	0.50
2	0	0.903	0.810	0.723	0.640	0.563	0.490	0.423	0.360	0.303	0.250
	1	0.998	0.990	0.978	0.960	0.938	0.910	0.878	0.840	0.798	0.750
3	0	0.857	0.729	0.614	0.512	0.422	0.343	0.275	0.216	0.166	0.125
	1	0.993	0.972	0.939	0.896	0.844	0.784	0.718	0.648	0.575	0.500
	2	1.000	0.999	0.997	0.992	0.984	0.973	0.957	0.936	0.909	0.875
4	0	0.815	0.656	0.522	0.410	0.316	0.240	0.179	0.130	0.092	0.063
	1	0.986	0.948	0.890	0.819	0.738	0.652	0.563	0.475	0.391	0.313
	2	1.000	0.996	0.988	0.973	0.949	0.916	0.874	0.821	0.759	0.688
	3		1.000	0.999	0.998	0.996	0.992	0.985	0.974	0.959	0.938
5	0	0.774	0.590	0.444	0.328	0.237	0.168	0.116	0.078	0.050	0.031
	1	0.977	0.919	0.835	0.737	0.633	0.528	0.428	0.337	0.256	0.188
	2	0.999	0.991	0.973	0.942	0.896	0.837	0.765	0.683	0.593	0.500
	3	1.000	1.000	0.998	0.993	0.984	0.969	0.946	0.913	0.869	0.813
	4			1.000	1.000	0.999	0.998	0.995	0.990	0.982	0.969
6	0	0.735	0.531	0.377	0.262	0.178	0.118	0.075	0.047	0.028	0.016
	1	0.967	0.886	0.776	0.655	0.534	0.420	0.319	0.233	0.164	0.109
	2	0.998	0.984	0.953	0.901	0.831	0.744	0.647	0.544	0.442	0.344
	3	1.000	0.999	0.994	0.983	0.962	0.930	0.883	0.821	0.745	0.656
	4		1.000	1.000	0.998	0.995	0.989	0.978	0.959	0.931	0.891
	5				1.000	1.000	0.999	0.998	0.996	0.992	0.984

Fundamentals of Quality Control and Improvement, Third Edition, By Amitava Mitra
Copyright © 2008 John Wiley & Sons, Inc.

APPENDIX A-1 (continued)

n	X	\multicolumn{10}{c}{p = Probability of Occurrence}
		0.05	0.10	0.15	0.20	0.25	0.30	0.35	0.40	0.45	0.50
7	0	0.698	0.478	0.321	0.210	0.133	0.082	0.049	0.028	0.015	0.008
	1	0.956	0.850	0.717	0.577	0.445	0.329	0.234	0.159	0.102	0.063
	2	0.996	0.974	0.926	0.852	0.756	0.647	0.532	0.420	0.316	0.227
	3	1.000	0.997	0.988	0.967	0.929	0.874	0.800	0.710	0.608	0.500
	4		1.000	0.999	0.995	0.987	0.971	0.944	0.904	0.847	0.773
	5			1.000	1.000	0.999	0.996	0.991	0.981	0.964	0.938
	6					1.000	1.000	0.999	0.998	0.996	0.992
8	0	0.663	0.430	0.272	0.168	0.100	0.058	0.032	0.017	0.008	0.004
	1	0.943	0.813	0.657	0.503	0.367	0.255	0.169	0.106	0.063	0.035
	2	0.994	0.962	0.895	0.797	0.679	0.552	0.428	0.315	0.220	0.145
	3	1.000	0.995	0.979	0.944	0.886	0.806	0.706	0.594	0.477	0.363
	4		1.000	0.997	0.990	0.973	0.942	0.894	0.826	0.740	0.637
	5			1.000	0.999	0.996	0.989	0.975	0.950	0.912	0.855
	6				1.000	1.000	0.999	0.996	0.991	0.982	0.965
	7						1.000	1.000	0.999	0.998	0.996
9	0	0.630	0.387	0.232	0.134	0.075	0.040	0.021	0.010	0.005	0.002
	1	0.929	0.775	0.599	0.436	0.300	0.196	0.121	0.071	0.039	0.020
	2	0.992	0.947	0.859	0.738	0.601	0.463	0.337	0.232	0.150	0.090
	3	0.999	0.992	0.966	0.914	0.834	0.730	0.609	0.483	0.361	0.254
	4	1.000	0.999	0.994	0.980	0.951	0.901	0.828	0.733	0.621	0.500
	5		1.000	0.999	0.997	0.990	0.975	0.946	0.901	0.834	0.746
	6			1.000	1.000	0.999	0.996	0.989	0.975	0.950	0.910
	7					1.000	1.000	0.999	0.996	0.991	0.980
	8							1.000	1.000	0.999	0.998
10	0	0.599	0.349	0.197	0.107	0.056	0.028	0.013	0.006	0.003	0.001
	1	0.914	0.736	0.544	0.376	0.244	0.149	0.086	0.046	0.023	0.011
	2	0.988	0.930	0.820	0.678	0.526	0.383	0.262	0.167	0.100	0.055
	3	0.999	0.987	0.950	0.879	0.776	0.650	0.514	0.382	0.266	0.172
	4	1.000	0.998	0.990	0.967	0.922	0.850	0.751	0.633	0.504	0.377
	5		1.000	0.999	0.994	0.980	0.953	0.905	0.834	0.738	0.623
	6			1.000	0.999	0.996	0.989	0.974	0.945	0.898	0.828
	7				1.000	1.000	0.998	0.995	0.988	0.973	0.945
	8						1.000	0.999	0.998	0.995	0.989
	9							1.000	1.000	1.000	0.999
11	0	0.569	0.314	0.167	0.086	0.042	0.020	0.009	0.004	0.001	0.000
	1	0.898	0.697	0.492	0.322	0.197	0.113	0.061	0.030	0.014	0.006
	2	0.985	0.910	0.779	0.617	0.455	0.313	0.200	0.119	0.065	0.033
	3	0.998	0.981	0.931	0.839	0.713	0.570	0.426	0.296	0.191	0.113
	4	1.000	0.997	0.984	0.950	0.885	0.790	0.668	0.533	0.397	0.274
	5		1.000	0.997	0.988	0.966	0.922	0.851	0.753	0.633	0.500
	6			1.000	0.998	0.992	0.978	0.950	0.901	0.826	0.726
	7				1.000	0.999	0.996	0.988	0.971	0.939	0.887
	8					1.000	0.999	0.998	0.994	0.985	0.967
	9						1.000	1.000	0.999	0.998	0.994
	10								1.000	1.000	1.000
12	0	0.540	0.282	0.142	0.069	0.032	0.014	0.006	0.002	0.001	0.000
	1	0.882	0.659	0.443	0.275	0.158	0.085	0.042	0.020	0.008	0.003
	2	0.980	0.889	0.736	0.558	0.391	0.253	0.151	0.083	0.042	0.019
	3	0.998	0.974	0.908	0.795	0.649	0.493	0.347	0.225	0.134	0.073

APPENDIX A-1 *(continued)*

		p = Probability of Occurrence									
n	X	0.05	0.10	0.15	0.20	0.25	0.30	0.35	0.40	0.45	0.50
	4	1.000	0.996	0.976	0.927	0.842	0.724	0.583	0.438	0.304	0.194
	5		0.999	0.995	0.981	0.946	0.882	0.787	0.665	0.527	0.387
	6		1.000	0.999	0.996	0.986	0.961	0.915	0.842	0.739	0.613
	7			1.000	0.999	0.997	0.991	0.974	0.943	0.888	0.806
	8				1.000	1.000	0.998	0.994	0.985	0.964	0.927
	9						1.000	0.999	0.997	0.992	0.981
	10							1.000	1.000	0.999	0.997
	11									1.000	1.000
13	0	0.513	0.254	0.121	0.055	0.024	0.010	0.004	0.001	0.000	0.000
	1	0.865	0.621	0.398	0.234	0.127	0.064	0.030	0.013	0.005	0.002
	2	0.975	0.866	0.692	0.502	0.333	0.202	0.113	0.058	0.027	0.011
	3	0.997	0.966	0.882	0.747	0.584	0.421	0.278	0.169	0.093	0.046
	4	1.000	0.994	0.966	0.901	0.794	0.654	0.501	0.353	0.228	0.133
	5		0.999	0.992	0.970	0.920	0.835	0.716	0.574	0.427	0.291
	6		1.000	0.999	0.993	0.976	0.938	0.871	0.771	0.644	0.500
	7			1.000	0.999	0.994	0.982	0.954	0.902	0.821	0.709
	8				1.000	0.999	0.996	0.987	0.968	0.930	0.867
	9					1.000	0.999	0.997	0.992	0.980	0.954
	10						1.000	1.000	0.999	0.996	0.989
	11								1.000	0.999	0.998
	12									1.000	1.000
14	0	0.488	0.229	0.103	0.044	0.018	0.007	0.002	0.001	0.000	0.000
	1	0.847	0.585	0.357	0.198	0.101	0.047	0.021	0.008	0.003	0.001
	2	0.970	0.842	0.648	0.448	0.281	0.161	0.084	0.040	0.017	0.006
	3	0.996	0.956	0.853	0.698	0.521	0.355	0.220	0.124	0.063	0.029
	4	1.000	0.991	0.953	0.870	0.742	0.584	0.423	0.279	0.167	0.090
	5		0.999	0.988	0.956	0.888	0.781	0.641	0.486	0.337	0.212
	6		1.000	0.998	0.988	0.962	0.907	0.816	0.692	0.546	0.395
	7			1.000	0.998	0.990	0.969	0.925	0.850	0.741	0.605
	8				1.000	0.998	0.992	0.976	0.942	0.881	0.788
	9					1.000	0.998	0.994	0.982	0.957	0.910
	10						1.000	0.999	0.996	0.989	0.971
	11							1.000	0.999	0.998	0.994
	12								1.000	1.000	0.999
	13										1.000
15	0	0.463	0.206	0.087	0.035	0.013	0.005	0.002	0.000	0.000	0.000
	1	0.829	0.549	0.319	0.167	0.080	0.035	0.014	0.005	0.002	0.000
	2	0.964	0.816	0.604	0.398	0.236	0.127	0.062	0.027	0.011	0.004
	3	0.995	0.944	0.823	0.648	0.461	0.297	0.173	0.091	0.042	0.018
	4	0.999	0.987	0.938	0.836	0.686	0.515	0.352	0.217	0.120	0.059
	5	1.000	0.998	0.983	0.939	0.852	0.722	0.564	0.403	0.261	0.151
	6		1.000	0.996	0.982	0.943	0.869	0.755	0.610	0.452	0.304
	7			0.999	0.996	0.983	0.950	0.887	0.787	0.654	0.500
	8			1.000	0.999	0.996	0.985	0.958	0.905	0.818	0.696
	9				1.000	0.999	0.996	0.988	0.966	0.923	0.849
	10					1.000	0.999	0.997	0.991	0.975	0.941
	11						1.000	1.000	0.998	0.994	0.982
	12								1.000	0.999	0.996
	13									1.000	1.000

(continued)

APPENDIX A-1 (*continued*)

						$p = $ Probability of Occurrence					
n	X	0.05	0.10	0.15	0.20	0.25	0.30	0.35	0.40	0.45	0.50
16	0	0.440	0.185	0.074	0.028	0.010	0.003	0.001	0.000	0.000	0.000
	1	0.811	0.515	0.284	0.141	0.063	0.026	0.010	0.003	0.001	0.000
	2	0.957	0.789	0.561	0.352	0.197	0.099	0.045	0.018	0.007	0.002
	3	0.993	0.932	0.790	0.598	0.405	0.246	0.134	0.065	0.028	0.011
	4	0.999	0.983	0.921	0.798	0.630	0.450	0.289	0.167	0.085	0.038
	5	1.000	0.997	0.976	0.918	0.810	0.660	0.490	0.329	0.198	0.105
	6		0.999	0.994	0.973	0.920	0.825	0.688	0.527	0.366	0.227
	7		1.000	0.999	0.993	0.973	0.926	0.841	0.716	0.563	0.402
	8			1.000	0.999	0.993	0.974	0.933	0.858	0.744	0.598
	9				1.000	0.998	0.993	0.977	0.942	0.876	0.773
	10					1.000	0.998	0.994	0.981	0.951	0.895
	11						1.000	0.999	0.995	0.985	0.962
	12							1.000	0.999	0.997	0.989
	13								1.000	0.999	0.998
	14									1.000	1.000
17	0	0.418	0.167	0.063	0.023	0.008	0.002	0.001	0.000	0.000	0.000
	1	0.792	0.482	0.252	0.118	0.050	0.019	0.007	0.002	0.001	0.000
	2	0.950	0.762	0.520	0.310	0.164	0.077	0.033	0.012	0.004	0.001
	3	0.991	0.917	0.756	0.549	0.353	0.202	0.103	0.046	0.018	0.006
	4	0.999	0.978	0.901	0.758	0.574	0.389	0.235	0.126	0.060	0.025
	5	1.000	0.995	0.968	0.894	0.765	0.597	0.420	0.264	0.147	0.072
	6		0.999	0.992	0.962	0.893	0.775	0.619	0.448	0.290	0.166
	7		1.000	0.998	0.989	0.960	0.895	0.787	0.641	0.474	0.315
	8			1.000	0.997	0.988	0.960	0.901	0.801	0.663	0.500
	9				1.000	0.997	0.987	0.962	0.908	0.817	0.685
	10					0.999	0.997	0.988	0.965	0.917	0.834
	11					1.000	0.999	0.997	0.989	0.970	0.928
	12						1.000	0.999	0.997	0.991	0.975
	13							1.000	1.000	0.998	0.994
	14									1.000	0.999
	15										1.000
18	0	0.397	0.150	0.054	0.018	0.006	0.002	0.000	0.000	0.000	0.000
	1	0.774	0.450	0.224	0.099	0.039	0.014	0.005	0.001	0.000	0.000
	2	0.942	0.734	0.480	0.271	0.135	0.060	0.024	0.008	0.003	0.001
	3	0.989	0.902	0.720	0.501	0.306	0.165	0.078	0.033	0.012	0.004
	4	0.998	0.972	0.879	0.716	0.519	0.333	0.189	0.094	0.041	0.015
	5	1.000	0.994	0.958	0.867	0.717	0.534	0.355	0.209	0.108	0.048
	6		0.999	0.988	0.949	0.861	0.722	0.549	0.374	0.226	0.119
	7		1.000	0.997	0.984	0.943	0.859	0.728	0.563	0.391	0.240
	8			0.999	0.996	0.981	0.940	0.861	0.737	0.578	0.407
	9			1.000	0.999	0.995	0.979	0.940	0.865	0.747	0.593
	10				1.000	0.999	0.994	0.979	0.942	0.872	0.760
	11					1.000	0.999	0.994	0.980	0.946	0.881
	12						1.000	0.999	0.994	0.982	0.952
	13							1.000	0.999	0.995	0.985
	14								1.000	0.999	0.996
	15									1.000	0.999
	16										1.000

APPENDIX A-1 (*continued*)

						$p =$ Probability of Occurrence					
n	X	0.05	0.10	0.15	0.20	0.25	0.30	0.35	0.40	0.45	0.50
19	0	0.377	0.135	0.046	0.014	0.004	0.001	0.000	0.000	0.000	0.000
	1	0.755	0.420	0.198	0.083	0.031	0.010	0.003	0.001	0.000	0.000
	2	0.933	0.705	0.441	0.237	0.111	0.046	0.017	0.005	0.002	0.000
	3	0.987	0.885	0.684	0.455	0.263	0.133	0.059	0.023	0.008	0.002
	4	0.998	0.965	0.856	0.673	0.465	0.282	0.150	0.070	0.028	0.010
	5	1.000	0.991	0.946	0.837	0.668	0.474	0.297	0.163	0.078	0.032
	6		0.998	0.984	0.932	0.825	0.666	0.481	0.308	0.173	0.084
	7		1.000	0.996	0.977	0.923	0.818	0.666	0.488	0.317	0.180
	8			0.999	0.993	0.971	0.916	0.815	0.667	0.494	0.324
	9			1.000	0.998	0.991	0.967	0.913	0.814	0.671	0.500
	10				1.000	0.998	0.989	0.965	0.912	0.816	0.676
	11					1.000	0.997	0.989	0.965	0.913	0.820
	12						0.999	0.997	0.988	0.966	0.916
	13						1.000	0.999	0.997	0.989	0.968
	14							1.000	0.999	0.997	0.990
	15								1.000	0.999	0.998
	16									1.000	1.000
20	0	0.358	0.122	0.039	0.012	0.003	0.001	0.000	0.000	0.000	0.000
	1	0.736	0.392	0.176	0.069	0.024	0.008	0.002	0.001	0.000	0.000
	2	0.925	0.677	0.405	0.206	0.091	0.035	0.012	0.004	0.001	0.000
	3	0.984	0.867	0.648	0.411	0.225	0.107	0.044	0.016	0.005	0.001
	4	0.997	0.957	0.830	0.630	0.415	0.238	0.118	0.051	0.019	0.006
	5	1.000	0.989	0.933	0.804	0.617	0.416	0.245	0.126	0.055	0.021
	6		0.998	0.978	0.913	0.786	0.608	0.417	0.250	0.130	0.058
	7		1.000	0.994	0.968	0.898	0.772	0.601	0.416	0.252	0.132
	8			0.999	0.990	0.959	0.887	0.762	0.596	0.414	0.252
	9			1.000	0.997	0.986	0.952	0.878	0.755	0.591	0.412
	10				0.999	0.996	0.983	0.947	0.872	0.751	0.588
	11				1.000	0.999	0.995	0.980	0.943	0.869	0.748
	12					1.000	0.999	0.994	0.979	0.942	0.868
	13						1.000	0.998	0.994	0.979	0.942
	14							1.000	0.998	0.994	0.979
	15								1.000	0.998	0.994
	16									1.000	0.999
	17										1.000

APPENDIX A-2 Cumulative Poisson Distribution

X	0.01	0.05	0.10	0.20	0.30	0.40	0.50	0.60	0.70	0.80	0.90
						$\lambda = $ Mean					
0	0.990	0.951	0.905	0.819	0.741	0.670	0.607	0.549	0.497	0.449	0.407
1	1.000	0.999	0.995	0.982	0.963	0.938	0.910	0.878	0.844	0.809	0.772
2		1.000	1.000	0.999	0.996	0.992	0.986	0.977	0.966	0.953	0.937
3				1.000	1.000	0.999	0.998	0.997	0.994	0.991	0.987
4						1.000	1.000	1.000	0.999	0.999	0.998
5									1.000	1.000	1.000

X	1.0	1.1	1.2	1.3	1.4	1.5	1.6	1.7	1.8	1.9	2.0
0	0.368	0.333	0.301	0.273	0.247	0.223	0.202	0.183	0.165	0.150	0.135
1	0.736	0.699	0.663	0.627	0.592	0.558	0.525	0.493	0.463	0.434	0.406
2	0.920	0.900	0.879	0.857	0.833	0.809	0.783	0.757	0.731	0.704	0.677
3	0.981	0.974	0.966	0.957	0.946	0.934	0.921	0.907	0.891	0.875	0.857
4	0.996	0.995	0.992	0.989	0.986	0.981	0.976	0.970	0.964	0.956	0.947
5	0.999	0.999	0.998	0.998	0.997	0.996	0.994	0.992	0.990	0.987	0.983
6	1.000	1.000	1.000	1.000	0.999	0.999	0.999	0.998	0.997	0.997	0.995
7					1.000	1.000	1.000	1.000	0.999	0.999	0.999
8									1.000	1.000	1.000

X	2.2	2.4	2.6	2.8	3.0	3.5	4.0	4.5	5.0	5.5	6.0
0	0.111	0.091	0.074	0.061	0.050	0.030	0.018	0.011	0.007	0.004	0.002
1	0.355	0.308	0.267	0.231	0.199	0.136	0.092	0.061	0.040	0.027	0.017
2	0.623	0.570	0.518	0.469	0.423	0.321	0.238	0.174	0.125	0.088	0.062
3	0.819	0.779	0.736	0.692	0.647	0.537	0.433	0.342	0.265	0.202	0.151
4	0.928	0.904	0.877	0.848	0.815	0.725	0.629	0.532	0.440	0.358	0.285
5	0.975	0.964	0.951	0.935	0.916	0.858	0.785	0.703	0.616	0.529	0.446
6	0.993	0.988	0.983	0.976	0.966	0.935	0.889	0.831	0.762	0.686	0.606
7	0.998	0.997	0.995	0.992	0.988	0.973	0.949	0.913	0.867	0.809	0.744
8	1.000	0.999	0.999	0.998	0.996	0.990	0.979	0.960	0.932	0.894	0.847
9		1.000	1.000	0.999	0.999	0.997	0.992	0.983	0.968	0.946	0.916
10				1.000	1.000	0.999	0.997	0.993	0.986	0.975	0.957
11						1.000	0.999	0.998	0.995	0.989	0.980
12							1.000	0.999	0.998	0.996	0.991
13								1.000	0.999	0.998	0.996
14									1.000	0.999	0.999
15										1.000	0.999
16											1.000

X	6.5	7.0	7.5	8.0	9.0	10.0	12.0	14.0	16.0	18.0	20.0
0	0.002	0.001	0.001	0.000	0.000						
1	0.011	0.007	0.005	0.003	0.001	0.000	0.000				
2	0.043	0.030	0.020	0.014	0.006	0.003	0.001				
3	0.112	0.082	0.059	0.042	0.021	0.010	0.002	0.000			
4	0.224	0.173	0.132	0.100	0.055	0.029	0.008	0.002	0.000		
5	0.369	0.301	0.241	0.191	0.116	0.067	0.020	0.006	0.001	0.000	
6	0.527	0.450	0.378	0.313	0.207	0.130	0.046	0.014	0.004	0.001	0.000
7	0.673	0.599	0.525	0.453	0.324	0.220	0.090	0.032	0.010	0.003	0.001
8	0.792	0.729	0.662	0.593	0.456	0.333	0.155	0.062	0.022	0.007	0.002
9	0.877	0.830	0.776	0.717	0.587	0.458	0.242	0.109	0.043	0.015	0.005
10	0.933	0.901	0.862	0.816	0.706	0.583	0.347	0.176	0.077	0.030	0.011
11	0.966	0.947	0.921	0.888	0.803	0.697	0.462	0.260	0.127	0.055	0.021

(continued)

APPENDIX A-2   (*continued*)

	6.5	7.0	7.5	8.0	9.0	10.0	12.0	14.0	16.0	18.0	20.0
						$\lambda = $ Mean					
12	0.984	0.973	0.957	0.936	0.876	0.792	0.576	0.358	0.193	0.092	0.039
13	0.993	0.987	0.978	0.966	0.926	0.864	0.682	0.464	0.275	0.143	0.066
14	0.997	0.994	0.990	0.983	0.959	0.917	0.772	0.570	0.368	0.208	0.105
15	0.999	0.998	0.995	0.992	0.978	0.951	0.844	0.669	0.467	0.287	0.157
16	1.000	0.999	0.998	0.996	0.989	0.973	0.899	0.756	0.566	0.375	0.221
17		1.000	0.999	0.998	0.995	0.986	0.937	0.827	0.659	0.469	0.297
18			1.000	0.999	0.998	0.993	0.963	0.883	0.742	0.562	0.381
19				1.000	0.999	0.997	0.979	0.923	0.812	0.651	0.470
20					1.000	0.998	0.988	0.952	0.868	0.731	0.559
21						0.999	0.994	0.971	0.911	0.799	0.644
22						1.000	0.997	0.983	0.942	0.855	0.721
23							0.999	0.991	0.963	0.899	0.787
24							0.999	0.995	0.978	0.932	0.843
25							1.000	0.997	0.987	0.955	0.888
26								0.999	0.993	0.972	0.922
27								0.999	0.996	0.983	0.948
28								1.000	0.998	0.990	0.966
29									0.999	0.994	0.978
30									0.999	0.997	0.987
31									1.000	0.998	0.992
32										0.999	0.995
33										1.000	0.997
34											0.999
35											0.999
36											1.000

APPENDIX A-3   Cumulative Standard Normal Distribution

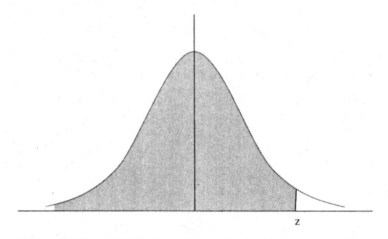

Z	0.00	0.01	0.02	0.03	0.04	0.05	0.06	0.07	0.08	0.09
−3.40	0.0003	0.0003	0.0003	0.0003	0.0003	0.0003	0.0003	0.0003	0.0003	0.0002
−3.30	0.0005	0.0005	0.0005	0.0004	0.0004	0.0004	0.0004	0.0004	0.0004	0.0003
−3.20	0.0007	0.0007	0.0006	0.0006	0.0006	0.0006	0.0006	0.0005	0.0005	0.0005
−3.10	0.0010	0.0009	0.0009	0.0009	0.0008	0.0008	0.0008	0.0008	0.0007	0.0007
−3.00	0.0013	0.0013	0.0013	0.0012	0.0012	0.0011	0.0011	0.0011	0.0010	0.0010
−2.90	0.0019	0.0018	0.0018	0.0017	0.0016	0.0016	0.0015	0.0015	0.0014	0.0014
−2.80	0.0026	0.0025	0.0024	0.0023	0.0023	0.0022	0.0021	0.0021	0.0020	0.0019
−2.70	0.0035	0.0034	0.0033	0.0032	0.0031	0.0030	0.0029	0.0028	0.0027	0.0026
−2.60	0.0047	0.0045	0.0044	0.0043	0.0041	0.0040	0.0039	0.0038	0.0037	0.0036
−2.50	0.0062	0.0060	0.0059	0.0057	0.0055	0.0054	0.0052	0.0051	0.0049	0.0048
−2.40	0.0082	0.0080	0.0078	0.0075	0.0073	0.0071	0.0069	0.0068	0.0066	0.0064
−2.30	0.0107	0.0104	0.0102	0.0099	0.0096	0.0094	0.0091	0.0089	0.0087	0.0084
−2.20	0.0139	0.0136	0.0132	0.0129	0.0125	0.0122	0.0119	0.0116	0.0113	0.0110
−2.10	0.0179	0.0174	0.0170	0.0166	0.0162	0.0158	0.0154	0.0150	0.0146	0.0143
−2.00	0.0228	0.0222	0.0217	0.0212	0.0207	0.0202	0.0197	0.0192	0.0188	0.0183
−1.90	0.0287	0.0281	0.0274	0.0268	0.0262	0.0256	0.0250	0.0244	0.0239	0.0233
−1.80	0.0359	0.0351	0.0344	0.0336	0.0329	0.0322	0.0314	0.0307	0.0301	0.0294
−1.70	0.0446	0.0436	0.0427	0.0418	0.0409	0.0401	0.0392	0.0384	0.0375	0.0367
−1.60	0.0548	0.0537	0.0526	0.0516	0.0505	0.0495	0.0485	0.0475	0.0465	0.0455
−1.50	0.0668	0.0655	0.0643	0.0630	0.0618	0.0606	0.0594	0.0582	0.0571	0.0559
−1.40	0.0808	0.0793	0.0778	0.0764	0.0749	0.0735	0.0721	0.0708	0.0694	0.0681
−1.30	0.0968	0.0951	0.0934	0.0918	0.0901	0.0885	0.0869	0.0853	0.0838	0.0823
−1.20	0.1151	0.1131	0.1112	0.1093	0.1075	0.1056	0.1038	0.1020	0.1003	0.0985
−1.10	0.1357	0.1335	0.1314	0.1292	0.1271	0.1251	0.1230	0.1210	0.1190	0.1170
−1.00	0.1587	0.1562	0.1539	0.1515	0.1492	0.1469	0.1446	0.1423	0.1401	0.1379
−0.90	0.1841	0.1814	0.1788	0.1762	0.1736	0.1711	0.1685	0.1660	0.1635	0.1611
−0.80	0.2119	0.2090	0.2061	0.2033	0.2005	0.1977	0.1949	0.1922	0.1894	0.1867
−0.70	0.2420	0.2389	0.2358	0.2327	0.2296	0.2266	0.2236	0.2206	0.2177	0.2148
−0.60	0.2743	0.2709	0.2676	0.2643	0.2611	0.2578	0.2546	0.2514	0.2483	0.2451
−0.50	0.3085	0.3050	0.3015	0.2981	0.2946	0.2912	0.2877	0.2843	0.2810	0.2776
−0.40	0.3446	0.3409	0.3372	0.3336	0.3300	0.3264	0.3228	0.3192	0.3156	0.3121
−0.30	0.3821	0.3783	0.3745	0.3707	0.3669	0.3632	0.3594	0.3557	0.3520	0.3483
−0.20	0.4207	0.4168	0.4129	0.4090	0.4052	0.4013	0.3974	0.3936	0.3897	0.3859

APPENDIX A-3 (*continued*)

Z	0.00	0.01	0.02	0.03	0.04	0.05	0.06	0.07	0.08	0.09
−0.10	0.4602	0.4562	0.4522	0.4483	0.4443	0.4404	0.4364	0.4325	0.4286	0.4247
−0.00	0.5000	0.4960	0.4920	0.4880	0.4840	0.4801	0.4761	0.4721	0.4681	0.4641
0.00	0.5000	0.5040	0.5080	0.5120	0.5160	0.5199	0.5239	0.5279	0.5319	0.5359
0.10	0.5398	0.5438	0.5478	0.5517	0.5557	0.5596	0.5636	0.5675	0.5714	0.5753
0.20	0.5793	0.5832	0.5871	0.5910	0.5948	0.5987	0.6026	0.6064	0.6103	0.6141
0.30	0.6179	0.6217	0.6255	0.6293	0.6331	0.6368	0.6406	0.6443	0.6480	0.6517
0.40	0.6554	0.6591	0.6628	0.6664	0.6700	0.6736	0.6772	0.6808	0.6844	0.6879
0.50	0.6915	0.6950	0.6985	0.7019	0.7054	0.7088	0.7123	0.7157	0.7190	0.7224
0.60	0.7257	0.7291	0.7324	0.7357	0.7389	0.7422	0.7454	0.7486	0.7517	0.7549
0.70	0.7580	0.7611	0.7642	0.7673	0.7704	0.7734	0.7764	0.7794	0.7823	0.7852
0.80	0.7881	0.7910	0.7939	0.7967	0.7995	0.8023	0.8051	0.8078	0.8106	0.8133
0.90	0.8159	0.8186	0.8212	0.8238	0.8264	0.8289	0.8315	0.8340	0.8365	0.8389
1.00	0.8413	0.8438	0.8461	0.8485	0.8508	0.8531	0.8554	0.8577	0.8599	0.8621
1.10	0.8643	0.8665	0.8686	0.8708	0.8729	0.8749	0.8770	0.8790	0.8810	0.8830
1.20	0.8849	0.8869	0.8888	0.8907	0.8925	0.8944	0.8962	0.8980	0.8997	0.9015
1.30	0.9032	0.9049	0.9066	0.9082	0.9099	0.9115	0.9131	0.9147	0.9162	0.9177
1.40	0.9192	0.9207	0.9222	0.9236	0.9251	0.9265	0.9279	0.9292	0.9306	0.9319
1.50	0.9332	0.9345	0.9357	0.9370	0.9382	0.9394	0.9406	0.9418	0.9429	0.9441
1.60	0.9452	0.9463	0.9474	0.9484	0.9495	0.9505	0.9515	0.9525	0.9535	0.9545
1.70	0.9554	0.9564	0.9573	0.9582	0.9591	0.9599	0.9608	0.9616	0.9625	0.9633
1.80	0.9641	0.9649	0.9656	0.9664	0.9671	0.9678	0.9686	0.9693	0.9699	0.9706
1.90	0.9713	0.9719	0.9726	0.9732	0.9738	0.9744	0.9750	0.9756	0.9761	0.9767
2.00	0.9772	0.9778	0.9783	0.9788	0.9793	0.9798	0.9803	0.9808	0.9812	0.9817
2.10	0.9821	0.9826	0.9830	0.9834	0.9838	0.9842	0.9846	0.9850	0.9854	0.9857
2.20	0.9861	0.9864	0.9868	0.9871	0.9875	0.9878	0.9881	0.9884	0.9887	0.9890
2.30	0.9893	0.9896	0.9898	0.9901	0.9904	0.9906	0.9909	0.9911	0.9913	0.9916
2.40	0.9918	0.9920	0.9922	0.9925	0.9927	0.9929	0.9931	0.9932	0.9934	0.9936
2.50	0.9938	0.9940	0.9941	0.9943	0.9945	0.9946	0.9948	0.9949	0.9951	0.9952
2.60	0.9953	0.9955	0.9956	0.9957	0.9959	0.9960	0.9961	0.9962	0.9963	0.9964
2.70	0.9965	0.9966	0.9967	0.9968	0.9969	0.9970	0.9971	0.9972	0.9973	0.9974
2.80	0.9974	0.9975	0.9976	0.9977	0.9977	0.9978	0.9979	0.9979	0.9980	0.9981
2.90	0.9981	0.9982	0.9982	0.9983	0.9984	0.9984	0.9985	0.9985	0.9986	0.9986
3.00	0.9987	0.9987	0.9987	0.9988	0.9988	0.9989	0.9989	0.9989	0.9990	0.9990
3.10	0.9990	0.9991	0.9991	0.9991	0.9992	0.9992	0.9992	0.9992	0.9993	0.9993
3.20	0.9993	0.9993	0.9994	0.9994	0.9994	0.9994	0.9994	0.9995	0.9995	0.9995
3.30	0.9995	0.9995	0.9995	0.9996	0.9996	0.9996	0.9996	0.9996	0.9996	0.9997
3.40	0.9997	0.9997	0.9997	0.9997	0.9997	0.9997	0.9997	0.9997	0.9997	0.9998

Z	F(z)	Z	F(z)	Z	F(z)
3.50	0.99976 73709	4.35	0.99999 31931	5.20	0.99999 99003
3.55	0.99980 73844	4.40	0.99999 45875	5.25	0.99999 99239
3.60	0.99984 08914	4.45	0.99999 57065	5.30	0.99999 99420
3.65	0.99986 88798	4.50	0.99999 66023	5.35	0.99999 99560
3.70	0.99989 22003	4.55	0.99999 73177	5.40	0.99999 99666
3.75	0.99991 15827	4.60	0.99999 78875	5.45	0.99999 99748
3.80	0.99992 76520	4.65	0.99999 83403	5.50	0.99999 99810
3.85	0.99994 09411	4.70	0.99999 86992	5.55	0.99999 99857
3.90	0.99995 19037	4.75	0.99999 89829	5.60	0.99999 99892

(*Continued*)

APPENDIX A-3 (*continued*)

3.95	0.99996 09244	4.80	0.99999 92067	5.65	0.99999 99919
4.00	0.99996 83288	4.85	0.99999 93827	5.70	0.99999 99940
4.05	0.99997 43912	4.90	0.99999 95208	5.75	0.99999 99955
4.10	0.99997 93425	4.95	0.99999 96289	5.80	0.99999 99966
4.15	0.99998 33762	5.00	0.99999 97133	5.85	0.99999 99975
4.20	0.99998 66543	5.05	0.99999 97790	5.90	0.99999 99981
4.25	0.99998 93115	5.10	0.99999 98301	5.95	0.99999 99986
4.30	0.99999 14601	5.15	0.99999 98697	6.00	0.99999 99990

APPENDIX A-4   Values of t for a Specified Right-Tail Area

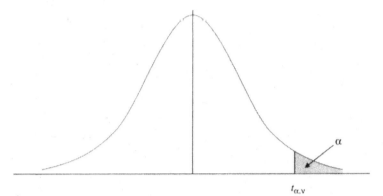

$t_{\alpha,\nu}$

df	α = Right-Hand Tail Area							
ν	0.250	0.200	0.100	0.050	0.025	0.010	0.005	0.001
1	1.000	1.376	3.078	6.314	12.706	31.821	63.657	318.309
2	0.816	1.061	1.886	2.920	4.303	6.965	9.925	22.327
3	0.765	0.978	1.638	2.353	3.182	4.541	5.841	10.215
4	0.741	0.941	1.533	2.132	2.776	3.747	4.604	7.173
5	0.727	0.920	1.476	2.015	2.571	3.365	4.032	5.893
6	0.718	0.906	1.440	1.943	2.447	3.143	3.707	5.208
7	0.711	0.896	1.415	1.895	2.365	2.998	3.499	4.785
8	0.706	0.889	1.397	1.860	2.306	2.896	3.355	4.501
9	0.703	0.883	1.383	1.833	2.262	2.821	3.250	4.297
10	0.700	0.879	1.372	1.812	2.228	2.764	3.169	4.144
11	0.697	0.876	1.363	1.796	2.201	2.718	3.106	4.025
12	0.695	0.873	1.356	1.782	2.179	2.681	3.055	3.930
13	0.694	0.870	1.350	1.771	2.160	2.650	3.012	3.852
14	0.692	0.868	1.345	1.761	2.145	2.624	2.977	3.787
15	0.691	0.866	1.341	1.753	2.131	2.602	2.947	3.733
16	0.690	0.865	1.337	1.746	2.120	2.583	2.921	3.686
17	0.689	0.863	1.333	1.740	2.110	2.567	2.898	3.646
18	0.688	0.862	1.330	1.734	2.101	2.552	2.878	3.610
19	0.688	0.861	1.328	1.729	2.093	2.539	2.861	3.579
20	0.687	0.860	1.325	1.725	2.086	2.528	2.845	3.552
21	0.686	0.859	1.323	1.721	2.080	2.518	2.831	3.527
22	0.686	0.858	1.321	1.717	2.074	2.508	2.819	3.505
23	0.685	0.858	1.319	1.714	2.069	2.500	2.807	3.485
24	0.685	0.857	1.318	1.711	2.064	2.492	2.797	3.467
25	0.684	0.856	1.316	1.708	2.060	2.485	2.787	3.450
26	0.684	0.856	1.315	1.706	2.056	2.479	2.779	3.435
27	0.684	0.855	1.314	1.703	2.052	2.473	2.771	3.421
28	0.683	0.855	1.313	1.701	2.048	2.467	2.763	3.408
29	0.683	0.854	1.311	1.699	2.045	2.462	2.756	3.396
30	0.683	0.854	1.310	1.697	2.042	2.457	2.750	3.385
35	0.682	0.852	1.306	1.690	2.030	2.438	2.724	3.340
40	0.681	0.851	1.303	1.684	2.021	2.423	2.704	3.307
45	0.680	0.850	1.301	1.679	2.014	2.412	2.690	3.281

(*continued*)

APPENDIX A-4 *(continued)*

df ν	α = Right-Hand Tail Area							
	0.250	0.200	0.100	0.050	0.025	0.010	0.005	0.001
50	0.679	0.849	1.299	1.676	2.009	2.403	2.678	3.261
55	0.679	0.848	1.297	1.673	2.004	2.396	2.668	3.245
60	0.679	0.848	1.296	1.671	2.000	2.390	2.660	3.232
65	0.678	0.847	1.295	1.669	1.997	2.385	2.654	3.220
70	0.678	0.847	1.294	1.667	1.994	2.381	2.648	3.211
80	0.678	0.846	1.292	1.664	1.990	2.374	2.639	3.195
90	0.677	0.846	1.291	1.662	1.987	2.368	2.632	3.183
100	0.677	0.845	1.290	1.660	1.984	2.364	2.626	3.174
110	0.677	0.845	1.289	1.659	1.982	2.361	2.621	3.166
120	0.677	0.845	1.289	1.658	1.980	2.358	2.617	3.160
∞	0.674	0.842	1.282	1.645	1.960	2.326	2.576	3.090

APPENDIX A-5   Chi-Squared Values for a Specified Right-Tail Area

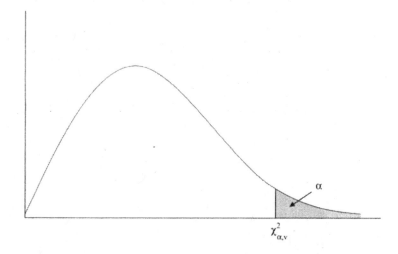

$$\chi^2_{\alpha,\nu}$$

df	α = Right-Hand-Tail Area											
ν	0.999	0.995	0.990	0.975	0.950	0.900	0.100	0.050	0.025	0.010	0.005	0.001
1	0.00	0.00	0.00	0.00	0.00	0.02	2.71	3.84	5.02	6.63	7.88	10.83
2	0.00	0.01	0.02	0.05	0.10	0.21	4.61	5.99	7.38	9.21	10.60	13.82
3	0.02	0.07	0.11	0.22	0.35	0.58	6.25	7.81	9.35	11.34	12.84	16.27
4	0.09	0.21	0.30	0.48	0.71	1.06	7.78	9.49	11.14	13.28	14.86	18.47
5	0.21	0.41	0.55	0.83	1.15	1.61	9.24	11.07	12.83	15.09	16.75	20.52
6	0.38	0.68	0.87	1.24	1.64	2.20	10.64	12.59	14.45	16.81	18.55	22.46
7	0.60	0.99	1.24	1.69	2.17	2.83	12.02	14.07	16.01	18.48	20.28	24.32
8	0.86	1.34	1.65	2.18	2.73	3.49	13.36	15.51	17.53	20.09	21.95	26.12
9	1.15	1.73	2.09	2.70	3.33	4.17	14.68	16.92	19.02	21.67	23.59	27.88
10	1.48	2.16	2.56	3.25	3.94	4.87	15.99	18.31	20.48	23.21	25.19	29.59
11	1.83	2.60	3.05	3.82	4.57	5.58	17.28	19.68	21.92	24.72	26.76	31.26
12	2.21	3.07	3.57	4.40	5.23	6.30	18.55	21.03	23.34	26.22	28.30	32.91
13	2.62	3.57	4.11	5.01	5.89	7.04	19.81	22.36	24.74	27.69	29.82	34.53
14	3.04	4.07	4.66	5.63	6.57	7.79	21.06	23.68	26.12	29.14	31.32	36.12
15	3.48	4.60	5.23	6.26	7.26	8.55	22.31	25.00	27.49	30.58	32.80	37.70
16	3.94	5.14	5.81	6.91	7.96	9.31	23.54	26.30	28.85	32.00	34.27	39.25
17	4.42	5.70	6.41	7.56	8.67	10.09	24.77	27.59	30.19	33.41	35.72	40.79
18	4.90	6.26	7.01	8.23	9.39	10.86	25.99	28.87	31.53	34.81	37.16	42.31
19	5.41	6.84	7.63	8.91	10.12	11.65	27.20	30.14	32.85	36.19	38.58	43.82
20	5.92	7.43	8.26	9.59	10.85	12.44	28.41	31.41	34.17	37.57	40.00	45.31
21	6.45	8.03	8.90	10.28	11.59	13.24	29.62	32.67	35.48	38.93	41.40	46.80
22	6.98	8.64	9.54	10.98	12.34	14.04	30.81	33.92	36.78	40.29	42.80	48.27
23	7.53	9.26	10.20	11.69	13.09	14.85	32.01	35.17	38.08	41.64	44.18	49.73
24	8.08	9.89	10.86	12.40	13.85	15.66	33.20	36.42	39.36	42.98	45.56	51.18
25	8.65	10.52	11.52	13.12	14.61	16.47	34.38	37.65	40.65	44.31	46.93	52.62
26	9.22	11.16	12.20	13.84	15.38	17.29	35.56	38.89	41.92	45.64	48.29	54.05
27	9.80	11.81	12.88	14.57	16.15	18.11	36.74	40.11	43.19	46.96	49.64	55.48
28	10.39	12.46	13.56	15.31	16.93	18.94	37.92	41.34	44.46	48.28	50.99	56.89
29	10.99	13.12	14.26	16.05	17.71	19.77	39.09	42.56	45.72	49.59	52.34	58.30
30	11.59	13.79	14.95	16.79	18.49	20.60	40.26	43.77	46.98	50.89	53.67	59.70

(continued)

APPENDIX A-5 (*continued*)

df						α = Right-Hand-Tail Area						
ν	0.999	0.995	0.990	0.975	0.950	0.900	0.100	0.050	0.025	0.010	0.005	0.001
32	12.81	15.13	16.36	18.29	20.07	22.27	42.58	46.19	49.48	53.49	56.33	62.49
34	14.06	16.50	17.79	19.81	21.66	23.95	44.90	48.60	51.97	56.06	58.96	65.25
36	15.32	17.89	19.23	21.34	23.27	25.64	47.21	51.00	54.44	58.62	61.58	67.99
38	16.61	19.29	20.69	22.88	24.88	27.34	49.51	53.38	56.90	61.16	64.18	70.70
40	17.92	20.71	22.16	24.43	26.51	29.05	51.81	55.76	59.34	63.69	66.77	73.40
42	19.24	22.14	23.65	26.00	28.14	30.77	54.09	58.12	61.78	66.21	69.34	76.08
44	20.58	23.58	25.15	27.57	29.79	32.49	56.37	60.48	64.20	68.71	71.89	78.75
46	21.93	25.04	26.66	29.16	31.44	34.22	58.64	62.83	66.62	71.20	74.44	81.40
48	23.29	26.51	28.18	30.75	33.10	35.95	60.91	65.17	69.02	73.68	76.97	84.04
50	24.67	27.99	29.71	32.36	34.76	37.69	63.17	67.50	71.42	76.15	79.49	86.66
55	28.17	31.73	33.57	36.40	38.96	42.06	68.80	73.31	77.38	82.29	85.75	93.17
60	31.74	35.53	37.48	40.48	43.19	46.46	74.40	79.08	83.30	88.38	91.95	99.61
65	35.36	39.38	41.44	44.60	47.45	50.88	79.97	84.82	89.18	94.42	98.11	105.99
70	39.04	43.28	45.44	48.76	51.74	55.33	85.53	90.53	95.02	100.43	104.21	112.32
75	42.76	47.21	49.48	52.94	56.05	59.79	91.06	96.22	100.84	106.39	110.29	118.60
80	46.52	51.17	53.54	57.15	60.39	64.28	96.58	101.88	106.63	112.33	116.32	124.84
85	50.32	55.17	57.63	61.39	64.75	68.78	102.08	107.52	112.39	118.24	122.32	131.04
90	54.16	59.20	61.75	65.65	69.13	73.29	107.57	113.15	118.14	124.12	128.30	137.21
95	58.02	63.25	65.90	69.92	73.52	77.82	113.04	118.75	123.86	129.97	134.25	143.34
100	61.92	67.33	70.06	74.22	77.93	82.36	118.50	124.34	129.56	135.81	140.17	149.45

APPENDIX A-6 Values of *F* for a Specified Right-Tail Area

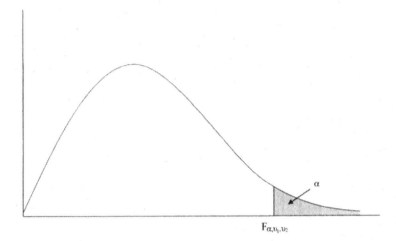

$$F_{\alpha, v_1, v_2}$$

		$v_1 = $ Degrees of Freedom for Numerator									
v_2	α	1	2	3	4	5	6	7	8	9	10	11
	0.100	39.9	49.5	53.6	55.8	57.2	58.2	58.9	59.4	59.9	60.2	60.5
	0.050	161.4	199.5	215.7	224.6	230.2	234.0	236.8	238.9	240.5	241.9	243.0
1	0.025	647.8	799.5	864.2	899.6	921.8	937.1	948.2	956.7	963.3	968.6	973.0
	0.010	4052.2	4999.5	5403.4	5624.6	5763.6	5859.0	5928.4	5981.1	6022.5	6055.8	6083.3
	0.005	16210.7	19999.5	21614.7	22499.6	23055.8	23437.1	23714.6	23925.4	24091.0	24224.5	24334.4
	0.001	405284.1	499999.5	540379.2	562499.6	576404.6	585937.1	592873.3	598144.2	602284.0	605621.0	608367.7
	0.100	8.53	9.00	9.16	9.24	9.29	9.33	9.35	9.37	9.38	9.39	9.40
	0.050	18.51	19.00	19.16	19.25	19.30	19.33	19.35	19.37	19.38	19.40	19.40
2	0.025	38.51	39.00	39.17	39.25	39.30	39.33	39.36	39.37	39.39	39.40	39.41
	0.010	98.50	99.00	99.17	99.25	99.30	99.33	99.36	99.37	99.39	99.40	99.41
	0.005	198.50	199.00	199.17	199.25	199.30	199.33	199.36	199.37	199.39	199.40	199.41
	0.001	998.50	999.00	999.17	999.25	999.30	999.33	999.36	999.37	999.39	999.40	999.41
	0.100	5.54	5.46	5.39	5.34	5.31	5.28	5.27	5.25	5.24	5.23	5.22
	0.050	10.13	9.55	9.28	9.12	9.01	8.94	8.89	8.85	8.81	8.79	8.76
3	0.025	17.44	16.04	15.44	15.10	14.88	14.73	14.62	14.54	14.47	14.42	14.37
	0.010	34.12	30.82	29.46	28.71	28.24	27.91	27.67	27.49	27.35	27.23	27.13
	0.005	55.55	49.80	47.47	46.19	45.39	44.84	44.43	44.13	43.88	43.69	43.52
	0.001	167.03	148.50	141.11	137.10	134.58	132.85	131.58	130.62	129.86	129.25	128.74
	0.100	4.54	4.32	4.19	4.11	4.05	4.01	3.98	3.95	3.94	3.92	3.91
	0.050	7.71	6.94	6.59	6.39	6.26	6.16	6.09	6.04	6.00	5.96	5.94
4	0.025	12.22	10.65	9.98	9.60	9.36	9.20	9.07	8.98	8.90	8.84	8.79
	0.010	21.20	18.00	16.69	15.98	15.52	15.21	14.98	14.80	14.66	14.55	14.45
	0.005	31.33	26.28	24.26	23.15	22.46	21.97	21.62	21.35	21.14	20.97	20.82
	0.001	74.14	61.25	56.18	53.44	51.71	50.53	49.66	49.00	48.47	48.05	47.70
	0.100	4.06	3.78	3.62	3.52	3.45	3.40	3.37	3.34	3.32	3.30	3.28
	0.050	6.61	5.79	5.41	5.19	5.05	4.95	4.88	4.82	4.77	4.74	4.70
5	0.025	10.01	8.43	7.76	7.39	7.15	6.98	6.85	6.76	6.68	6.62	6.57
	0.010	16.26	13.27	12.06	11.39	10.97	10.67	10.46	10.29	10.16	10.05	9.96
	0.005	22.78	18.31	16.53	15.56	14.94	14.51	14.20	13.96	13.77	13.62	13.49
	0.001	47.18	37.12	33.20	31.09	29.75	28.83	28.16	27.65	27.24	26.92	26.65

(continued)

APPENDIX A-6 (*continued*)

v_2	α	1	2	3	4	5	6	7	8	9	10	11

v_2	α	\multicolumn{11}{c}{$v_1 =$ Degrees of Freedom for Numerator}
		1	2	3	4	5	6	7	8	9	10	11
	0.100	3.78	3.46	3.29	3.18	3.11	3.05	3.01	2.98	2.96	2.94	2.92
	0.050	5.99	5.14	4.76	4.53	4.39	4.28	4.21	4.15	4.10	4.06	4.03
6	0.025	8.81	7.26	6.60	6.23	5.99	5.82	5.70	5.60	5.52	5.46	5.41
	0.010	13.75	10.92	9.78	9.15	8.75	8.47	8.26	8.10	7.98	7.87	7.79
	0.005	18.63	14.54	12.92	12.03	11.46	11.07	10.79	10.57	10.39	10.25	10.13
	0.001	35.51	27.00	23.70	21.92	20.80	20.03	19.46	19.03	18.69	18.41	18.18
	0.100	3.59	3.26	3.07	2.96	2.88	2.83	2.78	2.75	2.72	2.70	2.68
	0.050	5.59	4.74	4.35	4.12	3.97	3.87	3.79	3.73	3.68	3.64	3.60
7	0.025	8.07	6.54	5.89	5.52	5.29	5.12	4.99	4.90	4.82	4.76	4.71
	0.010	12.25	9.55	8.45	7.85	7.46	7.19	6.99	6.84	6.72	6.62	6.54
	0.005	16.24	12.40	10.88	10.05	9.52	9.16	8.89	8.68	8.51	8.38	8.27
	0.001	29.25	21.69	18.77	17.20	16.21	15.52	15.02	14.63	14.33	14.08	13.88
	0.100	3.46	3.11	2.92	2.81	2.73	2.67	2.62	2.59	2.56	2.54	2.52
	0.050	5.32	4.46	4.07	3.84	3.69	3.58	3.50	3.44	3.39	3.35	3.31
8	0.025	7.57	6.06	5.42	5.05	4.82	4.65	4.53	4.43	4.36	4.30	4.24
	0.010	11.26	8.65	7.59	7.01	6.63	6.37	6.18	6.03	5.91	5.81	5.73
	0.005	14.69	11.04	9.60	8.81	8.30	7.95	7.69	7.50	7.34	7.21	7.10
	0.001	25.41	18.49	15.83	14.39	13.48	12.86	12.40	12.05	11.77	11.54	11.35
	0.100	3.36	3.01	2.81	2.69	2.61	2.55	2.51	2.47	2.44	2.42	2.40
	0.050	5.12	4.26	3.86	3.63	3.48	3.37	3.29	3.23	3.18	3.14	3.10
9	0.025	7.21	5.71	5.08	4.72	4.48	4.32	4.20	4.10	4.03	3.96	3.91
	0.010	10.56	8.02	6.99	6.42	6.06	5.80	5.61	5.47	5.35	5.26	5.18
	0.005	13.61	10.11	8.72	7.96	7.47	7.13	6.88	6.69	6.54	6.42	6.31
	0.001	22.86	16.39	13.90	12.56	11.71	11.13	10.70	10.37	10.11	9.89	9.72
	0.100	3.29	2.92	2.73	2.61	2.52	2.46	2.41	2.38	2.35	2.32	2.30
	0.050	4.96	4.10	3.71	3.48	3.33	3.22	3.14	3.07	3.02	2.98	2.94
10	0.025	6.94	5.46	4.83	4.47	4.24	4.07	3.95	3.85	3.78	3.72	3.66
	0.010	10.04	7.56	6.55	5.99	5.64	5.39	5.20	5.06	4.94	4.85	4.77
	0.005	12.83	9.43	8.08	7.34	6.87	6.54	6.30	6.12	5.97	5.85	5.75
	0.001	21.04	14.91	12.55	11.28	10.48	9.93	9.52	9.20	8.96	8.75	8.59
	0.100	3.23	2.86	2.66	2.54	2.45	2.39	2.34	2.30	2.27	2.25	2.23
	0.050	4.84	3.98	3.59	3.36	3.20	3.09	3.01	2.95	2.90	2.85	2.82
11	0.025	6.72	5.26	4.63	4.28	4.04	3.88	3.76	3.66	3.59	3.53	3.47
	0.010	9.65	7.21	6.22	5.67	5.32	5.07	4.89	4.74	4.63	4.54	4.46
	0.005	12.23	8.91	7.60	6.88	6.42	6.10	5.86	5.68	5.54	5.42	5.32
	0.001	19.69	13.81	11.56	10.35	9.58	9.05	8.66	8.35	8.12	7.92	7.76
	0.100	3.18	2.81	2.61	2.48	2.39	2.33	2.28	2.24	2.21	2.19	2.17
	0.050	4.75	3.89	3.49	3.26	3.11	3.00	2.91	2.85	2.80	2.75	2.72
12	0.025	6.55	5.10	4.47	4.12	3.89	3.73	3.61	3.51	3.44	3.37	3.32
	0.010	9.33	6.93	5.95	5.41	5.06	4.82	4.64	4.50	4.39	4.30	4.22
	0.005	11.75	8.51	7.23	6.52	6.07	5.76	5.52	5.35	5.20	5.09	4.99
	0.001	18.64	12.97	10.80	9.63	8.89	8.38	8.00	7.71	7.48	7.29	7.14
	0.100	3.07	2.70	2.49	2.36	2.27	2.21	2.16	2.12	2.09	2.06	2.04
	0.050	4.54	3.68	3.29	3.06	2.90	2.79	2.71	2.64	2.59	2.54	2.51
15	0.025	6.20	4.77	4.15	3.80	3.58	3.41	3.29	3.20	3.12	3.06	3.01
	0.010	8.68	6.36	5.42	4.89	4.56	4.32	4.14	4.00	3.89	3.80	3.73
	0.005	10.80	7.70	6.48	5.80	5.37	5.07	4.85	4.67	4.54	4.42	4.33
	0.001	16.59	11.34	9.34	8.25	7.57	7.09	6.74	6.47	6.26	6.08	5.94

APPENDIX A-6 *(continued)*

v_2	α	$v_1 =$ Degrees of Freedom for Numerator										
		1	2	3	4	5	6	7	8	9	10	11
	0.100	3.01	2.62	2.42	2.29	2.20	2.13	2.08	2.04	2.00	1.98	1.95
	0.050	4.41	3.55	3.16	2.93	2.77	2.66	2.58	2.51	2.46	2.41	2.37
18	0.025	5.98	4.56	3.95	3.61	3.38	3.22	3.10	3.01	2.93	2.87	2.81
	0.010	8.29	6.01	5.09	4.58	4.25	4.01	3.84	3.71	3.60	3.51	3.43
	0.005	10.22	7.21	6.03	5.37	4.96	4.66	4.44	4.28	4.14	4.03	3.94
	0.001	15.38	10.39	8.49	7.46	6.81	6.35	6.02	5.76	5.56	5.39	5.25
	0.100	2.97	2.59	2.38	2.25	2.16	2.09	2.04	2.00	1.96	1.94	1.91
	0.050	4.35	3.49	3.10	2.87	2.71	2.60	2.51	2.45	2.39	2.35	2.31
20	0.025	5.87	4.46	3.86	3.51	3.29	3.13	3.01	2.91	2.84	2.77	2.72
	0.010	8.10	5.85	4.94	4.43	4.10	3.87	3.70	3.56	3.46	3.37	3.29
	0.005	9.94	6.99	5.82	5.17	4.76	4.47	4.26	4.09	3.96	3.85	3.76
	0.001	14.82	9.95	8.10	7.10	6.46	6.02	5.69	5.44	5.24	5.08	4.94
	0.100	2.92	2.53	2.32	2.18	2.09	2.02	1.97	1.93	1.89	1.87	1.84
	0.050	4.24	3.39	2.99	2.76	2.60	2.49	2.40	2.34	2.28	2.24	2.20
25	0.025	5.69	4.29	3.69	3.35	3.13	2.97	2.85	2.75	2.68	2.61	2.56
	0.010	7.77	5.57	4.68	4.18	3.85	3.63	3.46	3.32	3.22	3.13	3.06
	0.005	9.48	6.60	5.46	4.84	4.43	4.15	3.94	3.78	3.64	3.54	3.45
	0.001	13.88	9.22	7.45	6.49	5.89	5.46	5.15	4.91	4.71	4.56	4.42
	0.100	2.88	2.49	2.28	2.14	2.05	1.98	1.93	1.88	1.85	1.82	1.79
	0.050	4.17	3.32	2.92	2.69	2.53	2.42	2.33	2.27	2.21	2.16	2.13
30	0.025	5.57	4.18	3.59	3.25	3.03	2.87	2.75	2.65	2.57	2.51	2.46
	0.010	7.56	5.39	4.51	4.02	3.70	3.47	3.30	3.17	3.07	2.98	2.91
	0.005	9.18	6.35	5.24	4.62	4.23	3.95	3.74	3.58	3.45	3.34	3.25
	0.001	13.29	8.77	7.05	6.12	5.53	5.12	4.82	4.58	4.39	4.24	4.11
	0.100	2.84	2.44	2.23	2.09	2.00	1.93	1.87	1.83	1.79	1.76	1.74
	0.050	4.08	3.23	2.84	2.61	2.45	2.34	2.25	2.18	2.12	2.08	2.04
40	0.025	5.42	4.05	3.46	3.13	2.90	2.74	2.62	2.53	2.45	2.39	2.33
	0.010	7.31	5.18	4.31	3.83	3.51	3.29	3.12	2.99	2.89	2.80	2.73
	0.005	8.83	6.07	4.98	4.37	3.99	3.71	3.51	3.35	3.22	3.12	3.03
	0.001	12.61	8.25	6.59	5.70	5.13	4.73	4.44	4.21	4.02	3.87	3.75
	0.100	2.81	2.41	2.20	2.06	1.97	1.90	1.84	1.80	1.76	1.73	1.70
	0.050	4.03	3.18	2.79	2.56	2.40	2.29	2.20	2.13	2.07	2.03	1.99
50	0.025	5.34	3.97	3.39	3.05	2.83	2.67	2.55	2.46	2.38	2.32	2.26
	0.010	7.17	5.06	4.20	3.72	3.41	3.19	3.02	2.89	2.78	2.70	2.63
	0.005	8.63	5.90	4.83	4.23	3.85	3.58	3.38	3.22	3.09	2.99	2.90
	0.001	12.22	7.96	6.34	5.46	4.90	4.51	4.22	4.00	3.82	3.67	3.55
	0.100	2.79	2.39	2.18	2.04	1.95	1.87	1.82	1.77	1.74	1.71	1.68
	0.050	4.00	3.15	2.76	2.53	2.37	2.25	2.17	2.10	2.04	1.99	1.95
60	0.025	5.29	3.93	3.34	3.01	2.79	2.63	2.51	2.41	2.33	2.27	2.22
	0.010	7.08	4.98	4.13	3.65	3.34	3.12	2.95	2.82	2.72	2.63	2.56
	0.005	8.49	5.79	4.73	4.14	3.76	3.49	3.29	3.13	3.01	2.90	2.82
	0.001	11.97	7.77	6.17	5.31	4.76	4.37	4.09	3.86	3.69	3.54	3.42
	0.100	2.76	2.36	2.14	2.00	1.91	1.83	1.78	1.73	1.69	1.66	1.64
	0.050	3.94	3.09	2.70	2.46	2.31	2.19	2.10	2.03	1.97	1.93	1.89
100	0.025	5.18	3.83	3.25	2.92	2.70	2.54	2.42	2.32	2.24	2.18	2.12
	0.010	6.90	4.82	3.98	3.51	3.21	2.99	2.82	2.69	2.59	2.50	2.43
	0.005	8.24	5.59	4.54	3.96	3.59	3.33	3.13	2.97	2.85	2.74	2.66
	0.001	11.50	7.41	5.86	5.02	4.48	4.11	3.83	3.61	3.44	3.30	3.18

(continued)

APPENDIX A-6 (*continued*)

v_2	α	\multicolumn{11}{c}{$v_1 =$ Degrees of Freedom for Numerator}
		1	2	3	4	5	6	7	8	9	10	11
∞	0.100	2.71	2.30	2.08	1.95	1.85	1.77	1.72	1.67	1.63	1.60	1.57
	0.050	3.84	3.00	2.61	2.37	2.21	2.10	2.01	1.94	1.88	1.83	1.79
	0.025	5.03	3.69	3.12	2.79	2.57	2.41	2.29	2.19	2.11	2.05	1.99
	0.010	6.64	4.61	3.78	3.32	3.02	2.80	2.64	2.51	2.41	2.32	2.25
	0.005	7.88	5.30	4.28	3.72	3.35	3.09	2.90	2.75	2.62	2.52	2.43
	0.001	10.83	6.91	5.43	4.62	4.11	3.75	3.48	3.27	3.10	2.96	2.85

v_2	α	\multicolumn{10}{c}{$v_1 =$ Degrees of Freedom for Numerator}
		12	15	20	40	50	60	75	100	120	∞
1	0.100	60.7	61.2	61.7	62.5	62.7	62.8	62.9	63.0	63.1	63.32
	0.050	243.9	245.9	248.0	251.1	251.8	252.2	252.6	253.0	253.3	254.30
	0.025	976.7	984.9	993.1	1005.6	1008.1	1009.8	1011.5	1013.2	1014.0	1018.21
	0.010	6106.3	6157.3	6208.7	6286.8	6302.5	6313.0	6323.6	6334.1	6339.4	6365.55
	0.005	24426.4	24630.2	24836.0	25148.2	25211.1	25253.1	25295.3	25337.5	25358.6	25463.18
	0.001	610667.8	615763.7	620907.7	628712.0	630285.4	631336.6	632389.5	633444.3	633972.4	636587.61
2	0.100	9.41	9.42	9.44	9.47	9.47	9.47	9.48	9.48	9.48	9.49
	0.050	19.41	19.43	19.45	19.47	19.48	19.48	19.48	19.49	19.49	19.50
	0.025	39.41	39.43	39.45	39.47	39.48	39.48	39.48	39.49	39.49	39.50
	0.010	99.42	99.43	99.45	99.47	99.48	99.48	99.49	99.49	99.49	99.50
	0.005	199.42	199.43	199.45	199.47	199.48	199.48	199.49	199.49	199.49	199.50
	0.001	999.42	999.43	999.45	999.47	999.48	999.48	999.49	999.49	999.49	999.50
3	0.100	5.22	5.20	5.18	5.16	5.15	5.15	5.15	5.14	5.14	5.13
	0.050	8.74	8.70	8.66	8.59	8.58	8.57	8.56	8.55	8.55	8.53
	0.025	14.34	14.25	14.17	14.04	14.01	13.99	13.97	13.96	13.95	13.90
	0.010	27.05	26.87	26.69	26.41	26.35	26.32	26.28	26.24	26.22	26.13
	0.005	43.39	43.08	42.78	42.31	42.21	42.15	42.09	42.02	41.99	41.83
	0.001	128.32	127.37	126.42	124.96	124.66	124.47	124.27	124.07	123.97	123.48
4	0.100	3.90	3.87	3.84	3.80	3.80	3.79	3.78	3.78	3.78	3.76
	0.050	5.91	5.86	5.80	5.72	5.70	5.69	5.68	5.66	5.66	5.63
	0.025	8.75	8.66	8.56	8.41	8.38	8.36	8.34	8.32	8.31	8.26
	0.010	14.37	14.20	14.02	13.75	13.69	13.65	13.61	13.58	13.56	13.46
	0.005	20.70	20.44	20.17	19.75	19.67	19.61	19.55	19.50	19.47	19.33
	0.001	47.41	46.76	46.10	45.09	44.88	44.75	44.61	44.47	44.40	44.06
5	0.100	3.27	3.24	3.21	3.16	3.15	3.14	3.13	3.13	3.12	3.11
	0.050	4.68	4.62	4.56	4.46	4.44	4.43	4.42	4.41	4.40	4.37
	0.025	6.52	6.43	6.33	6.18	6.14	6.12	6.10	6.08	6.07	6.02
	0.010	9.89	9.72	9.55	9.29	9.24	9.20	9.17	9.13	9.11	9.02
	0.005	13.38	13.15	12.90	12.53	12.45	12.40	12.35	12.30	12.27	12.15
	0.001	26.42	25.91	25.39	24.60	24.44	24.33	24.22	24.12	24.06	23.79
6	0.100	2.90	2.87	2.84	2.78	2.77	2.76	2.75	2.75	2.74	2.72
	0.050	4.00	3.94	3.87	3.77	3.75	3.74	3.73	3.71	3.70	3.67
	0.025	5.37	5.27	5.17	5.01	4.98	4.96	4.94	4.92	4.90	4.85
	0.010	7.72	7.56	7.40	7.14	7.09	7.06	7.02	6.99	6.97	6.88
	0.005	10.03	9.81	9.59	9.24	9.17	9.12	9.07	9.03	9.00	8.88
	0.001	17.99	17.56	17.12	16.44	16.31	16.21	16.12	16.03	15.98	15.75

APPENDIX A-6 (*continued*)

v_2	α	12	15	20	40	50	60	75	100	120	∞
						$v_1 =$ Degrees of Freedom for Numerator					
7	0.100	2.67	2.63	2.59	2.54	2.52	2.51	2.51	2.50	2.49	2.47
	0.050	3.57	3.51	3.44	3.34	3.32	3.30	3.29	3.27	3.27	3.23
	0.025	4.67	4.57	4.47	4.31	4.28	4.25	4.23	4.21	4.20	4.14
	0.010	6.47	6.31	6.16	5.91	5.86	5.82	5.79	5.75	5.74	5.65
	0.005	8.18	7.97	7.75	7.42	7.35	7.31	7.26	7.22	7.19	7.08
	0.001	13.71	13.32	12.93	12.33	12.20	12.12	12.04	11.95	11.91	11.70
8	0.100	2.50	2.46	2.42	2.36	2.35	2.34	2.33	2.32	2.32	2.29
	0.050	3.28	3.22	3.15	3.04	3.02	3.01	2.99	2.97	2.97	2.93
	0.025	4.20	4.10	4.00	3.84	3.81	3.78	3.76	3.74	3.73	3.67
	0.010	5.67	5.52	5.36	5.12	5.07	5.03	5.00	4.96	4.95	4.86
	0.005	7.01	6.81	6.61	6.29	6.22	6.18	6.13	6.09	6.06	5.95
	0.001	11.19	10.84	10.48	9.92	9.80	9.73	9.65	9.57	9.53	9.34
9	0.100	2.38	2.34	2.30	2.23	2.22	2.21	2.20	2.19	2.18	2.16
	0.050	3.07	3.01	2.94	2.83	2.80	2.79	2.77	2.76	2.75	2.71
	0.025	3.87	3.77	3.67	3.51	3.47	3.45	3.43	3.40	3.39	3.33
	0.010	5.11	4.96	4.81	4.57	4.52	4.48	4.45	4.41	4.40	4.31
	0.005	6.23	6.03	5.83	5.52	5.45	5.41	5.37	5.32	5.30	5.19
	0.001	9.57	9.24	8.90	8.37	8.26	8.19	8.11	8.04	8.00	7.82
10	0.100	2.28	2.24	2.20	2.13	2.12	2.11	2.10	2.09	2.08	2.06
	0.050	2.91	2.85	2.77	2.66	2.64	2.62	2.60	2.59	2.58	2.54
	0.025	3.62	3.52	3.42	3.26	3.22	3.20	3.18	3.15	3.14	3.08
	0.010	4.71	4.56	4.41	4.17	4.12	4.08	4.05	4.01	4.00	3.91
	0.005	5.66	5.47	5.27	4.97	4.90	4.86	4.82	4.77	4.75	4.64
	0.001	8.45	8.13	7.80	7.30	7.19	7.12	7.05	6.98	6.94	6.76
11	0.100	2.21	2.17	2.12	2.05	2.04	2.03	2.02	2.01	2.00	1.97
	0.050	2.79	2.72	2.65	2.53	2.51	2.49	2.47	2.46	2.45	2.41
	0.025	3.43	3.33	3.23	3.06	3.03	3.00	2.98	2.96	2.94	2.88
	0.010	4.40	4.25	4.10	3.86	3.81	3.78	3.74	3.71	3.69	3.60
	0.005	5.24	5.05	4.86	4.55	4.49	4.45	4.40	4.36	4.34	4.23
	0.001	7.63	7.32	7.01	6.52	6.42	6.35	6.28	6.21	6.18	6.00
12	0.100	2.15	2.10	2.06	1.99	1.97	1.96	1.95	1.94	1.93	1.90
	0.050	2.69	2.62	2.54	2.43	2.40	2.38	2.37	2.35	2.34	2.30
	0.025	3.28	3.18	3.07	2.91	2.87	2.85	2.82	2.80	2.79	2.73
	0.010	4.16	4.01	3.86	3.62	3.57	3.54	3.50	3.47	3.45	3.36
	0.005	4.91	4.72	4.53	4.23	4.17	4.12	4.08	4.04	4.01	3.91
	0.001	7.00	6.71	6.40	5.93	5.83	5.76	5.70	5.63	5.59	5.42
15	0.100	2.02	1.97	1.92	1.85	1.83	1.82	1.80	1.79	1.79	1.76
	0.050	2.48	2.40	2.33	2.20	2.18	2.16	2.14	2.12	2.11	2.07
	0.025	2.96	2.86	2.76	2.59	2.55	2.52	2.50	2.47	2.46	2.40
	0.010	3.67	3.52	3.37	3.13	3.08	3.05	3.01	2.98	2.96	2.87
	0.005	4.25	4.07	3.88	3.58	3.52	3.48	3.44	3.39	3.37	3.26
	0.001	5.81	5.54	5.25	4.80	4.70	4.64	4.57	4.51	4.47	4.31
18	0.100	1.93	1.89	1.84	1.75	1.74	1.72	1.71	1.70	1.69	1.66
	0.050	2.34	2.27	2.19	2.06	2.04	2.02	2.00	1.98	1.97	1.92
	0.025	2.77	2.67	2.56	2.38	2.35	2.32	2.30	2.27	2.26	2.19
	0.010	3.37	3.23	3.08	2.84	2.78	2.75	2.71	2.68	2.66	2.57
	0.005	3.86	3.68	3.50	3.20	3.14	3.10	3.05	3.01	2.99	2.87
	0.001	5.13	4.87	4.59	4.15	4.06	4.00	3.93	3.87	3.84	3.67

(*continued*)

APPENDIX A-6 (*continued*)

v_2	α	12	15	20	40	50	60	75	100	120	∞
					v_1 = Degrees of Freedom for Numerator						
20	0.100	1.89	1.84	1.79	1.71	1.69	1.68	1.66	1.65	1.64	1.61
	0.050	2.28	2.20	2.12	1.99	1.97	1.95	1.93	1.91	1.90	1.84
	0.025	2.68	2.57	2.46	2.29	2.25	2.22	2.20	2.17	2.16	2.09
	0.010	3.23	3.09	2.94	2.69	2.64	2.61	2.57	2.54	2.52	2.42
	0.005	3.68	3.50	3.32	3.02	2.96	2.92	2.87	2.83	2.81	2.69
	0.001	4.82	4.56	4.29	3.86	3.77	3.70	3.64	3.58	3.54	3.38
25	0.100	1.82	1.77	1.72	1.63	1.61	1.59	1.58	1.56	1.56	1.52
	0.050	2.16	2.09	2.01	1.87	1.84	1.82	1.80	1.78	1.77	1.71
	0.025	2.51	2.41	2.30	2.12	2.08	2.05	2.02	2.00	1.98	1.91
	0.010	2.99	2.85	2.70	2.45	2.40	2.36	2.33	2.29	2.27	2.17
	0.005	3.37	3.20	3.01	2.72	2.65	2.61	2.56	2.52	2.50	2.38
	0.001	4.31	4.06	3.79	3.37	3.28	3.22	3.15	3.09	3.06	2.89
30	0.100	1.77	1.72	1.67	1.57	1.55	1.54	1.52	1.51	1.50	1.46
	0.050	2.09	2.01	1.93	1.79	1.76	1.74	1.72	1.70	1.68	1.62
	0.025	2.41	2.31	2.20	2.01	1.97	1.94	1.91	1.88	1.87	1.79
	0.010	2.84	2.70	2.55	2.30	2.25	2.21	2.17	2.13	2.11	2.01
	0.005	3.18	3.01	2.82	2.52	2.46	2.42	2.37	2.32	2.30	2.18
	0.001	4.00	3.75	3.49	3.07	2.98	2.92	2.86	2.79	2.76	2.59
40	0.100	1.71	1.66	1.61	1.51	1.48	1.47	1.45	1.43	1.42	1.38
	0.050	2.00	1.92	1.84	1.69	1.66	1.64	1.61	1.59	1.58	1.51
	0.025	2.29	2.18	2.07	1.88	1.83	1.80	1.77	1.74	1.72	1.64
	0.010	2.66	2.52	2.37	2.11	2.06	2.02	1.98	1.94	1.92	1.81
	0.005	2.95	2.78	2.60	2.30	2.23	2.18	2.14	2.09	2.06	1.93
	0.001	3.64	3.40	3.14	2.73	2.64	2.57	2.51	2.44	2.41	2.23
50	0.100	1.68	1.63	1.57	1.46	1.44	1.42	1.41	1.39	1.38	1.33
	0.050	1.95	1.87	1.78	1.63	1.60	1.58	1.55	1.52	1.51	1.44
	0.025	2.22	2.11	1.99	1.80	1.75	1.72	1.69	1.66	1.64	1.55
	0.010	2.56	2.42	2.27	2.01	1.95	1.91	1.87	1.82	1.80	1.68
	0.005	2.82	2.65	2.47	2.16	2.10	2.05	2.00	1.95	1.93	1.79
	0.001	3.44	3.20	2.95	2.53	2.44	2.38	2.31	2.25	2.21	2.03
60	0.100	1.66	1.60	1.54	1.44	1.41	1.40	1.38	1.36	1.35	1.29
	0.050	1.92	1.84	1.75	1.59	1.56	1.53	1.51	1.48	1.47	1.39
	0.025	2.17	2.06	1.94	1.74	1.70	1.67	1.63	1.60	1.58	1.48
	0.010	2.50	2.35	2.20	1.94	1.88	1.84	1.79	1.75	1.73	1.60
	0.005	2.74	2.57	2.39	2.08	2.01	1.96	1.91	1.86	1.83	1.69
	0.001	3.32	3.08	2.83	2.41	2.32	2.25	2.19	2.12	2.08	1.89
100	0.100	1.61	1.56	1.49	1.38	1.35	1.34	1.32	1.29	1.28	1.22
	0.050	1.85	1.77	1.68	1.52	1.48	1.45	1.42	1.39	1.38	1.28
	0.025	2.08	1.97	1.85	1.64	1.59	1.56	1.52	1.48	1.46	1.35
	0.010	2.37	2.22	2.07	1.80	1.74	1.69	1.65	1.60	1.57	1.43
	0.005	2.58	2.41	2.23	1.91	1.84	1.79	1.74	1.68	1.65	1.49
	0.001	3.07	2.84	2.59	2.17	2.08	2.01	1.94	1.87	1.83	1.62
∞	0.100	1.55	1.49	1.42	1.30	1.26	1.24	1.22	1.19	1.17	1.00
	0.050	1.75	1.67	1.57	1.40	1.35	1.32	1.28	1.25	1.22	1.00
	0.025	1.95	1.83	1.71	1.49	1.43	1.39	1.35	1.30	1.27	1.00
	0.010	2.19	2.04	1.88	1.59	1.53	1.48	1.42	1.36	1.33	1.00
	0.005	2.36	2.19	2.00	1.67	1.59	1.54	1.47	1.40	1.37	1.00
	0.001	2.75	2.52	2.27	1.84	1.74	1.66	1.58	1.50	1.45	1.00

APPENDIX A-7 Factors for Computing Center line and Three-Sigma Control Limits

Observations in Sample, n	$\bar{X}$-Charts Factors for Control Limits			s-Charts Factors for Center line		s-Charts Factors for Control Limits				R-Charts Factors for Center line			R-Charts Factors for Control Limits			
	A	A_2	A_3	c_4	$1/c_4$	B_3	B_4	B_5	B_6	d_2	$1/d_2$	d_3	D_1	D_2	D_3	D_4
2	2.121	1.880	2.659	0.7979	1.2533	0	3.267	0	2.606	1.128	0.8865	0.853	0	3.686	0	3.267
3	1.732	1.023	1.954	0.8862	1.1284	0	2.568	0	2.276	1.693	0.5907	0.888	0	4.358	0	2.574
4	1.500	0.729	1.628	0.9213	1.0854	0	2.266	0	2.088	2.059	0.4857	0.880	0	4.698	0	2.282
5	1.342	0.577	1.427	0.9400	1.0638	0	2.089	0	1.964	2.326	0.4299	0.864	0	4.918	0	2.114
6	1.225	0.483	1.287	0.9515	1.0510	0.030	1.970	0.029	1.874	2.534	0.3946	0.848	0	5.078	0	2.004
7	1.134	0.419	1.182	0.9594	1.0423	0.118	1.882	0.113	1.806	2.704	0.3698	0.833	0.204	5.204	0.076	1.924
8	1.061	0.373	1.099	0.9650	1.0363	0.185	1.815	0.179	1.751	2.847	0.3512	0.820	0.388	5.306	0.136	1.864
9	1.000	0.337	1.032	0.9693	1.0317	0.239	1.761	0.232	1.707	2.970	0.3367	0.808	0.547	5.393	0.184	1.816
10	0.949	0.308	0.975	0.9727	1.0281	0.284	1.716	0.276	1.669	3.078	0.3249	0.797	0.687	5.469	0.223	1.777
11	0.905	0.285	0.927	0.9754	1.0252	0.321	1.679	0.313	1.637	3.173	0.3152	0.787	0.811	5.535	0.256	1.744
12	0.866	0.266	0.886	0.9776	1.0229	0.354	1.646	0.346	1.610	3.258	0.3069	0.778	0.922	5.594	0.283	1.717
13	0.832	0.249	0.850	0.9794	1.0210	0.382	1.618	0.374	1.585	3.336	0.2998	0.770	1.025	5.647	0.307	1.693
14	0.802	0.235	0.817	0.9810	1.0194	0.406	1.594	0.399	1.563	3.407	0.2935	0.763	1.118	5.696	0.328	1.672
15	0.775	0.223	0.789	0.9823	1.0180	0.428	1.572	0.421	1.544	3.472	0.2880	0.756	1.203	5.741	0.347	1.653
16	0.750	0.212	0.763	0.9835	1.0168	0.448	1.552	0.440	1.526	3.532	0.2831	0.750	1.282	5.782	0.363	1.637
17	0.728	0.203	0.739	0.9845	1.0157	0.466	1.534	0.458	1.511	3.588	0.2787	0.744	1.356	5.820	0.378	1.622
18	0.707	0.194	0.718	0.9854	1.0148	0.482	1.518	0.475	1.496	3.640	0.2747	0.739	1.424	5.856	0.391	1.608
19	0.688	0.187	0.698	0.9862	1.0140	0.497	1.503	0.490	1.483	3.689	0.2711	0.734	1.487	5.891	0.403	1.597
20	0.671	0.180	0.680	0.9869	1.0133	0.510	1.490	0.504	1.470	3.735	0.2677	0.729	1.549	5.921	0.415	1.585
21	0.655	0.173	0.663	0.9876	1.0126	0.523	1.477	0.516	1.459	3.778	0.2647	0.724	1.605	5.951	0.425	1.575
22	0.640	0.167	0.647	0.9882	1.0119	0.534	1.466	0.528	1.448	3.819	0.2618	0.720	1.659	5.979	0.434	1.566
23	0.626	0.162	0.633	0.9887	1.0114	0.545	1.455	0.539	1.438	3.858	0.2592	0.716	1.710	6.006	0.443	1.557
24	0.612	0.157	0.619	0.9892	1.0109	0.555	1.445	0.549	1.429	3.895	0.2567	0.712	1.759	6.031	0.451	1.548
25	0.600	0.153	0.606	0.9896	1.0105	0.565	1.435	0.559	1.420	3.931	0.2544	0.708	1.806	6.056	0.459	1.541

Source: © ASTM. Reprinted with permission.

APPENDIX A-8 Uniform Random Numbers

97878	44645	60468	86596	29743	98439	64428	50357
64600	47935	55776	38732	70498	61832	11372	30484
21734	98488	94734	99531	25282	24016	50366	11021
71402	60964	28456	72686	45225	00076	81351	82365
00358	30750	23487	57473	60720	34874	64186	80531
40782	28908	67197	86933	87919	65522	30539	71547
76722	69512	95964	74114	44095	12630	81913	13102
11137	92332	34009	04099	92740	95264	04667	40145
23984	99243	90979	95199	25357	84703	07202	51970
30262	22720	64621	91122	00940	07670	24647	58469
41609	32500	37060	75444	99182	59388	38806	14520
17469	07308	37637	75591	67256	74415	97339	84763
85130	96392	70015	42639	36885	35159	20170	98134
04204	59281	44421	93374	42647	75392	25164	45359
02387	34404	31376	52748	41546	70173	03409	96409
98770	39674	03575	97601	91398	39995	10671	61442
68709	98636	78218	25617	46255	22234	88613	31217
56293	30564	01867	87371	50834	67311	96809	72744
35017	39984	13007	16757	37348	09247	22734	01217
03193	76349	97895	26047	80563	29319	70426	64120
67759	40380	74450	91825	03074	66039	28096	22809
18755	25573	14639	38260	47489	58234	50219	32596
49269	80057	24228	09605	34931	01224	76877	87911
97434	82611	58899	30042	16356	01293	31830	69230
68837	56094	82048	93441	72467	62565	38400	99459
03797	28132	17109	57402	62259	98531	14472	35450
57828	30374	98465	13151	43132	32193	78184	50939
32393	43266	85401	41622	45396	80588	53661	83531
54176	40988	97983	10019	97341	14550	47511	18987
82120	01302	88694	94271	22290	39296	63110	02916
25874	68357	81583	69948	24461	25326	60400	54006
76479	94047	52701	61100	21783	11980	48124	30173
29314	93695	41337	67648	70136	83831	54796	56998
08217	12006	91741	94542	68309	73687	66076	51625
83677	66495	97931	36805	92243	35255	38746	20177
59042	24587	47017	23384	29948	40591	17259	78456
27293	16782	69543	20522	15442	37452	62532	86340
44154	11467	26355	64544	00741	63428	49008	79631
56887	08762	65555	94922	56064	72304	13685	13303
02523	15422	14344	90718	98416	79818	15141	90939
74141	88274	34566	57489	34419	85325	64590	71890
28175	40295	48870	65330	13591	31724	96311	09956
99599	73883	62419	89694	87842	72571	09171	47657
84925	48510	18369	35883	88947	33572	51302	43004
46361	94992	31390	72792	18506	42730	72923	20226
91086	96394	97533	42656	20758	94888	83053	80529
60435	33374	18112	25029	62553	32646	84162	21814
30737	01996	59146	76739	63951	31707	04183	04168
33045	97426	48039	92940	47647	19067	75199	72413
06223	02998	98687	40526	32807	89534	02039	76278

INDEX

Fundamentals of Quality Control and Improvement, Third Edition, By Amitava Mitra
Copyright © 2008 John Wiley & Sons, Inc.